T0406295

Lecture Notes in Electrical Engineering 277

For further volumes:
http://www.springer.com/series/7818

W. Eric Wong • Tingshao Zhu
Editors

Computer Engineering and Networking

Proceedings of the 2013 International Conference on Computer Engineering and Network (CENet2013)

Volume 1

Editors
W. Eric Wong
University of Texas at Dallas
Richardson, Texas
USA

Tingshao Zhu
Chinese Academy of Sciences
Beijing, China, People's Republic

ISSN 1876-1100 ISSN 1876-1119 (electronic)
ISBN 978-3-319-01765-5 ISBN 978-3-319-01766-2 (eBook)
DOI 10.1007/978-3-319-01766-2
Springer Cham Heidelberg New York Dordrecht London

Library of Congress Control Number: 2014931403

Printed on acid-free paper

Springer is part of Springer Science+Business Media (www.springer.com)

Preface

This book is a collection of the papers accepted by CENet 2013—the third International Conference on *Computer Engineering and Network* (CENet), which was held from 20 to 21 July, 2013 in Shanghai, China. It has two volumes and three parts in each. Part I focuses on Algorithm Design with 29 papers over 232 pages; Part II emphasizes Data Processing containing 184 pages divided among 22 papers; Part III Pattern Recognition includes 29 papers in 234 pages; Part IV has 22 papers and 187 pages devoted to one of the most exciting technologies currently surging in popularity—Cloud Computing; Part V covers recent advances in Embedded Systems with 28 papers in 228 pages; and finally Part VI has 28 papers spanning 234 pages dedicated to Network Optimization.

Each part can be used as an excellent reference by industry practitioners, university faculty, and undergraduate as well as graduate students who need to build a knowledge base of the most current advances and state of practice in the topics covered by this book. This will enable them to produce, maintain, and manage systems with high levels of trustworthiness and complexity that provide critical services in a variety of applications.

Thanks go to the authors for their hard work and dedication as well as the reviewers for ensuring the selection of only the highest quality papers; their efforts made this book possible. Invaluable assistance with the publication was also provided by the editorial staff at Springer, especially Mr. Brett Kurzman and Miss Rebecca Hytowitz.

Richardson, Texas, USA — W. Eric Wong
Beijing, China — Tingshao Zhu

Contents

Volume 1

Part I Algorithm Design

1 **Simulation Algorithm of Adaptive Scheduling in Missile-Borne Phased Array Radar** 3
Qizhong Li, Shanshan Sun, and Jianfei Zhao

2 **A Second-Order Algorithm for Curve Parallel Projection on Parametric Surfaces** 9
Xiongbing Fang and Hai-Yin Xu

3 **Computation Method of Processing Time Based on BP Neural Network and Genetic Algorithm** 21
Danchen Zhou and Chao Guo

4 **Integral Sliding Mode Controller for an Uncertain Network Control System with Delay** 31
Zhenbin Gao

5 **Synthesis of Linear Antenna Array Using Genetic Algorithm to Control Side Lobe Level** 39
Zhigang Zhang, Ting Li, Feng Yuan, and Li Yin

6 **Wavelet Analysis Combined with Artificial Neural Network for Predicting Protein–Protein Interactions** 47
Juanjuan Li, Yuehui Chen, and Fenglin Wang

7 **Application Analysis of Slot Allocation Algorithm for Link16** 55
Hui Zeng, Qiang Chen, Xiaoqiang Li, and Jianguo Shen

8 **An Improved Cluster Head Algorithm for Wireless Sensor Network** 65
Feng Yu, Wei Liu, and Gang Li

9 An Ant Colony System for Dynamic Voltage Scaling Problem in Heterogeneous System 73
Yan Kang, Ying Lin, Yifan Zhang, and He Lu

10 An Improved Ant Colony System for Task Scheduling Problem in Heterogeneous Distributed System 83
Yan Kang, Yifan Zhang, Ying Lin, and He Lu

11 Optimization of Green Agri-Food Supply Chain Network Using Particle Swarm Optimization Algorithm 91
Qian Tao, Zhexue Huang, Chunqin Gu, and Chenxin Zhang

12 A New Model for Short-Term Power System Load Forecasting Using Wavelet Transform Fuzzy RBF Neural Network 99
Jingduan Dong, Changhao Xia, and Wei Zhang

13 Energy-Effective Frequency-Based Adaptive Sampling Algorithm for Clustered Wireless Sensor Network 107
Meiyan Zhang, Wenyu Cai, Liping Zhou, and Jilai Liu

14 An Indoor Three-Dimensional Positioning Algorithm Based on Difference Received Signal Strength in WiFi 115
Yibo Li and Xiting Liu

15 The Universal Approximation Capability of Double Flexible Approximate Identity Neural Networks 125
Saeed Panahian Fard and Zarita Zainuddin

16 A Novel and Real-Time Hand Tracking Algorithm for Gesture Manipulation 135
Zhiqin Zhang

17 A Transforming Quantum-Inspired Genetic Algorithm for Optimization of Green Agricultural Products Supply Chain Network 145
Chunqin Gu and Qian Tao

18 A Shortest Path Algorithm Suitable for Navigation Software 153
Peng Luo, Qizhi Qiu, Wenyan Zhou, and Pei Fang

19 An Energy-Balanced Clustering Routing Algorithm for Wireless Sensor Networks 163
Mingqiang Chen and Xianhai Tan

20 Simulation and Analysis of Binary Frequency Shift Keying Noise Cancel Adaptive Filter Based on Least Mean Square Error Algorithm 171
Zhongping Chen and Jinding Gao

21 Density-Sensitive Semi-supervised Affinity Propagation Clustering 177
Kunlun Li, Qi Meng, Shangzong Luo, Hexin Li, and Qian Wang

22 The Implementation of a Hybrid Particle Swarm Optimization Algorithm Based on Three-Level Parallel Model 185
Yi Xiao and Yu Liu

23 Optimization of Inverse Planning Based on an Improved Non-dominated Neighbor-Based Selection in Intensity Modulated Radiation Therapy 193
Xiao Zhang, Guoli Li, and Zhizhong Li

24 A Recommendation System for Paper Submission Based on Vertical Search Engine 201
Zhen Xu, Yi Yang, Fei Wang, Jiao Xu, Zhong Li, Fuqiang Mu, and Lian Li

25 Analysis and Improvement of SPRINT Algorithm Based on Hadoop 209
Shanshan Fei, Qiaoyan Wen, and Zhengping Jin

26 Prediction Model for Trend of Web Sentiment Using Extension Neural Network and Nonparametric Auto-regression Method 219
Haitao Zhang, Binjun Wang, and Guangxuan Chen

27 K-Optimal Chaos Ant Colony Algorithm and Its Application on Dynamic Route Guidance System 227
Hai Yang

28 A Certainty-Based Active Learning Framework of Meeting Speech Summarization 235
Jian Zhang and Huaqiang Yuan

29 Application of Improved BP Neural Network in the Frequency Identification of Piano Tone 243
Xu Chen and Jun Tang

Part II Data Processing

30 Implicit Factoring with Shared Middle Discrete Bits 255
Meng Shi, Xianghui Liu, and Wenbao Han

31 Loading Data into HBase 265
Juan Yang and Xiaopu Feng

32 Incomplete Decision-Theoretic Rough Set Model Based on Improved Complete Tolerance Relation 273
Xia Wang

33 A New Association Rule Mining Algorithm Based on Compression Matrix 281
Sihui Shu

34 Decoupling Interrupts from the Internet in Markov Models 291
Jinwen Ma, Jingchun Zhang, and Jinrong Guo

35 Parallel Feature Selection Based on MapReduce 299
Zhanquan Sun

36 Initial State Modeling of Interlocking System Using Maude 307
Rui Ma, Zhongwei Xu, Zuxi Chen, and Shuqing Zhang

37 Semi-supervised Learning Using Nonnegative Matrix Factorization and Harmonic Functions 321
Lin Li, Zhenyu Zhao, Chenping Hou, and Yi Wu

38 Exploring Data Communication at System Level Through Reverse Engineering: A Case Study on USB Device Driver 329
Leela Sedaghat, Brad Duerling, Xiaoxi Huang, and Ziying Tang

39 Using Spatial Analysis to Identify Tuberculosis Transmission and Surveillance 337
Jinrong Bai, Guozhong Zou, Shiguang Mu, and Yu Ma

40 Construction Method of Exception Control Flow Graph for Business Process Execution Language Process 345
Caoqing Jiang, Shi Ying, Shanming Hu, and Hua Guan

41 P300 Detection in Electroencephalographic Signals for Brain–Computer Interface Systems: A Neural Networks Approach 355
Seyed Aliakbar Mousavi, Muhammad Rafie Hj. Mohd. Arshad, Hasimah Hj. Mohamed, Putra Sumari, and Saeed Panahian Fard

42 Web Content Extraction Technology 365
Zhenyu Jiao, Xiaoben Yan, Jinjin Sun, Yuchen Wang, and Jiangbin Chen

43 A New Data-Intensive Parallel Processing Framework for Spatial Data 375
Dong Zhao, Yang Gu, and Zhenchun Huang

44 The Approach of Graphical User Interface Testing Guided by Bayesian Model 385
Zhifang Yang, Zhongxing Yu, and Chenggang Bai

45 A Model for Reverse Logistics with Collection Sites Based on Heuristic Algorithm 395
Xiaoqing Geng and Yu Wang

46 The Storage of Wind Turbine Mass Data Based on MongoDB 403
Qile Wang, Zhu Shen, Long Ma, and Shi Yin

47 Improvement of Extraction Method of Correlation Time Delay Based on Connected-Element Interferometry 411
Fei Wang, Zhenfei Wang, Dun Li, and Bingjie Yang

48 Modeling and Evaluation of the Performance of Parallel/Distributed File System 421
Tiezhu Zhao, Xin Ao, and Huaqiang Yuan

49 CoCell: A Low-Diameter, High-Performance Data Center Network Architecture 429
Peng Wang, Huaxi Gu, Yan Zhao, and Xiaoshan Yu

50 Simulation Investigation of Counterwork Between Anti-radiation Missile and Active Decoy System 437
Huaqiang Hu and Dandan Wen

51 Simulation Jamming Technique on Binary Phase-Coded Pulse Compression Radar 445
Yulin Yang and Lijuan Qiu

Part III Pattern Recognition

52 Personalized Information Service Recommendation System Based on Clustering and Classification 455
Yu Wang

53 Palmprint Recognition Based on Subclass Discriminant Analysis 465
Pengfei Yu, Haiyan Li, Hao Zhou, and Dan Xu

54 A Process Quality Monitoring Approach of Automatic Aircraft Component Docking 473
Guowei Yang, Chengjing Zhang, and Xiaofeng Zhang

55 Overhead Transmission Lines Sag Measurement Based on Image Processing 481
Wengang Cheng and Long Chen

56 Chinese Domain Ontology Learning Based on Semantic Dependency and Formal Concept Analysis 489
Lixin Hou, Shanhong Zheng, Haitao He, and Xinyi Peng

57 Text Classification Algorithm Based on Rough Set 499
Zhiyong Hong

58 Robust Fragment-Based Tracking with Online Selection of Discriminative Features 507
Yongqiang Huang and Long Zhao

59 Extraction Method of Gait Feature Based on Human Centroid Trajectory 515
Xin Chen and Tianqi Yang

60 An Algorithm for Bayesian Network Structure Learning Based on Simulated Annealing with Adaptive Selection Operator 525
Ao Lin, Bing Xiao, and Yi Zhu

61 Static Image Segmentation Using Polar Space Transformation Technique 533
Xuan Luo, Tiancai Liang, and Weifeng Wang

62 Image Restoration via Nonlocal *P*-Laplace Regularization 541
Chen Yao, Lijuan Hong, and Yunfei Cheng

63 Analysis and Application of Computer Technology on Architectural Space Lighting Visual Design 549
Yiwen Cao

64 Improving Online Gesture Recognition with WarpingLCSS by Multi-Sensor Fusion 559
Chao Chen and Haibin Shen

65 The Lane Mark Identifying and Tracking in Intense Illumination 567
Yanyun Xing, Bo Yu, and Fangqun Yang

66 Classification Modeling of Multi-Featured Remote Sensing Images Based on Sparse Representation 577
Xiaoting Hao, Chunmei Zhang, Jing Bai, Mo Dai, Wenxing Bao, and Wei Feng

67 A Parallel and Convergent Support Vector Machine Based on MapReduce 585
Yingying Ma, Liming Wang, and Longpu Li

68 Vehicle Classification Based on Hierarchical Support Vector Machine 593
Mengwan Jiang and Haoliang Li

69 Image Splicing Detection Based on Machine Learning Algorithm 601
Yan Xiao

70 A Lane Detection Algorithm Based on Hyperbola Model 609
Chaobo Chen, Bofeng Zhang, and Song Gao

71 Comparisons and Analyses of Image Softproofing Under Different Profile Rendering Intents 617
Qingxue Yu, Yunhui Luo, Maohai Lin, and Quantao Liu

72 An Improved Dense Matching Algorithm for Face Based on Region Growing 625
Xin Xia and Shaoyan Gai

73 An Improved Feature Selection Method for Chinese Short Texts Clustering Based on HowNet 635
Xin Chen, Yuqing Zhang, Long Cao, and Donghui Li

74 Internet Worm Detection and Classification Based on Support Vector Machine 643
Huihui Liang, Min Li, and Jiwen Chai

75 Real-Time Fall Detection Based on Global Orientation and Human Shape 653
Shuangcheng Wang, Yepeng Guan, and Ruiyue Xu

76 The Classification of Synthetic Aperture Radar Oil Spill Images Based on the Texture Features and Deep Belief Network 661
Xixi Huang and Xiaofeng Wang

77 The Ground Objects Identification for Digital Remote Sensing Image Based on the BP Neural Network 671
Shengkui Cao, Guangchao Cao, Kelong Chen, Chengyong Wu, Tao Zhang, and Jie Yuan

78 Detection of Image Forgery Based on Improved PCA-SIFT 679
Kunlun Li, Hexin Li, Bo Yang, Qi Meng, and Shangzong Luo

79 A Thinning Model for Handwriting-Like Image Skeleton 687
Shijiao Zhu, Jun Yang, and Xue-fang Zhu

80 Discrimination of the White Wine Based on Sparse Principal Component Analysis and Support Vector Machine 695
Rong Wang, Wu Zeng, and Jiao Ming

Volume 2

Part IV Cloud Computing

81 Design of Mobile Electronic Payment System 705
Ting Huang

82 Power Saving-Based Radio Resource Scheduling in Long-Term Evolution Advanced Network 713
Yen-Yin Chu, I-Hsuan Peng, Yen-Wen Chen, Chi-Fu Yi, and Addison Y.S. Su

83 Dispatching and Management Model Based on Safe Performance Interface for Improving Cloud Efficiency 723
Bin Chen, Zhijian Wang, and Yu Wang

84 A Proposed Methodology for an E-Health Monitoring System Based on a Fault-Tolerant Smart Mobile 731
Ahmed Alahmadi and Ben Soh

85 Design and Application of Indoor Geographical Information System 739
Yongfeng Suo, Tianhe Chi, and Tianyue Liu

86 Constructing Cloud Computing Infrastructure Platform of the Digital Library Base on Virtualization Technology 747
Tingbo Fu, Jinsheng Yang, Yu Gao, and Guang Yu

87 A New Single Sign-on Solution in Cloud 755
Guangxuan Chen, Yanhui Du, Panke Qin, Lei Zhang, and Jin Du

88 A Collaborative Load Control Scheme for Hierarchical Mobile IPv6 Network 763
Yi Yang, QingShan Man, and PingLiang Rui

89 A High Efficient Selective Content Encryption Method Suitable for Satellite Communication System 775
Yanyan Xu, Bo Yang, Zhengquan Xu, and Tengyue Mao

90 Network Design of a Low-Power Parking Guidance System 783
Ming Xia, Yabo Dong, Qingzhang Chen, Kai Wang, and Rongjie Wu

91 Strategy of Domain and Cross-Domain Access Control Based on Trust in Cloud Computing Environment 791
Bo Li, Ming Tian, Yongsheng Zhang, and Shenjuan Lv

92 Detecting Unhealthy Cloud System Status 799
Zhidong Chen, Buyang Cao, and Yuanyuan Liu

93 Scoring System of Simulation Training Platform Based on Expert System 809
Wei Nie, Ying Wu, and Dabin Hu

94 Analysis of Distributed File Systems on Virtualized Cloud Computing Environment 817
Tiezhu Zhao, Zusheng Zhang, and Huaqiang Yuan

95 A Decision Support System with Dynamic Probability Adjustment for Fault Diagnosis in Critical Systems 825
Qiang Chen and Yun Xue

96 Design and Implementation of an SD Interface to Multiple-Target Interface Bridge 835
Guoyong Li, Leibo Liu, Shouyi Yin, Dajiang Liu, and Shaojun Wei

97 Cloud Storage Management Technology for Small File Based on Two-Dimensional Packing Algorithm 847
Zhiyun Zheng, Shaofeng Zhao, Xingjin Zhang, Zhenfei Wang, and Liping Lu

98 Advertising Media Selection and Delivery Decision-Making Using Influence Diagram 855
Xiaoxuan Hu and Fan Jiang

99 The Application of Trusted Computing Technology in the Cloud Security 865
Bo Li, Shenjuan Lv, Yongsheng Zhang, and Ming Tian

100 The Application Level of E-commerce in Enterprises in China 873
Yinghan H. Tang

101 Toward a Trinity Model of Digital Education Resources Construction and Management 883
Yong Huang and Qingchun Hu

102 Geographic Information System in the Cloud Computing Environment 893
Yichun Peng and Yunpeng Wang

Part V Embedded Systems

103 Memory Controller Design Based on Quadruple Modular Redundant Architecture 905
Yuanyuan Cui, Wei Li, and Xunying Zhang

104 Computer Power Management System Based on the Face Detection 913
Li Xie, Yong He, Yanfang Tian, and Tinghong Yang

105 Twist Rotation Deformation of Titanium Sheet Metal in Laser Curve Bending Based on Finite Element Analysis 921
Peng Zhang, Qian Su, and Dong Luan

106 Voltage Transient Stability Analysis by Changing the Control Modes of the Wind Generator 929
Yu Shao, Feng Shi, and Xiang Li

107 The Generator Stator Fault Analysis Based on the Multi-loop Theory 939
Yu Shao, Feng Shi, and Xiang Li

108 An Improved Edge Flag Algorithm Suitable for Hardware Implementation 947
Lixiang Wang and Tiejun Xiao

109 A Handheld Controller with Embedded Real-Time Video Transmission Based on TCP/IP Protocol 955
Mingjie Dong, Wusheng Chou, and Yihan Liu

110 Evaluating the Energy Consumption of InfiniBand Switch Based on Time Series 963
Huifeng Wang, Zhanhuai Li, Xiaonan Zhao, Qinlu He, and Jian Sun

111 Real-Time Filtering Method Based on Neuron Filtering Mechanism and Its Application on Robot Speed Signals 971
Wa Gao, Fusheng Zha, Baoyu Song, Mantian Li, Pengfei Wang, Zhenyu Jiang, and Wei Guo

112 Multiple-View Spectral Embedded Clustering Using a Co-training Approach 979
Hong Tao, Chenping Hou, and Dongyun Yi

113 Feedback Earliest Deadline First Exploiting Hardware Assisted Voltage Scaling 989
Chuansheng Wu

114 Design and Realization of General Interface Based on Object Linking and Embedding for Process Control 997
Jiguang Liu, Jianbing Wu, and Zhiguo He

115 A Stateful and Stateless IPv4/IPv6 Translator Based on Embedded System 1007
Yanlin Yin and Dalin Jiang

116 A Novel Collaborative Filtering Approach by Using Tags and Field Authorities 1017
Zhi Xue, Yaoxue Zhang, Yuezhi Zhou, and Wei Hu

117 Characteristics of Impedance for Plasma Antenna 1027
Bo Yin and Feng Yang

118 A Low-Voltage 5.8-GHz Complementary Metal Oxide Semiconductor Transceiver Front-End Chip Design for Dedicated Short-Range Communication Application 1035
Jhin-Fang Huang, Jiun-Yu Wen, and Yong-Jhen Jiangn

119 A 5.8-GHz Frequency Synthesizer with Dynamic Current-Matching Charge Pump Linearization Technique and an Average Varactor Circuit 1045
Jhin-Fang Huang, Jia-Lun Yang, and Kuo-Lung Chen

120 Full-Wave Design of Wireless Charging System for Electronic Vehicle 1055
Yongxiang Liu, Yi Ren, and Yi Wang

121 A Hierarchical Local-Interconnection Structure for Reconfigurable Processing Unit 1063
Yujia Zou, Leibo Liu, Shouyi Yin, Min Zhu, and Shaojun Wei

122 High Impedance Fault Location in Distribution System Based on Nonlinear Frequency Analysis 1073
Jinqian Zhai, Di Su, Wenjian Li, Feng Li, and Guohong Zhang

123 Early Fault Detection of Distribution Network Based on High-Frequency Component of Residual Current 1083
Jinqian Zhai, Di Su, Wenjian Li, Feng Li, and Guohong Zhang

124 A Complementary Metal Oxide Semiconductor D/A Converter with R-2R Ladder Based on T-Type Weighted Current Network 1091
Junshen Jiao

125 Detecting Repackaged Android Applications 1099
Zhongyuan Qin, Zhongyun Yang, Yuxing Di, Qunfang Zhang, Xinshuai Zhang, and Zhiwei Zhang

126 Design of Wireless Local Area Network Security Program Based on Near Field Communication Technology 1109
Pengfei Hu and Leizhen Wang

127 A Mechanism of Transforming Architecture Analysis and Design Language into Modelica 1117
Shuguang Feng and Lichen Zhang

128 Aspect-Oriented QoS Modeling of Cyber-Physical Systems by the Extension of Architecture Analysis and Design Language 1125
Lichen Zhang and Shuguang Feng

129 Using RC4-BHF to Construct One-way Hash Chains 1133
Qian Yu and Chang N. Zhang

130 Leakage Power Reduction of Instruction Cache Based on Tag Prediction and Drowsy Cache 1143
Wei Li and Jianqing Xiao

Part VI Network Optimization

131 The Human Role Model of Cyber Counterwork 1155
Fang Zhou

132 A Service Channel Assignment Scheme for IEEE 802.11p Vehicular Ad Hoc Network 1165
Yao Zhang, Licai Yang, Haiqing Liu, and Lei Wu

133 An Exception Handling Framework for Web Service 1173
Hua Guan, Shi Ying, and Caoqing Jiang

134 Resource Congestion Based on SDH Network Static Resource Allocation 1181
Fuyong Liu, Jianghe Yao, Gang Wu, and Huanhuan Wu

135 Multilayered Reinforcement Learning Approach for Radio Resource Management 1191
Kevin Collados, Juan-Luis Gorricho, Joan Serrat, and Hu Zheng

136 A Network Access Security Scheme for Virtual Machine 1201
Mingkun Xu, Wenyuan Dong, and Cheng Shuo

137 Light Protocols in Chain Network 1209
Ying Wang, Yifang Chen, and Lenan Wu

138 Research and Implementation of a Peripheral Environment Simulation Tool with Domain-Specific Languages 1217
Maodi Zhang, Zili Wang, Ping Xu, and Yi Li

139 Probability Model for Information Dissemination on Complex Networks 1225
Juan Li and Xueguang Zhou

140 Verification of UML Sequence Diagrams in Coq 1233
Liang Dou, Lunjin Lu, Ying Zuo, and Zongyuan Yang

141 Quantitative Verification of the Bounded Retransmission Protocol 1245
Xu Guo, Ming Xu, and Zongyuan Yang

142 A Cluster-Based and Range-Free Multidimensional Scaling-MAP Localization Scheme in WSN 1253
Ke Xu, Yuhua Liu, Cui Xu, and Kaihua Xu

143 A Resource Information Organization Method Based on Node Encoding for Resource Discovering 1263
Zhuang Miao, Qianqian Zhang, Songqing Wang, Yang Li, Weiguang Xu, and Jiang Xiao

144 The Implementation of Electronic Product Code System Based on Internet of Things Applications for Trade Enterprises 1271
Huiqun Zhao and Biao Shi

145 The Characteristic and Verification of Length of Vertex-Degree Sequence in Scale-Free Network 1281
Yanxia Liu, Wenjun Xiao, and Jianqing Xi

146 A Preemptive Model for Asynchronous Persistent Carrier Sense Multiple Access 1289
Lin Gao and Zhijun Wu

147 Extended Petri Net-Based Advanced Persistent Threat Analysis Model 1297
Wentao Zhao, Pengfei Wang, and Fan Zhang

148 Energy-Efficient Routing Protocol Based on Probability of Wireless Sensor Network 1307
Kaiguo Qian

149 A Dynamic Routing Protocols Switching Scheme in Wireless Sensor Networks 1315
Zusheng Zhang, Tiezhu Zhao, and Huaqiang Yuan

150 Incipient Fault Diagnosis in the Distribution Network Based on S-Transform and Polarity of Magnitude Difference 1323
Jinqian Zhai and Xin Chen

151 Network Communication Forming Coalition S4n-Knowledge Model Case 1331
Takashi Matsuhisa

152 An Optimization Model of the Layout of Public Bike Rental Stations Based on B+R Mode 1341
Liu He, Xuhong Li, and Dawei Chen

153 Modeling of Train Control Systems Using Formal Techniques 1349
Bingqing Xu and Lichen Zhang

154 A Clock-Based Specification of Cyber-Physical Systems 1357
Bingqing Xu and Lichen Zhang

155 Polymorphic Worm Detection Using Position-Relation Signature 1365
Huihui Liang, Jiwen Chai, and Yong Tang

156 Application of the Wavelet-ANFIS Model 1373
Rijun Zhang, Caishui Hou, Hui Lin, Meiyan Zhuo, Meixin Zhang, Zhongsheng Li, Liwu Sun, and Fengqin Lin

157 Visualization of Clustered Network Graphs Based on Constrained Optimization Partition Layout 1381
Fang Huang, Wenjie Xiao, and Hao Zhang

158 An Ultra-Wideband Cooperative Communication Method Based on Transmitted Cooperative Reference 1395
Tiefeng Li, Ou Li, and Zewen Zhou

Author Index for Volume 1 1407

Subject Index for Volume 1 1411

Author Index for Volume 2 1417

Subject Index for Volume 2 1421

CENet 2013 Committee

RADJEF Mohammed Said	University of Bejaia, Algeria
Rui Chen	Xi'an University of Electronic Science and Technology, China
S. EL-Rabaie	Faculty of Electronic Engineering, 32952 Menouf, Egypt
Stephan Chalup	Heidelberg University, Germany
Sunil Kumar Khatri	Amity University, Uttar Pradesh, India
Tao Zhou	North China Electric Power University, China
Vong Chi-Man	University of Macau, Macau, China
Xianzhong Yi	Yangtze University, China
Yan Pei	Kyushu University, Japan
Yuan Gao	Tsinghua University, China
Yuduo Wang	Beijing Information Science and Technology University, China
Zhongtian Jia	University of Jinan, China

Part I
Algorithm Design

Chapter 1
Simulation Algorithm of Adaptive Scheduling in Missile-Borne Phased Array Radar

Qizhong Li, Shanshan Sun, and Jianfei Zhao

Abstract In contrast to conventional radars, phased array radars have the capability to switch the direction of the radar beam very quickly without inertia. The measurements from phased array radar can contribute to many application fields such as data and intelligence process and radar performance evaluation. However, it often costs more than we can bear to obtain phased array radar measurements. It is necessary to model and simulate the phased array radar, especially for missile-borne phased array radar. This chapter lays a strong emphasis on the search and simulation technology of missile-borne phased array radar. According to operational theory of phased array radar, this chapter focuses on the functional modeling and simulation techniques. It contains three parts: beam arrangement of phased array radar in sin coordinate, parameter optimization of missile-borne phased array radar, and a function simulation model of missile-borne phased array radar.

1.1 Introduction

In ballistic missile defense system, the beam width of tracking radar is usually very narrow, and the dwell time is longer, so the implementation of routine large airspace search is not realistic. Only under the guide information of early warning system, the radar can intercept the target in small space [1]. The flight time of ballistic missile is usually short, and the lethality is enormous, which requires that the tracking radar can intercept the target as early as possible. Therefore, the missile-borne radar search strategy, especially on the wave search order, is very important.

Based on the previous research, considering the constraints of radar, we complete the simulation of adaptive scheduling method and make the constraints

---

Q. Li (✉) • S. Sun • J. Zhao
North China Institute of Science and Technology, Beijing 101601, China
e-mail: li_qz@126.com.cn

W.E. Wong and T. Zhu (eds.), *Computer Engineering and Networking*, Lecture Notes in Electrical Engineering 277, DOI 10.1007/978-3-319-01766-2_1,

modeling for recursive form, through adjusting the scheduling of periodic inspection the constraint conditions [2]. This chapter presents the realization method and process of the algorithm in detail, through the scenario simulation, and verifies the effectiveness and practicability of the method.

1.2 Scheduling Model

1.2.1 Single Beam Search Simulation

In the distance occlusion problem under a single frequency, pulse Doppler waveform exists, which is mainly due to pulse Doppler radar seeker using the same antenna for transmitting and receiving. Transmitting and receiving is timeshare; when the transmitter sends electromagnetic wave, the receiver is in a closed state, which can avoid the leakage of the transmitter signal to the receiving system and burn the frequency receiver [3]. The high-frequency receiver uses gating switch to guarantee isolation in certain, if the target echo arrival seeker at the time of the transmitter sends pulse. Because the receiving system is in the closed state, the return signal cannot be received by target seeker in the "shelter area." And while the missile is close to the target, the delay time of the target echo pulse with respect to the transmission pulse is fluxing. Thus, the blocking periodic phenomenon appears, which has entered the "transparent area," "translucent area," "shelter area," "translucent area," and then the "transparent area." In the "translucent area," the return signal is not affected; in the "shelter area," it cannot receive the return signal completely; and in the "translucent area," the width of echo pulse which is received by seeker gradually increases to the width of transmission pulse or from the width of transmitted pulse which decreases to zero [4].

We assume that the occlusion period is T_{zr}, width of "shelter area" is τ_z, width of "transparent area" is τ_m, and width of "translucent area" is τ_{bm}; they were calculated by the following equation:

$$T_{zr} = \frac{c}{2v_r} T_r \tag{1.1}$$

$$\tau_z = \frac{c}{2v_r} (T_r - \tau_r - \tau_l) \tag{1.2}$$

$$\tau_m = \frac{c}{2v_r} (\tau_r - \tau_l) \tag{1.3}$$

$$\tau_{bm} = \frac{c}{2v_r} \tau_l \tag{1.4}$$

In formula (1.1) to (1.4), C is the speed of light, v_r is the missile-target relative velocity, T_r is pulse width, τ_r is received pulse width, and τ_l is transmitted pulse width.

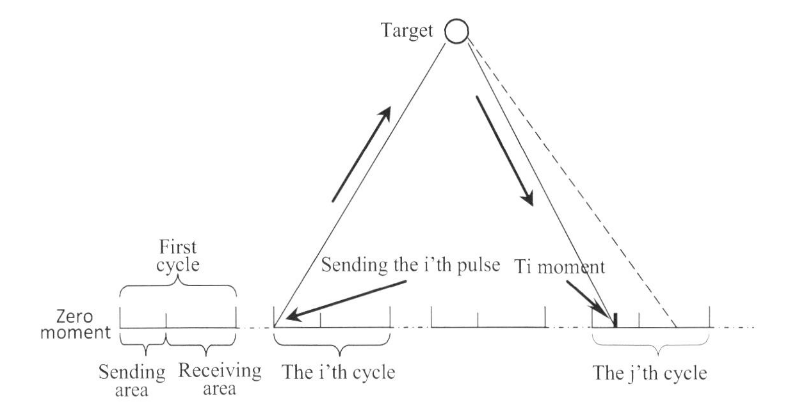

Fig. 1.1 The radar "eclipse" phenomenon mathematical model diagram

1.2.2 Establish Mathematical Model

According to the basic principle of the radar "eclipse" phenomenon, we can start from a simple mathematical model as shown in Fig. 1.1 and establish the idealized mathematical model of "eclipse" phenomenon step by step.

Assume that the target can be detected by radar, and the distance to the target is D, the radar frequency is f, the speed of light is C, pulse width of radar is T_r, the return moments of the i beam radar is T_i, and $V_1, V_2 \ldots V_i$ is the speed of the target in different intervals.

$$T_i = (i-1) \times T_r + \frac{2(D - (V_1 + V_2 \ldots + V_{i-1})) \times T_r}{C + v_i} \tag{1.5}$$

1.3 Multi-beam Search Simulation Modeling and Algorithm

1.3.1 Beam Position Arrangement Method

The antenna of phased array radar is fixed while scanning. When the scanning angle deviates from the normal direction, the beam will change. So the beam arrangement was usually complete in the sinusoidal spherical coordinates.

First step: determine the scanning space which was required in the radar station coordinates and change the space into the front spherical coordinates through coordinate transformation.

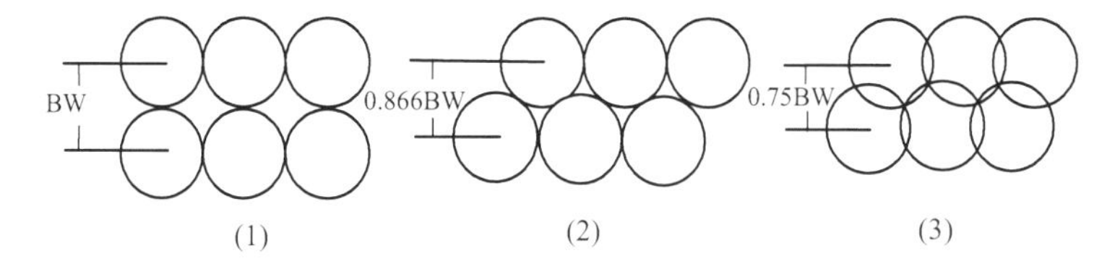

Fig. 1.2 Beam accumulation method

Table 1.1 Three waves of a style of the coverage and overlap rate

Arrangement	The wave number	Fraction of coverage (%)	Coincidence rate (%)
Vertical and horizontal arrangement wave	645	86.4	0
Staggered wave	783	98.7	0
Crisscross wave	920	100	3.56

Second step: change the space of the front spherical coordinates into front sine coordinate system and complete the beam arrangement in the sine coordinate system.

Third step: after the completion of beam arrangement in sine coordinate system, we can get the beam distribution in radar station coordinates by coordinate transformation.

The optimal beam position arrangement aims at making full use of the prior information and the phased array radar system resources and makes the average time of finding target as short as possible. So we can think it is an optimization problem under constraints [5], as shown in Fig. 1.2 (Table 1.1).

From the table above, we know that three kinds of beam position arrangement each have advantages. The first arrangement needs the minimum number of wave, but the coverage rate is only 86.4 %. When the resource is limited or in the small target distribution density area, we can use it. Although the coverage rate of the third arrangement can reach 100 %, it is easy to cause the redundancy detection, which limits its application. In the second arrangement, the wave number and the coverage rate achieve a better balance, so it is a common beam position arrangement style. To make a long story short, the specific choice of what kind of arrangement style cannot lump together; we should think about the actual background and phased array radar system resources to make a reasonable choice.

1.3.2 *Mathematical Model*

According to the phased array radar beam position arrangement theory and calculation, we use the above three kinds of beam position arrangement. If the number of beam is N in the first forms, according to the theory, we need 1.15N beam in the second forms and 1.54N beam for the third forms.

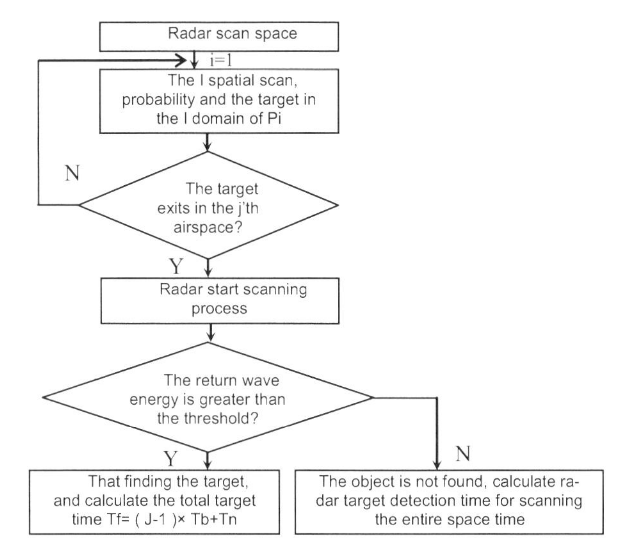

Fig. 1.3 Radar scanning airspace flow chart

Assuming a certain air-to-air missile radar scan area for $[10^0, 10^0]$, $[10^0, 10^0]$, radar beam width is 2^0, then the first form needs 50 beams in the area, the second form needs 58 beams, and the third form needs 77 beams.

So we can design a mode which has three different scanning forms. Assume that the target must be in the airspace that the radar will scan; to find the target or not, we will use the basic model of radar scanning to determine by target's frequency, radar's duty ratio, radar return wave threshold, and other factors. At the same time, we assume that the probability of the target appears in a wave of the space is different.

To take the first mode, for example, in the 50 wave, we will begin numbering from the largest probability of targets appearing as no. 1, 2 . . . N and the probability as P1, P2 . . . Pn (n is the wave number). We design an algorithm of target emerging randomly by the probability and make probability of the target emerging in the wave I as Pi. We set radar to start scanning from the maximum probability of waves until it finds the target. If it has scanned j wave from the start of radar scanning, the time the target is found is $Tf = (J-1) \times Tb + Tn$, where Tb is the time of scanning a wave and Tn is the time of finding the target in the jth wave, as shown in Fig. 1.3.

The simulation program uses C/C++ language to compile the simulation model of the function such as source code, establishes user interface in the VC++6.0 environment, tests results under various conditions through the simulation calculation, and uses the program for calibration and correction of simulation model [6]. We select Access as the background data management and ODBC for accessing dynamic target echo simulation database.

1.4 Conclusion

Using the above method, we can do the simulation for each beam position arrangement. For each method, we tested the time of radar target detection as Tf for three groups of 100 times, and then we calculated the average time as Ta. We can analyze the data to find the shortest average time which we take as the optimal model. Of course it should consider radar's frequency, target distance, duty ratio, and other factors. We need to complete simulation and calculation under the conditions of different target distance, frequency, and duty ratio and analyze large amounts of data to derive the optimal searching adaptive algorithm that can automatically adjust the radar's frequency and duty ratio in order to find the target at the shortest time under the changing target distance.

Acknowledgments A lot of thanks to the fundamental research funds for the Central Universities Fund (DX2013B01).

References

1. Skolnik, M. I. (1990). *Radar handbook* (2nd ed., pp. 231–233). New York: McGraw-Hill.
2. Horne, G., & Meyer, T. (2004). Data farming: Discovering surprise. In *Proceedings of the 2004 Winter Simulation Conference* (pp. 807–813). Huntsville, AL: Society for Computer Simulation International.
3. Liao, S. Y., Lu, H. W., Chen, J., et al. (2006). Research on conceptual framework for agent-based modeling and simulation. *Journal of System Simulation (S1004-731X), 18*(S2), 616–620 (In Chinese).
4. Wang, X., Li, D., Li, B., & Wang, X. S. (2010). Coherent video modeling and simulation method of phased array radar. *Journal of System Simulation (S1004-731X), 27*(3), 741–747 (In Chinese).
5. Yang, L. B., Wang, X. S., Dan, M., & Xiao, S. P. (2004). Simulation and evaluation of phased array radar's tactical ballistic missile defense capability. *Journal of System Simulation (S1004-731X), 16*(7), 1417–1426 (In Chinese).
6. Brandstein, A., & Horne, G. (2008). *Maneuver warfare science 2001 [EB/OL]*. Retrieved from http://www.projectalbert.org/files/MWS2001On-line.pdf

Chapter 2
A Second-Order Algorithm for Curve Parallel Projection on Parametric Surfaces

Xiongbing Fang and Hai-Yin Xu

Abstract A second-order algorithm is presented to calculate the parallel projection of a parametric curve onto a parametric surface in this chapter. The essence of our approach is to transform the problem of computing parallel projection curve on the parametric surface into that of computing parametric projection curve in the two-dimensional parametric domain of the surface. First- and second-order differential geometric characteristics of the parametric projection curve in the parametric domain of the surface are firstly analyzed. A marching method based on second-order Taylor Approximation is formulated to calculate the parametric projection curve. A first-order correction technique is developed to depress the error caused by the truncated higher order terms in the marching method. Several examples are finally implemented to demonstrate the effectiveness of the proposed scheme. Experimental results indicate that both the computational efficiency and accuracy of the presented method have dominant performance as compared with the first-order differential equation method.

2.1 Introduction

Curves on a surface have a wide range of applications in the fields of Computer Graphics, Computer-Aided Geometric Design, Computer Animation, CNC, etc. For instance, curves on a surface can be used for surface trimming [1], surface blending [2], NC tool path generation [3, 4], and so on. According to the designing manner, curves on a surface can be the intersection curve of two surfaces [4],

X. Fang (✉)
China Ship Development and Design Center, Wuhan 430064, China
e-mail: fangxb2013@sina.cn

H.-Y. Xu
School of Computer Science and Technology, Huazhong University of Science and Technology, Wuhan 430074, China

W.E. Wong and T. Zhu (eds.), *Computer Engineering and Networking*, Lecture Notes in Electrical Engineering 277, DOI 10.1007/978-3-319-01766-2_2,

the offset of a given curve on a surface [3], the projection curve of a spatial curve onto a surface [5–9], the image of a curve in the parametric domain of a parametric surface [1, 2], or the fitting curve of a sequence of points lying on a surface [10].

In this chapter, we focus on computation of parallel projection of a parametric curve onto a parametric surface. Presently, there are two ways to do this problem. One is the first-order differential equation method [6] and the other is discrete method.

For the problem of calculating the parallel projection of parametric curves onto parametric surfaces, Wang, et al. transformed the condition of parallel projection into a system of differential equations and then formulated the problem as a first-order initial value problem. Numerical methods such as Runge–Kutta and Adams-Bashforth can be utilized to solve the initial value problem to generate a sequence of points. Under this transformation, difficulties lie in the choice of an accurate initial value and the stability of the adopted numerical method. For the discrete method, the parametric curve should first be discretized into a series of points. Parallel projections of these separated points can be calculated through the technique of intersection of a line with a surface. Usually, the computational efficiency of the discrete method depends on that of the intersection algorithm. Though the current projected point can be taken as the initial value of the next iteration, efficiency of the discrete method is generally slow as it does not fully utilize the differential geometric properties of both the parametric curve and the projection curve. The projection curve can be constructed by fitting the projected points generated by the two aforementioned types of approaches.

A second-order algorithm is put forward for tracing the parallel projection of a parametric curve onto a parametric surface. Experimental results show that the proposed scheme has dominant performance in both efficiency and computational accuracy as compared with Wang's approach [6]. The rest of the chapter is organized as follows. An overview for our approach is presented in the next section. A second-order technique with error adjustment for tracing the projection curve is given in Sect. 2.3. Implementation and experimental results of our approach are carried out in Sect. 2.4 and we conclude the chapter in Sect. 2.5.

2.2 Overview

A 3D parametric curve $\boldsymbol{p}(t)$ and a parametric surface $\boldsymbol{S}(u,w)$ given in Fig. 2.1a are represented as

$$\boldsymbol{p}(t) = [x(t)\ y(t)\ z(t)] \tag{2.1}$$

and

$$\boldsymbol{S}(u,w) = [x(u,w)\ y(u,w)\ z(u,w)] \tag{2.2}$$

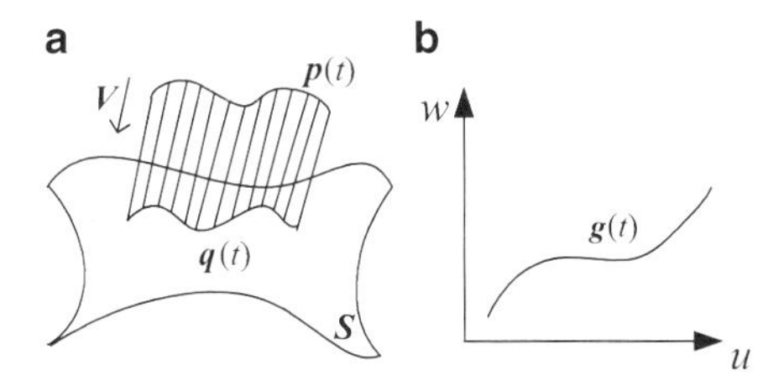

Fig. 2.1 Parallel projection of 3D curve onto surface

respectively. Suppose $\boldsymbol{p}$ is a point on the curve $\boldsymbol{p}(t)$ and the $\boldsymbol{q}$ is the corresponding point generated by projecting the point $\boldsymbol{p}$ along with the direction $\boldsymbol{V}$ onto the surface $\boldsymbol{S}$. While the point $\boldsymbol{p}$ moving along the curve $\boldsymbol{p}(t)$, its parallel projection point $\boldsymbol{q}$ also moves along a curve on the surface, which is termed as the parallel projection curve of $\boldsymbol{p}(t)$ along with the direction $\boldsymbol{V}$ onto the surface $\boldsymbol{S}$. According to the definition of parallel projection, one has

$$(\boldsymbol{q}(t) - \boldsymbol{p}(t)) \times \boldsymbol{V} = \boldsymbol{0} \tag{2.3}$$

where "×" denotes the cross product of two vectors. As the point lies on the parametric surface $\boldsymbol{S}(u,w)$, the curve $\boldsymbol{q}(t)$ can be represented as $\boldsymbol{q}(t) = [x(u(t), w(t))\ y(u(t), w(t))\ z(u(t), w(t))]$.

From the above analysis, there is a curve $\boldsymbol{g}(t) = [u(t), w(t)]$ in the parametric domain of the surface (please see Fig. 2.1b), which has a one-to-one corresponding relationship with the parallel projection curve on the surface. Thus a one-to-one corresponding relationship exists between the 3D curve $\boldsymbol{p}(t)$ and the 2D curve $\boldsymbol{g}(t)$. For the convenience of description, we call $\boldsymbol{g}(t)$ as parametric projection curve in the remainder.

In this chapter, we transform the problem of computing parallel projection curve on parametric surface into the one of tracing parametric projection curve in 2D u–w parametric domain. In Sect. 2.3.1, the first- and second-order differential quantities of the parametric projection curve are analyzed. A second-order iteration method for marching a series of points on the parametric projection curve based on Taylor Approximation is established and two methods for choosing the iterative step size are developed in Sect. 2.3.2. Considering the error caused by truncated higher order terms, a simple and efficient way to decrease the error is put forward in Sect. 2.3.3.

2.3 Parallel Projection Marching

2.3.1 The First- and Second-Order Differential Quantities of the Parametric Projection Curve

In order to calculate the first-order differential quantities, i.e., u_t and w_t, of the parameters u and w with respect to the parameter t of the 3D curve, differentiating Eq. (2.3) with respect to t produces

$$(\boldsymbol{S}_u \times \boldsymbol{V})u_u + (\boldsymbol{S}_w \times \boldsymbol{V})w_t = \boldsymbol{p}_t \times \boldsymbol{V} \tag{2.4}$$

Taking the cross product of the $\boldsymbol{S}_w$ and both sides of the Eq. (2.4), one has

$$([\boldsymbol{S}_u \times \boldsymbol{V}] \cdot \boldsymbol{S}_w)u_t = [\boldsymbol{p}_t \times \boldsymbol{V}] \cdot \boldsymbol{S}_w \tag{2.5}$$

Substituting $[\boldsymbol{S}_u \times \boldsymbol{V}] \cdot \boldsymbol{S}_w = -(\boldsymbol{S}_u \times \boldsymbol{S}_w) \cdot \boldsymbol{V}$ into Eq. (2.5), we have

$$u_t = \frac{-\boldsymbol{L} \cdot \boldsymbol{S}_w}{[\boldsymbol{S}_u \times \boldsymbol{S}_w] \cdot \boldsymbol{V}} \tag{2.6}$$

where $\boldsymbol{L} = \boldsymbol{p}_t \times \boldsymbol{V}$. Similarly, dot-multiplying Eq. (2.4) by $\boldsymbol{S}_u$ gives

$$w_t = \frac{\boldsymbol{L} \cdot \boldsymbol{S}_u}{[\boldsymbol{S}_u \times \boldsymbol{S}_w] \cdot \boldsymbol{V}} \tag{2.7}$$

One can continue to differentiate Eq. (2.3) with respect to the parameter t

$$\left(\frac{d\boldsymbol{S}_u}{dt} \times \boldsymbol{V}\right)u_t + (\boldsymbol{S}_u \times \boldsymbol{V})u_{tt} + \left(\frac{d\boldsymbol{S}_w}{dt} \times \boldsymbol{V}\right)w_t + (\boldsymbol{S}_w \times \boldsymbol{V})w_{tt} = \boldsymbol{p}_{tt} \times \boldsymbol{V} \tag{2.8}$$

Since $\frac{d\boldsymbol{S}_u}{dt} = \boldsymbol{S}_{uu}u_t + \boldsymbol{S}_{uw}w_t$ and $\frac{d\boldsymbol{S}_w}{dt} = \boldsymbol{S}_{uw}u_t + \boldsymbol{S}_{ww}w_t$, Eq. (2.8) can be rewritten as

$$\begin{aligned}&\left[\boldsymbol{S}_{uu}(u_t)^2 + 2\boldsymbol{S}_{uw}u_t w_t + \boldsymbol{S}_{ww}(w_t)^2\right] \times \boldsymbol{V} + (\boldsymbol{S}_u \times \boldsymbol{V})u_{tt} + (\boldsymbol{S}_w \times \boldsymbol{V})w_{tt}\\ &\quad = \boldsymbol{p}_{tt} \times \boldsymbol{V}\end{aligned} \tag{2.9}$$

Dot-multiplying Eq. (2.9) by $\boldsymbol{S}_w$ and $\boldsymbol{S}_u$, respectively, gives

$$\begin{cases} u_{tt} = \dfrac{-\boldsymbol{J} \cdot \boldsymbol{S}_w}{(\boldsymbol{S}_u \times \boldsymbol{S}_w) \cdot \boldsymbol{V}} \\ w_{tt} = \dfrac{\boldsymbol{J} \cdot \boldsymbol{S}_u}{(\boldsymbol{S}_u \times \boldsymbol{S}_w) \cdot \boldsymbol{V}} \end{cases} \tag{2.10}$$

where $\boldsymbol{J} = [\boldsymbol{p}_{tt} - \boldsymbol{S}_{uu}(u_t)^2 - 2\boldsymbol{S}_{uw}u_t w_t - \boldsymbol{S}_{ww}(w_t)^2] \times \boldsymbol{V}$, and u_t, w_t are computed by Eqs. (2.6) and (2.7), respectively.

2.3.2 Parametric Projection Curve Marching

In this section, we will use a second-order Taylor Approximation method to trace the parametric projection curve $\boldsymbol{g}(t)$ to generate a series of points.

Let $\boldsymbol{g}_i = (u_i, w_i)$ and $\boldsymbol{g}_{i+1} = (u_{i+1}, w_{i+1})$ be the current and the next parametric projected point in the u–w domain respectively. Then the position of point $\boldsymbol{g}_{i+1}$ can be iteratively calculated through the current projected point and the first- and second-order differential quantities u_t, w_t, u_{tt}, w_{tt} at the point as follows

$$[u_{i+1}, w_{i+1}] = [u_i, w_i] + [u_t, w_t][k + 1/2]u_{tt}, w_{tt}k^2 \tag{2.11}$$

where k is a constant iteration step size.

If $k = \Delta t$ is constant, one can utilize the Eq. (2.11) to get a sequence of points (u_i, w_i) on the parametric projection curve $\boldsymbol{g}(t)$, where $i = 1, 2, \ldots$. However, the distribution of the projected points may not be well-proportioned. In order to fully use the differential geometric characteristics of $\boldsymbol{g}(t)$ and $\boldsymbol{p}(t)$, we will give two ways to choose the iteration step size.

One can consider marching along the parametric projection curve based on a constant step size v_p along the 3D parametric curve. Suppose r is the arc-length parameter of the 3D parametric curve, one has

$$dr = \|x_t \ \ y_t \ \ z_t\| dt = \|\boldsymbol{p}_t\| dt \tag{2.12}$$

Marching along the 3D parametric curve uses

$$t_{i+1} = t_i + k \ \text{ with } \ k = \frac{1}{\|\boldsymbol{p}_t\|} v_p \tag{2.13}$$

in addition to using Eqs. (2.6), (2.7), (2.10) and (2.11).

For marching along the parametric projection curve based on constant step size v_q along the projection curve, one can write

$$ds = \|\boldsymbol{q}_t\| dt \tag{2.14}$$

where s is the arc-length parameter of the parallel projection curve and $\boldsymbol{q}_t = \boldsymbol{S}_u u_t + \boldsymbol{S}_w w_t$. Marching along the parallel projection curve uses

$$t_{i+1} = t_i + k \ \text{ with } \ k = \frac{1}{\|\boldsymbol{S}_u u_t + \boldsymbol{S}_w w_t\|} v_q \tag{2.15}$$

in addition to using Eqs. (2.6), (2.7), (2.10) and (2.11).

2.3.3 Error Adjustment

Considering the truncated higher order terms, the position of point computed through Eq. (2.11) may depart from the parametric projection curve $\boldsymbol{g}(t)$ in u–w domain. Hence, a first-order adjustment technique is presented to reduce the error.

Suppose $\hat{\boldsymbol{g}}_{i+1} = (\hat{u}_{i+1}, \hat{w}_{i+1})$ be the parametric projection point computed by Eq. (2.11)

$$[\hat{u}_{i+1}, \hat{w}_{i+1}] = [u_i, w_i] + [u_t, w_t]k + 1/2[u_{tt}, w_{tt}]k^2$$

As the above iteration formula omits higher order terms, the image point of $\hat{\boldsymbol{g}}_{i+1}$ may deviate from the parallel projection curve on the parametric surface $\boldsymbol{S}$, i.e.,

$$\left(\hat{\boldsymbol{q}}_{i+1} - \boldsymbol{p}_{i+1}\right) \times \boldsymbol{V} = \boldsymbol{\varepsilon} \neq \boldsymbol{0} \tag{2.16}$$

where $\hat{\boldsymbol{q}}_{i+1} = \boldsymbol{S}(\hat{u}_{i+1}, \hat{w}_{i+1}), \boldsymbol{p}_{i+1} = \boldsymbol{p}(t_{i+1})$.

Let $[\Delta u, \Delta w]$ be the correction vector, the corrected parametric projection point and parallel projection point be $\boldsymbol{g}_{i+1} = (u_{i+1}, w_{i+1}) = (\hat{u}_{i+1} + \Delta u, \hat{w}_{i+1} + \Delta w)$ and $\boldsymbol{q}_{i+1} = \boldsymbol{S}(u_{i+1}, w_{i+1})$, respectively. According to the definition of parallel projection, one has

$$\left(\boldsymbol{q}_{i+1} - \boldsymbol{p}_{i+1}\right) \times \boldsymbol{V} = \boldsymbol{0} \tag{2.17}$$

In order to calculate the error adjustment vector $[\Delta u, \Delta w]$, we use a first-order Taylor formula to expand the point $\boldsymbol{q}_{i+1}$ at $\hat{\boldsymbol{g}}_{i+1}$, i.e., $\boldsymbol{q}_{i+1} = \boldsymbol{S}(\hat{u}_{i+1}, \hat{w}_{i+1}) + \boldsymbol{S}_u \Delta u + \boldsymbol{S}_w \Delta w$. Substituting $\boldsymbol{q}_{i+1}$ into Eq. (2.17) produces

$$(\boldsymbol{S}_u \Delta u + \boldsymbol{S}_w \Delta w) \times \boldsymbol{V} = -\boldsymbol{\varepsilon} \tag{2.18}$$

Dot-multiplying Eq. (2.18) by $\boldsymbol{S}_w$ and $\boldsymbol{S}_u$ respectively, gives

$$\begin{cases} \Delta u = \dfrac{\boldsymbol{\varepsilon} \cdot \boldsymbol{S}_w}{(\boldsymbol{S}_u \times \boldsymbol{S}_w) \cdot \boldsymbol{V}} \\ \Delta w = \dfrac{-\boldsymbol{\varepsilon} \cdot \boldsymbol{S}_u}{(\boldsymbol{S}_u \times \boldsymbol{S}_w) \cdot \boldsymbol{V}} \end{cases} \tag{2.19}$$

Note that all variables in Eq. (2.19) are evaluated at point $\hat{\boldsymbol{g}}_{i+1}$ and t_{i+1}. When the deviation $\|\boldsymbol{\varepsilon}\|$ calculated by Eq. (2.16) is greater than a given threshold α, i.e., $\|\boldsymbol{\varepsilon}\| > \alpha$, one can adopt the error adjustment vector calculated by Eq. (2.19) to adjust the parametric projection point $\hat{\boldsymbol{g}}_{i+1}$. In the following, the quantities $\|\boldsymbol{\varepsilon}\|$ and α are called as projection error and projection error threshold, respectively.

Suppose $\boldsymbol{g}_i$ and $\boldsymbol{g}_{i+1}$ are two adjacent points after error correction lying on the parametric projection curve $\boldsymbol{g}(t) = [u(t)\ w(t)]$. One can use a line segment to connect the two points $\boldsymbol{g}_i$ and $\boldsymbol{g}_{i+1}$ as follows:

$$\begin{cases} u = t \ \ (u_i \leq t \leq u_{i+1}) \\ w = w_i + l(t - u_i) \end{cases} \tag{2.20}$$

where $l = (w_{i+1} - w_i)/(u_{i+1} - u_i)$. A G^0 continuous parallel projection curve is acquired by substituting Eq. (2.20) into Eq. (2.2). One can also use the G^1 approximation method [10] to fit the sequence of points $S(u_i,w_i)$, $i = 1, 2, \ldots$, to get a parallel projection curve with G^1 or higher continuity.

2.4 Demonstrations

Our parallel projection algorithm framework is outlined in Sect. 2.4.1 and a number of examples are given to demonstrate the validity of our proposed method in Sect. 2.4.2. The examples make use of cubic NURBS curves and bicubic NURBS surfaces. In this context, these are viewed as parametric curves and surfaces, respectively.

2.4.1 Outline of Our Parallel Projection Algorithm

Given the control points and knot vectors for a NURBS curve and a NURBS surface, we need to compute a series of parallel projection points on the surface.

Algorithm 1. Parallel projection of parametric curves onto parametric surfaces

Input: Parametric curve $\boldsymbol{p}(t)$, parametric surface $\boldsymbol{S}(u, w)$, the initial projection point $\boldsymbol{p}_0=\boldsymbol{p}(t_0)$ and its parallel projection point $\boldsymbol{q}_0=\boldsymbol{S}(u_0, w_0)$ on surface $\boldsymbol{S}$, projection error threshold α

Output: A series of parallel projection points $\boldsymbol{q}_i=\boldsymbol{S}(u_i, w_i)$ $(i=1,2,..)$ on surface $\boldsymbol{S}$.

Algorithm Description:

1. $t_i= t_0$, $(u_{i+1}, w_{i+1})= (u_0, w_0)$, $\boldsymbol{p}_i = \boldsymbol{p}_0$, $\boldsymbol{q}_i = \boldsymbol{q}_0$;
2. **do**{
3. Compute step size k , t_{i+1}: $t_{i+1}=t_i+k$ and $\boldsymbol{p}(t_{i+1})$;
4. Compute (u_{i+1}, w_{i+1}) and point $\boldsymbol{q}_{i+1}=\boldsymbol{S}(u_{i+1}, w_{i+1})$;
5. Compute the projection error $\boldsymbol{\varepsilon}$;
6. **while**($\|\boldsymbol{\varepsilon}\| > \alpha$) **do{**
7. Compute the adjustment vector $[\Delta u, \Delta w]$ with equation (19);
8. Compute $(\hat{u}_{i+1}, \hat{w}_{i+1}) = (u_{i+1}, w_{i+1}) + (\Delta u, \Delta w)$ and $\hat{\boldsymbol{q}}_{i+1} = \boldsymbol{S}(\hat{u}_{i+1}, \hat{w}_{i+1})$;
9. Renew (u_{i+1}, w_{i+1}): $(u_{i+1}, w_{i+1})= (\hat{u}_{i+1}, \hat{w}_{i+1})$ and $\boldsymbol{q}_{i+1}$: $\boldsymbol{q}_{i+1} = \hat{\boldsymbol{q}}_{i+1}$;
10. Recalculate the projection error $\boldsymbol{\varepsilon}$ with new parameters (u_{i+1}, w_{i+1}) and $\boldsymbol{q}_{i+1}$;
11. }**end while**
12. Renew t_i: $t_i= t_{i+1}$, (u_i, w_i): $(u_i, w_i)=(u_{i+1}, w_{i+1})$ and $\boldsymbol{q}_i$: $\boldsymbol{q}_i = \boldsymbol{q}_{i+1}$;
13. Output the parallel projection point $\boldsymbol{q}_i$;
14. }**while**($\boldsymbol{p}(t_i)$ is at the end of the 3D parametric curve $\boldsymbol{p}(t)$)

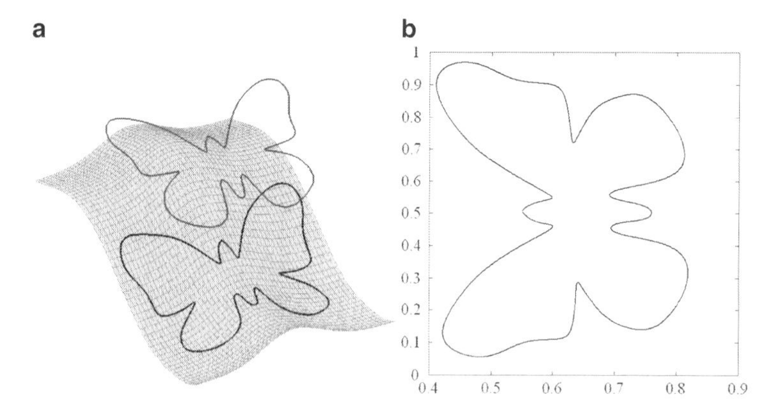

Fig. 2.2 Parallel projection example 1. (**a**) A butterfly curve onto an undee surface, and (**b**) parametric projection curve in u–w domain

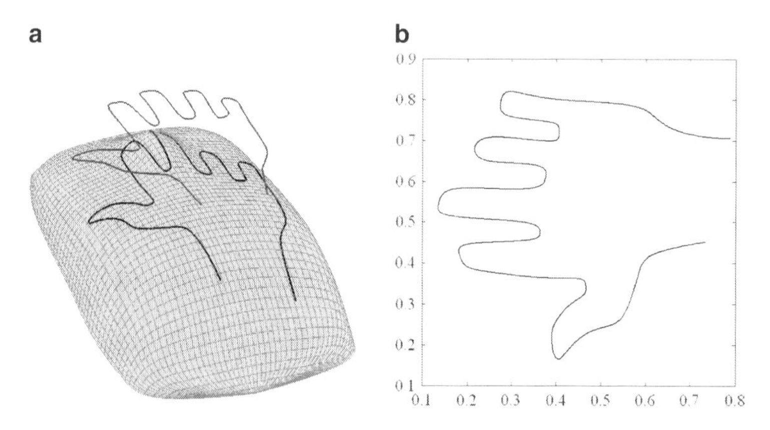

Fig. 2.3 Parallel projection example 2. (**a**) A hand-form curve onto a mouse surface, and (**b**) parametric projection curve in u–w domain

2.4.2 Examples

In this section we firstly give four examples (please see Figs. 2.2, 2.3, 2.4, and 2.5) obtained by the **Algorithm 1**. Further detailed comparisons of our scheme with Wang's first-order differential equation method [6] are carried out to testify the computational efficiency and accuracy of our methods (please see Tables 2.1, 2.2, 2.3, and 2.4). All the examples are implemented with Matlab, and run on a PC with 2.80 GHz CPU and 1 GB memory.

Figure 2.2a shows an example of projecting a closed butterfly curve onto an undee surface. The corresponding parametric projection curve in the u–w parametric domain is shown in Fig. 2.2b. The second example shows the projection of a small hand-form curve onto a mouse surface. The hand-form curve and its parallel

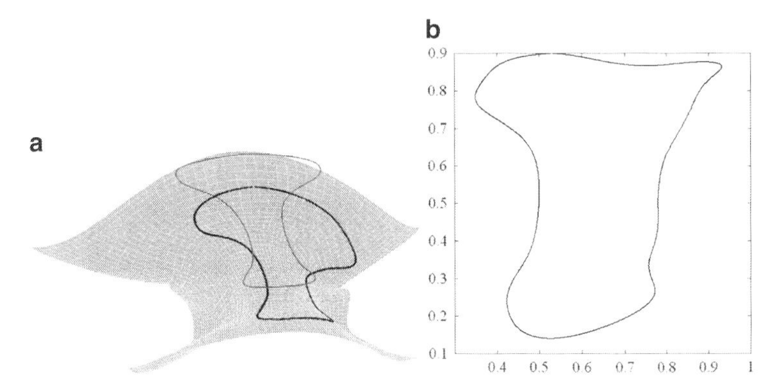

Fig. 2.4 Parallel projection example 3. (**a**) A closed NURBS curve onto a ridge surface, and (**b**) parametric projection curve in u–w domain

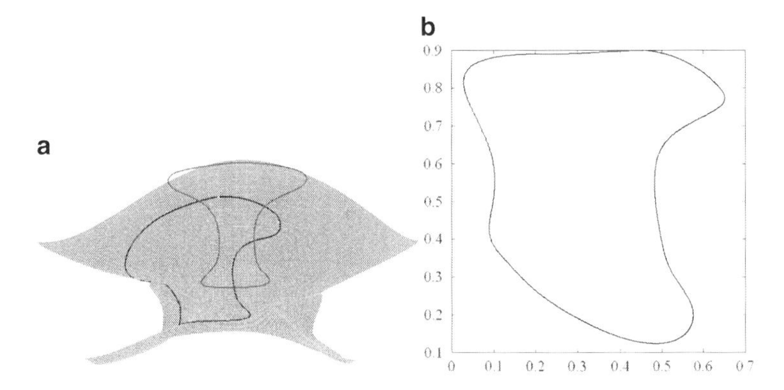

Fig. 2.5 Parallel projection example 4. (**a**) A closed NURBS curve onto a ridge surface, and (**b**) parametric projection curve in u–w domain

projection curve on the NURBS surface is show in Fig. 2.3a. Figure 2.3b denotes the parametric projection curve in the u–w parametric domain.

Figures 2.3a and 2.4a are examples of projecting a same closed NURBS curve onto a same complex ridge NURBS surface along two different directions. The corresponding parametric projection curves in the u–w parametric domain are shown in Figs. 2.3b and 2.4b.

For the parallel projection examples 1 and 2, we compared our method with Wang's first-order differential equation method [6] focusing on computational accuracy and efficiency (please see Tables 2.1, 2.2, 2.3, and 2.4). For the sake of fairness, the step size is specified by constant parametric increment Δt and the tolerance α of the parallel projection error ε is set as constant (α is set as 1.0e-009 for all the comparisons). Moreover, the initial values for the two methods are the same. As the parallel projection error of Wang's method may overrun the error threshold α, it is solved with the classical fourth-order Runge–Kutta method and our error adjustment technique deduced in Sect. 2.3.3.

Table 2.1 Comparison of accuracy: projection of a butterfly curve onto an undee surface

Step size, Δt	APE of Wang's method (no error adjustment)	APE of our method (no error adjustment)	(error adjustment)
0.05	2.2598e-009	1.3133e-004	1.8662e-012
0.04	4.7598e-010	8.3957e-005	4.8905e-013
0.03	2.0282e-007	4.9389e-005	8.6769e-014
0.02	6.7689e-011	2.0942e-005	7.6519e-015

Table 2.2 Comparison of accuracy: projection of a hand-form curve onto a mouse surface

Step size, Δt	APE of Wang's method (no error adjustment)	APE of our method (no error adjustment)	(error adjustment)
0.05	2.8146e-004	4.1585e-002	7.7697e-011
0.04	1.2407e-004	2.4439e-002	2.6262e-011
0.03	4.1559e-005	1.2371e-002	5.4997e-012
0.02	8.4125e-006	4.8313e-003	1.4454e-012
0.01	5.0854e-007	1.0423e-003	4.5111e-013

Table 2.3 Comparison of efficiency: projection of a butterfly curve onto an undee surface

	Wang's method (error adjustment)		Our method (error adjustment)	
Step size, Δt	APE	CPU time (s)	APE	CPU time (s)
0.05	3.0445e-010	30.1094	1.8662e-012	27.2031
0.04	2.1330e-010	37.2500	4.8905e-013	33.0781
0.02	6.7689e-011	73.0781	7.6519e-015	66.1406

For comparison of accuracy of the two methods, the first-order method is realized just by using the classical four-order Runge–Kutta method, while our method is implemented in two ways that one is carried out with error adjustment and the other is without correction. From the results in Tables 2.1 and 2.2 (Note that the symbol "APE" in Tables 2.1, 2.2, 2.3, and 2.4 denotes average projection error), we can see that our method is much lower than Wang's in precision when implemented without error correction. After the error adjustment, the accuracy of our method is much finer than Wang's. From the results of large numbers of experiments, we found that the CPU time of our method with error adjustment is less than that of Wang's as projecting the same number of points onto a surface, while the precision of ours still has predominant performance.

Tables 2.3 and 2.4 are the comparison results of accuracy of our method with Wang's, which are gained under the conditions of same initial values, iteration step size, and error threshold. Both of the two methods adopted the error correction technique given in Sect. 2.3.3. From the column of CPU time in Tables 2.3 and 2.4, we can see that the efficiency of our method is about 1.1 times of Wang's. On the

Table 2.4 Comparison of efficiency: projection of a hand-form curve onto a mouse surface

	Wang's method (error adjustment)		Our method (error adjustment)	
Step size, Δt	APE	CPU time (s)	APE	CPU time (s)
0.05	4.5873e-010	30.5469	7.7697e-011	26.9219
0.03	1.7461e-010	49.7031	5.4997e-012	43.7813
0.01	3.9048e-010	145.3438	4.5111e-013	131.0313

other hand, from the comparisons of Tables 2.1, 2.2 and Tables 2.3, 2.4, one can also find that our first-order error adjustment approach could efficiently improve the computational precision of the Wang's method.

2.5 Conclusion

A second-order algorithm based on Taylor Approximation is proposed to compute the parallel projection of a parametric curve onto a parametric surface. Several examples are presented to demonstrate the effectiveness of the presented approach. Experimental results and comparisons indicate that the computational efficiency of our method is about 1.1 times of that of the first-order differential equation method and our method has superior performance in the computational accuracy.

References

1. Sederberg, T. W., Finnigan, G. T., Li, X., Lin, H. W., & Ipson, H. (2008). Watertight trimmed NURBS. *ACM Transaction on Graphics, 27*(3), 79:1–79:8.
2. Chuang, J. H., Lin, C. H., & Hwang, W. C. (1995). Variable-radius blending of parametric surfaces. *The Visual Computer, 11*(10), 513–525.
3. Tam, H.-Y., Law, H. W., & Xu, H.-Y. (2004). A geometric approach to the offsetting of profiles on the three-dimensional surfaces. *Computer-Aided Design, 36*(10), 887–902.
4. Xu, H.-Y., Tam, H.-Y., Fang, X., & Hu, L. (2009). Quart-parametric interpolations for intersecting paths. *Computer-Aided Design, 41*(6), 432–440.
5. Pegna, J., & Wolter, F. E. (1996). Surface curve design by orthogonal projection of space curves onto free-form surface. *Journal of Mechanical Design, 118*(1), 45–52.
6. Wang, X., Wei, W., & Zhang, W.-Z. (2010). Projecting curves onto free-form surfaces. *International Journal of Computer Applications in Technology, 37*(2), 153–159.
7. Wang, X., An, L. L., Zhou L. S., & Zhang, L. (2010). Constructing G^2 continuous curve on freeform surface with normal projection. *Chinese Journal of Aeronautics, 23*(1), 137–144.
8. Xu, H.-Y., Fang, X., Hu, L., Xiao F., & Li, D. (2010). An algorithm for curve orthogonal projections onto implicit surfaces. *Journal of Computer-Aided Design and Computer Graphics, 22*(12), 2103–2110.
9. Xu, H.-Y., Fang, X., Tam, H.-Y., Wu, X., & Hu, L. (2012). A second-order algorithm for curve orthogonal projection onto parametric surface. *International Journal of Computer Mathematics, 89*(1), 98–111.
10. Yang, Y.-J., Zeng, W., Yang, C.-L., Meng, X.-X., Yong J.-H., & Deng, B. (2012). G^1 continuous approximate curves on NURBS surfaces. *Computer-Aided Design, 44*(9), 824–834.

Chapter 3
Computation Method of Processing Time Based on BP Neural Network and Genetic Algorithm

Danchen Zhou and Chao Guo

Abstract Looking-up standard processing time table is a commonly used and important determination method of processing time. However, the large error in nonstandard nodes brings adverse effect on its accuracy. In view of the problem, a computation method of processing time based on back propagation neural network (BPNN) and genetic algorithm (GA) is proposed. Several key technologies of BPNN based on Matlab, including computation of the number of neurons in hidden layer, determination of training algorithm, and affecting factors of generalization ability, are researched in depth. In order to improve the training efficiency of BPNN, GA is used to optimize its connection weights and thresholds. The encoding method, selection operation, crossover, and mutation operation of GA are discussed in detail. The higher computation precision and faster operation speed of the proposed method is demonstrated through application cases.

3.1 Introduction

Processing time is the time consumption of per unit product or per unit work under the condition of certain production technology and organization manner. The main methods for determining processing time include experience estimation, probability estimation, statistic analysis, analogism comparison, and standard looking-up. On the basis of systematical and whole-set processing time standards, the method of standard looking-up obtains processing time of job, operation, or step in terms of decomposition of work elements and confirmation of one-to-one corresponding factor. As the criterion for determining processing time, processing time standards

D. Zhou (✉) • C. Guo
Institute of Machinery Manufacturing Technology, China Academy of Engineering Physics, Mianyang 621900, Sichuan, China
e-mail: zdc69@163.com

W.E. Wong and T. Zhu (eds.), *Computer Engineering and Networking*, Lecture Notes in Electrical Engineering 277, DOI 10.1007/978-3-319-01766-2_3,

Table 3.1 An example of standard processing time table (min)

		Length (L/mm)								
Width (B/mm)	Depth (t/mm)	20	30	43	64	94	138	204	300	426
14	3	2.0	2.1	2.3	2.5	2.9	3.4	4.2	5.3	6.8
22	5	2.1	2.2	2.4	2.7	3.2	3.8	4.8	6.2	8.1
30	7	2.2	2.4	2.7	3.1	3.7	4.5	5.8	7.6	10.0
38	9	2.4	2.7	3.0	3.5	4.3	4.4	7.1	9.5	12.7
50	11	2.8	3.2	3.7	4.5	5.6	7.3	9.8	13.4	18.2
Working condition		Cutting tool: high-speed steel end mill								

generally are established based on multiple measurements by taking processing factors (surface roughness, cutting tool, length, width, depth, height, diameter, radius, volume, area, modulus, etc.) into full consideration under the premise of the same process conditions, such as machine, working environment, and personnel skill. Processing time standard is usually represented into the tabular form, which is called standard processing time table (SPTT). An example of SPTT with three factors (length, width, and depth) is shown in Table 3.1.

As for factors in standard nodes, processing time value can be obtained directly from SPTT. However, as for factors in nonstandard nodes, processing time value can only be obtained according to the principle of taking value of the closest to a standard node. Obviously, it is difficult to satisfy customers' requirements due to a large error. Aiming at the problem, for SPTT with only one factor, data fitting method can be used to transform SPTT into mathematical model. Unfortunately, the great majority of SPTT have more than one factor, each of whose influencing regularity on processing time is very difficult to be mastered at present. As a result, it is almost impossible to implement data fitting under the condition of indefinite curve equation.

Nowadays, artificial neural network (ANN) has been widely applied in function approach, data fitting, and structural optimization. Compared with general data fitting method, it has greater advantage and higher precision in training and computing data with no distinct regularity and multiple parameters. As a kind of multilayer feed-forward neural network based on error back propagation algorithm, back propagation neural network (BPNN) is the most popular neural network learning algorithm [1–3]. However, BPNN has some disadvantages such as over long training time and convergence difficulties, thanks to easily getting stuck in local minimum caused by randomicity of connection weights and thresholds [4, 5]. Thus, this paper attempts to find an accurate and fast computation method of processing time based on the combination of BPNN and genetic algorithm (GA) according to the characteristics of SPTT.

3.2 GA–BPNN Algorithm

GA–BPNN algorithm is the combination of BPNN and GA. Based on the evolutionary ideas of natural selection and genetics, GA is a stochastic parallel search algorithm with high robust performance and strong global search ability [6–8]. In order to improve training efficiency and convergence rate of BPNN by overcoming the randomicity of its initialization, GA is used to optimize the connection weights and thresholds of BPNN.

The main ideas of GA–BPNN algorithm are as follows. Firstly, population size and chromosome length of GA are determined in accordance with the structure of BPNN. Secondly, chromosome with the maximal fitness value is decoded as initial connection weights and thresholds of BPNN by means of the selection, crossover, and mutation operation of GA. Eventually, the satisfactory model of BPNN is determined through repeated training, by which the processing time can be accurately computed.

According to Kolmogorov theorem, a three-layer feed-forward neural network can exactly approach sample sets with regularity [5]. Accordingly, BPNN involved in this paper belongs to three-layer neural network, including input layer, hidden layer, and output layer, and research works are based on neural network toolbox and genetic algorithm toolbox in Matlab. In addition, all sample sets (SPTT) used in this paper are from processing time of petroleum machinery (SY/T 5179–93) in Chinese oil and natural gas industry standards.

3.3 BPNN-Based Computation Method of Processing Time

3.3.1 Computation of the Number of Neurons in Hidden Layer

For three-layer neural network, the number of neurons in input layer and output layer neurons is determined by sample sets. As for SPTT, the former is the number of factors in a table, and the latter is always one (processing time value). Hence, as long as the number of neurons in hidden layer has been decided, the structure of BPNN will be fixed. However, there is no theoretical guideline for its determination at present [8]. Consequently, some empirical formulas are used to determine its approximate range at first, and then, the model of BPNN meeting the error requirement is found out by gradual adjustment of the number of neurons in hidden layer and repeated training. Five tested empirical formulas for the number of neurons in hidden layer are listed as below.

$$h = \sqrt{0.43mn + 2.54m + 0.77n + 0.35 + 0.12n^2} + 0.51 \tag{3.1}$$

$$h = 2m + 1 \tag{3.2}$$

$$h = \sqrt{m + n} + a \tag{3.3}$$

$$h = \frac{m + n + 0.5n(m^2 + m)}{2} + \sqrt{p} + b \tag{3.4}$$

$$h = round\left(\frac{n(p-1)}{m+n+1}\right) \tag{3.5}$$

where h is the number of neurons in hidden layer, m is the number of neurons in input layer , n is the number of neurons in output layer (here, $n = 1$), a is the integer between 0 and 10, p is the quantity of samples, b is 1 or -1, and *round( )* is integral function.

The experimental results show that, for five empirical formulas, Eq. (3.3) has the highest predicting accuracy for SPTT. However, because value obtained by Eq. (3.3) is only an estimation range, it will spend more training times to determine the most appropriate structure of BPNN. Therefore, Eq. (3.2) is selected as preferred formula because its accuracy is secondary to Eq. (3.3) and the value is definite. At the same time, if the structure of BPNN determined by Eq. (3.2) does not converge, or has poor generalization ability in the process of training, Eq. (3.3) will be selected.

3.3.2 Determination of Training Algorithm

Multiple training algorithms are provided by neural network toolbox in Matlab [9]. For selecting the appropriate training algorithms, an experiment scheme for evaluating training performance of every algorithm oriented to SPTT is designed. It is divided into two parts. The first part is to examine the training epochs (maximum is 5,000) and time of every algorithm under the condition of reaching specified convergence error (0.00001). The second part is to examine minimal convergence error and time of every algorithm under the condition of setting fixed training epochs (10,000). The comparisons of two parts are shown in Tables 3.2 and 3.3, respectively.

It can be seen that trainlm algorithm is the most excellent in training speed and convergence error among these algorithms. However, it is also discovered during the experiment that, when there are a large number of training samples, trainlm algorithm will take a large amount of resources of computer. By comprehensive consideration, trainlm, traingdx, and trainrp algorithms are selected as training algorithm of BPNN.

Table 3.2 Comparison of training performance for different algorithms through first part of experiment

Algorithm name	Training epochs	Whether reaching specified convergence error	Time (s)
Traingd	5,000	No	29
Traingdm	5,000	No	31
Traingda	5,000	No	31.2
Traingdx	3,776	Yes	23.6
Trainrp	549	Yes	3.9
Trainlm	8	Yes	0.9
Trainbr	64	Yes	1.6

Table 3.3 Comparison of training performance for different algorithms through the second part of experiment

Algorithm name	Training epochs	Minimal convergence error	Time (s)
Traingd	10,000	3.5E-3	56.5
Traingdm	10,000	4.6E-3	59.1
Traingda	10,000	2.7E-4	65.4
Traingdx	10,000	9.1E-5	59.6
Trainrp	10,000	1.6E-5	60.4
Trainlm	1,019	3.2E-28	10.9
Trainbr	6,965	5.6E-20	96

3.3.3 *Affecting Factors of Generalization Ability*

3.3.3.1 Quality of Samples

Quality of samples denotes whether data in sample sets are accurate and their distributions are uniform. In other words, bad quality of samples will directly lead to poor generalization ability. Because all sample sets used in this paper are from SPTT, in which data have been verified by practice and have better regularity, it can be believed that quality of samples has less effect on generalization ability.

3.3.3.2 Structure of BPNN

The structure of BPNN has greater impact on generalization ability, which is mainly reflected in the number of neurons in hidden layer. Namely, the appropriate number of neurons in hidden layer will make BPNN reach the specified convergence error with the least training epochs and meanwhile easily meet generalization error requirement. Through extensive experiments for SPTT, relations of the number of neurons in hidden layer, convergence error and generalization error are studied. An experimental result is shown in Fig. 3.1.

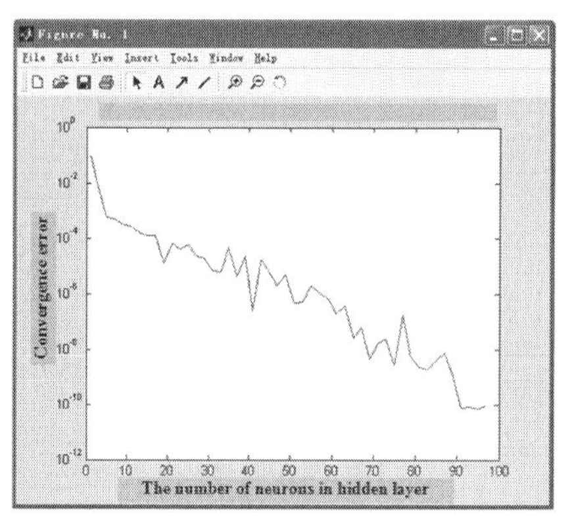

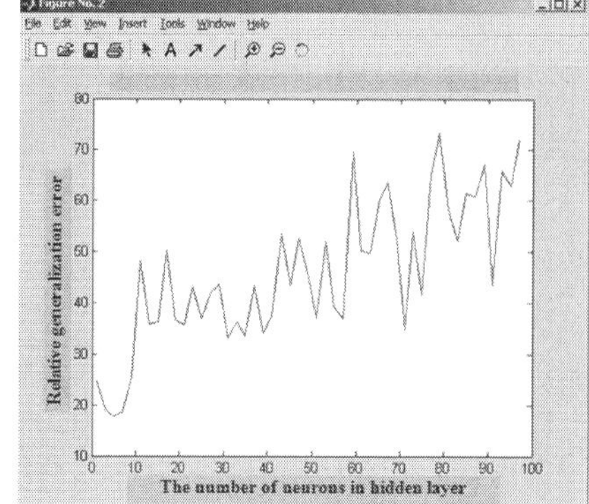

Fig. 3.1 Experimental result for structure of BPNN

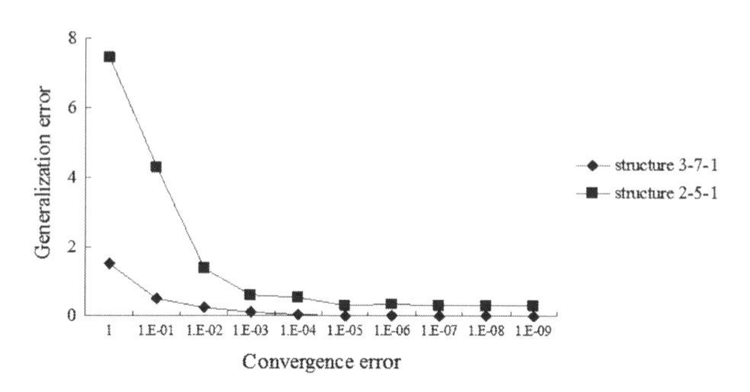

Fig. 3.2 Relation between generalization error and convergence error

It can be seen from Fig. 3.1 that with the increase of number of neurons in hidden layer, the tendency is gradual reduction of convergence error and gradual increase of generalization error, but in the beginning, both of them reduce simultaneously.

3.3.3.3 Convergence Error

Experimental result of relation between generalization error and convergence error for two different structures of BPNN is shown in Fig. 3.2.

It can be seen from Fig. 3.2 that for a fixed structure of BPNN, generalization error also reduces correspondingly with the reduction of convergence error, which is maybe because the hidden regularity in training samples is learned in depth step-by-step. Due to higher correctness of the regularity guaranteed by higher reliability of data in SPTT, predicting results are more and more accurate.

3.4 GA-Based Optimization for BPNN

3.4.1 Chromosome Encoding Method

For GA-based optimization for BPNN, binary encoding, real encoding, and Gray code encoding are dominant chromosome encoding methods. To test the performance of three encoding methods, a comparative experiment is conducted. Therefore, three SPTTs with 216, 60, and 72 data are selected as sample sets. The structures of BPNN are 4-9-1, 3-7-1, and 2-5-1, respectively. In the experiment, their fitness function is defined as reciprocal of the mean squared error (MSE), which is expressed in Eq. (3.6).

$$fitness = \frac{1}{1 + \sqrt{\frac{1}{N}\sum_{i=1}^{N}\left(y_i - d_i\right)^2}}, \tag{3.6}$$

where N represents the number of training samples and y_i and d_i represent the computing and expected processing time value, respectively. The time consumption and maximum of fitness values of three encoding methods are listed in Table 3.4.

It can be found from experimental result that three encoding methods have the approximate maximum of fitness value, but real encoding method consumes the least running time of program. Therefore, the real encoding is chosen as encoding method.

3.4.2 Selection Operation

An improved ranking selection method is adopted in this paper. The steps are as below.

Step 1: Rank all chromosomes of initial parent population in descending order according to their own fitness values.

Table 3.4 Performance comparison among three encoding methods

Comparison item	Sample set	Binary encoding	Real encoding	Gray code encoding
Time consumption (s)	SPTT1	809.940	718.000	3,562.4
	SPTT2	359.391	318.015	1,725.6
	SPTT3	329.830	299.580	1,124.7
Maximum of fitness values	SPTT1	0.998 65	0.998 75	0.994 92
	SPTT2	0.998 54	0.999 66	0.866 17
	SPTT3	0.999 74	0.999 50	0.942 84

Step 2: Choose the top N chromosomes as initial child population for crossover and mutation operation.
Step 3: Generate new child population through carrying out crossover and mutation operation for initial child population.
Step 4: Generate new parent population through mixing new child population into initial parent population.
Step 5: Rank all chromosomes of new parent population in descending order according to their own fitness values.
Step 6: Choose the top N chromosomes as next-generation population.

3.4.3 Crossover and Mutation Operation

In this paper, single-point crossover method is adopted due to its low possibility of destroying structure of chromosome in population. At the same time, uniform mutation method is adopted because, on the one hand, it is suitable for real encoding, and on the other hand, it accords with the actual states of nature for the reason that the number in mutation range is randomly generated with the same possibility.

3.5 Application Cases

To verify the validity of the proposed method, two standard processing time tables with three factors, which have 65 and 60 data, respectively, are selected as sample sets. The structure of BPNN is 3-10-1 and 3-7-1, respectively. Parameters setting of BPNN and GA are shown in Table 3.5. Comparison between computing values and actual values of testing samples is shown in Table 3.6. The convergence curves of training error for two SPTTs are shown in Fig. 3.3.

It can be calculated from Table 3.6 that average relative errors are 0.376 % and 0.620 % and the MSE are 0.016 and 0.048, respectively. The result proves the

Table 3.5 Parameters setting of BPNN and GA

BPNN		GA	
Parameter	Value	Parameter	Value
Required generalization error	0.05	Population size	50
Training epochs	2,000	Maximum of generation	500
Specified convergence error	0.0001	Generation gap	0.7
Learning rate	0.7	Crossover probability	0.75
Momentum	0.5	Mutation probability	0.1
Training algorithm	Traingdx	Maximum and minimum of encoding	(2, −2)

Table 3.6 Comparison between computing values and actual values of testing samples

	SPTT1			SPTT2		
Sample	Actual value	Computing value	Relative error	Actual value	Computing value	Relative error
1	3.34	3.3313	0.26	3.5	3.4945	0.16
2	3.39	3.3758	0.42	3.7	3.6882	0.32
3	3.43	3.4229	0.21	3.8	3.8675	1.78
4	3.44	3.4500	0.29	3.9	3.8520	1.23
5	3.45	3.4358	0.41	4.0	3.9800	0.50
6	3.48	3.4690	0.32	4.1	4.1051	0.12
7	3.52	3.5154	0.13	4.2	4.1863	0.33
8	3.54	3.5570	0.48	4.4	4.3905	0.22
9	3.63	3.6439	0.38	4.5	4.4620	0.84
10	3.64	3.6552	0.42	4.7	4.7640	0.98
11	3.67	3.6943	0.66	4.8	4.8068	0.14
12	3.74	3.7472	0.19	5.2	5.2194	0.37
13	3.83	3.8357	0.15	5.3	5.2637	0.68
14	3.93	3.9230	0.18	5.8	5.7651	0.6
15	4.03	4.0112	0.47	6.5	6.5149	0.23
16	4.13	4.1023	0.67	6.8	6.6884	1.64
17	4.28	4.2483	0.74	7.4	7.3693	0.42
18	4.38	4.3542	0.59	7.5	7.5163	0.22

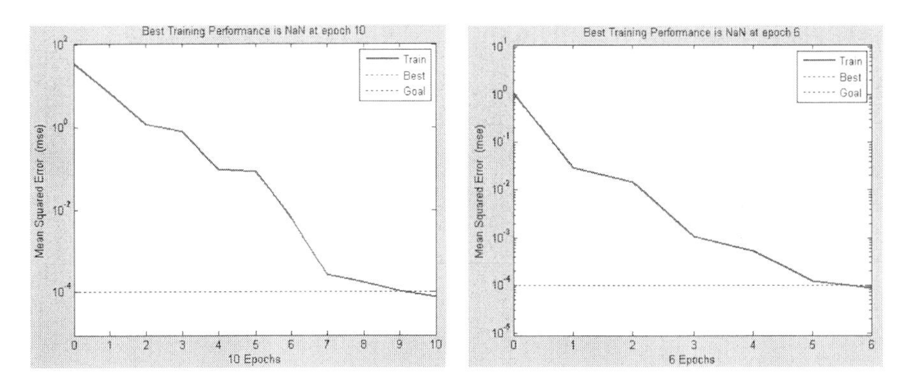

Fig. 3.3 Convergence curves of training error for SPTT1 and SPTT2

models of GA–BPNN have better generalization ability. Meanwhile, Fig. 3.3 shows that the GA–BPNN models can converge to the specified accuracy with much fewer training epochs. Therefore, application cases demonstrate that the computation method proposed in this paper has higher computation precision and faster operation speed and is capable of accurately determining processing time.

3.6 Conclusion

In view of the characteristics of SPTT, a computation method of processing time based on BPNN and GA is proposed, and depending on the optimization of GA for connection weights and thresholds of BPNN, training efficiency and convergence rate are obviously improved because the probability of getting stuck in local minimum is greatly decreased. Application cases have proved its higher computation precision and faster operation speed by means of in-depth studies on some key technologies.

References

1. Ding, S., Su, C., Yu, J. (2011). An optimizing BP neural network algorithm based on genetic algorithm. *Artificial Intelligence Review, 36*(2), 153–162.
2. Wang, H., & Liu, M. (2012). Design of robotic visual servo control based on neural network and genetic algorithm. *International Journal of Automation and Computing, 9*(1), 24–29.
3. Wang, Z., Zhao, Z., Wang, L., Wang, K. (2011). Study on the law of short fatigue crack using genetic algorithm-BP neural networks. *Lecture Notes in Computer Science, 6677*(1), 586–593.
4. Long, J., Lan, F., Chen, J., Yu, P. (2009). Mechanical properties prediction of the mechanical clinching joints based on genetic algorithm and BP neural network. *Chinese Journal of Mechanical Engineering, 22*(1), 36–40.
5. Zhang, J., Xu, C., Yi, M., Fang, B. (2012). Design of nano-micro-composite ceramic tool and die material with back propagation neural network and genetic algorithm. *Journal of Materials Engineering and Performance, 21*(4), 463–470.
6. Lin, J. (2012). A systematic estimation model for fraction nonconforming of a wafer in semiconductor manufacturing research. *Applied Soft Computing Journal, 12*(6), 1733–1740.
7. Zemin, F., & Mo, J. (2011). Springback prediction of high-strength sheet metal under air bending forming and tool design based on GA-BPNN. *The International Journal of Advanced Manufacturing Technology, 53*(5–8), 473–483.
8. Xu, M., Jin, B., Yu, Y., Shen, H., Li, W. (2010). Using artificial neural networks for energy regulation based variable-speed electrohydraulic drive. *Chinese Journal of Mechanical Engineering, 23*(3), 327–335.
9. Zain, A. M., Haron, H., Sharif, S. (2010). Prediction of surface roughness in the end milling machining using Artificial Neural Network. *Expert Systems with Applications, 37*(2), 1755–1768.

Chapter 4
Integral Sliding Mode Controller for an Uncertain Network Control System with Delay

Zhenbin Gao

Abstract Integral sliding mode control is formulated with respect to an uncertain continuous network control system with the state delay. The parameter uncertainty is assumed to be the norm-bounded and satisfy the sliding mode matching requirements. The switching function is presented which include the integral term of the state feedback gain and the sliding mode compensator. The sliding mode controller is designed which is divided into the equivalent controller and switching controller, so the reachability of the sliding surface is ensured. A sufficient condition is derived by means of linear matrix inequality such that the asymptotical stability of the closed-loop system is guaranteed. The validity and feasibility of the proposed approach is investigated via the corresponding numerical simulation.

4.1 Introduction

A networked control system (NCS) is a feedback control system with network channels used for the communications between spatially distributed system components, such as sensors, actuators, and controllers. Since the signals are transmitted over a communication network of limited bandwidth, there exist network-induced delays, which may deteriorate the performance and stability of the closed-loop control system [1, 2].

Owing to the fact that the conventional control methods do not take into account the uncertainties of the network environment, the study of new control strategies dealing with this problem is of practical importance.

Sliding mode control (SMC) has been successfully used in controlling many uncertain systems; it is preferred because it's robust character and superior performance [3, 4]. During the last decades, numerous contributions to SMC theory have

Z. Gao (✉)
Department of Applied Mathematics, School of Statistic, Xi'an University of Finance and Economics, Xi'an 710100, China
e-mail: gaozb2700@sina.com

W.E. Wong and T. Zhu (eds.), *Computer Engineering and Networking*, Lecture Notes in Electrical Engineering 277, DOI 10.1007/978-3-319-01766-2_4,

been made. The switching schemes, putting the differential equations into canonical forms, and generating simple SMC strategies are considered in detail. The application of SMC scheme to robotic manipulator is studied, and the quality of the scheme is discussed from the point of robustness [5]. The performance of SMC scheme is proven to be satisfactory in the face of external disturbances and uncertainties in the system model representation.

A new sliding mode, called integral sliding mode, is proposed [6]. Having this feature, integral sliding mode control has been widely applied to various systems [7, 8].

In this paper, we design a novel integral switching function to tackle a continuous NCS which is comprised of the parameter uncertainties and the state delay. The sliding mode controller is achieved and the reachability of SMC is analyzed. The example is given to verify the feasibility and effectiveness.

4.2 Problem Formulation

In an NCS, the controlled plant with parameter uncertainty is a continuous system that can be described as follows:

$$\begin{aligned} \dot{x}(t) &= (A + \Delta A)x(t) + (A_d + \Delta A_d)x(t-d) + (B + \Delta B)u(t) \\ x(t) &= \phi(t), t \in [-d, 0] \end{aligned} \tag{4.1}$$

where $x(t) \in R^n$ is the state vector; $u(t) \in R^m$ is the control input; $d > 0$ is the delay constant; $\phi(t)$ is the original state vector defined in $[-d, 0]$; A, A_d, B are matrices of the appropriate dimension; and ΔA, $\Delta A_d \Delta B$ are uncertain matrices, which are satisfied with the condition of the norm-bounded, that is,

$$[\Delta A, \Delta A_d, \Delta B] = DF[E_1, E_d, E_2]$$

where D, E_1, E_d, E_2 are the known matrices of the appropriate dimension and F is the unknown matrix which satisfies $F^T F \leq I$.

The sliding mode matching requirements are satisfied with

$$\begin{bmatrix} \Delta A & \Delta A_d & \Delta B \end{bmatrix} = B \begin{bmatrix} \Delta \tilde{A} & \Delta \tilde{A}_d & \Delta \tilde{B} \end{bmatrix}$$

Then, the system (4.1) can be formed as follows:

$$\dot{x}(t) = Ax(t) + A_d x(t-d) + B[u(t) + W(t)] \tag{4.2}$$

where $W(t) = B^+[\Delta A x(t) + \Delta A_d x(t-d) + \Delta B u(t)]$ and B^+ is the pseudo inverse, namely, $B^+ = (B^T B)^{-1} B^T$.

4.3 Main Results

The main idea is to find a sliding mode controller for the abovementioned systems (4.1) and (4.2); the corresponding normal system is

$$\dot{x}(t) = Ax(t) + A_d x(t-d) + Bu(t) \tag{4.3}$$

Define the integral switching function as

$$s(t) = C\left[x(t) - \int_0^t (A + BK)x(t)dt\right] + \mathrm{T} \tag{4.4}$$

where the state feedback gain K is unknown, C is the matrix with the appropriate dimension and satisfy $CB > 0$, and T is the sliding mode compensator which satisfies

$$\dot{\mathrm{T}} = -CA_d x(t-d) \tag{4.5}$$

During sliding surface $s(t) = \dot{s}(t) = 0$, from Eqs. (4.3), (4.4), and (4.5), we have

$$\dot{s}(t) = C\dot{x}(t) - C(A + BK)x(t) - CA_d x(t-d) = CBu(t) - CBKx(t) = 0 \tag{4.6}$$

Then, the equivalent controller is

$$u_{eq}(t) = Kx(t) \tag{4.7}$$

4.3.1 SMC Design

For the system defined by Eq. (4.2), the SMC law is written as

$$u(t) = u_{eq}(t) + u_N(t) \tag{4.8}$$

where $u_N(t)$ is the switching controller which is used to overcome the system uncertainties; the form is selected as

$$u_N(t) = -\gamma s(t) - f\mathrm{sgn}(s(t)) \tag{4.9}$$

where $f \geq |W(t)|$, $\gamma > 0$ and sgn(•) is the sign function.

Theorem 1 *Consider the system (4.2), design the switching function as Eq. (4.4), if the SMC law is given by Eq. (4.8), then, the system (4.2) can be asymptotically stable.*

Proof Suppose Lyapunov function is $V(t) = \frac{1}{2}s^2(t)$, $s(t)$ is described as Eq. (4.4), then

$$\begin{aligned}\dot{V}(t) &= s\dot{s} = s\big[C\dot{x} - C(A + BK)x - CA_d x(t - d)\big]\\ &= s\big[CB(u(t) + W(t)) - CBKx(t)\big]\\ &= s\big[CB(u_{eq}(t) + u_N(t) + W(t)) - CBKx(t)\big]\\ &= s\big[CB(-\gamma s - f\mathrm{sgn}(s) + W(t))\big] = -CB\big(\gamma|s|^2 + f|s| - sW\big) \leq 0\end{aligned}$$

Then, the SMC law is considered to guarantee the reachability of the sliding surface. Thus, the proof is complete.

4.3.2 Stability of the Sliding Surface

Lemma *Given a scalar, if there exist matrices and satisfying, where, is an identity matrix, then the following matrix inequality holds:*

$$DFE + E^T F^T D^T \leq \varepsilon DD^T + \varepsilon^{-1} E^T E$$

Put the equivalent controller Eq. (4.7) into system (4.1), we have the sliding mode dynamics equation as follows:

$$\dot{x}(t) = [(A + \Delta A) + (B + \Delta B)K]x(t) + (A_d + \Delta A_d)x(t - d) \tag{4.10}$$

Theorem 2 *For system (4.1), given constant $\varepsilon > 0$, select switching function (4.4). If exist symmetric definite matrices X, V and matrix W, satisfy linear matrix inequality (LMI) as follows:*

$$\begin{bmatrix} \Theta & A_d V & (E_1 X + E_2 W)^T & X \\ * & -V & VE_d^T & 0 \\ * & * & -\varepsilon I & 0 \\ * & * & * & -V \end{bmatrix} < 0 \tag{4.11}$$

where $\Theta = AX + XA^T + BW + W^T B^T + \varepsilon DD^T$. The asterisk denotes the transpose of the corresponding block above the main diagonal, and I denote the identity matrix of appropriate dimension. Then, the equivalent controller is $u_{eq}(t) = WX^{-1}x(t)$, and the sliding mode dynamics system (4.10) is globally asymptotically stable.

Proof Suppose the Lyapunov function is

$$V(t) = x^T(t)Px(t) + \int_{t-d}^{t} x^T(\tau)Rx(\tau)d\tau$$

where P, R are symmetry definite matrices. Calculate $\dot{V}(t)$ as follows:

$$\begin{aligned}
\dot{V}(t) &= 2x^T(t)P\dot{x}(t) + x^T(t)Rx(t) - x^T(t-d)Rx(t-d) \\
&= 2x^T(t)P\left[(A + DFE_1) + (B + DFE_2)K\right]x(t) + 2x^T(t)P(A_d + DFE_d)x(t-d) \\
&\quad + x^T(t)Rx(t) - x^T(t-d)Rx(t-d) \\
&= x^T(t)(PA + A^TP + PBK + K^TB^TP + R)x(t) + x^T(t)(PA_d + A_d^TP)x(t-d) \\
&\quad - x^T(t-d)Rx(t-d) + x^T(t)(PDFE_1 + PDFE_2K + E_1^TF^TD^TP + K^TE_2^TF^TD^TP)x(t) \\
&\quad + x^T(t)PDFE_dx(t-d) + x^T(t-d)E_d^TF^TD^TPx(t)
\end{aligned}$$

Using Lemma, we get

$$\begin{aligned}
&x^T(t)(PDFE_1 + PDFE_2K + E_1^TF^TD^TP + K^TE_2^TF^TD^TP)x(t) \\
&\le x^T(t)\left[\varepsilon PDD^TP + \varepsilon^{-1}(E_1 + E_2K)^T(E_1 + E_2K)\right]x(t) \\
&\quad x^T(t)PDFE_dx(t-d) + x^T(t-d)E_d^TF^TD^TPx(t) \\
&\quad \le x^T(t)\varepsilon PDD^TPx(t) + \varepsilon^{-1}x^T(t-d)E_d^TE_dx(t-d)
\end{aligned}$$

Consequently, we obtain

$$\dot{V}(t) \le \xi^T(t)\Phi\xi(t)$$

where $\xi(t) = \left[x(t) \quad x(t-d)\right]$, and

$$\Phi = \begin{bmatrix} Z & PA_d \\ A_d^TP & -R + \varepsilon^{-1}E_d^TE_d \end{bmatrix} \tag{4.12}$$

and $Z = PA + A^TP + PBK + K^TB^TP + R + \varepsilon PDD^TP + \varepsilon^{-1}(E_1 + E_2K)^T(E_1 + E_2K)$

Thus, $\dot{V}(t) < 0$ is satisfied if $\Phi < 0$. From Eq. (4.12) pre-multiply and post-multiply the matrix diag$\{P^{-1}, P^{-1}\}$; the nest LMI can be derived:

$$\begin{bmatrix} \Sigma & A_d & X(E_1 + E_2K)^T & X \\ * & -R & E_d^T & 0 \\ * & * & -\varepsilon I & 0 \\ * & * & * & -R^{-1} \end{bmatrix} < 0$$

where $\Sigma = AX + XA^T + BKX + XK^TB^T + \varepsilon DD^T$. Let $W = KX$ and pre-multiply and post-multiply the diag$\{I, R^{-1}, I, I\}$, and set $V = R^{-1}$, we have the LMI (4.11). Thus, the proof is complete.

4.4 Experiment Studying

For the system (4.1), we have the system parameters as follows:

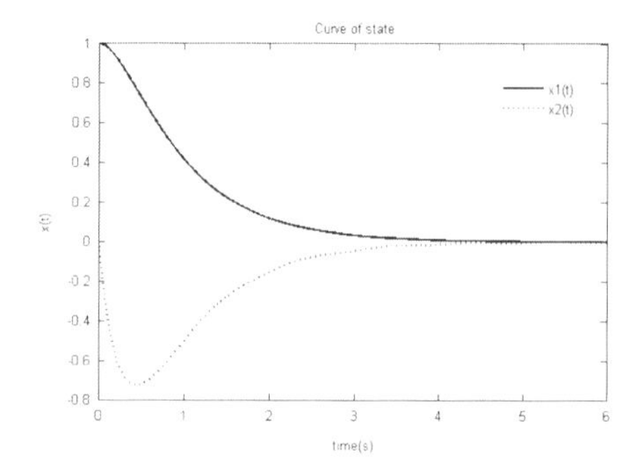

Fig. 4.1 The curve of state

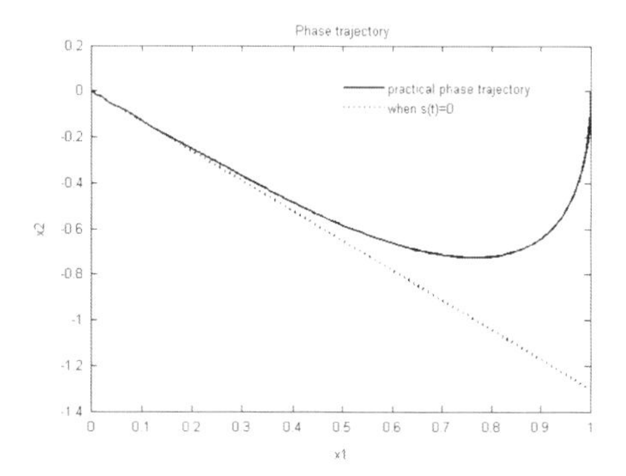

Fig. 4.2 The phase trajectory

$$A = \begin{bmatrix} 0 & 1 \\ 0 & -1 \end{bmatrix}, B = \begin{bmatrix} 0 \\ 5 \end{bmatrix}, A_d = \begin{bmatrix} 0 & 0 \\ 0.1 & 0.1 \end{bmatrix}, C = \begin{bmatrix} 1.3 & 1 \end{bmatrix}, D = \begin{bmatrix} 0.5 \\ 0.1 \end{bmatrix}, E_1 = \begin{bmatrix} 0 & 0.1 \end{bmatrix},$$
$$E_d = \begin{bmatrix} 0 & 0.1 \end{bmatrix}, E_2 = 0.1$$

the delay constant $d = 3$; original state is $X_0 = \begin{bmatrix} 1 & 0 \end{bmatrix}^T$.

From the Eqs. (4.8) and (4.9), we have the SMC law $u(t)$. Set $f = 0.01$, $\gamma = 0.6$ and the perturbation $W(t) = 0.005\ \sin(2\pi t)$. According to Theorem 2, using Matlab LMI toolbox, we have

$$X = \begin{bmatrix} 0.4002 & -0.2891 \\ -0.2891 & 0.3968 \end{bmatrix}, V = \begin{bmatrix} 1.1928 & -0.0078 \\ -0.0078 & 1.1850 \end{bmatrix},$$
$$W = \begin{bmatrix} -0.1470 & -0.0418 \end{bmatrix}$$

then, the equivalent controller gain is $K = WX^{-1} = -\begin{bmatrix} 0.9358 & 0.7870 \end{bmatrix}$.

The simulation results are shown in Figs. 4.1, 4.2, 4.3, and 4.4. The evolutions of two state variables $x_1(t)$, $x_2(t)$ are depicted in Fig. 4.1.

The motion in the phase plane is illustrated in Fig. 4.2. It shows that after a fast reaching mode, a sliding mode is maintained on the sliding surface $s(t) = 0$ by the

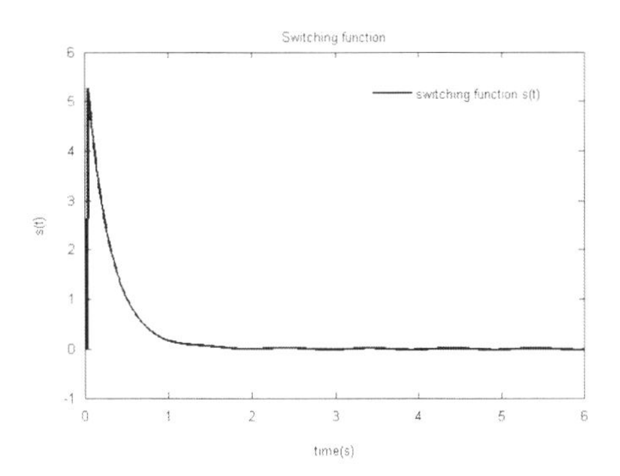

Fig. 4.3 Switching function

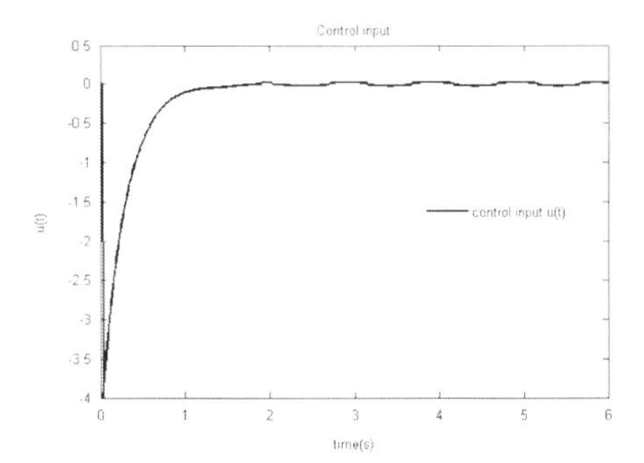

Fig. 4.4 Control input

suitable control input. It can efficiently eliminate the chatter due to apply the integral sliding mode controller.

The evolution of the switching function is depicted in Fig. 4.3. In Fig. 4.4, it is seen that the control signal slightly vibrates at first time and then tends to zero in the finite time.

4.5 Conclusion

In this paper, the proposed integral sliding mode controller is applied for an uncertain continuous NCS with the state delay. The parameter uncertainties satisfy the norm-bounded, and the sliding modes comply with the matching requirements. The main contribution of this work is to design the new integral switching function which includes the state feedback control gain and the sliding mode compensator. Furthermore, according to Lyapunov stability condition and in basis of the LMI convex optimal technology, a sufficient condition is derived so that the asymptotical stability of the closed-loop system is guaranteed. The numerical simulation is provided to verify the correctness and superiority of the approach.

Acknowledgements This paper is supported by Science Research Term of the Education Department of Shaanxi province of China (No. 12JK0548).

References

1. Dai, S.L., Lin, H., Ge, S.S. (2010). Scheduling-and-control code-sign for a collection of networked control systems with uncertain delays. *IEEE Transactions on Control Systems Technology, 18*(1), 66–78.
2. Gao, H.J., Meng, X.Y., Chen, T.W. (2008). Stabilization of networked control system with a new delay characterization. *IEEE Transactions on Automatic Control, 53*(9), 2142–2148.
3. Utkin, V.I. (1992). *Sliding modes in control optimization* (pp. 145–154). New York: Springer.
4. Efe, M.O., Kaynak, O., Yu, X.H. (2000). Sliding mode control of a three degrees of freedom anthropoid robot by driving the controller parameters to an equivalent regime. *Transactions of the ASME Journal of Dynamic Systems, Measurement and Control, 122*(4), 632–640.
5. Gao, W.B., Hung, J.C. (1993). Variable structure control of nonlinear systems: A new approach. *IEEE Transactions on Industrial Electronics, 40*(1), 45–55.
6. Utkin, V.I., Shi, J. (1996). Integral sliding mode in systems operating under uncertainty conditions. In *Proceedings of the 35th IEEE Conference Decision Control* (pp. 4591–4596). New York: IEEE Press.
7. Chou, C., & Cheng, C. (2003). A decentralized model reference adaptive variable structure controller for large-scale time-varying delay systems. *IEEE Transactions on Automatic Control, 48*(7), 1213–1217.
8. Poznyak, A., Fridman, L., & Bejarano, F.J. (2004). Mini-max integral sliding mode control for multimodel linear uncertain systems. *IEEE Transactions on Automatic Control, 49*(1), 97–102.

Chapter 5
Synthesis of Linear Antenna Array Using Genetic Algorithm to Control Side Lobe Level

Zhigang Zhang, Ting Li, Feng Yuan, and Li Yin

Abstract In array pattern synthesis, it is often designed to achieve the desired radiation pattern. In this paper, real-coded genetic algorithm (RGA) optimization method is presented to optimize the value of weights of each antenna element to minimize side lobe level of the uniform spaced linear array geometries with a certain main beam width. The optimization program is done by using MATLAB. It compared with the conventional analytical methods such as Chebyshev and Taylor through radiation patterns with different number of elements and intervals of each element. The simulation results show that the optimization results are of little difference when $d \geq \lambda/2$, but GA can get a more optimal result when $d < \lambda/2$. The application of genetic algorithm for pattern synthesis is found to be useful.

5.1 Introduction

Synthesis of array antennas is very important to get the desired pattern, and how to achieve low side lobe in the condition of a fixed main beam width has been considered since a long period [1]. Conventional analytical methods such as Chebyshev and Taylor are widely used for uniformly spaced linear arrays with isotropic elements [2, 3]. However, these methods cannot synthesize antenna array with complicated geometry layout. In recent years, numerical approach has become more popular in synthesis of antenna arrays such as Powell's method, memetic algorithm (MA), tabu search (TS), particle swarm optimization (PSO), and genetic algorithm (GA) [4–6].

Z. Zhang • T. Li (✉) • L. Yin
Electronic Engineering College, University of Naval Engineering,
Wuhan 430033, Hubei, China
e-mail: liting89720@126.com

F. Yuan
Department of Information, Command of East China Sea Fleet,
Ningbo 315122, Zhejiang, China

W.E. Wong and T. Zhu (eds.), *Computer Engineering and Networking*, Lecture Notes in Electrical Engineering 277, DOI 10.1007/978-3-319-01766-2_5,

As a general optimization algorithm, genetic algorithm is simple in coding and genetic operations. Optimization is not bounded by other constraint. Because of its global search ability, GA does not have to rely on good initial values to obtain the optimal solution. J. Michael Johnson and Yahya Rahmat-Samii applied genetic algorithm to antenna synthesis firstly, illustrated the basic theory of array synthesis by using genetic algorithm, and have proved the probability of synthesized one and two dimensions of uniform and nonuniform antenna array by genetic algorithm [7]. Randy L. Haupt used genetic algorithm to optimize the 200 symmetric linear array and plane array, respectively, to obtain the minimum side lobe level, got less than 20 db side lobe level, and provided a general method of synthesized array by genetic algorithm [8]. R. L. Haupt presented a genetic algorithm which could deal with real number coding and binary coded simultaneously [9]. V. R. Lakshmi and G. S. N Raju compared GA with synthesis of Taylor by different scan angles.

In this paper, it is assumed that the array is uniform, where all the antenna elements are identical and equally spaced. The design criterion here considered is to minimize the side lobe level with a fixed main beam width. In these conditions, GA compared with the methods of Chebyshev and Taylor through radiation patterns with different number of elements and intervals of each element.

5.2 Uniform Linear Antenna Array

In linear antenna array, all the antenna elements of mutually uncoupled isotropic radiator are arranged along the x-axis with equal spacing between them, N is the total number of elements in the antenna array, and d is the distance between two consecutive elements [10, 11]. A_m is the amplitude of each element, A symmetric linear array with even number is shown in Fig. 5.1.

The far-field radiation pattern in the XY—plane of the array in the free space—can be obtained with the array factor (AF) as given in Eq. (5.1):

$$AF(\theta) = 2\sum_{n=1}^{M} A_n \cos\left(kz_n \cos\theta + \alpha_n\right) \tag{5.1}$$

where A_n and α_n are excitation magnitude and phase, respectively, the wave number of the carrier signal is $k = 2\pi/\lambda$, θ is the angle from broadside, and z_n is the distance between nth element with origin (the array center) as given in Eq. (5.2):

$$z_n = \begin{cases} \pm(n-1)d, n = 1, 2, \cdots, M+1 (N = 2M+1) \\ \pm\dfrac{(2n-1)}{2}d, n = 1, 2, \cdots, M (N = 2M) \end{cases} \tag{5.2}$$

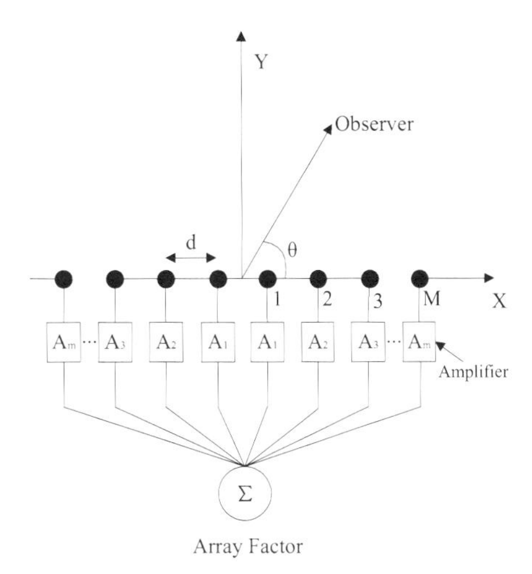

Fig. 5.1 Geometry of 2M elements array

In this paper, only amplitude is considered in pattern synthesis, uniform the phase of excitation ($\alpha_n = 0$), and formula (5.1) can be translated as given in Eq. (5.3):

$$AF'(\theta) = \begin{cases} 2\sum_{n=1}^{M} A_n \cos\left(\frac{2\pi}{\lambda}(n-1)d\cos\theta\right), (N = 2M+1) \\ 2\sum_{n=1}^{M} A_n \cos\left(\frac{\pi}{\lambda}(2n-1)d\cos\theta\right), (N = 2M) \end{cases} \quad (5.3)$$

5.3 Optimizing by GA

Genetic algorithm is a stochastic global search algorithm based on the natural selection and genetic. GA is different with the traditional optimum algorithms, which are used to generate a deterministic test solution sequence merely based on the gradient calculation of evaluation function [12, 13]. Genetic algorithm searched the optimal solution through imitating the natural evolutionary process. It provides a generic framework for solving complex system optimization problems. It is not dependent on the problem of specific fields with a strong robustness. Genetic algorithm has a strong vitality in nonlinear numerical optimization. It is very useful for many electromagnetic optimization problems and adapted to beams forming of antenna array.

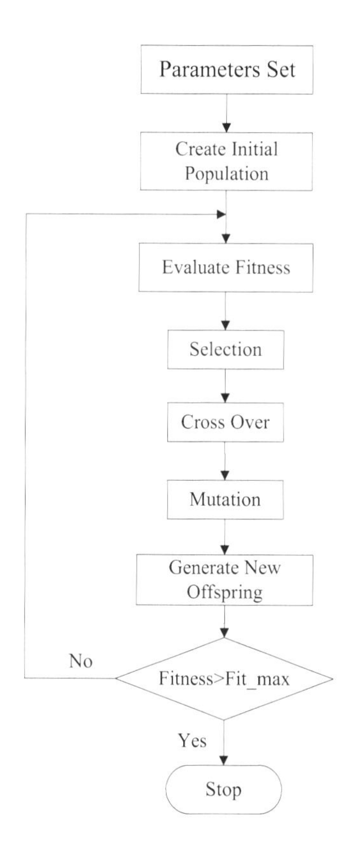

Fig. 5.2 Flowchart of GA

GA produced the first generation randomly and calculated the fitness value of each individual based on the fitness function. Fitness criterion suggested the advantages and disadvantages of each individual. The fittest individual has the biggest fitness value. The best will be survived to the next generation, and the worst will be eliminated directly. The rest are selected to produce new generation by recombination. Recombination consists of two genetic operators named crossover and mutation. And it will be recycled for a certain number of generations to get the optimal solution. The flowchart is shown in Fig. 5.2.

The important operators of GA can be summarized as follows:

1. Selection: The purpose of selection is to get good individuals from present population and then make them have chances to generate descendants as father generation. The selection probability is dependent on the fitness value.
2. Crossover: The new individuals can be generated through the operator of crossover and combined the characters of two individuals of their father generation. The operator happened based on the crossover rate.
3. Mutation: The change in a single individual depended on the mutation rate. The operator provides a chance to generate new individuals.

In amplitude-only synthesis, the main concern is to optimize the amplitude levels to produce the lowest side lobe level. The objective function associated with this array is the maximum side lobe level of its radiation field pattern to be minimized with a certain main beam width. The general form of the objective function is given by Eq. (5.4)

$$Objective = 20\ \log_{10}\left(\frac{\max|AF'(\theta)|}{\max|AF'(\theta_0)|}\right) \tag{5.4}$$

where $AF'(\theta)$ is array factor as given in Eq. (5.3), and $0 \leq \theta_0 \leq \pi$, θ belong a certain bands. Fitness function is ordered the values of objective function of every generation from small to big by nonlinear sequence. The fitness value decreased follow with the increasement of objective value. The range of fitness value is from 0 to 1.

5.4 Experimental Results

Here we have used genetic algorithm for linear antenna array design. The results have been compared with that of Chebyshev optimization and Taylor optimization. The antenna model consists of certain number elements and equally spaced along the x-axis. Voltage sources are at the center segment of each element, and the amplitude of the voltage level is the antenna element weight. Only amplitude distribution is changed to find the optimum result; the array geometry and elements remain constant. GA with population size of 40, generation gap of 0.9, crossover rate of 0.8, and mutation rate of 0.2 runs for a total of 100 generations by using MATLAB. The optimal results will be founded each iteration. The fitness function is minimized the side lobe level of antenna pattern. Figure 5.3 shows that the antenna array with $N = 12$ elements and interval of each element is $d = \lambda/2$ has been designed for minimum side lobe level in bands [0, 80] and [100, 180]. The maximum of SLL calculated by GA is -18.49 dB, approximated with the method of Chebyshev and Taylor. Figure 5.4 shows the maximum side lobe level value of every generation. The evolutionary tendency suggested that the optimal solution tends to be converged with the increase of generation and there is no premature convergence.

The synthesis of Chebyshev and Taylor can achieve the minimum side lobe level in a given main beam width based on the classical array theory. However, the optimal is conditional. It required array element spaced $d \geq \lambda/2$. When the array element spaced $d < \lambda/2$, Chebyshev and Taylor were not the optimal, but genetic algorithm was still valid. Figure 5.5 showed the radiation pattern with array element spaced $d = 3\lambda/4$ and the minimum side lobe level in bands [0, 82.5] and [97.5, 180].

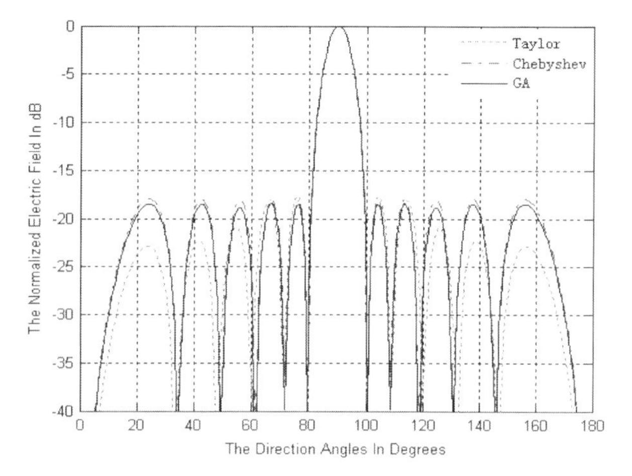

Fig. 5.3 Normalized patterns with $d = \lambda/2$

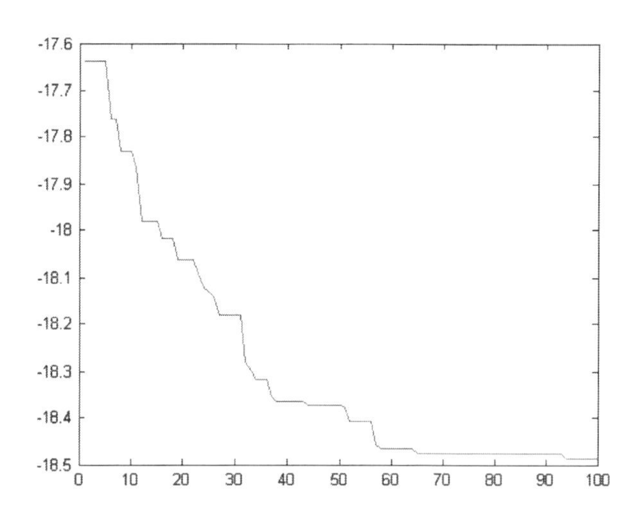

Fig. 5.4 Convergence of side lobe level of every generation

Figure 5.6 showed the radiation pattern with array element spaced $d = \lambda/4$ and the minimum side lobe level in bands [0, 70] and [110, 180]. The maximum side lobe level calculated by GA is -19 dB, which is lower about 3 dB than the result by Chebyshev and Taylor.

Figure 5.7 showed the radiation pattern with 24 elements spaced $d = \lambda/4$ and the minimum side lobe level in bands [0, 80] and [100, 180]. The maximum side lobe level calculated by GA is -19.5 dB, which is lower about 2 dB than the result by Chebyshev and Taylor.

Fig. 5.5 Normalized patterns with $d = 3\lambda/4$

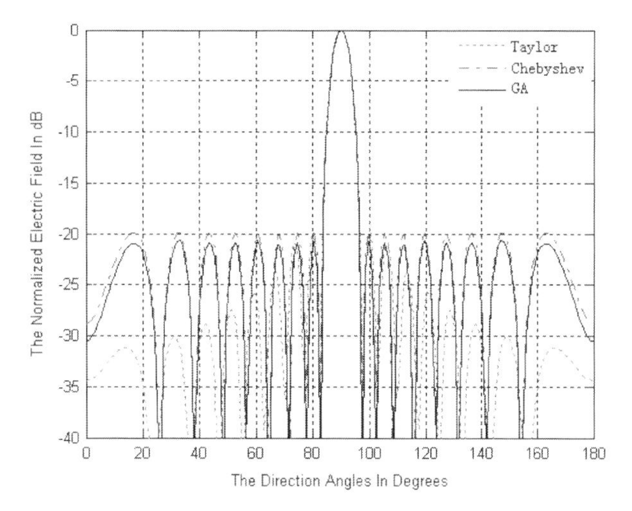

Fig. 5.6 Normalized patterns with $d = \lambda/4$

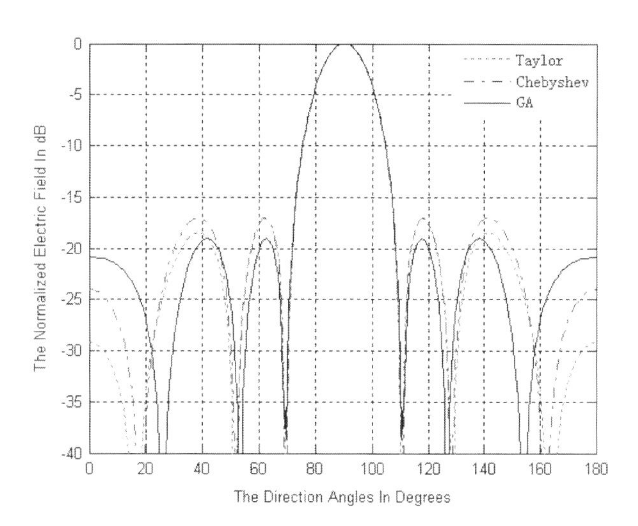

Fig. 5.7 Normalized patterns with $d = \lambda/4$ for 24 elements array

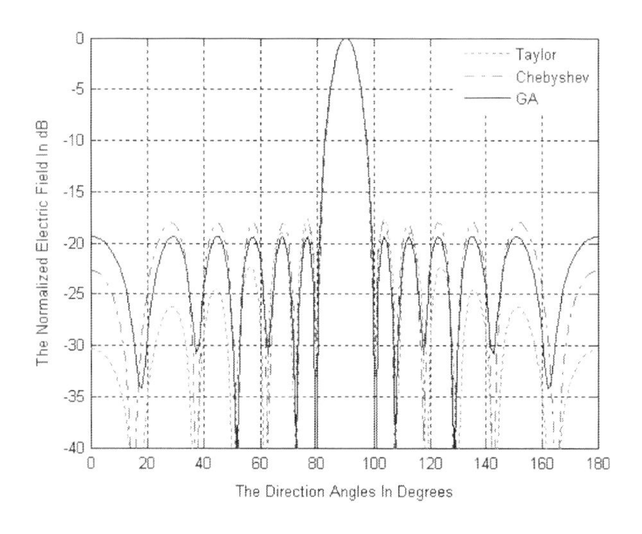

5.5 Conclusion

From the results, it is clear that genetic algorithm performs better than the method of Chebyshev and Taylor synthesis. As calculated in the previous section, the three methods performed almost the same when $d \geq \lambda/2$. However, GA can achieve about 3 dB lower than Chebyshev and Taylor in radiation pattern with the same main beam width when $d < \lambda/2$ for 12 elements array. The application of genetic algorithm for pattern synthesis is found to be useful.

References

1. Rajya Lakshmi, V., & Raju, G. S. N. (2011). Synthesis of linear antenna arrays using array designer and genetic algorithm. *IJAEST, 9*(1), 44–48.
2. Raju, G. S. N. (2005). *Antennas and wave propagation* (pp. 132–139). India: Pearson Education.
3. Elliot, R. S. (1981). *Antenna theory and design* (pp. 49–54). New Jersey: Prentice-Hall.
4. Jiao, Y. C., Wei, W. Y., Huang, L. W., & Wu, H. S. (1993). A new low side lobe pattern synthesis technique for conformal arrays. *Antenna Propagation, 41*(8), 824–831.
5. Er, M. H., Sim, S. L., & Koh, S. N. (1993). Application of constrained optimization techniques to array pattern synthesis. *Signal Processing, 34*(5), 323–334.
6. Chen, T. B., Chen, Y. B., Jiao, Y. C., & Zhang, E. S. (2005). Synthesis of antenna array using particle swarm optimization. In Microwave conference proceedings 2005. Asia-Pacific conference proceedings, China (pp. 4–9).
7. Johnson, J. M., & Rahmat-Samii, Y. (1994). Genetic algorithm optimization and its application to antenna design. In *Antennas and propagation society international symposium, 1994, USA*, AP-S. Digest Volume 1 (pp. 326–329).
8. Haupt, R. L. (1994). Thinned arrays using genetic algorithms. *Antennas and Propagation, 12*(7), 993–999.
9. Haupt, R. L. (2007). Antenna design with a mixed integer genetic algorithm. *Antennas and Propagation, 55*(5), 577–582.
10. Shrivastava, S., & Cecil, K. (2012). Performance analysis of linear antenna array using genetic algorithm. *IJEIT, 2*(5), 84–88.
11. Laseetha, T. S. J., & Sukanesh, R. (2011). Synthesis of linear antenna array using genetic algorithm to maximize side lobe level reduction. *International Journal of Computer Applications, 20*(7), 27–33.
12. Mandal, D., & Chandra, A. (2010). Side lobe reduction of a concentric circular antenna array using genetic algorithm. *Serbian Journal of Electrical Engineering, 7*(2), 214–218.
13. Johnson, J. M., & Samii, Y. R. (1997). Genetic algorithm in engineering electromagnetic. *Antennas and Propagation, 39*(4), 7–21.

Chapter 6
Wavelet Analysis Combined with Artificial Neural Network for Predicting Protein–Protein Interactions

Juanjuan Li, Yuehui Chen, and Fenglin Wang

Abstract In order to solve the prediction problem of interaction between proteins, we use a wavelet coefficient combined with artificial neural network method, improving the prediction accuracy of the problem of protein–protein interactions. By introducing the Biorthogonal Wavelet 3.3 coefficients as the feature extraction method and the three-layer feedforward neural network as a classifier, we solve the problem of protein interaction effectively. Using the Human dataset verifies the validity of this method. Through testing the Human dataset, using Biorthogonal Wavelet 3.3 coefficient combined with the three-layer feedforward neural network, solve the prediction problem of protein interactions with well results. This combination of wavelet coefficients and the three-layer feedforward neural network to predict protein interaction problem is an effective method. At the same time, compared with other prediction methods, this method performs at least 4 % higher accuracy than the better accuracy of auto-covariance (11) combined with PNN on the same dataset.

6.1 Introduction

Proteins are major component of organism and play a very important role in organism. Proteins play biological function through the interactions in organism. The study of interaction between proteins starts from biological experiment, and commonly biological experimental methods have the following: yeast two-hybrid (Y2H) [1], a biological method aiming at the yeast; mass spectrometry protein complex identification (MS-PCI) [2], which is from the protein molecular and atomic micro, based on which forecast is carried on; and protein chip technology [3], a technology that solidifies some known proteins and chips are used to predict

J. Li (✉) • Y. Chen • F. Wang
Computational Intelligence Laboratory, University of Jinan, Jinan 201305, China
e-mail: juanjuannuli@126.com

W.E. Wong and T. Zhu (eds.), *Computer Engineering and Networking*, Lecture Notes in Electrical Engineering 277, DOI 10.1007/978-3-319-01766-2_6,

the interactions of solidified proteins. Using methods of biological experiment for studying the interactions between protein prediction problems has its own advantages. Biological experiments operate simply and can get intuitive and reliable results. But with the development and use of high-throughput biological experiments, a large number of biological data are produced. Using biological experiment for high throughput data is a very large project and also time consuming. So, intelligent calculation methods are introduced to solve biological problems. Intelligent calculation method is a technology combining computer technology with biology. It is generated through the rapid development of computer and can be a very good solution in solving biological problems of high throughput calculation [4].

At present, many intelligent computation methods are used for the study of protein–protein interaction (PPI) problem. Among them, typical feature extraction methods have amino acid composition, pseudo-amino acid composition, and physicochemical properties. Classification methods have kernel-nearest neighbor, support vector machine [5], and probabilistic neural network. In this chapter, we proposed the resonant recognition model combined with artificial neural network to predict interaction between proteins. Experiment results show that our proposed method performs at least 4 % higher accuracy than the best of other related works.

6.2 Technical Introduction

The term wavelet means small wave; the so-called small means it is capable of decaying; and wave refers to the volatility, the amplitude of positive and negative form of shocks. Compared with the Fourier transform, the wavelet transform is a time frequency analysis (spatial) localization, signals (functions) multi-scale refinement through telescopic translation operations, and ultimately achieves frequency time segments and low frequency subdivision. The wavelet transform can automatically adapt to the time-frequency signal analysis requirements, which can focus on any signal details, solve the difficult problem of Fourier transform, and become a major breakthrough in the scientific methods of Fourier transform since [6, 7].

Neural network is an algorithm of the simulation to process human neurons. The input data are combined to a single output node through the nonlinear transformation. The typical structure of the neural network has three layers, and it has strong robustness and fault tolerance; at the same time, it can learn and adapt uncertain systems. The neural network needs to set the number of neurons of the input layer, hidden layer, and output layer. The neural network classification process has two steps: first, training the neural network parameters by continuous input training data; and second, inputting test data to get the results of the test data used in the optimization of neural network. Using artificial neural network (ANN) classification has the advantages of speed and potential super speed and good fault tolerance ability. It is suitable for solving some problem. Compared to some clustering ideological classification, for example, support vector machine (SVM) and nuclear nearest neighbor, it has better classification results [8, 9].

6.3 Research Programs

The dataset used in this study is Human dataset with more difficulty. Human biological structure is very complex, and the human body to maintain its own physiological function needs a lot of proteins to participate, so the interaction between the human proteins is more complex, and then the prediction of this dataset has some difficulty. The human data contain 914 positive protein pairs and 941 negative protein pairs for a total of 1,882 protein pairs. This dataset is used in this study to verify the effectiveness of the method we chose [10].

Feature extraction. The following are the three main steps to extraction feature of the protein interaction sequence pairs and finally into a one-dimensional vector of the neural network input:

1. Assign the two test proteins to respective IC value, which will be converted to two numerical sequences. The IC values are shown in Table 6.1.
2. Transform the two protein sequences using the discrete wavelet transform of DWT, using Biorthogonal Wavelet 3.3 (see Fig. 6.1). The discrete wavelet transform is performed at three levels (see Fig. 6.2).The original numerical signal is decomposed into A3 (approximation at level 3), D3 (detail at level 3), D2 (detail at level 2), and D1 (detail at level 1) signals of four scales. D3, D2, and D1 contain details of the signal in the original signal at different levels, but A3 only retained a few low-frequency signals [11].
3. Wavelet coefficients are expressed by $CD_n^{K,p}$, and wavelet coefficients were normalized; see Eq. (6.1)

$$NOM_n^{K,p} = \frac{\left|CD_n^{K,p}\right|}{\max_n\left(\left|CD_n^{K,p}\right|\right)} \tag{6.1}$$

where n = 1,2,. . .N; p = 1,2; K = A3, D3, D2, D1.

Among them, P is protein serial number.

So we can get the wavelet coefficient, using the wavelet coefficients as the feature for protein–protein prediction. But because the sequence length is too long, the result in numerical sequences of wavelet coefficients is also very long; as such, the input of the classifier is not reasonable, so we choose 1 of the 20 largest values as the feature representation input to the classifier prediction. But we save these maximum values with their original order. Therefore, our neural network input is 20D.

Classify. Using artificial neural network to classify PPIs has two steps. Firstly, we need to use the training set to train parameters of neural network. Secondly, we need to use the test set to test the trained classifier. We choose particle swarm

Table 6.1 IC value of the amino acid

Amino acid	L	A	N	C	I	P	W	T
IC value	2.40	2.30	2.20	1.96	2.40	2.00	2.37	2.09
Amino acid	Y	H	Q	F	H	K	M	R
IC value	2.20	2.30	2.06	1.98	2.46	2.20	2.17	1.82
Amino acid	V	E	S	D				
IC value	2.35	2.30	2.10	1.88				

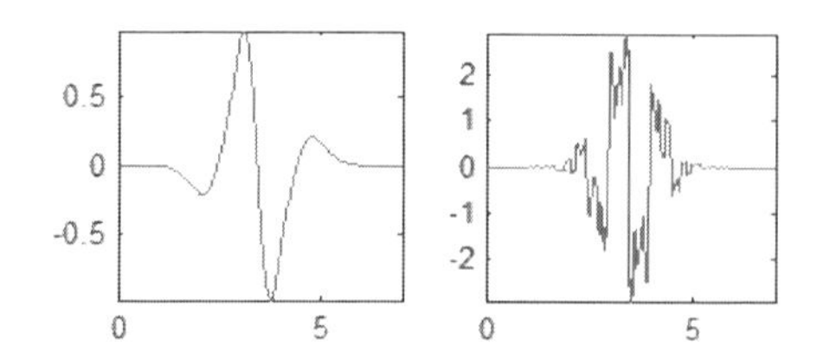

Fig. 6.1 Biorthogonal wavelet 3.3

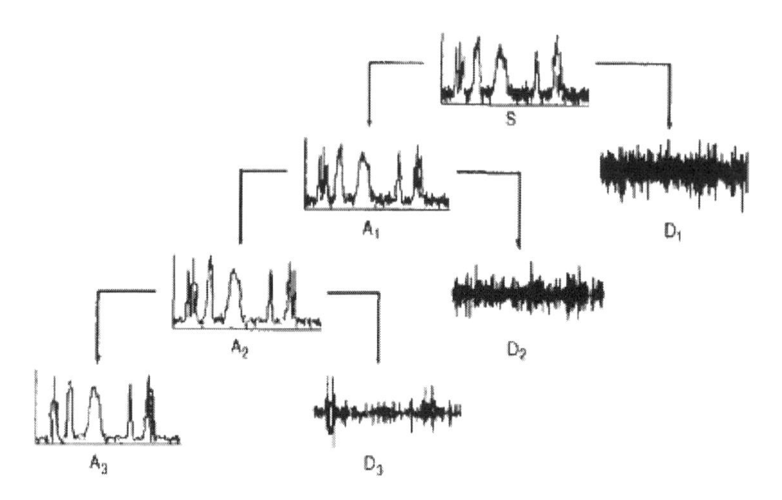

Fig. 6.2 Discrete wavelet transform three levels

algorithm as parameter optimization algorithm to optimize parameters of neural network. It is a random search algorithm developed by simulating the foraging behavior of bird flocks based on group cooperation. It is generally considered swarm intelligence. Particle swarm optimization parameters mainly consist of two steps, one is the change of speed, and other is the change of position, using formulas (6.2) and (6.3):

$$V_{id} = w * V_{id} + c_1 * rand() * (P_{id} - X_{id}) + c_2 * rand() * \left(P_{gd} - X_{id}\right) \tag{6.2}$$

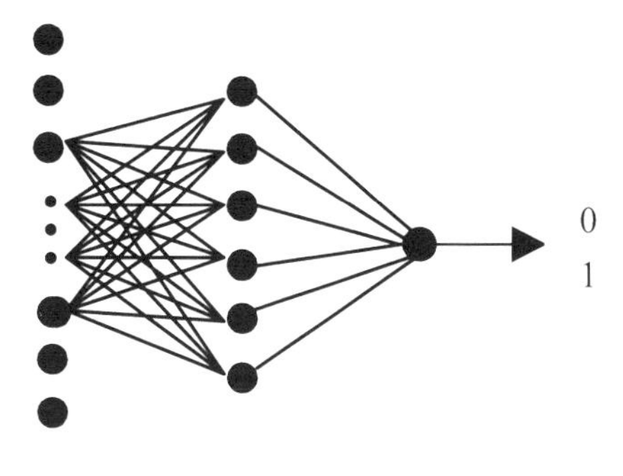

Fig. 6.3 The model of the artificial neural network

$$X_{id} = X_{id} + V_{id} \tag{6.3}$$

Through the feature dimension obtained, we design the neural network structure as shown in Fig. 6.3. The input dimension is 20, and after trying, hidden layer is set to 6. Because this is a two-classification problem, the output is one dimensional.

6.4 Analysis of Results

This chapter uses 10-fold cross-validation test to verify this method performance. The 10-fold cross-validation divides the dataset into 10 groups, each with nine datasets to do the training set and the rest of the group to do the test set, turns to a total of 10 times, and then takes the average value as a result of the test. Each test set has 94 positive samples and 94 negative samples, a total of 188; each train set has 867 positive samples and 867 negative samples, a total of 1,694. Classification index using the total classification accuracy (Acc) and sensitivity (S) and precision (P) [12] is defined as follows:

$$Acc = \frac{PT + NT}{PT + PF + NT + NF} \tag{6.4}$$

$$S = \frac{PT}{PT + NF} \tag{6.5}$$

$$P = \frac{PT}{PT + PF} \tag{6.6}$$

in which PT is the correct number of positive samples, PF is the error number of positive samples, NT is the correct number of negative samples, and NF is error number of negative samples. The positive samples here refer to the interaction of the protein pairs, while the negative samples refer to the non-interaction protein pairs (Table 6.2).

The chapter is experimented many times to verify the selected methods' effectiveness; the following tables show the results of ten pairs of train set and test set and the best result of the integrated methods (Table 6.3).

Table 6.2 Best result of the experiment

Dataset	P	S	Acc
Human	78.723	84.09	81.9149

Table 6.3 Best results of 10-fold cross-validation

P	P1	P 2	P 3	P 4	P 5	
R	80.8511	80.8511	76.0638	77.6596	81.9149	
P	P 6	P 7	P 8	P 9	P 10	Best
R	80.8511	78.1915	80.8511	79.7872	78.1915	81.9149

Table 6.4 Results of different methods used

Data	Classifier	Feature	Code	Result
Human	SVM	KMC + KNN + BIO [10]	Link	73.10
Human	KNN	ACC [9]	Summation	73.90
Human	PNN	AC (11)	Link	78.37
Human	ANN	Wavelet	Summation	81.9149

List the results of the different methods, where AC (11) presents calculating AC of 11 physicochemical properties (Table 6.4).

Through this table, we can achieve a conclusion: the wavelet coefficients combined with artificial neural network perform at least 4 % higher accuracy than the AC (11) combined with PNN on the same dataset.

6.5 Conclusion

This chapter uses ANN as a classifier to predict protein interactions. Through experiment of a variety of physicochemical properties of the protein, we find that the wavelet analysis combined with ANN reached good results. This feature extraction method combined with ANN achieves more high precision. It can obviously be seen that the accuracy is improved. Through confirmation, this method we used is effective.

Acknowledgments This research was partially supported by the NSFC grant (61201428), the Natural Science Foundation of China (61070130), the Key Project of Natural Science Foundation of Shandong Province (ZR2011FZ003), the Key Subject Research Foundation of Shandong Province, and the Shandong Provincial Key Laboratory of Network Based Intelligent Computing.

References

1. Ito, T., Chiba, T., Ozawa, R., Yoshida, M., et al. (2001). A comprehensive two-hybrid analysis to explore the yeast protein interaction. *Proceedings of the National Academy of Sciences U S A, 98*(8), 4569–4574.

2. Ho, Y., Gruhler, A., Heibut, A., et al. (2002). Systematic identification of protein complexes in Saccharomyces cerevisiae by mass spectrometry. *Nature, 415*, 180–183.
3. Zhu, H., Bilgin, M., Bangham, R., et al. (2001). Global analysis of protein activities using proteome chips. *Science, 293*, 2101–2105.
4. Zhou, J. L., & Yi, M. C. (2008). Study of protein interactions in a review of calculation methods. *Journal of Computer Research and Development, 45*(12), 2129–2137.
5. Chang, C. C., & Lin, C. J. (2011). LIBSVM: A library for support vector machines. *ACM Transactions on Intelligent Systems and Technology, 2*(3), 27.
6. Zhuo, M., Zhang, Y. Q., Li, M. L., et al. (2011). Based on the integrated learning approach for predicting protein-protein interactions. *Journal of Sichuan University (Engineering Science Edition), 43*(3), 324–328.
7. Feng, T. N., Da, L., Jin, D. L., et al. (2011). Using lifting wavelet extracting protein-protein interaction characteristics. *Communications in Applied Mathematics and Computational Science, 25*(2), 12.
8. Guo, Y., Yu, L., Wen, Z., & Li, M. (2008). Using support vector machine combined with auto covariance to predict protein-protein interactions from protein sequences. *Nucleic Acids Research, 36*(9), 3025–3030.
9. Song, J. (2009). Prediction of protein-protein interaction using kernel nearest neighbor algorithm. *Application Research of Computers, 26*(11), 124–130.
10. Guo, Y., Li, M., Pu, X., et al. (2010). PRED-PPI: A server for predicting protein-protein interactions based on sequence data with probability assignment. *BMC Research Notes, 3*(1), 145–151.
11. Feng, T. N., Zhang, H., & Wang, Y. F. (2008). Prediction of protein SNR based on interaction. *Journal of Shanghai University (Natural Science), 14*(6), 12.
12. Zhou, Z. R., Song, X. F., & Wang, M. B. (2010). Using a combination of classifiers for predicting protein-protein interactions. *Acta Electronica Sinica, 38*(6), 275–282.

Chapter 7
Application Analysis of Slot Allocation Algorithm for Link16

Hui Zeng, Qiang Chen, Xiaoqiang Li, and Jianguo Shen

Abstract In order to improve the efficiency of slot utilization of Link16, the traditional binary tree model of timeslot allocation is improved, and the performances of three kinds of timeslot assignment algorithm are compared and analyzed in the average transmission delay and message sending failure rate. When the user is determined and the message arrival rate is lower, the fixed timeslot allocation is simple and effective. When there are a lot of users or the user is uncertain, and the message arrival rate is very low, competitive slot allocation should be more efficient. When the number of users is small, the user message arrives sudden, and arrival rate is high, the performance of the dynamic slot allocation has a very distinct advantage. Finally, based on the characters of different timeslot assignment algorithms, the overall timeslot allocation scheme is presented for Link16 network.

7.1 Introduction

In the traction of modern information warfare demand, the data link has been significant developed in the field of military information, and Link16 is one of the most widely used tactical data link which uses the TDMA protocol. Because the platforms which need to access data link networks become more and more, timeslot allocation is very important for the efficiency to the Link16 network. In this chapter, the binary tree slot allocation model is analyzed. According to uniformity principle,

H. Zeng (✉) • J. Shen
National Defense Information Academy, WuHan 430019, China
e-mail: zenghui_1974@163.com

Q. Chen
Institute of China Electronic System Engineering Corporation, Beijing 100039, China

X. Li
Communications Training Base of the General Staff Headquarters, Xuanhua 075100, China

W.E. Wong and T. Zhu (eds.), *Computer Engineering and Networking*, Lecture Notes in Electrical Engineering 277, DOI 10.1007/978-3-319-01766-2_7,

a method of slot allocation based on binary tree model [1] is proposed when the number of timeslots is not 2^n.

Timeslot allocation algorithm of Link16 includes fixed allocation (FA) algorithm, competitive allocation (CA) algorithm, and dynamic allocation (DA) algorithm [2]. There are some researches in recent years of the performance of various algorithms, but their performance indicators are inconsistent, such as queuing delay of FA algorithm [3], collision probability of CA algorithm [4], and the throughput of DA algorithm [5], which cannot be an effective evaluation or meet the characteristic of Link16 application. In this chapter, the relations between the delay and the failure rate of message transmission and the load rate are analyzed for the three kinds of slot allocation algorithm. Each algorithm has advantages and disadvantages. According to the characteristics of the network participation group (NPG), different algorithms should be used for different NPG of Link16 network.

7.2 The Model of Timeslot Resource

One day is divided into 112.5 time units in Link16. Each unit is divided into 64 time frames, and the time frame is divided into 1,536 application timeslots. There is 7.8125 ms for every timeslot. Because the time unit is too long to describe the real-time demand of Link16, usually, the time frame is analyzed as a slot cycle in the process of timeslot allocation.

For the stability of the system work, the timeslot resource of every NPG should be uniformly distributed in the slot cycle. Binary tree model is commonly used in the slot block allocation of resources. 1,536 application timeslots of a time frame is divided into three groups (A, B, and C); each group consists of 512 timeslots. The timeslots of a group are staggered arranged as A-0, B-0, C-0, A-1, B-1, C-1,. . ., A-511, B-511, and C-511. In order to simplify the identification and assignment, a slot block which contains 2^n timeslots can be defined by the group of timeslots (A, B, or C), index (0~511), and recurrence rate number (RNN). As shown in Fig. 7.1, A-0-9 which contains all 512 slots in the A group is divided into A-0-8

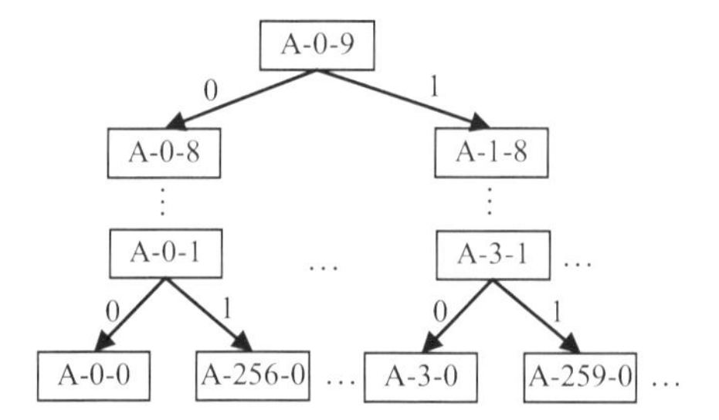

Fig. 7.1 The model of binary tree slot

and A-1-8. A-0-8 contains A-0, A-2,. . ., A-510, and A-1-8 contains A-1, A-3,. . ., A-511. Finally, the timeslots A-0-0, A-1-0, A-2-0,. . ., A-511-0, respectively, denote A-0, A-1, A-2,. . ., A-511.

As can be seen from the binary tree model diagram, if the timeslot block assigned to the NPG just by the power of 2 slots, it ensures that the timeslots belonging to a NPG are uniformly distributed in the slot ring, and the distance between the timeslots is $3 \times 2^{9\text{-RNN}}$. In the decomposition process of the model, if encoding the left represents 0 and the right represents 1, the tree node index number of the slots is the binary values of the coding which consists of a string from the leaf node to the root node of the direction of the binary code, and the number of timeslots of the slot block is 2^{RNN}.

If the timeslot block assigned to the NPG is not a power of 2, then it should be decomposed into a group of timeslot subblocks which are the power of 2. Because the slots of timeslot block are not a power of 2, the slot cannot be uniformly distributed in the slot ring. In order to achieve the slot distribution as uniform as possible, the choice of timeslot subblock should comply with the principle that the common ancestor node is closest.

7.3 Performance Analysis of the TDMA Slot Allocation Algorithm

7.3.1 Fixed Slot Allocation Algorithm for Link16

Based on the estimated amount of the user's business, fixed timeslot resources are assigned to each user by fixed timeslot allocation, and then the user can only occupy its own dedicated slot to send a message. Assuming that the messages arriving obey Poisson distribution with the parameter λ, the message transmission rate (μ) of the user is directly related to the amount of the dedicated timeslot; therefore, the message processing model is a M/D/1 queuing models. The message have to wait for the arrival of the transmit timeslot to send, so the arrival of the first message is the equivalent to the second message of the M/D/1 queuing models. Considering the change of the amount of messages in the send buffer $\{X(t), t \geq 0\}$, it is a typical birth and death process. State space defined as $E = \{1, 2, 3, \ldots\}$, and the state transition matrix is

$$Q = \begin{bmatrix} -\lambda & \lambda & 0 & 0 & \cdots \\ \mu & -\lambda-\mu & \lambda & 0 & \cdots \\ 0 & \mu & -\lambda-\mu & \lambda & \cdots \\ \vdots & \vdots & \vdots & \vdots & \ddots \end{bmatrix} \tag{7.1}$$

The stationary distribution is obtained as

$$\begin{cases} \pi_1 = \left[\sum_{k=1}^{\infty} \left(\frac{\lambda}{\mu} \right)^{\mathrm{k}} \right] = 1 - \rho, \quad \left(\rho = \frac{\lambda}{\mu} < 1 \right) \\ \pi_{\mathrm{k}} = \left(\frac{\lambda}{\mu} \right)^{\mathrm{k}-1} \pi_1 = \rho^{k-1}(1-\rho), \quad (k = 2, 3, \cdots) \end{cases} \tag{7.2}$$

ρ is called load rate. Therefore, the average number of the message to wait for transmission is obtained as

$$L = \sum_{k=2}^{\infty} (k-2)\pi_k = \frac{\rho^2}{1-\rho} \tag{7.3}$$

The average transmission delay is equal to

$$W_f = \frac{L}{\lambda} = \frac{\rho}{(1-\rho)\mu} \tag{7.4}$$

When the message in the transmission queue is more than N_g, it is assumed that the message will be discarded for the real-time requirements of sending a message, so message sending failure rate should be obtained as follows:

$$F_f = 1 - \sum_{k=1}^{N_g+1} \pi_k = \rho^{N_g+1} \tag{7.5}$$

Figures 7.2 and 7.3 illustrate the relation between the system average waiting delay, message transmission failure rate, and user load rate, respectively.

As indicated in Fig. 7.2, the average waiting time is increased as the load rate increases. In the lower interval of ρ, the change of W_f is slow and gentle. However, when ρ increases to a certain extent, the value of W_f sharply becomes high.

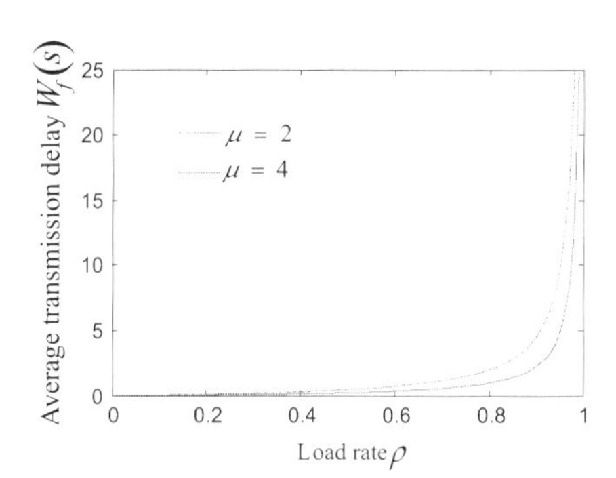

Fig. 7.2 Average transmission delay W_f vs load rate ρ

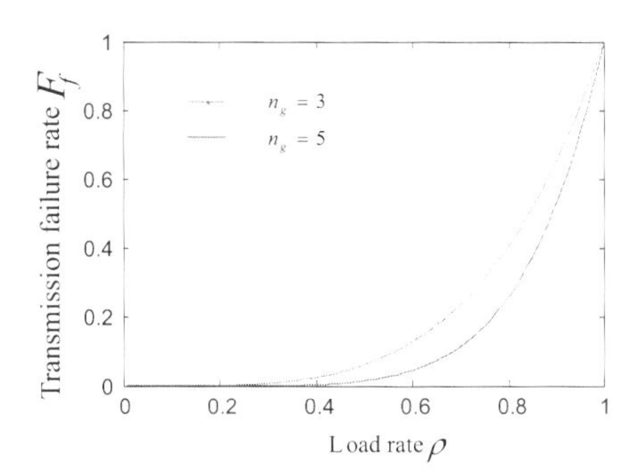

Fig. 7.3 Transmission failure rate F_f vs load rate ρ

As shown in Fig. 7.3, F_f increase with the increasing of ρ. The major factor to reduce F_f is to decrease ρ, and the increasing of the transmit buffer can also be appropriate to reduce the F_f.

7.3.2 Competition Slot Allocation Algorithm for Link16

All users of the NPG randomly grab slot to send messages. Assumed all users have the same priority, which sent the message as soon as possible, so there may be a plurality of users competing for the same timeslot. Then the slot conflict will occur, which will result in a communication failure.

Assuming that the NPG number of users is N, the messages arriving obey Poisson distribution with the parameter λ and are independent of each other. The available slot resource of NPG is evenly distributed in the slot ring with the interval T. Then, the probability of n messages arriving in NPG is obtained as

$$P(n) = \frac{(N\lambda T)^n e^{-N\lambda T}}{n!} \tag{7.6}$$

The probability of two or more messages arriving is much smaller than one for a user at time T, so we can ignore the probability that a user received two or more messages at time T, then the probability of the slot competition conflict which will result in communication failure can be obtained as

$$F_c = 1 - e^{-N\lambda t} - N\lambda t e^{-N\lambda t} = 1 - (1 + C)e^{-C}, \quad (C = N\lambda t) \tag{7.7}$$

$C = N\lambda t$ is defined as the competitive strength of user timeslot. The number of users, the user message arrival rate, and the slot interval have the equivalent effects on C. The relationship between F_c and C is shown in Fig. 7.4.

In this case,the message transmission delay also can be obtained by the formula (7.3). Because the load rate ρ is a user relative to the entire NPG and μ is message transmission rate of the entire NPG, the average waiting delay is greatly reduced.

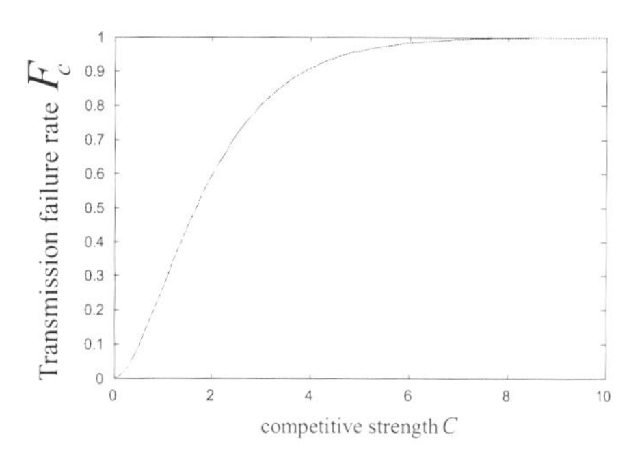

Fig. 7.4 Transmission failure rate F_c vs competitive strength C

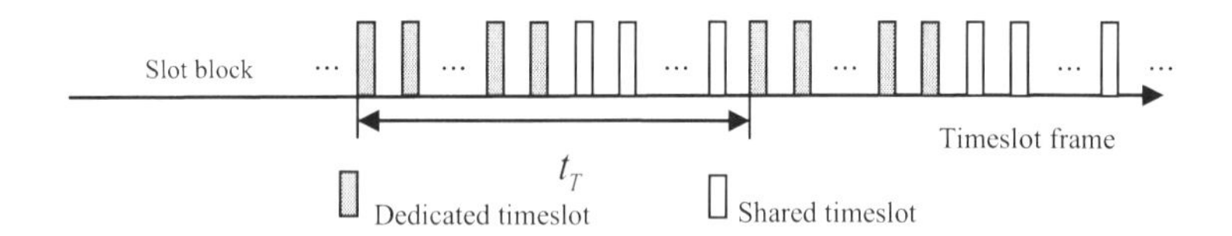

Fig. 7.5 Timeslot distribution of the dynamic slot allocation algorithm

7.3.3 Dynamic Slot Allocation Algorithm for Link16

The slot block is divided into a dedicated timeslot and shared timeslot. Dedicated timeslot is assigned to the specified user to send a message or make an appointment, and the shared timeslot is that all users can make an appointment to use. The distribution of timeslots in the timeslot frame is shown in Fig. 7.5.

Assuming message source can use ON/OFF model to describe, there are active period T_{on} and quiet period T_{off}, the message is generated. The messages arriving obey the Poisson distribution with parameter λ in active period, and no message arrives in the silent period. Active factor r is defined as

$$r = \frac{T_{on}}{T_{on} + T_{off}} \tag{7.8}$$

In this case, assuming that the length of T_{on} and T_{off} obeys negative exponential distribution, in a reservation period t_T, the probability that n active period is triggered by the all N independent users can be obtained as

$$P(n) = \frac{(Nrt_T)^n e^{-Nrt_T}}{n!} \tag{7.9}$$

Shared slot averaging assigned to the user that enters the active phase, the average transmission delay of the message can be calculated by the formula (7.4) as follows:

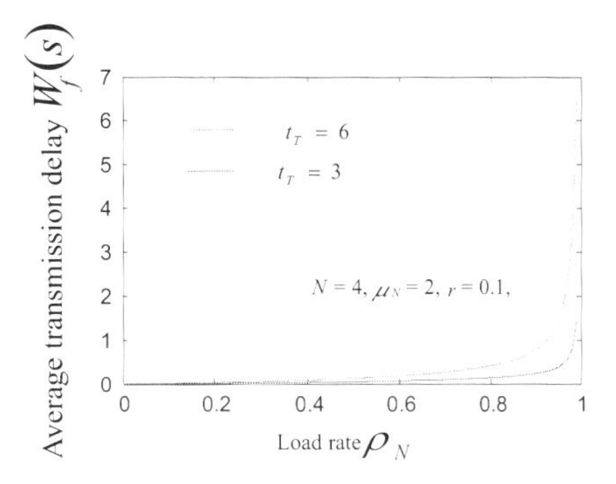

Fig. 7.6 Average transmission delay W_f vs load rate ρ_N

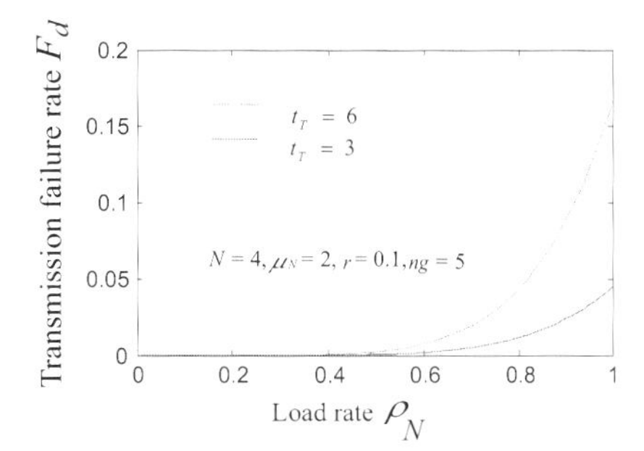

Fig. 7.7 Transmission failure rate F_f vs load rate ρ_N

$$W_d = \sum_{n=1}^{N} P(n)\frac{\rho_n}{(1-\rho_n)\mu_n} = \sum_{n=1}^{N} \frac{(Nrt_T)^n e^{-Nrt_T}}{n!}\frac{\frac{n\rho_N}{N}}{\left(1-\frac{n\rho_N}{N}\right)\frac{N\mu_N}{n}} \tag{7.10}$$

ρ_n and μ_n are the load ratio and the message sending rate of n users that is triggered, respectively, wherein the condition of $\mu_N > \lambda$ should be complied with.

The message sending failure rate can be calculated by the formula (7.5) as follows:

$$F_d = \sum_{n=1}^{N} P(n){\rho_n}^{n_g+1} = \sum_{n=1}^{N} \frac{(Nrt_T)^n e^{-Nrt_T}}{n!}\left(\frac{n\rho_N}{N}\right)^{n_g+1} \tag{7.11}$$

The average transmission waiting delay and message transmission failure rate with the variation of the load ratio are shown in Figs. 7.6 and 7.7, respectively.

Since the dynamic slot assignment takes account of the burst of the message arrival and the sharing of resources, performance in both the average transmission waiting delay or message transmission failure rate is improved significantly. The analysis here, however, does not consider the impact of the dedicated timeslot occupying the timeslot resources. The amount of the timeslot resources occupied by dedicated timeslot is determined by the number of users and the appointment cycle length. Dynamic slot allocation should give full consideration to the impact of the increase in the number of users and the shorter of the appointment cycle length. Specially, when the dedicated timeslots occupy the all timeslots of the slot block, it becomes fixed slot allocation algorithm. On the other hand, it becomes competition slot allocation algorithm if the all timeslots is defined as shared timeslots.

7.4 Link16 Slot Allocation Scheme

The network of Link16 is divided into different NPG. Firstly, the resources of the timeslot frame allocated to the NPG and then reallocated to the users of NPG.

The purpose of NPG division is to reduce the mutual influence of the communication of the different functional domains, so the resources of timeslot frame divide into timeslot blocks for NPG by fixed timeslot allocation algorithm.

There are about 20 major NPGs of Link16 [6]. Because the number and type of users and the characteristics of message are different from NPG, the timeslot block of NPG should be assigned to the user in different ways. From the above analysis, the slot allocation scheme for Link16 NPG is shown in Table 7.1.

Table 7.1 The slot allocation scheme for the NPG of Link16

NPG-index	NPG-function	Timeslot allocation algorithm
1	Initial entry	FA CA
2	RTT-A	FA
3	RTT-B	CA
4	Network management	FA CA
5	PPLI-A	FA
6	PPLI-B	CA
7	Surveillance	FA DA
8	Mission management	FA CA
9	Air control	FA CA
10	Electronic warfare	FA
12	Voice A	CA
13	Voice B	CA
14	I-PPLI	FA DA
18	Weapons coordination	FA
19	Fighter-to-fighter net	FA CA
27	Joint PPLI	FA
28	Distributed network management	FA
29	Residual message	FA CA
30	IJMS P message	FA DA
31	IJMS T message	FA DA

7.5 Conclusion

When the user is determined and the message arrival rate is lower, the fixed timeslot allocation is a simple and effective allocation method. When there are a lot of users or the user is uncertain, and the message arrival rate is very low, competitive slot allocation should be more efficient. When the user message arrives sudden, the message arrival rate is high, and the number of users is small, the performance of the dynamic slot allocation has a very distinct advantage. Accordingly, the slot allocation method of the different NPG of Link16 should be adapted to its features, which help to reduce the delay of message transmission, improve the slot utilization, and reduce the probability of the message failed to be sent.

References

1. Hou, X. (2008). *Data link dynamic slot allocation and its network planning technology research*. Chengdu: University of Electronic Science and Technology of China (In Chinese).
2. Commanding Officer Navy Centre for Tactical Systems Interoperability (NCTSI). (2001). *Understanding Link-16: A guidebook for new users*. San Diego: NCTSI.
3. He, Z.-x., Luo, X.-h. (2011). Towards fixed allocation timeslot number prediction method for TDMA tactical data link based on operation requirements. *International Journal of Advancements in Computing Technology, 3*(6), 57–64 (In Chinese).
4. Xia, B., & Li, H. (2010). Collision analysis of time slot contention access in data-link. *Ship Electronic Engineering, 30*(6), 66–69 (In Chinese).
5. Zhang, L., Yin, H. (2008). *The Link16 throughput simulation based on DTDMA* (pp. 464–466). China-Ireland International Conference on Information and Communications Technologies, Beijing, China.
6. Mei, W.-h., & Cai, S.-f. (2004). *JTIDS/Link16 data link* (p. 103). Beijing: Higher Education Press (In Chinese).

Chapter 8
An Improved Cluster Head Algorithm for Wireless Sensor Network

Feng Yu, Wei Liu, and Gang Li

Abstract Routing algorithm is one of the key technologies of wireless sensor network. Based on the LEACH algorithm, the second choice of cluster head algorithm for WSN is proposed. It takes into account the cluster head as residual energy and distance to BS, and then it chooses a senior cluster head. The improved algorithm avoids direct communication between BS and the cluster head which has low energy and is far away from BS, which prolongs the lifetime of network and enhances the ability of data collection. The experiments show that this technique of event-driven can cut down the data transmission and further extend the lifetime of network.

8.1 Introduction

Wireless sensor network (WSN) is a convergence technology of computer network, communications, and sensor. WSN usually is widely applied in the military operations, environmental monitoring, medical rescue, traffic control, home automation, and other commercial applications [1, 2]. Study at home and abroad of WSN is not yet mature. Carrying out this cutting-edge technology research timely will bring great value and strategic significance to the development of the country. As the bright future in their application, a main goal in designing such WSNs is optimizing energy consumption to maximize the network lifetime. However, the influencing factors are various in impacting lifetime of network [3]. Actually, after deploying sensor nodes, effective routing control protocol is the key to improve the efficiency of energy. In the current study, the clustering algorithm is considered to be one of the most effective ways of efficient energy management. Typically, LEACH (Low Energy Adaptive Clustering Hierarchy) is the earliest cluster-organized

F. Yu (✉) • W. Liu • G. Li
Network Center of Shenyang Jianzhu University, Shenyang 100168, China
e-mail: wind@sjzu.edu.cn

W.E. Wong and T. Zhu (eds.), *Computer Engineering and Networking*, Lecture Notes in Electrical Engineering 277, DOI 10.1007/978-3-319-01766-2_8,

routing protocol in clustering algorithm [4, 5]. Inspired from LEACH, many of clustering algorithms are created, such as LEACH-C [6, 7], PEGASIS [8], and TEEN [9]. In this paper, contraposing the deficiencies of LEACH, we present an improved routing protocol LEACH-SC, which put forward the cluster head of secondary selection algorithm and single-multi-hop routing algorithm to further reduce energy costs.

8.2 Related Works

8.2.1 LEACH Protocol

LEACH protocol is based on the hypothesis that all sensor nodes can communicate with sink node directly and run with many rounds. Each round begins with a setup phase when the clusters are organized, followed by a steady-state phase when data are transferred directly from the ordinary nodes to the cluster heads and then transmit to sink node or base station (BS). In the setup phase, each node chooses a random number between 0 and 1, if the number is lower than the threshold $T(n)$, then the node becomes a cluster head for current round. The threshold limit is defined by the following formula (8.1).

$$T(n) = \begin{cases} \dfrac{p}{1 - p * [r * \mathrm{mod}(1/p)]}, & \text{if } n = G \\ 0\ , & \text{otherwise} \end{cases} \tag{8.1}$$

Where p denotes the percentage of cluster head in the total number of nodes, r is current rounds, G is a node set which have not been elected in the past $1/p$ rounds. When one node is selected as cluster head, it will inform other nodes, or else it will choose a cluster to join in deciding as the distance to cluster heads.

8.2.2 Energy Model

LEACH employs the familiar radio model discussed in Lv's method [10], which is the first order radio model. If transmit an l-bit message over distance d, the energy consumption of the node is

$$E_{Tx}(k,d) = lE_{Tx-elec} + E_{Tx-amp}(l,d) = \begin{cases} kE_{elec} + k\varepsilon_{fs}d^2, & d < d_0 \\ kE_{elec} + k\varepsilon_{amp}d^4, & d \geq d_0 \end{cases} \tag{8.2}$$

kE_{elec} is the energy consumption of radio dissipation, while the second item presents the energy consumption for amplifying radio. Depending on the transmission distance, both the free space $k\varepsilon_{fs}$ and the multipath fading channel models $k\varepsilon_{amp}$ are used. When receiving this data, the radio expends:

$$E_{Rx}(l,d) = E_{Rx-elec}(l) = l \cdot E_{elec} \tag{8.3}$$

8.2.3 *Problem Description*

Cluster heads are selected randomly in LEACH. It is possible that the number of node joining the same cluster head could be too much or too little, which will eventually lead to cluster heads with the excessive number of nodes energy exhausted and die too fast [11]. But only a few nodes of the cluster heads energy will be surplus. Accordingly, the entire network load becomes imbalanced. In addition, because in LEACH protocol the cluster heads communicate with base stations by single-hop manner, its energy consuming and expandability is limited so it could not adapt to large network. In the following, we will discuss how to improve LEACH algorithm from two points: optimizing the method of cluster heads choosing and cluster heads routing.

8.3 The Second Choice of Cluster Head Algorithm

8.3.1 *The Second Choice of Cluster Heads*

Because the cluster heads of LEACH protocol are randomly generated by ordinary nodes in the network, from this way, there are some defects that cannot ensure the uniform distribution of cluster heads and guarantee the reasonable scale of the clusters. So the major scheme of LEACH-SC is to maintain the formula (8.1) without any changes and to propose the thought of second choice for cluster heads. After the clusters formed, the large scale of clusters will be broken, and then reselect cluster heads within the cluster; accordingly utilize multi-cluster heads to share the cluster together. Meanwhile, break up the small-scale clusters and cancel the cluster heads, and finally let the members of the cluster join other clusters. This can optimize the distribution of cluster heads, balance the scale of the clusters, and, to some certain extent, solve the problem of uneven distribution of cluster heads. However, we need to choose the number of cluster head and the scale of cluster reasonably, bringing them close or equal to the optimal number of cluster head. As is referred in Lv's paper [10], Eq. (8.4) was proposed.

$$k_{opt} = \sqrt{\frac{N}{2\pi}}\sqrt{\frac{\varepsilon_{fs}}{\varepsilon_{amp}}}\frac{M}{d_{bs}^2} \tag{8.4}$$

Where N is numbers of nodes in our experiment, ε_{fs} and ε_{amp} are amplifying coefficient. M is side length of square region, and d_{bs}^2 denotes square of distance from cluster heads to base station. We choose $\varepsilon_{fs} = 10$ pJ/bit/m^2, $N = 100, M = 100$, $\varepsilon_{amp} = 0.0013$ pJ/bit/m^2, $50 \leq d_{bs} \leq 71$, next by the Eq. (8.4) we can get $7 \leq kopt \leq 14$. Accordingly, in the theory, the scale of each cluster is [N/14, N/7], and that approximately means [7,14].So we can stipulate that when the cluster size is greater than 14 will be rebuilt and when its less than 7 will be broken.

8.3.2 Clusters Formation

After choice of the cluster heads have been completed, the message "joining cluster" will be broadcast within their communication range, to tell other nodes they are cluster heads. Typically, when receiving the broadcast message, the ordinary nodes will select a cluster joining in as the member of it.

In the LEACH algorithm, ordinary nodes choose which cluster to join in judging by their distance to the cluster heads. There being a shortcoming by this way. When the ordinary nodes distance to the sink node is far less than to the any cluster heads, they will do not join to the sink node, instead of the cluster heads, then the energy demand could be larger to communicate with the cluster heads than to communicate with the sink node directly. In the improved algorithm LEACH-SC, therefore, to join in a cluster for ordinary nodes, not only take into account the cluster heads, but think over the sink node, based on distance and energy consuming, and decide to join a cluster or as a stand-alone node communicating with the sink node directly.

8.3.3 Cluster Heads Routing

When without range of the sink nodes radio, the cluster heads cannot communicate with them directly, even if within the scope, resultingly far away from the sink nodes, it would consume greater energy to communicate between them. In the improved algorithm LEACH-SC, the cluster heads adopt the communication of single-multi-hop routing, which take the method of single-hop routing when close to the sink nodes, meanwhile, when the cluster heads farther from the sink nodes, they will use multi-hop manner to pass data to the sink nodes through other cluster heads.

8.4 Simulation and Analysis

In order to evaluate the capability of the improved routing protocol, this paper makes a simulation for LEACH and LEACH-SC. For our experiments, in a zone with dimensions of $\omega \times \omega$ and the distribution of N homogeneous sensor nodes are randomly simulated by MATLAB simulator. All nodes have the same initial energy E_0 and energy consumption of radio E_{elec}. Table 8.1 shows the simulation parameters.

Figure 8.1a shows the building of clusters for the LEACH. From Fig. 8.1a, we can see the network appearing as many small clusters, which increase the number of cluster heads and cannot get them a reasonable distribution. If this kind of circumstance is appearing frequently, it will affect the overall energy distribution, so as to shorten the whole network of survival cycle. Figure 8.1b shows the building of clusters for the LEACH-SC. Compared to Fig. 8.1(a), Fig. 8.1(b) reveals that the number of cluster head is more reasonable and its cluster size is more balanced, which overcomes the disadvantage of LEACH clustering algorithm.

Figure 8.2 shows the total number of dead nodes over time, where the solid line represents LEACH-SC, and the forked line expresses LEACH. We assume the

Table 8.1 Table of parameter value

Description	Parameter	Value
Initial energy	E_0	0.5 J
Radio electronics energy	E_{elec}	50 nJ/bit
Radio amplifier energy for free space model	ε_{fs}	10 pJ/bit/m^2
Radio amplifier energy for multipath fading model	ε_{amp}	0.0013 pJ/bit/m^2
Desired probability of cluster heads	P	0.07
Data size	l	500 × 50 bit
Network size	$\omega \times \omega$	100 m× 100 m
Nodes	N	100

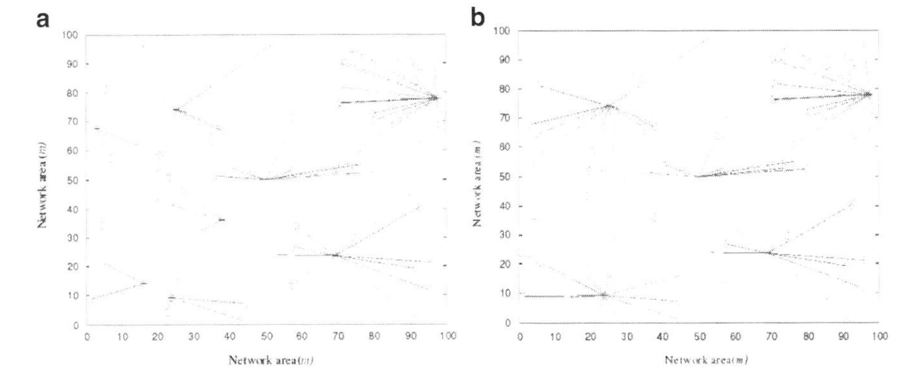

Fig. 8.1 The cluster head choosing of LEACH (**a**) and LEACH-SC (**b**)

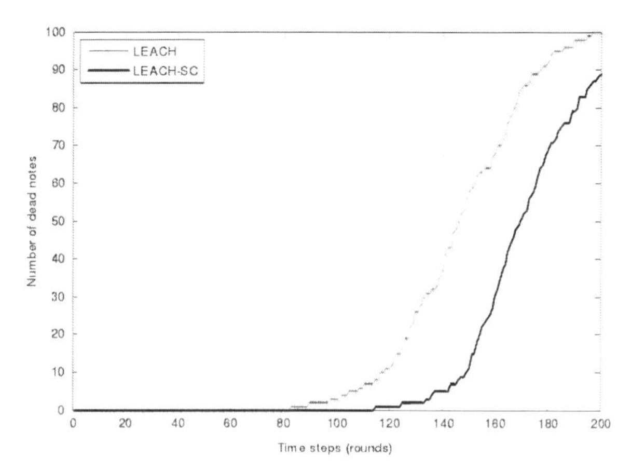

Fig. 8.2 Number of dead nodes per rounds

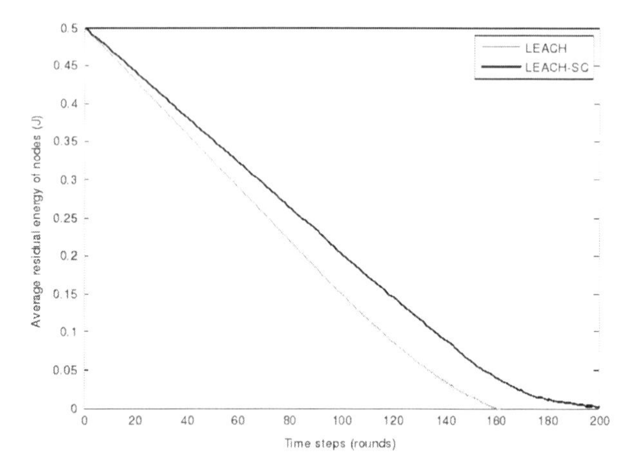

Fig. 8.3 Average residual energy of nodes per round

interval of First Node Dies (FND) as a steady period of network. Simulation results appear that FND = 85 rounds for LEACH, whereas FND = 115 for LEACH-SC. We can see that the improved algorithm increases 15 % of network stability on average comparing to LEACH. After running 200 rounds, the whole of network life increases 12 %.

Figure 8.3 is the result of average residual energy simulation. Judging from the figure, we can see the improved protocol have the more average residual energy of nodes than LEACH at any time, reflecting the greater energy-saving advantage. After running 160 rounds, the LEACH network of energy exhausted in advance, but the LEACH-SC still has a surplus energy to keep the network running and to extend the lifetime of the entire network up to 20 %. Thus, the improved protocol adopts a more reasonable means of cluster head election and intercluster communication, ultimately, effectively reducing the overall consumption of the network.

8.5 Conclusion

Cluster head is a chief undertaker of maximum energy consumption, and it is important to consider the energy consumption balance among cluster heads for WSNs. This paper introduced the improved routing algorithm LEACH-SC, which optimized the method of cluster heads choosing and cluster heads routing, realizing efficient energy distribution and prolonging the lifetime of network.

References

1. Wu, C., & Hu, Y. (2009). Improvement of LEACH in wireless sensor networks. *Computer Technology and Development, 19*(3), 270–274.
2. Martirosyan, A., Boukerche, A., & Pazzi, R. W. N. (2008). Energy-aware and quality of service-based routing in wireless sensor networks and vehicular ad hoc networks. *Annals of Telecommunications, 35*(11), 669–681.
3. Wei, X., Wang, Y., & Yu, X. (2009). Cluster-heads multi-hop algorithm of LEACH protocol in wireless sensor network. *Journal of Guangxi Academy of Sciences, 21*(4), 89–92.
4. Wang, H.-Y., & Liu, S. (2010). A clustering algorithm based on fusion and transmission cost for WSN. *Dalian Ligong Daxue Xuebao in Chinese, 50*(4), 591–596.
5. Mao, S., & Zhao, C.-L. (2011). Unequal clustering algorithm for WSN based on fuzzy logic and improved ACO. *Journal of China Universities of Posts and Telecommunications in Chinese, 18*(6), 89–97.
6. Heinzelman, W. B. (2002). An application-specific architectures for wireless networks. *IEEE Transactions On Wireless Communications, 1*(4), 660–670.
7. Siv D. Murugana., Daniel C.F. Ma., Rolly I. Bhasin., Abraham O. Fapojuwo. (2005). A centralized energy-efficient routing protocol for wireless sensor networks. *IEEE Communications Magazine, 43*(3), 8–13.
8. Lindsey, S., & Raghavendra, C. S. (2002). PEGASIS: Power efficient gathering in sensor information systems. In *Proceedings of IEEE aerospace conference* (pp. 1125–1130). New York: The Aerospace Corporation.
9. Manjeshwar, A., & Agrawal, D. P. (2001). TEEN: A routing protocol for enhanced efficiency in wireless sensor networks. In *Proceedings of the 15th parallel and distributed processing symposium* (pp. 2009–2015). San Francisco: IEEE Computer Society.
10. Lv, T., Zhu, Q.-x., & Zhang, L.-q. (2011). An improved LEACH algorithm in wireless sensor network. *Acta Electronica Sinica Journal, 39*(6), 1405–1409.
11. Owais, A., Ahthsham, S., & Aamir, M. M. (2011). Comparison of routing protocols to assess network lifetime of WSN. *International Journal of Computer Science Issues, 8*(6), 220–224.

Chapter 9
An Ant Colony System for Dynamic Voltage Scaling Problem in Heterogeneous System

Yan Kang, Ying Lin, Yifan Zhang, and He Lu

Abstract Dynamic voltage scaling is an effective energy minimization technique by conjointly changing the supply voltage and the operational frequency during run-time. In this chapter, an improved ant colony system is presented for distributed systems consisting dynamic voltage scalable processing elements. The energy saving can be obtained by using the DVS algorithm on the schedule obtained by the presented scheduling algorithm. The pheromone information of the ants and the heuristic information inspired by the list heuristic rule and energy consumption are combined together to guide the ants search. The parameter value of heuristic is varied from higher value to lower value to lessen its impact on ants search, while the parameter value of pheromone information is increased during the run of ant algorithm. And the elitist solution is discarded if it cannot be improved from generation to generation. By cooperating several generations of artificial ants, the ants search for the path with a minimum energy consumption cost, and the quality of the solution can be improved for minimizing the energy consumption. Experiments are implemented to demonstrate the performance of the algorithm.

Y. Kang (✉) • Y. Zhang
Department of Software Engineering, School of Software, Yunnan University, Kunming 650091, Yunnan, China
e-mail: kangyan@ynu.edu.cn

Y. Lin
Department of Information Security, School of Software, Yunnan University, Kunming 650091, Yunnan, China

H. Lu
School of Software, Yunnan University, Kunming 650091, Yunnan, China

W.E. Wong and T. Zhu (eds.), *Computer Engineering and Networking*, Lecture Notes in Electrical Engineering 277, DOI 10.1007/978-3-319-01766-2_9,

9.1 Introduction

Energy efficiency has become an important issue of the distributed system especially the embedded system which is composed of processing elements (PES) normally powered by batteries. The power consumption depends on function executed, resulting in the various processing elements characteristics if gated clocks switch off unused circuit parts during idle periods. Energy consumption is at least a quadratic function of the supply voltage (hence CPU speed). Modern processors can reduce the power consumption by using dynamic voltage scaling (DVS) to utilize the idle intervals between the deadline and the real finishing time. It is shown that further energy consumption can be reduced by using the process element power characteristic during the voltage selection [1]. The voltage scales is investigated that DVS can decrease the power consumption by up to 10 times when executing real-life applications by taking full use of the intervals to process tasks as slowly as possible [2].

Yao presented one of the earliest theoretical models for DVS and presented an $O(n^3)$ algorithm for computing a characterization of the minimum energy DVS schedule [3]. This optimal schedule did not make any special assumption on the power consumption function except convexity, and it has been referenced widely as it gave a main benchmark for evaluating other scheduling algorithms in both theoretical and simulation work.

There are static and dynamic techniques to using dynamic voltage scaling to decide appropriate operating voltage/speed. Schmitz researched static techniques by taking advantage of off-line parameters, such as periods and worst-case execution cycles [4]. Dynamic techniques (based on slack reclamation) exploit early completions of tasks to further decrease the speed and energy consumption. Here, we investigate static one. Scordino and Lipari used dynamic method to reduce the speed based on these predicted values and achieve more energy saving than static ones [5]. But the interest in static techniques is still high since static approaches can be enhanced to develop dynamic ones.

Pillai and Shin computed a single-optimal speed off-line and obtain the minimal speed to make a task set schedulable under earliest deadline first, and proposed a near-optimal method under rate monotonous [6]. Without fixing the processor speed, Saewong and Rajkumar assumed that the speed of the processor can be changed continuously in a given range and presented an algorithm to find the optimal speed value based on fixed priority assignments [7]. Based on the task parameters, Aydin proposed some methods to decide statically processor speed and assign different speed to every task before system execution [8, 9]. Liu and Mok gave a more general scheme to choose the speed switching instants more freely during the activation/deadline of some jobs [10].

Recently, there are several DVS-based algorithms proposed for slack allocation for a multiprocessor real-time system, while extensive researches on DVS scheduling algorithms have been proposed for independent tasks in a single processor real-time system.

In this chapter, we combine the schedule algorithm and dynamic scaling voltage strategy together to save the energy consumption effectively. Complicate task sets are concurrently processed on the distributed system with different degrees of parallelism. The presented scheduling algorithm wants to utilize the nonuniform workloads of processors and generates promising schedule pattern to reduce the energy consumption by using DVS strategy. Dynamic scaling voltage can reduce the energy consumption by exploiting the idle or slack time among the schedule pattern without obeying the precedence restricts and concurrence situation. By using ant colony optimization and heuristic strategy together to accumulate the pheromone information on path, the searches move towards the path with high trail, and find the solution with smaller energy consumption by iteratively ant generations. The algorithm includes three important co-synthesis steps: (a) assigning, determining the assignment of computational tasks to machines; (b) allocating, determining the execution order (sequencing) of tasks mapped to machines and communications cost; and (c) evaluation, determining the quality of the implementation candidate (energy consumption). The experiments are implemented to verify the performance of the algorithm.

This chapter is organized as follows: The next section briefly describe the problem model which includes task model and energy model. In Sect. 9.3, an ACO algorithm is presented while its performance is studied in Sect. 9.4. In Sect. 9.5, a conclusion is presented.

9.2 Task and Energy Mode

In the task scheduling problem, a deterministic set of tasks $ST = \{st_1, st_2, \ldots, st_n\}$ that are divided from an application is processed on a available set of processors $SP = \{sp_1, sp_2, \ldots, sp_m\}$ with fixed computing times $SC = \{sc_{11}, sc_{12}, \ldots, sc_{nm}\}$. A directed acyclic graph (DAG) is given to demonstrate the set of tasks and the precedence restriction among them. The precedence constraints mean the successor task can be assigned after all its processor tasks have been completed. The communication cost $cm_{i,j}$ represents the transfer time from precedence task t_i to its successor task t_j if they assigned on the different processors.

The task model is a fixed model, i.e., the execution time and the precedence constraint among the tasks are deterministic. Each task can be run on different processors with different processing time and cannot be interrupted before it finishes. All the processors can execute a task at a time, and they are continuously available and can process at most one task at a time. A feasible solution to the task scheduling problem satisfies the precedence constraint among the tasks.

It is shown that the power of a processing element is the sum of each CMOS circuit power which is composed of the static power and the dynamic one [4]. Most research groups have focused on reducing the dynamic power while the static power is negligible with respect to the dynamic power. The power consumption of a processor is calculated as

Table 9.1 Computation time of Fig. 9.1

1	2	3	4	5	6
8 (32)	10 (30)	19 (47)	7 (19)	15 (42)	24 (61)
11 (41)	9 (27)	25 (52)	6 (43)	19 (36)	13 (41)

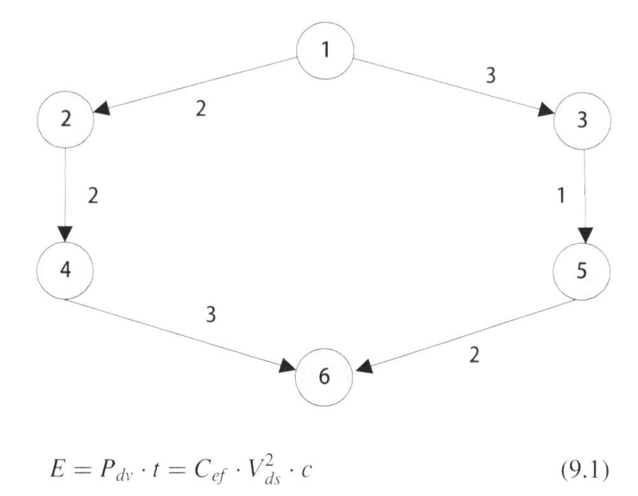

Fig. 9.1 Example of a task graph with six tasks

$$E = P_{dv} \cdot t = C_{ef} \cdot V_{ds}^2 \cdot c \tag{9.1}$$

where P_{dy}, C_{ef}, V_{ds} are the dynamic power, the workload capacitance, and the supply voltage, respectively.

It is noted that various task scheduling patterns generate different initial energy consumptions and the slack and idle times. While a lot of researches focus on how to scale the tasks' execution speed to appropriate operating voltage/speed, we research the impact of task scheduling algorithm on the energy consumption.

First row in the table shows the tasks, second row shows the information about the first machine, and third row shows the information about the second machine. Table 9.1 gives the processing time (initial energy consumption) of the tasks in Fig. 9.1. Tables 9.2 and 9.3 show the normal execution time and scaling execution time of each task with the deadline 90, and the system consumption decreases from 2,482.56 to 945.19. It is demonstrated that energy saving obtained by DVS strategy is different based on the various schedule patterns.

The objective function of the scheduling problem by using DVS strategy is to find a feasible schedule with minimum energy consumption.

9.3 ACO–DVS Algorithm

Huang presented an ACO (ant colony optimization) algorithm combined with taboo search for the job shop scheduling problem [11]. In this chapter, a real number matrix (PT_{nxm}) is used to imitate the pheromone trail of the real ants and is updated during the run of the algorithm. When a good solution is found by assigning task st_i

Table 9.2 Energy consumption of Fig. 9.1 on one specific scheduling

1	2	3	4	5	6
0–8.45	8.45–18.72	18.72–41.41		41.41–58.7	58.7–90
(0–8.0)	(8.0–18.0)	(18.0–37.0)		(37.0–52.0)	(52.0–76.0)
			20.72–55.7		
			(20.0–26.0)		

Table 9.3 Energy consumption of Fig. 9.1

1	2	3	4	5	6
0–11.96		11.96–43.11		43.11–67.02	
(0–8.0)		(8.0–27.0)		(27.0–42.0)	
	13.96–44.83		44.83–69.02		69.02–90.0
	(10.0–19.0)		(19.0–25.0)		(44.0–57.0)

on processor sp_j, pheromone is added to an element of the pheromone matrix T_{ixj}. The following ants of the next generation directly use the value of T_{ixj} to evaluate the promise of placing task on the processor sp_j. T_{ixj} is actually simplified by omitting the iteration counter k from $T_{ixj}(k)$ which the amount of pheromone depends on previously found solution and will be changed after the following iteration. In our ACO, each task is considered as an ant, and the pheromone trail of ants is imitated by the solution obtained by the list heuristic strategy.

9.3.1 Initialization Phase

In ACO, the generated schedules by artificial ants may be so coarse that they should be improved by some complementary local search method. The reason that the earliest application of ACO to the problem generates unsatisfactory results may be due to the lack of an appropriate local search.

A set of artificial ants is initially created according to initial schedule and the pheromone information obtained by using the list scheduling heuristics. The energy consumption of each task and the total energy consumption of the schedule are used to generate the initial pheromones to all ants.

$$pst_i = \begin{cases} c + a/ec_{ij} + b/initec & \text{if} \quad sch(i,j) = 1 \\ c & \text{else} \end{cases} \tag{9.2}$$

where $sch(i,j) = 1$ means that task st_i is processing on processor sp_j, and ec_{ij} is the initial energy consumption of task st_i, and $initec$ is the total energy consumption of the schedule obtained by the DVS algorithm.

9.3.2 Construction and DVS Phase

As the DVS slow down certain tasks by using available idle and slack time obtained following, the execution sequence of tasks is optimized by list scheduling strategy. The list scheduling heuristic is preferable than other heuristics in terms of the solution quality and time performance. The order of the tasks is obtained by using heuristic strategy, and then leads the ant to explore the new neighborhood based on the pheromone.

The execution sequence of the tasks is obtained by ordering the priority of tasks. The priority of tasks t_i is obtained as follows:

$$pst_i = se_{ij} + \max_{t_k \in tsu(t_i)} \left(ES_{ij}, se_{kj} + sc_{ij} \right) \tag{9.3}$$

where $tsu(t_i)$ is the conjunctive and disjunctive successors of task st_i, ES_{ij} is the earliest start time of task st_i if it is executed on processor sp_j.

The schedule is constructed according to the execution order obtained by the heuristic strategy, and then the processor is select based on the pheromone trail of the ants. The probability pr, a real number uniformly distributed in [0,1], is generated randomly. The pheromone trails for task st_i is calculated as follows:

$$pt_{ij} = \begin{cases} \max\left\{ [pt_{ij}]^{\alpha} \left[ech_{ij}\right]^{\mathrm{ij}}\beta \right\} & \text{if} \quad pr < thre \\ [pt_{ij}]^{\alpha} \left[ech_{ij}\right]^{\mathrm{ij}}\beta / \sum_{j=1}^{m} [pt_{ij}]^{\alpha} \left[ph_{ij}\right]^{\mathrm{ij}}\beta & \text{else} \end{cases} \tag{9.4}$$

where the heuristic information ech_{ij} is the sum of earliest finish time and energy consumption for assigning task st_i on processor sp_j. The value of heuristic is decreased from generation to generation and is increased the pheromone values to further improve the solution quality it has found so far.

When extended by a time quantum which is a slice of the slack, the energy consumption of task t_i during the iterative j is computed as:

$$E_i^k(e_i + \Delta e_i) = E_i^k(e_i) \cdot V_{ids}^2 / div \cdot V_{iUp}^2 \tag{9.5}$$

where div, the number of the processors which run the direct successor of the task t_i, is used to evaluate the extend of energy saving. V_{ids} is the scaling voltage of task t_i and is given as:

$$V_{ids} = V_{it} + V_{iC}/rd_i + \sqrt{(V_{it} + V_{iC}/rd)^2 - V_{it}^2} \tag{9.6}$$

The strategy allocates the slack or idle time to the task with most perspective to optimize the power consumption of the tasks. And then it can reduce the processors power consumption more effectively by unevenly dividing the slack or idle interval among the tasks. In the construction phase, each task is considered as an artificial ant, and the heuristic value is generated by some problem-dependent heuristic.

For lessening the influence of the old generations of ants, the corresponding pheromone information is updated by applying the local pheromone update rule as follows:

$$pst_{ij} = (1 - r) \cdot pst_{ij} + r \cdot C0 \tag{9.7}$$

Based on the best solution found so far and the evaporation of some portion of pheromone, the global pheromone information is updated as follows:

$$pst_{ij} = (1 - k)pst_{ij} + k/Btec \tag{9.8}$$

$Btec$ is the best found energy consumption of the obtained schedule, and r and k are the pheromone evaporation rate.

9.4 Experiments

The algorithm was implemented in eclipse and run on Intel Xeon processors with 1 GHz speed with 1 GB of memory. The performance of ACO–DVS algorithm is compared with and genetic algorithm. For this purpose, we consider randomly generated tests and some specific sets by using various graph sizes and the energy consumptions. Figure 9.2 demonstrates the comparison of the energy consumption obtained by ACO–DVS and genetic algorithm. The horizontal line represents the number of the tasks, and the vertical line represents the average energy consumption. The average energy consumption obtained by ACO–DVS is smaller than that obtained by genetic algorithm for graph size from 5 to 20.

The results show that the ACO–DVS outperforms the general genetic algorithm. And the energy saving is larger when the number of tasks is increased.

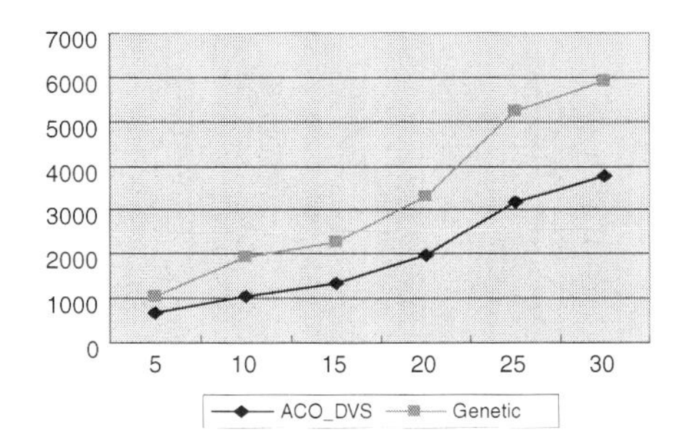

Fig. 9.2 Comparison of the energy consumption obtained by ACO–DVS and genetic algorithm

9.5 Conclusion

This chapter presented a combination of ACO algorithm and the DVS strategy for minimizing the energy consumption of the set of tasks executed on the heterogeneous distributed system. The heuristic information is obtained according to the energy consumption of the schedule system. The pheromone information and the list heuristic strategy are used to guide the search of the ants and generate the scheduling pattern. The factor of the heuristic and pheromone information is the variable based on the execution of the ant algorithm. Parameters of the algorithm are assigned by various values in the procedure of the ACO–DVS algorithm as parameter determines the convergence speed of the algorithm. At the beginning of the algorithm, parameter values are set to explore different regions of the search space and do not focus the search too early on a small region. At the end of the algorithm, parameter values are set to focus search near the best solution that has been found so far. The computational results on a set of random and specific instances testify the performance of the presented approach.

Acknowledgments This work has been supported by the Open Foundation of Key Laboratory in Software Engineering of Yunnan Province under Grant No. 2011SE03, Digital media technology and visualization of innovative communication platform under Grant No. 2012EI05, "CDIO-based software system modelling and design research and implementation" (Grant No. Rj14), and National Natural Science Foundation of China (Grant No. 60763008).

References

1. Yun, H. S., & Kim, J. (2003). On energy-optimal voltage scheduling for fixed priority hard real-time systems. *ACM Transactions on Embedded Computing Systems, 2*(3), 393–430.
2. Burd, T. D., Pering, T. A., Stratakos, A. J., & Brodersen, R. W. (2000). A dynamic voltage scaled microprocessor system. *IEEE Solid-State Circuits, 35*(11), 1571–1580.
3. Li, M. N., & Yao, F. (2005). An efficient algorithm for computing optimal discrete voltage schedules. *SIAM Journal on Computing, 35*(3), 658–671.
4. Schmitz, M. T., Al-Hashimi, B. M., & Eles, P. (2004). Iterative schedule optimization for voltage scalable distributed embedded system. *ACM Transactions on Embedded Computing Systems, 3*(1), 182–217.
5. Scordino, C., & Lipari, G. (2006). A resource reservation algorithm for power-aware scheduling of periodic and aperiodic real-time tasks. *IEEE Transactions on Computers, 12*(55), 1509–1522.
6. Pillai, P., & Shin, K. G. (2001). Real-time dynamic voltage scaling for low-power embedded operating systems. In *Proceedings of the 18th ACM symposium on operating system principles* (pp. 89–201). New York, NY: ACM.
7. Saewong, S., & Rajkumar, R. (2003). Practical voltage-scaling for fixed-priority RT-systems. In *Proceedings of the 9th IEEE real-time and embedded technology and applications symposium* (pp. 106–115). Washington, DC: IEEE CS Press.
8. Aydin, H., Devadas, V., & Zhu, D. (2006). System-level energy management for periodic real time tasks. In *Proceedings of the 27th IEEE international real-time systems symposium* (pp. 313–322). Washington, DC: IEEE CS Press.

9. Aydin, H., Melhem, R., Mossé, D., & Mejia-Alvarez, P. (2004). Power-aware scheduling for periodic real-time tasks. *IEEE Transaction on Computers, 53*(5), 584–600.
10. Liu, Y., & Mok, A.K. (2003). An integrated approach for applying dynamic voltage scaling to hard real-time systems. In *Proceedings of the 9th IEEE real-time and embedded technology and applications symposium* (pp. 116–123). Washington, DC: IEEE CS Press.
11. Huang, K. L., & Liao, C. J. (2008). Ant colony optimization combined with taboo search for the job shop scheduling problem. *Computers and Operation Research, 35*(1), 1030–1046.

Chapter 10
An Improved Ant Colony System for Task Scheduling Problem in Heterogeneous Distributed System

Yan Kang, Yifan Zhang, Ying Lin, and He Lu

Abstract Task scheduling problem is a major issue of distributed system. An improved ant colony algorithm is presented to solve the scheduling problem in heterogeneous distributed system whose complexity is known to be NP-complete in general cases. To speed up the converging rate of the algorithm, elite initial solutions are generated by using an adaptable list heuristic algorithm which is a good tradeoff between the computation complexity and solution quality. A novel representation of pheromone can make effectively use of the task fitness value to accumulate pheromone in ACO (ant colony optimization) algorithm. To improve the self-adaptability of the algorithm, the ACO algorithm and the heuristic rule are combined together to adjust the searching space on the progress of the algorithm. Finally, local and global neighborhood searching are performed on the best solution obtained in iterations. Simulation results show that the performance of the improve ACO algorithm is better on finding optimal or near-optimal solutions than general genetic algorithm.

Y. Kang (✉) • Y. Zhang
Department of Software Engineering, School of Software, Yunnan University, Kunming 650091, Yunnan, China
e-mail: kangyan@ynu.edu.cn

Y. Lin
Department of Information Security, School of Software, Yunnan University, Kunming 650091, Yunnan, China

H. Lu
School of Software, Yunnan University, Kunming 650091, Yunnan, China

W.E. Wong and T. Zhu (eds.), *Computer Engineering and Networking*, Lecture Notes in Electrical Engineering 277, DOI 10.1007/978-3-319-01766-2_10,

10.1 Introduction

Scheduling for the tasks of an application represented as a directed acyclic graph (DAG) is very important in both fields of scientific computation and commercial application. The task scheduling problem is a branch of scheduling problem, which is complex combinatorial optimization problem and is very difficult to solve. It is well known that task scheduling problem is NP-complete, and its optimal solutions are hard to be achieved with traditional optimization approaches owing to the high computational complexity [1]. The classical task scheduling problem consists in scheduling a set of tasks on a set of processors with the objective to minimize the makespan of the schedule, subject to the precedence constraints among the tasks.

Various approaches have been proposed to solve task scheduling problem in the homogeneous system as this problem has been widely studied in the literature. Recently, many researches are focus on the application scheduling for achieving high performance in heterogeneous systems with nonuniform task processing time [2]. Since task scheduling problem in heterogeneous systems cannot be solved to guarantee optimality even under simplified assumption, many heuristics have been proposed for giving a suboptimal in polynomial time.

In this chapter, we improve the effectiveness and efficiency of the algorithm by proposing a hybrid algorithm for the task scheduling problem. The proposed algorithm combines a new list heuristic rule and ACO approach for the task scheduling problem. The list scheduling heuristic generates good initial solution by using characteristics of the heterogeneous multiprocessor scheduling problem, and ACO enhances it iteratively. The proposed hybrid algorithm has high performance in terms of both performance metrics (schedule length ratio, speedup, efficiency, and frequency of best results) and a cost metric (scheduling time).

This chapter is organized as follows: the next sections briefly describe the scheduling problem and a presented list scheduling algorithms, respectively. In Sect. 10.4, we propose an improved ant algorithm. In Sect. 10.5, experiments are given to investigate the performance of the algorithm which Sect. 10.6 concludes the chapter.

10.2 Task Scheduling Problem Model

In the task scheduling problem, a finite set of tasks $T = \{t_1, t_2,\ldots, t_n\}$ that are divided from an application is processed on a finite set of machines $N = \{n_1, n_2,\ldots, n_m\}$ with deterministic processing times $E = \{e_{11}, e_{12},\ldots, e_{nm}\}$. $e_{i,j}$ is the execution time of task t_i on machine n_j that may be different on different processor depending on the processors computational capability. All tasks and their precedence constraints are represented as a DAG, and the precedence restriction means the successor task cannot be scheduled until all its processor tasks have been finished. The communication cost $CM = \{cm_{11}, cm_{12},\ldots, cm_{nn}\}$, $cm_{i,j}$ represents the cost of the data transfer from task t_i to its successor task t_j.

10.3 An Adaptable Scheduling Heuristic

10.3.1 Related Work

These heuristics are classified into a variety of categories such as list scheduling algorithms, clustering algorithms [3], genetic algorithms [4, 5], and task duplication-based algorithms [6, 7].

In list scheduling algorithms, the task in a list is constructed by assigning priority to it, and each task is assigned to the processor based on its priority. Several variant list scheduling algorithms have been proposed to deal with heterogeneous system, for example, Mapping Heuristic (MH), Levelized MinTime (LMT), Dynamic-Level Scheduling (DLS), Heterogeneous Earliest Finish Time (HEFT), and Critical Path On a processor (CPOP) [8, 9].

Ant colony optimization (ACO), one of the population-based metaheuristics dedicated to combinatorial optimization problems, has been successfully applied to a large number of discrete optimization problems, such as the traveling salesman problem and the vehicle routing problem, for which ACO was shown to be very competitive to other metaheuristics. Also, ACO has been applied successfully to task scheduling problems [10, 11].

With respect to list scheduling algorithms, it has been shown that minimizing the makespan of the tasks throughout the schedule is preferable to sequence the tasks in the descend order of their previously computed priorities, and then assign tasks on machines according to their sequence.

10.3.2 An Adaptable Scheduling Heuristic

If a task can be assigned to more than one predecessor and the computation cost of the task varies in the system, we think that the performance of the scheduling algorithm will be improved by considering the change ratio of the computation cost of the tasks. We present an elastic factor to imply the elastic of the task, i.e., the flexibility of the task performance. The priority of task t_i is given as

$$pt_i = \left(w_i + \max_{v_j \in succ(v_i)} (pt_j + cm_{ij}) \right)^* a + ef_i^* b, \qquad (10.1)$$

where $succ(t_i)$ is the set of immediate successors of task t_i, w_i, the time-weight of task t_i, is the average computation cost of task t_i. a, b are coefficients whose values are smaller than 1, and the sum of a,b is 1. And the value of the elastic factor ef_i is the computation time difference between the maximum computation time and the minimum computation time of task t_i.

The execution order (sequencing) of tasks is based on the priorities of all tasks. To assign priorities to all tasks, the upward rank of a task is computed as the critical path of that task.

Each task is scheduled onto the processor that gives the earliest completion time for the task. For the tasks to be scheduled, the value is computed recursively as shown in Eq. (10.2); all immediate predecessor tasks of t_i must have been scheduled.

$$EC_i = \min_{j \in n_i} \left(te_{ij} + \max_{t_j \in succ(v_i)} \left(AV_{ij}, e_{jk} + cm_{ij} \right) \right), \tag{10.2}$$

where AV_{ij} is the earliest time that processor n_k completed the execution of the last assigned task or the idle slot between the assigned tasks enough to schedule task t_i as early as possible. The inner max block in the EC equation returns the ready time, i.e., the time when all the data needed by t_i has arrived at processor n_j. The elastic factor of a task is given as a valid heuristic to minimize the schedule length. The solution is obtained by scheduling the tasks t_i onto processor n_j that gives the earliest finish time for the task.

10.4 Improved Ant Colony Optimization Algorithm

The general principle of ant algorithms is that the pheromone information reflects the outcomes of the decisions which have been made by former ants that found good solutions. In case of task scheduling problems, the decisions are which task to put on which place in the schedule. In our ACO, each task is considered as an ant, and the pheromone trail of ants is imitated by the solution obtained by the adaptable heuristic strategy. This pheromone trail, the executing situation of the processor, is updated during the run of the algorithm. The task is selected according to the order given by task prioritizing phase of the adaptable heuristic strategy. And ants select the processor based on its pheromone information and heuristic information. This is done to make explicit that the amount of pheromone depends on the current iteration and changes during the run of the algorithm. Yet, for simplicity, we omit the iteration counter.

10.4.1 Initialization Phase and Heuristic Rule

In ACO, the generated schedules by artificial ants may be so coarse that they should be improved by some complementary local search method. The reason that the earliest application of ACO to the problem generates unsatisfactory results may be due to the lack of an appropriate local search.

A set of artificial ants is initially created according to initial schedule and the pheromone information obtained by using the adaptable list scheduling heuristics given in Sect. 10.2. The execution situations of tasks give the pheromones to all ants, especially the execution situation of all tasks scheduled onto the same processor.

As in the task scheduling problem, the absolute position of a task is of importance; we use the following list scheduling strategy to generate the positions of tasks that represent the desire of setting task at the position in the sequence. And then ants build up new schedule by extending the already fixed prefix of the schedule and profit from existing list scheduling heuristics. The priority of task t_i is computed according to the former scheduling pattern as

$$pt_i = e_{ij} + \max_{t_k \in asu(v_i)} \left(AV_{ij}, e_{kj} + cm_{ij} \right), \tag{10.3}$$

where $asu(t_i)$ is the set of immediate successors of task t_i and the immediate task running after task t_i on the same processor n_j, cm_{ij} is the communication time between task t_i and its successor task t_j.

10.4.2 Construction Phase and Update Pheromone Rule

We construct a schedule by choosing the tasks based on the order given by the adaptable list heuristic strategy, and then a task can only be selected if all its predecessors are already in the partial schedule in each construction step. To assign each task onto a particular machine, an important issue is to define the pheromone trails for the machines to be assigned. When the random probability rp, a real number uniformly distributed in [0,1], is less than the given probability gp, an ant chooses the machine with maximal pheromone for the task to be scheduled.

$$vt_i = \max\left\{ [phr(i,j)][heu(i,j)]^{\beta} \right\} \tag{10.4}$$

When the random probability rp is larger than the given probability gp, a processor is chosen according to the following probability distribution.

$$pt_i = \frac{[phr(i,j)][heu(i,j)]^{\beta}}{\sum_{j=1}^{n} [phr(i,j)][heu(i,j)]^{\beta}}, \tag{10.5}$$

where the pheromone information $phr(i,j)$ is the last completion time for the processor n_j. And the heuristic information $heu(i,j)$ is the earliest start time for the task t_i on the processor n_j, which is given as

$$heu(i,j) = e_{ij} + AV_{ij} + \max_{t_k \in succ(t_i)} (e_{kj} + cm_{ik}), \tag{10.6}$$

where AV_{ij} is the earliest start time that processor n_j can process task t_i.

The heuristic and pheromone information are indicators of how good it seems to put task t_i on the processor n_j. Parameters β determine the relative influence of the pheromone values and the heuristic values on the decision of the ant.

After an artificial ant has been assigned to a machine, the corresponding pheromone information is updated by applying the local pheromone update rule as follows:

$$phr(i,j) = (1-\rho)phr(i,j) + \rho \cdot C, \tag{10.7}$$

where $0 < \rho < 1$ is the pheromone evaporation rate, C is the initial pheromone level. The effect of the local pheromone update rule is to make the choice of putting t_i on the processor n_j less desirable for other ants to achieve diversification. Consequently, this mechanism favors the exploration of different schedules.

10.5 Experiment and Results

The algorithms described in Sect. 10.4 were implemented in eclipse-jee-indigo-SR1-win32 and run on Intel Xeon processors with 1 GHz speed with 1 GB of memory. Figure 10.1 demonstrates a DAG with 10 tasks and 15 edges on which the communication times of each task are labeled and assumed to be the same on the whole network.

Table 10.1 shows the processing times of each task on three available processors in the heterogeneous computing system. Table 10.2 shows the minimum results obtained by the IACO algorithm, i.e., begin time and completion time of all tasks. Table 10.3 shows the minimum results obtained by the genetic algorithm. Comparison between the genetic algorithm and IACO algorithm is shown in Tables 10.2 and 10.3 under the used number of ants is 52 and initial genetic individuals is 52 after 10 iterative. The schedule length obtained by IACO algorithm is 47 which is less than the schedule length obtained by genetic algorithm is 66. The improvement ratio between the difference between two results and the larger result is 28.8 %.

The performances and cost of the algorithms were compared with respect to set of experiments with various graph characteristics. The results show that the IACO algorithm is more effective than genetic algorithm with improvement is larger than 10 %.

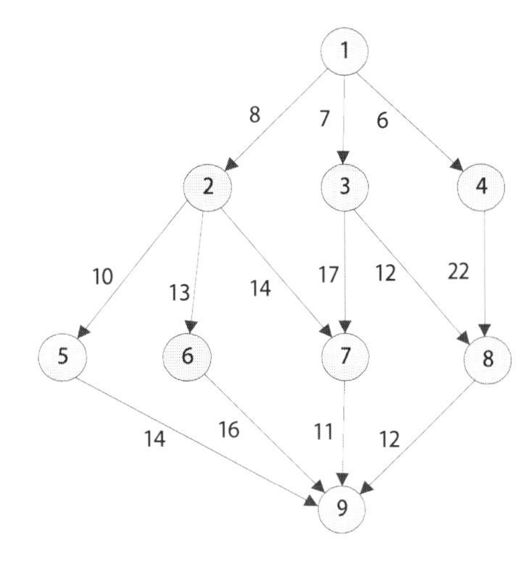

Fig. 10.1 Example of a task graph with 10 tasks

Table 10.1 Communication and computation time of Fig. 10.1

V	1	2	3	4	5	6	7	8	9
P1	8	16	9	15	8	8	11	9	12
P2	7	17	11	12	9	14	17	13	14
P3	10	12	15	14	7	16	13	15	19

Table 10.2 Schedule of Fig. 10.1 by IACO algorithm

V	1	2	3	4	5	6	7	8	9
P1			7–16		27–35	19–27			12
P2	0–7			7–19				19–32	14
P3		7–19					19–32		19

Table 10.3 Schedule of Fig. 10.1 by genetic algorithm

V	1	2	3	4	5	6	7	8	9
P1	0–8			8–23			36–47		12
P2		8–25	25–36		36–45				14
P3						38–54		23–38	19

10.6 Conclusion

This chapter presented a combination of ACO algorithm and the list heuristic algorithm for the task scheduling algorithm in the heterogeneous multiprocessor system. We present a heuristic strategy to generate the initial solution for the ACO algorithm. And then new ant can generate by using pheromone and heuristic information based on the characteristic of the problem. The feasible solution is obtained by sequence the task based on the priority given by the heuristic rule. The computational results on a set of random and specific instances testify the performance of the approach. And it is also demonstrated that the percentage of final schedule length is less than the initial one and the average improvement ratio are both sensitive to the graph structure and the initial one.

Acknowledgments This work has been supported by the Open Foundation of Key Laboratory in Software Engineering of Yunnan Province under Grant No. 2011SE03. "Digital media technology and visualization of innovative communication platform" under Grant No. 2012EI05, and National Natural Science Foundation of China (Grant No. 60763008).

References

1. Cassavant, T., & Kuhl, J. A. (1988). A taxonomy of scheduling in general purpose distributed memory systems. *IEEE Transactions on Software Engineering, 14*(2), 141–154.
2. Ilavarasan, E., & Thambidurai, P. (2007). Low complexity performance effective task scheduling algorithm for heterogeneous computing environments. *Journal of Computer Sciences, 3* (2), 94–103.
3. Kafil, M., & Ahmed, I. (1998). Optimal task assignment in heterogeneous distributed computing systems. *IEEE Concurrency, 6*(3), 42–51.
4. Wu, A. S., & Jin, S. Y. (2004). An incremental genetic algorithm approach to multiprocessor scheduling. *IEEE Transactions on Parallel and Distributed Systems, 15*(9), 824–834.
5. Kaur, K., Chhabra, A., & Singh, G. (2010). Heuristics based genetic algorithm for scheduling static tasks in homogeneous parallel system. *International Journal of Computer Science and Security, 4*(2), 183–189.
6. Ahmed, I., & Kwok, Y. (1998). On exploiting task duplication in parallel program scheduling. *IEEE Transactions on Parallel and Distributed Systems, 9*(9), 872–892.
7. Bajaj, R., & Agrawal, D. P. (2004). Improving scheduling of tasks in a heterogeneous environments. *IEEE Transactions on Parallel and Distributed Systems, 15*(7), 107–118.
8. Iverson, M., Ozguner, F., & Follen, G. (1995). Parallelizing existing applications in a distributed heterogeneous environments. In *Proceedings of heterogeneous computing workshop* (pp. 93–100). Washington, DC: IEEE CS Press.
9. Topcuoglu, H., Hariri, S., & Wu, M. Y. (2002). Performance effective and low-complexity task scheduling for heterogeneous computing. *IEEE Transactions on Parallel and Distributed Systems, 13*(3), 260–274.
10. Chen, L., & Pan, Q. (2011). Improved ant colony algorithm to solve identical paraHel machine task scheduling problem. *Computer Engineering and Applications, 47*(6), 44–48.
11. Deng, R., Cheng, H. Z., Wang, B., Wang, X. M., & Li, C. (2010). Permutation ant colony system for heterogeneous DAG scheduling problem. *Journal of Computer Science, 12*(37), 193–196.

Chapter 11
Optimization of Green Agri-Food Supply Chain Network Using Particle Swarm Optimization Algorithm

Qian Tao, Zhexue Huang, Chunqin Gu, and Chenxin Zhang

Abstract The green agri-food supply chain network (GASCN) design is critical to reduce the total transportation cost for efficient and effective supply chain management. This paper proposes a new solution based on particle swarm optimization (PSO) to find optimal solution for GASCN problem. PSO adopts transforming operator to modify particles in the population. The novelty of the transforming operator is that it can avoid applying the penalty function so that the diversity of populations is decreased. To show the efficacy of the algorithm, PSO is also tested on three cases. Results show that the proposed algorithm is promising and outperforms GA by both optimization speed and solution quality, especially when the scale of problem is large.

Q. Tao (✉)
Department of Computer Science, Guangdong University of Education, Guangzhou 201305, China

Shenzhen Institutes of Advanced Technology, Chinese Academy of Sciences, Shenzhen 201305, China
e-mail: taoalex66@gmail.com

Z. Huang
Shenzhen Institutes of Advanced Technology, Chinese Academy of Sciences, Shenzhen 201305, China

C. Gu
Department of Computer Science, Zhongkai University of Agriculture and Engineering, Guangzhou 201305, China

C. Zhang
School of Software Engineering, Sun Yat-Sen University, Guangzhou 201304, China

W.E. Wong and T. Zhu (eds.), *Computer Engineering and Networking*, Lecture Notes in Electrical Engineering 277, DOI 10.1007/978-3-319-01766-2_11,

11.1 Introduction

With the increasing demand for green agri-food, the green agri-food supply chain network (GASCN) design problem has been gaining focus of more and more researchers recently. Due to increasing competitiveness, the logistics firms are obliged to maintain high customer service levels at the same time they are forced to reduce total transportation cost for profit maximization.

As any other SCN, the GASCN is a network of organizations in order to bring agri-food to customers. The network involves planting bases (PB), distribution centers (DC), and points of sales (POS) with purpose of satisfying POS's demands, beginning with the PB of green agri-food, following with DC and ending with POS such as vegetable markets and supermarkets. Traditionally, manufacturing, distributing, and marketing along the supply chain operated independently. These organizations have their own objectives, and these objectives are often conflicting. But there is a need for a mechanism through which these different functions can be integrated together [1].

As for the SCN problem, the models and solutions for industrial products are relatively massive, but the solutions of the SCN about green agri-food are limited. In literature, there are many different studies on the design problem of supply networks, and these studies have been reviewed by Erenguc et al. [2] and Pontrandolfo et al. [3]. Amiri [4] has presented a Lagrangian relaxation approach to minimize the total cost of two-stage supply chain. Costa et al. [5] have worked on three stages of SCN optimization problem. Most of the researchers have concentrated on the improvement of the supply chain performance, and very few have considered the performance improvement of the algorithm concurrently [6].

Due to the NP hardness of SCN problems [7] and the large-sized problems in the real world, intelligent algorithms such as genetic algorithms (GA) [8] have been proposed to solve the SCN problems. Particle swarm optimization (PSO) [9], which is an important optimization tool as well as a continuous optimization method, was proposed by Kennedy and Eberhart in 1995. As a swarm intelligence algorithm, PSO simulates preying behaviors of bird flocking and fish schooling and searches for the optimal solution iteratively. Since PSO was put forward, it has attracted much attention from many scholars especially in the research fields of multi-objective problems [10]. The standard PSO has great advantage in searching for non-inferior solutions in a multidimensional complex space.

In this paper, an enhanced PSO is proposed in order to solve the GASCN problem for optimization. The proposed algorithm can be applied to SCN problems. The distinct feature of PSO is that it adopts transforming operator to modify chromosomes in the population and uses effective genetic operations. The transforming operator can assure that the solutions are always feasible. The performance of the proposed PSO has been compared with the state-of-the-art GA [8]. Results show that the proposed algorithm can achieve high-quality solutions with a much faster optimization speed.

The remainder of this paper is organized as follows. Section 11.2 presents the statement of the optimization problem. Section 11.3 describes the implementation of the proposed algorithm in detail, including the encoding method of particles, the design of transforming operator during the process of initialization of chromosomes, the design of fitness function, the identification of the local best position and global best position, and updating the particle's velocity and position. In Sect. 11.4, a series of experiments are conducted, and the results are analyzed to illustrate the performance of the proposed algorithm. Finally, in Sect. 11.5, this paper is concluded, and suggestions are given for future research.

11.2 Problem Statement

In order to optimize performance of GASCN, we should design the distribution network strategy that will satisfy demand requirement for the products imposed by points of sales. The problem is a SCN design problem for single green agri-food. The main objective to solve such an optimization problem of GASCN is to evaluate the selection of different planting bases or set of the planting bases and different distribution centers or set of the distribution centers, whereas the performance criteria are the minimization of the total transportation cost. Simultaneously, the demand for green agri-food from each point of sales must be satisfied. The assumptions used in this problem are as follows.

Planting bases, distribution centers, and points of sales are known. The requirement for green agri-food from each point of sales is known. Green agri-food and distribution centers are enough for distribution. The transportation costs of green agri-food on the path from each planting base pb_i to each distribution center dc_j and from each dc_j to each point of sales pos_k are known. Green agri-food distributed to pos_k are supplied by a dc_j or multi-dc_j, and green agri-food distributed to dc_j are supplied by a pb_i or multi-pb_i.

Suppose there are planting bases $PB = \{pb_1, pb_2, \ldots, pb_{|PB|}\}$; distribution centers $DC = \{dc_1, dc_2, \ldots, dc_{|DC|}\}$, points of sales $POS = \{pos_1, pos_2, \ldots, pos_{|POS|}\}$; a green agri-food SCN $GASCN \subseteq PB \times DC \cup DC \times POS$, where "×" is Cartesian product; and demand of POS $Req_{POS} = \left\{Req_{pos_1}, Req_{pos_2}, \ldots, Req_{|POS|}\right\}$, the objective of optimization of GASCN is to find $PB_i = \left\{pb_{i1}, pb_{i2}, \ldots, pb_{i|PB_i|}\right\} \subseteq PB$, $DC_j = \left\{dc_{j1}, dc_{j2}, \ldots, dc_{j|DC_j|}\right\} \subseteq DC$, $GASCN = \{(pb_i, dc_j), (dc_j, pos_k))|pb_i \in PB_i, dc_j \in DC_j, pos_k \in POS\}$, satisfying

$$\min f = TQ_i^j \times Cost_{pb_i}^{dc_j} + TQ_j^k Cost_{dc_j}^{pos_k}$$

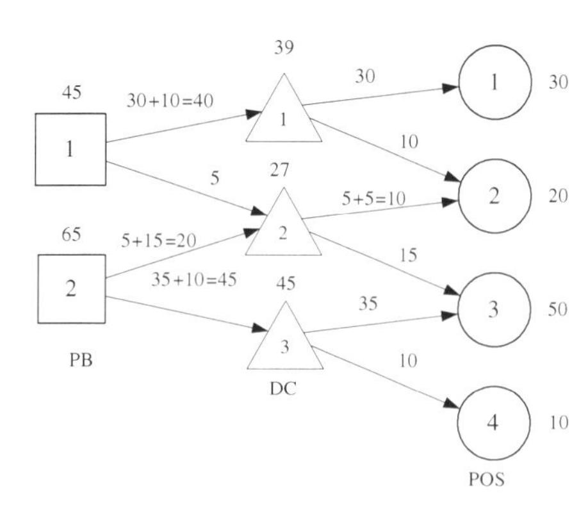

Fig. 11.1 A GASCN application used in our experiments

Table 11.1 Notations

Symbol	Descriptions
PB	Set of planting bases
pb_i	Planting base i
DC	Set of distribution centers
dc_j	Distribution centers j
POS	Set of points of sales
pos_k	Points of sales k
TQ_i^j	Transportation quantity from node i to node j
X_i	Particle i
x_i^n	The position of dimension n particle i
M	Number of particles in the population
N	Total number of impossible paths
n	The index of paths, $n \in [1,N]$
Req_{pos_k}	The requirement of point of sales k
$f\_pb(n)$	The map from n to a planting base
$g\_dc(n)$	The map from n to a distribution center
$h\_pos(n)$	The map from n to a point of sales
$Cost_i^j$	Unit transportation cost from node i to node j

$$Req_{pos_k} = TQ_{dc_j}^{pos_k}$$

A GASCN is shown in Fig. 11.1. Table 11.1 lists the notations used in this paper.

11.3 Proposed PSO Algorithm

This paper proposes a PSO algorithm for optimization of green agri-food supply chain networks. In PSO, a novel transforming operator is applied. In addition, the method of representation is the critical to solve the optimization problem. In this

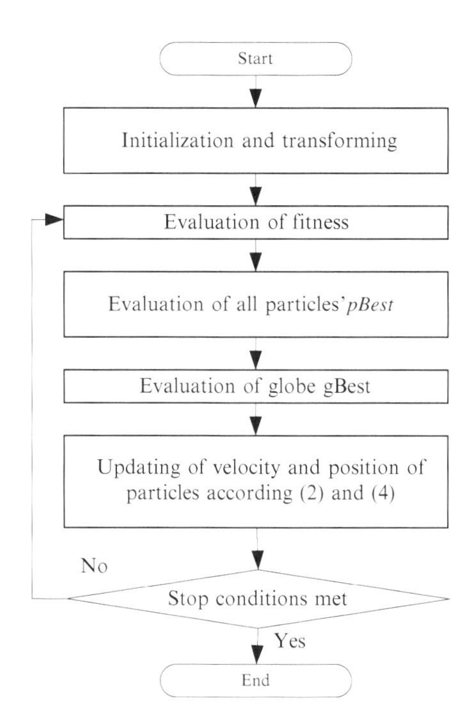

Fig. 11.2 Flowchart of the proposed PSO

section, the particle's representation is firstly described. Then the particle's initialization, the evaluation of fitness, and the identification of the local and global best position, updating the particle's velocity and position, are presented. A complete flowchart of the proposed PSO is shown in Fig. 11.2.

11.3.1 Encoding Method of a Particle

The encoding of particles is the premise of using particle swarm to search for an optimal solution in a GASCN problem. In the PSO, the number of particle dimensions is equal to the number of paths from *PB* to *POS* by *DC*, and each dimension corresponds to the transportation quantity of green agri-food on each path. In Fig. 11.2, the GASCN has 2 *PB*, 3 *DC*, and 4 *POS*, and the particles are encoded by 24 ($2 \times 3 \times 4 = 24$) dimensions. The number of solution in the searching space of a particles is $|PB| \times |DC| \times |POS|$. Figure 11.3 gives an illustration of a searching of particle swarm in a 24-dimensional space.

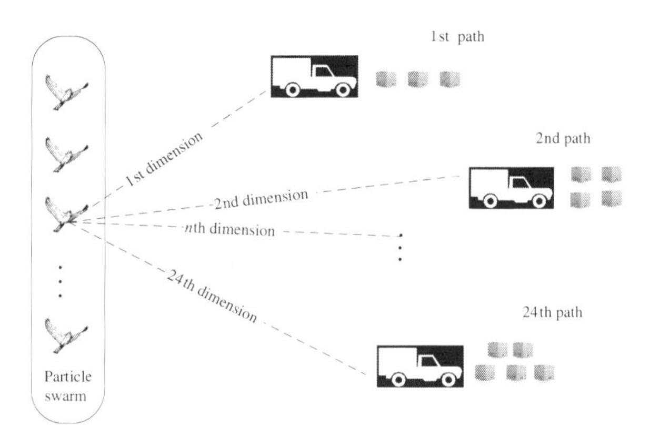

Fig. 11.3 An illustration of searching of particle swarm in a 24-dimensional space

11.3.2 Evaluation of Fitness

The fitness function of a particle X_i in the population is defined as

$$fitness_i = \sum_{n=1}^{N} x_i^n \times \left(Cost_{f\_pb(n)}^{g\_dc(n)} + Cost_{g\_dc(n)}^{h\_pos(n)} \right) \tag{11.1}$$

where $f\_pb(n) = \frac{n}{|DC| \times |POS|}$, $g\_dc(n) = \frac{n}{|POS|} \% |DC|$, $h\_pos(n) = n\%|POS|$, $Cost_{f_p b(n)}^{g_p b(n)}$ is the transportation cost of green agri-food from $PB_{f\_pb(n)}$ to $DC_{g\_dc(n)}$, and $Cost_{g_d c(n)}^{h_p os(n)}$ is the transportation cost from $DC_{g\_dc(n)}$ to $POS_{h\_pos(n)}$. Moreover, every particle is evaluated according to Eq. (11.1).

11.3.3 Updating the Particle's Velocity and Position

$$\begin{aligned} v_i^n(t+1) = \omega v_i^n(t) + c_1 \cdot r_1 \cdot \left(pBest_i^n(t) - x_i^n(t)\right) \\ + c_2 \cdot r_2\left(gBest_n(t) - x_i^n(t)\right) \end{aligned} \tag{11.2}$$

$$\omega = 1.0 - gen^{*}0.6/499 \tag{11.3}$$

$$x_i^n(t+1) = x_i^n(t) + v_i^n(t+1) \tag{11.4}$$

The position and velocity of the particle are updated by Eqs. (11.2) and (11.4), respectively. In Eq. (11.2), ω is the inertia weight and is calculated by Eq. (11.3), c_1, c_2 are acceleration constants (also known as cognitive parameters and social parameters), and r_1, r_2 are the independent random variables evenly distributed within [0, 1]. If the new particle does not satisfy the requirement of points of sales, the transforming operator is applied.

11.4 Experimental Study and Discussions

The proposed PSO for optimization of green agri-food supply chain network was performed on a machine with Pentium D 3.00GHz CPU and 1024MB of RAM. The operating system was MS Windows XP, and the compiler was VC++ 6.0. We tested the PSO on Case_1, Case_2, and Case_3 from small scale to large scale (Table 11.2).

The following parameters were used in the PSO: $M = 20$, $c_1 + c_2 = 1.8 + 1.8 < 4$. To validate the proposed PSO, we compared the PSO with GA [10] to tackle the various problems in GASCN. Experiments were independently simulated 30 times. Table 11.3 presents the comparison of GA and PSO. To validate the reliability and robustness of PSO, we tested two algorithms with a longer running time and 10,000 iterations. The mean fitness, best fitness, the average time in second used for obtaining the best result in each run of the two algorithms are compared with each other. The statistical results are shown in Table 11.3. From Table 11.3, GA outperforms PSO on Case_1 in speed, but as the increasing of cases scale, PSO outperforms GA both in the solution quality and in the optimization speed.

Table 11.2 Test cases

Test cases	$\lvert PB \rvert$	$\lvert LC \rvert$	$\lvert ST \rvert$	Scale	Req_{st_k}	$Cost_i^j$
Case_1	2	3	4	24	[10, 50]	[1, 10]
Case_2	10	15	16	2,400	[10, 50]	[1, 10]
Case_3	15	20	25	7,500	[100, 500]	[1, 10]

Table 11.3 Convergence speed and the solution accuracy comparisons

	GA			PSO		
Test cases	Mean fitness	Best fitness	Time (s)	Mean fitness	Best fitness	Time (s)
Case_1	758.978	750.229	17.656	590	590	24.757
Case_2	2,878.92	2,874.12	421.563	1,253.598	1,110	171.829
Case_3	78,239.3	76,546.6	2,867.44	39,694.79	36,667.6	689.686

[a]Convergence speed being measured on the mean CPU time needed to reach an acceptable solution; solution accuracy being measured on the best fitness and the mean fitness on 30 independent trials

11.5 Conclusion

In this paper, we have proposed a PSO for optimization of green agri-food supply chain network. PSO uses float-point representation to encode the SCN problems. In PSO, a transforming operator is used to keep solutions in feasible ones when total transportation number does not meet the demands of each POS. Experiments were conducted on three cases, and the experimental results demonstrate the superiority and effectiveness of the proposed algorithm for green agri-food supply chain network. In our future research, to the best of our knowledge, we will study more effective and efficient solution methodology based PSO algorithm.

Acknowledgments This work is supported by the Special Funds for Doctoral Research Project of Guangdong University of Education under Grant No. 2012ARF05, the Natural Science Foundation of Guangdong Province of China under Grant No. S2012020011067 and the National High Technology Research and Development Program of China (863) under Grant No. 2012AA101701.

References

1. Altiparmak, F., & Gen, M. (2006). A genetic algorithm approach for multi-objective optimization of supply chain networks. *Computers and Industrial Engineering, 51*(1), 197–216.
2. Erenguc, S. S., Simpson, N. C., & Vakharia, A. J. (1999). Integrated production/distribution planning in supply chains: an invited review. *European Journal of Operational Research, 115*(2), 219–236.
3. Pontrandolfo, P., & Okogbaa, O. G. (1999). Global manufacturing: A review and a framework for planning in a global corporation. *International Journal of Production Economics, 37*(1), 1–19.
4. Amiri, A. (2006). Designing a distribution network in a supply chain system: Formulation and efficient solution procedure. *European Journal of Operational Research, 171*(2), 567–576.
5. Costa, A., Celano, G., Fichera, S., & Trovato, E. (2010). A new efficient encoding/decoding procedure for the design of a supply chain network with genetic algorithms. *Computers and Industrial Engineering, 59*(4), 986–999.
6. Prakash, A., Chan, F. T. S., Liao, H., & Deshmukh, S. G. (2012). Network optimization in supply chain: A KBGA approach. *Decision Support Systems, 52*(2), 528–538.
7. Altiparmak, F., Gen, M., Lin, L., et al. (2006). A genetic algorithm approach for multi-objective optimization of supply chain networks. *Computers and Industrial Engineering, 51*(1), 196–215.
8. Henry, C. W. L. (2009). Cost optimization of the supply chain network using Genetic Algorithms. *IEEE Transactions on Knowledge and Data Engineering, 99*(1), 1–36.
9. Kennedy, J., & Eberhart, R. C. (1995). Particle swarm optimization. In *Proceedings of IEEE international conference on neural networks* (pp. 1942–1948). Piscataway, NJ: IEEE Press.
10. Heo, J., Lee, K., & Garduno-Ramirez, R. (2006). Multiobjective control of power plants using particle swarm optimization techniques. *IEEE Transactions on Energy Conversion, 21*(2), 552–561.

Chapter 12
A New Model for Short-Term Power System Load Forecasting Using Wavelet Transform Fuzzy RBF Neural Network

Jingduan Dong, Changhao Xia, and Wei Zhang

Abstract Power load changes periodically. And the effects of climatic (precipitation, relative humidity, temperature, wind speed) on the load should be fuzzy. In order to solve the problem, this chapter presents a method combining wavelet transform, fuzzy set concept, and neural networks for short-term load forecasting. Through the wavelet transform, the load sequence decomposes into subsequences consisting of different wavelet coefficients. On the other side, by the fuzzy neural network, the samples of five meteorological factors influencing power load are transformed into fuzzy input with the subsequences, and then, the suitable RBF neural networks for the forecasting are selected. Finally, the load forecasting sequence is obtained by the reconstruction of the forecasted results from the subsequences. The simulation results demonstrate the proposed method possesses validity and practicability with a mean absolute error below 1.5 %.

12.1 Introduction

Power load forecasting is to study the change rule and variation trend of the electric load. Accurate power load forecasting can improve the security and stability as well as economy of the power systems. Because of the factors such as climate, economy, price, and policy many more efforts have been made towards the application and improvement of power load forecasting techniques in order to adapt to the impact of these factors. Short-term load forecasting is an essential part of power system operation mainly used in optimal unit commitment, generation schedule, hydro-thermal coordination, power exchange schedule, and so on [1]. Accurate prediction can forecast the load variation and lower power loss for power grid in advance.

J. Dong (✉) • C. Xia • W. Zhang
College of Electrical Engineering and New Energy, China Three Gorges University, Yichang 443002, China
e-mail: 76471786@qq.com

W.E. Wong and T. Zhu (eds.), *Computer Engineering and Networking*, Lecture Notes in Electrical Engineering 277, DOI 10.1007/978-3-319-01766-2_12,

However, the power load is uncertain and changeable by external factors. The study shows that the impact of different climatic factors on the electricity load is also different from each other. Climate change is one of the important reasons of load fluctuations which make further research on improving the prediction methods [2].

There are various methods used in load forecasting over the years. The traditional short-term load forecasting methods are time series, autoregressive average, exponential smoothing, etc. The prediction models in these mathematical expressions cannot meet the requirements of power systems nowadays. The modern prediction algorithm artificial neural network (ANN) will be introduced in this chapter for the reason that the network can approximate any nonlinear function then model the performance and learn characteristics. ANN has been widely applied in short-term load forecasting in power system.

This chapter presents a wavelet transform fuzzy neural network model. It uses the subsequences of the load and the transformed climate fuzzy variable as the input of the neural network. The subsequences of the load are decomposed by wavelet transform, and the climate variables are transformed by membership functions. Then, the predicted results are obtained by the reconstruction through inverse wavelet changes to get the load sequence of the forecast day. The simulated results demonstrated that the accuracy meets the demand.

12.2 Wavelet Decomposition and Reconstruction

To process data scientifically and rationally is the most basic part to improve the load forecasting precision. For the short-term load forecasting, load data also need the process in an appropriate way. It is because of the power load changes periodically and will also fluctuate with the users' contingency change and the variety of climate conditions (precipitation, relative humidity, temperature, wind speed) at the same time that the various types of load signal generally performs as a continuous spectrum. This load type can be transformed into nonperiodic components, and these load components have different frequency characteristics including low and high frequencies [3]. By this way, the load transforms into different scales of wavelet coefficients. Then, the feature vector composed by the wavelet coefficients replaces the load data for neural network training. Finally, the load forecasting sequence was obtained by the reconstruction of the forecasted results from the output.

The historical power load data can be treated as time series signals. An original signal can be expressed by a multi-resolution representation. In order to give a brief introduction on multi-resolution decomposition, we need to take a three-tier multi-resolution decomposition, for example. The signal decomposition processes is shown in Fig. 12.1, where S means the original signal, c is low-frequency component, and d is high-frequency component.

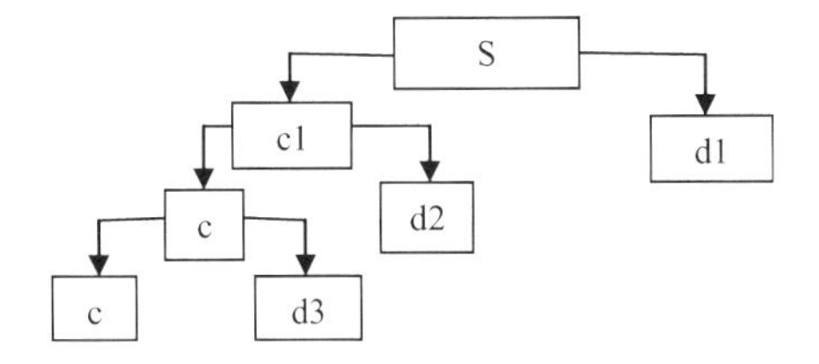

Fig. 12.1 The decomposition process

As we can see from the figure, multi-resolution analysis is just the decomposition of low-frequency part, and the high-frequency part will not be considered further. What is more, the multi-resolution analysis is essentially an iterative process of signal analysis between the two adjacent resolutions [4].

Mallat algorithm is employed in this chapter for wavelet decomposition and reconstruction. Using Mallat algorithm, the known signal can be decomposed into different scales. Mallat decomposition algorithm uses the extraction method that the subsequence is shortened to half of the original sequence. So it greatly reduces the amount of computation of the wavelet transform that improves computing speed which is conducive to processing large amounts of information. Briefly, in the space of discrete sequence, under the prerequisite of Shannon sampling theorem, the Mallat algorithms use a low-pass filter $h(k)$ and a high-pass filter $g(k)$ to the sampling signal binary band division. Through observing the signal in different frequency bands, the signal feature can be extracted. Besides, Daubechies function is selected as the mother wavelet.

The decomposition algorithm has the form of

$$c_{j+1,k} = \sum_{l \in Z} c_{j.l} \overline{h_{l-2k}} \tag{12.1}$$

$$d_{j+1,k} = \sum_{l \in Z} c_{j.l} \overline{g_{l-2k}} \tag{12.2}$$

On the contrary, the algorithm also provides the simple way to reconstruct the original single:

$$c_{j,k} = \sum_{l \in Z} h_{k-2l} c_{j+1,l} + \sum_{j \in Z} g_{k-2l} d_{j+1,l} \tag{12.3}$$

12.3 Wavelet Transform Fuzzy Neural Network Forecasting Model

The neural network model based on the method of this chapter is shown in Fig. 12.2 where $x_1 \sim x_m$ are original load inputs and $x_{m+1} \sim x_n$ are original climatic factors inputs. The way to deal with load sequence via wavelet transform will not be given

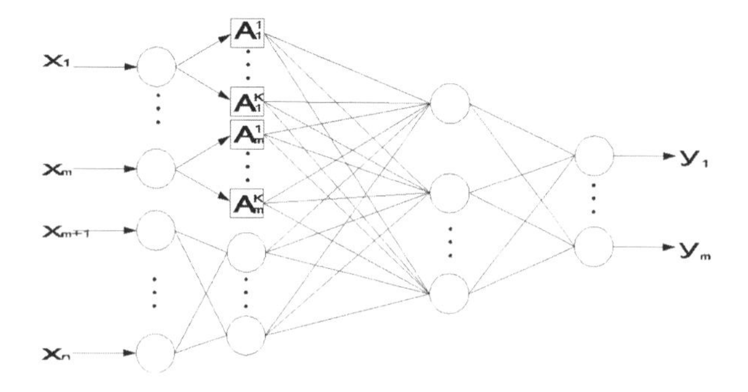

Fig. 12.2 Wavelet transform combined with fuzzy RBF neural network forecasting model

unnecessary details here. As for climate data, they are turned into fuzzy input via membership functions. The learning network chooses radial basis function (RBF) neural network which is similar to nonlinear continuous function with uniform approximation so as to speed learning process.

This new model considers the climatic factors affecting load, the analysis of the uncertainty factors, load data processing, and the basic forecast model. It combines these algorithms in an optimal way which ensures the accuracy of the prediction.

12.3.1 Fuzzification of Meteorological Factors

Fuzzy neural network combines fuzzy theory with neural network. The fuzzy logic is inserted into the fuzzy neural network instead of Boolean logic, so that the network can obtain the ability of fuzzy inference to improve the learning speed. Fuzzy neural network just blurs some of elements and retains the basic properties and structure.

The fuzzy set is the collection of elements that with some specific characters in some degree. The characteristic function which describes the elements' degree of membership in a fuzzy set is membership function. Furthermore, fuzzy set theory is the method of mapping from region to region while ANN mapping from point to point. Through combining the merits of the two, the network can be applied to different power systems.

Natural factors (typical micro factors such as temperature, precipitation, and wind speed) will lead to short-term relative volatility of the load [5]. Lots of load forecasting models just consider the influence of temperature on load. Obviously, temperature is an important indicator of the load fluctuation [6]. However, some other natural factors will affect human comfort so that cause fluctuations of power load. The fact is people tie between meteorological factors and load change. Specifically, meteorological factors affect the comfort of the human and thereby affect human behavior which will lead the load change [7].

There are five primary factors (i.e., daily maximum temperature, daily minimum temperature, precipitation, relative humidity, wind speed) which will be considered as the fuzzy input. The selection of the membership functions is based on the output and input functions linear or not. Due to the inputs above are mainly linear factors, we elect 2~3 membership functions for simplified precondition here. For these inputs, use fuzzy number of triangular or trapezoidal distribution to express the input spatial distribution. Therefore, we can simultaneously obtain multiple fuzzy membership degree of the input if a specific input value has been given.

12.3.2 Radial Basis Function Neural Network for Training

The RBF neural network involves three layers with only one hidden layer which applies a nonlinear transformation form input space to hidden space in its most basic form. Comparing to BP network, it has a higher learning efficiency and function approximation.

The input in this chapter is a feature vector constituted by the load coefficients of wavelet and fuzzy meteorological factors. And the input-output mapping of RBF network has the following form:

$$f(x) = \sum_{j=1}^{m} w_j \varphi\left(\left\|x - x_j\right\|\right) \tag{12.4}$$

where m is the number of hidden nodes, w_j is the adjustment weight, $\varphi(\|x - x_j\|)$ is a set of m arbitrary functions, and $\|.\|$ refers to Euclidean norm.

For regularization RBF network itself, while it is in the process of learning, the hidden layer's activation functions evolve slowly in accordance with some nonlinear optimizations strategies, and output layer's weights adjust themselves rapidly through a linear optimization strategy. Here, take the method of fixed center selected at random for learning and employ an isotropic Gaussian function whose standard deviation is fixed according to the spread of the centers. In effect, the standard deviation of the Gaussian radial basis functions is fixed at

$$\delta = \frac{d_{\max}}{\sqrt{2P}} \tag{12.5}$$

where $d_{\max}$ is the maximum distance between the chosen centers and P is the number of centers. This formula ensures that the individual RBF is not too peaked or too flat. Both of these two extreme conditions should be avoided.

Then, the weights of output layer adopt least mean square algorithm (LMS), and the weight adjustment takes the following form:

$$\Delta W_k = \eta\left(d_k - W_k{}^T\Phi\right)\Phi \tag{12.6}$$

In formula (12.6), W_k is the linear weight in the output layer of the network, d_k is the desired respond, and Φ is an m-by-m matrix with elements.

Only a small number of connection weights need to be adjusted by training. It is because of this characteristic that the RBF neural network has a relatively higher learning speed. And RBF network proves to be effective in power load forecasting [8].

12.4 Practical Example

In this practical example, the method above is used to deal with 24-h-ahead load forecasting. In order to verify the reliability of the algorithm, we select the vagaries of climate on season transition. Hubei Yichang is just the city where the weather is changeable and rainy. Choose the historical hourly load data (24 points a day, a total of 720 points, load measured in MW) and corresponding climate data from October 1, 2005, to October 30, 2005, of Yichang, Hubei, China. The previous 29 days' data are taken as the training samples into the forecasting model for learning, and then, the data sequence of October 29 is used to forecast the power load data of October 30.

The load sequences adopt Mallat algorithm with layer 3 scale decomposition. Besides, take the method of periodic extension to get biorthogonal compactly supported wavelets db4 (Daubechies function as the mother wavelet) thus obtaining the wavelet coefficients. At the same time, select five climate factors (daily maximum temperature, daily minimum temperature, precipitation, relative humidity, and wind speed) related to the load fluctuation and choose respectively appropriate trapezoidal membership functions or triangular membership functions to obtain the fuzzy inputs. This step can be achieved easily by Matlab programming language. The neural network model is built via Matlab R2010a simulation software for simulation training and prediction.

The training error curve is shown in Fig. 12.3. As we can see from the figure, the network has a high convergence speed, and the number of iteration is just 27. It is a short time before the target error is met. And this is one of the reasons why we choose the RBF neural network. After the simulating, the mean absolute percentage error (MAPE) is used to appraise the performance of the network on the test. From the simulation result we can easily get that the network has a mean absolute error of 1.4775 % whose forecast value is almost identical with the actual value. This forecast accuracy meets the power system operation needs. The specific experiment results are shown in Table 12.1 and Fig. 12.4.

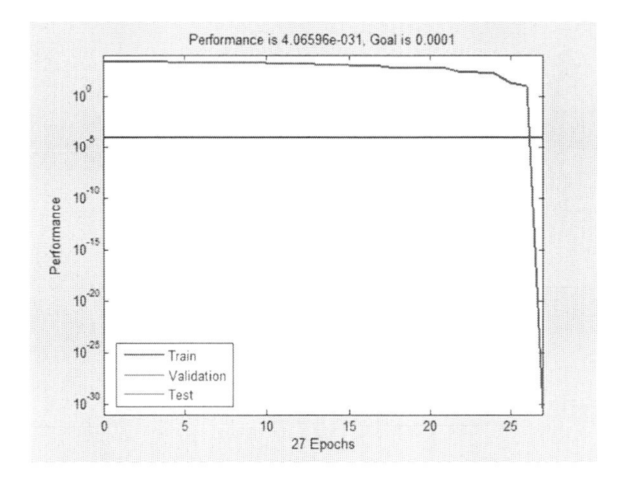

Fig. 12.3 The error curve of the network

Table 12.1 Comparison between the forecast load and actual load

Hours	Actual load	Forecasting load	Error (%)	Hours	Actual load	Forecasting load	Error (%)
1	737	748.1	1.44	13	766.2	764.3	−0.25
2	734.5	747.3	1.74	14	790.3	782.8	−0.95
3	740.1	753.8	1.85	15	797.5	788.6	−1.12
4	720.8	732.4	1.61	16	811.4	788.3	−2.85
5	715.5	729.2	1.91	17	814.9	793.9	−2.58
6	718.5	728.7	1.42	18	833.0	841.6	1.03
7	749.1	759.6	1.40	19	845.6	829.4	−1.92
8	806.0	810.9	0.61	20	807.8	793.1	−1.82
9	798.3	800.3	0.25	21	774.7	769.1	0.72
10	764.5	750.4	−1.84	22	740.1	755.5	2.08
11	747.0	729.5	−3.27	23	763.3	753.8	1.24
12	766.2	741.7	−0.71	24	755.2	748.8	0.85
MAPE = 1.4775							

12.5 Conclusion

The predictive model established in this chapter for short-term load forecasting combines wavelet transform with fuzzy neural network. The short-term power system load is periodic and nonlinear that the chapter puts forward a model considering different scales of load as well as the impact made by climate. This method highlights the relevance of the power systems and outside environment impact.

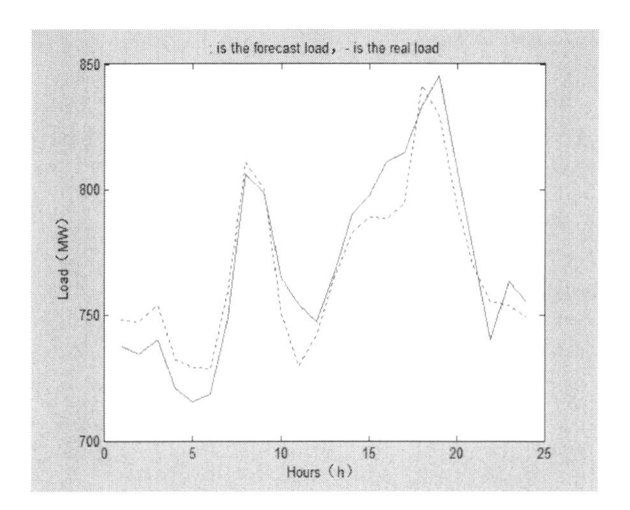

Fig. 12.4 Curves of the forecast and actual load

It proves to be viable and applicable through the practical example. The forecasting precision meets the demand for power system operation. Although the impacts of climate have been considered in this model, the degrees of influences vary from each other. Future work is to deal with the diversity to improve the prediction accuracy of the forecasting.

References

1. Kang, C., Xia, Q., & Zhang, B. (2004). Review of power system load forecasting and its development. *Automation of Electric Power Systems, 28*(17), 1–11 (in Chinese).
2. Khotanzad, A., Zhou, E., & Elragal, H. (2002). A neuro-fuzzy approach to short-term load forecasting in a price-sensitive environment. *IEEE Transactions on Power System, 17*(4), 1273–1282.
3. Tai, N., & Hou, Z. (2003). New short-term load principle with the wavelet transform fuzzy neural network for the power systems. *Proceedings of the CSEE, 24*(1), 24–29 (in Chinese).
4. Hu, C., Li, G., & Zhou, T. (2008). *Systems analysis and design based on MATLAB7.x* (pp. 25–27). Xi'an: University of Electronic Science and Technology Press (in Chinese).
5. Guiqin, F., & Li, Y. (2008). Influence analysis of meteorological conditions on electric loads. *Meteorological Science and Technology, 36*(6), 795–800 (in Chinese).
6. Soozanchi-K, Z. (2010). Modeling and forecasting short-term electricity load based on adaptive neural-fuzzy inference system by using temperature. *IEEE Signal processing systems*, 18–22.
7. Du, Y., & Lin, L. (2006). Analysis of the effect of compositive meteorology index on power load. *Journal of Chongqing University, 29*(12), 56–60 (in Chinese).
8. Han, M., Xu, Z., & Yu, Y. (1994). Electric load-forecasting using RBF neural networks. *Journal of North China Electric Power University, 21*(4), 1–6 (in Chinese).

Chapter 13
Energy-Effective Frequency-Based Adaptive Sampling Algorithm for Clustered Wireless Sensor Network

Meiyan Zhang, Wenyu Cai, Liping Zhou, and Jilai Liu

Abstract The objective of wireless sensor networks is to extract the synoptic structures (spatiotemporal sequence) of the phenomena of ROI (region of interest) in order to make effective predictive and analytical characterizations. Energy limitation is one of the main obstacles to the universal application of wireless sensor networks. Recently, adaptive sampling strategy is regarded as a much promising method for improving energy efficiency. In this paper, we dedicate to investigating how to regulate sampling frequency of sensor nodes in different clusters dynamically following the change of signal frequency. The adaptive frequency-based sampling (FAS) algorithm proposed in this literature is to measure periodic signal frequency online in different clustered region, afterwards regulate signal sampling frequency following with minimal necessary frequency criterion; as a result, the previous desired level of accuracy is achieved, and the energy consumption is decreased. The simulation results are compared with that of fixed sampling rate approach with respect to energy conservation.

13.1 Introduction

Wireless sensor networks have received considerable academia research attention in present years [1]. Wireless sensor networks consist of a large number of tiny sensor nodes deployed over a geographical area; each node is a low-power device that integrates computing, communication, and sensing abilities. The key application of

M. Zhang • L. Zhou • J. Liu
Electric Engineering Department, Zhejiang University of Water Resources and Electric Power, Hangzhou 310018, China
e-mail: Meiyan19831109@163.com

W. Cai (✉)
School of Information Engineering, Hangzhou Dianzi University, Hangzhou 310018, China
e-mail: dreampp2000@163.com

W.E. Wong and T. Zhu (eds.), *Computer Engineering and Networking*, Lecture Notes in Electrical Engineering 277, DOI 10.1007/978-3-319-01766-2_13,

wireless sensor networks is monitoring physical phenomena and acquiring environment information. Typically, each sensor node collects raw sensory data to the users through network interconnection for further analysis. The simplest way is to permit each sensor node to deliver its raw sensory data to the base station periodically, where the data can be assembled for subsequent analysis. However, this approach results in excessive communication, and the energy is wasteful. Energy limitation is one of main obstacles to the universal application of wireless sensor networks. In recent years, several energy management schemes have been proposed to reduce power consumption in the literatures, the novel one is adaptive sampling. It is emphasized that reducing the amount of acquired data by using adaptive sampling techniques also reduces the energy consumption for communication. A detailed survey shows that data acquisition and processing have energy consumption, which is significantly lower than that of communication [2].

Generally, data acquisition and processing consume energy that is significantly lower than that of communication. Therefore, traditional researches are concerned with conserving energy as much as possible by reducing transmission capability. Unfortunately, this assumption does not always hold in a number of practical applications. Therefore, this paper proposes a frequency-based adaptive sampling (FAS) algorithm to solve the problem of how to regulate sampling frequency of sensor nodes dynamically following the change of signal frequency. The key idea of this paper is to measure periodic signal frequency online, afterwards adjust signal sampling frequency to the real needs of the physical phenomena under observation. As a result, these two complementary requirements can be achieved: the previous desired level of accuracy is achieved through guaranteeing certain minimal sampling rate for reconstruction, e.g., Nyquist sampling frequency; moreover, the energy consumption is decreased through reducing sampling frequency rate of sensor nodes as much as possible. The FAS algorithm does not assume any hypothesis with regard to the observed signal; therefore, it is of more general applicability. The sensory signal evolution rates of different clustering regions are different for large-scale wireless sensor networks with many clusters; therefore, signal sampling rate ought to adjust subsequently following signal evolution tendency of different clusters.

13.2 Related Works

The problem of energy-efficient transmission has been investigated with certain technology such as mathematical optimization in the current papers [3]. However, most researches are concerned with energy-efficient transmission, not energy-efficient sampling. Unfortunately, data acquisition or sampling will consume much more energy than data transmission in a number of practical applications because acquisition times are typically longer than transmission or some sophisticated acquisition process such as multimedia sensor networks. Therefore, some researchers have been investigating energy-efficient sampling scheme. Temporal correlation is used in an adaptive sampling algorithm for minimizing the energy consumption of a snow sensor [4]. A similar approach has been suggested that the sampling rate is adapted based on the outcome of a Kalman filter [5]. Adaptive sampling is also proposed that a flood alerting system is presented. The system includes a flood predictor that is used to adjust the reporting rate of individual node [6]. Other researchers are discussing how to perform energy-efficient sensory data sampling [7].

Generally, adaptive sampling schemes can be divided in two categories: one is adaptive spatial sampling scheme and the other is adaptive temporal sampling scheme. Adaptive spatial sampling schemes assure monitor accuracy using region location adjustment or wake-up state scheduling. Adaptive temporal sampling schemes assure monitoring accuracy using sampling frequency adjustment or online model estimation of signal tendency. Certainly, there are a few mutational adaptive sampling schemes such as multi-scale adaptive sampling, which provides multi-resolution sensory information as possible as required.

There are only a few researches concerned to temporal sampling schemes for wireless sensor networks. Lygouras presented a velocity-adaptive measurement system for closed-loop position control that relies on the adaptation of the sampling frequency to improve the response time [8]. Aplippi proposed an adaptive sampling algorithm (ASA) that adapted the sampling frequencies of the sensors to the evolving dynamics of the process. The ASA reduced the power consumption of the measurement phase by adapting the real needs of the physical phenomena under observation; however, it is a centralized approach but not adaptable for large-scale wireless sensor networks [9]. In this paper, we dedicate to studying adaptive temporal sampling by regulating sampling rate adaptively and proposing a novel adaptive frequency-based sampling algorithm. To the best of our knowledge, this is the first novel approach for clustered adaptive frequency-based technology. By adaptively sampling ROI of different clusters, the sampling rate decreases energy consumption; moreover, the necessary sensory accuracy is guaranteed.

13.3 FAS Algorithm

As we know, following the most famous Nyquist law in signal processing field, if the highest frequency $F_{\max}$ of signal is given, the minimum sampling frequency F_N is more than at least twice the highest frequency so that signal reconstruction is guaranteed [10]. In particular, it is common to pick a sampling frequency 3–5 times higher than the signal maximum frequency, i.e., $F_N = cF_{\max}$, where $c = 3$–5 generally [11].

First of all, the principle of adaptive frequency sampling algorithm proposed in this paper is explained with Fig. 13.1, where a specific signal with amplitude shifting with time is figured. Fixed sampling rate by Nyquist law is illustrated in Fig. 13.1a, and over-sampling and under-sampling cases are illustrated in

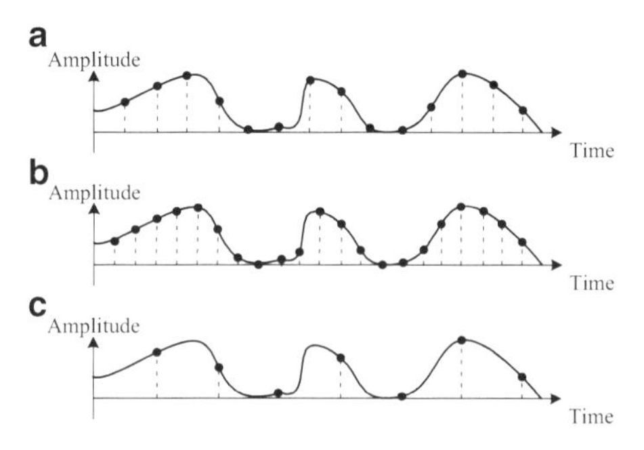

Fig. 13.1 Over-sampling and under-sampling

Fig. 13.1b, c. Obviously, over-sampling denotes that the sampling rate is much higher than necessary Nyquist sampling rate, and under-sampling denotes that the sampling rate is low than necessary Nyquist sampling rate. Consequently, design appropriate and necessary sampling rate to satisfy certain accuracy and minimal energy consumption is a promising but difficult mission.

Unfortunately, signal maximum frequency is not available a priori and changes over time in a nonstationary process. Consequently, the sampling frequency has to change adaptively following change tendency. For large-scale wireless sensor networks with many clusters, signal evolution rates of different clustering regions are different; therefore, signal sampling rate ought to adjust subsequently following signal evolution tendency of different clusters. Therefore, our proposed FAS algorithm must be operated on different clustering regions. We will illustrate the details of three procedures that form FAS approach.

13.3.1 Clusters Construction

In the first phase, sensor nodes establish different clusters autonomously and elect cluster head (CH) in a fully distributed fashion. Our design proposes a simple distributed clustering algorithm named maximum energy and minimum distance (MEMD) clustering algorithm to establish one-hop clusters, which is described below. First of all, the weight value of sensor node to elect CHs is defined as

$$\omega(i) > \omega(j) \Leftrightarrow E(i) > E(j) \text{ or } E(i) = E(j) \ \&\& \ id(i) > id(j) \tag{13.1}$$

where $E(i)$ and $E(j)$ denote the residual energy of sensor node i and j, respectively, and $id(i)$ and $id(j)$ denote ID number of sensor node i and j, respectively. It is obvious that the defined weight is positive correlation with node residual energy and the ID value of sensor node is the second factor. More specially, sensor nodes with more residual energy within all the neighbor nodes should be chosen to be cluster heads with higher probability, thus implementing maximum energy first principle. After cluster heads election, each cluster member sensor node will select the nearest CH and join that cluster. It is required that the distance between each cluster member and its nearest CH must be smaller than maximal transmission power radius R_{max}.

13.3.2 Intra-cluster Sampling Frequency Regulation

Within each cluster, sensor nodes are divided into two categories: one particular sensor node is elected as CH, which is responsible for data aggregation, and the other sensor nodes as cluster members (CMs) that are responsible for data acquisition. The most important assumption of this paper is the region of interest (ROI) within each cluster is data correlated and variation identical. Therefore, the sampling frequencies of sensor nodes in the same cluster ought to be same. The distributed nature of FAS algorithm is that different sampling frequency regulation processes are operated in different clusters at the same time; therefore, the FAS algorithm is appropriate even for large-scale sensor networks.

The primary and difficult problem is to estimate the maximal frequency of phenomena signal within certain cluster. Unfortunately, precise estimation of signal maximal frequency is hard, so we use a heuristic method for sampling frequency modification in this paper. In each iteration, CH broadcasts a certain sampling frequency to all CMs which is derived by current sensory accuracy. If the acquired sensory accuracy is lower than the threshold value θ, then increases sampling rate $f_c = f_c + \beta$ ($\beta = (Th_{max} - Th_{min})/M$). On the contrary, if the acquired sensory accuracy is larger than the threshold value θ, then decreases the sampling rate $f_c = \alpha \times f_c$ ($\alpha = 0.5 - 0.8$). The sampling frequency regulation scheme obeys multiplicative increase additive decrease (MIAD) rule. The details are illustrated in Fig. 13.2. The parameters used in this algorithm comprise of α, β, Th_{max}, Th_{min}, M, and T. The sensory accuracy value θ is a prefixed threshold and calculated by model-based correlation mechanism, which will be illustrated in our further paper.

In order to improve transfer reliability in wireless sensor networks, several retransmission manners such as ACK are used. Therefore, the impact of communication unreliability will influence the performance of FAS algorithm; nevertheless, the influence of the communication reliability has not been studied in this paper.

13.3.3 Inter-clusters Data Delivery

The final phase is inter-clusters data delivery from CHs to the base station. Although sampling frequencies of different clusters are different, CHs transmit aggregated data to the base station with the same frequency T. Actually, the delivery ratios of different CHs can be different with different regions, because the signal evolution over time in each region is different. The relay method between sensor nodes is the same with traditional routing schemes.

Fig. 13.2 Pseudo-code of intra-cluster sampling frequency regulation

```
#Cluster i Sampling Frequency Regulation procedure
begin
Given initial sampling frequency f_i;
CH broadcast f_i;
while(each Time interval T)
{
CH receive data and aggregation;
CH calculate sensory accuracy;
if(sensory accuracy ≥ θ)
   f_c = α × f_c        (α = 0.5 − 0.8)
else
   f_c = f_c + β        (β = (Th_max − Th_min) / M)
if(f_c ≥ Th_max)   f_c = Th_max; if(f_c ≤ Th_min)   f_c = Th_min
CH broadcast f_c;
}
end
```

13.4 Performance Evaluation

In this section, we describe the evaluation settings and the metrics we have chosen for the evaluation. In order to verify the performance of FAS algorithm, some simulations are carried out with a series of fancied data with mutable and limited signal frequency. The evaluation metrics comprise of sampling frequency f_c and energy consumption rate comparing with fixed sampling rate. In our simulations, some important parameters of FAS algorithm are defined in Table 13.1.

The half value of actual sampling frequency $f_c/2$ and the signal frequency are described in Fig. 13.3. The original sampling frequency is defined as $(Th_{\max} + Th_{\min})/2 = 50$ Hz. It is obvious that the change of sampling frequency drops behind that of signal maximal frequency with windows $M = 10$ samples from Fig. 13.3. Moreover, the actual sampling frequency approaches the maximal frequency of signal. Therefore, the sampling frequency derived from FAS algorithm is better than fixed sampling frequency.

The energy consumption ratio is defined as the rate of actual sampling frequency with maximal frequency threshold, which is set as 100 Hz in our simulations. The results are illustrated in Fig. 13.4. As we know, if the actual signal frequency increases, the sampling rate ought to increase, thus consuming much more energy; therefore, energy consumption ratio will increase because of the fixed maximal frequency threshold $Th_{\max}$.

Table 13.1 FAS parameters for simulations

Parameter	Value
Th_{max}	80 Hz
Th_{min}	20 Hz
M	10
α	0.5–0.8
T	10 samples

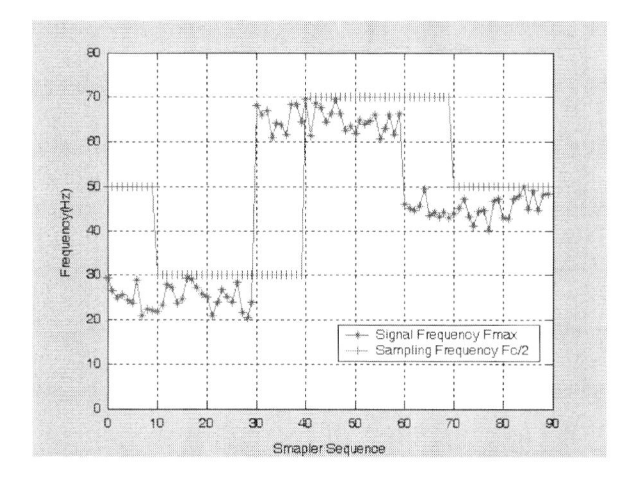

Fig. 13.3 Sampling frequency following signal frequency evaluation

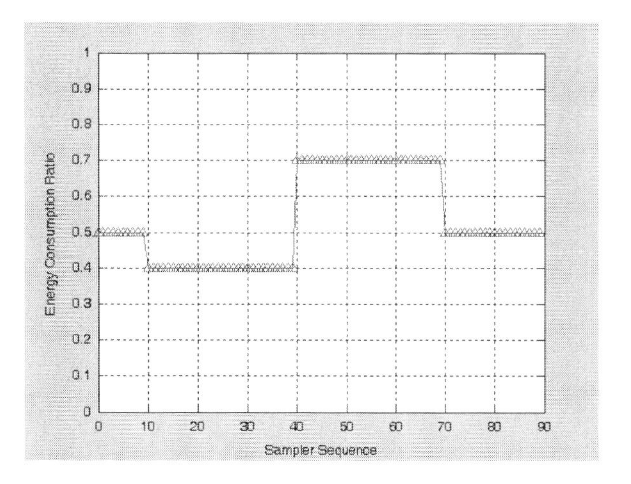

Fig. 13.4 Energy consumption ratio

13.5 Conclusion

In this paper, we proposed an energy-efficient adaptive sampling algorithm based on adaptive sampling frequency, which regulates signal sampling frequency of related sensor nodes so as to reduce energy consumption intelligently. However, most of the proposed solutions are limited to either temporal or spatial correlation.

Our future work is dedicated to finding more energy-efficient approach using the spatiotemporal correlation of wireless sensor networks. Moreover, in our next simulations, we will use real sensory data, which is derived from Intel Berkeley Research lab.

Acknowledgements This research was supported by the National Natural Science Foundation of China (No. 61102067) and the Natural Science Foundation of Zhejiang Province (No. LQ12F03006).

References

1. Akyildiz, I. F., Weilian, S., Sankarasubramaniam, Y., & Cayirci, E. (2002). A survey on sensor networks. *IEEE Communication Magazine, 40*(8), 102–114.
2. Anastasi, G., Conti, M., Di Francesco, M., & Passarella, A. (2009). Energy conservation in wireless sensor networks. *Ad Hoc Networks, 7*(3), 537–568.
3. Ordonez, F., & Krishnamachari, B. (2004). Optimal information extraction in energy-limited wireless sensor networks. *IEEE Journal on Selected Areas in Communications, 22*(6), 1121–1129. Special Issue Fundamental Performance Limits Wireless Sensor Networks.
4. Alippi, C., Anastasi, G., Galperti, C., Mancini, F., & Roveri, M. (2007). Adaptive sampling for energy conservation in wireless sensor networks for snow monitoring applications. In *Proceedings of the MASS 2007* (pp. 1–6). Pisa, Italy: IEEE.
5. Gedik, B., Lu, L., & Yu, P. S. (2004). ASAP: An adaptive sampling approach to data collection in sensor networks. *IEEE Transactions on Parallel and Distributed Systems, 18*(12), 10–16.
6. Zhou, J., & De Roure, D. (2007). FloodNet: Coupling adaptive sampling with energy aware routing in a flood warning system. *Computer Science Technology, 22*(6), 121–130.
7. Gedik, B., Liu, L., & Yu, P. S. (2007). ASAP: An adaptive sampling approach to data collection in sensor networks. *IEEE Transactions on Parallel Distributed Systems, 18*(12), 1766–1783.
8. Lygouras, J. N., Lalakos, K. A., & Tsalides, P. G. (1998). High-performance position detection and velocity adaptive measurement for closed loop position control. *IEEE Transactions on Instrumentation and Measurement, 47*(4), 978–985.
9. Aplippi, C., Anastasi, G., Francesco, M. D., & Roveri, M. (2010). An adaptive sampling algorithm for effective energy management in wireless sensor networks with energy-hungry sensors networks. *IEEE Transactions on Instrumentation and Measurement, 59*(2), 335–344.
10. Jerri, A. J. (1977). The Shannon sampling theorem-its various extensions and applications: A tutorial review. *Proceedings of the IEEE, 65*(11), 1565–1595.
11. Rauth, D. A., & Randal, V. T. (2005). Analog-to-digital conversion part 5. *IEEE Instrumentation and Measurement Magazine, 8*(4), 44–54.

Chapter 14
An Indoor Three-Dimensional Positioning Algorithm Based on Difference Received Signal Strength in WiFi

Yibo Li and Xiting Liu

Abstract To further solve the problem that using the positioning algorithm directly based on received signal strength (RSS) in WiFi technology has lower positioning accuracy because the characteristics of the wireless channel affect the signal attenuation largely and randomly, an indoor three-dimensional positioning algorithm based on difference received signal strength (DRSS) is proposed through the analysis that two close propagation paths have similar interference. It can reduce the large positioning error caused by time-varying interference and directly use the fluctuant values of the interfered received signal. Meanwhile, the wireless signal attenuation model with a parameter of time-varying environment factor is used. A method of real-timely estimating and modifying the parameters by least square estimation (LS) and the way of average is proposed to solve the problem that the model cannot describe the real-time changes of signal attenuation accurately. The environmental test results show that this method not only can obtain a more accurate model but also has higher positioning accuracy in three-dimensional multi-interference environment.

14.1 Introduction

Indoor positioning in WiFi technology has became a hot research with the growing popularity of WiFi access points. It can be used in many fields such as indoor airship positioning and personnel positioning [1]. The walls, pillars, personnel walking, and humidity can influence the signal attenuation [2]. The difficulty of indoor positioning is how to locate accurately based on large fluctuations of received signal strength values.

Y. Li (✉) • X. Liu
College of Automation, Shenyang Aerospace University, Shenyang 110136, China
e-mail: lyb20040612@aliyun.com

W.E. Wong and T. Zhu (eds.), *Computer Engineering and Networking*, Lecture Notes in Electrical Engineering 277, DOI 10.1007/978-3-319-01766-2_14,

The positioning based on received signal strength (RSS) in WiFi includes model-based method and model-free method. The model-free method mainly uses fingerprint database and ray-tracing technology [3, 4]. The model-based method is used widely and mostly improved by separating section or adding different compensation factors to describe the signal attenuation more accurately [1, 5]. The estimation of the parameters in the model is very important. Usually, the empirical values and the fixed values which are estimated before positioning are used [6]. In recent years, parameters are dynamically revised by mean value method [7, 8].

To further improve the positioning accuracy, the algorithm based on difference received signal strength (DRSS) is proposed through the analysis that two close propagation paths have similar interference. It can reduce the large positioning error caused by directly using the interfered received signal values. A method of real-timely modifying the parameters by least square estimation (LS) and the way of average is also proposed to make the model describe the real-time changes of signal accurately. The environmental test results show that this method has higher positioning accuracy in three-dimensional multi-interference environment.

14.2 Analysis of the Model

The signal attenuation model with an environmental factor is used, as below.

$$P_r(d) = P_r(d_0) - 10n\log_{10}\left(\frac{d}{d_0}\right) + h \tag{14.1}$$

Thereinto, $P_r(d)$ and $P_r(d_0)$ express the received signal strength at the distance of d and d_0 from the transmitter, and they are measured in decibel. d_0 expresses the distance of the fixed receiver to the transmitter, and the value is generally set to 1 m. h is environment factor which compensates the real-time changes of the environment. The exponent n is path loss exponent which means the degree of the signal strength attenuating with the propagation distance. The value is two in ideal free space, but it increases in practical environment. Meanwhile, there are extra losses because of the uncertain multi-interference factors such as the temperature, humidity, and personnel walking. So the values of n and h are time-varying and influence the positioning accuracy largely.

14.3 Proposed Approach

14.3.1 Real-Time Estimation and Correction of Parameters

Multigroup signal strength values received by the fixed reference receivers from the fixed transmitters with known position are used to estimate the values of n and h by least square estimation (LS) and the way of average. Then, use the present parameter values in the model to calculate the current position. To describe conveniently, a access points and b fixed receivers are set with known positions, and a mobile receiver is used as the object to be located. The access points are divided into m ($m \geq 4$) groups, and each group has two access points with little distance. Use AP_{ij} with the coordinate (x_{ij},y_{ij},z_{ij}) to express the ith group and jth access point ($i = 1, 2, \ldots, m$; $j = 1, 2$). FR_λ with the coordinate (x_λ,y_λ,z_λ) expresses the λth ($\lambda = 1, 2, \ldots, b$, $\lambda \geq m$) fixed receiver, and each fixed receiver has a group of access points at a distance of approximately 1 m at least. MR expresses the mobile receiver. The structure of the positioning is shown in Fig. 14.1.

The steps of estimating the model parameters are introduced as follows.

(1) The nearest group of access points to the FR_λ is found out and set to AP_i which consists of AP_{i1} and AP_{i2}.
(2) According to the above model, the signal strength value received from AP_{i1} by FR_λ is used as $P_r(d_0)$, and the distance between FR_λ and AP_{i1} is used as d_0. The values received from other $2(m-1)$ groups of access points by FR_λ are used as $P_r(d)$, and the corresponding distance is used as d. Then, $C_{2(m-1)}^2$ values of n and h can be obtained through the model by least square estimation, and they can be expressed as $\phi_{\lambda 1} = \left\{ n_1 \quad n_2 \quad \cdots \quad n_{C_{2(m-1)}^2} \right\}$ and $\psi_{\lambda 1} = \left\{ h_1 \quad h_2 \quad \cdots \quad h_{C_{2(m-1)}^2} \right\}$. In the same way, other $C_{2(m-1)}^2$ values of

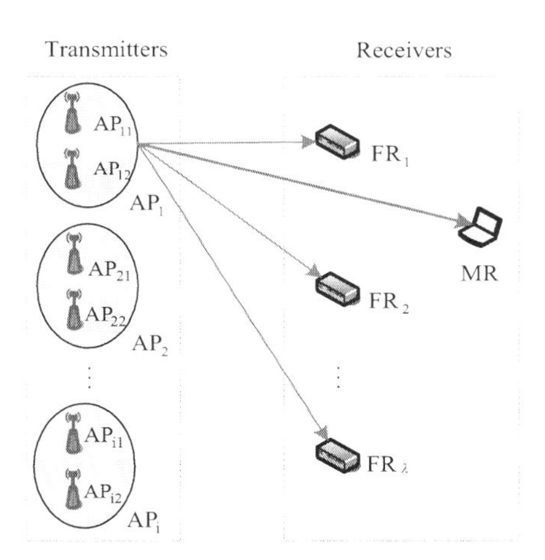

Fig. 14.1 The structure of positioning

n and h can be obtained by using the values received from AP_{i2} by FR_{λ} as $P_r(d_0)$. They can be expressed as $\phi_{\lambda 2} = \left\{ n_1 \quad n_2 \quad \ldots \quad n_{C^2_{2(m-1)}} \right\}$ and $\psi_{\lambda 2} = \left\{ h_1 \quad h_2 \quad \ldots \quad h_{C^2_{2(m-1)}} \right\}$. Then, the averages of n and h can be calculated based on only one fixed receiver FR_{λ} by Eq. (14.2).

$$n_{\lambda} = \frac{\sum \phi_{\lambda 1} + \sum \phi_{\lambda 2}}{2C^2_{2(m-1)}}, \quad h_{\lambda} = \frac{\sum \psi_{\lambda 1} + \sum \psi_{\lambda 2}}{2C^2_{2(m-1)}} \tag{14.2}$$

(3) Because b fixed receivers are used, the current values of the path loss exponent and the environment factor can be estimated by Eq. (14.3).

$$n = \frac{\sum_{\lambda=1}^{b} n_{\lambda}}{b}, \quad \zeta = \frac{\sum_{\lambda=1}^{b} \zeta_{\lambda}}{b} \tag{14.3}$$

Then, the model with time-varying environment factor and path loss exponent can be obtained by the substitution of the estimated values in real time.

14.3.2 The Positioning Algorithm Based on Difference Received Signal Strength

Directly using the signal strength value can cause large error because the value includes theoretical value and interference value which is difficult to estimate, as below.

$$P_r(d) = P_{rth}(d) + P_{rin}(d) \tag{14.4}$$

Thereinto, $P_{rth}(d)$ expresses the theoretical value, and $P_{rin}(d)$ expresses the interference value. In general indoor positioning, the wireless signal transmitted by a transmitter will arrive at a receiver through some paths. So the wireless signal transmitted by two very close transmitters will arrive at the same receiver through nearly the same paths, and they will have almost the same interferences and the interference values, as shown in Fig. 14.2.

Based on this idea, the difference received signal strength is presented to reduce the time-varying and random influences, as below.

$$P_r(d_{i1}) - P_r(d_{i2}) = [P_{rth}(d_{i1}) - P_{rth}(d_{i2})] + [(P_{rin}(d_{i1}) - P_{rin}(d_{i2})] \tag{14.5}$$

Thereinto, $P_r(d_{i1})$ and $P_r(d_{i2})$ express the signal strength values received by mobile receiver from two transmitters AP_{i1} and AP_{i2}. d_{i1} and d_{i2} express the

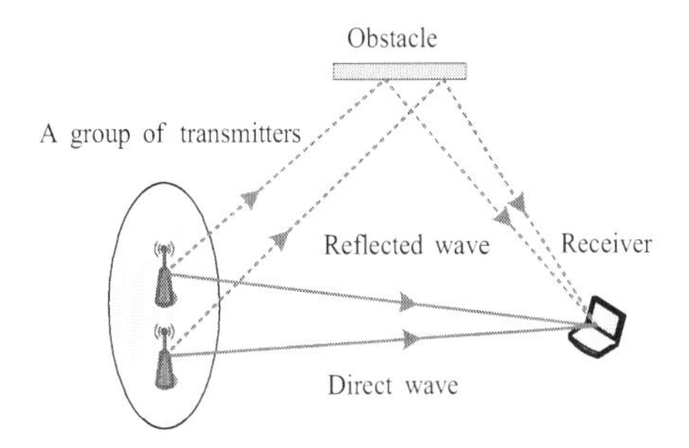

Fig. 14.2 The signal propagation paths of two close transmitters

distance between mobile receiver and AP_{i1} , AP_{i2}. So, combining with the model, Eq. (14.6) can be obtained.

$$\frac{d_{i2}}{d_{i1}} = \frac{1}{\eta_i} \cdot 10^{\frac{RSSIe_{i12}}{10n}} \tag{14.6}$$

Thereinto, $\eta_i = \frac{d_{i01}}{d_{i02}}$. d_{i01} and d_{i02} separately express the distance between the nearest fixed receiver and AP_{i1} and AP_{i2}. $RSSIe_{i12} = RSSI_{i12} - RSSI_{i012} - \zeta_{i12}$, $RSSI_{i12} = P_r(d_{i1}) - P_r(d_{i2})$, $RSSI_{i012} = P_r(d_{i01}) - P_r(d_{i02})$, $\zeta_{i12} = \zeta_{i1} - \zeta_{i2}$. $P_r(d_{i01})$ and $P_r(d_{i02})$ express the signal strength values received by a fixed receiver from AP_{i1} and AP_{i2}. ζ_{i1} and ζ_{i2} express the environment factors of the signal received from AP_{i1} and AP_{i2}.

According to the relation between distance and coordinates shown in Eq. (14.7),

$$d_{ij} = \sqrt{\left(x_{ij} - x\right)^2 + \left(y_{ij} - y\right)^2 + \left(z_{ij} - z\right)^2} \tag{14.7}$$

Equation (14.8) is obtained through a further derivation of Eq. (14.6).

$$\begin{aligned} R^2 &+ \frac{2(x_{i2} - \gamma_i^2 x_{i1})}{(\gamma_i^2 - 1)} x + \frac{2(y_{i2} - \gamma_i^2 y_{i1})}{(\gamma_i^2 - 1)} y + \frac{2(z_{i2} - \gamma_i^2 z_{i1})}{(\gamma_i^2 - 1)} z \\ &= \frac{x_{i2}^2 + y_{i2}^2 + z_{i2}^2 - \gamma_i^2 (x_{i1}^2 + y_{i1}^2 + z_{i1}^2)}{(\gamma_i^2 - 1)} \end{aligned} \tag{14.8}$$

Thereinto, (x,y,z) expresses the coordinate of mobile receiver. $R^2 = x^2 + y^2 + z^2$. $\gamma_i = \frac{d_{i2}}{d_{i1}}$.

Because this algorithm needs four groups of access points at least, for convenience, four groups are set and substituted in Eq. (14.8), as below.

$$\begin{cases} R^2 + \dfrac{2(x_{12}-\gamma_1{}^2x_{11})}{(\gamma_1{}^2-1)}x + \dfrac{2(y_{12}-\gamma_1{}^2y_{11})}{(\gamma_1{}^2-1)}y + \dfrac{2(z_{12}-\gamma_1{}^2z_{11})}{(\gamma_1{}^2-1)}z = \dfrac{x_{12}{}^2+y_{12}{}^2+z_{12}{}^2-\gamma_1{}^2(x_{11}{}^2+y_{11}{}^2+z_{11}{}^2)}{(\gamma_1{}^2-1)} \\ R^2 + \dfrac{2(x_{22}-\gamma_2{}^2x_{21})}{(\gamma_2{}^2-1)}x + \dfrac{2(y_{22}-\gamma_2{}^2y_{21})}{(\gamma_2{}^2-1)}y + \dfrac{2(z_{22}-\gamma_2{}^2z_{21})}{(\gamma_2{}^2-1)}z = \dfrac{x_{22}{}^2+y_{22}{}^2+z_{22}{}^2-\gamma_2{}^2(x_{21}{}^2+y_{21}{}^2+z_{21}{}^2)}{(\gamma_2{}^2-1)} \\ R^2 + \dfrac{2(x_{32}-\gamma_3{}^2x_{31})}{(\gamma_3{}^2-1)}x + \dfrac{2(y_{32}-\gamma_3{}^2y_{31})}{(\gamma_3{}^2-1)}y + \dfrac{2(z_{32}-\gamma_3{}^2z_{31})}{(\gamma_3{}^2-1)}z = \dfrac{x_{32}{}^2+y_{32}{}^2+z_{32}{}^2-\gamma_3{}^2(x_{31}{}^2+y_{31}{}^2+z_{31}{}^2)}{(\gamma_3{}^2-1)} \\ R^2 + \dfrac{2(x_{42}-\gamma_4{}^2x_{41})}{(\gamma_4{}^2-1)}x + \dfrac{2(y_{42}-\gamma_4{}^2y_{41})}{(\gamma_4{}^2-1)}y + \dfrac{2(z_{42}-\gamma_4{}^2z_{41})}{(\gamma_4{}^2-1)}z = \dfrac{x_{42}{}^2+y_{42}{}^2+z_{42}{}^2-\gamma_1{}^2(x_{41}{}^2+y_{41}{}^2+z_{41}{}^2)}{(\gamma_4{}^2-1)} \end{cases} \tag{14.9}$$

Using the first formula to subtract other ones separately can get another equation including three formulas, and it can be expressed in the form of matrix equation, as below.

$$A\theta = \beta \tag{14.10}$$

Thereinto,

$$A = \begin{bmatrix} \dfrac{(x_{12}-\gamma_1{}^2x_{11})}{(\gamma_1{}^2-1)} - \dfrac{(x_{22}-\gamma_2{}^2x_{21})}{(\gamma_2{}^2-1)} & \dfrac{(y_{12}-\gamma_1{}^2y_{11})}{(\gamma_1{}^2-1)} - \dfrac{(y_{22}-\gamma_2{}^2y_{21})}{(\gamma_2{}^2-1)} & \dfrac{(z_{12}-\gamma_1{}^2z_{11})}{(\gamma_1{}^2-1)} - \dfrac{(z_{22}-\gamma_2{}^2z_{21})}{(\gamma_2{}^2-1)} \\ \dfrac{(x_{12}-\gamma_1{}^2x_{11})}{(\gamma_1{}^2-1)} - \dfrac{(x_{32}-\gamma_3{}^2x_{31})}{(\gamma_3{}^2-1)} & \dfrac{(y_{12}-\gamma_1{}^2y_{11})}{(\gamma_1{}^2-1)} - \dfrac{(y_{32}-\gamma_3{}^2y_{31})}{(\gamma_3{}^2-1)} & \dfrac{(z_{12}-\gamma_1{}^2z_{11})}{(\gamma_1{}^2-1)} - \dfrac{(z_{32}-\gamma_3{}^2z_{31})}{(\gamma_3{}^2-1)} \\ \dfrac{(x_{12}-\gamma_1{}^2x_{11})}{(\gamma_1{}^2-1)} - \dfrac{(x_{42}-\gamma_4{}^2x_{41})}{(\gamma_4{}^2-1)} & \dfrac{(y_{12}-\gamma_1{}^2y_{11})}{(\gamma_1{}^2-1)} - \dfrac{(y_{42}-\gamma_4{}^2y_{41})}{(\gamma_4{}^2-1)} & \dfrac{(z_{12}-\gamma_1{}^2z_{11})}{(\gamma_1{}^2-1)} - \dfrac{(z_{42}-\gamma_4{}^2z_{41})}{(\gamma_4{}^2-1)} \end{bmatrix}$$

$$\beta = \begin{bmatrix} \dfrac{x_{12}{}^2+y_{12}{}^2+z_{12}{}^2-\gamma_1{}^2(x_{11}{}^2+y_{11}{}^2+z_{11}{}^2)}{2(\gamma_1{}^2-1)} - \dfrac{x_{22}{}^2+y_{22}{}^2+z_{22}{}^2-\gamma_2{}^2(x_{21}{}^2+y_{21}{}^2+z_{21}{}^2)}{2(\gamma_2{}^2-1)} \\ \dfrac{x_{12}{}^2+y_{12}{}^2+z_{12}{}^2-\gamma_1{}^2(x_{11}{}^2+y_{11}{}^2+z_{11}{}^2)}{2(\gamma_1{}^2-1)} - \dfrac{x_{32}{}^2+y_{32}{}^2+z_{32}{}^2-\gamma_3{}^2(x_{31}{}^2+y_{31}{}^2+z_{31}{}^2)}{2(\gamma_3{}^2-1)} \\ \dfrac{x_{12}{}^2+y_{12}{}^2+z_{12}{}^2-\gamma_1{}^2(x_{11}{}^2+y_{11}{}^2+z_{11}{}^2)}{2(\gamma_1{}^2-1)} - \dfrac{x_{42}{}^2+y_{42}{}^2+z_{42}{}^2-\gamma_4{}^2(x_{41}{}^2+y_{41}{}^2+z_{41}{}^2)}{2(\gamma_4{}^2-1)} \end{bmatrix},$$

$$\theta = \begin{bmatrix} x \\ y \\ z \end{bmatrix}$$

Then, the value of the real-time coordinate of the mobile receiver can be estimated by least squares estimation, as below.

$$\widehat{\theta_{LS}} = \left(A^T A\right)^{-1} A^T \beta \tag{14.11}$$

14.4 Experimental Test and Validation

To evaluate the performance of the proposed method, the test environment which is about 28 m × 19 m × 8 m is set up at first floor lobby of laboratory building, as shown in Fig. 14.3. Four groups of access points are placed near the middle of four walls. The distance between the two access points in one group is set to 0.2 m. Computers are used as the mobile receiver and the fixed receivers. The fixed receivers have to send the real-time received signal to the mobile receiver to modify the parameters and evaluate the coordinates. Twenty coordinates are selected to analyze the results. Figure 14.4 shows

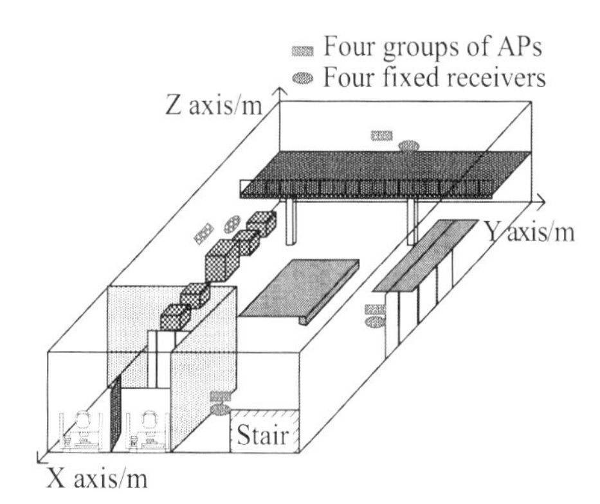

Fig. 14.3 The schematic plot of test environment

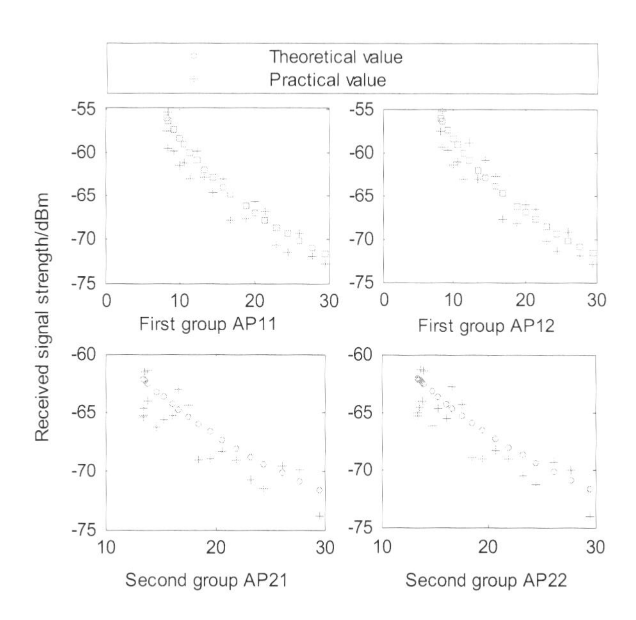

Fig. 14.4 The comparison of practical received signal strength values with theoretical values

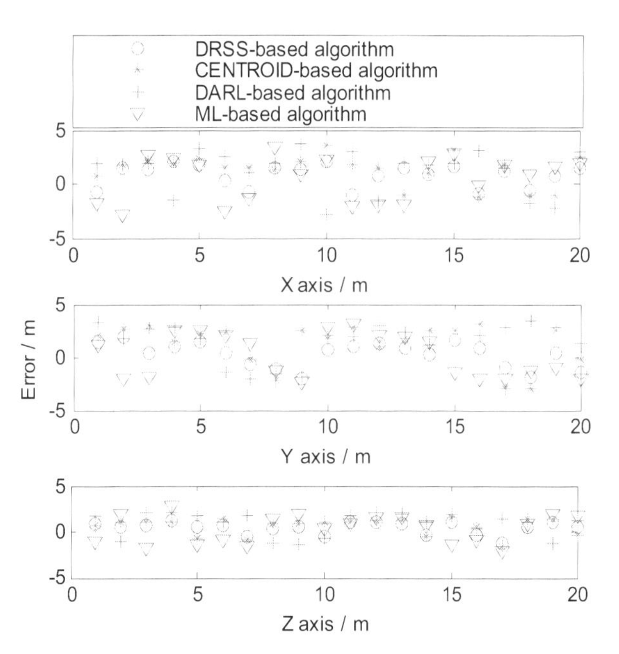

Fig. 14.5 The positioning error on X, Y, Z coordinate axis

the comparison of the theoretical signal strength values with the practical ones which are received from two groups of access points. It can be seen that the practical values fluctuate largely and unregularly around theoretical ones and this will result in large positioning error. But the values received from one group of access points have almost the same fluctuations.

To know the positioning performance in the experiment, the proposed algorithm is compared with other three received signal strength-based algorithms including dynamic correction and centroid-based algorithm, DARL (dominant access point RSS location)-based algorithm, and ML (maximum likelihood)-based algorithm [6, 8, 9]. The positioning errors which are analyzed on X, Y, Z coordinate axis are shown in Fig. 14.5. We can see that most errors of proposed algorithm are less than 2.6 m and smaller than that of other algorithms. At present, the positioning error is generally about 3–4 m in similar research (based on WiFi technology), and the better ones can control the most errors within 3 m.

Another experiment is made to know the improvement degree of using new proposed method of estimating model parameters. Other three methods, weighted average, MVU (minimum variance unbiased), and fixed empirical value [1, 7, 8], are also used to compare with it, and the comparison results are shown in Fig. 14.6. The positioning accuracy is obviously improved when the real-time values are used and the new proposed method has better effect than other three ones.

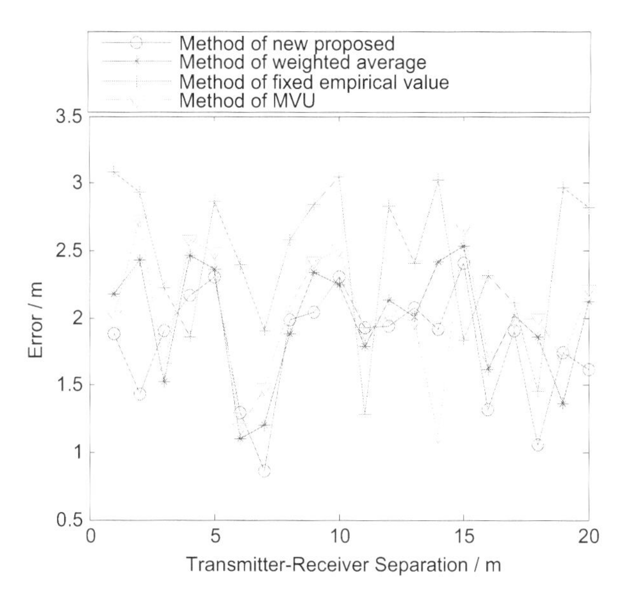

Fig. 14.6 The positioning error with different estimation methods of parameter

14.5 Conclusion

The algorithm of indoor three-dimensional positioning based on difference received signal strength (DRSS) is presented in this paper, and a new method of the real-time estimation and correction of parameters of the model with an environment factor is also used at the same time. As we have seen from the environmental test, the effect of the positioning based on this method is more sound. The multipath propagation, multi-interference factors, nonlinearity, and time variation of the wireless channel bring great influence on the positioning. So this method can use the difference received signal strength to weaken it to some degree, and the indoor propagation model of wireless signal can describe the real-time changes of the attenuation more accurately through the correction. Further research will be made to improve the performance of indoor positioning based on WiFi access points.

References

1. Shi, W. R., Xiong, Z. G., & Xu, L. (2010). In-building RSSI-based user localization algorithm. *Computer Engineering and Applications, 46*(17), 232–235 (In Chinese).
2. Kaemarungsi, K., & Krishnamurthy, P. (2012). Analysis of WLAN's received signal strength indication for indoor location fingerprinting. *Pervasive and Mobile Computing, 8*(2), 292–316.
3. Zaruba, G. V., Huber, M., Kamangar, F. A., & Chlamtac, I. (2007). Indoor location tracking using RSSI readings from a single Wi-Fi access point. *Wireless Networks, 13*(2), 221–235.

4. Swangmuang, N., & Krishnamurthy, P. (2008). An effective location fingerprint model for wireless indoor localization. *Pervasive and Mobile Computing, 4*(6), 836–850.
5. Narzullaev, A., Park, Y., Yoo, K., & Yu, J. (2011). A fast and accurate calibration algorithm for real-time locating systems based on the received signal strength indication. *AEU-International Journal of Electronics and Communications, 65*(4), 305–311.
6. Park, D. W., & Park, J. G. (2011). An enhanced ranging scheme using WiFi RSSI measurements for ubiquitous location. *IEEE CNSI*, 296–301. doi:10.1109/CNSI.2011.29
7. Xu, F. Y., Shan, H. G., & Wang, Z. X. (2008). An indoor location algorithm based on received signal strength with parameter estimation. *Journal of Microwaves, 24*(2), 67–72 (In Chinese).
8. Xu, R. M., Zhuang, C. Y., & Yu, B. (2010). Dynamic correction algorithm for indoor wireless location based on RSSI. *Computer Knowledge and Technology, 6*(3), 686–688 (In Chinese).
9. Ni, W., & Wang, Z. X. (2006). Indoor location algorithm based on the measurement of the received signal strength. *Frontiers of Electrical and Electronic Engineering in China, 1*(1), 48–52.

Chapter 15
The Universal Approximation Capability of Double Flexible Approximate Identity Neural Networks

Saeed Panahian Fard and Zarita Zainuddin

Abstract This study investigates the universal approximation capability of three-layer feedforward double flexible approximate identity neural networks in the space of continuous functions with two variables. First, we propose double flexible approximate identity functions, which are a combination of double approximate identity functions and flexible approximate identity functions as investigated in our previous studies. Then, we prove that any continuous function f with two variables will converge to itself if it convolves with double flexible approximate identity. Finally, we prove a main theorem by using the obtained results.

15.1 Introduction

Universal approximation capability is the first question in the approximation theory of neural networks. That is, neural networks can approximate any continuous function in the topological space w.r.t. to a proper norm to any accuracy. Thus, the most fundamental theoretical studies for neural networks are related to the universal approximation capability of a new class of neural networks.

In this study, the main results of the universal approximation capability of artificial neural networks are surveyed in the space of continuous functions. It has been shown that any continuous function can be approximated on a compact set with uniform topology by sigmoidal neural networks [1, 2]. It has been proved that Gaussian radial basis function neural networks are universal approximators [3]. Moreover, it has been obtained that radial basis function neural networks can uniformly approximate any continuous function on a compact set provided that the activation function is continuous almost everywhere, locally

S.P. Fard (✉) • Z. Zainuddin
School of Mathematical Sciences, Universiti Sains Malaysia,
11800 USM Pulau Pinang, Malaysia
e-mail: saeedpanahian@yahoo.com; zarita@cs.usm.my; http://math.usm.my/

W.E. Wong and T. Zhu (eds.), *Computer Engineering and Networking*, Lecture Notes in Electrical Engineering 277, DOI 10.1007/978-3-319-01766-2_15,

essentially bounded and not a polynomial [4]. Furthermore, it has been surveyed the universal approximation capability of feedforward neural networks [5].

On the other hand, we have introduced double approximate identity neural networks [6]. Double approximate identity neural networks are the generalization of Plane-Gaussian neural networks which were introduced in [7]. Then, we have proved that flexible approximate identity neural networks are universal approximators [8, 9]. Flexible approximate identity neural networks are the generalization of generalized Gaussian RBF neural networks which were presented in [10]. In addition, in our latest result, it has been shown that Mellin approximate identity neural networks are universal approximators [11].

The motivation of this study is to combine double approximate identity functions and flexible approximate identity functions in order to obtain an extension class of approximate identity. This extension is called double flexible approximate identity functions. Thus, a new class of neural networks is obtained by using double flexible approximate identity as activation functions.

The approach of this study is illustrated below. First, a theorem is proved by using the fundamental concept of the double flexible approximate identity functions. The proof of this result is based on the double convolution linear operator in the space of the continuous functions with two variables. Then, by using this result, a main theorem is obtained. The proof of the main theorem is in the framework of the theory of ε-net. This theorem shows the universal approximation capability of a three-layer feedforward double approximate identity artificial neural networks in the space of the continuous functions with two variables.

The remainder of this study is organized as follows: in Sect. 15.2, some basic definitions are studied which will be used in the following sections. In Sect. 15.3, a theoretical result is presented which will be constructed the fundamental structure of the next section. In Sect. 15.4, a main theoretical result is given. Finally, in Sect. 15.5, conclusions are derived.

15.2 Basic Definitions

In this section, a certain number of definitions are reviewed which will be used in the succeeding sections. As the fundamental concept, the definition of the double flexible approximate identity functions is presented.

Definition 1 Let $A = A(a_1, \ldots, a_m), a_i \in \mathbb{R}, i = 1, \ldots, m$ be any parameters, $\{\phi_n(x, y, A)\}_{n=1}^{\infty}$, $\phi_n : \mathbb{R}^2 \to \mathbb{R}$, be a sequence of two-dimensional flexible functions. The sequence is called a double flexible approximate identity if it satisfies the following conditions:

1) $\int_{\mathbb{R}} \int_{\mathbb{R}} \phi_n(x,y,A) dx dy = 1;$
2) for any $\varepsilon > 0$ and $\delta > 0$, there exists a number N such that if $n \geq N$ it results:

(i) $\int_{|y|<\delta} \int_{|x|\geq\delta} |\phi_n(x, y, A)| dxdy \leq \epsilon$,
(ii) $\int_{|y|\geq\delta} \int_{|x|<\delta} |\phi_n(x, y, A)| dxdy \leq \epsilon$,
(iii) $\int_{|y|\geq\delta} \int_{|x|\geq\delta} |\phi_n(x, y, A)| dxdy \leq \epsilon$.

The above definition will be used in Theorems 2 and 4. In the following definition, double convolution is reviewed which will be used in Theorem 2.

Definition 2 ([12]) A double convolution of two integrable functions f and g is defined as

$$f(x, y) *^x *^y g(x, y) = \int_{\mathbb{R}} \int_{\mathbb{R}} f(u, v) g(x - u, y - v) dudv.$$

Now, the definitions of ε-net and finite ε-net will be studied, respectively. These definitions will be used in Theorems 3 and 4.

Definition 3 ([13]) Let $\varepsilon > 0$. A set $V_\varepsilon \subset C(I, J)$ is called ε-net of a set V, if $\tilde{f} \in V_\epsilon$ can be found for $\forall f \in V$ such that $\|f - \tilde{f}\|_{C(I,J)} < \epsilon$.

Definition 4 ([13]) The ε-net is said to be finite if it is a finite set of elements.

In the next section, a theoretical result in the space of continuous functions with two variables will be given.

15.3 Theoretical Result for Continuous Functions with Two Variables

In this section, the following theorem is studied which will be used in the proof of Theorem 2.

Theorem 1 ([14]) *Let $I, J \subseteq \mathbb{R}$ be closed and bounded intervals and let $f : I \times J \to \mathbb{R}$ be a continuous function with two variables. Then f is uniformly continuous.*

Now, Theorem 2 is presented. This theorem shows a very important property of double approximate identity functions in the space of continuous functions with two variables.

Theorem 2 *Let $A = A(a_1, \ldots, a_m), a_i \in \mathbb{R}, i = 1, \ldots, m$ be any parameters, $\{\phi_n(x, y, A)\}_{n\in\mathbb{N}}$, $\phi_n(x, y, A) : \mathbb{R}^2 \to \mathbb{R}$ be a double approximate identity. Let f be a function on $C(I,J)$. Then $\phi_n *^x *^y f$ uniformly converges to f on $C(I, J)$.*

Proof. Let $x \in I$, $y \in J$ and $\varepsilon > 0$. Based on Theorem 1, there exists a $\delta > 0$ such that $|f(x, y) - f(u, v)| < \frac{\epsilon}{2\|\phi\|_{L^1(\mathbb{R}^2)}}$ for all u, v, $|x - u| < \delta$, $|y - v| < \delta$. Let us define $\{\phi_n *^x *^y f\}_{n\in\mathbb{N}}$ by $\phi_n(x, y, A) = n^2\phi(n x, n y, A)$. We consider

$$
\begin{aligned}
&\phi_n *^x *^y f(x,y) - f(x,y) \\
&= \int_{\mathbb{R}} \int_{\mathbb{R}} \phi_n(u,v,A)\{f(x-u,y-v) - f(x,y)\}dudv \\
&= \left(\int_{|y|<\delta} \int_{|x|<\delta} + \int_{|y|<\delta} \int_{|x|\geq\delta} + \int_{|y|\geq\delta} \int_{|x|<\delta} + \int_{|y|\geq\delta} \int_{|x|\geq\delta} \right) n^2 . \\
&\quad \phi(nu,nv,A)\{f(x-u,y-v) - f(x,y)\}dudv \\
&= I_1 + I_2 + I_3 + I_4.
\end{aligned}
$$

We calculate I_1, I_2, I_3, and I_4 as follows:

$$
\begin{aligned}
|I_1| &\leq \int_{|y|<\delta} \int_{|x|<\delta} n^2 |\phi(nu,nv,A)\{f(x-u,y-v) - f(x,y)\}|dudv \\
&< \frac{\epsilon}{4\|\phi\|_{L^1(\mathbb{R}^2)}} \int_{|y|<\delta} \int_{|x|<\delta} n^2 |\phi(nu,nv,A)|dudv \\
&= \frac{\epsilon}{4\|\phi\|_{L^1(\mathbb{R}^2)}} \int_{|t|<n\delta} \int_{|s|<n\delta} |\phi(s,t,A)|dsdt \\
&\leq \frac{\epsilon}{4\|\phi\|_{L^1(\mathbb{R}^2)}} \int_{\mathbb{R}} \int_{\mathbb{R}} |\phi(s,t,A)|dsdt = \frac{\epsilon}{4}.
\end{aligned}
$$

For I_2, we have

$$
\begin{aligned}
|I_2| &\leq 2\|f\|_{C(I,J)} \int_{|y|<\delta} \int_{|x|\geq\delta} n^2 |\phi(nu,nv,A)|dudv \\
&= 2\|f\|_{C(I,J)} \int_{|t|<n\delta} \int_{|s|\geq n\delta} |\phi(s,t,A)|dsdt.
\end{aligned}
$$

Since

$$
\lim_{n\to\infty} \int_{|t|<n\delta} \int_{|s|\geq n\delta} |\phi(s,t,A)|dsdt = 0,
$$

there exist $n_0 \in \mathbb{N}$ such that for all $n \geq n_0$,

$$
\int_{|t|<n\delta} \int_{|s|\geq n\delta} |\phi(s,t,A)|dsdt < \frac{\epsilon}{8\|f\|_{C(I,J)}}.
$$

For I_3, we have

$$
\begin{aligned}
|I_3| &\leq 2\|f\|_{C(I,J)} \int_{|y|\geq\delta} \int_{|x|<\delta} n^2 |\phi(nu,nv,A)|dudv \\
&= 2\|f\|_{C(I,J)} \int_{|t|\geq n\delta} \int_{|s|<n\delta} |\phi(s,t,A)|dsdt.
\end{aligned}
$$

Since

$$\lim_{n\to\infty}\int_{|t|\geq n\delta}\int_{|s|\delta}|\phi(s,t,A)|dsdt=0,$$

there exist $n_0 \in \mathbb{N}$ such that for all $n \geq n_0$,

$$\int_{|t|\geq n\delta}\int_{|s|<n\delta}|\phi(s,t,A)|dsdt<\frac{\epsilon}{8\|f\|_{C(I,J)}}.$$

For I_4, we have

$$\begin{aligned}|I_4| &\leq 2\|f\|_{C(I,J)}\int_{|y|\geq\delta}\int_{|x|\geq\delta}n^2|\phi(nu,nv,A)|dudv\\ &=2\|f\|_{C(I,J)}\int_{|t|\geq n\delta}\int_{|s|\geq n\delta}|\phi(s,t,A)|dsdt.\end{aligned}$$

Since

$$\lim_{n\to\infty}\int_{|t|\geq n\delta}\int_{|s|\geq n\delta}|\phi(s,t,A)|dsdt=0,$$

there exists an $n_0 \in \mathbb{N}$ such that for all $n \geq n_0$,

$$\int_{|t|\geq n\delta}\int_{|s|\geq n\delta}|\phi(s,t,A)|dsdt<\frac{\epsilon}{8\|f\|_{C(I,J)}}.$$

Combining I_1, I_2, I_3, and I_4 for $n \geq n_0$, we have

$$\|\phi_n *^x *^y f(x,y)-f(x,y)\|_{C(I,J)}<\epsilon.$$

The above theorem will be used in the next section. The next section shows a main result for continuous functions with two variables

15.4 Main Result for Continuous Functions with Two Variables

In this section, Theorem 3 is reviewed which will be used in the proof of the Theorem 4.

Theorem 3 ([13]) *A set V in $C(I,J)$ is compact iff $\forall\varepsilon>0$ in $\mathbb{R}$ there exists a finite ε-net.*

Now, Theorem 4 is presented. This main result shows the universal approximation capability of double flexible approximate identity neural networks in the space of continuous functions with two variables.

Theorem 4 *Let $I, J \subseteq \mathbb{R}$ be closed and bounded intervals. Let $C(I,J)$ be the real linear space of all continuous functions on any compact subset of real two-dimensional space, and $V \subset C(I,J)$ a compact set. Let $A = A(a_1, \ldots, a_m)$, $a_i \in \mathbb{R}, i = 1, \ldots, m$ be any parameters, $\{\phi_n(x, y, A)\}_{n \in \mathbb{N}}, \phi_n : \mathbb{R}^2 \to \mathbb{R}$ be a double flexible approximate identity. Let the family of functions $\{\sum_{j=1}^{M} \lambda_j \phi_j(x, y) | \lambda_j \in \mathbb{R}, x \in \mathbb{R}, y \in \mathbb{R}, M \in \mathbb{N}\}$, be dense in $C(I,J)$, and given $\varepsilon > 0$. Then there exists $N \in \mathbb{N}$ which depends on V and ε but not on f, such that for any $f \in V$, there exist weights $c_k = c_k(f,V,\varepsilon)$ satisfying*

$$\left\| f(x, y) - \sum_{k=1}^{N} c_k \phi_k(x, y, A) \right\|_{C(I,J)} < \epsilon.$$

Moreover, every c_k is a continuous function of $f \in V$.

Proof Since V is compact, for any $\varepsilon > 0$, there is a finite $\frac{\epsilon}{2}$-net $\{f^1, \ldots, f^M\}$ for V. This implies that for any $f \in V$, there is an f^j such that $\| f - f^j \|_{C(I,J)} < \frac{\epsilon}{2}$. For any f^j, by assumption of the theorem, there are $\lambda_i^j \in \mathbb{R}, N_j \in \mathbb{N}$, and $\phi_i^j(x, y, A)$ such that

$$\left\| f^j(x) - \sum_{i=1}^{N_j} \lambda_i^j \phi_i^j(x, y, A) \right\|_{C(I,J)} < \frac{\epsilon}{2}. \tag{15.1}$$

For any $f \in V$, we define

$$F_-(f) = \left\{ j | \; \| f - f^j \|_{C(I,J)} < \frac{\epsilon}{2} \right\},$$

$$F_0(f) = \left\{ j | \; \| f - f^j \|_{C(I,J)} = \frac{\epsilon}{2} \right\},$$

$$F_+(f) = \left\{ j | \; \| f - f^j \|_{C(I,J)} > \frac{\epsilon}{2} \right\}.$$

Therefore, $F_-(f)$ is not empty according to the definition of $\frac{\epsilon}{2}$-net. If $\tilde{f} \in V$ approaches f such that $\| \tilde{f} - f \|_{C(I,J)}$ is small enough, then we have $F_-(f) \subset F_-(\tilde{f})$ and $F_+(f) \subset F_+(\tilde{f})$. Thus $F_-(\tilde{f}) \bigcap F_+(f) \subset F_-(\tilde{f}) \bigcap F_+(\tilde{f}) = \emptyset$, which implies $F_-(\tilde{f}) \subset F_-(f) \cup F_0(f)$. We conclude the following:

$$F_-(f) \subset F_-(\tilde{f}) \subset F_-(f) \cup F_0(f). \tag{15.2}$$

Define

$$d(f) = \left[\sum_{j \in F_-(f)} \left(\frac{\epsilon}{2} - \| f - f^j \|_{C(I,J)} \right) \right]^{-1}$$

and

$$f_h = \sum_{j \in F_-(f)} \sum_{i=1}^{N_j} d(f) \left(\frac{\epsilon}{2} - \| f - f^j \|_{C(I,J)} \right) \lambda_i^j \phi_i^j(x, y, A) \tag{15.3}$$

then $f_h \in \left\{ \sum_{j=1}^{M} \lambda_j \phi_j(x, y, A) \right\}$ approximates f with accuracy ε :

$$\begin{aligned}
&\| f - f_h \|_{C(I,J)} \\
&= \left\| \sum_{j \in F_-(f)} d(f) \left(\frac{\epsilon}{2} - \| f - f^j \|_{C(I,J)} \right) \right. \\
&\quad \left. \left(f - \sum_{i=1}^{N_j} \lambda_i^j \phi_i^j(x, y, A) \right) \right\|_{C(I,J)} \\
&= \left\| \sum_{j \in F_-(f)} d(f) \left(\frac{\epsilon}{2} - \| f - f^j \|_{C(I,J)} \right) \right. \\
&\quad \left. \left(f - f^j + f^j - \sum_{i=1}^{N_j} \lambda_i^j \phi_i^j(x, y, A) \right) \right\|_{C(I,J)} \\
&\le \sum_{j \in F_-(f)} d(f) \left(\frac{\epsilon}{2} - \| f - f^j \|_{C(I,J)} \right) \\
&\quad \left(\| f - f^j \|_{C(I,J)} + \left\| f_j - \sum_{i=1}^{N_j} \lambda_i^j \phi_i^j(x, y, A) \right\|_{C(I,J)} \right) \\
&\le \sum_{j \in F_-(f)} d(f) \left(\frac{\epsilon}{2} - \| f - f^j \|_{C(I,J)} \right) \left(\frac{\epsilon}{2} + \frac{\epsilon}{2} \right) = \epsilon.
\end{aligned} \tag{15.4}$$

In the next step, we prove the continuity of c_k. For the proof, we use Eq. (15.2) to obtain

$$\begin{aligned}
&\sum_{j \in F_-(f)} \left(\frac{\epsilon}{2} - \| \tilde{f} - f^j \|_{C(I,J)} \right) \\
&\le \sum_{j \in F_-(\tilde{f})} \left(\frac{\epsilon}{2} - \| \tilde{f} - f \|_{C(I,J)} \right) \\
&\le \sum_{j \in F_-(\tilde{f})} \left(\frac{\epsilon}{2} - \| \tilde{f} - f^j \|_{C(I,J)} \right) + \sum_{j \in F_0(f)} \left(\frac{\epsilon}{2} - \| \tilde{f} - f^j \|_{C(I,J)} \right).
\end{aligned} \tag{15.5}$$

Let $\tilde{f} \to f$ in Eq. (15.5), then we have

$$\sum_{j \in F_{-}(\tilde{f})} \left(\frac{\epsilon}{2} - \| \tilde{f} - f^j \|_{C(I,J)} \right) \to \sum_{j \in F_{-}(f)} \left(\frac{\epsilon}{2} - \| f - f^j \|_{C(I,J)} \right). \tag{15.6}$$

This obviously demonstrates that $d(\tilde{f}) \to d(f)$. Thus, $\tilde{f} \to f$ results

$$d(\tilde{f}) \left(\frac{\epsilon}{2} - \| \tilde{f} - f^j \|_{C(I,J)} \right) \lambda_i^j \to d(f) \left(\frac{\epsilon}{2} - \| f - f^j \|_{C(I,J)} \right) \lambda_i^j. \tag{15.7}$$

Let $N = \sum_{j \in F_{-}(f)} N_j$ and define c_k in terms of

$$\begin{aligned} f_h &= \sum_{j \in F_{-}(f)} \sum_{i=1}^{N_j} d(f) \left(\frac{\epsilon}{2} - \| f - f^j \|_{C(I,J)} \right) \lambda_i^j \phi_i^j(x, y, A) \\ &\equiv \sum_{k=1}^{N} c_k \phi_k(x, y, A) \end{aligned}$$

From Eq. (15.7), c_k is a continuous functional of f. Thus the approximation result follows.

15.5 Conclusions

In this study, the universal approximation capability of three-layer feedforward double approximate identity neural networks was investigated in the space of continuous functions with two variables. First, the definition of double flexible approximate identity functions was proposed. Then, Theorem 2 was obtained through this definition. Theorem 2 showed that any continuous function f with two variables convolved with double flexible approximate identity functions converges to itself. Through the obtained result, Theorem 4 was proved as the main result. The proof of Theorem 4 was in the framework of theory of ε-net and similar to the proof of Theorem 1 [15]. Theorem 4 showed that three-layer feedforward neural networks were capable of uniformly approximating any continuous function f with two variables on a compact set to arbitrary accuracy.

References

1. Cybenko, G. (1989). Approximation by superpositions of sigmoidal function. *Mathematics of Control, Signals and Systems*, *2*(4), 303–314.
2. Funahashi, K. (1989). On the approximate realization of continuous mappings by neural network. *Neural Networks*, *2*(3), 183–192.
3. Park, J., & Sandberg, I. W. (1991). Universal approximation using radial-basis-function networks. *Neural Computation*, *3*(2), 246–257.
4. Liao, Y., Fang, S. C., & Nuttle, H. L. W. (2003). Relaxed conditions for radial-basis function networks to be universal approximators. *Neural Networks*, *16*(7), 1019–1028.
5. Sanguineti, M. (2008). Universal approximation by ridge Computational models and neural networks: a survey. *The Open Applied Mathematics Journal*, *2*(1), 31–58.
6. Zainuddin, Z., & Panahian Fard, S. (2012). Double approximate identity neural networks universal approximation in real Lebesgue spaces. *Neural information processing*. Lecture Notes in Computer Science, vol. 7663, (pp. 409–415). Berlin, Heidelberg: Springer.
7. Yang, X., Chen, S., & Chen, B. (2012). Plane-Gaussian artificial neural network. *Neural Computing and Applications*, *21*(2), 305–317.
8. Panahian Fard, S., & Zainuddin, Z. (2013). On the universal approximation capability of flexible approximate identity neural networks. *Emerging technologies for information systems, computing, and management*. Lecture Notes in Electrical Engineering, vol. 236, (pp. 201–27). New York: Springer.
9. Panahian Fard, S., & Zainuddin, Z. (2013). Analyses for L^P [a, b]-norm approximation capability of flexible approximate identity neural networks. *Neural Computing and Applications*, DOI 10.1007/s00521-013-1493-9
10. Fernández, N. F., Hervás, M. C., Sanchez, M. J., & Gutiírrez, P. A. (2011). MELM-GRBF: A modified version of the extreme learning machine for generalized radial basis function neural networks. *Neurocomputing*, *74*(16), 2502–2510.
11. Panahian Fard, S., & Zainuddin, Z. (2013). The universal approximation capabilities of Mellin approximate identity neural networks. *Advances in Neural Networks- ISNN 2013*. Lecture Notes in Computer Science, vol. 7951, (pp. 205–213). Berlin, Heidelberg: Springer.
12. Ditkin, V. A., & Prudnikov, A. P. (1962). *Operation calculus in two variable and its applications*. New York: Pergamon press.
13. Lebedev, V. (1997). *An introduction to functional analysis and computational mathematics*. Boston: Birkhäuser.
14. Jones, F. (1997). *Lebesgue integration on Euclidean space*. Boston: Jones and Bartlett.
15. Wu, W., Nan, D., Li, Z., & Long, J. (August 2007). Approximation to compact set of functions by feedforward neural networks. In 20th International Joint Conference on Neural Networks, Orlando, Fl, USA, pp. 1222–1225.

Chapter 16
A Novel and Real-Time Hand Tracking Algorithm for Gesture Manipulation

Zhiqin Zhang

Abstract Direct use of the hand as an input device is an attractive method for providing natural human–computer interaction (HCI). Computer vision (CV) has the potential to provide more natural, noncontact solutions. As a result, there have been considerable research efforts to use the hand as an input device for HCI in recent years. Hand tracking is the most important procedure for HCI. This chapter presents a novel hand tracking algorithm which can track a hand stable and is real time, and we review on the latter hand tracking research direction, which is a very challenging problem in the context of HCI. Our algorithm is based on mean-shift and we improved it to fit for robust hand tracking by using integrated GIH and skin color mask, our improved algorithm can track hand reliably even in clutter environments. Finally, we demonstrate the benefits of our approach in contrast to existing methods.

16.1 Introduction

There has been a great emphasis lately in HCI research to create easier to use interfaces by directly employing natural communication and manipulation skills of humans. Adopting direct sensing in HCI will allow the deployment of a wide range of applications in more sophisticated computing environments such as Virtual Environments (VEs) or Augmented Reality (AR) systems. The development of these systems involves addressing challenging research problems including effective input/output techniques, interaction styles, and evaluation methods [1, 2]. In the input domain, the direct sensing approach requires capturing and interpreting the motion of head, eye gaze, face, hand, arms, or even the whole body [3].

Z. Zhang (✉)
School of Computer Science, Wuhan Donghu University, Wuhan 430212, China
e-mail: zzq9908@sohu.com

W.E. Wong and T. Zhu (eds.), *Computer Engineering and Networking*, Lecture Notes in Electrical Engineering 277, DOI 10.1007/978-3-319-01766-2_16,

Among different body parts, the hand is the most effective, general-purpose interaction tool due to its dexterous functionality in communication and manipulation. Object manipulation interfaces [4–6] utilize the hand for navigation, selection, and manipulation tasks in VEs. A reliable hand tacking method can make the manipulation tasks easy-to-use and stable.

Target tracking is a difficult problem which can be affected by environments, occlusion, movement speed, scaling, and rotation. A reliable hand tracking should solve several pivotal problems: deformation, background disturbance, and scaling. Our approach has been inspired by an established trend in hand tracking, mean-shift is a good tracking method which can track deformation targets, but the traditional mean-shift is not reliable in complicated background. In this study, we presents a novel algorithm which is a significant improvement comparing with existing methods, experiments show that our algorithm can achieve similar or greater performance at a lower computational cost in comparison to existing methods, and the hand tracking system can be real time.

16.2 Background

There exist many reviews on target tracking [7–10], but the reviews involved in hand tracking is seldom, latest of which only covers studies up to 100.

A recent survey includes a very interesting overview of the use of color for face (and, therefore skin color) detection. A major decision towards providing a model of skin color is the selection of the color space to be employed. Several color spaces have been proposed including RGB [11], normalized RGB, HSV, YCrCb, YUV, etc. Color spaces efficiently separating the chrominance from the luminance components of color are typically considered preferable. This is due to the fact that by employing chrominance-dependent components of color only, some degree of robustness to illumination changes can be achieved. Terrillon et al. [12] review different skin chrominance models and evaluate their performance.

Another approach is to assume that the probabilities of skin colors follow a distribution that can be learned either offline or by employing an online iterative method [13].

The skin colors information is sensitive to the background and local luminance, the presented method aims to solve this disturbance by using both color information and shape information.

The remainder of this chapter is organized as follows: Sect. 16.3 describes the traditional mean-shift tracking algorithm, Sect. 16.4 describes our method details, Sect. 16.5 shows our experiments results, and the final section makes a conclusion.

16.3 Mean-Shift Tracker

The heart of the mean-shift algorithm is computation of an offset from location vector x to a new location according to the mean-shift vector

$$\Delta_x = \frac{\sum_a K(a-x)w(a)a}{\sum_a K(a-x)w(a)} - x = \frac{\sum_a K(a-x)w(a)(a-x)}{\sum_a K(a-x)w(a)} \tag{16.1}$$

where K is a suitable kernel function and the summations are performed over a local window of pixels a around the current location x. A "suitable" kernel K is one that can be written in terms of a profile function k such that

$$K(x) = k\left(||x||^2\right) \tag{16.2}$$

and profile k is nonnegative, nonincreasing, piecewise continuous, and $\int_0^\infty k(r)dr < \infty$.

An important theoretical property of the mean-shift algorithm is that the local mean-shift offset x computed at position x using kernel K points opposite to the gradient direction of the convolution surface.

$$C(x) = \sum_a H(a-x)w(a) \tag{16.3}$$

Kernels K and shadow kernels H must satisfy the relationship

$$h'(r) = -ck(r) \tag{16.4}$$

where h and k are the respective profiles of H and K, $r = \| a - x \|^2$, and $c > 0$ is some constant.

Note that the mean-shift vector computed by Eq. (16.1) is invariant to scaling of the sample weights w(a) by a positive constant c, that is, if each w(a) is replaced by cw(a) for $c > 0$. Note that the mean-shift vector is not invariant to a constant offset w(a) + c. However, all mean-shift kernels explored to date are symmetric about the origin, such that K(x) = K(−x), in which case

$$\Delta_x = \frac{\sum K(a-x)(w(a)+c)(a-x)}{\sum K(a-x)(w(a)+c)} = \frac{\sum K(a-x)w(a)(a-x)+c\sum K(a-x)(a-x)}{\sum K(a-x)(w(a)+c)}$$
$$= \frac{\sum K(a-x)w(a)(a-x)}{\sum K(a-x)(w(a)+c)} \tag{16.5}$$

showing that the direction of the mean-shift vector is invariant and just the step size changes [14].

The mean-shift target tracking procedure as follows:

1. Initialize the location of the target, in the current frame with $\hat{\mathbf{y}}_0$, compute the distribution $\{\hat{p}(\hat{\mathbf{y}}_0)\}_{u=1\ldots m}$ and evaluate

$$\rho\left[\hat{\mathrm{p}}(\hat{\mathrm{y}}_0), \hat{\mathrm{q}}\right] = \sum_{u=1}^{m} \sqrt{\hat{\mathrm{p}}_u(\mathrm{y}_0)\hat{\mathrm{q}}_u} \tag{16.6}$$

2. Derive the weights $\{w_i\}_{i=1\ldots n_h}$ according to Eq. (16.7)

$$\mathrm{w}_i = \sum_{u=1}^{m} \delta\left[\mathrm{b}(\mathrm{x}_i) - \mathrm{u}\right]\sqrt{\frac{\hat{\mathrm{q}}_u}{\hat{\mathrm{p}}_u(\hat{\mathrm{y}}_0)}} \tag{16.7}$$

3. Based on the mean-shift vector, derive the new location of the target

$$\hat{\mathrm{y}}_1 = \frac{\sum_{i=1}^{n_h} \mathrm{x}_i \mathrm{w}_i \mathrm{g}\left(\left\|\frac{\hat{\mathrm{y}}_0 - \mathrm{x}_i}{\mathrm{h}}\right\|^2\right)}{\sum_{i=1}^{n_h} \mathrm{w}_i \mathrm{g}\left(\left\|\frac{\hat{\mathrm{y}}_0 - \mathrm{x}_i}{\mathrm{h}}\right\|^2\right)} \tag{16.8}$$

   Update $\{\hat{\mathrm{p}}_u(\hat{\mathrm{y}}_1)\}_{u=1\ldots m}$ and evaluate $\rho\left[\hat{\mathrm{p}}(\hat{\mathrm{y}}_1), \mathrm{q}\right] = \sum_{u=1}^{m} \sqrt{\hat{\mathrm{p}}_u(\mathrm{y}_1)\hat{\mathrm{q}}_u}$

4. While $\rho\left[\hat{\mathbf{p}}(\hat{\mathbf{y}}_1), \hat{\mathbf{q}}\right] < \rho\left[\hat{\mathbf{p}}(\hat{\mathbf{y}}_0), \hat{\mathbf{q}}\right]$ Do $\hat{\mathrm{y}}_1 \leftarrow \frac{1}{2}(\hat{\mathrm{y}}_0 + \hat{\mathrm{y}}_1)$
5. If $\left\|(\hat{\mathbf{y}}_1 - \hat{\mathbf{y}}_0\right\| < \epsilon$ Stop.

Otherwise Set $\hat{\mathrm{y}}_0 \leftarrow \hat{\mathrm{y}}_1$ and go to step 1. The tracking consists in running for each frame of the optimization algorithm described above. Thus, given the target model, the new location of the target in the current frame minimizes the distance in the neighborhood of the previous location estimate [15].

16.4 Our Hand Tracking Method

16.4.1 Features Selecting

Traditional mean-shift algorithm s not reliable in complicated environments and is easy to produce track offset error because of falling into local optimum. Using the skin color detection to track hand is also easy to track wrong targets because of local luminance and background similar to skin color. The kernel reason of producing track error is that the selected features is not sufficient to identify the hand and the other fault objects, so we extract the skin color features and extended histogram features to represent the hand. HSV is a good color space which can distinguish efficiently hand color from background color [13], so we extract the integrated features in HSV.

16.4.2 Features Extracting

The three channel color features is computed in HSV, every channel responds to two thresholds which can be used to distinguish hand color and background color:

$$f_i(x) = k(x - {}_i)\{i = 1, 2, 3\} \tag{16.9}$$

k(x) is a signal function and if (x−Φi) > 0, k(x) = 1, else k(x) = 0. Φi represents for three color channel thresholds and the thresholds were calculated from experiment results. After the procedure we can get the mask image which represents for the most probability candidate hand regions, then the extended histograms can be calculated based on the three mask images. The single channel image histogram is not sufficient to represents for the hand shape information, which inspires us to use GIH (gradient-integrated histogram) features to extend the mean-shift tracking.

16.4.3 Algorithm Procedure

The algorithm can be described as follows:

1. Calculating the three channels mask images using Eq. (16.8).
2. Using the channel mask images and original HSV images to extract the hand candidate region images.
3. Weighting the three mask image using Eq. (16.10) and get the singe integrated result mask image:

$$F(x) = w1 * h(x) + w2 * s(x) + w3 * v(x); \tag{16.10}$$

where $w3 = 0.6$, $w2 = 0.3$, $w1 = 0.1$, and G(x) is the result mask image, $x = \{x,y\}$ is image coordinates, and h(x), s(x), v(x) correspond to the three channel mask images.

4. Normalized the result mask image F(x) using Eq. (16.11)

$$F_N = \max(0, F(x)/(\max(F(x) + 1))); \tag{16.11}$$

5. Get the gradient image G(x), where $G(x) = F_N'(x)$. According to the hand shape distribution, we calculate the gradient features using only vertical and two 45 degree diagonal gradient features: Gv, Gd1, Gd2, then G(x) can be calculated using the following equation:

$$G(x) = w3 * Gv + w4 * Gd1 + w5 * Gd2 + w6 * F_N; \tag{16.12}$$

where $w3 = 0.3$, $w4 = 0.15$, $w5 = 0.15$, $w6 = 0.4$.

6. Using the G(x) as the input image to calculate the mean-shift iterative as the above Sect. 16.3 mentioned.

16.5 Experimental Results

In this section, representative results from a prototype implementation of the proposed tracker are provided. The reported experiment consists of five long (total 12,815 frames) sequences that has been acquired and processed on-line and in real-time on a Pentium 4 laptop computer running MS Windows at 2.00 GHz. A camera with an usb2.0 interface has been used for this experiment.

Figure 16.1 provides a few characteristic snapshots of the experiment. For visualization purposes, the bounding rectangle of tracked object is shown. The first column shows our method tracking results, the second column shows the traditional mean-shift tracking results, and the third column shows the skin color method tracking results.

The tracker is initialized using a hand detector such as haar detector or some others color detectors, then the tracker can work well on the whole tracking procedure even encountering the clutter background and drastic hand deformation.

As it can be verified from the snapshots, the labeling of the object hypotheses is consistent throughout the whole sequence, which indicates that they are correctly tracked. Thus, the proposed tracker performs very well in all the above cases, some of which are challenging.

Besides the reported example, the proposed tracker has also been extensively tested with different cameras and in different settings involving different scenes and humans.

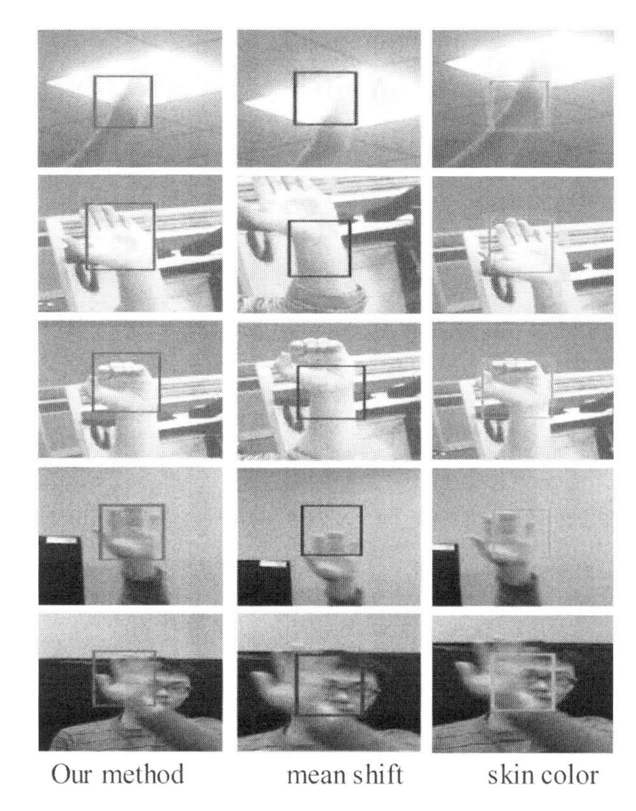

Fig. 16.1 Comparing of several tracking results

The second column and the third column in Fig. 16.1 shows the snapshots of the traditional mean-shift tracking and skin color tracker, respectively, results using the same cameral recording video, the two trackers are lost under clutter environments. From the results we can make a conclusion that our method can achieve good performance improvement comparing to the existing mean-shift tracker and skin color tracker even under clutter background.

Table 16.1 shows the cost of our algorithm and existing methods, the cost is almost the same and real time. Moreover, the cost can be reduced greatly by using parallel algorithm design [16].

16.6 Conclusion

In this chapter a new method for tracking hand has been presented. The proposed method can cope successfully with hands moving in complex patterns and local illumination environments using a still camera. The experiments verified that the proposed method is fast and reliable comparing to the existing methods. Since the tracker is not based on explicit background modeling and subtraction, it may

Table 16.1 The cost comparison table of several methods

Video no.	Mean-shift (traditional)	Proposed method	Skin color
1.	8 ms	10 ms	20 ms
2.	10 ms	13 ms	28 ms

operate even with images acquired by a moving camera. Ongoing research efforts are currently focused on improving the reliability of tracking with a moving camera.

References

1. Oikonomidis, I., Kyriazis, N., & Argyros, A. (2011). Efficient model-based 3d tracking of hand articulations using Kinect. In *Proceedings of BMVC'2011* (pp. 101.1–101.11). BMVA Press.
2. Oikonomidis, I., Kyriazis, N., & Argyros, A. A. (2012). Tracking the articulated motion of two strongly interacting hands. In *Proceedings of CVPR'2012* (pp. 1862–1869). NJ: IEEE Conference Publications.
3. Erol, A., & Bebis, G. (2007). Vision-based hand pose estimation: A review. *Computer Vision and Image Understanding, 108*(1), 52–73.
4. Bowman, D. (2002). Principles for the design of performance-oriented interaction techniques. In K. M. Stanney (Ed.), *Handbook of virtual environments: Design, implementation, and applications* (pp. 201–207). Hillsdale, NJ: Lawrence Erlbaum Associates.
5. Gabbard, J. (1997). *A taxonomy of usability characteristics in virtual environments*. Master Dissertation of University of Western Australia, Australia.
6. Buchmann, V., Violich, S., Billinghurst, M., & Cockburn, A. (2004). FingARtips: Gesture based direct manipulation in augmented reality. In *Proceedings of GRAPHITE '04: 2nd International Conference on Computer Graphics and Interactive Techniques in Australasia and South East Asia* (pp. 212–221). New York, NY: ACM Press.
7. Comaniciu, D., & Meer, P. (2002). Mean shift: A robust approach toward feature space analysis. *IEEE Transactions on Pattern Analysis and Machine Intelligence, 24*(5), 603–619.
8. Levinshtein, A., Stere, A., Kutulakos, K., Fleet, D., Dickinson, S., & Siddiqi, K. (2009). Turbopixels: Fast superpixels using geometric flows. *IEEE Transactions on Pattern Analysis and Machine Intelligence, 31*(12), 2290–2297.
9. Paragios, N., & Deriche, R. (2000). Geodesic active contours and level sets for the detection and tracking of moving objects. *IEEE Transactions on Pattern Analysis and Machine Intelligence, 22*(3), 266–280.
10. Kwon, J., & Lee, K. M. (2010). Visual tracking decomposition. In *Proceedings of CVPR'2010* (pp. 1269–1276). NJ: IEEE Conference Publications.
11. Jebara, T. S., & Pentland, A. (1997). Parameterized structure from motion for 3d adaptive feedback tracking of faces. In *Proceedings of CVPR'97* (pp. 144–150). NJ: IEEE Conference Publications.
12. Terrillon, J. C., Shirazi, M. N., Fukamachi, H., & Akamatsu, S. (2000). Comparative performance of different skin chrominance models and chrominance spaces for the automatic detection of human faces in color images. In *Proceedings of Automatic Face and Gesture Recognition'2000* (pp. 54–61). NJ: IEEE Conference Publications.
13. Saxe, D., & Foulds, R. (1996). Toward robust skin identification in video images. In *Proceedings of Automatic Face and Gesture Recognition'1996* (pp. 379–384). NJ: IEEE Conference Publications.

14. Collins, R. T. (2003). Mean-shift blob tracking through scale space. In *Proceedings of CVPR'2003* (pp. 234–240). NJ: IEEE Conference Publications.
15. Comaniciu, D., Ramesh, V., & Meer, P. (2000). Real-time tracking of non-rigid objects using mean shift. In *Proceedings of CVPR'2000* (pp. 142–149). NJ: IEEE Conference Publications.
16. Vijayanarasimhan, S., & Kapoor, A. (2010). Visual recognition and detection under bounded computational resources. In *Proceedings of CVPR'2010* (pp. 1006–1013). NJ: IEEE Conference Publications.

Chapter 17
A Transforming Quantum-Inspired Genetic Algorithm for Optimization of Green Agricultural Products Supply Chain Network

Chunqin Gu and Qian Tao

Abstract The green agricultural products supply chain network (GAP-SCN) design provides an optimal platform for efficient and effective supply chain management. This chapter proposes a new solution based on transforming quantum-inspired genetic algorithm (TQGA) to find optimal solution for the GAP-SCN problem. TQGA adopts transforming representation to convert the Q-bit representation to float-point number, and the float-point number to Q-bit representation, using transforming operator to modify chromosomes. The novelty of the transforming operator, that it can avoid the diversity of populations, is decreased. To show the efficacy of the algorithm, TQGA is tested on two cases. Results show that the proposed algorithm is promising and outperforms the classic GA by both optimization speed and solution quality.

17.1 Introduction

In recent years, the demand for green agricultural products is increasing. The green agricultural products supply chain network (GAP-SCN) design problem has been gaining importance due to increasing competitiveness. The logistics firms are obliged to maintain high customer service levels while at the same time they are forced to reduce total transportation cost.

The GAP-SCN, as any other supply chain, is a network of organizations. The network involves production bases (PB), logistics centers (LC), and sale terminals (ST) with purpose of satisfying sale terminals' demands, beginning with the PB of

C. Gu (✉)
Department of Computer Science, Zhongkai University of Agriculture and Engineering, Guangzhou 201305, China
e-mail: guchunqin@gmail.com

Q. Tao
Department of Computer Science,
Guangdong University of Education, Guangzhou 201305, China

W.E. Wong and T. Zhu (eds.), *Computer Engineering and Networking*, Lecture Notes in Electrical Engineering 277, DOI 10.1007/978-3-319-01766-2_17,

green agricultural products, following with LC, and ending with ST such as supermarkets and vegetable markets. Traditionally, manufacturing, distribution, and marketing along the supply chain operated independently. These organizations have their own objectives, and these objectives are often conflicting. But there is a need for a mechanism through which these different functions can be integrated together [1].

As for the SCN problem, the models and solutions for industrial products are relatively massive, but the solutions of the SCN about green agricultural products are limited. In literature, there are many different studies on the design problem of supply networks, and these studies have been reviewed by Erenguc et al. [2] and Pontrandolfo et al. [3]. Amiri [4] has presented a Lagrangian relaxation approach to minimize the total cost of two-stage supply chain. Costa et al. [5] have worked on three stages of SCN optimization problem. Moreover, there are also some intelligent algorithms for SCN problem, such as genetic algorithms (GA) [6, 7]. Most of the researchers have concentrated on the improvement of the supply chain performance, and very few have considered the performance improvement of the algorithm concurrently. Recently, Han and Kim [8] proposed several quantum-inspired GAs (QGAs). Obviously, it is a challenge to apply the algorithm for other different problems. Since the solution representation of the GAP-SCN problem is completely different from that of the knapsack problem, the original QGAs cannot be directly applied to the GAP-SCN problem. The main contribution of the study is to propose a novel transforming QGA in order to solve GAP-SCN problem for performance optimization.

The remainder of this chapter is organized as follows. Section 17.2 presents the statement of the optimization problem. Section 17.3 describes the preliminary knowledge of QGA. Section 17.4 gives the implementation of the proposed algorithm in detail. In Sect. 17.5, a series of experiments are conducted, and the results are analyzed to illustrate the performance of the proposed algorithm. Finally, in Sect. 17.6, this chapter is concluded.

17.2 Problem Statement

In order to optimize the performance of GAP-SCN, we should design the distribution network strategy that will satisfy demand requirement for the products imposed by sale terminals. The main objective to solve such a performance optimization problem of GAP-SCN is to evaluate the selection of different production bases or set of the production bases and different logistics centers or set of the logistics centers, whereas the performance criteria are the minimization of the total transportation cost. Simultaneously, the demand for GAP from each sale terminal must be satisfied. The assumptions used in this problem are as follows:

1. Production bases, logistics centers, and sale terminals are known.
2. The requirement for GAP from each sale terminal is known.

Table 17.1 Notations

Symbol	Descriptions	Symbol	Descriptions
PB	Set of production bases	p_t^l	The lth individual in the gth generation with the Q-bit representation
pb_i	Production base i	M	Number of chromosomes in the population
LC	Set of logistics centers	m	Length of Q-bit gene
lc_j	Logistics center j	N	Total number of impossible paths
ST	Set of sale terminals	n	The index of paths, $n \in [1,N]$
st_k	Sale terminal k	Req_{st_k}	The requirement of sale terminal k
TQ_i^j	Transportation quantity from node i to node j	$f\_pb(n)$	The map from n to a production base
C_t	Chromosome t	$g\_lc(n)$	The map from n to a logistics center
QC_t	Quantum chromosome t	$h\_st(n)$	The map from n to a sale terminal
x_i^n	nth gene of chromosome i	$Cost_i^j$	Unit transportation cost from node i to node j

3. GAP and logistics centers are enough for distribution.
4. The transportation costs of green agricultural products on the path from each production base pb_i to each logistics center lc_j and from each lc_j to each sale terminal st_k are known.
5. GAP distributed to st_k are supplied by a lc_j or multi-lc_j, and GAP distributed to lc_j are supplied by a pb_i or multi-pb_i.

Suppose there is a set $PB = \{pb_1, pb_2, \ldots, pb_{|PB|}\}$ of production bases, a set $LC = \{lc_1, lc_2, \ldots, lc_{|LC|}\}$ of logistics centers, a set $ST = \{st_1, st_2, \ldots, st_{|ST|}\}$ of sale terminals, a $GAP - SCN \subseteq PB \times LC \cup LC \times ST$, where "×" is a Cartesian product, and a set $Req_{ST} = \left\{Req_{st_1}, Req_{st_2}, \ldots, Req_{|ST|}\right\}$ of requirement of sale terminal ST; the objective of performance optimization of GAP-SCN is to find $PB_i = \left\{pb_{i1}, pb_{i2}, \ldots, pb_{i|PB_i|}\right\} \subseteq PB$, $LC_j = \left\{lc_{j1}, lc_{j2}, \ldots, lc_{j|LC_j|}\right\} \subseteq LC$, ST, and $SCN = \{(pb_i, lc_j), (lc_j, st_k))|pb_i \in PB_i, lc_j \in LC_j, st_k \in ST\}$, satisfying $\min f = TQ_i^j \times Cost_{pb_i}^{lc_j} + TQ_j^k Cost_{lc_j}^{st_k}, Req_{st_k} = TQ_{lc_j}^{st_k}$ (Table 17.1).

17.3 Preliminary Knowledge of QGA

The smallest unit in a two-state quantum computer is called a Q-bit. The Q-bit may be in the "1" state or in the "0" state, or in any superposition of the two. The state of a Q-bit can be represented as

$$|\Psi\rangle = \alpha|0\rangle + \beta|1\rangle, \tag{17.1}$$

where α and β are complex numbers that specify the probability amplitudes of the corresponding states. $|\alpha|^2$ and $|\beta|^2$ denote the probability that the Q-bit will be found in the "0" state and "1" state, respectively. Normalization of the state to the unity guarantees $|\alpha|^2 + |\beta|^2 = 1$.

A Q-bit individual as a string of m Q-bit is defined as

$$\left[\begin{array}{c|c|c|c} \alpha_1 & \alpha_2 & \cdots & \alpha_m \\ \beta_1 & \beta_2 & \cdots & \beta_m \end{array}\right], \tag{17.2}$$

where $|\alpha_i|^2 + |\beta_i|^2 = 1, i = 1, 2, \ldots, m$.

Q-bit representation has the advantage that it can represent a linear superposition of states. Because the Q-bit representation and represent linear superposition of the state's probability, evolutionary computing with Q-bit representation has a better characteristic of population diversity than other representation. The state of a Q-bit can be updated by the quantum gate operation U.

In QGA, rotation quantum gate is selected. The quantum gate is adjusted as follows:

$$\begin{bmatrix} \alpha_i' \\ \beta_i' \end{bmatrix} = U(\theta_i)\begin{bmatrix} \alpha_i \\ \beta_i \end{bmatrix} = \begin{bmatrix} \cos(\theta_i) & -\sin(\theta_i) \\ \sin(\theta_i) & \cos(\theta_i) \end{bmatrix}\begin{bmatrix} \alpha_i' \\ \beta_i' \end{bmatrix}, \tag{17.3}$$

where (α_i, β_i) is the ith Q-bit and θ_i is the rotation angle of each Q-bit toward either 0 or 1 state depending on its sign.

Inspired by the concept of quantum computing, QGA applies the probabilistic amplitude representation of Q-bit to the coding of chromosome, based on the representation of the quantum vector, so that one chromosome can represent any linear superposition.

17.4 Proposed Transforming Quantum-Inspired Genetic Algorithm

This chapter proposes a TQGA for performance optimization of green agricultural products supply chain networks. In TQGA, the transforming representation and transforming operator are applied. In addition, the chromosome representation is critical to solve the optimization problem. In this section, the chromosome representation is firstly described. Then, the chromosome initialization, the evaluation of fitness, and updating operator by rotary quantum gate are presented.

17.4.1 Representation of Chromosomes

Each gene in the chromosome is mapped to a path from PB to ST by LC. The gene value indicates transportation quantity of green agricultural products on a path.

Based on this idea, we use float-point representation to encode the SCN problems. Each chromosome C_t in the population is represented as

$$C_t = \left(x_t^1, x_t^2, \ldots, x_t^n, x_t^{n+1}, \ldots, x_t^N\right), \tag{17.4}$$

where $x_t^n \in [0, Req_{h\_st(n)}]$ represents the transportation quantity of green agricultural products on path n, $n = 1, 2, \ldots, N$, $N = |PB| \times |LC| \times |ST|$ is the total maximum number of impossible paths, $t = 1, 2, \ldots, M$. $Req_{h\_st(n)}$ is the requirement of sale terminal $h\_st(n)$ and is applied to clamp the maximum gene value.

Each quantum chromosome QC_t in the population is represented as

$$QC_t = \left(p_t^{11}, p_t^{12}, \ldots, p_t^{1m}, p_t^{21}, p_t^{22}, \ldots, p_t^{2m}, \ldots, p_t^{N1}, p_t^{N2}, \ldots, p_t^{Nm}\right)$$

17.4.2 Transforming of the Encoding

After QC_t is created, we should transform it into float-number representation for evaluating the fitness. x_t^n is calculated by Eq. (17.5).

$$x_t^n = U_{\min} + \left(\sum_{i=1}^{m} b_t^l \times 2^{i-1}\right) \times \frac{U_{\max} - U_{\min}}{2^m - 1}, \tag{17.5}$$

where $U_{\min} = 0$, $U_{\max} = Req_{h\_st(n)}$.

Before applying rotation quantum gate to updating quantum chromosome, the float-number representation chromosome also should be transformed into quantum chromosome. $\sum_{i=1}^{m} b_t^l \times 2^{i-1}$ is calculated by Eq. (17.6); then, b_t^l is calculated by method of successive division

$$\sum_{i=1}^{m} b_t^l \times 2^{i-1} = \left(x_t^n - U_{\min}\right) \times \frac{2^m - 1}{U_{\max} - U_{\min}} \tag{17.6}$$

17.4.3 Initialization of Chromosomes

Initially, a population with M Q-chromosomes is created. α_t^l and β_t^l, $l = 1, 2, \ldots, m$, are initialized with $1/\sqrt{2}$. If the gene value of some chromosome is more or less than the maximum requirement, an additional operator, called transforming, is applied on chromosomes. After transforming operator, the initial population C_1, C_2, ..., C_M is generated.

17.4.4 Evaluation of Fitness

The fitness function of a chromosome C_t in the population is defined as

$$fitness_i = \sum_{n=1}^{N} x_i^n \times \left(Cost_{f\_pb(n)}^{g\_lc(n)} + Cost_{g\_lc(n)}^{h\_st(n)} \right), \tag{17.7}$$

where $f\_pb(n) = \frac{n}{|LC| \times |ST|}$, $g\_lc(n) = \frac{n}{|ST|} \% |LC|$, $h\_st(n) = n\% |ST|$, $Cost_{f\_pb(n)}^{g\_lc(n)}$ is the transportation cost of green agricultural products from $PB_{f\_pb(n)}$ to $LC_{g\_lc(n)}$, and $Cost_{g\_lc(n)}^{h\_st(n)}$ is the transportation cost from $LC_{g\_lc(n)}$ to $ST_{h\_st(n)}$.

17.4.5 Rotation Operation

The rotation operation is only used in QGA to adjust the probability amplitudes of each quantum gene. According to the rotation operation in Eq. (17.3), a quantum gate $U(\theta_i)$ is a function of $\theta_i = s(\alpha_i,\beta_i) \cdot \Delta\theta_i$, where $s(\alpha_i,\beta_i)$ is the sign of θ_i that determines the direction and $\Delta\theta_i$ is the magnitude of rotation angle. The lookup table of $\Delta\theta_i$ is proposed by [8].

17.5 Experimental Study and Discussions

The proposed TQGA for performance optimization of green agricultural products supply chain logistics network was performed on a machine with Pentium D 3.00GHz CPU and 1024MB of RAM. The operating system was MS Windows XP, and the compiler was VC++ 6.0. We tested the TQGA on Case_1 and Case_2 as given in Table 17.2.

The following parameters were use in the GA: $M = 50$; in the TQGA, $M = 20$. To validate the proposed TQGA, we compared the TQGA with GA to tackle the various problems in GAP-SCN. Experiments were independently simulated 30 times. Figure 17.1 presents the comparison of GA and TQGA. To validate the reliability and robustness of TQGA, we tested two GAs with a longer running time and 10,000 iterations. It can be seen from Fig. 17.1 that TQGA has higher solution accuracy and quicker convergence speed than GA. The advantage of optimization speed of TQGA is outstanding. TQGA outperforms GA both in the solution quality and in the optimization speed, especially in the optimization speed.

Table 17.2 Test cases

Test cases	$\|PB\|$	$\|LC\|$	$\|ST\|$	Scale	Req_{st_k}	$Cost_i^j$
Case_1	2	3	3	18	[10,50]	[1,10]
Case_2	10	15	16	2400	[10,50]	[1,10]

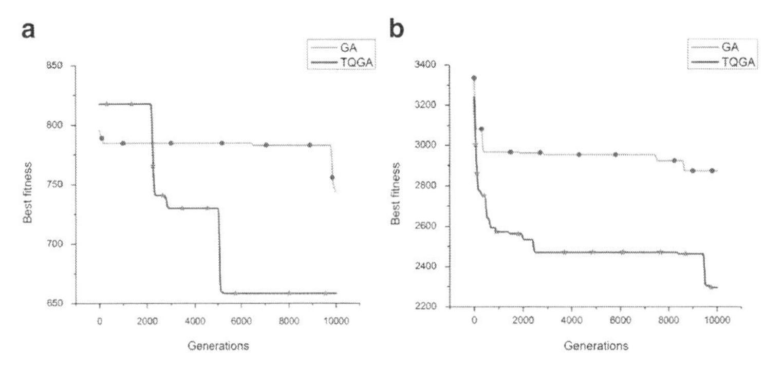

Fig. 17.1 Comparison of GA and TQGA. (**a**) Case_1, (**b**) Case_2

17.6 Conclusion

In this chapter, we have proposed a TQGA for performance optimization of GAP-SCN. In TQGA, a transforming operator is used to keep solutions in feasible ones when total transportation number does not meet the demands of each ST. A transforming of encoding between float-number representation and quantum representation is described. Experimental results demonstrate the superiority and effectiveness of the proposed algorithm for green agricultural products supply chain network.

Acknowledgments This work is supported by the Natural Science Foundation of Guangdong Province of China under Grant No. S2012020011067, the National High Technology Research and Development Program of China (863) under Grant No. 2012AA101701, and the Natural Science Foundation of Zhongkai University of Agriculture and Engineering under Grant No. G3100013.

References

1. Altiparmak, F., & Gen, M. (2006). A genetic algorithm approach for multi-objective optimization of supply chain networks. *Computers and Industrial Engineering, 51*(1), 197–216.
2. Erenguc, S. S., Simpson, N. C., & Vakharia, A. J. (1999). Integrated production/distribution planning in supply chains: An invited review. *European Journal of Operational Research, 115* (2), 219–236.
3. Pontrandolfo, P., & Okogbaa, O. G. (1999). Global manufacturing: A review and a framework for planning in a global corporation. *International Journal of Production Economics, 37*(1), 1–19.

4. Amiri, A. (2006). Designing a distribution network in a supply chain system: Formulation and efficient solution procedure. *European Journal of Operational Research, 171*(2), 567–576.
5. Costa, A., Celano, G., Fichera, S., & Trovato, E. (2010). A new efficient encoding/decoding procedure for the design of a supply chain network with genetic algorithms. *Computers and Industrial Engineering, 59*(4), 986–999.
6. Altiparmak, F., Gen, M., Lin, L., & Paksoy, T. (2006). A genetic algorithm approach for multi-objective optimization of supply chain networks. *Computers and Industrial Engineering, 51*(1), 196–215.
7. Xu, J., Liu, Q., & Wang, R. (2008). A class of multi-objective supply chain networks optimal model under random fuzzy environment and its application to the industry of Chinese liquor. *Information Sciences, 178*(8), 2022–2043.
8. Han, K. H., & Kim, J. H. (2002). Quantum-inspired evolutionary algorithm for a class of combinatorial optimization. *IEEE Transactions on Evolutionary Computation, 6*(6), 580–593.

Chapter 18
A Shortest Path Algorithm Suitable for Navigation Software

Peng Luo, Qizhi Qiu, Wenyan Zhou, and Pei Fang

Abstract In allusion to the shortage of hardware configuration in the mobile devices and high time-complexity of Dijkstra algorithm, the chapter comes up with a shortest path algorithm based on cut-corner for restricted searching area. This algorithm aims at shrinking the smallest searching area quickly and considers the advantages of Ellipse algorithm and Rectangle algorithm. When tested in the simulator, we find that the time-complexity of Cut-corner algorithm is reduced by 5–20 % compared with that of other conventional algorithms. Thus, it has better effect when used in navigation software of low-end mobile device.

18.1 Introduction

The shortest path algorithm plays an important role in the research of intelligent transportation systems [1], and with the high development of mobile devices, it can be used more widely in navigation software.

The Dijkstra algorithm [2] is a conventional single-source shortest path algorithm. Dijkstra algorithm uses blind search from the source S nearly with concentric circles. It does not stop until the radius reaches the destination D [3]. It is no use to search the nodes, which are deviated far from the destination orientation in the algorithm. Navigation software has high requirements of time. Hence Dijkstra algorithm is not suitable.

A large amount of studies focused on searching strategy, data structure, and restricted searching area to improve the efficiency of searching. But the improvement of data structure is very complicated. By contrast, the improved algorithm for restricted searching area is more mature.

P. Luo • Q. Qiu • W. Zhou (✉) • P. Fang
Department of Computer Science and Technology, Wuhan University of Technology, Wuhan 430000, China
e-mail: nancychouwhut@gmail.com

W.E. Wong and T. Zhu (eds.), *Computer Engineering and Networking*, Lecture Notes in Electrical Engineering 277, DOI 10.1007/978-3-319-01766-2_18,

Nordbeck [4] first came up with the idea of shortest path algorithm based on ellipse for restricted searching area (hereafter referred as Ellipse algorithm). Then many scholars studied on the Ellipse algorithm and made some progress. Lu [5] proposed shortest path algorithm based on rectangle for restricted searching area (hereafter referred as Rectangle algorithm). In succession, shortest path algorithm based on orientation rectangle for restricted searching area [6] (hereafter referred as Orientation Rectangle algorithm) was proposed. Wang [7] delivered Ellipse-based shortest path algorithm for typical urban road networks to concentrate on the relationship between the Euclidean distance and the restricted searching area.

The main idea of restricted searching area is to shrink the area to reduce the total time-complexity.

In this chapter, we come up with a shortest path algorithm based on cut-corner for restricted searching area (hereafter referred as Cut-corner algorithm). The experiment results show that the time-complexity of Cut-corner algorithm is obviously less than other algorithms. So, it is more suitable for navigation software whose capacity of hardware configuration is limited.

18.2 Relevant Work

18.2.1 Hardware Configuration of Mobile Device

Mobile device is used more widely in our daily life because of its portability, and it also promotes the spread of navigation software. Nowadays, navigation software is widely applied in Android mobile device. We have investigated 57 kinds of Android mobile phones. Among them, 32 phones are with the memory of 512 MB and 31 phones are the clock speed of 1 GHz.

18.2.2 Conventional Shortest Path Algorithm for Restricted Searching Area

In order to explicate the algorithms, symbolic interpretation is listed.

For the given source S and destination D:

$|SD|$: the Euclidean distance between S and D. d: the shortest path length between S and D. r: the ratio of shortest path length to Euclidean distance, $r=d/|SD|$.

The main idea of Ellipse algorithm is as follows: each node in the shortest path ought to fall in the ellipse (or on the boundary) whose focal points are S, D and long axis is $r \times |SD|$.

Thus for the given source and destination at random, the key of Ellipse algorithm is to determine the value of r. Current idea is to calculate r by statistics and the method is as follows. For the given road network G, through testing a certain number of samples at random, calculate the ratio r_i of shortest path length to Euclidean distance of each sample and then determine a value of r whose confidence interval is 95 % when r is greater to r_i, that is, the r corresponding to G.

Rectangle algorithm aims at avoiding the high time-complexity of nonlinear operations in Ellipse algorithm. Ellipse is a nonlinear conic section; thus, it needs a lot of involution and evolution operations [8] when judging whether a new added node is inside the ellipse or not. Hence, Lu [5] came up with Rectangle algorithm which constructed minimum circumscribed rectangle contained ellipse. Rectangle algorithm expands the searching area leading to more time of searching the shortest path compared to Ellipse algorithm while its linear operation reduces the time of constructing searching area.

Orientation Rectangle algorithm was proposed to ulteriorly shrink the searching area of Rectangle algorithm. The orientation from the source to the destination roughly represents the orientation of the shortest path. Rectangle algorithm pays no attention to the orientation of the source and the destination, thus Han [6] put forward the Orientation Rectangle algorithm. It also only needs simple linear operations just as Rectangle algorithm does. In one hand, it spends more time to rotate the coordinate system. In the other hand, it reduces the time of searching the shortest path because of fewer nodes inside. In all, the time-complexity of Orientation Rectangle algorithm may be lower than that of Rectangle algorithm.

18.2.3 Time-Complexity of Conventional Algorithms by Comparison

In order to evaluate the performance of the mentioned algorithms, we construct a virtual road network area of 4,115 nodes and 7,743 roads, which are even distribution. We test Dijkstra algorithm, Ellipse algorithm, Rectangle algorithm, and Orientation Rectangle algorithm in this area. The experiment environment is Eclipse 3.7.2 and Android simulator. The memory of simulator sets as 512 MB. Adjacency list is adopted to store road network.

Table 18.1 shows the experiment results of the shortest path from the source ID 1023 to the destination ID 1193, and Table 18.2 reveals that from 1830 to 2102. T_1 (ms) is the time of constructing searching area, T_2 (ms) is the time of searching shortest path, L is the length of shortest path, and N is the number of nodes in the searching area.

Tables 18.1 and 18.2 reveal the following facts:

- Each algorithm can work efficiently. All these three restricted searching area algorithms can find the same accurate shortest path.
- Each algorithm can improve the performance greatly when compared with Dijkstra algorithm. The method to shrink searching area can obviously reduce the time-complexity of algorithms. The searching area of Ellipse algorithm is the

Table 18.1 Experiment result of the source ID 1023 and destination ID 1193

	Dijkstra algorithm	Ellipse algorithm	Rectangle algorithm	Orientation rectangle algorithm
T_1	0	51.5	18.6	28.7
T_2	3,223.4	43.6	57.6	54.8
L	370.2	370.2	370.2	370.2
N	–	81	107	99

Table 18.2 Experiment result of the source ID 1830 and destination ID 2102

	Dijkstra algorithm	Ellipse algorithm	Rectangle algorithm	Orientation rectangle algorithm
T_1	0	53.5	22.0	30.1
T_2	8,099.5	171.9	313.5	270.7
L	541.2	541.2	541.2	541.2
N	–	175	257	228

smallest and is markedly superior to that of other two algorithms. The number of the searching nodes can embody the fact. The searching node number of Ellipse algorithm is reduced by 24.3 % and 18.2 %, respectively, when compared with Rectangle algorithm and Orientation Rectangle algorithm in Table 18.1. In Table 18.2, the rate is 31.9 % and 23.2 %.

- Linear operations can greatly reduce the time of constructing searching area. In Table 18.1, Rectangle algorithm reduces the time by 63.8 % compared with Ellipse algorithm; meanwhile, Orientation Rectangle algorithm reduces 44.3 %. The reduction is 58.9 % and 43.7 % in Table 18.2.

18.3 A Shortest Path Algorithm Based on Cut-Corner for Restricted Searching Area

18.3.1 The Idea of Algorithm

The conclusions in Sect. 18.2.3 throw light on the key to the algorithm for navigation software. It is more important to use lower-complicated operations. The time-complexity of Dijkstra algorithm is proportional to the square of the number of the nodes. Both Rectangle algorithm and Orientation Rectangle algorithm expand the searching area leading to more searching time than Ellipse algorithm. With the increase of the length between S and D leading to the larger constructed area, the time reduced by simple linear operations is less than that increased by path searching, shown in Table 18.2. Therefore, the chapter focuses on the following two criteria to judge whether a restricted searching area algorithm is suitable when applied in navigation software.

- The time-complexity of the algorithm should be much lower. Tables 18.1 and 18.2 show that Rectangle algorithm and Orientation Rectangle algorithm spend

less time than Ellipse algorithm, because they only execute the linear operations to compare the node coordinate with the boundary of rectangle when judging whether a new added node is inside the restricted area.
- The searching area should be constructed as small as possible in the supposed even-distribution area. Through the results conveyed by Tables 18.1 and 18.2, we can find that the searching area of Ellipse algorithm is the smallest. The number of searching nodes also accounts for this.

According to the above two points, we propose Cut-corner algorithm so as to construct polygon whose area is approaching ellipse as restricted searching area to the utmost extent. Its main idea is to cut-corner based on Orientation Rectangle to make the restricted area close to that of ellipse. Cut-corner algorithm uses the advantages of both the minimum restricted searching area and simple linear operations.

18.3.2 Algorithm Description

The procedure of Cut-corner algorithm is as follows:

- To rotate the origin coordinate system until the new X-axis parallels with the concatenated line between S and D. The new rotation coordinate is called X0Y.
- To define two containers *listN* and *listR* in order to respectively store the nodes and roads in restricted searching area.
- To calculate the Euclidean distance $|SD|$ between the source S and destination D, half major axis of ellipse a, half minor axis of ellipse b, and the value of cut-angle β to construct the searching area.
- To judge whether all the nodes in road network is traversed, if it is, go to step 6.
- To get next coordinate of node, change it into the corresponding coordinate in X0Y; judge whether this node is inside the cut-corner area. If it is, add it into *listN* or go to step 4.
- To judge whether all the roads in road network is traversed, if it is, go to step 8.
- To get the next road, judge whether it is in the cut-corner area. If it is, put the road into *listR* or go to step 6.
- To use Dijkstra algorithm to search the shortest path and print it out.

18.3.3 Algorithm Implementation

Figure 18.1 illustrates the Cut-corner algorithm. In Fig. 18.1, S is the source and D is the destination, x0y is the original coordinate system, and the angle between the line SD and X-axis is θ. Then rotate the x0y coordinate system to X0Y by θ, and the concatenated line of S, D parallels with the X-axis in the rotated coordinate system. P is the midpoint of S and D, and A, B, C, E are the four tangent

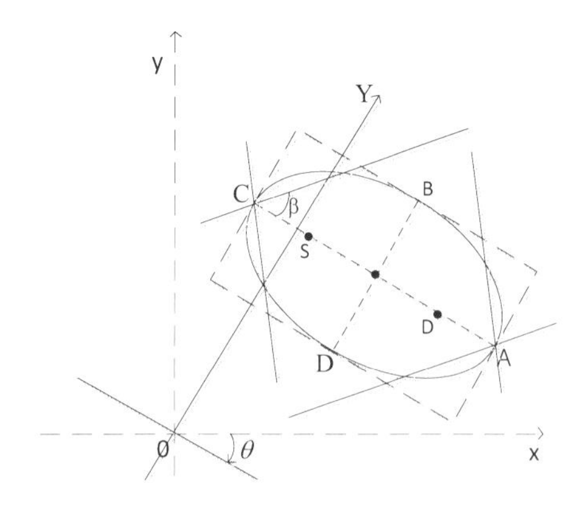

Fig. 18.1 The sketch of constructing searching area based on Cut-corner algorithm

intersections of orientation rectangle and ellipse. β is the value of cut-angle in Cut-corner algorithm.

In the x0y coordinate system, assume that the coordinate of the source S is (x_1, y_1) and that of the destination D is (x_2,y_2) and then come to the conclusion:

The Euclidean distance of S and D is

$$|SD| = \sqrt{(x_1 - x_2) \times (x_1 - x_2) + (y_1 - y_2) \times (y_1 - y_2)} \tag{18.1}$$

The rotated angle of coordinate system θ is

$$\theta = \arctan((y2 - y1)/(x2 - x1)), \theta \in (-\pi/2, \pi/2) \tag{18.2}$$

According to the rotation equation of coordinate system (18.3),

$$\begin{aligned} X &= x\cos(-\theta) - y\sin(-\theta) = x\cos\theta + y\sin\theta \\ Y &= x\sin(-\theta) + y\cos(-\theta) = y\cos\theta - x\sin\theta \end{aligned} \tag{18.3}$$

We can gain the coordinates of the source S and destination D in the X0Y coordinate system:

$$S = (x_1\cos\theta + y_1 sin\theta, y_1\cos\theta - x_1\sin\theta) \tag{18.4}$$

$$D = (x_2\cos\theta + y_2\sin\theta, y_2\cos\theta - x_2\sin\theta) \tag{18.5}$$

Then, the coordinate of midpoint P of S and D:

$$\begin{aligned} P = (m, n) = \big(&(x_1\cos\theta + y_1\sin\theta + x_2\cos\theta + y_2\sin\theta)/2 \\ &(y_1\cos\theta - x_1\sin\theta + y_2\cos\theta - x_2\sin\theta)/2\big) \end{aligned} \tag{18.6}$$

Assume that a is the half major axis of ellipse, b is the half minor axis of ellipse, and c is the half focal length of ellipse.

Define the proportional coefficient $r = a/c$, and r can be gained by statistics [7]. Thus,

$$a = r \times c = r \times |SD|/2 \tag{18.7}$$

According to equation of ellipse $a^2 = b^2 + c^2$, combined with Eq. (18.7), we can get

$$b = \sqrt{a^2 - c^2} = \sqrt{r^2 - 1} \times c = \sqrt{r^2 - 1} \times |SD|/2 \tag{18.8}$$

The searching area of Cut-corner algorithm is supposed to equal to that of Ellipse algorithm. Thus we can get the relationship of cut-angle β and proportional coefficient r. The derivation refers to Zhou et al. ("An improved shortest path algorithm based on orientation rectangle for restricted searching area," Unpublished) which concentrates on the theoretical research. But this chapter focuses more on practical application.

$$\beta = \arctan\left(\left(2 - \sqrt{r^2 - 1}\right)/((4 - \pi) \times r)\right) \tag{18.9}$$

The horizontal coordinate of A is m+a. The vertical coordinate of B is n+b. The horizontal coordinate of C is m−a. The vertical coordinate of E is n−b. Thus the searching area of cut-corner is

$$Z = \{(x, y)|m - a \leq x \leq m + a, n - b \leq y \leq n + b,$$
$$\arctan(|(y - n)/(x - (m - a))|) \leq \beta, \arctan(|(y - n)/(x - (m + a))|) \leq \beta\} \tag{18.10}$$

According to Eqs. (18.1), (18.2), (18.6), (18.7), (18.8), (18.9), and (18.10), the searching area of Cut-corner algorithm can be solved.

18.4 The Analysis of Experiment Results

In the same experiment environment as Sect. 18.2.3, we test Cut-corner algorithm. Table 18.3 shows the experiment result; the meaning of each parameter is the same as Sect. 18.2.3. The second column represents the experiment result from the source ID 1023 to the destination ID 1193, and the third column represents that from ID 1830 to ID 2102.

We can learn from Tables 18.1, 18.2, and 18.3.

- The feasibility of algorithms: as mentioned in Sect. 18.2.3, all the algorithms can find the accurate shortest path, so does the Cut-corner algorithm.

Table 18.3 The experiment results of Cut-corner algorithm

	ID 1023 to ID 1193	ID 1830 to ID 2102
T_1	32.6	39.5
T_2	40.1	174.1
L	370.22	541.2
N	82	176

- The time of constructing searching area: the tables show the time–cost relationship of all the discussed algorithms, which is simplified as:

Rectangle algorithm $\le$Orientation Rectangle algorithm $\le$Cut-corner algorithm $\le$Ellipse algorithm.

For Rectangle algorithm, all the operations are linear and simple, so it spends the least time. While Orientation Rectangle algorithm needs more time to rotate the coordinate system, based on orientation rectangle, Cut-corner algorithm requires additional comparison. Therefore, the time reveals the incremental relationship. However, the three algorithms are all linear operations, which are simpler than the nonlinear operations of Ellipse algorithm. Thus the constructing time is lower than that of Ellipse. For example, in the case of the source ID 1023 and destination ID 1193, Cut-corner algorithm reduces the time by 36.7 % compared with Ellipse algorithm. On the contrary, it increases the time by 75.3 %, 13.6 % compared with Rectangle algorithm and Orientation Rectangle algorithm.

- The time of path searching: the time of Cut-corner algorithm and that of Ellipse algorithm are nearly the same. However, they are both lower than that of Rectangle algorithm or Orientation Rectangle algorithm. So, it costs less time in Cut-corner algorithm. For example, in the case of the source ID 1023 and the destination ID 1193, Cut-corner algorithm reduces the time of searching shortest path by 8.0 %, 30.4 %, and 26.8 %, respectively, compared with Ellipse algorithm, Rectangle algorithm, and Orientation algorithm.
- Among all the algorithms, the total time of Cut-corner algorithm is the least. In the case of the source ID 1023 and destination ID 1193, it reduces the total time by 23.6 %, 4.6 %, and 12.9 %, respectively, compared with Ellipse algorithm, Rectangle algorithm, and Orientation algorithm.

From the above results, when the Euclidean distance between the source and the destination is relatively short, Rectangle algorithm and Orientation Rectangle algorithm are better than Ellipse algorithm. But when the Euclidean distance becomes longer, those two algorithms cannot reflect superiority compared with Ellipse algorithm. Cut-corner algorithm is a combination of advantages of Ellipse algorithm, which has a relatively small searching area, and Rectangle algorithm and Orientation Rectangle algorithm that have the simple linear operations. The time in each situation is the least, which coincides with the main idea, proposed in Cut-corner algorithm in Sect. 18.3.1. Navigation software has high requirement of time. So it values the time-complexity of algorithm more than the shortest path.

In fact, the confidence interval for Ellipse algorithm to search the shortest path is 95 %, which completely satisfy the veracity requirements of navigation software. So Cut-corner algorithm is more suitable for navigation software.

18.5 Conclusion

Searching the shortest path is an extremely important function in the navigation software. The fast development of mobile device makes the navigation software more popular. However, because of lacking in hardware resource, the navigation software in mobile device needs more effective algorithm. The chapter aims at solving the shortage of conventional restricted searching area algorithm and proposes the Cut-corner algorithm. It constructs a simulated area and experiments several shortest path algorithm for restricted searching area in the Android simulator. The experiment results show that the Cut-corner algorithm combines the advantage of small searching area and simple linear operations. Its time-complexity is lower than that of Ellipse, Rectangle, and Orientation Rectangle algorithms. Thus it is more suitable for navigation software in mobile device.

References

1. Wang, Y. B. (2007). Efficient implementation to Dijkstra arithmetic in ITS. *Computer Engineering, 33*(6), 256–258 (In Chinese).
2. Dijkstra, E. W. (1959). A note on two problems in connection with graphs. *Numerische Mathematik, 1*, 269–271.
3. Zhou, Y., Cao, H., & Li, J. X. (2007). A shortest route-planning algorithm within a restricted area. *Microelectronics and Computer, 24*(8), 110–112 (In Chinese).
4. Nordbeck, S., & Rystedt, B. (1969). *Computer cartography shortest route programs* (pp. 28–32). Sweden: The Royal University of Lund.
5. Lu, F., Lu, D. M., & Cui, W. H. (1999). Time shortest path algorithm for restricted searching area in transportation networks. *Journal of Image and Graphics, 4A*(10), 849–853 (In Chinese).
6. Han, G., & Jiang, J. (2003). *The optimal path algorithm with orientation rectangle restriction in vehicle navigation system* (pp. 66–70). In The 7th Proceeding of China GPS Association, Shenzhen (In Chinese).
7. Wang, S. M., Xing, J. P., Zhang, Y. T., & Bai, B. H. (2011). Ellipse-based shortest path algorithm for typical urban road networks. *Systems Engineering – Theory and Practice, 31* (6), 1158–1164 (In Chinese).
8. Fu, M. Y., Li, J., & Deng, Z. H. (2004). A practical route planning algorithm for vehicle navigation system. In *Proceedings of the 5th world congress on intelligent control and automation* (pp. 5326–5329). New York: Institute of Electrical and Electronics Engineers Inc.

Chapter 19
An Energy-Balanced Clustering Routing Algorithm for Wireless Sensor Networks

Mingqiang Chen and Xianhai Tan

Abstract Aimed at the problem of nodes energy imbalance, which is caused by the heavy burden of cluster heads in clustering wireless sensor networks, an uneven clustering routing algorithm based on multihop communication has been proposed for wireless sensor networks. An election algorithm is used for reasonable selection of cluster heads based on candidate threshold and time driven, the independent nodes are introduced to reduce burden of the cluster heads, and the multihop routing based on angle is applied to optimize the intercluster routing algorithm. Simulation results show that the algorithm can save the network energy effectively and balance the energy consumption.

19.1 Introduction

Wireless sensor network (WSN) is a self-organizing and distributed network which is comprised of a large number of microsensor nodes and some base stations (BS) owning wireless communication and computing capacity. WSN has been widely used in many fields, such as environmental monitoring, traffic management, health care, and national defense [1]. Sensor nodes generally come with a disposable battery powered, and their energy is very limited; therefore, the design of routing protocol is the key. In recent years, many researchers have proposed a variety of WSN routing algorithms. Among them, the clustering protocols are concerned by the researchers. LEACH is a clustering routing algorithm [2], in which each node has a certain probability of becoming a cluster head per round, and data deliver directly to BS by single-hop. Previous research has proved that

M. Chen (✉) • X. Tan
School of Information Science and Technology, Southwest Jiaotong University,
Chengdu 610000, China
e-mail: newseem@qq.com

W.E. Wong and T. Zhu (eds.), *Computer Engineering and Networking*, Lecture Notes in Electrical Engineering 277, DOI 10.1007/978-3-319-01766-2_19,

multihop routing communication can save more energy [3], but it easily leads to dissipation of energy and the hot spots problem [4]. An energy-efficient unequal clustering (EEUC) algorithm was proposed [5], but it still has the following problems (1) the candidate cluster heads are selected by a preset probability, sensors with low residual energy can still become candidate cluster heads, and their energy is consumed in the election of final cluster heads. The energy of nodes is more imbalanced in the network. (2) The ordinary nodes join the cluster whose signal is the strongest, but it does not consider that some nodes near BS which send directly to BS will consume smaller energy than indirectly. (3) The method of multihop routing is unreasonable, resulting in waste of energy.

In view of the above problems, this chapter presents an energy-balanced uneven clustering routing algorithm for wireless sensor networks (EBUC). It owns the following advantages (1) differentiated from randomly producing candidate cluster heads, the EBUC selects candidate cluster head whose residual energy is not less than the cluster average energy, and the residual energy of nodes is quantified as the time for driving the cluster heads election. The selected cluster heads own high energy, and the message complexity of the cluster formation algorithm is reduced. (2) By defining the independent nodes, the loads of some cluster heads are reduced nearby BS. (3) The routing algorithm based on angle is proposed, reducing the intercluster energy consumption.

19.2 Network Model

The network consists of N sensor nodes, which are randomly deployed over a vast field. Each node has a unique ID in the network, and we assume it has the following characteristics.

1. All sensor nodes are energy heterogeneous, whose positions are fixed, and have the capability of sensing signal emission angle.
2. BS has endless energy, a strong computing and storage capacity.
3. If transmission power is known, the distance between the sender and the receiver can be calculated [6].
4. Sensors can use power control to vary the amount of transmission power which depends on the receiver.

We use a simplified model in which the radio hardware energy dissipation consists of the sending circuit loss and the power amplifier loss [7]. As shown in Eq. (19.1), both the free space (d^2 power loss) and the multipath fading (d^4 power loss) channel models are used in the model, depending on the distance between the transmitter and receiver. The energy spent for transmission of an l-bit packet over distance d is

$$E_{Tx}(l,d) = \begin{cases} lE_{elec} + l\varepsilon_{fs}d^2, d < d_o \\ lE_{elec} + l\varepsilon_{mp}d^4, d \geq d_o \end{cases} \tag{19.1}$$

Equation (19.2) shows energy consumption, when the node receives this packet.

$$E_{Rx}(l) = lE_{elec} \tag{19.2}$$

A cluster head consumes E_{DA} amount of energy for data aggregation. We also assume that the sensed information is highly correlated, so the cluster heads can always aggregate the data received from its members into a single length-fixed packet.

19.3 The EBUC Mechanism

After the deployment of sensor nodes, BS broadcasts BS_MSG to the entire network with the determined power. Each node calculates the distance to BS by the receiving signal strength. We need to control the range of competition radius in the network and set R_c of s_i as Eq. (19.3).

$$s_i.R_c = \left(1 - c \cdot \frac{d_{\max} - d_{s_i-BS}}{d_{\max} - d_{\min}}\right) \cdot R_c^0 \tag{19.3}$$

c is a constant coefficient between 0 and 1. d_{max} and d_{min} represent the maximum and minimum distance between sensor nodes to BS. d_{s_i-BS} is the distance from s_i to the BS. R_c^0 is the maximum competition radius which is predefined. According to Eq. (19.3), R_c varies from $(1 - c)R_c^0$ to R_c^0. As an example, if c is set to ⅓, R_c varies from ⅔ R_c^0 to R_c^0 according to its distance to BS.

19.3.1 Election of Cluster Heads

In many WSNs applications, the network node density is large, so it is not necessary for all nodes to become the candidate cluster heads. E_{ave} is the energy threshold, which is the cluster average remaining energy in the last round. If the residual energy of some node is not less than E_{ave}, it will become a candidate cluster head, and its competition time t is calculated in Eq. (19.4).

$$t = \alpha \cdot T_{CH} \cdot \frac{E_{\max} - E_i}{E_{\max} - E_{ave}}, E_i \geq E_{ave} \tag{19.4}$$

T_{CH} is a period of time for the cluster head election. E_i represents the residual energy of s_i. E_{max} means the maximum residual energy of the intercluster nodes in the last round. $\alpha \in (0.9,1)$ is a random number which is used to avoid the collisions of HEAD_MSGs which are caused by the nodes owning same residual energy.

In the election, the candidate cluster heads broadcast HEAD_MSGs (containing the node ID, the competition radius R_c, and competition time t) by radius R_c^0 at their competition time t, in order to declare that they win. The adjacent candidate cluster head set of s_i is defined as $s_i.N_c = \{s_j | d_{s_i - s_j} < \max(s_i.R_c, s_j.R_c)\}$. When the sensors receive HEAD_MSGs from N_c, they will quit their own election process. After T_{CH}, all cluster heads are selected, and then they broadcast CH_ADV_MSGs.

19.3.2 The Formation of Clusters

After all cluster heads are selected, the nodes are divided into the ordinary nodes and the cluster heads. T_r is a threshold for distinguishing the ordinary nodes. If the distance between an ordinary node and BS is less than T_r, the ordinary node is divided into the first class nodes. Otherwise, it is divided into the second class nodes. We assume that s_i is one of the first class nodes, CH is the nearest cluster head to s_i, and the distance between the two is recorded as $d_{CH\text{-}BS}$. According to Eqs. (19.1) and (19.2), when k-bit packet is transmitted to BS by single-hop, the energy consumption is shown in Eq. (19.5).

$$s_i.E_{direct} = k \cdot E_{elec} + k \cdot \varepsilon_{fs} \cdot d_{s_i - BS}^2 \tag{19.5}$$

If s_i sends k-bit packet to BS via CH, the energy consumption is shown in Eq. (19.6).

$$s_i.E_{CH} = k \cdot \left(E_{elec} + \varepsilon_{fs} \cdot d_{s_i - CH}^2\right) + k \cdot E_{elec} + k \cdot E_{DA} \tag{19.6}$$

Comparing E_{direct} with E_{CH}, we know that by which energy consumption is lower.

$$s_i.E_{CH} - s_i.E_{direct} = k \cdot \left(E_{elec} + E_{DA} + \varepsilon_{fs} \cdot d_{s_i - CH}^2 - \varepsilon_{fs} \cdot d_{s_i - BS}^2\right) \tag{19.7}$$

Order $H = (E_{elec} + E_{DA})/\varepsilon_{fs}$, then:

1. If $d_{s_i - BS}^2 - d_{s_i - CH}^2 \leq H$, that is $s_i.\ E_{CH} - s_i.\ E_{direct} \geq 0$. Energy consumption that s_i sends data directly to BS is smaller than indirectly. s_i is defined as an independent node and sends a JOIN_BS_MSG to BS.
2. If $d_{s_i - BS}^2 - d_{s_i - CH}^2 > H$, that is $s_i.\ E_{CH} - s_i.\ E_{direct} < 0$. Energy consumption that s_i sends packet indirectly to BS is lower than directly, and s_i joins the nearest cluster.

For far from BS, the energy consumption is extremely high, when the second class nodes send packet directly to BS, so they send JOIN_CH_MSGs to join the nearest cluster. When all clusters are established, a Voronoi diagram [5] is formed in the network.

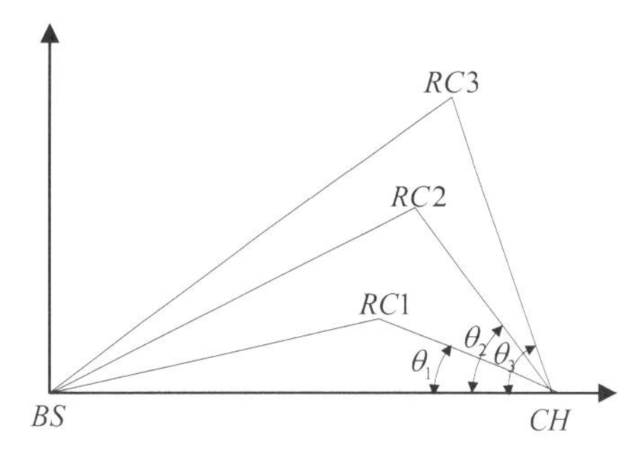

Fig. 19.1 The relay node model

19.3.3 Data Transmission

From the adjacent cluster head set of CH, the cluster heads which satisfy the condition of $0 \leq \theta < \pi/2$ within the range of $n \times R_c^0$ are selected to constitute N_{RC} which is the candidate relay set of CH. n is the smallest positive integral, which makes that N_{RC} non-null. θ is an angle and measured by CH. As shown in Fig. 19.1, we assume N_{RC} of CH contains RC1, RC2, and RC3 and $\theta_1 < \theta_2 < \theta_3$.

If CH chooses RC1 as the relay node, we assume a free space propagation channel model, and RC1 communicates with BS directly. To deliver an l-length packet to BS, the energy consumed by RC1 and CH is

$$\begin{aligned} E_{2hop} &= E_{Tx}(l, d_{CH-RC1}) + E_{Rx} + E_{Tx}(l, d_{RC1-BS}) \\ &= 3lE_{elec} + l\varepsilon_{fs}\left(d_{CH-RC1}^2 + d_{RC1-BS}^2\right) \\ &= 3lE_{elec} + l\varepsilon_{fs}\left(d_{CH-BS}^2 + 2d_{CH-RC1}^2 - 2d_{CH-BS}d_{CH-RC1}\cos\theta_1\right) \end{aligned} \quad (19.8)$$

According to Eqs. (19.1) and (19.2), thus we define

$$\cos t1 = d_{CH-RC1}^2 - d_{CH-BS}d_{CH-RC1}\cos\theta_1 \quad (19.9)$$

Among them, d_{CH-RC1} means the distance between CH and RC1. After CH is placed in position, d_{CH-BS} is a constant, which means the distance between CH and BS. Similarly, we can calculate cost2 of RC2 and cost3 of RC3. Because of cost1 < cost2 < cost3, RC1 is selected as the relay node of CH.

19.4 Algorithm Analysis and Simulation

19.4.1 Message Complexity

Lemma 1 *The message complexity of the cluster formation algorithm is O(N) in the network.*

Table 19.1 Simulation parameters

Parameter	Value	Parameter	Value
Network coverage	(0,0)~(200,200) m	E_{elec}	50 nJ/bit
BS location	(0,250)	ε_{fs}	10 pJ/(bit·m^2)
N	400	ε_{mp}	0.0013 pJ/(bit·m^4)
Initial energy	0.50 ± 0.01 J	d_o	87 m
Data packet size	4,000 bits	E_{DA}	5 nJ/(bit·signal)

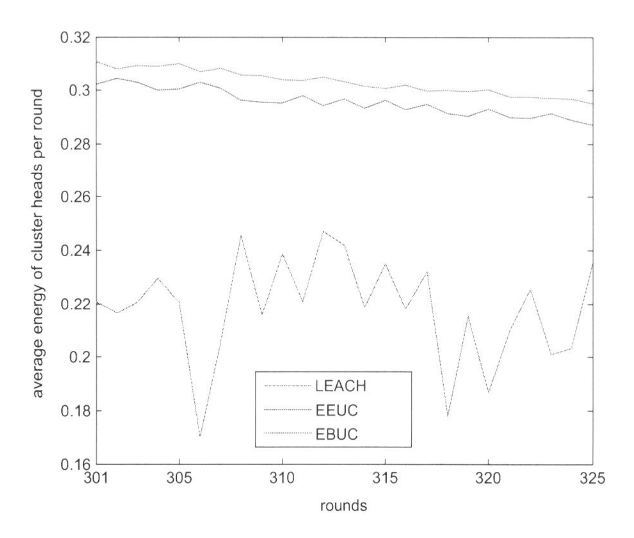

Fig. 19.2 Cluster heads average energy

Proof At the beginning of the cluster head selection phase, $N \times T$ candidate cluster heads are produced, and each of them broadcasts a HEAD_MSG. If it has x cluster heads and y independent nodes, it will send out x CH_ADV_MSGs and y JOIN_BS_MSGs, and then the ordinary nodes will send $N - x - y$ JOIN_CH_MSGs. So the messages add up to $N \times T + x + y + N - x - y = N (T + 1)$ per round. The message overhead of EBUC is $N \times T$ smaller than EEUC.

19.4.2 Simulation Analysis

To illustrate the energy efficiency of EBUC, we use MATLAB to simulate LEACH, EEUC, and EBUC under an ideal condition, and packet loss and delays are ignored. The energy consumption of data aggregation and transmission is measured in the experiment. Firstly, survival time of the network is measured, and the energy efficiency of EBUC is analyzed. Secondly, the residual energy and energy consumption characteristic of cluster heads are analyzed. The simulation parameters are shown in Table 19.1.

Figure 19.2 is the cluster heads average energy curve of three protocols in 25 randomly selected rounds, and it shows that the cluster heads average energy

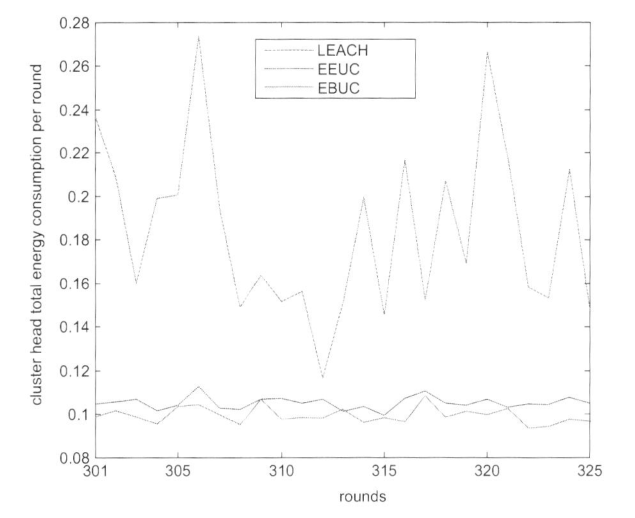

Fig. 19.3 Cluster heads energy consumption

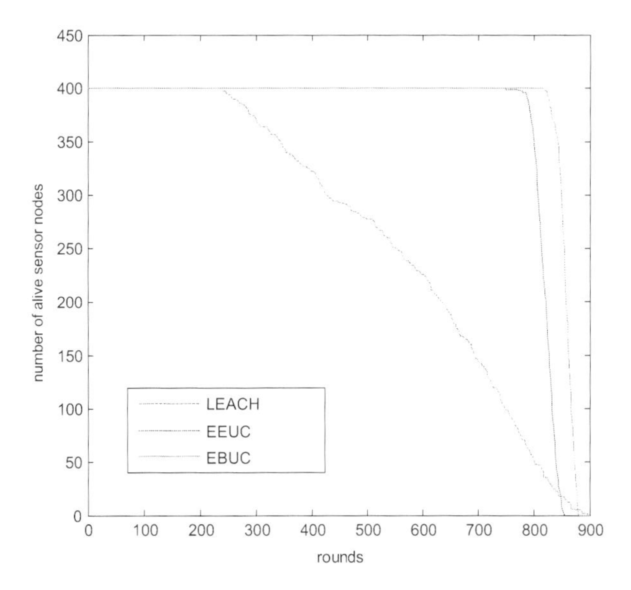

Fig. 19.4 Number of survival nodes curve

of EBUC is the highest of the three algorithms. Figure 19.3 is the comparison of cluster heads total energy consumption of three protocols in 25 randomly selected rounds. It shows that the total energy consumption of EBUC is the lowest of the three algorithms.

Finally, we examine the network lifetime of three algorithms. Figure 19.4 shows that EBUC clearly improves the network lifetime over LEACH and EEUC. In EEUC, candidate cluster heads are randomly selected. Therefore, sensors with low residual energy can still become candidate cluster heads, and their energy is consumed in the election of final cluster heads. Such is avoided in EBUC. Multihop routing based on angle is used in EBUC, so its energy consumption of EBUC is lower than EEUC as illustrated in Fig. 19.3. Comparing to EEUC, stable period of EBUC is increased by 10.6 %. Simulation results show (1) EBUC is able to effectively balance the energy consumption of nodes in the network and extend the network survival time. (2) The cluster heads of EBUC own the highest average energy and capacity of data transmission of the three algorithms. (3) The energy consumption of EBUC cluster heads is the smallest of the three algorithms.

19.5 Conclusion

The article proposes an energy-balanced uneven clustering WSN routing algorithm EBUC. The main idea is that the candidate cluster heads are produced by the candidate threshold and their own competition time, so it avoids that the candidate cluster heads are produced randomly and at the same time the message complexity is reduced. Then, it introduces the concept of the independent nodes to reduce some cluster head loads, and the intercluster route is optimized by the factor of multihop energy consumption for saving energy of cluster heads. Simulation results show that the cluster heads energy of EBUC is the highest of the three algorithms, and its energy consumption is the lowest, and it can efficiently prolong the network lifetime.

References

1. Su, I. J., Tsai, C. C., & Sung, W. T. (2012). Area temperature system monitoring and computing based on adaptive fuzzy logic in wireless sensor networks. *Applied Soft Computing, 12*(5), 1532–1541.
2. Heinzelman, W., Chandrakasan, A., & Balakrishnan, H. (2000). Energy-efficient communication protocol for wireless microsensor networks. *IEEE System Sciences*, 3005–3014. doi:10.1109/HICSS.2000.926982
3. Mhatre, V., & Rosenberg, C. (2004). Design guidelines for wireless sensor networks: Communication, clustering and aggregation. *Ad Hoc Networks, 2*(1), 45–63.
4. Perillo, M., Cheng, Z., & Heinzelman, W. (2004). *On the problem of unbalanced load distribution in wireless sensor networks* (pp. 74–79). Global Telecommunications Conference Workshops 2004, IEEE, Texas.
5. Li, C., Mao, Y., Chen, G., & Wu, J. (2005). *An energy-efficient unequal clustering mechanism for wireless sensor networks* (Vol. 8, p. 604). Mobile Ad Hoc and Sensor Systems Conference, IEEE, Washington, DC.
6. Pei, Z. M., Deng, Z. D., Xu, S., & Xu, X. (2010). A new localization method for wireless sensor network nodes based on N-best rank sequence. *Acta Automatica Sinica, 36*(2), 119–207.
7. Heinzelman, W., Chandrakasan, A., & Balakrishnan, H. (2002). An application specific protocol architecture for wireless microsensor networks. *IEEE Transactions on Wireless Communications, 1*(4), 660–670.

`Chapter 20
Simulation and Analysis of Binary Frequency Shift Keying Noise Cancel Adaptive Filter Based on Least Mean Square Error Algorithm

Zhongping Chen and Jinding Gao

Abstract Pseudorandom binary frequency shift keying (2FSK) sine wave signals with the frequency of 1,200 bit/s and 2,200 bit/s are produced by using the rand function in MATLAB. A noise cancel adaptive filter based on least mean square error LMS algorithm is designed and simulated on MATLAB. The relationships between the adaptive parameters (filter taps N and iterations step length μ) and the convergence speed and precision are analyzed. The optimized adaptive parameters which are not only sufficed for filtering performance but also suited for FPGA hardware implementation are found out, that is, filter taps N = 22 and iteration step length $\mu = 0.006$. It may provide a good foundation for FPGA hardware implementation.

20.1 Introduction

Filtering is one of the most basic and important techniques in the field of signal and information processing. In the process of signal transmission, the signal usually will be disturbed such that signal transmission due to multipath delay caused by the inter-code interference and additive noise affects the communication quality and is the main cause of bit errors. The digital filter can change the relative proportion of the frequency component contained in the signal by a numerical calculation, to the purposes of suppressing or eliminating interference [1].

The adaptive filter belongs to the category of modern filter; when the statistical characteristics of the input signal and the noise are unknown or the statistical characteristics of the input process change, the adaptive filter can automatically

Z. Chen
Hunan Engineering Polytechnic, Changsha 410151, China
e-mail: czpmcu@126.com

J. Gao (✉)
Hunan International Economics University, Changsha 410205, China
e-mail: jdgao@qq.com

W.E. Wong and T. Zhu (eds.), *Computer Engineering and Networking*, Lecture Notes in Electrical Engineering 277, DOI 10.1007/978-3-319-01766-2_20,

adjust its parameters to meet the requirements of certain criteria, its research mainly including adaptive algorithm and hardware implementation. Adaptive noise cancellation is one of the four classic applications [2, 3].

Binary frequency shift keying (2FSK) is one of the most widely used digital communication modulations; the International Telegraph and Telephone Consultative Committee (CCITT) recommends a data rate of lower than 1,200 bit/s with 2FSK modulation way in the voice band. In order to find out the optimal adaptive parameters that not only satisfy the filtering performance but also are suitable for FPGA hardware implementation, in this chapter, take the binary frequency shift keying signal (2FSK), for example, design and simulation of the 2FSK signal denoising adaptive filter based on LMS algorithm on MATLAB, analyze the performance of adaptive filter, obtain the optimum adaptive parameters that both can satisfy the requirement of filtering performance and can suit hardware implementation, and provide reference basis for the FPGA hardware implementation [4, 5].

20.2 Summary of the LMS Algorithm

There are two main adaptive algorithm: least mean square error (LMS) algorithm and the recursive least squares (RLS) algorithm. The least mean square error (LMS) algorithm proposed by Widow and Hoff is a fast search algorithm using gradient estimated value instead of the gradient vector, because it has a small amount of calculation and easy-to-realize benefits that are widely used in practice. The basic idea is to adjust the filter weighting parameter to figure out the mean square error between the filter output signal and the least desired signal. As shown in Fig. 20.1, using the direct form FIR structure, x(k) is the input signal; y(k) is the output of the filter; d(k) is the desired signal, i.e., the reference input; and e(k) is the error signal W(k) for the variable filter coefficient.

The least mean square error algorithm iterative formula based on the steepest descent algorithm is as follows:

$$\mathrm{y(k)} = \sum_{\mathrm{i}=0}^{\mathrm{N}} \mathrm{w_i}(k)x(k-i) \tag{20.1}$$

$$e(k) = d(k) - y(k) \tag{20.2}$$

$$w(k+1) = w(k) + 2\mu e(k)x(k) \tag{20.3}$$

where N is the filter order and μ is the iterative step. The LMS algorithm convergence speed and accuracy mainly depend on these two parameters. In order to ensure the convergence of the algorithm, the iterative step μ requirements in $0 < \mu < 1/\lambda_{max}$, λ_{max} are the maximum eigenvalue of the input signal autocorrelation matrix and the step size μ that influences the convergence speed of the filter. The higher the order of the filter, the greater the accuracy; however, the higher

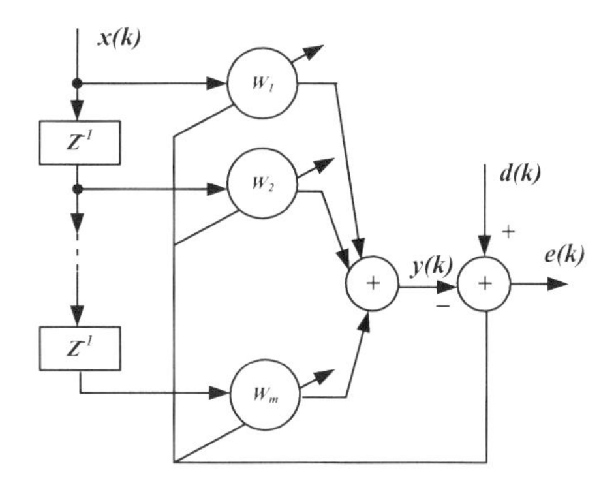

Fig. 20.1 Structure of LMS adaptive filter

the order number, the more hardware implementation cost of the resources. In this chapter, by MATLAB simulation, find out the optimum adaptive parameters μ and N that both can satisfy the requirement of filtering performance and can suit hardware implementation, to provide a reference for the FPGA hardware implementation.

20.3 Simulations and Analysis

20.3.1 2FSK Signal Generation

Rand (N) is the MATLAB built-in pseudorandom number generating function, the function value within (0.0 1.0), and obeys uniform distribution.

According to Rand (N), if the value is greater than or less than the mean value 0.5, then take 1 cycle of frequency 1,200 Hz for the sine signal or take one cycle of frequency 2,200 Hz for the sine wave; random 2FSK sequence generated from 2FSK pseudorandom signal is shown in Fig. 20.2.

20.3.2 LMS Adaptive Filter Simulation and Analysis

Taking the 2FSK pseudorandom sequence as the desired signal d(k), the input signal x(k) of filter is the 2FSK pseudorandom sequence superimposed with white Gaussian noise. e(k) is the error signal and y(k) is the filter output signal. According to the LMS iterative algorithm, written the m file in MATLAB, to reduce the FPGA implementation cost of hardware resources, the main thing is to meet the minimum requirements of the filtering performance of the adaptive filter order N. In this chapter, by changing the iteration step size μ and order N, a series of simulation was conducted. Shown in Figure 20.3 is a filter input signal x(k). Figures 20.4 and 20.5 compare how to select a few key parameters of the filter error simulation output;

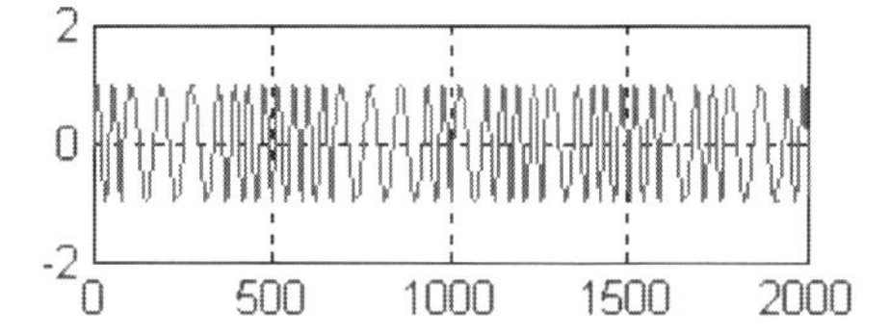

Fig. 20.2 Random 2FSK sequence

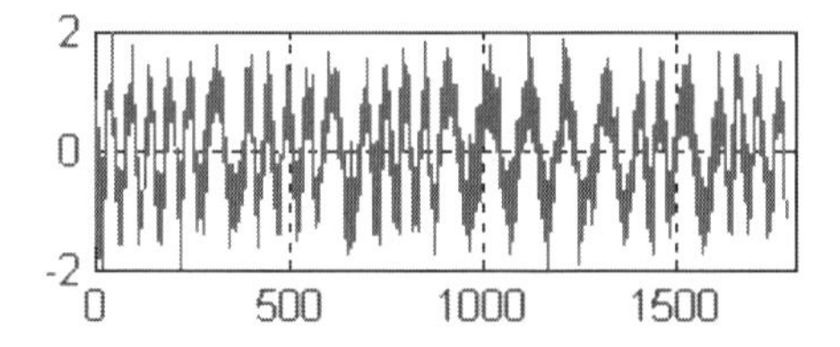

Fig. 20.3 Filter input signal

the abscissa is the number of sampling points, and the ordinates are the relative amplitudes of the size.

Shown in Figure 20.4 is the step size mu = 0.006, the error output image of the various orders adaptive filter. It can be seen from the figure that when the filter step size μ is fixed, with the increase of the filter order N, error output of the filter is smaller, filtering performance is better, and the accuracy of the adaptive filter is higher; N = 12, the larger error output of the filtering performance cannot meet the requirements; N = 22 and N = 32, the relatively small error output can meet the performance requirements; and N = 32, error output smaller than N = 22; however, taking into account the more orders of FPGA hardware implementation and the greater cost of hardware resources, so as long as they can meet the filter performance requirements, N = 22 is more ideal.

Shown in Figure 20.5 is a filter order N = 22, the error output image of the adaptive filter of the various iteration step. It can be seen in Fig. 20.5 that the iteration step size μ has great influence of the convergence speed of the filter. μ value is small; when mu = 0.001, the error convergence speed is very slow and the filter cannot meet the filtering requirements; the greater the value of mu, the faster the convergence speed. When mu = 0.006, it can meet the filtering requirements, but as mentioned above, in order to ensure the convergence of the algorithm, the step size mu requires a $0 < mu < 1/\lambda_{max}$ range. When the iteration step length mu = 0.01 h, the filter output is unstable and even divergence.

After a series of simulation and analysis, the best adaptive parameters are identified, which meet not only the filtering requirements but also those for FPGA hardware implementation: filter order N = 22 and the step size mu = 0.006.

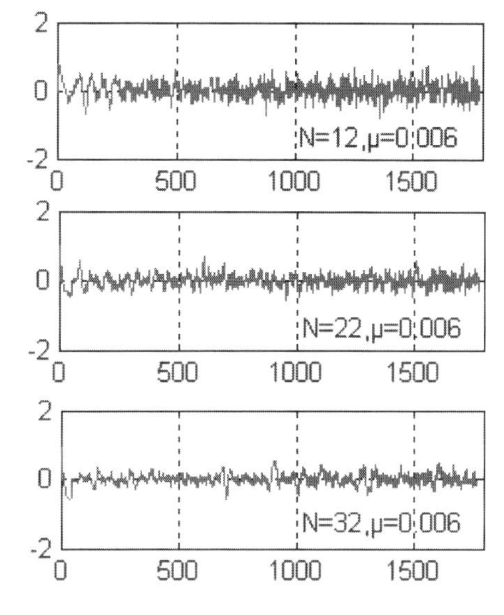

Fig. 20.4 The step size μ = 0.006, various orders of the adaptive filter error output

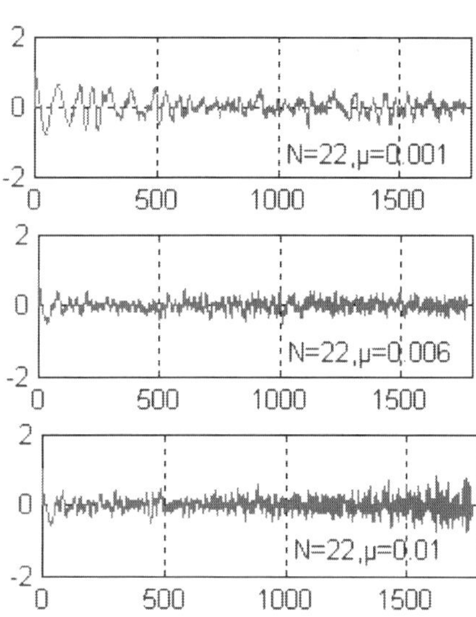

Fig. 20.5 Order of N = 22 different iterations of step adaptive filter error output

20.4 Conclusion

FPGA hardware implementation adaptive filter is an important way to meet the high-speed real-time signal processing requirements. LMS adaptive de-noising filter about pseudorandom binary frequency shift keying (2FSK) signal is designed and simulated on MATLAB. The relationships between the adaptive parameters (filter taps N and iterations step length μ) and the convergence speed and precision are analyzed in this paper. The convergence of the algorithm to identify the filtering performance that meets the requirements for hardware is achieved. The best adaptive parameters are filter order N = 22 and the step size mu = 0.006; it provides a reference for the FPGA hardware implementation.

Acknowledgements This work is partially supported by the Research Fund of Excellent Youth Project of Hunan Provincial Education Department (09B058), Innovative Special Projects of Research Conditions of Hunan Province (2013TT2037), and the Research Fund of Key Project of Hunan Provincial Education Department (10A068). The authors also gratefully acknowledge the helpful comments and suggestions of the reviewers, which have improved the presentation.

References

1. Pan, S., & Wang, G. (2000). *VHDL practical tutorial* (pp. 52–56). Chengdu: University of Electronic Science and Technology of China Press.
2. Diniz, P. S. R. (2004). *Adaptive filtering: Algorithms and practical implementation* (2nd ed., pp. 12–159). Beijing: Publishing House of Electronics Industry.
3. Haykin, S. (2003). *Adaptive filter theory* (4th ed., pp. 1–169). Beijing: Publishing House of Electronics Industry.
4. Windrow, B., & Hoff, M. E. (1960). Adaptive switching circuits. *Proceedings of Wescon Convention Record, Part 4, 12*(1), 96–140.
5. Gao, J. (2013). An improved delayed structure of least mean square adaptive filter and its field programmable gate array implementation. *ICIC Express Letters, Part B: Applications, 4*(1), 69–73.

Chapter 21
Density-Sensitive Semi-supervised Affinity Propagation Clustering

Kunlun Li, Qi Meng, Shangzong Luo, Hexin Li, and Qian Wang

Abstract A density-sensitive semi-supervised affinity propagation clustering algorithm (DS-SAP) is proposed in this chapter. The DS-SAP uses supervised information of the pairwise constraints for adjusting data points distance matrix. Then we introduce a novelty similarity metric based on the characteristics of global and local data distribution. This metric can effectively reflect the reality of data distribution. The DS-SAP clustering algorithm is based on the frame of the traditional AP algorithm and has extended data processing capacity compared to the traditional AP algorithm. Experimental results show that the new algorithm is outperforming traditional AP clustering algorithm.

21.1 Introduction

Traditional clustering algorithms are sensitive to the choice of initial clustering center. Frey and others put forward a new clustering algorithm based on factor graph theory, the affinity propagation algorithm, which can avoid the above problems [1]. Different from many classic clustering algorithms, AP makes all the data points as the potential exemplars, which could avoid the initial clustering center influence of clustering results. It can offer a better performance with a higher efficiency of data work with no requirement of being symmetry made by the similarity matrices of data points [2]. Since 2007 AP algorithms have been proposed, many methods have been introduced to improve the clustering accuracy of AP. AP is applied in a wider range with the improved methods introduced.

Examples are the following: DS-AP algorithm by Prof. Lu makes dense date sparse and parallel based on AP clustering algorithm [3]. KMNC-AP by Prof. Xing increases the efficiency and accuracy of clustering by using the idea of K-mutual

K. Li (✉) • Q. Meng (✉) • S. Luo • H. Li • Q. Wang
College of Electronic and Information Engineering, Hebei University, Baoding 071002, China
e-mail: likunlun@hbu.edu.cn; meng88219@163.com

W.E. Wong and T. Zhu (eds.), *Computer Engineering and Networking*, Lecture Notes in Electrical Engineering 277, DOI 10.1007/978-3-319-01766-2_21,

nearest neighbor consistency to adjust the similarity between data points [4]. Dong proposed a new algorithm, which had the ability of describing the characters of data clustering effectively, and this method has extended data processing capacity compared with traditional AP [5]. Ahmad Akl applied the AP algorithm to the gesture recognition system, to achieve a high identification accuracy [6].

We propose a new similarity metric to improve AP in this chapter. With the introduction of semi-supervised learning strategy to improve similarity matrices, we take the pairwise constraints and a kind of density-sensitive metric into AP, to introduce the density-sensitive semi-supervised affinity propagation clustering algorithm.

21.2 Affinity Propagation

AP algorithm [1] is a new algorithm by B. Frey from Toronto University. The main characteristic of AP is that it makes all the data points as the initial exemplars, so that it can avoid the situation that clustering result is influenced by the choice of the initial selection of exemplars.

AP sets up a collection of real-valued similarities for all the points on how to attract others, which is the similarity metric between any of the two points x_i and x_k; each similarity metric is set to a negative squared error (Euclidean distance): $s(i,k) = -\ \|x_i - x_k\|^2$. AP at first assumes that the possibility of all data points choosing to be the exemplar is the same, which means to set the same P for all $s(i,i)$. Meanwhile, Preference eventually decides number of clusters.

AP algorithm propagates two kinds of information between each two data points: the "responsibility" $r(i,k)$ sent from data point x_i to data point x_k, which reflects how well x_k serves as the exemplar of x_i considering other potential exemplars for x_i, and the "availability" $a(i,k)$ sent from data point x_k to data point x_i, which reflects how appropriate x_i chooses x_k as its exemplar considering other potential points that may choose x_k as their exemplars. The information is updated in an iterative way as

$$r(i,k) \leftarrow s(i,k) - \max_{k s.t k' \neq k} \left\{a\left(i,k'\right) + s\left(i,k'\right)\right\} \tag{21.1}$$

$$\text{if}, i \neq k, \mathrm{a(i,k)} \leftarrow \min\left\{0, r(k,k) + \sum_{i' s.t i' \notin \{i,k\}} \max\left\{0, \mathrm{r}\left(i',k\right)\right\}\right\} \tag{21.2}$$

$$a(k,k) \leftarrow \sum_{i' s.t i' \neq k} \max\left\{0, r\left(i',k\right)\right\} \tag{21.3}$$

The parameter λ, known as the damping factor, is also introduced in the information updating of AP, $\lambda \in [0,1)$. The default value of λ is 0.5. Damping factor could help AP to avoid shocking. Shocking can be avoided by enlarging the damping factor.

21.3 Density-Sensitive Measurement

In the processing of data propagation, two types of consistency can be found [7], which are also in accordance with the semi-supervised learning consistency of the data in a priori assumptions:

1. Global consistency, the high similarity between the data points on the same manifold
2. Local consistency, the high similarity between the neighboring data points

According to the analysis of data's space distribution, the data from the same clustering tend to be in a dense manifold area; however, the data from different clustering tend to be in the sparse area [8]. Generally, we can design a new similarity measure based on local data density characteristics. From two aspects, we introduced a kind of density-sensitive measure to improve similarity measure of AP algorithm.

According to the global consistency, data points in the center or around are more concentrated and more likely to be grouped into one type when clustering, so shortening their distance is needed. On the contrary, enlarging the distance should be applied to the data points which are far from the center. So we accordingly devise new measures, defined as follows:

$$DS\left(x_i, x_j\right) = \rho^{\left[D\left(x_i, x_j\right)/\theta_{ij}\right]} - 1 \tag{21.4}$$

Among the types above, $D\left(x_i, x_j\right) = \frac{d\left(x_i, x_j\right)}{\max\left(d\left(x_{i'}, x_{j'}\right)\right)}$ is the normalized Euclidean distance. This distance makes evaluation results of different scales unified to the same scale, $d(x_i, x_j)$ is the Euclidean distance of data points, $\max\left(d\left(x_{i'}, x_{j'}\right)\right)$ is the MAX distance of data points, and we can set the θ_{ij} and adjust the $D(x_i, x_j)$ amplitude to more effectively reflect the reality of data distribution, $\rho > 1$, known as the damping factor.

According to local manifold properties, through the appropriate transforming of local comparability measurement between data points, we can convert the points on the same manifold into hyperellipsoidal or hypersphere convex data, so that the distribution characteristics of the data can be more precisely expressed. As a result, the clustering accuracy of AP is improved. Thus we can define a local manifold distance metric as follows:

$$DS\left(x_i, x_j\right) = \rho^{\left[\frac{\varepsilon}{\theta_{ij} \bullet \max\left(d\left(x_i, x_j\right)\right)}\right]} - 1 \tag{21.5}$$

The data points x_i and x_j are about ε density that are linked together and on the same manifold in the data set X and $x_i \neq x_j$. When the value of ε is small, which leads the class number a few more, then we can adjust ρ to achieve better clustering results. So we can transform data points on the same manifold through the transformation to the approximation for the center with x_i hypersphere; of course, it is also easier to AP clustering algorithm identification to improve the algorithm accuracy.

According to the space distribution of data, data belonging to the same propagation group stay in a high dense area; however, data belonging to different propagation groups stay in sparse areas. Measurement of density sensitiveness can enlarge the distance of data points in different areas and reduce the distance of data points in the same area.

21.4 Density-Sensitive Semi-supervised AP

In the reality, it is easier to get pairwise constraints compared with labeling the unlabelled patterns. We can take some labeled data for semi-supervised clustering. Semi-supervised AP adjusts the similarity matrix using the priori known labeled data or pairwise constraints. The principle of pairwise constraints is that when the constraint points $\{(x_i,x_j)\} \in M$, we think the two points have high similarity, so we adjust the $s(i,j) = 0$ and they belong to a same cluster; when the constraint points $\{(x_i,x_j)\} \in C$, we think two points have low similarity, then we adjust the $s(i,j) = -\infty$ and divided them into different cluster. The details are as follows:

$$(x_i, x_j) \in M \Rightarrow s(i,j) = 0 \ \& \ s(j,i) = 0 \tag{21.6}$$

$$\begin{aligned} &(x_i, x_k) \notin M \& (x_i, x_j) \in M \ \& \ (x_j, x_k) \in M \Rightarrow \\ &\quad s(i,k) = 0 \ \& \ s(k,i) = 0 \ \& \ M = (x_i, x_k) \cup M \end{aligned} \tag{21.7}$$

$$(x_i, x_j) \in C \Rightarrow s(i,j) = -\infty \ \& \ s(j,i) = -\infty \tag{21.8}$$

$$\begin{aligned} &(x_i, x_j) \notin \{M \cup C\} \ \& \ (x_i, x_k) \in C \ \& \ (x_k, x_j) \in M \Rightarrow \\ &\quad s(i,j) = -\infty \ \& \ s(j,i) = -\infty \end{aligned} \tag{21.9}$$

The capacity for pairwise constraints helping the unsupervised clustering algorithms is limited; we should also utilize the priori knowledge belonging to data set to assist clustering. In this chapter, we use semi-supervised pairwise constraints to adjust data points to optimize the similarity matrix. Furthermore, we introduce density-sensitive measure to improve the data metric matrix between the points; therefore, the more closely the distribution of the various points within the same category, the more decentralized for various points between the different classes; so as to achieve better clustering results, an improved algorithm is proposed as follows:

Algorithm: Density-sensitive semi-supervised affinity propagation clustering algorithm (DS-SAP)

1. $\forall\, x_i,\, x_j \in X$, calculate the Euclidean distance between two points:

$$D_{ij} = \left(\left\|x_i - x_j\right\|^2\right)^{\frac{1}{2}} \tag{21.10}$$

2. Add pairs limit information based on a priori pairwise constraints:

$$\begin{cases} D_{ij}, D_{ij} = 0, if\left(x_i, x_j\right) \in ML \\ D_{ij}, D_{ij} = -\infty, if\left(x_i, x_j\right) \in CL \end{cases} \tag{21.11}$$

3. $\forall\, x_i,\, x_j \in X$, calculated the density-sensitive measure between two points:

$$DS_{ij} = \begin{cases} \rho^{\left[D\left(x_i, x_j\right)/\theta_{ij}\right]} - 1 & x_1, x_j \;\; cannot \;\; link \;\; to \;\; \varepsilon \\ \rho^{\left[\frac{\varepsilon}{\theta_{ij} \bullet \max\left(d\left(x_i, x_j\right)\right)}\right]} - 1 & x_i, x_j \;\; link \;\; to \;\; \varepsilon \end{cases} \tag{21.12}$$

4. The similarity matrix is configured according to the similarity measure of the density-sensitive $S \in R^{n \times n}$:

$$S_{ij} = \frac{1}{DS_{ij} + 1}, S_{ii} = 0 \tag{21.13}$$

5. Based on the similarity matrix S of the above configuration, completed clustering by AP algorithm.

   DS-SAP algorithm directly modifies the distance measure based on semi-supervised pairwise constraint a priori information, so the similar relationship will change with the constraint condition. The spatial consistency a priori information adjusts the distance between data points automatically; it plays an indirect role to modify the similarity matrix. Making full use of the two types of a priori information can maximize the guiding role of clustering search.

21.5 Experiments

The data sets used in this experiment are from UCI, including Iris, Ionosphere, Glass, and Sonar. We compare DS-SAP with ad-AP and AP clustering effects on the above data sets. Meanwhile, the running time of DS-SAP and AP is also included with Circles, Spirals, and Face Image.

Experimental environment: Processor Core2 2.2GHz, memory 2 GB, hard-drive 500G, Windows 7, parameters in the experiment are as: Maxits = 3,000; Convits = 100; Lam = 0.8; Table 21.1 shows the properties of the dataset:

Table 21.1 The experiment dataset

	Iris	Ionosphere	Glass	Sonar	Circles	Spirals	Face image
Instance	150	351	214	208	600	1,000	900
Dimension	4	34	9	60	2	2	50 × 50
Class	3	2	6	2	3	2	100

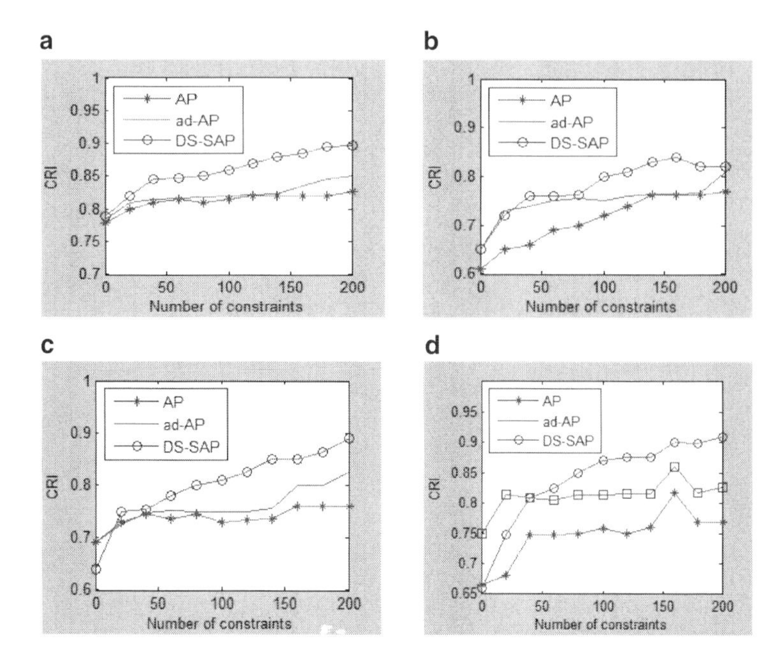

Fig. 21.1 The CRI index of the three algorithms in the UCI data sets. (**a**) Iris, (**b**) Ionosphere, (**c**) Glass, and (**d**) Sonar

Table 21.2 Run-time list unit/s

Algorithm	Circles	Spirals	Face image
DS-SAP	3.06	12.05	16.57
AP	3.25	11.38	17.69

During the experiment, the clustering result is assessed according to CRI:

$$CRI = \frac{correct\ \ free\ \ decisions}{total\ \ free\ \ decisions}$$

Among the type above, *total free decisions* $= (n \times (n - 1))/2 - C_n$, where n is the number of data points, C_n is the number of the pairwise constraints, and correct free decisions are divided in the correct data to minus the number of constraints to the division for the number of correct data. The experimental results are shown (Fig. 21.1; Table 21.2) as follows:

The experimental results show the DS-SAP clustering accuracy is improved. DS-SAP adjust the similarity matrix using pairwise constraints of semi-supervised and the density-sensitive distance measure; it can express the nature of the internal structure of the data, so that it spontaneously improves performance of the algorithm. Meanwhile, the time complexity of the proposed DS-SAP algorithm is mainly depended on the similarity matrix and AP algorithm clustering iterative times. Although the building time of the similarity matrix is relatively increased, the similarity matrix is closer to the ideal after optimization. It can greatly reduce the number of iterations of the clustering, so the total time of our algorithm is not increased and the experiments have verified it. In summary, due to the optimization of the similarity matrix, DS-SAP has higher accuracy than AP and ad-AP, the running time of the algorithm generally flat or even less than the AP.

21.6 Conclusion

This chapter put forward an improved framework of affinity propagation based on semi-supervised (DS-SAP). In this algorithm we used semi-supervised pairwise constraints and introduced density-sensitive measure to improve the data metric matrix between the points, so as to achieve better clustering results. The experimental results on the four data sets of UCI and other three data sets showed that the algorithm clustering performance was improved.

Acknowledgements The project is supported by the National Natural Science Foundation of China (No. 61073121), National Science and Technology Support Plan Project (No. 2013BAK07B04), Natural Science Foundation of Hebei Province. China (No. F2013201170) and Medical Engineering Alternate Research Center Open Foundation of Hebei University (No. BM201102).

References

1. Frey, B. J., & Dueck, D. (2007). Clustering by passing messages between data points. *Science, 315*(5814), 972–976.
2. Givoni, I. E., & Frey, B. J. (2009). A binary variable model for affinity propagation. *Neural Computation, 21*(6), 1589–1600.
3. Weiming, L., Chenyang, D., Baogang, W., Chunhui, S., & Zhenchao, Y. (2012). Distributed affinity propagation clustering based on map reduce. *Journal of Computer Research and Development, 49*(8), 1762–1772.
4. Yan, X., & Yong, Z. (2012). Affinity propagation based on K-mutual nearest neighbor consistency. *Application Research of Computers, 29*(7), 2524–2526.
5. Jun, D., Suoping, W., & Fanlun, X. (2010). Affinity propagation clustering based on variable-similarity measure. *Journal of Electronics and Information Technology, 32*(3), 509–514.
6. Akl, A., & Valaee, S. (2010). *Accelerometer-based gesture recognition via dynamic-time warping, affinity propagation, and compressive sensing* (pp. 2270–2273). IEEE International Conference on Acoustics Speech and Signal Processing, Dallas, TX.

7. Zhai, D., Chang, H., Shan, S., Chen, X., & Gao, W. (2012). Multiview metric learning with global consistency and local smoothness. *ACM Transactions on Intelligent Systems and Technology (TIST), 3*(3), 53.1–53.22.
8. Gui, J., Zhao, Z., Hu, R., & Jia, W. (2013). Semi-supervised learning with local and global consistency by geodesic distance and sparse representation. In *Intelligent science and intelligent data engineering* (Vol. 7751, pp. 125–132). Berlin: Springer.

Chapter 22
The Implementation of a Hybrid Particle Swarm Optimization Algorithm Based on Three-Level Parallel Model

Yi Xiao and Yu Liu

Abstract In order to improve the efficiency of hybrid particle swarm optimization (PSO) algorithm, a PSO merging simulated annealing and hill climbing (SAHCPSO) is implemented based on a three-level parallel model to increase its convergence speed and to decrease the operation time. SAHCPSO can enhance the diversity of the population and avoid population premature convergence. By analyzing and optimizing the SAHCPSO, we complete the task mapping on the model and make full use of CPU/GPU heterogeneous cluster resources. Optimization for parallel accessing further improves the efficiency of the algorithm. The parallel SAHCPSO implements the coarse-grained parallelism between computation nodes and fine-grained parallelism within each node, greatly reducing the operation time. The experimental results show that with the increase of particle scale, higher speedup can be obtained. The high efficiency of the parallel strategy of the model makes the parallel SAHCPSO more easily to solve large-scale problems.

22.1 Introduction

The particle swarm optimization algorithm (PSO) is an evolutionary intelligent algorithm that was first proposed by Kennedy and Eberhart [1]. The idea comes from birds' predation. The advantages of PSO are less parameters and easy to implement. Many significant problems in the field of scientific research and engineering applications can be solved by using PSO. But PSO is easy to premature

Y. Xiao (✉)
College of Information Science and Engineering, Guilin University of Technology, Guilin 541004, China
e-mail: louisxcode@yahoo.com

Y. Liu
College of Mechanical and Control Engineering, Guilin University of Technology, Guilin 541004, China

W.E. Wong and T. Zhu (eds.), *Computer Engineering and Networking*, Lecture Notes in Electrical Engineering 277, DOI 10.1007/978-3-319-01766-2_22,

convergence, and it lacks local search capability. A particle swarm algorithm based on dynamic inertia weight vector and dimension mutation was proposed to balance the local and global searching of particle swarm and increase the speed of convergence [2], but it still takes a long time to compute when solving large-scale optimization problems. In the field of scientific and engineering computing, parallel computing has become an important method to solve large-scale complex problems. With the development of science and technology, CPU/GPU heterogeneous architecture is gradually becoming an important parallel computing platform [3]. We apply a three-level parallel computing model to a hybrid particle swarm optimization algorithm merging simulated annealing and hill climbing (SAHCPSO), which can enhance the diversity of the population, avoid population premature convergence, and improve the efficiency of operations [4].

22.2 Three-Level Parallel Computing Model

22.2.1 MPI, OpenMP, and CUDA

MPI (Message Passing Interface) is a standard of message-passing system. It is a library and a language-independent communication protocol used to write parallel programs, and in fact, it has become a current mainstream programming model in distributed storage architecture. OpenMP (Open Multi-Processing) defines a collection which contains a set of compile-guide statements, runtime library functions, and environment variables. OpenMP completely describes the parallelism architecture and fully uses the advantage of shared memory architecture. It avoids the overhead of message passing and provides runtime scheduling, fine-grained operation mechanism, and coarse-grained operation mechanism. CUDA (Compute Unified Device Architecture) is a general purpose parallel computing architecture—with a new parallel programming model and instruction set architecture which leverages the parallel compute engine in both CPU and GPU to solve many complex computational problems in a more efficient way [5].

22.2.2 Mixed Parallel Computing

The three-level parallel computing model is based on TC (Thread Communication) mixed mode of MPI + OpenMP parallel model. In this model, CPU creates threads to control GPU running kernel function [6, 7]. The parallel model is shown in Fig. 22.1.

According to the parallel model, parallel algorithm use GPU as the central computing component. On the cluster, program processes are divided into subtasks by main thread. Main thread using message-passing function sends subtasks to all computing nodes. After each of the computing nodes receiving the task computing threads for logic operation, communication threads and control threads are created

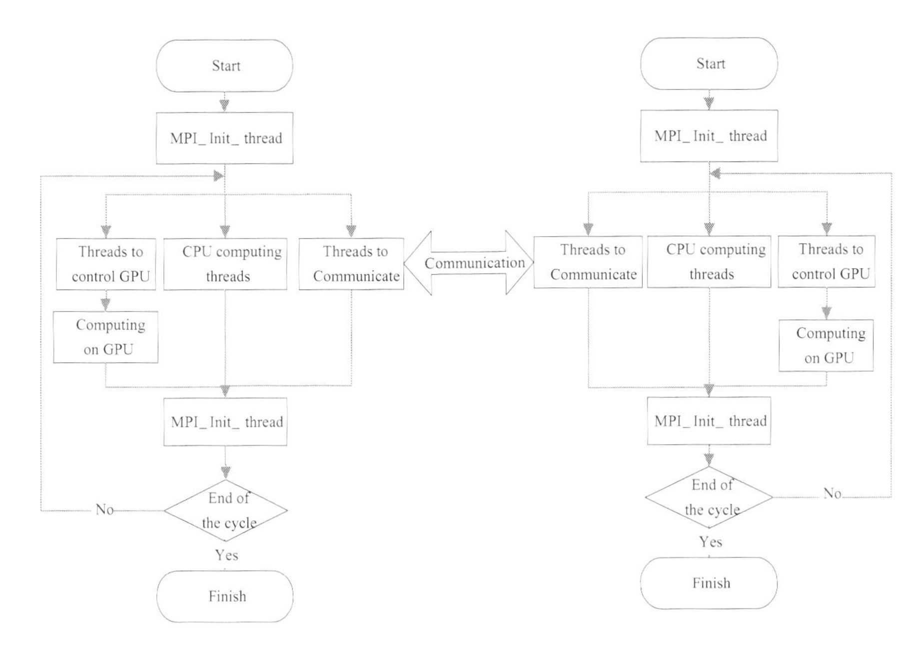

Fig. 22.1 Three-level parallel computing model

by OpenMP. Communication threads are used to communicate with other nodes through MPI function. Control threads are created to run CUDA kernel function [8].

22.3 Algorithm Parallelization Solving

22.3.1 Analysis of SAHCPSO

SAHCPSO program will repeatedly execute the algorithm and calculate the correct result which is equal to the optimal value. The number of correct results determines the ability of convergence. SAHCPSO is divided into four stages: initialization of particles information, mutation of particles information, updating of particles information, and searching for the group best fitness.

Stage 1: SAHCPSO initializes the position (*pos*) and velocity (v) of all particles. For each particle, fitness can be calculated by using particle position information (x_i) as follow:

$$fitness = \sum_{i=1}^{n} \left[x_i^2 - 10\cos(2\pi x_i) + 10\right] \tag{22.1}$$

Each particle calculates their fitness value and saves as personal best value (*p-Best*). Then SAHCPSO compares all particles' p-Best to find the optimal value which saves as group best value (*g-Best*). When SAHCPSO update the

Table 22.1 The proportion of each part of the serial algorithm running time

Stage	Initialization	Mutation	Updating	Searching
Proportion	1.3 %	39.39 %	43.56 %	4.2 %

value of *p-Best* and *g-Best*, corresponding particle position information will be saved. In order to avoid aimless search, position and velocity information will be limited to a definite range.

Stage 2: In this stage SAHCPSO create mutational position information based on the current particle position information according to Eq. (22.2). Where pos' is mutational position information, *Gaussian* is a Gaussian random number.

$$pos' = pos + (1 + Gaussian) \tag{22.2}$$

The mutation probability of each dimension of position information is 10 %. For each particle, fitness can be calculated by using mutational position information according to Eq. (22.1). If the fitness is better than the corresponding particle current fitness value, update the particle fitness and *p-Best*. Otherwise, the particle current fitness will be forced to updating the fitness and *g-Best* with the probability of 1 and 5 %.

Stage 3: SAHCPSO starts updating the particles information. The position and velocity information of all particles are updated according to Eqs. (22.3) and (22.4). Where $w, r1, r2, c1, c2$ is algorithm parameters, k is the current number of iterations, *p-BestPos* and *g-BestPos* is the position information of current *p-Best* and *g-Best*.

$$v_i^{k+1} = wv_i^k + c_1 r_1 \left(p - BestPos_i^k - x_i^k\right) + c_2 r_2 \left(g - BestPos^k - x_i^k\right) \tag{22.3}$$

$$x_i^{k+1} = x_i^k + v_i^{k+1} \tag{22.4}$$

Stage 4: When *p-Best* has been updated, SAHCPSO compares all particles' *p-Best* and searches the optimal value saving as group best value (*g-Best*).

According to Amdahl's law, the impact of the overall system performance depends on proportion and acceleration efficiency of the program which can be optimized [9]. Table 22.1 shows the test of the original serial SAHCPSO for different stages of the average of the proportion of time in different iterations. Therefore, the stages we need to parallelize are particles information mutation and updating.

22.3.2 Parallel Implementation of SAHCPSO

For the characteristics of the procedure, there are two parallel strategies in the parallel computing of SAHCPSO on three-level parallel computing model:

1. Parallelization between the repeated algorithm processes of program
2. Parallelization between the calculations of each particle

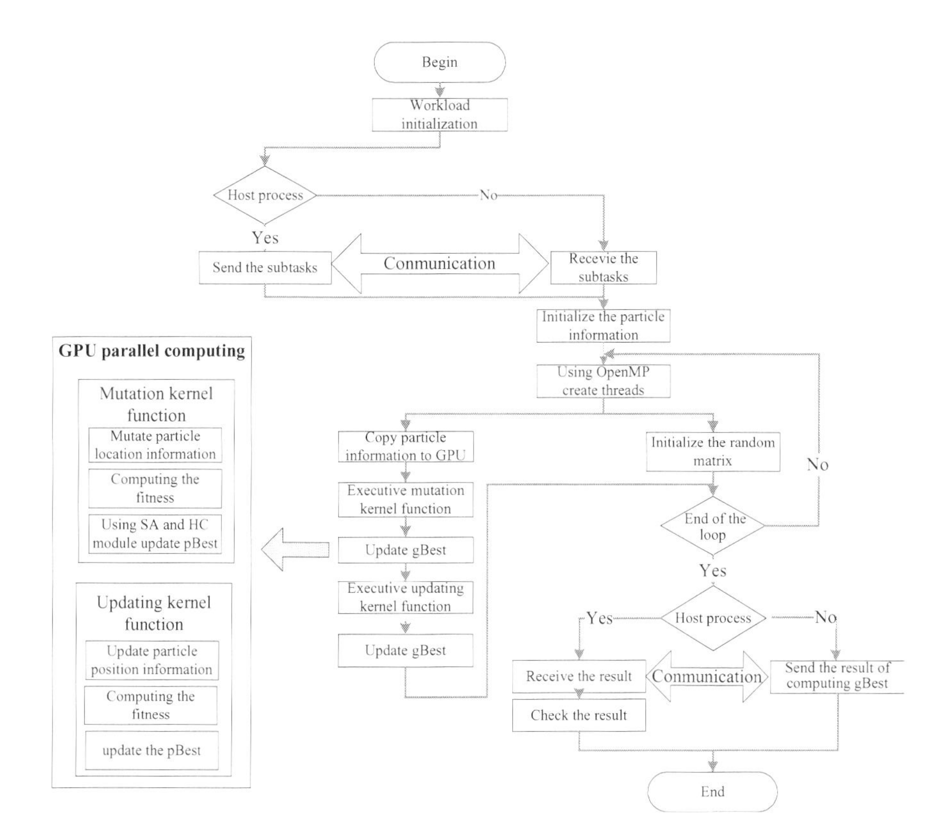

Fig. 22.2 The flow chart of the parallel SAHCPSO

According to the three-level parallel computing model, the repeated algorithm processes of program are mapped to cluster nodes which are coarse-grained parallelism; the processes of particle information calculation are mapped to a single node which uses the CPU and GPU to achieve fine-grained parallel computing. GPU threads are responsible for calculating mutation of particles information and updating of particle information. The random number which GPU required is calculated by CPU threads. For the lack of global synchronization mechanism, g-Best cannot be updated on GPU and the stages of mutation and updating have to be divided into two kernel functions.

The parallel SAHCPSO flow chart is shown in Fig. 22.2. In order to improve the speed of GPU parallel accessing, data structure of the parallel algorithm consists of a series of particles information matrixes and random matrixes which can provide data accessing patterns for memory coalescing [10]. Take particle position information matrix, for example, the number of particles is *p-Count* and the dimension is *dim*. Then we can get the particle position information matrix: *PSO* [*p-Count*] [*dim*].

The contents of random number matrixes are prejudge on CPU according to the high speed of calculation for integer data on GPU. When a single-precision float random number is generated, CPU compares the random number with the corresponding probability parameters and stores a specific integer number to random number matrix.

22.4 Experimental Results

The experiments are performed on cluster which has four nodes with one Intel Core i3 3.3GHz processor connecting to windows XP operating system (2 GB of memory). Each computing node is equipped with a NVIDIA GT405 card. The bandwidth of cluster network is 100 Mbps.

We execute the original serial SAHCPSO and improve parallel algorithm under different particle scale (p-Count) and iterations to test the impact of different conditions on execution time. Figure 22.3 shows the performance of serial algorithm and parallel algorithm computing on 2000 and 5000 *p-Count*. With the expansion of the particles scale, the growth of the time also gradually expands. It is obvious that the parallel algorithm can greatly reduce the growth of time.

Speedup refers to the ratio of the time required by the serial algorithm running on the single node and the parallel algorithm running on the cluster. Figure 22.4 shows the test of speedup under different *p-Count* in the case of 2,000 iterations. With the expansion of particles scale, the growth in the time of the serial algorithm is greater than the parallel algorithm. Therefore, the speedup increases gradually.

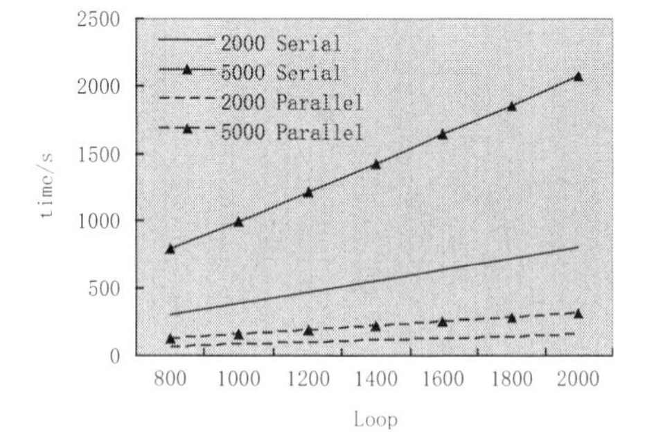

Fig. 22.3 The effect of different particle scale on execution time

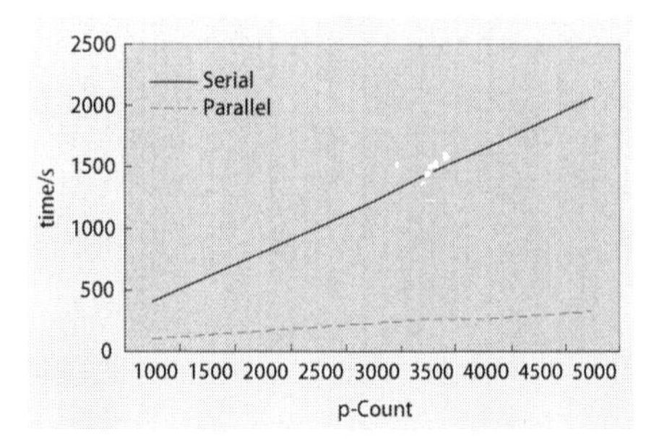

Fig. 22.4 Speedup of the different particle scale

22.5 Conclusion

The paper implemented the SAHCPSO parallel computation on a three-level parallel computing model. Practice proved that this hybrid model can make full use of CPU/GPU heterogeneous cluster recourses. By analyzing the characteristics of the serial algorithm, parallelism embedded in SAHCPSO can be excavated. The calculation of particles' fitness can be mapped to GPU and CPU threads. The repeat of calculation result can be mapped to computing node processes. Therefore, most of SAHCPSO calculations can be executed in parallel. Optimization for parallel accessing further improves the efficiency of the algorithm. The results showed that the parallel algorithm is correct and efficient. The speedup we have achieved is up to 6.6. In addition, because the model has a good scalability, larger particles scale can be computed.

Acknowledgements This work has been supported by The National Natural Science Foundation of China under research project 41264005 and also supported by The Guangxi department of education under the research project 201102ZD018. We also thank master degree candidate Jiaxing You for his generous help to our work especially in SAHCPSO programming.

References

1. Kennedy, J., & Eberhart R. (1995). *Particle swarm optimization* (pp. 1942–1948). IEEE International Conference on Neural Networks, IEEE, Perth.
2. Liang, X. M., Dong, S. H., Long, W., & Xiao X. F. (2011). PSO algorithm with dynamical inertial weight vector and dimension mutation. *Computer Engineering and Applications, 47*(5), 29–31 (In Chinese).
3. Tian, G., & Lu. F. S. (2011). Measuring MPI/OpenMP+CUDA high-performance computing environment configuration and application. *Silicon Valley, 17*(9), 118–119 (In Chinese).
4. Li, J. M., Wan, D. L., Chi, Z. X., & Hu X. P. (2006). A parallel particle swarm optimization algorithm based on fine-grained model with GPU-accelerating. *Journal of Harbin Institute of Technology, 12*(38), 2162–2836 (In Chinese).
5. Xu, Q. Y., & Chen Q. K. (2010). Research and implementation of MPI+CUDA model based on SMP clusters. *Computer Engineering and Design, 15*(31), 3408–3412 (In Chinese).
6. Liu, Q. K., Ma, M. W., & Yan W. C. (2011). Parallel matrix multiplication based on MPI + CUDA asynchronous model. *Journal of Computer Applications, 12*(31), 3327–3330 (In Chinese).
7. Teng, R. D., & Liu Q. K. (2010). Mixed CUDA, MPI and OpenMP in three mode parallel programming. *Microcomputer Applications, 9*(31), 63–69 (In Chinese).
8. Sanders, J., & Kandrot E. (2011). *CUDA by example: An introduction to general-purpose GPU programming* (pp. 162–165). NJ: Addison-Wesley.
9. Pacheco P. S. (2011). *An introduction to parallel programming* (pp. 60–62). Burlington: Morgan Kaufmann.
10. Kirk, D. B., & Hwu W.W. (2012). *Programming massively parallel processors: A hands-on approach* (2nd ed., pp. 135–137). Burlington: Morgan Kaufmann.

Chapter 23
Optimization of Inverse Planning Based on an Improved Non-dominated Neighbor-Based Selection in Intensity Modulated Radiation Therapy

Xiao Zhang, Guoli Li, and Zhizhong Li

Abstract Intensity modulated radiation therapy (IMRT) is a principal cancer treatment at present, and the optimization of inverse planning is the core to realize the IMRT treatment planning, so it is important to study how to optimize the inverse planning as much as possible. The optimization of inverse planning in IMRT refers to a number of parameters and requires rapid calculation speed clinically, so an improved NNIA for multi-objective is adopted in this paper. At first, according to the IMRT dose constraints of multiple targets, an average dose-based function is used, and then the optimization results compared with the SAGA algorithm under the water phantom show the feasibility and efficiency of the improved NNIA algorithm.

23.1 Introduction

Cancer has become the main disease in the world [1]. Intensity modulated radiation therapy (IMRT) as a precise radiotherapy technology is an effective method to treat cancer. The key step for IMRT is the inverse planning, and the purpose of IMRT inverse planning is to determine the ray type and energy, beam orientation and number, and so on [2]. Because of the complexity of the implementation for IMRT inverse planning, it usually needs computer-aided inverse planning to solve the problem. The optimization of inverse planning is the core to realize the IMRT treatment planning, so it has important meaning to research on this problem.

The optimization of inverse planning in IMRT refers to a number of parameters; therefore, it has multi-objective characteristic. But the heart of multi-objective optimization is to solve the contradiction between convergence speed and population diversity. According to the study of the optimization algorithms in IMRT

X. Zhang (✉) • G. Li • Z. Li
College of Information Engineering, Zhejiang University of Technology,
Hangzhou 310023, China
e-mail: tmny@zjut.edu.cn

W.E. Wong and T. Zhu (eds.), *Computer Engineering and Networking*, Lecture Notes in Electrical Engineering 277, DOI 10.1007/978-3-319-01766-2_23,

inverse planning, the optimization algorithms can be classified as deterministic algorithms and multi-objective evolutionary algorithms. The deterministic algorithms dominated by gradient method are with the advantage of rapid convergence, but may fall into the local minimum for non-convex problems. The multi-objective evolutionary algorithms can escape local minimum theoretically, but their calculating speed usually is relatively slower than deterministic algorithms.

In our previous study, several algorithms are used for the optimization of IMRT inverse planning, such as hybrid multi-objective gradient algorithm [3], multi-objective hybrid genetic approach based on simulated annealing (SAGA) [4], and improved fast non-dominated sorting genetic algorithm (NSGA-II) [5].

In this paper, in order to solve the problem of convergence speed for better, an improved multi-objective immune algorithm with non-dominated neighbor-based selection (NNIA) is applied to the optimization of IMRT inverse planning; the improved NNIA adopts the nonuniform mutation operator making the search step length adjusted adaptively during the different stages of evolution. Finally, an irregular field of segment at deep of 1.5 cm under the condition of 30 cm × 30 cm × 30 cm water phantom is used as the simulation case, through the comparison with SAGA algorithm; the feasibility and efficiency of the improved NNIA algorithm for IMRT inverse panning is to be proved.

23.2 Optimization Algorithm

Immune algorithm (IA) is a multi-objective evolutionary algorithm based on artificial immune system (AIS), in which the antigen and antibody are respectively defined as the optimization problem and the solution of problem; through immune operations, the antibodies will evolve; after evaluating the affinity between antigens and antibodies, the Pareto solutions are obtained.

In this paper, an improved NNIA has been adopted, which has rapid convergence speed and can effectively maintain the diversity of population.

This improved NNIA is implemented by using the non-dominated neighbor-based selection, proportional cloning operator [6], nonuniform mutation operator, and elitism.

For the non-dominated neighbor-based selection, the calculation of crowding distance is a key problem. Through calculating the crowding distance of an individual, its probability of being selected will be determined. The crowding distance of an individual d belonged to the population D can be expressed as

$$I(d,D) = \sum_{i=1}^{m} \frac{I_i(d,D)}{f_i^{\max} - f_i^{\min}} \tag{23.1}$$

where $f_i^{\max}$ and $f_i^{\max}$ are the maximum and minimum for the ith subgoal.

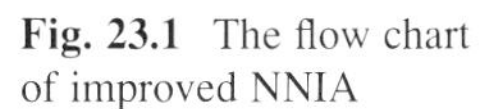

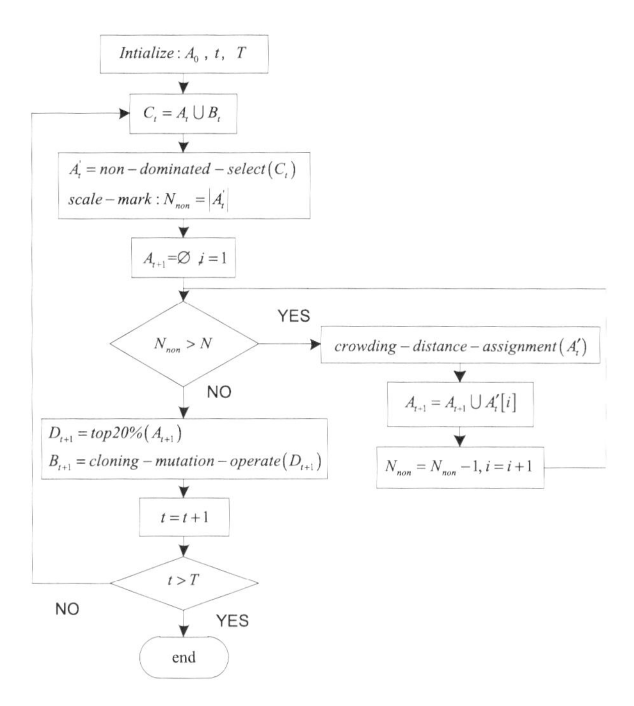

Fig. 23.1 The flow chart of improved NNIA

For the nonuniform mutation operator, the results of mutation operator are associated with its iteration t. The range of variation is relatively large in an early phase, and with the advance of the evolution, the range of variation becomes smaller and smaller, mainly for local search. The variable $x_k \in [a_k, b_k]$ after mutation operation can be determined by following expressions:

$$x_k^{'} = \begin{cases} x_k + \Delta(t, b_k - x_k), & rnd(0,1) = 0 \\ x_k + \Delta(t, x_k - a_k), & rnd(0,1) = 1 \end{cases} \tag{23.2}$$

where $\Delta(t, y) = y\left(1 - r^{(1-t/T)^{\lambda}}\right)$; λ is a unified parameter, it determines the dependence of random disturbance on the iteration t; $r \in [0,y]$. The flow chart of improved NNIA algorithm is shown in Fig. 23.1.

23.3 The Objective Function

There are two main categories of objective functions: the model based on dose constraint and the model based on dose-volume constraint. For the objective function based on dose constraint, it has a simple concept of optimization and is easier to be used; for the objective function based on dose-volume constraint, it has definite physical significances and its application is more flexible. But the form of the objective function can have a significant impact on the speed and effect of optimization.

In our study, an average dose-based function for the planning target volume (PTV) and the surrounding normal tissue (NT) are used in this paper; the mathematical expression is as follows:

$$y_1 = f_{PTV} = \frac{1}{N_{PTV}} \sum_{i=1}^{N_{PTV}} \left(d_i^{PTV} - D_{ref}^{PTV} \right)^2$$
$$y_2 = f_{NT} = \frac{1}{N_{NT}} \sum_{i=1}^{N_{NT}} \left(d_i^{NT} - D_{ref}^{NT} \right)^2 \tag{23.3}$$

where N^{PTV} and N^{NT} are the corresponding numbers of sampling points for PTV and NT, respectively. d_i is the calculated dose value at the ith sampling point. D_{ref}^{PTV} and D_{ref}^{NT} are the average prescription doses for the PTV and NT, respectively.

The fitness functions are expressed as follows:

$$fitness_i = \frac{1}{1 + y_i}, i = 1, 2 \tag{23.4}$$

The dose calculation method among inverse planning is a finite-size pencil beam (fsPB) model based on Monte Carlo (MC) [7], which comes from the previous research of our group.

23.4 Optimization Result

The simulation case is an irregular field of segment at deep of 1.5 cm under the condition of 30 cm × 30 cm × 30 cm water phantom. The slice of the irregular field is shown in Fig. 23.2. The part of shadow is PTV, and the other is NT, in which $N_{PTV} = 13$ and $N_{NT} = 87$.

The optimization parameters are set as Table 23.1.

We also adopt SAGA algorithm which has studied in early stage for comparison; it uses the similar input condition, and the annealing conditions are set as initial temperature is 7 and random seed is 0.7.

The results of these two algorithms are shown as Table 23.2.

The intensity distribution maps of improved NNIA and SAGA are shown as Figs. 23.3 and 23.4.

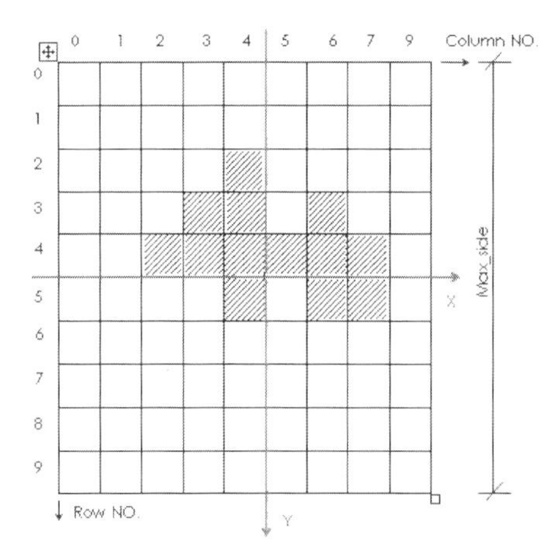

Fig. 23.2 The slice of the irregular field

Table 23.1 Optimization parameters

Parameter name	Parameter value
D_{ref}^{PTV}	0.922
D_{ref}^{NT}	0.013
$fitness_i(i = 1, 2)$	>0.982
Cloning scale	100
Population	100
Generations	500

Table 23.2 Optimization results

Algorithm	$fitness_1$	$fitness_2$	Time of calculation
Improved NNIA	0.9972	0.9893	64s857 ms
SAGA	0.9899	0.9656	87s253 ms

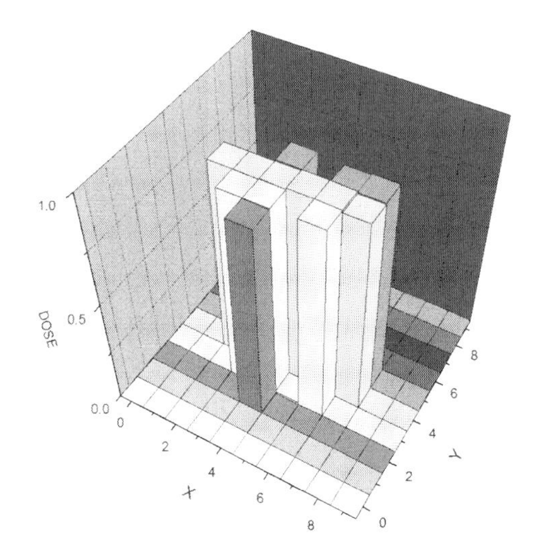

Fig. 23.3 Improved NNIA

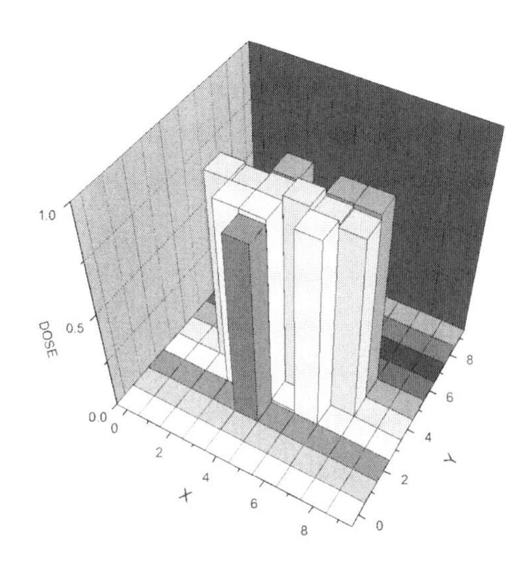

Fig. 23.4 SAGA

23.5 Conclusion

Through comparing the intensity distribution maps of Fig. 23.3 with that of Fig. 23.4 and judging the optimization results of two algorithms, it can be found that the improved NNIA can obtain solutions in a relatively short time, and the conformity of intensity distribution map is better. Therefore, a conclusion can be drawn: the improved NNIA is an efficient optimization algorithm for IMRT inverse planning. But when the algorithm is used in clinic, many practical problems need to be considered, such as the choice of treatment angle. Therefore the focus of our future work is how to apply this algorithm to clinic.

References

1. Zheng, J., Chen, D., Yu, X., & Tian, X. (2013). Investigation of the cancer-related fatigue and quality of life in cancer patients with chemotherapy. *Shanxi Medical Journal, 42*(2), 141–143.
2. Hu, Y. (2005). Advances of intensity modulation therapy. *China Medical Device Information, 11*(2), 1–5.
3. Wang, Z. (2008). *Application of optimization with scaled conjugate gradient algorithm in IMRT* (pp. 751–754). In Proceedings of the 2nd International Conference on Bioinformatics and Biomedical Engineering, Shanghai.
4. Li, G., Li, Z., & Lin, L. (2009). *Multi-objective optimization for irregular field in IMRT based on hybrid SA algorithm* (pp. 1–4). In Proceedings of the International Conference on Information Engineering and Computer Science, Wuhan.
5. Lin, L. (2009). *An improved NSGA-II algorithm for the optimization of IMRT inverse planning* (pp. 1–3). In Proceedings of the 2nd International Conference on Biomedical Engineering and Informatics, Hangzhou.

6. Gong, M., & Jiao, L. (2008). Multi-objective immune algorithm with non-dominated neighbor-based selection. *Evolutionary Computation, 16*(2), 225–255.
7. Zhou, J. (2009). A fast size pencil beam algorithm based on Monte Carlo simulation. *Journal of University Science and Technology of China, 39*(5), 515–519.

Chapter 24
A Recommendation System for Paper Submission Based on Vertical Search Engine

Zhen Xu, Yi Yang, Fei Wang, Jiao Xu, Zhong Li, Fuqiang Mu, and Lian Li

Abstract In this work, the proposed orchestrating and sharing system for online paper aims at managing papers from information collecting, paper editing, paper type-setting, and paper submitting to paper sharing. In the five aspects above, there are many available tools which help science researchers write papers, but these tools work separately not cooperatively. Orchestrating and sharing system for online paper integrates functions of these tools, which offers one-stop service. As an important part of this system, the recommendation for paper submission is to provide valuable information about the latest international conferences and journal for paper publication. When papers are written, our system, a context-aware solution for paper, automatically obtains the keywords from context. Given that the recommendation for paper submission is subject-oriented search, we design a recommendation system for paper submission based on vertical search engine, which enhances the search accuracy by the improved URL-based filtering algorithm and the improved content-based filtering algorithm.

24.1 Introduction

PapersCloud [1] is an online paper orchestrating and sharing system that supports the whole life cycle of science papers (briefly called paper in this paper below). The basic meaning of the life cycle can be popularly understood as "the whole process from the cradle to the grave" (Cradle-to-Grave). In accordance with the definition of the life cycle, we propose a paper life cycle approach. Paper life cycle

Z. Xu • Y. Yang (✉) • F. Wang • J. Xu • L. Li
School of Information Science and Engineering, Lanzhou University,
Lanzhou 730000, China
e-mail: xuzhen1109@163.com; yy@lzu.edu.cn

Z. Li • F. Mu
Gansu Wanwei Company, Lanzhou 730000, China

W.E. Wong and T. Zhu (eds.), *Computer Engineering and Networking*, Lecture Notes in Electrical Engineering 277, DOI 10.1007/978-3-319-01766-2_24,

is the process that from collection information, paper editing, paper type-setting, paper submission to paper sharing. In our system, paper life cycle includes references recommendation, paper editing, paper type-setting, paper submission recommendation, and paper sharing management. References recommendation is to offer some related papers for the interesting topics. Paper editing and type-setting get the input of the paper and type-setting according to a certain format. Paper submission recommendation gives a list of institutes for submitting the paper for publication. The last is to manage the available papers remotely. This chapter mainly aims at papers submission recommendation.

Paper submission recommendation gives information about paper publication for the authors after their accomplishment of their papers. How to always get better result is the main part of this paper. We can get the research field and the keywords from the paper, and we search the most related information for the author from our recommendation library. According to the preference of every author, we give out the result in different ways: for new authors we give out the most related and for old users we just give out the recently information such as the main dates of their preferences.

24.2 System Overview

This system is mainly designed to provide information of conferences and periodicals when users want to submit their papers. And the system continues to use the classical framework, it is consists of web spider module, dumper module, index module, and query module. The main process of the system is shown as following Fig. 24.1. For instance, when user finishes a paper about cloud computing, our system will automatically obtains the keywords from context. When system obtains the keywords, web spiders crawl the web pages form Internet, after that we filter some useless pages and becoming the web page database. Then web page go through information extraction, calculation of the PageRank value, etc., and after these steps, data can be indexing. The last step is similarity calculation, and the query result will be shown to the user. In next, we will introduce the four modules in detail.

The first module of search engine is web spider module. It is the foundation of the search engine. It is in charge of downloading useful web pages from the World Wide Web. All search data are derived from the work of web spider module.

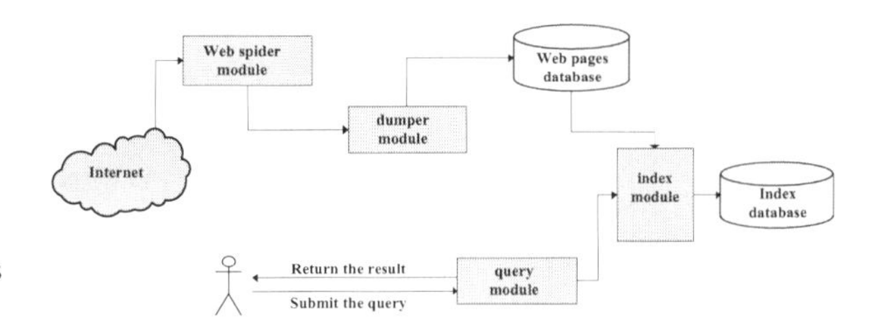

Fig. 24.1 The main process of the system

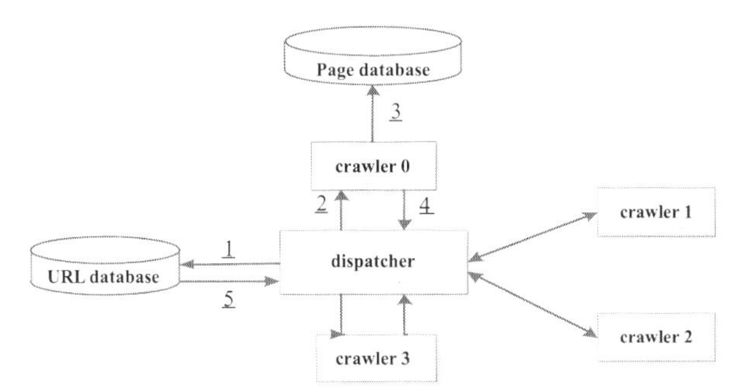

Fig. 24.2 The crawlers working process

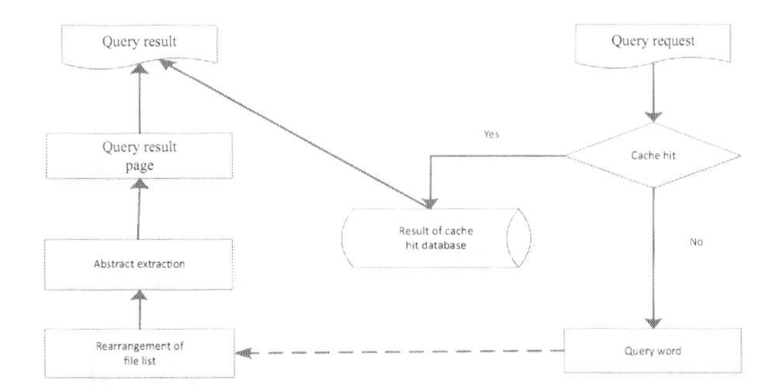

Fig. 24.3 The process of query module

This is a huge project because there are tens of thousands of web pages [2] on the Internet. World Wide Web has bow-tie structure. Because of this, we select directory-type Website as the crawler start page. Figure 24.2 shows how crawlers work.

In Fig. 24.3, we can see crawler get URL from the URL database and then download the web page, and after the filter, useless pages form the page database.

In this module, our main consideration is the selection of the URL library. The choice of text search engine is portals and directory-type site. As for this, it get much more information about subsites or related sites of the portal. Thus, there is much more unconcerned information it get. Because the selection of URL library influences the search result to a large extent and we pay more attention on the search for international conferences and journals, we use both method of URL setting and portals for the crawler. To set the certain sites, we can get more accurate information than we get from the portals; on the contrary, we can get entire information from the method of portals.

The second module is dumper module. For dumper module, it basic and primary work is categorized extract [3] valuable information from the semi-structured page. And this information can represent the attributes of the page, such as anchor text, title, and content.

Because the information we get through the crawler are complicated and uncertain, we need to eliminate the unconcerned information and put the rest to our recommendation library. As to dumper module, it does the work to search results according to the conditions of the authors such as deadline of paper submission, journals or conferences, SCI or EI, and the rest from our recommendation library.

The third one is index module. The index module is the search engine's data warehouse; it stores and indexes millions of thousands of web pages. The purpose of indexing the web page is convenient for querying in the next stage. We need to index the page crawl from the web that can accelerate the speed of query.

In this module, we will first progress the Chinese word segmentation. It is mainly to segment the sentence into a collection with suitable word. Then, we will calculate the PageRank value. The result of offline calculation will be returned a list of PageRank, including a PageRank value of every page and will be easily retrieved in the query module. At last, we will index the pages.

The last module is query module. Query module directly faces the users. It receives the query request by online users and gives the user result in accordance with the calculation by retrieving, sorting, and abstracting extract, and so on.

Figure 24.3 shows the process of query module. Firstly, the system receives the query request from the user and then compares the request with the cache hit. If the request is in it, we directly show the result to the user, and if not we query the word from index module. When the index module returns the result, rearrange the file and extract the abstract, form the result page to user. The whole query requirements not only are faster but also are able to provide users with available results.

24.3 Detailed Design

Our system provides the function that can offer some information about recommended conferences or periodicals for users, which requires the system to consider how to control the search results that will not be offset and filter the useless query information. These issues will be exhaustively described in next context.

24.3.1 URL Filter Analysis and Implementation

Search engine crawlers work mechanism's priority is to grab the web pages which are the highly relevant to the subject. Web pages were sorted according to the page ranking, and only keep themes that are above the URL threshold.

Currently widely used URL filtering algorithm consists of two classes, PageRank algorithm and HITS algorithm [4]. The basic idea of PageRank algorithm [5, 6] is that, web pages from a number of high-quality web links must have the high quality web pages. HITS algorithm bases on the idea that the really value of the page is highly relevant to the theme of the user's search content. HITS algorithm easily occurs that pages deviate from the core theme and irrelevant results returned.

Compared with the two algorithms, PageRank algorithm is in dominant position. So, we selected the PageRank algorithm and improved it. PageRank algorithm is simple, but it ignores the user's understanding of web page. It gives different web pages with the same weight; this will search high value but have little relationship about the subject.

In order to improve the results, we improved PageRank algorithm by using the user's visiting navigation path diagram to modify the traditional PageRank value.

Different web page has different probability to be visited by the users, so you can through the web page's visiting probability express the initial PageRank value. The simple way is Eq. (24.1). A_P is the number that page p was visited; $\sum A_P$ is the number that all pages were visited.

$$PR_1(P) = \frac{A_P}{\sum A_P} \tag{24.1}$$

We should consider the actual situation that different users access to different pages and unbalanced to set the current page's ability to recommend the outlink web page. And combined with actual visiting condition, the PageRank value expresses as below.

$$PR_{n+1}(\mathrm{P}) = (1 - \mathrm{d}) * 1 + \mathrm{d} * \sum \frac{PR_{n+1}(\mathrm{T_i}) * \mathrm{W_p}(\mathrm{T_i})}{\mathrm{C}(\mathrm{T_i})} \tag{24.2}$$

In formula (24.2), $\mathrm{W_p}(\mathrm{T_i})$ is the weight that is obtained from page $\mathrm{T_i}$ visiting page A, and this weight is proportional to the number page $\mathrm{T_i}$ outlink page A, inverse proportion to the page $\mathrm{T_i}$ outlink all the pages. So $\mathrm{W_p}(\mathrm{T_i})$ is

$$\mathrm{W_p}(\mathrm{T_i}) = \frac{A_P^{\mathrm{T}_i}}{A_{B(T_i)}} \tag{24.3}$$

In formula (24.3), $A_P^{\mathrm{T}_i}$ is the number that user through page $\mathrm{T_i}$ visited page P; $A_{B(Ti)}$ is the number that user through page $\mathrm{T_i}$ all outlinks, $B(\mathrm{T_i})$ is the page $\mathrm{T_i}$'s all outlinks.

According to the formula (24.1), the PageRank value expresses is

$$PR_{n+1}(\mathrm{P}) = (1 - \mathrm{d}) * \mathrm{W_P} + \mathrm{d} * \sum \frac{PR_{n+1}(\mathrm{T_i}) * \mathrm{W_p}(\mathrm{T_i})}{\mathrm{C}(\mathrm{T_i})} \tag{24.4}$$

In formula (24.4), $\mathrm{W_p}$ is the weight that user random visit some page; the weight is proportional to the number that user not through other link visit page A and inversely proportional to the number that user visit page A; the expression is

$$W_P = \frac{A'_P}{A_P} \tag{24.5}$$

In formula (24.5), A'_P is the number that user not through other link visit page A, and A_P is the number that user visit page A.

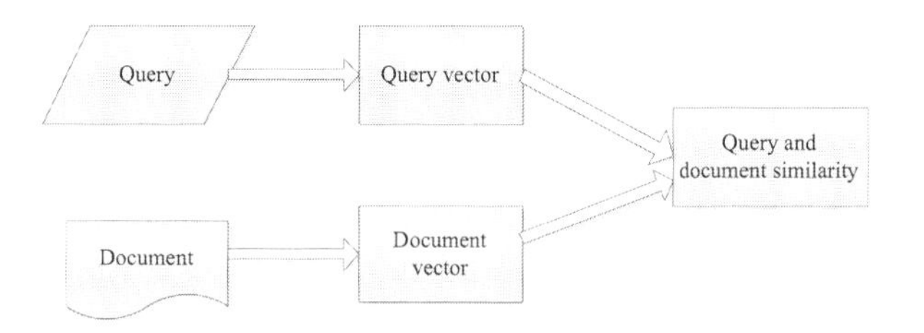

Fig. 24.4 The calculation method of vector space retrieval model

At last, we got the modified PageRank value:

$$PR_{n+1}(\mathrm{P}) = (1 - \mathrm{d}) * \frac{A'_P}{A_P} + \mathrm{d} * \sum \frac{PR_{n+1}(\mathrm{T_i}) * \mathrm{W_p}(\mathrm{T_i})}{\mathrm{C}(\mathrm{T_i})} \tag{24.6}$$

By improving the PageRank algorithm, the URL's filtering accuracy is improved. And through that we can get more useful page that related to the user's requirements.

24.3.2 Content Filter Analysis and Implementation

Content filtering needs to make sure the content filtering algorithm with contextual and real-time result; therefore, filtration precision and filtration velocity become a key content filtering criterion. Current algorithms for matching model include Boolean model, vector space model, and analysis semantic model. Boolean model is a strict matching model; its fast speed is very convenient to realize and suitable for structured information. Vector space model [7] is expressed as a vector to page document, and it will be submitted to customers with the search content. Semantic analysis model based on keyword matching will search for the link between the search item and the actual content to build for a semantic model.

In content-based filtering algorithm, we need to evaluate the similar degree between page and the search subject, that is to say the keywords [8] and the web page will sort by correlation. We would improve the algorithm to increase the search relevance of content. Specific ideas of the improve algorithm:

- Combined the keywords in the text with the frequency and location to determine the relative weights
- Web crawler is to analyze and collect the network data that have higher relative weights and filter irrelevant information
- Using vector space model to collected text for the N dimensional vector and calculate the similarity

Therefore, we simplified the user's query keyword and network resources to a vector which means weight. The method of vector space model works as Fig. 24.4.

Vector similarity algorithm is as follows: Suppose we have two text data which have relation to the author's paper and are expressed as D1, D2. w_{1k}, w_{2k} indicate text weight. Similarity sim(D1, D2) can be expressed with the distance between the vector:

$$sim(D_1, D_2) = \frac{\sum_{k=1}^{n} w_{1k} * w_{2k}}{\sqrt{\sum_{k=1}^{n} w_{1k}^2 \sum_{k=1}^{n} w_{2k}^2}} \tag{24.7}$$

Through the improved algorithm we can further filter out content which is not related to web pages or the related degree is not high, increase the precision of the system.

24.4 Experiment Studying

In this section, we describe an experiment to test our vertical search engine. We crawl about 26,756 web pages on the Internet by open source search engine Nutch. After that we transported the crawling web pages to database. The keywords, "cloud computing," "data base," and "mobile computing," respectively, are chosen for the experiments. And we compute the accuracy. The comparison of the two results is as follows in Fig. 24.5. Obviously the accuracy of the improved one is higher to the original one.

24.5 Conclusion

In this chapter, one vertical search engine for paper submitted was designed, which can give users a help to search available conference or periodicals. We improved the accuracy through the selection of URL library, data filtering, better PageRank, and better similarity algorithm. The key problem, in detailed design, filtering

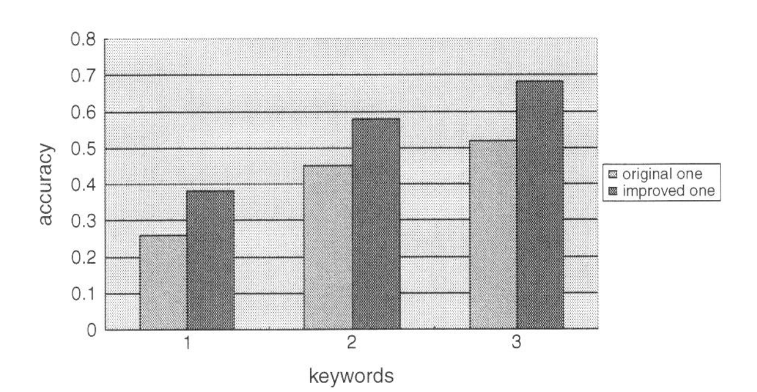

Fig. 24.5 The contrast result

algorithm was discussed in detail. With the improved algorithm, the search engine's query result is comprehensive and its precision is high and it gives users more available results.

There are also some further problems to solve, such as better user interface to display the result, more reasonable Chinese word segmentation and site revisit time, and so on. All above-mentioned issues deserve further research.

Acknowledgments This work was supported by the Natural Science Foundation of P. R. of China (90912003, 90812001 and 61073193), the Key Science and Technology Foundation of Gansu Province (1102FKDA010), Natural Science Foundation of Gansu Province (1107RJZA188), and the Fundamental Research Funds for the Central Universities (lzujbky-2012-47, lzujbky-2012-48).

References

1. Zhang, L., Yang, L., & Xu, Z. (2012). *Paperscloud: A composing-free, collaborative editing platform for scientific papers* (pp. 286–291). The Proceeding of the 2012 IET International Conference on Frontier Computing-Theory, Technologies and Applications (IETFC 2012), ISBN 978-1-849-19604-8.
2. Yang, J. C., & Ling, P. L. (2009). Improvement of PageRank algorithm for search engine. *Computer Engineering, 35*, 35–37.
3. Chau, M., & Chen, H. C. (2008). A machine learning approach to web page filtering using content and structure analysis. *Elsevier Science Direct, 44*(2), 482–494.
4. Kleinberg, J. M. (1999). Authoritative sources in a hyperlinked environment. *Journal of the ACM (JACM), 46*(5), 604–632.
5. Hjghjf, T. (2002). *Topic-sensitive PageRank* (pp. 517–526). Proceedings of the 11th International Conference on World Wide Web (WWW02), Honolulu, Hawaii.
6. L. Page, S. Brin, R. Motwani. (1998). *The page rank citation ranking: Bringing order to the web*. Standford Digital Library Technologies Project 1998.
7. Zhang, Y. M., & Zhou, J. F. (2000). A train able method for extracting chinese entity names and their relations. *Proceedings of the Second Chinese Language Processing Workshop, 12*, 66–72.
8. Amit, C., & Jaewoo, K. (2007). *Selective approach to handing topic oriented tasks on the World Wide Web* (pp. 343–348). Proceeding of the 2007 I.E. Symposium on Computational Intellingence and Data Ming (CIDM2007), ISBN 1-4244-0705-2/07.

Chapter 25
Analysis and Improvement of SPRINT Algorithm Based on Hadoop

Shanshan Fei, Qiaoyan Wen, and Zhengping Jin

Abstract With the rapid development of computers and networks, the growth of data causes the data mining increasingly difficult. To solve this problem, this paper proposes an improved SPRINT algorithm based on the Hadoop platform. By analyzing the traditional SPRINT algorithm, we improve it in three aspects: eliminate unnecessary and repetitive calculations in the processing of discrete attributes; none presort of continuous attributes and split by line directly when splitting; and add the node field for attributes list in the data structure. For illustration, a performance test of acceleration and accuracy is executed to prove the effectiveness of the improved SPRINT algorithm. Compared to the original SPRINT algorithm, experimental result shows that the improved SPRINT algorithm guarantees the accuracy and reduces the computing time for the best split point thus accelerates the speed of decision-tree construction.

25.1 Introduction

The rapid popularity of the Internet has brought a wealth of information for people, but followed by a massive data explosion struck, that is, to say "information explosion." Nowadays, the storage and process of large data sets has become the new challenge faced by many enterprises. It becomes a new target for data mining that how to mining valuable and understandable information from massive data in a fast, efficient, and cost-effective way [1].

The emergence and development of cloud computing has brought new opportunities and challenges for data mining. Cloud computing achieves the process for large data sets by evenly distributing storage and calculation to a multiple of storage

S. Fei (✉) • Q. Wen • Z. Jin
Network Security Laboratory, Beijing University of Posts and Telecommunications, Beijing 100876, China
e-mail: shanshanfei@163.com

W.E. Wong and T. Zhu (eds.), *Computer Engineering and Networking*, Lecture Notes in Electrical Engineering 277, DOI 10.1007/978-3-319-01766-2_25,

computing nodes in a cluster [2, 3]. It greatly reduces the cost owing to the low-cost computer clusters instead of the expensive servers. Data mining is going into the era of cloud based with the help of superior computing power of cloud computing.

Decision tree is the most basic and common classification algorithm in data mining [4]. The basic idea of decision-tree induction is the greedy algorithm, which takes the top-down way and crashes one by one. The construction of the decision tree usually consists of two stages: construction stage and pruning stage. In the construction stage, a fully grown decision tree is built by recursively calling the algorithm. This paper will focus on tree construction stage.

This paper will focus on the SPRINT algorithm that is a kind of decision-tree algorithms [5]. We propose an improved SPRINT algorithm based on the Hadoop platform and make a performance test. The performance test includes the writing of program code, the building of Hadoop cluster, the submitting of program, and the comparative analysis of results.

25.2 Hadoop and MapReduce

Our study is based on the Hadoop platform, while the writing of program code uses the MapReduce programming model [6]. Hadoop is an Apache open source project used to build the cloud platform. It allows users more easily to write and run applications for processing huge amounts of data. Because the cluster has high access speed and a good backup, we set up a Hadoop cluster to verify the effectiveness of the improved SPRINT algorithm.

MapReduce is Google's core computing model. It is a simple and linearly scalable model for data processing. MapReduce highly abstracts the complex parallel computing process running on large-scale clusters to two functions: Map and Reduce [7, 8]. Each of them defines a mapping from one set of key-value pairs to another. Map is responsible for breaking down tasks, while Reduce is responsible for merging the decomposed tasks. Programmers only need to specify the Map function and the Reduce function to write distributed parallel programs.

25.3 Basic Idea of SPRINT Algorithm

The selection of test attributes and division of the sample set are the key links in constructing a decision tree. Gini index can effectively search for the best split point [9]. The point that has minimum Gini index is chosen as the best split point with the maximum information gain. It is very beneficial to generate a good decision tree.

The Gini index is described as:

1. If the set S contains M records belonging to N categories, the Gini index is

$$Gini(S) = 1 - \sum_{i=1}^{n} P_i^2$$

where P_i is the frequency of class i.

2. If the set S is divided into S1 and S2, respectively, corresponding to M1 records and M2 records, the Gini index is

$$Gini_{split}(S) = \frac{m_1}{m} Gini(S1) + \frac{m2}{m} Gini(S2)$$

SPRINT algorithm uses two data structures: the attributes list and histograms [10]. When the data set is too large to fit entirely in memory, SPRINT algorithm can save the rest of the attributes list to the hard disk and only put the current attributes list into memory. The attributes list is split with the division of the node. It includes three fields: attributes value, attributes class label, and row index. Histogram is shown in two forms based on the attribute is continuous attribute or discrete attribute. When the attributes list is generated for the first time, it will be sorted firstly.

For continuous attributes, firstly presorted, assume that the results are v_1, v_2, ..., v_i, ...v_n. Because the split point is between two adjacent points, so there are n−1 possibilities. The split forms are two parts: $V \leq v_i$ and $V > v_i$; it usually takes the midpoint $\frac{v_i + v_{i+1}}{2}$ as the candidate split point. Scanning the attributes list from top to bottom and calculating the Gini index of all candidate points, the best splitting point is the point that has the minimum Gini index. For discrete attributes, there are m mutually different values, the segmented nature is divided into two sets, and there are 2^m possible divisions. We need to calculate the Gini index for each division and look for the best candidate split point. Ultimately, the best splitting point is the point with minimum Gini index and the corresponding attribute is the test attribute.

For example, in the "car insurance" case, the attributes list and the Gini index are shown in Fig. 25.1.

The process of SPRINT algorithm is shown below:

1. Generate initial attributes list: A. Create root node N. Presort A with continuous attribute. Attach A to N.
2. Create the decision tree for node N as following BuildTree algorithm.

The process of BuildTree is shown below:

1. If all attributes in A are of the same class, return; otherwise, go to 2.
2. Scan A and update histograms.
3. Compute the minimum Gini index for candidate split points to get the best split point.

rid	*age*	*carType*	*risk*
0	23	familyCar	high
1	17	racingCar	high
2	43	racingCar	high
3	68	familyCar	low
4	32	truck	low
5	20	familyCar	high

⇨

rid	*age*	*risk*
0	23	high
1	17	high
2	43	high
3	68	low
4	32	low
5	20	high

+

rid	*carType*	*risk*
0	familyCar	high
1	racingCar	high
2	racingCar	high
3	familyCar	low
4	truck	low
5	familyCar	high

S1	*S2*	*gini*
age $\leq$ 18.5	age>18.5	0.4
age $\leq$ 21.5	age>21.5	0.333
age $\leq$ 27.5	age>27.5	0.222
age $\leq$ 37.5	age>37.5	0.417
age $\leq$ 55.5	age>5.5	0.267
age $\leq$ 55.5	age>5.5	0.267

S1	*S2*	*gini*
{ }	{ familyCar , racingCar, truck }	0.444
{ familyCar }	{ racingCar, truck }	0.444
{ racingCar }	{ familyCar, truck }	0.333
{ truck }	{ familyCar , racingCar }	0.267
{ familyCar , racingCar }	{ truck }	0.267
{ familyCar, truck }	{ racingCar }	0.333
{ racingCar, truck }	{ familyCar }	0.444
{ familyCar , racingCar, truck }	{ }	0.444

Fig. 25.1 Attributes list generate and Gini index

4. Use the best split point to split A into A1 and A2 and associate them with N1 and N2.
5. Execute BuildTree recursively to create tree for N1 and N2.

25.4 Analysis and Improvement of SPRINT Algorithm

25.4.1 The Analysis of SPRINT Algorithm

SPRINT algorithm is completely free from memory limit, easy to parallel, and better scalability, acceleration, and expansion, and the resulting decision tree is more compact and accurate.

Through in-depth study of SPRINT algorithm, its shortcomings are as follows:

1. There are repetitive and unnecessary calculations when calculating the Gini index of discrete attributes.

   For discrete attributes, assuming it has m values different from each other, the nature of the split is that m values are divided into two sets and there are 2^m possible divisions. This means we must calculate 2^m Gini values. From Fig. 25.2,

```
Algorithm SPRINT
Input: training sample S
Output: decision-tree
generate attributes list A according S
create node N and attach it to A
BuildTree(A,N)
{
    if(each attribute in A belong to the same class) then
        return
    for( A in A)
    if A is continuous then
        take the midpoint of adjacent points as candidate split point
        obtain the minimum Gini index
    else if A is discrete then
        compute each Gini index of half the possible divisions
        obtain the minimum Gini index
    obtain the best split point of A and split A into A_Left and A_Right
    create nodes N_Left and N_Right
    attach N_Left and N_Right with A_Left and A_Right
    BuildTree(A_Left, N_Left)
    BuildTree(A_Right, N_Right)
}
```

Fig. 25.2 Process of the improved SPRINT algorithm

we can find that some divisions have the same Gini index. For example, the divisions S1 = {family car}, S2 = {racing car, truck} and S1 = {racing car, truck}, S2 = {family car} have the same value 0.444. In fact, when the set S is divided into S1 and S2, exchanging data in S1 and S2, the Gini index is unchanged according to the definition of Gini index. Therefore, the Gini index of nearly half of the candidate split points with that of the other half are same. Such calculations are repeated.

In addition, from the basic idea of SPRINT algorithm, we can see that when a subset is belonging to the same class, the division is terminated. If a set S needs to split, it means that the set S does not belong to the same class. So, the sets which contain 0 value and M values have no sense and the Gini index of them must have the maximum value. Therefore, such calculations are unnecessary.

2. The split of nondividing attributes becomes difficult after the continuous attributes perform presorted.

SPRINT algorithm will perform presorted for continuous attributes, and the discrete attributes do not need to be sorted when the attributes list is generated for the first time. If it utilizes presorted, when split by a continuous attribute, you need to loop scan to determine whether the row is in the left sub tree or in the right sub tree for the nondividing attributes. It increases the scan of attributes list and causes the large amount of computation.

3. The traditional data structure cannot meet the mapping function on the MapReduce framework.

   The function of Map is mapping. By defining different petitioner functions, we can map different <key, value> pairs to different Reducers for processing. The attributes list on each node uses Map to be mapped to different Reducers for parallel processing. The original attributes list includes attributes value, class label, and the row index, which obviously does not meet the requirement. This requires a new data structure to solve this problem.

25.4.2 The Improvement of SPRINT Algorithm

The original SPRINT algorithm is improved in the following three aspects:

1. Eliminate repetitive and unnecessary calculations of discrete attributes.

   The repetitive and unnecessary calculations should be eliminated in the processing of discrete attributes. In general, you need to calculate N Gini index with the unimproved algorithm, but you only need to calculate $\frac{N-2}{2}$ Gini index with the improved algorithm. Obviously, the improved algorithm reduces nearly half of the amount of computation for the best split point.

2. None presort of continuous attributes.

   Continuous attributes do not perform presorted. We just mark the split index so the nondividing attributes can be split by line directly that saves split time spent on the nondividing especially when the data set is large. The cost of this improvement is that the time spent on the calculation of Gini index for continuous attributes is more than that using presorting technology. Overall, the effect is still relatively good. Maybe the input sample is not typical enough which requires further verification.

3. Add node field in attributes list.

   The node field is added in the attributes list to record the current processing node that is used to perform the mapping. However, the improvements have not yet reached the purpose of reducing disk scans. The improvement of the data structure requires further study.

In addition, the current split attribute is deleted immediately after each split because the candidate split attributes are the attributes that never used to split. This saves space and time. The process of the improved SPRINT algorithm is shown in Fig. 25.2.

The detail procedure of improved SPRINT algorithm is shown as follows:

1. Generate initial attributes list: A. Create root node N. Attach A to N.
2. Create the decision tree for node N as following BuildTree algorithm.

The process of BuildTree is described as follows:

1. If all attributes in A belong to the same class, return.
2. For (A_i in A).
   If A_i is continuous, the midpoint of the adjacent points is taken as the candidate split point and the Gini index is calculated. If A_i is discrete, the Gini index of half the possible divisions are calculated.
   Compute the minimum Gini index for candidate split points to get the best split point. Use it to split A into A_Left and A_Right and associate them with N_Left and N_Right.
3. Recursively execute BuildTree(A_Left, N_Left) and BuildTree(A_Right, N_Right) until a decision tree is generated.

25.4.3 *Performance Test*

In this experiment, we implement the original and the improved SPRINT algorithm in MapReduce framework based on the Hadoop. In this part, we focus on analyzing the performance such as accuracy and acceleration of the improved SPRINT algorithm. The experiment uses the virtual machine to build a three-node Hadoop cluster [11].

The cluster includes three nodes: one Master and two Slaves. The Master machine is responsible for the implementation of the data distribution and task decomposition as NameNode and JobTracker. These two Slave machines are responsible for distributed data storage and the execution of tasks as DataNode and TaskTracker. The IP addresses of each node are shown in Table 25.1.

The comparison of results is shown in Table 25.2.

From Figs. 25.3 and 25.4, we can see the time spent on dealing with the same number of records decreased for the improved SPRINT algorithm and still achieves almost the same accuracy that proves the effectiveness of the improvement.

25.5 Conclusion

This paper proposes an improved SPRINT algorithm that improves in three aspects: eliminate unnecessary and repetitive calculations in the processing of discrete attributes; none presort of continuous attributes and split by line directly when splitting; and add the node field for attributes list in the data structure. Through the performance test, we prove that compared to the original SPRINT algorithm, the improved SPRINT algorithm reduces the computing time for the best split point

Table 25.1 Hosts in cluster

Host name	IP address	Experimental environment
Master	192.168.1.133	Operating system: CentOS 6.0; Hadoop: Hadoop 1.0.4; IDE platform: Eclipse 3.3.0; Java runtime environment: JDK1.6.0.19
Slave1	192.168.1.134	
Slave2	192.168.1.135	

Table 25.2 Comparison between unimproved and improved algorithm

	Unimproved		Improved		
Recorder/1,000,000	Accuracy (%)	Time (s)	Accuracy (%)	Time (s)	Save (s)
1	88.59	202	88.59	130	72
2	89.77	248	89.77	165	83
3	90.12	298	90.12	199	99
4	90.33	341	90.34	236	105
5	90.67	387	90.68	274	113
6	91.04	441	91.06	315	126
7	91.32	502	91.35	367	135
8	91.61	544	91.65	399	145

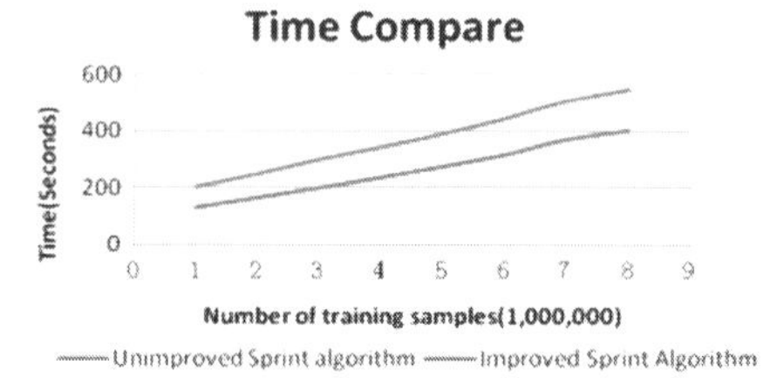

Fig. 25.3 Time compares

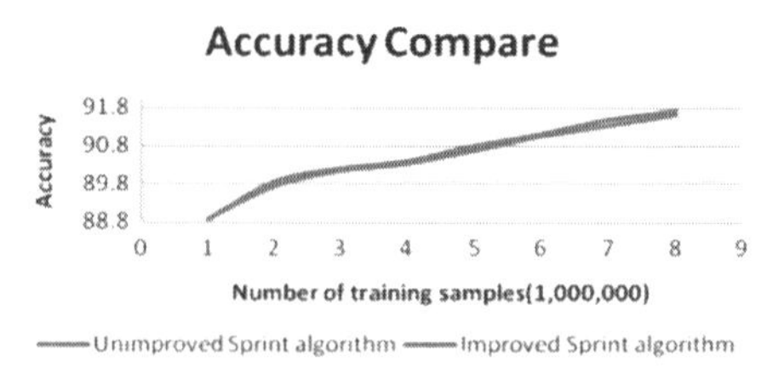

Fig. 25.4 Accuracy compares

thus accelerates the speed of decision-tree construction and still guarantees the accuracy. Of course, there are still some defects of the improved SPRINT algorithm, such as the attributes list is scanned many times and a better data structure is needed in order to save more time and space. This is also the next further study for us.

Acknowledgements National Natural Science Foundation of China (Grant Nos. 61170270, 61100203, 60903152, 61003286, 61121061) and Fundamental Research Funds for the Central Universities (Grant Nos. support this work BUPT2011YB01, BUPT2011RC0505, 2011PTB-00-29, 2011RCZJ15).

References

1. Fayyad, U., Piatetsky-Shapiro, G., & Smyth, P. (1996). From data mining to knowledge discovery in databases. *AI Magazine, 17*(3), 37.
2. Armbrust, M., Fox, A., Griffith, R., Joseph, A.D., Katz, R., Konwinski, A., & Zaharia, M. (2010). A view of cloud computing. *Communications of the ACM, 53*(4), 50–58.
3. Chen, Q., & Deng, Q.-n. (2009). Cloud computing and its key techniques. *Journal of Computer Applications, 29*(9), 2565 (In Chinese).
4. Durkin, J., Jingfeng, C., & Zixing, C. (2005). Decision tree technique and its current research. *Control Engineering, 12*(1), 15–21 (In Chinese).
5. Liu, H., Chen, J., & Chen, G. (2002). Review of classification algorithms for data mining. *Journal of Tsinghua University: Science and Technology, 42*(6), 727–730 (In Chinese).
6. White, T. (2012). *Hadoop: The definitive guide* (pp. 15–38). Sebastopol, CA: O'Reilly Media.
7. Dean, J., & Ghemawat, S. (2008). MapReduce: Simplified date processing on large clusters. *Communications of the ACM, 51*(1), 107–113.
8. Dean, J. (2006). Experiences with MapReduce, an abstraction for large-scale computation. In *PACT: Proceedings of the 15th international conference on parallel architectures and compilation techniques* (Vol. 16, Issue 20, p. 1). Washington, DC: IEEE Computer Society.
9. Raileanu, L. E., & Kilian, S. (2004). Theoretical comparison between the gini index and information gain criteria. *Annals of Mathematics and Artificial Intelligence, 41*(1), 77–93.
10. Ganti, V., Gehrke, J., & Ramakrishnan, R. (1999). Mining very large database. *Computer, 32* (8), 38–45.
11. Lu, Q., & Cheng, X. (2012). The research of decision tree mining based on Hadoop. In *9th international conference on fuzzy systems and knowledge discovery (FSKD) 2012* (pp. 798–801). Piscataway, NJ: IEEE.

Chapter 26
Prediction Model for Trend of Web Sentiment Using Extension Neural Network and Nonparametric Auto-regression Method

Haitao Zhang, Binjun Wang, and Guangxuan Chen

Abstract In order to solve the problem of prediction for long-term web sentiment, a prediction model is built using the proposed method in this paper. First, a novel clustering method based on the extension neural network (ENN) is introduced to recognize the types of subclass of web sentiment. For each class of social events, the class model library of the development trend of web sentiment is established by cycle analysis and ENN clustering combined with nonparametric auto-regression analysis (NAR) method. Then the adaptive transformation is applied to the already known development trend of a new social event, and the min-sum of mean square error (MSE) from the library is selected to predict the future development trend of web sentiment. Empirical findings indicated that compared with the traditional methods, such as the GM (1,1) and least squares estimation (LS) method, the approach presented in this paper yields a higher correlation value in predicting the long-term development trend of web sentiment and can predict the turning points of the development trend more effectively. The ENN- and NAR-based prediction model can effectively solve the problem of prediction for long-term web sentiment.

26.1 Introduction

The World Wide Web and other textual databases provide a convenient platform for exchanging opinions. Reviews and blogs written with the purpose of conveying a particular opinion or sentiment are more and more popular in China. Therefore, prediction of web sentiment development is one of the most important subjects for plan and operation in the field of mass media and sociology. An accurate sentiment prediction is the basis of making decision on the trend of public security and public

H. Zhang (✉) • B. Wang • G. Chen
People's Public Security University of China, Beijing 100038, China
e-mail: okhaitao@126.com

W.E. Wong and T. Zhu (eds.), *Computer Engineering and Networking*, Lecture Notes in Electrical Engineering 277, DOI 10.1007/978-3-319-01766-2_26,

opinions flow for the sociology researchers. The preliminary research in this paper has been carried out by literature survey and observation. Various models for web sentiment prediction have been reported in the literature. Detailed methods include statistical methods [1–3], the regression functions [4], and grey theory [5], neural network method [6]. Artificial neural networks are currently established as a promising approach to sentiment prediction since they are able to learn the functional relationship between system inputs and outputs through a training process. However, a limitation of the neural network (NN) approach is difficult to understand; the content of network memory and the traditional neural network approaches need large amounts of training data and thus require heavy computational efforts to create satisfactory models. Grey prediction models do not need large amounts of data [7]. They have been successfully applied in many fields, but the accuracy of grey prediction models still needs further improvement. Autoregression method is difficult to solve nonlinear problems such as population growth forecasting due to the form of linear parametric. This paper proposed a novel prediction method; we use a new neural network topology, called the extension neural network (ENN) proposed in Wang's paper [8], to recognize the trend types of web sentiment. Then, according to the actual data of every type, we use the nonparametric auto-regression analysis method (NAR) to build the prediction models of every type.

26.2 Extension Neural Network

Extension neural network is a combination of the neural network and the extension theory. The extension theory proves a novel distance measurement for classification processes, and the neural network can embed the salient features of parallel computation power and learning capability. The ENN permits classification of problems with range features, continuous input, and discrete output [9].

The schematic structure of the ENN is depicted in Fig. 26.1. It comprises of both the input layer and the output layer. The nodes in the input layer receive an input feature pattern and use a set of weighted parameters to generate an image of the input pattern. In this network, there are two connection values (weights) between input nodes and output nodes; one connection represents the lower bound for this classical domain of the features and the other connection represents the upper bound. The connections between the jth input node and the kth output node are w_{kj}^{L} and w_{kj}^{U}. This image is further enhanced in the process characterized by the output layer. The output layer is a competitive layer. There is one node in the output layer for each prototype pattern and only one output node with nonzero output to indicate the prototype pattern that is closest to the input vector.

The learning of the ENN can be seen as supervised learning; before the learning, several variables have to be defined. Let training set be $X = \{X_1, X_2, \ldots, X_{Np}\}$,

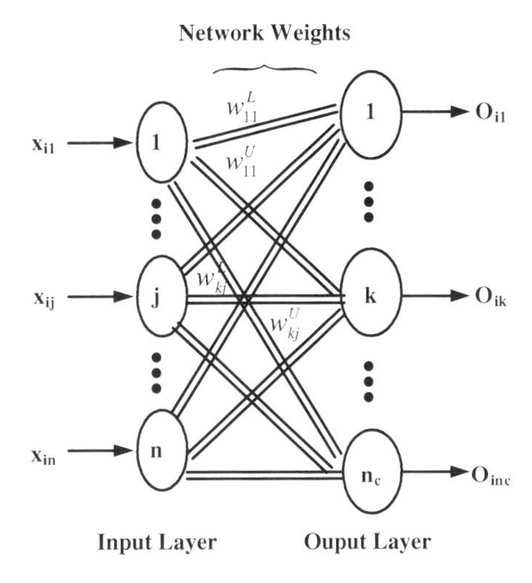

Fig. 26.1 The structure of extension neural network (ENN)

where the total number of training patterns is N_p.The ith input vector is $X_i^p = \{x_{i1}^p, x_{i2}^p, \ldots, x_{in}^p\}$, where n is the total number of the features. To evaluate the clustering performance, the total error number is set as N_m. The total number of the input pattern is set as N_p and the total error rate E_T is defined below:

$$E_T = \frac{N_m}{N_p} \tag{26.1}$$

The detailed supervised learning algorithm can be described as follows:

Step 1: Set the connection weights between input nodes and output nodes according to the range of classical domains. The range of classical domains can be directly obtained from previous experience or be determined from training data as follows:

$$w_{kj}^{L} = \min_{T_i \in k} \{x_{ij}\}; w_{kj}^{U} = \max_{T_i \in k} \{x_{ij}\} \tag{26.2}$$

Step 2: Read ith training pattern and its cluster number p:

$$X_i = \{x_{i1}, x_{i2}, \ldots, x_{in}\} \tag{26.3}$$

Step 3: Use the extension distance (ED) to calculate the distance between the input pattern X_i and the kth cluster as follows:

$$ED_{ik} = \sum_{j=1}^{n} \left(\frac{\left|x_{ij} - \left(w_{kj}^{U} + w_{kj}^{L}\right)/2\right| - \left(w_{kj}^{U} - w_{kj}^{L}\right)/2}{\left(w_{kj}^{U} - w_{kj}^{L}\right)/2} + 1 \right) \tag{26.4}$$

The proposed distance in this paper is a modification of ED [10]. It can describe the distance between the x and a range $<W^L, W^U>$. Different ranges of classical domains can arrive at different distances due to different sensitivities. This is a significant advantage in classification applications. Usually, if the feature covers a large range, the requirement of data is fuzzy or low in sensitivity to distance. On the other hand, if the feature covers a small range, the requirement of data is precision or high sensitivity to distance.

Step 4: Find the m, such that $ED_{im} = \text{Min}\{ED_{ik}\}$. If $m = p$, then go to step 6; otherwise step 5.

Step 5: Update the weights of the pth and the mth clusters as follows:

$$\begin{cases} w_{pj}^{L(new)} = w_{pj}^{L(old)} + \eta\left(x_{ij} - \dfrac{w_{pj}^{L(old)} + w_{pj}^{U(old)}}{2}\right) \\ w_{pj}^{U(new)} = w_{pj}^{U(old)} + \eta\left(x_{ij} - \dfrac{w_{pj}^{L(old)} + w_{pj}^{U(old)}}{2}\right) \end{cases} \tag{26.5}$$

$$\begin{cases} w_{mj}^{L(new)} = w_{mj}^{L(old)} - \eta\left(x_{ij} - \dfrac{w_{mj}^{L(old)} + w_{mj}^{U(old)}}{2}\right) \\ w_{mj}^{U(new)} = w_{mj}^{U(old)} - \eta\left(x_{ij} - \dfrac{w_{mj}^{L(old)} + w_{mj}^{U(old)}}{2}\right) \end{cases} \tag{26.6}$$

where η is a learning rate. In this step, we can clearly see that the learning process is only to adjust the weights of the pth and the mth clusters.

Step 6: Repeat step 2 to step 5; if all patterns have been classified, then a learning epoch is finished.

Step 7: Stop if the clustering process has converged or the total error has arrived at a preset value; otherwise, return to step 3.

26.3 The Proposed Method

Sentiment prediction is different from traditional text categorization because sentiments are ordinal variables in contrast to the categorical nature of topics; lots of contradicting opinions might coexist on the same event, which interact with each other to produce the global sentiment. Indeed, sentiment prediction is a much harder task than topic of text classification tasks. For many historic events or other great issues that happened in society, studies found that not only the same type of event development trend has a high similarity but also the development of the same event experiences the same cycle of history. In order to improve the fitting accuracy of the model and prediction ability, a series of procedures must be done.

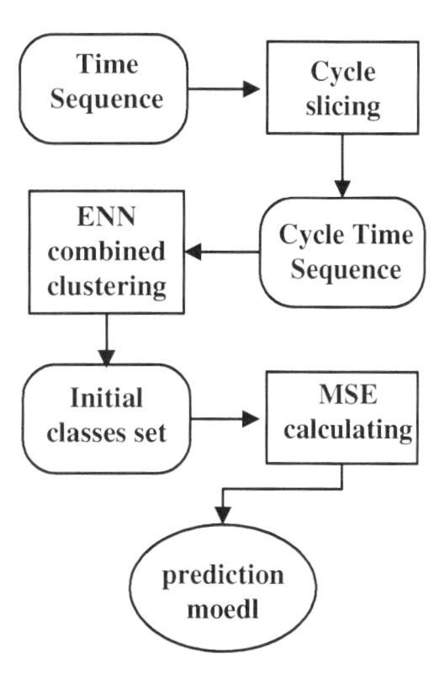

Fig. 26.2 Flow chart of prediction model for web sentiment

As showed in Fig. 26.2, firstly, this paper will make a classification of events and cycle slicing. Then establish the cycle class model recognized by ENN and NAR method for each type of event and create class model libraries. After that, pick up the class model, which has the minimum mean square error (MSE) for long-term prediction. Due to the nonlinear relation between input factors and output values, the prediction of sentiment is not an easy work. Past value of the time sequence will affect the future value.

In short, the main ideas of the proposed method are divided into two parts. First, we use a novel clustering method based on the ENN to recognize the type of every subclass of time sequences. When the dates are given, the changed ranges will be obtained from the proposed models. Second, using the nonparametric auto-regression analysis method to build the prediction model of web sentiment development of long term, when the dates are given, the prediction values can be calculated according to the calculation of MSE. Use the ENN to learn the clustering models of every cycle time sequence derived from the statistical data of Google trends. The detail learning method can refer to Sect. 26.2; the prediction model can be calculated by the nonparametric auto-regression analysis method as in Sect. 26.3; the typical prediction model in our problem can be written as a NAR model will be found on the basis of sequence $\{\Delta Y_t\}$, $\Delta Y_t = m(\Delta Y_{t-1}, \Delta Y_{t-2}, \ldots, \Delta Y_{t-p}) + \varepsilon_t$, and random error sequence $\{\varepsilon_t\}$ is independent and identically distributed. $E(\varepsilon_t) = 0, E\left(\varepsilon_t^2\right) = \sigma^2$, and ε_t are independent with $\Delta Y_{t-1}, \Delta Y_{t-2}, \ldots, \Delta Y_{t-p}$. The parameter p is defined according to cross-validation (CV) method. The value of CV (p) is obtained though calculating using MATLAB programming. An orthogonal sequence estimation method will be operated on the model [11]. After abovementioned procedures, the prediction data should be input, and the growth type of the prediction will be determined by the ENN-based clustering method. Use the prediction model of this type as NAR method to calculate the detailed values of the prediction data.

26.4 Case Studies and Discussions

It is clear that the relations between the input factors and output values are highly nonlinear curves. Therefore, if the prediction models are directly to use the NAR, the prediction error will be large, which is the main reason in the proposed method to delimit some growth types (or some sections) in the related curves. When the subtypes are delimited, every section can be seen as approximately linear, thus the sectioned prediction models will provide higher accuracy. This paper uses the historical data from Google trends and forums to build the prediction model to predict the bird flu event. In this case, we could make conclusion that when k = 1, the value of CV (k) is minimum, so, the optimal selection of lag is 1, in every recognized type the corresponding NAR model is $\Delta Y_t = m(\Delta Y_{t-1}) + \varepsilon_t$. Comparisons of prediction results using the grey GM (1,1) model and LS method are also conducted . In this paper, the accuracy of the prediction model is calculated with the value of correlation degree analysis. The compared prediction results with the grey GM (1,1) model and least square (LS) method show that the proposed method has better accuracy. Selecting the first 10 days of data for six different events, the compared results of modeling accuracy using different methods are shown in Table 26.1. The comparison between actual and prediction curve for the event of bird flu is shown in Fig. 26.3. Obviously, the compared results indicate that the proposed prediction method has the higher accuracy than both grey GM (1,1) and LS method.

Table 26.1 Comparison between different methods on correlation degree

Event	Proposed method	LS	GM (1,1)
Bird flu	0.7910	0.5496	0.5189
Li Gang	0.8103	0.4834	0.3987
Needle shot	0.7637	0.5313	0.4897
Maiden suicide	0.8123	0.4720	0.5108
Dust lungs	0.8527	0.5435	0.4412
Drag racing	0.7879	0.5104	0.3925

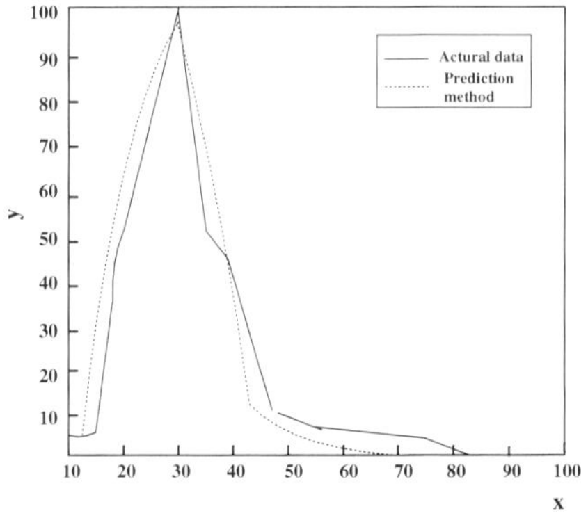

Fig. 26.3 Comparison between the actual data and long-term trend prediction curve

26.5 Conclusion

Web sentiment prediction has always been a great issue for the society. The sentiment of public reflects the attitude towards society and government at that time. In this paper, we address the ENN concept in details. As existing models are inadequate for a variety of reasons, we introduce the ENN- and NAR-based model that is suited to predict the trend of web sentiment. The proposed method not only can be used to predict the values of web sentiment, it also can be used with multi-input factors in nonlinear prediction problems. The main advantage of the proposed method is that it can give the range and values at the same time. Moreover, because of the simple structure of ENN, the computing time of the proposed method is also short. We also demonstrate the usefulness of the web sentiment representation for long-term prediction. Comparisons with the grey GM (1,1) model and LS method show that the proposed prediction method has better accuracy and provides more information that is scientific to sociology researchers.

References

1. Hardle, W., & Chen, R. (1997). Nonparametric time analysis, a selective review with examples. *International Statistical Review, 65*(1), 49–72.
2. Smith, B. L., & Williams, B. M. (2002). Comparison of parametric and nonparametric models for traffic flow forecasting. *Transportation Research, 10*(4), 15–19.
3. Wang, L. (2007). *Public opinion study—theory, method and reality hotspot* (pp. 27–41). Tianjin: Tianjin Social Sciences Press.
4. Masry, E., & Fan, J. (1997). Local polynomial estimation of regression functions for mixing processes. *Scandinavian Journal of Statistics, 24*(2), 165–179.
5. Kayacan, E., & Ulutas, B. (2010). Grey system theory based models in time series prediction. *Expert System with Applications, 37*(2), 1784–1789.
6. Kaur, H., & Raghava, G. P. (2004). A neural network method for prediction of β turn types in proteins using evolutionary information. *Bioinformatics, 20*(16), 2751–2758.
7. Xie, N. M., & Liu, S. F. (2005). Discrete GM(1,1) and mechanism of grey forecasting model. *Systems and Engineering Theory and Practices, 25*(1), 95–97.
8. Wang, M. H., & Hung, C. P. (2005). Extension neural network-type2 and its applications. *IEEE Transactions on Neural Networks, 16*(6), 1352–1361.
9. Wang, M. H., Tseng, Y. F., Chen, H. C., & Chao, K. H. (2009). A novel clustering algorithm based on the extension theory and genetic algorithm. *Expert Systems with Applications, 36*(4), 8269–8276.
10. Cai, W. (1983). The extension set and incompatibility problem. Journal of Scientific Exploration, 1, 81–93.
11. Xia, Y., & Li, W. K. (2002). Asymptotic behavior of band width selected by the cross-validation method for local polynomial fitting. *Journal of Multivariate Analysis, 83*(2), 265–287.

Chapter 27
K-Optimal Chaos Ant Colony Algorithm and Its Application on Dynamic Route Guidance System

Hai Yang

Abstract Dynamic route guidance system is an important part of the intelligent transportation system; the core part of which is optimal path algorithm. This paper has analyzed the main influencing factors on the choice of optimal path, then provided an improved K-optimal chaos ant colony algorithm (K-CACA). The road impedance factor in K-CACA is based on the length, crowdedness, condition, and traffic load of the road sections. The optimizing procedure of the algorithm is speeded up by introducing the included angle threshold of direction. The chaos perturbation effectively refrains the algorithm from trapping into local optima. The results of simulation experiment show that K-CACA is effective and has much higher capacity of global optimization than Dijkstra algorithm and basic ant colony algorithm for optimal route choice.

27.1 Introduction

Dynamic route guidance system (DRGS) is an important part of the intelligent transportation system [1]. It offers the useful optimal route guidance information to the driver according to starting and destination point. The aim of DRGS is to achieve the goals of improving traffic system, voiding traffic jam, reducing the vehicles' travel time, and realizing the rational distribution of traffic flow on each road section by guiding the driver's travel decision [2]. So the core content of DRGS is the detection of optimal path in the traffic network.

With the constant enlargement of urban road network, the graph theoretic algorithms such as Dijkstra algorithm and mathematical programming methods are difficult to satisfy, the real-time requirements of DRGS. Initially proposed by

---

H. Yang (✉)
College of Information Science and Electricity Engineering, Shandong Jiaotong University, Jinan 250023, Shandong, China
e-mail: yanghai_sdjtu@163.com

W.E. Wong and T. Zhu (eds.), *Computer Engineering and Networking*, Lecture Notes in Electrical Engineering 277, DOI 10.1007/978-3-319-01766-2_27,

Marco Dorigo in 1991, the ant colony algorithm (ACO) was aiming to search for an optimal path in a graph based on the behavior of ants seeking a path between ant nest and a source of food. ACO has been successfully applied to solve different combinatorial optimization problems such as TSP [3] and also is suitable for DRGS because of its robustness, positive feedback, and distributed computing.

27.2 The Influencing Factors of the Optimal Path Selection

27.2.1 The Length of Road Section

The length of road section is the most common weight during the procedure of finding the optimal path because the length is easy to access and presents the travel cost directly [4]. But the optimal path does not only mean the shortest path in dynamic route guidance system. The travel time, crowdedness, conditions, and traffic loadings of the roads also affect the driver's decision of optimal path.

27.2.2 The Travel Time and Crowdedness

The travel time is more important than the length of road section in the real urban traffic network when drivers decide which path they should go. The travel time factor is dynamic according to the vehicle number per unit time. So we can compute the crowdedness of road by the travel time factor [5].

The crowdedness of road refers to the situation that the vehicles on the road cannot run at the normal speed so that both the travel time and parking delay become longer. The crowdedness factor can be expressed as $[T(i,j) - T_0(i,j)]/T_0(i, j)$, where $T(i,j)$ is the real travel time of road section (i,j) and $T_0(i,j)$ is the normal travel time of road section (i,j) without jam. In general, the smaller $T(i,j)$ is, the lower the crowdedness factor is, and vice versa.

27.2.3 Traffic Load

The traffic load refers to the number of vehicle staying on some road section for a moment. Particularly, the traffic load per kilometer presents the traffic density [6]. So the traffic load factor presents the change of traffic situation over time and its equation of state is as follows:

$$\frac{dx_{(i,j)}(t)}{dt} = I_{(i,j)}(t) - O_{(i,j)}(t) \tag{27.1}$$

where $x_{(i,j)}(t)$ is the vehicle number on road section (i,j) at time t. $I_{(i,j)}(t)$ is the number of vehicles coming into road section (i,j) per unit time. $O_{(i,j)}(t)$ is the number of vehicles leaving from road section (i,j) per unit time.

27.3 K-Optimal Chaos Ant Colony Algorithm

The urban traffic network is usually abstracted as a weighted, directed graph G = (V,A,R), with node set V, road section set A, and R = $\{r_{ij}|(i,j) \in A\}$; r_{ij} denotes the road impedance factor, that is, heuristic information of road sections [7].

27.3.1 The Selection of Subsequent Node Based on the Direction

Because the dynamic route guidance system provides optimal path information according to the specific starting point and destination point, the procedure of finding optimal path is obviously directional [8]. When the ant selects next node, the probability of succeeding in finding optimal path is larger, if the ant's moving direction is close to the direction of destination point. Otherwise, the probability of succeeding in finding optimal path is lower, if the ant's moving direction is far away from the direction of destination point.

Let v_s denote the starting node with the coordinate (x_s, y_s); v_d denotes the destination node with the coordinate (x_d, y_d); v_i denotes the current node where the ant stays with the coordinate (x_i, y_i), and v_j denotes some node adjacent to v_i with the coordinate (x_j, y_j). Let $\overrightarrow{v_i v_d}$ denote the direction from current node v_i to destination node v_d, called the approximate destination direction. Thus, the included angle θ_{ij} of road section (v_i, v_j) and $\overrightarrow{v_i v_d}$ is as follows:

$$\theta_{ij} = \arctan\frac{k_d - k_{next}}{1 + k_d \cdot k_{next}} \tag{27.2}$$

where $k_d = \dfrac{y_d - y_i}{x_d - x_i}$ denotes the slope of the approximate destination direction $\overrightarrow{v_i v_d}$.
$k_{next} = \dfrac{y_j - y_i}{x_j - x_i}$ denotes the slope of road section (v_i, v_j).

In order to enlarge the probability of succeeding in finding optimal path, we can set an included angle threshold ϕ . The subsequent nodes whose included angle is greater than the threshold ϕ will be omitted. The convergence speed of K-optimal chaos ant colony algorithm (K-CACA) will be accelerated by decreasing ϕ, and the solutions diversity will be enlarged by increasing ϕ.

27.3.2 The Heuristic Information of Road Section

In the new algorithm, the road impedance factor, which is gained by using weighted summation of the length, crowdedness, condition, and traffic load of the road sections [9], presents the heuristic information of road sections.

Let r_{ij} denote the road impedance factor of road section (v_i, v_j):

$$r_{ij} = \frac{\omega_1 \cdot \frac{1}{L_{ij}} + \omega_2 \cdot \frac{1}{C_{ij}} + \omega_3 \cdot Q_{ij} + \omega_4 \cdot S_{ij}}{\sum_{k=1}^{4} \omega_k} \tag{27.3}$$

where L_{ij} denotes the length of road section (v_i, v_j); C_{ij} denotes the crowdedness of (v_i, v_j); Q_{ij} denotes the condition of (v_i, v_j); S_{ij} denotes the traffic load of (v_i, v_j), and ω_k($k = 1, 2, 3, 4$) is weight factor.

27.3.3 Pheromone Updating Strategy

The pheromone updating strategy of the new algorithm is as follow:

$$\tau_{ij}(t+1) = (1-\rho) \cdot \tau_{ij}(t) + \sum_{l=1}^{L} \Delta\tau_{ij}^{l} \quad i = 1, \ldots, N \quad j = 1, \ldots, K \tag{27.4}$$

where ρ is the evaporation rate of pheromone.

27.3.4 Chaos Selection Strategy

When ant colony algorithm initializes, the strength of pheromones on every road section is equal so that the possibility of every road section is equal. It is difficult for ant colony to find an optimization path; besides, the convergence speed of algorithm is slow. So using chaos operator is necessary because it can increase the searching efficiency by the random and ergodic of chaos [10].

Take the logistics mapping, for example, iterative formula is as follows:

$$z_{i+1} = \mu \cdot z_i \cdot (1 - z_i), \quad i = 0, 1, 2, \cdots, \quad \mu \in (2, 4] \tag{27.5}$$

where μ is the control parameter, with the domain between (2, 4]. When $\mu = 4$, $0 \leq z_0 \leq 1$, logistic function is the full mapping in (0, 1), which is at a totally chaos status. A chaos sequence will be created by the iteration and then be converted to a chaos ergodic parameter when solving optimization problems in space. After introducing chaos perturbation, the new status transition strategy becomes

$$p_{ij}^{k}(t) = \begin{cases} \dfrac{\left[\tau_{ij}^{s}(t)\right]^{\alpha} \cdot \left(r_{ij}^{s}\right)^{\beta} \cdot (1+Z_{ij})^{\gamma}}{\sum\limits_{j \in allowed_k} \left[\tau_{ij}^{s}(t)\right]^{\alpha} \cdot \left(r_{ij}^{s}\right)^{\beta} \cdot (1+Z_{ij})^{\gamma}}, j \in allowed_w \quad and \quad \theta_{ij} \leq \phi \\ 0, otherwise \end{cases} \tag{27.6}$$

In the formula above, θ_{ij} is the included angle of road section (v_i, v_j) and the approximate destination direction $\overrightarrow{v_i v_d}$. ϕ is the included angle threshold.

27.3.5 K-Optimal Paths Guidance

The K-optimal chaos ant colony algorithm proposed in this paper provides drivers with K-optimal paths, which are almost exactly the same. Drivers decide which path they want to adopt by themselves, according to the real conditions of roads and the travel preference. The value of parameter K is referred to the scale of the traffic network, generally, K = 3. The newly improved algorithm can relax the conflict of traffic jam due to that many drivers choose the same optimal path.

27.3.6 The Realization Steps of K-CACA

The realization steps of K-CACA are proposed as follows:

Step 1: Set the iteration number n = 0, initialize the pheromone of all road section, and let K = 0.

Step 2: Put the starting node into $tabu_k(s)$, compute the road impedance factors of every road, and then choose one subsequent node j to move according to p_{ij}^{k} and the included angle threshold, finally put node j into $tabu_k(s)$.

Step 3: Update the pheromone of roads the ants passed by. Take record the current optimal path and let K = K + 1.

Step 4: If K equals to the specific value, then stop and go to step 6.

Step 5: Adjust the initial parameters according to the latest traffic situations, run this algorithm again in order to obtain new optimal paths. Go to step 2.

Step 6: Output K optimal paths.

27.4 Experimental Results

We successfully achieve this arithmetic and carry out simulation experiments using the road network of Jinan City shown in Fig. 27.1 from Google Earth. The topology of the road network shown in Fig. 27.1 is demonstrated in Fig. 27.2. The road impedance factors demonstrated in Table 27.1 are obtained by calculating the weighted summation of the length, crowdedness, condition, and traffic load of the road sections at 8:30 in the morning. Let m denote the number of ants and set $m = 20$, $\alpha = 1$, $\beta = 3$, and $\rho = 0.3$, and use logistic mapping. The aim of

Fig. 27.1 The road network of Jinan City

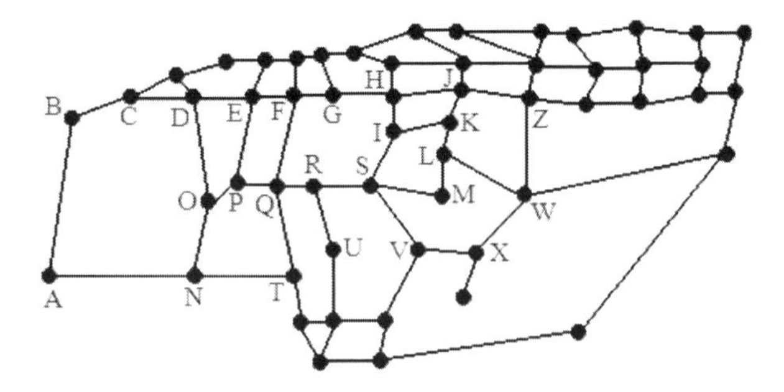

Fig. 27.2 The topology of road network of Jinan City

Table 27.1 The road impedance factor r_{ij} of road sections

Road section	r_{ij}	Road section	r_{ij}	Road section	r_{ij}	Road section	r_{ij}
(A, B)	4.96	(B, C)	2.16	(C, D)	2.02	(D, E)	3.13
(E, F)	2.87	(F, G)	3.01	(G, H)	3.08	(H, J)	3.97
(J, Z)	3.99	(Z, W)	4.01	(A, N)	4.86	(N, O)	3.22
(O, P)	2.13	(D, O)	3.99	(P, E)	3.91	(P, O)	2.71
(P, Q)	2.21	(E, P)	3.63	(Q, F)	3.70	(Q, R)	1.86
(N, T)	3.72	(T, Q)	3.91	(X, W)	3.77	(R, S)	3.64
(S, V)	3.71	(S, M)	3.57	(M, L)	2.28	(H, I)	1.13
(K, L)	1.97	(F, Q)	4.11	(L, W)	3.11	(V, X)	2.99

Table 27.2 The results of three algorithms

Algorithm	Optimal path	Distance (km)	Average arrival time
K-CACA	A-N-T-Q-R-S-M-L-W	13.67	22 min and 7 s
Dijkstra	A-N-O-P-Q-R-S-M-L-W	14.75	26 min and 3 s
Basic ACO	A-B-C-D-E-F-G-H-J-Z-W	14.14	23 min and 21 s

simulation experiment is to find the optimal path from node A to node W using K-CACA, Dijkstra, and basic ant colony algorithm (ACO) separately. The results are shown in Table 27.2.

The results of simulation experiment show that K-CACA is effective and has much higher capacity of global optimization than Dijkstra algorithm and basic ant colony algorithm for optimal path choice.

27.5 Conclusion

In this article, K-optimal chaos ant colony algorithm is proposed for finding optimal path in dynamic route guidance system. The road impedance factor is introduced into K-CACA based on the length, crowdedness, condition, and traffic load of the road sections. The included angle threshold both accelerates the convergence speed and enlarges the solutions diversity. The chaos perturbation effectively refrains the algorithm from trapping into local optima. The experiment results show K-CACA is much more suitable for DRGS.

Acknowledgements The research is supported by Chinese Natural Science Foundation (61103022) and Scientific Research Fund Project of Shandong Jiaotong University (Z201213).

References

1. Boyce, D. E., Ran, B., & Leblanc, L. J. (1995). Solving an instantaneous dynamic user optimal route choice model. *Transportation Science, 29*(2), 128–142.
2. Kuwahara, M., & Akamatsu, T. (1997). Dynamic user optimal assignment with physical queues for a many-to-many origin–destination pattern. *Transportation Research B, 31*(1), 1–10.
3. Dorigo, M. (1997). Ant colonies for the traveling salesman problem. *Biosystems, 43*(2), 73–81.
4. Ben-Akiva, M., Palama, A., & Kaysi, I. (1991). Dynamic network models and driver information systems. *Transportation Research A, 25*(5), 251–266.
5. Deek, H. L., & Kanafani, A. (1993). Modeling the benefits of advanced traveler information systems in corridors with incidents. *Transportation Research C, 1*(4), 303–324.
6. Daganzo, C. F. (1998). Queue spillovers in transportation networks with a route choice. *Transportation Science, 32*(1), 3–11.
7. Kim, J., Mitchell, J. S. B., & Polishchuk, V. (2012). Routing multi-class traffic flows in the plane. *Computational Geometry-Theory and Applications, 45*(3), 99–114.

8. Cordeau, J. F., & Maischberger, M. (2012). A parallel iterated tabu search heuristic for vehicle routing problems. *Computers and Operations Research, 39*(9), 2033–2050.
9. Blum, C., & Dorigo, M. (2004). The hyper-cube framework for ant colony optimization. *IEEE Transactions on Systems, Man and Cybernetics B, 34*(2), 1161–1172.
10. Yunwu, W. (2009). Application of chaos ant colony algorithm in web service composition based on QoS. *International Forum on Information Technology and Applications, 172*(2), 225–227.

Chapter 28
A Certainty-Based Active Learning Framework of Meeting Speech Summarization

Jian Zhang and Huaqiang Yuan

Abstract This paper proposes using a certainty-based active learning framework for extractive meeting speech summarization in order to reduce human effort in generating reference summaries. Active learning chooses a selective set of samples to be labeled by annotators. A combination of informativeness and representativeness criteria for sample selection is proposed. The results of summarizing parliamentary meeting speech show that the amount of labeled data needed for a given summarization accuracy can be reduced by more than 40 % compared to random sampling. The certainty-based active learning framework can effectively reduce the need of labeling samples for training. Furthermore, compared with lecture speech summarization task, the experiments show that the proposed active learning method of meeting speech summarization is obviously more affected by choice of different kinds of classifiers.

28.1 Introduction

The need for the summarization of spontaneous speech, such as classroom lectures, conference speeches, and parliamentary speeches, is ever increasing with the advent of remote learning, distributed collaboration, and electronic archiving. Short abstracts cannot sufficiently meet these user needs. State-of-the-art summarization systems are built by the extractive summarization method in a passive supervised learning framework, which compiles a summary from sentences or segments chosen from the transcribed document using some saliency criteria [1–3]. However, these supervised learning extractive summarization systems require a large amount of training data of reference summaries [4, 5]. There is no

J. Zhang (✉) • H. Yuan
Engineering and Technology Institute, Dongguan University of Technology,
Dongguan 523808, China
e-mail: zjian03@gmail.com

W.E. Wong and T. Zhu (eds.), *Computer Engineering and Networking*, Lecture Notes in Electrical Engineering 277, DOI 10.1007/978-3-319-01766-2_28,

clear guideline for compiling stable and reproducible reference summaries. To minimize the human annotation efforts, yet still producing the same level of performance as a supervised learning approach, we study how to apply active learning approach for training extractive speech summarizer. We then propose three criteria and strategies in our certainty-based active learning framework of meeting speech summarization.

Being the first piece of work on active learning for lecture speech summarization task, we produced stable and reproducible reference summary and minimized the need for human annotation efforts, yet still producing the same level of performance as a supervised learning approach [6]. In this article, we further describe our certainty-based active learning framework in detail and verify its effectiveness on different genres of spontaneous speech.

The rest of this article is organized as follows: Sect. 28.2 describes our certainty-based active learning framework, the criteria, and the strategies for selecting samples. Section 28.3 first describes the parliamentary meeting speech corpus, for our experiments, and then outlines the acoustic/prosodic and lexical feature sets for representing each sentence of the transcriptions and then briefly depicts the probabilistic SVM classifier and naive Bayesian classifier as our extractive summarizers. Our experimental setup and the evaluation results are described in Sect. 28.4. Our conclusion follows in the end of this article.

28.2 Active Learning Framework and Sample Selection Strategies

28.2.1 Active Learning Algorithm

Human annotators label a small set of training data with summary labels. The training data is used to learn the initial model of the classifier. Then the classifier is used to predict labels for the sentences belonging to all transcribed documents from an unlabeled pool. Because human annotators usually annotate sentences of each document with summary labels by taking the context of the document into account, at each iteration they choose several unlabeled transcribed documents from the pool according to some sample selection strategies. Next the chosen transcribed documents are labeled by human annotators. These annotated transcribed documents are then added for retraining the classifier. This approach is described as follows:

Initialization

For an unlabeled data set: U_{all}, $i = 0$

(1) Randomly choose a small set of data X{i} from Uall; $U\{i\} = U_{all} - X\{i\}$.

(2) Label each sentence in X{i} as summary or non-summary using the RDTW-based semiautomatic annotation procedure proposed by Zhang et al. [5], and save these sentences and their labels in L{i}.

Active Learning Process

(3) X{i} = null.
(4) Train the classifier M{i} using L{i}.
(5) Test U{i} by M{i} and select the most useful documents from Ufig based on informativeness/representativeness strategies described in Sect. 28.2.3.
(6) Save selected documents into X{i}.
(7) Label each sentence in X{i} as summary or non-summary using the RDTW-based semiautomatic annotation procedure.
(8) L{i + 1} = L{i} + X{i}, U{i + 1} = U{i} − X{i}.
(9) Evaluate M{i} on the testing set E.
(10) i = i + 1, and repeat from (3) until U{i} is empty or M{i} obtains expected performance.
(11) M{i} is produced and the process ends.

We propose the following two criteria for choosing the most useful samples X from the unlabeled data U for annotation.

28.2.2 Sampling Criteria

28.2.2.1 Criterion 1: Informativeness

All active learning scenarios involve evaluating the informativeness of unlabeled samples. The simplest and most commonly used query framework is uncertainty sampling [7]. The basic idea of informativeness criterion is that samples that the current model is most uncertain about are selected for annotation. This criterion is often straightforward for probabilistic learning models. When we use a probabilistic model for binary classification as the summarizer, an informativeness-based uncertainty sampling simply selects the samples whose posterior probability of being summary sentences (positive examples) are close to the classification hyperplane. This intuition is justified by D. Lewis and J. Catlett [8] and G. Schohn and D. Cohn [9] based on a version space analysis. They claim that labeling a sample that lies on or close to the hyperplane is guaranteed to have an effect on the model construction.

We measure the informativeness score of unlabeled transcribed document $D = \{s_1, s_2, \ldots, s_n, \ldots, s_N\}$ by $Score_{inf}(D) = (1/N)\Sigma$ informative(s_n), which indicates the ratio of the number of informative sentences to that of all sentences in the document D.

If the sentence s_n satisfies the informativeness criterion,

$$P\left(c\left(\vec{s}_n\right) = 1 \middle| D\right) \in [(1-\beta)^* T, (1+\beta)^* T]$$

where $c\left(\vec{s}_n\right) = 1$ means the sentence s_n is a summary sentence, **then** informative (sn) is equal to 1; **if not**, informative(s_n) is equal to 0.

We denote $T = P\left(c\left(\vec{s}_n\right) = 1\right)$ as the prior probability of the sentence s_n which is a summary sentence without considering any information on document D. We have tried different values for T when we evaluate our method on the development set. We found T as 0.12 is suitable empirically for meeting compression tasks. β is tuned by evaluating the development set sentences to optimize the active learning algorithm. Sentences that satisfy the informativeness criterion are close to the classification hyperplane. Misclassification implies that the model is most uncertain about the samples.

28.2.2.2 Criterion 2: Representativeness

Another general active learning criterion [10] is to query the sample that would impart the greatest change to the current model if we knew its label. This criterion has also been applied to probabilistic sequence models like conditional random fields (CRFs) [10]. The intuition behind this criterion is that it prefers samples that are likely to most influence the model regardless of the resulting query label. This criterion has been shown to work well in empirical studies, though it will be computationally expensive if both the feature space and set of labels are very large [10].

We consider that the model will become more robust if it can be retrained by the samples selected by this representativeness criterion. The basic idea of representativeness criterion is that samples that are most generalized and are likely to be margin points are selected.

We measure the representativeness score of unlabeled transcribed document $D = \{s_1, s_2, \ldots, s_n, \ldots, s_N\}$ by $Score_{repre}(D) = (1/N)\Sigma$ representative(s_n), which indicates the ratio of the number of representative sentences to that of all sentences in the document D.

If the sentence s_n satisfies the representativeness criterion,

$$P\left(c\left(\vec{s}_n\right) = 1 \middle| D\right) \in [0, \beta * T] \cup [(1 - \beta * T), 1],$$

then representative(s_n) is equal to 1; **if not**, representative(s_n) is equal to 0.

Sentences with classification scores in the range [0, β*T] are believed to be most likely non-summary sentences. Sentences with scores in the range [(1−β*T), 1] are believed to belong most likely to the summary-sentence class. It means that if $P\left(c\left(\vec{s}_n\right) = 1 \middle| D\right)$ is close to 1, the sentence s_n is most likely summary sentence. If the samples from the above range are misclassified, those samples would impart the greatest change to the current model. Adding those samples to the training data set for labeling will improve the robustness of the model.

28.2.3 Sample Selection Strategies

We apply the following sample selection strategies based on the above criteria for selecting the most useful unlabeled examples from the unlabeled pool at each active learning iteration.

Random Selection Baseline: A baseline sample selection strategy is to randomly choose K documents from U{i} to X{i}.

Length Selection Baseline: A baseline sample selection strategy is to choose K documents from U{i} to X{i} according to the average length of a sentence. The documents with larger average length of a sentence are chosen first.

Sample Selection Strategy 1 (Informativeness): We choose K documents with the highest value of $Score_{info}(D)$ from U{i} to X{i} for annotation.

Sample Selection Strategy 2 (Robustness): We choose K documents with the highest value of $Score_{repre}(D)$ from U{i} to X{i} for annotation.

Sample Selection Strategy 3 (Hybrid): We build a hybrid strategy by considering criterion 1 and criterion 2 together for striking a proper balance between the informativeness and representativeness criteria to reach the maximum effectiveness on speech summarization. We choose K documents with the highest value of Score (D), defined as $Score(D) = Scoreinfo(D) *[Score_{repre}(D)]^{\lambda}$ from U{i} to X{i} for annotation, where λ $(0 \leq \lambda \leq 1)$ is tuned by evaluating the development set sentences to optimize the active learning algorithm.

28.3 Meeting Corpus, Features, and Summarizers

Our parliamentary meeting speech corpus contains meeting audio files, the Hansard transcriptions, and the meeting minutes from the Hong Kong Legislative Council. For our experiments, we use all 70 ordinary session meeting data from the year 2008 and the year 2009, including audio files, Hansards, and minutes.

We represent each sentence by a feature vector which consists of acoustic features and linguistic features as follows.

Similar to text summarization, the linguistic information can help us predict the summary sentences. We extract eight linguistic features from transcribed documents: len I (the number of words in the sentence), len II/len III (the previous/next sentence's len I value), TFIDF (the summation of the tf *idf value of each word in the sentence), and cosine (cosine similarity measure between two sentence vectors).

We extract all linguistic features from the manual and ASR-transcribed documents respectively. We segment Chinese words of these transcribed documents for calculating length features. We use an off-the-shelf Chinese lexical analysis system, the open source HIT IR Lab Chinese Word Segmenter [5] to process our corpora.

We had a detailed study of linguistic features based on the word segmentation result. We find that the errors produced by the word segmentation process have little effect on the summarization process. We also find the number of words as a feature is better than the number of characters because word can convey more unambiguous information.

Acoustic/prosodic features in speech summarization system are usually extracted from audio data. Researchers commonly use acoustic/prosodic variation—changes in pitch, intensity, speaking rate—and duration of pause for tagging the important contents of their speeches. We also investigate these features for their efficiency in predicting summary sentences of presentation speech and meeting speech.

Our acoustic feature set contains 12 features: Duration I (time duration of the sentence), SpeakingRate (average syllable duration), F0 (I–V) (F0 min, max, mean, slope, range), and E (I–V) (energy min, max, mean, slope, range).

We calculate Duration I from the annotated manual transcriptions that align the audio documents. We then obtain SpeakingRate by phonetic forced alignment by HTK [6]. Next, we extract F0 features and energy features from audio data by using the Praat tool [6].

We then build a discriminative model—support vector machine (SVM) classifier as our summarizer based on these sentence feature vectors.

We apply a generative model—Naive Bayesian classifier implemented by ourselves for the summarizer to investigate whether our active learning method is affected by choice of different machine learning methods.

28.4 Experimental Results and Evaluation

We perform tenfold cross-validation experiments on manual transcriptions, one group for lecture speeches and another one for meeting speeches. First, we divide the 70 meeting speeches into ten subsets. Each subset has seven ones. We use nine subsets as the unlabeled data pool and use the remaining one for test.

We start our experiments by randomly choosing seven speeches from the unlabeled data pool as seeds for manual labeling. We then train the initial classifier using the seeds. We gradually increase the training data pool by choosing five more speeches from the unlabeled data pool each time for annotation. We separately carry out ten sets of experiments for each genre of speech for comparison according to the above three strategies. For each sample selection step, we also choose the unlabeled documents by random selection baseline strategy and length selection baseline strategy. In each group of experiment, we repeat the process of randomly choosing seven speeches as seeds, six times. We evaluate the summarizer by ROUGE-L (summary-level longest common subsequence) F-measure [5]. We calculate the mean value of ROUGE-L F-measure of each active learning step as the final performance of this step. For meeting speeches, all the evaluation results of the three active learning strategies and the two baseline strategies using the SVM

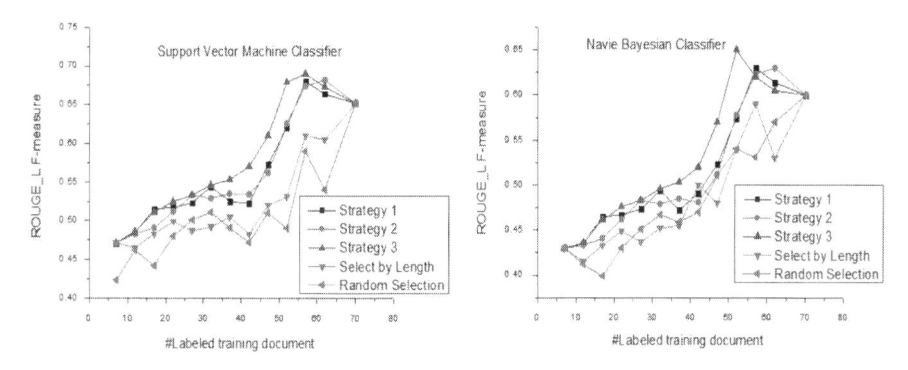

Fig. 28.1 Average performance of the SVM classifier and the Naive Bayesian classifier trained by using the combination of acoustic features and linguistic features on parliamentary meeting speech corpus

classifier and the Naive Bayesian classifier are respectively shown in Fig. 28.1. We train the classifiers by using a combination of linguistic features and acoustic features.

We consider the extractive summarization problem as a binary classification problem. From Fig. 28.1, we find that for classification performance, random sample selection strategy is always consistently and significantly outperformed by the active learning methods using our proposed sample selection strategies. By using only 37 documents for training, the performance of the SVM classifier achieved by strategy 3 which combines the informativeness criterion and representativeness criterion for selecting samples is better than that of the SVM classifier trained by random sample selection using all 70 presentations (ROUGE-L F-measure of 0.512 vs. that of 0.502). This shows that our active learning approach requires 40 % less training data. We also find that using strategy 3, our active learning approach requires less training data to obtain the same level of summarization performance compared to that using strategy 1. In other words, strategy 3 is better than strategy 1. When we apply Naive Bayesian classifier as our summarizer, we obtain the same finding as that shown in Fig. 28.1. Besides, we find that the length sample selection method produces worse performance than our active learning sample selection strategies. This indicates that the performance gains of the active learning methods are due to their ability to choose the unlabeled transcribed documents which contain more informative and representative samples which help find more accurate classification hyperplane and improve the summarization performance. However, the random sample selection strategy and the length sample selection strategy cannot guarantee selecting those documents with more informative and representative samples.

We compare a discriminative model—SVM classifier with a generative model—Naive Bayesian classifier as shown in Fig. 28.1 to investigate the effectiveness of active learning using different kinds of machine learning methods. We find that our active learning method of meeting speech summarization is obviously affected by choice of different classifiers. For meeting speech corpus, the best performance of SVM classifier is absolute a 4 % higher than that of the Naive Bayesian classifier (0.69 vs. 0.65).

28.5 Conclusion

This paper described an approach of active learning to reduce the need for human annotation for summarizing parliamentary meeting speech. We chose the unlabeled examples according to a combination of informativeness criterion and representativeness criterion. The summarization results showed an increasing learning curve, consistently higher than that by using randomly chosen training samples. Furthermore, the experiments showed that the proposed active learning method of meeting speech summarization was obviously affected by choice of different classifiers.

Acknowledgements Supported by the Natural Science Foundation of Guangdong Province of China (Grant No. S2012040007560), the Foundation of Guangdong Educational Committee (Grant No. 2012KJCX0099), and the National Natural Science Foundation of China (Grant No. 61300197).

References

1. Fujii, Y., Yamamoto, K., Kitaoka, N., & Nakagawa, S. (2008). *Class lecture summarization taking into account consecutiveness of important sentences* (pp. 2438–2441). In Proceedings of Interspeech, IEEE.
2. Hori, C., & Furui, S. (2001). *Advances in automatic speech summarization* (pp. 1771–1774). In Proceedings of Eurospeech 2001.
3. Mrozinski, J., Whittaker, E., Chatain, P., & Furui, S. (2005). *Automatic sentence segmentation of speech for automatic summarization* (Vol. 1, Issue 5, p. 12). In Proceedings of ICASSP.
4. Kawahara, T., Nanjo, H., & Furui, S. (2001). *Automatic transcription of spontaneous lecture speech* (pp. 186–189). In Proceedings of the IEEE Workshop on Automatic Speech Recognition and Understanding, IEEE.
5. Zhang, J., Huang, S., & Fung, P. (2008). *RSHMM++ for extractive lecture speech summarization* (pp. 161–164). In Proceedings of 2008 I.E. Workshop on Spoken Language Technology, IEEE.
6. Zhang, J., & Fung, P. (2009). *Active learning of extractive reference summaries for lecture speech summarization* (pp. 23–26). In Proceedings of the 2nd Workshop on Building and Using Comparable Corpora (BUCC), Association for Computational Linguistics.
7. Schohn, G., & Cohn, D. (2000). *Less is more: Active learning with support vector machines* (pp. 839–846). In Machine Learning-International Workshop THEN Conference.
8. Tong, S., & Koller, D. (2002). Support vector machine active learning with applications to text classification. *The Journal of Machine Learning Research, 2*(1), 45–66.
9. Lewis, D., & Catlett, J. (1994). Heterogeneous uncertainty sampling for supervised learning. In *Proceedings of the eleventh international conference on machine learning* (pp. 148–156). Morgan Kaufmann.
10. Settles, B. (2009). *Active learning literature survey* (pp. 1648–1715). University of Wisconsin-Madison. Computer Sciences Technical Report.

Chapter 29
Application of Improved BP Neural Network in the Frequency Identification of Piano Tone

Xu Chen and Jun Tang

Abstract For the problems existing in the identification process of piano tone, this paper puts forward an MFCC-based (Mel Frequency Cepstrum Coefficient) feature extraction algorithm and a new piano tone identification method with BP neural network as the matching model. Using the MFCC feature extraction algorithm to extract parameters is a good alternative, which could improve the identification rate. Regarding the improved BP neural network as the matching model of tone identification consumes moderate training time and owns high recognition rate. Simulation results show that the piano tone identification combining the BP neural network with the MFCC algorithm is simple, fast, and highly accurate.

29.1 Introduction

With the rapid development of multimedia technology, using computer and electronic technology to carry out feature extraction, tone recognition and auxiliary music creation have been given more and more attention [1]. If the results of computer and electronic technologies could be used in music appreciation, learning, and creative fields [2], using computer to automatically identify the music and complete automatic evaluation not only can reduce the work intensity of music educators but also is conducive to improve the scientificity, impartiality, and objectivity of music evaluation.

As a branch of speech recognition [3], tone recognition technology has been paid much attention to in recent years. From the current research status, to construct the tone signal recognition system, analog circuits or digital circuits are frequently used to conduct filtering, adding, Fourier transform, spectral analysis and autocorrelation analysis, and so on for signals. The identification rate has been the greatest

X. Chen (✉) • J. Tang
Hunan Urban Construction College, Xiangtan 411101, China
e-mail: xtchenxu@163.com; xttangjun@163.com

W.E. Wong and T. Zhu (eds.), *Computer Engineering and Networking*, Lecture Notes in Electrical Engineering 277, DOI 10.1007/978-3-319-01766-2_29,

impediment to limit its development. All the piano tones contain the information of 88 keys, while the speech recognition requires a very large amount of data for model training and match to ensure the recognition rate [4]. From every point of view, the theoretical property and practical operability of the tone recognition are better than speech recognition. Therefore, it is entirely possible to use speech recognition theory, combined with the characteristics of tone recognition, to study and process the tone recognition [5].

In this paper, MFCC (Mel Frequency Cepstrum Coefficient) [6] and BP (Back Propagation) neural network algorithm are regarded as the research basis to complete a set of feasible musical note recognition methods [7]. In MATLAB, the aforementioned algorithm is used to test the single key tone, and the simulation result is good; subsequently, validate the system by continuous tones. The results show that the algorithm can be used for a wide range of tone identification fields, which will promote the development of tone recognition.

29.2 Piano Tone Identification Methods

29.2.1 Physical Characteristics of Piano Tone

Set the tension of the piano string as T_0, the density as ρ, the cross-sectional area as A, and the length as l. The string is divided into $n + 1$ segments, and the number of the division points is n. Then we can obtain the free response function of the string, as shown in Eq. (29.1).

$$y(x,t) = \sum_{n=1}^{x} \left\{ \frac{2}{l} \left[\int_0^l f_1 \sin \frac{n\pi}{l} x dx \right] \cos P_n t + \frac{2}{nc\pi} \left[\int_0^l f_2(x) \sin \frac{n\pi}{l} x dx \right] \sin P_n t \right\} \sin \frac{P_n}{c} x \tag{29.1}$$

The natural frequency P_n of the string is as shown in Eq. (29.2).

$$P_n = \frac{n\pi}{l} \sqrt{\frac{T_0}{\rho A}} \quad (\mathrm{n} = 1, 2, 3, \ldots) \tag{29.2}$$

When $n = 1$, $P_1 = \frac{n\pi}{l} \sqrt{\frac{T_0}{\rho A}}$ is called the baseband of string which determines the pitch of the string, the main frequency to be identified in this paper; when $n = 2, 3, 4, \ldots$, the frequency P_n is called overtone and its value is an integer multiple of the frequency.

The frequency domain constitution of single notes (such as every piano key) is basically the same in the frequency segment, and the difference lies in the linear

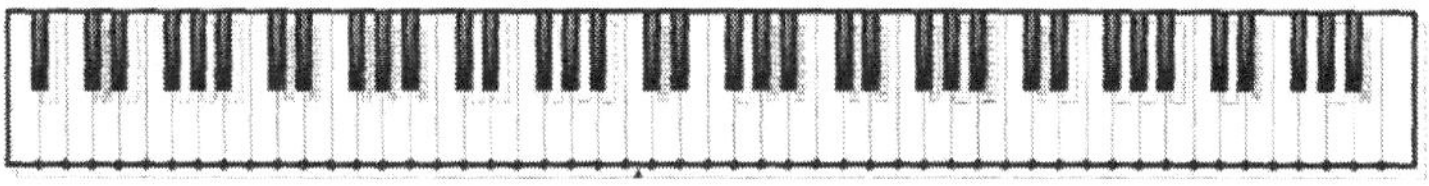

Fig. 29.1 Diagram of 88-key piano keyboard

combination between the harmonics. During the whole process from signal to signal generation to the signal ending, the physical form of harmonics and dominant frequency is smoothly weakened. That is, the composition is constant, and it is the direct amplitude of waveforms that takes changes, weakened from the maximum to zero [8].

29.2.2 Music Theory Analysis of Piano Tone

A piano consists of 88 keys, with 7 full octaves. The piano keyboard is shown in Fig. 29.1.

The piano musical alphabet is obtained in accordance with the law of averages of the 12-tone system [9]. Each octave is divided into a frequency doubling, and musical alphabet between each adjacent octave is a multiple relationship in which the frequency ratio of musical alphabet with the difference of one frequency doubling is 2. Each octave is divided into 12 half-degree syllables. The main frequencies of 88-key piano are shown in Table 29.1.

29.2.3 Identification Method of Piano Music

The recognition process of piano tone includes the acquisition and digitization, preprocessing, and feature extraction of musical tone signals and the study of tone recognition model.

A special microphone and sound capture card are used to acquire a collection of 88 keys of the piano five times, with 440 tone data saved in WAV format. Sound capture card with 16-bit accuracy and sampling frequency of 22,050 Hz is used to digitize the signal for the operating system [10].

Use the feature extraction algorithm of MFCC and methods based on BP neural network recognition model to analyze and design a system for tone recognition starting from the analysis of the unique performance of tones in combination with a variety of recognition models of speech recognition system. Departing from the physical and musical characteristics of the musical tone, the main frequency and the musical alphabet of the tone signal are identified to complete the entire recognition system design. The whole tone recognition process is shown in Fig. 29.2.

Table 29.1 Main frequency of 88-key piano

Sequence number	Main frequency	Frequency (Hz)
1	c^1	261.63
2	d^1	293.7
3	e^1	329.63
4	f^1	349.23
5	g^1	392.0
6	a^1	440
7	b^1	493.88

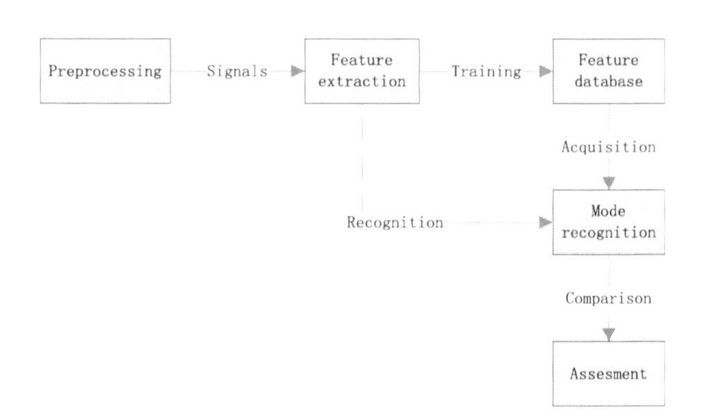

Fig. 29.2 The whole tone recognition process

29.3 Mel Frequency Cepstrum Coefficient

The MFCC is used to conduct transformation for the spectrum of the time domain by the nonlinear spectrum and carry out studies after it is ultimately transformed in cepstrum spectrum. The specific calculation processes are shown as follows:

1. Conduct windowing frame selection for the tone signal to change the tone signal to the short-term signal; then adopt the Fast Fourier Transform to transform the time domain signal $x(n)$ into a frequency domain signal $X(m)$, and calculate the short-time energy spectrum on this basis $P(f)$.
2. By the formula (29.3), the frequency spectrum of short-term energy spectrum $P(f)$ in the frequency spectrum axis is transformed into the $P(M)$ in the Mel axis.

$$F_{mel} = 2595\lg(1 + f_{HZ}/700) \tag{29.3}$$

3. Set M band-pass filters $H_m(k)$ in the spectral range of the tone, and calculate the input of band-pass filter with the energy spectrum $P(M)$ as input, as shown in Eq. (29.4).

$$\theta(M_k) = In\left[\sum_{k=1}^{N} |X(k)|^2 H_m(k)\right] \quad \mathrm{m} = 1, 2, \ldots \mathrm{M} \quad \mathrm{N} = 5,000 \tag{29.4}$$

In this equation, N is the width of the windowed pour election. Set the coverage area of each filter is 0–5,000 Hz and they have the same MEL characteristics, namely, the span of filter group in the Mel coordinates is equal. Set the center frequency as $f(m)$. Based on the corresponding relationship between Mel coordinates and the time field coordinates, the span of band-pass filters in the time domain coordinates increases with the rise of m, i.e., the spacing between adjacent $f(m)$ is continuously increasing. The transfer function of band-pass filter system is as shown in Eq. (29.5).

$$H_m(k) = \begin{cases} 0 & k < f(m-1) \\ \dfrac{2[k - f(m-1)]}{[f(m) - f(m-1)][f(m+1) - f(m-1)]} & f(m-1) \leq k \leq f(m) \\ \dfrac{2[f(m+1) - k]}{[f(m+1) - f(m)][f(m+1) - f(m-1)]} & f(m) \leq k \leq f(m+1) \\ 0 & k > f(m+1) \end{cases} \tag{29.5}$$

The definition of center frequency $f(m)$ is as shown in Eq. (29.6):

$$f(m) = \frac{N}{F_s} F_{mel}^{-1} \left[F_{mel}(f_l + m \frac{F_{mel}(f_h) = F_{mel}(f_l)}{M+1} \right] \tag{29.6}$$

wherein, f_l f_h, respectively, represents the lowest frequency and highest frequency in the filter frequency range, F_s is the sampling frequency, and F_{mel}^{-1} is the inverse function of F_{mel}, $F_{mel}^{-1}(b) = 700(e^{h/2595} - 1)$.

4. Output energy of the mth filter, noted as $\theta(M_k)$. Use the modified discrete cosine transform (IDCT) to obtain the Mel frequency cepstral $c(n)$ on the left side of the Mel, as shown in Eq. (29.7).

$$c(n) = \sum_{k=1}^{M} \theta(M_k) \cos\left(n(k - 0.5)\frac{\pi}{M}\right) \quad n = 1, 2, \ldots P \tag{29.7}$$

In this equation, P is the order of MFCC parameters, and in this system, it is set as 12 orders.

29.4 BP Neural Network Model

12 neurons are adopted in input layer to correspond to the 12-dimensional characteristic parameters of the tone signal; 1 neuron is adopted in output layer, corresponding to the main frequency of piano tone; the number of hidden layer neurons is determined through experimental comparison.

MLP mode predictor in BP network is selected as the system neuron. Its working principle is to use the neuron output $f(t - \tau), \ldots f(t - 1)$ before time t as the input of time t to give the output of feature parameter $\hat{f}(t)$ at this time.

The relationship between the feature input and prediction output is as shown in Eqs. (29.8) and (29.9).

$$h(t) = F\left(\sum_{s=1}^{r} \omega_s^{(1)} f(t-s)\right) \tag{29.8}$$

$$\hat{f}(t) = \omega_0^{(2)} h(t) \tag{29.9}$$

In the equation, $\omega_s^{(1)}$ and $\omega_0^{(2)}$ are weight matrixes between neurons, and $h(t)$ is used as the output of the hidden layer, while $f(t)$ refers to the sigmoid function.

Each neuron network node is replaced by an MLP forecast period. Assuming that there are N_n predictive models, and for musical tone recognition, we can split it into N_n sides and use the corresponding forecasting models to predict each piece of musical tone signals. Finally, the optimization of musical tone segmentation is achieved through the calculation of a minimum cumulative prediction residual, as shown in Eq. (29.10).

$$D(n) = \min_{\{i(t)\}} \sum_{t=1}^{T} \left\| \hat{f}(n, i(t)) - f(t) \right\|^2 \tag{29.10}$$

In the formula, $\hat{f} n, i(t)$ represents the feature component predicted by s the ith predictor. The recurrence of algorithm is as shown in Eqs. (29.11), (29.12), and (29.13).

$$g(t,i) = d(t,i) + \min\{g(t-1,i), g(t-1,i-1)\} \tag{29.11}$$

$$d(t,i) = \left\| \hat{f}(t,i) - f(t) \right\|^2 \tag{29.12}$$

$$D(n) = g(T, Nn) \tag{29.13}$$

$D(n)$ is the total distance between the input musical tone signal and the template, and the tone recognition will be completed if the minimum $D(n)$ is obtained.

The mean of accumulated prediction residual can be used to judge, as shown in Eq. (29.14).

$$\overline{D}(n) = \frac{1}{M} \sum_{m=1}^{M} D(n,m) \tag{29.14}$$

In the equation, M represents the number of samples in the training sample database.

29.5 Experimental Results

In light of the fact that the piano keys contained only 88 keys, the data is relatively less than normal, so we extracted the musical tone of each key in different environments five times to get a total of 440 sets of data constructing the input unit the neural network, while the output unit of training is the main frequency value corresponding to each note. After that, the input data is normalized, which enables the input of certain exceptions to fall within the expected range, enhancing the adaptability of the system to the musical tone data.

Through the setting of relevant decomposition filter bank in MATLAB, Daubechies4 order wavelet decomposition is conducted on musical signals. MFCC is used to extract the feature parameters of music, including the use of a 12-order extraction algorithm, 256 audio signal FFT transform length, 20,500 Hz sampling frequency, and the frame length of 256. First, read the preprocessed musical tone file mentioned above; and then conduct the normalization of Mel filter bank coefficients by the programmed melbankm function, followed by the solution to DCT parameters, the filtering of the preemphasis filter, as well as the reframing of the musical tone signal; and finally calculate the MFCC parameters of musical tone signals. The 8-order MFCC parameters of "A11.wav" are shown in Table 29.2.

After obtaining the trained BP neural network model, we randomly selected eight musical tone signals of keys to have them verified and corrected again.

As can be seen from Table 29.3, musical tone signals whose BP neural network matches are conducted using MFCC parameters as the feature parameters have the actual value of the signals consistent with the expected value, which shows a relatively high recognition rate.

Table 29.2 The 12-order MFCC parameters of "A11.wav"

Sequence number	Feature parameters					
1	50.214	8.991	21.365	1.5712	16.298	4.709
	20.287	4.4121	5.7611	2.6902	5.021	0.1187
2	51.098	9.6791	24.461	2.8612	16.675	3.1981
	17.2016	1.0015	9.1417	0.4098	4.1005	0.01941
3	49.4572	7.7453	26.7091	4.5546	16.4416	2.8717
	14.4908	4.0198	13.1121	1.7789	2.9103	0.7659
4	54.5476	13.1154	22.3224	2.7621	18.5532	1.9012
	11.3910	2.5846	10.2217	0.50012	4.4199	0.1876

Table 29.3 Recognition results of partial musical tones of 88-key piano

Actual musical alphabet	Expected value	Actual value	Error (%)	Expected musical alphabet
e^4	2,594	2,561	98.73	e^4
$\# A_2$	28.91	27.83	96.26	$\# A_2$
A_1	55	53.92	98.04	A_1
$\# A$	116	114.87	99.03	$\# A$
e^1	337	336.13	99.74	e^1
d	155	155.72	99.54	d
b^2	988	987.06	99.90	b^2
c^2	523	521.37	99.69	c^2

29.6 Conclusion

This paper proposed piano tone recognition methods including signal de-noising of high-frequency tone, endpoint detection, single note segmentation, feature parameters extraction, and recognition of model matching. The MFCC method with good recognition performance and noise resistance has been adopted to extract feature parameters of piano musical tone, improving the efficiency and accuracy of feature extraction. With an improved BP neural network model to identify the piano musical tone, a highly efficient neural network was built. The experimental analysis proves the correctness and feasibility of musical tone recognition system which improves the recognition rate.

Acknowledgement This paper is supported by Scientific Research Fund of Hunan Provincial Education Department (11C0231, 12C0995).

References

1. Abeyratne, U. R., Kinouchi, Y., Oki, H., Okada, J., Shichijo, F., & Matsumoto, K. (1991). Artificial neural networks for source localization in the human brain. *Brain Topography, 4*(1), 3–21.
2. Zhang, Q., Nagashino, H., & Kinouchi, Y. (2003). Accuracy of single dipole source localization by BP neural networks from 18-channel EEGs. *IEICE Transactions on Information and Systems, 86*(4), 1447–1455.
3. Anguita, D., Parodi, G., & Zunino, R. (1994). An efficient implementation of BP on RISC-based workstations. *Neurocomputing, 6*(2), 57–65.
4. Ye, H., & Young, S. (2006). Quality-enhanced voice morphing using maximum likelihood transformations. *IEEE Transactions on Audio, Speech, and Language Processing, 14*(4), 1301–1312.
5. Curry, B., & Morgan, P. (1997). Neural networks: A need for caution. *Omega, International Journal of Management Sciences, 25*(8), 123–133.
6. Chen, D., & Mohler, R. R. (2003). Neural-network-based load modeling and its use in voltage stability analysis. *IEEE Transactions on Control Systems Technology, 11*(4), 460–470.

7. Baldi, P. (1996). Gradient descent learning algorithm overview: A general dynamical systems perspective. *IEEE Transactions on Neural Networks, 6*(1), 182–195.
8. Moller, M. (1993). A scaled conjugate gradient algorithm for fast supervised learning. *Neural Networks, 6*(10), 525–533.
9. Hagan, M., & Menhaj, M. (1994). Training feed-forward networks with the Marquadt algorithm. *IEEE Transactions on Neural Networks, 5*(6), 152–157.
10. Storn, R., & Price, K. (1997). Differential evolution — a simple and efficient heuristic for global optimization over continuous spaces. *Journal of Global Optimization, 11*(6), 341–359.

Part II
Data Processing

Chapter 30
Implicit Factoring with Shared Middle Discrete Bits

Meng Shi, Xianghui Liu, and Wenbao Han

Abstract We study the problem of implicit factoring presented by May and Ritzenhofen in 2009 and apply it to more general settings, where prime factors of both integers are only known by implicit information of middle discrete bits. Consider two integers $N_1 = p_1q_1$ and $N_2 = p_2q_2$ where p_1, p_2, q_1, and q_2 are primes and $q_1, q_2 \approx N^{\alpha}$. In the case of $t\log_2 N$ bits shared in one consecutive middle block, we describe a novel lattice-based method that leads to the factorization of two integers in polynomial time as soon as $t > 4\alpha$. Moreover, we use much lower lattice dimensions and obtain a great speedup. Subsequently, we heuristically generalize the method to an arbitrary number n of shared blocks. The experimental results show that the constructed lattices work well in practical attacks.

30.1 Introduction

Efficient factoring of large integers is one of the most fundamental problems in algorithmic number theory and plays an essential role in RSA's security. Since RSA system was proposed in 1977, many researchers have been trying to solve large integer factoring problem. For classical computing model, the quadratic sieve, the elliptic curve method, and the number field sieve have been presented to optimize the factorization complexity. It is not known if there exist integer factorization algorithms whose running time is polynomial in the bit size of N.

In 2009, May and Ritzenhofen first presented the large integer factorization problem with implicit information [1]. Consider two integers $N1 = p1q1$ and $N2 = p2q2$, where $p1$, $p2$, $q1$, $q2$ are primes and $p1$, $p2$ share $t\log2N$ least significant bits (LSBs). They proved that when $q1$, $q2 \approx N\alpha$, $N1$, $N2$ can be factored in polynomial time if $t \geq 2\alpha$. These results can further be extended to

M. Shi (✉) • X. Liu • W. Han
Department of Applied Mathematics, Zhengzhou Information Science and Technology Institute, Zhengzhou 450002, China
e-mail: bagoubj@yahoo.cn

W.E. Wong and T. Zhu (eds.), *Computer Engineering and Networking*, Lecture Notes in Electrical Engineering 277, DOI 10.1007/978-3-319-01766-2_30,

the situation when $N1 = p1q1$, $N2 = p2q2, \ldots, Nk = pkqk$, and all the pi's share $t\log 2N$ LSBs. Note that the additional information here is only implicit: the attacker only knows that $p1$ and $p2$ share them. Subsequently, Sarkar and Maitra did further research applying Coppersmith and Gröbner-basis techniques [2]. Contrary to May's result, their method works when $p1$, $p2$ share either $t\log 2N$ LSBs or MSBs or bits in the middle. In 2010, Faugere, Marinier, and Renault presented a method different from previous one and improved their conclusions [3].

However, this attack works on a condition that pi is sharing the MSBs, LSBs, or bits in the middle. Unfortunately, the unknown part is in general not located in one consecutive bit block but widely spread over the whole bit string [4]. Hence, whether we can extend our method to the general scenario is a work worth researching.

Our contribution: Instead of the strategy of finding roots of multivariate polynomials applied in Sarkar's results [2], we transformed this problem into finding small modular roots to reduce lattice dimensions. Table 30.1 compares our results against the previous results [1–3].

What is more is that our theoretical results are supported by experiments. We have written the corresponding programs in VC++6.0 over WINXP SP3 on computer with Intel(R) Pentium(R) 4 3.00 GHz CPU, 512M RAM.

The paper is organized as follows. In Sect. 30.2, we briefly discuss some background information. Next, in Sect. 30.3 we present our strategy in the case when $p1$, $p2$ share a contiguous portion of bits at the middle and describe the complete experimental details with comparison of existing works. Then, in Sect. 30.4 we extend the method to the arbitrary contiguous portions of bits. In Sect. 30.5 we conclude the paper.

30.2 Preliminaries

In this section, we briefly present some basics on basis reduction in lattice [5–7].

Definition 1 Let $b_1, b_2, \ldots, b_m \in R^n$ be linearly independent (row) vectors, where n and m are integers such that $n \geq m$. A lattice L is described as the set of vectors in R^n that are integer linear combinations of the basis vectors $b_1, b_2, \ldots, b_m$. Formally,

$$L(b_1, b_2, \ldots, b_m) = \left\{ z = \sum_{i=1}^{m} \lambda_i b_i \middle| \lambda_i \in Z \right\}.$$

When given a lattice L, finding the shortest vector in a basis of L has been proved to be an NP-hard problem. In 1982, Lenstra, Lenstra, and Lovasz proposed the famous LLL algorithm for lattice basis reduction [8], which can produce a reduced basis in polynomial time. Here is the result:

Table 30.1 Comparison of our results against the previous results

	May's results [1]	Sarkar's results[2]	Faugere's results [3]	Our results
$k = 2$	share $t\log_2 N$ LSBs: $t \geq 2\alpha + 3$ using two-dimensional lattices	share $t\log_2 N$ LSBs or MSBs: better than $t \geq 2\alpha + 3$ when $\alpha \geq 0.266n$ share $t\log_2 N$ bits in the middle: $t \geq 4\alpha + 7$ but depending on the position of the shared bits using 46-dimentsional lattices	share $t\log_2 N$ MSBs: $t \geq 2\alpha + 3$ using two-dimensional lattices share $t\log_2 N$ bits in the middle: $t \geq 4\alpha + 7$ using three-dimensional lattices	share $t\log_2 N$ bits in the middle: $t > 4\alpha$ using eight-dimensional lattices share $t\log_2 N$ discrete bits in the middle: $t > 7\alpha$ using 12-dimensional lattices
$k > 2$	Can be applied	Cannot be applied	Can be applied	Cannot be applied

Lemma 1 *Let* L *be a lattice of dimension* m. *The reduced basis vectors* $\{v_1, v_2, \ldots, v_m\}$ *that the LLL algorithm outputs satisfy*

$$||v_1|| \leq ||v_2|| \leq \cdots \leq ||v_i|| \leq 2^{\frac{\omega(\omega-1)+(i-1)(i-2)}{4(\omega-i+1)}} \det(L)^{\frac{1}{\omega-i+1}}, \quad \text{for all } 1 \leq i \leq \omega.$$

Definition 2 Given a polynomial $|f(x_1, x_2, \ldots, x_n)| = \sum_{i_1, i_2, \ldots, i_n} a_{i_1, i_2, \ldots, i_n} x_1^{i_1} x_2^{i_2} \ldots x_n^{i_n}$, we define the norm as $\|f(x_1, x_2, \ldots, x_n)\| = \left(\sum_{i_1, i_2, \ldots, i_n} a^2_{i_1, i_2, \ldots, i_n} \right)^{1/2}$.

In 1998, Howgrave-Graham reformulated Coppersmith's method and proposed a theorem as follows [9]:

Lemma 2 *Let* $h(x_1, x_2, \ldots, x_n) \in Z[x_1, x_2, \ldots, x_n]$ *be an integer polynomial that consists of at most* w *monomials. Suppose that*
(1) $h(x_1^{(0)}, x_2^{(0)}, \ldots, x_n^{(0)}) = 0 \bmod N^m$ *for some* $|x_1^{(0)}| < X_1, \ldots, |x_n^{(0)}| < X_n$ *and*
(2) $\|h(X_1 x_1, X_2 x_2, \ldots, X_n x_n)\| < N^m/\sqrt{w}$

then $h(x_1^{(0)}, x_2^{(0)}, \ldots, x_n^{(0)}) = 0$ *holds over the integers.*

Lemma 2 plays an important role in this paper. A common root of *i* variables can be extracted efficiently if the *i* polynomials are algebraically independent. Therefore, we have encountered the following heuristic. This heuristic often works perfectly.

Assumption 1 *The resultant computations for the multivariate polynomials constructed in our approaches yield nonzero polynomials.*

30.3 Primes p_1, p_2 Share a Contiguous Portion of Bits at the Middle

Throughout this paper, we assume p_1, p_2 are primes of same bit size and q_1, q_2 are primes of same bit size. Thus $N_1 = p_1 q_1$ and $N_2 = p_2 q_2$ are also of same bit size. We use N to represent an integer of same bit size as of N_1, N_2. Assuming that we are given two different module $N_1 = p_1 q_1$ and $N_2 = p_2 q_2$, where contiguous portion of bits of p_1, p_2 are same at the middle leaving the $\gamma_1 \log_2 N$ many MSBs and $\gamma_2 \log_2 N$ many LSBs, i.e., $p_1 = N^{1-\alpha-\gamma_1} p_{10} + N^{\gamma_2} p_{11} + p_{12}$ and $p_2 = N^{1-\alpha-\gamma_1} p_{20} + N^{\gamma_2} p_{11} + p_{22}$ for some common p_{11} that is unknown to us. In this section, we will show that there is an algorithm that recovers the factorization of N_1 and N_2 in polynomial time provided that $t = 1 - \gamma_1 - \gamma_2 - \alpha$ is sufficiently large.

We start with

$$p_1 - p_2 = N^{1-\alpha-\gamma_1}(p_{10} - p_{20}) + (p_{12} - p_{22}) \tag{30.1}$$

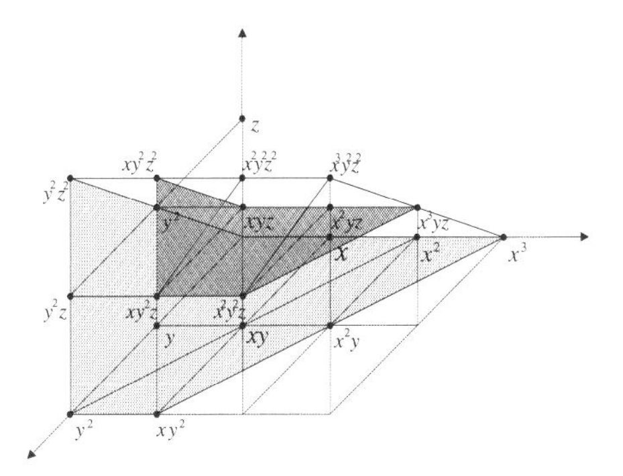

Fig. 30.1 Example of $M_k \backslash M_{k+1}$ when $m = 2$, $d = 1$

By reducing this equation modulo $N^{1-\alpha-\gamma_1}$, we can eliminate the two unknowns and get

$$p_1 - p_2 = p_{12} - p_{22} \text{mod} N^{1-\alpha-\gamma_1} \quad (30.2)$$

Since $p_1 = N_1/q_1$ and $p_2 = N_2/q_2$, we have

$$N_1q_2 - N_2q_1 - N^{1-\alpha-\gamma_1}(p_{10} - p_{20})q_1q_2 - (p_{12} - p_{22})q_1q_2 = 0 \quad (30.3)$$

Then we can rewrite the equation modulo $N^{1-\alpha-\gamma_1}$ to

$$N_1q_2 - N_2q_1 - (p_{12} - p_{22})q_1q_2 = 0 \text{mod} N^{1-\alpha-\gamma_1} \quad (30.4)$$

Thus we need to solve the following polynomial:

$$\begin{aligned} f_N(x, y, z) &= f'_N(x - 1, y, z) = N_1x - N_2y - xyz - N_1 - yz \\ &= 0 \text{mod} N^{1-\alpha-\gamma_1} \end{aligned} \quad (30.5)$$

The root (x_0, y_0, z_0) of f is $(q_2 - 1, q_1, p_{11} - p_{22})$. Let $X = N^{\alpha}, Y = N^{\alpha}, Z = N^{\gamma_2}$, then we can take X, Y, Z as the upper bound of x_0, y_0, z_0 respectively.

Hence, we can get Theorem 1.

Theorem 1 *Let* $N_1 = p_1q_1$ *and* $N_2 = p_2q_2$, *where* p_1, p_2, q_1, q_2 *are primes. Let* $q_1, q_2 \approx N^{\alpha}$. *If a contiguous portion of bits of* p_1, p_2 *are same at the middle leaving the* $\gamma_1 \log_2 N$ *many MSBs and* $\gamma_2 \log_2 N$ *many LSBs, we can factor both* N_1, N_2 *if* $t > 4\alpha$ *under Assumption 1.*

Proof Let us discuss in general how small a root (x_0, y_0, z_0) of a polynomial $f_N(x, y, z) = a_1x + a_2y + a_3xyz + a_4yz + a_5 = 0 \text{mod} N^{1-\alpha-\gamma_1}$ could be such that it can be found by Coppersmith method.

Jochemsz described the extended strategy [5]. So it can be prescribed as follows:

$$M_k = \bigcup_{0 \le j \le d} \left\{ x^{i_1+j} y^{i_2} z^{i_3} \middle| x^{i_1} y^{i_2} z^{i_3} \, monomial\, of\, f_N^m \, and \, \frac{x^{i_1} y^{i_2} z^{i_3}}{l^k} \, monomial\, of\, f_N^{m-k} \right\}$$

for $l = xyz$, $d = \tau m$, and $\tau > 0$.

Next, we define the following shift polynomials:

$$g_{i_1 i_2 i_3 k}(x, y, z) = \frac{x^{i_1} y^{i_2} z^{i_3}}{l^k} f_N^k N^{(1-\alpha-\gamma_1)(m-k)}$$

for $k = 0, \ldots, m$, and $x^{i_1} y^{i_2} z^{i_3} \in M_k \setminus M_{k+1}$.

Figure 30.1 shows that the new sets M_k when $m = 2$, $d = 1$. Points of the large shadow belong to $M_0 \setminus M_1$, and points of the small shadow belong to $M_1 \setminus M_2$.

Hence, if we reasonably order the matrix describing the lattice L, we are sure to get a lower triangular matrix.

We conclude that the condition $\det(L) < q^{m\omega - \varepsilon}$ reduces to

$$X^{3+8\tau+6\tau^2} Y^{5+8\tau} Z^{3+4\tau} < N^{(1-\alpha-\gamma_1)(2+4\tau)} \tag{30.6}$$

Substituting the values of $|x| \le X = N^{\alpha}, |y| \le Y = N^{\alpha}, |z| \le Z = N^{\gamma_2}$, if $1-5\alpha-\gamma_1-\gamma_2 > 0$, we can solve Eq. (30.6) for the maximal value of $\tau = \frac{1-\gamma_1-\gamma_2-5\alpha}{3\alpha}$. Let $t = 1-\alpha-\gamma_1-\gamma_2$, we get $t > 4\alpha$.

Next, we compare the conclusion of Theorem 1 in this paper and Theorem 2 in May's paper [1] via experimental data. In this experiment, we still consider 850-bit primes p_1 and p_2.

$p_1 =$ 60100632917456734116303555865209875275018541235077520316344103
38922261325200790844122487056278592666345958977017699817179683704 6
32052368934119368753632404396752810137898732502224812203770830561 7
873 96400563774598643554294853982575818856214151329271376535 73.
$p_2 =$ 59848256418709315858233822209629263442206705325544039333521 05
67557106667270360325973032351634730768233113437592153548409315545 3
66310986033574699693202605043239631638906892532523449394047385276
94071 49342403534694186411407834893900732303380146620012528421339.

Note that p_1, p_2 share middle 504 many bits (leaving 177 bits from the least significant side).

Further, q_1, q_2 are 150-bit primes

$q_1 =$ 1038476608131498405684472704928794724111541861 and
$q_2 =$ 1281887704228770097092001008195142506836912053.

Given N_1, N_2 with only the implicit information, we can factorize both of them efficiently. We use lattice of dimension 8 (parameters $m = 1, d = 1$) and the lattice reduction costs 15 ms.

From the test we obtain two conclusions. First, the experimental results are better than the theoretical results: the theoretical result of Theorem 1 is $t > 4\alpha$, that is, if p_1, p_2 are 850-bit prime numbers and q_1, q_2 are 150-bit prime numbers, then N_1, N_2 can be factored in polynomial time when p_1, p_2 share the middle of 600-bit. However, in our experiment, p_1, p_2 share only the middle 504-bit. Second, because we applied the method of finding small modular roots of multivariate polynomials, the number of variables of equation in Theorem 1 of this paper was smaller than the Theorem 2 in Sarkar's paper [2] under the same conditions. The experimental results show that compared with the 70-dimentional lattice in Sarkar's paper [2], N_1, N_2 can be factored more efficiently by constructing an eight-dimensional lattice according to Theorem 1 in this paper.

30.4 Sharing More Contiguous Portions of Bits at the Middle

In practice, the attacker can get the implicit information of the prime factors with shared middle discrete bits easily. In this section, we heuristically generalize the method of Theorem 1 to an arbitrary number n of shared blocks. In fact, the difference between Theorem 3 and the Theorem 2 is that the polynomials contain increased number of variables as well as monomials. For simplicity, let $n = 2$ first.

Theorem 2 *Let* $N_1 = p_1q_1$ *and* $N_2 = p_2q_2$, *where* p_1, p_2, q_1, q_2 *are primes. Let* $q_1, q_2 \approx N^{\alpha}$. *If a contiguous portion of bits of* p_1, p_2 *is the same at the middle leaving the* $\gamma_1 \log_2 N$ *many MSBs,* $\gamma_3 \log_2 N$ *many LSBs, and* $\gamma_2 \log_2 N$ *many bits in the middle, we assume that the first same part is* t_1*-bit and the second same paragraph is* t_2*-bit. Then under Assumption 1, we can factor both* N_1, N_2 *if* $t = t_1 + t_2 > 7\alpha$.

Proof We can write the following formulas similarly:

$$p_1 = N^{1-\alpha-\gamma_1}p_{10} + N^{1-\alpha-\gamma_1-t_1}p_{11} + N^{\gamma_3+t_2}p_{12} + N^{\gamma_3}p_{13} + p_{14} \quad (30.7)$$

$$p_2 = N^{1-\alpha-\gamma_1}p_{20} + N^{1-\alpha-\gamma_1-t_1}p_{11} + N^{\gamma_3+t_2}p_{22} + N^{\gamma_3}p_{13} + p_{24} \quad (30.8)$$

By reducing we can get

$$N_1q_2 - N_2q_1 - N^{\gamma_3+t_2}(p_{12} - p_{22})q_1q_2 - (p_{14} - p_{24})q_1q_2 = 0 \bmod N^{1-\alpha-\gamma_1} \quad (30.9)$$

Therefore, our goal is to find the small root $(q_2 - 1, q_1, p_{11} - p_{22}, p_{14} - p_{24})$ of the polynomial $f_N = a_1x + a_2y + a_3xyz + a_4yz + a_5wxy + a_6wy + a_7 = 0$ $\bmod N^{1-\alpha-\gamma_1}$. We can take $X = N^{\alpha}, Y = N^{\alpha}, Z = N^{\gamma_2}, W = N^{\gamma_3}$ as the upper bound of root respectively.

By calculating, the upper bounds X, Y, Z, and W have to satisfy that

$$X^{10+20\tau+10\tau^2+20\tau^4+20\tau^5}Y^{15+15\tau+20\tau^4+5\tau^5}Z^{5+5\tau+20\tau^4+15\tau^5}W^{5+5\tau-5\tau^5} < N^{(1-\alpha-\gamma_1)(4+5\tau+20\tau^4+9\tau^5)}. \quad (30.10)$$

Substituting the values $|x| \leq X = N^{\alpha}, |y| \leq Y = N^{\alpha}, |z| \leq Z = N^{\gamma_2}, |w| \leq W = N^{\gamma_3}$, if $1-8\alpha - \gamma_1 - \gamma_2 - \gamma_3 > 0$, we can solve Eq. (30.10) for the maximal value of $\tau = \frac{1-\gamma_1-\gamma_2-\gamma_3-8\alpha}{4\alpha}$. Let $t = 1-\alpha - \gamma_1 - \gamma_2$, we get $t > 7\alpha$.

Obviously, we can generalize the method of Theorem 2 to an arbitrary number n of shared blocks.

Theorem 3 *Let* $N_1 = p_1q_1$ *and* $N_2 = p_2q_2$, *where* p_1, p_2, q_1, q_2 *are primes. Let* $q_1, q_2 \approx N^{\alpha}$. *If a contiguous portion of bits of* p_1, p_2 *is same at the middle leaving the* $\gamma_1 log_2N$ *many MSBs,* $\gamma_{n+2} log_2N$ *many LSBs, and* $\gamma_2 log_2N, \ldots, \gamma_{n+1} log_2N$ *many bits in the middle. We assume that the first same part is* t_1*-bit, the second same part is* t_2*-bit, and the last same part is* t_{n+1}*-bit. Then under Assumption 1, we can factor both* N_1, N_2 *if* $t = \sum_i^{n+1} t_i > 7\alpha$.

The next example considers the case of primes p_1, p_2 of 900 bits and q_1, q_2 of 100 bits. This is to demonstrate how our method woks experimentally for discrete bits. Here we only consider 900-bit primes p_1 and p_2.

$p_1 =$ 7505061092492112375563523501603029893103360600496947681215898539028978880558579468898992711484001866062693594592629842369133358102966297068584054427273763300567492499617802017459636475620491510231406148324755843778363996138755147469460323318959606760968500389785083814343 and
$p_2 =$ 4226356249085321971634266775991800272784808798610270102561708704362527040205013775291568103094901198355262363763755300814608766443080087384559762394512580130940071069071767173778872370100880021226567400723767205747855563912712394927148504565667243123047573407832687510917.

Note that p_1, p_2 share the most bits at the middle expect 95 bits which are distributed from the 1-th to the 60-th bits, from the 322-th to the 396-th bits, and from the 836-th to the 900-th bits. Further, q_1, q_2 are 100-bit primes

$q_1 =$ 658018347648025712456491553623 and
$q_2 =$ 684627888228678540762517278787, respectively.

Given N_1, N_2 with only the implicit information, we can factorize both of them efficiently. We use lattice of dimensions 12 (parameter $m = 1$, $d = 1$) and the lattice reduction costs 16 ms.

30.5 Conclusion

In this article we have studied polynomial time factorization strategy when $t\log_2 N$ bits shared in one consecutive middle block of p_1 and p_2. We transformed the problem into finding the small modular roots of multivariate polynomials, and then the bound $t > 4\alpha$ is obtained by using eight-dimensional lattices. Moreover, we heuristically generalize the method to an arbitrary number n of shared blocks and obtain a bound of $t > 7\alpha$. Our results generalize the idea presented in May and Sarkar's papers. Furthermore, for the former case, we obtain better results than May's result under same conditions.

However, the techniques presented here cannot immediately be extended to the generalized problem in Sarkar's paper where $N_1, N_2, \ldots, N_k$ are considered. In this case, the problem has much more variables and our method cannot be directly applied. This problem would be remained as an open issue.

References

1. May, A., & Ritzenhofen M. (2009). Implicit factoring: On polynomial time factoring given only an implicit hint. *LNCS, 5443*, 1–14.
2. Sarkar, S., & Maitra S. (2009). Further results on implicit factoring in polynomial time. *Mathematics of Communications, 3*(2), 205–217.
3. Faugère J.-C., Marinier R., & Renault G. (2010). Implicit factoring with shared most significant and middle bits. *LNCS, 6056*, 70–87.
4. Herrman, M., & May, A. (2008). Solving linear equations modulo divisors: On factoring given any bits. *LNCS, 5350*, 406–424.
5. Jochemsz, E. (2007). *Cryptanalysis of RSA variants using small roots of polynomials*. Netherlands: Technische Universiteit Eindhoven.
6. Blömer, J., & May, A. (2003). New partial key exposure attacks on RSA. *LNCS, 2729*, 27–43.
7. Jochemsz, E., et al. (2006). A strategy for finding roots of multivariate polynomials with new applications in attacking RSA variants. *LNCS, 4284*, 267–282.
8. Lenstra, A. K., Lenstra Jr. H. W., & Lovász L. (1982). Factoring polynomials with rational coefficients. *Mathematiche Analen, 261*(4), 515–534.
9. Howgrave-Graham, N. (1997). Finding small roots of univariate modular equations revisited. *LNCS, 1355*, 131–142.

Chapter 31
Loading Data into HBase

Juan Yang and Xiaopu Feng

Abstract HBase is a top Apache open-source project that separated from Hadoop. As it has most of the features of Google's BigTable system and is implemented in Java, it is very popular in days of massive data. HBase's advantages are reflected in the massive data read and query. Loading huge amounts of data into HBase is the first step to use HBase. HBase itself has several methods to load data, and different methods have different application scenarios. This article made an exhaustive study and a performance testing of them. Also, this article achieved the custom loading data, and experiments show that it has good efficiency.

31.1 Introduction

With the increase of massive data, a large number of business scenarios began to consider expanding the level of data storage, so that storage services can be added or deleted easily. But the current relational databases focus on a machine. The storage of massive amount of data becomes the bottleneck, and a single machine is unable to load large amounts of data. The IO read and write of a single machine have become the bottleneck when it comes to the storage of massive data. When the amount of data is increasing with linear growth, how to solve the consistency of the data IO and how to combine MapReduce computing framework to analyze massive data have become big problems of traditional databases [1]. Traditional databases are unable to meet these needs, and the emergence of HBase solves these problems well. HBase is a top open-source project separated from Hadoop Apache. As it achieves most features of Google's BigTable system with Java, it is welcome in days of massive data [2]. The scenarios of HBase are mainly massive data queries and related operations. Therefore, the first step to use HBase is loading massive data

J. Yang (✉) • X. Feng
Beijing University of Posts and Telecommunications, Beijing 100876, China
e-mail: yangjuan@bupt.edu.cn

W.E. Wong and T. Zhu (eds.), *Computer Engineering and Networking*, Lecture Notes in Electrical Engineering 277, DOI 10.1007/978-3-319-01766-2_31,

into HBase. HBase itself has several methods to load data, and each method has its own characteristics and application scenarios. However, the methods of HBase are lack of flexibility, and cannot fully meet the personalized needs of users. Based on this feature, the author achieved custom loading data with the combination of HBase interface and MapReduce framework through learning the methods of HBase. The experiments show that the effect of this method is good [3].

31.2 HBase

31.2.1 HBase Architecture

HBase is the open-source version of BigTable and is based on HDFS (Hadoop Distributed File System). It is a real-time read and write database system with high reliability and high performance. It is scalable and column storage. HBase is between NoSQL (Not Only SQL) and RDBMS (Relational Database Management System), and it can only retrieve the data by the primary key (row key) and the range of the main keys. It supports single-line transaction (via Hive support to achieve complex operations such as multi-table join). HBase is mainly used to store loose of unstructured and semi-structured data. Just like Hadoop, HBase target mainly relies on horizontal expansion. With the increase of cheap commercial server, it can increase the computing and storage capacity. HBase table generally has the following characteristics:

1. Large: a table can have hundreds of millions of lines and millions of columns.
2. Column-oriented: column-oriented (family) storage and access control, independent, and retrieve of column (family).
3. Sparse: for the empty (null) column, it does not occupy storage space. Therefore, the table can be designed very sparse.

Figure 31.1 is a diagram that shows the HBase Hadoop Ecosystem [4].

HBase's data is stored in the form of table. Table consists of rows and columns. Column is divided into several column families (row family). Figure 31.2 is the logical view of HBase's storage.

Same with NoSQL databases, row key is the primary key used to retrieve the record. There are only three ways to access rows in HBase's table:

1. Key access through a single row
2. Through the row key range
3. Full table scan

Row key can be any string (maximum length is 64 KB and the length of the practical application is generally 10–100 bytes). In the internal of HBase, row key is saved as byte array. When the data is stored, it is sorted in accordance with the row key lexicographic sorting (byte order). When designing the key, it is better to

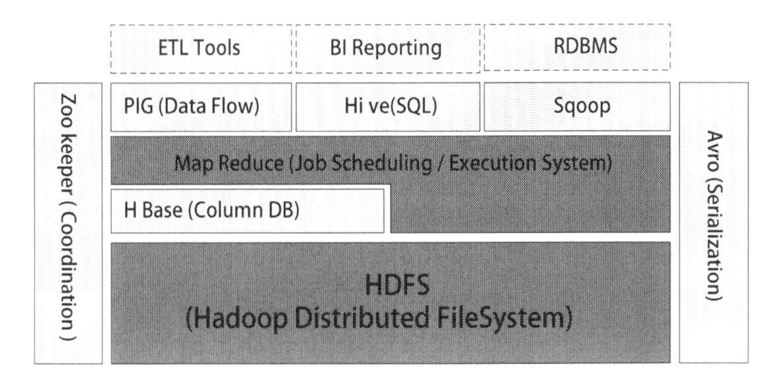

Fig. 31.1 HBase Hadoop ecosystem

Row Key	Time Stamp	Column "Contenst"	Column "anchor"		Column "mime"
"com.cnn. www"	t9		"anchor:cnnsi.com"	"CNN"	
	t8		"anchor:my.look.ca"	"CNN.com"	
	t6	"<html>..."			
	t5	"<html>..."			"text/html"
	t3	"<html>..."			

Fig. 31.2 HBase logical view

consider the characteristic of sorting storage and put the data together which are often read. The reading and writing of the line are atomic operations (regardless the number of columns to read and write). This design enables users to easily understand the program behavior when concurrent update operations are in the same line [5].

31.2.2 The Methods to Load Data

For the latest version of HBase (HBase-0.94.1), there are three methods to load data. In the following, we will explore and compare these three methods.

Method one: In the Hadoop environment, run the HBase own jar package to load specified data on HDFS into the specified data table. This method just needs Map function to complete the loading of data, without Reduce function. It uses the MapReduce interface of HBase to complete this operation. HBase has detailed instructions for using this method and the meaning of the parameters. It can be directly viewed in Hadoop environment with the related command.

Method two: In the Hadoop environment, run the HBase's related jar package. There are two steps to complete it. First, it will translate data in HDFS to HFile format. Then, run the related jar package to load HFile format data into specified

Table 31.1 Test data information

Data	pk1.data	pk2.data	pk3.data	pk4.data	pk5.data	pk6.data	pk7.data
Rows	50,000	600,000	3,000,000	6,000,000	20,000,000	40,000,000	300,000,000
Size	1.10 mb	13.74 mb	71.04 mb	143.14 mb	561.61 mb	1,133.81 mb	7,860.25 mb

data table. During the implementation process in first step, both Map function and Reduce function will be executed. The execution time of Reducing/reduction function is significantly longer than the Map function. The Reducing/reduction function consumes more system resources. The second step, it is just a simple data transfer, and it does not execute MapReduce function.

Method three: In the HBase environment, run the HBase's own method. It will load the sequential file on HDFS directly into the specified data table. This method is the same as method one, completing the data loading in the Map function and did not execute the Reduce function [6].

31.2.3 Performance Test Comparison

31.2.3.1 Test Environment and Data Description

The test cluster comprises three nodes, including a master node and two slave nodes. Each single node contains four hard disks, and each hard disk has 750G storage space, four CUP with 2.00 GHz frequency, and 8G memory. The Hadoop version is 1.0.3 and the HBase version is 0.94.1. The test data is a text file and each file has five columns. The data is randomly generated. Details of the data are shown in Table 31.1.

31.2.3.2 Test Result Contrast

The performance testing of method one and method two is shown in Fig. 31.3. The ordinate is the time to load data and the time unit is seconds. While using pk5.data and pk6.data to test method two, due to consuming too many resources, the hardware environment cannot meet the demand to complete.

As can be seen from Fig. 31.3, when the test data volume is small, method one is faster than method two. But when the data volume is big enough, method two is much faster than method one. However, method two consumes more system resources, and when the hardware system cannot meet the demand, it could not be completed. Method one is easy to operate, for it just needs one step to be completed. Therefore, when the hardware environment is not very well and the amount of data is small, it is better to use method one to complete loading data. When the amount of data is very big, method two is better to save time, but it needs more system resources.

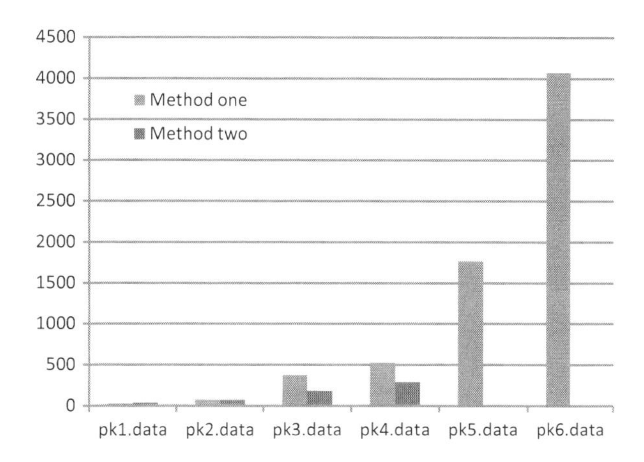

Fig. 31.3 HBase's loading data performance testing

31.3 Custom Loading Data

31.3.1 Design Description

HBase's interface allows users to customize program to load data into HBase. HBase's encapsulated MapReduce interface allows Hadoop and HBase to communicate with each other. It can define the HBase's interface in the Map function to load data into HBase. With the help of Hadoop MapReduce parallel framework, it will be easy to load massive data into HBase. Loading operation can be completed directly in the Map function, without writing the Reduce function. Firstly, set up connection with HBase in the run method of MapReduce and then in the Mapfunction calling HBase's interface to load data into HBase.

Additionally, we can optimize settings of HBase's relevant parameters for specific data and experimental environment. This chapter focuses on the following parameters for experimental cluster to optimize settings.

1. Delay log flushes
   HBase's mechanism is to refresh the log once you put a write. If we use a delay mechanism, log will be stored in memory until the log refresh cycle is over. It can delay the refresh cycle properly, thus each refresh can load more data into the data table.
2. Automatic reflash
   Once HBase executes a Put operation, it will send a request to the Region Server. Through htable.add (Put), it will add Put to the write buffer. If the autoFlush is set to false, it will have to wait until the write buffer is full to initiate the request. In order to explicitly initiate the request, you can also call the flushCommits.

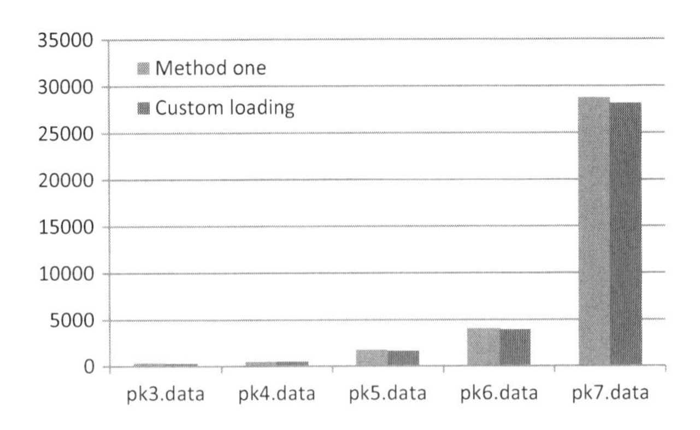

Fig. 31.4 Custom loading data performance testing

3. Close Puts on WAL
   When executing the Put operation, the data is firstly written to the log. Setting put.setWriteToWAL to false means that it will not be written to the log but directly write to the memory. So when the data is loaded, it will eliminate the time to write the log.
4. Cache size
   When the data size in the cache reaches a certain percentage, HBase will automatically refresh, and the data in the cache will be loaded into the data table. We can adjust the cache size for the experimental environment and take full advantage of the cache to load as much data into the cache as possible.

   In addition, we can adjust and optimize the number of regions, region size, and intermediate result storage format according to the experimental environment to achieve the best results [7].

31.3.2 Performance Testing

According to the characteristics of the test cluster, custom loading data optimizes the relevant parameters and, based on this, executing a performance test. Test environment and test data are the same as before. Figure 31.4 is a performance testing comparison of custom loading data and method one.

As we can see from the above results, custom loading method is flat or slightly better than method one. Additionally, you can specify a particular column and make preprocessing and other operations with custom loading method. Especially when the amount of data is large and the data contains illegal data, you can use custom loading method to filter data and do some other operations. So, custom loading method has a lot of flexibility and has advantages in terms of performance compared with the methods of HBase itself [8].

31.4 Conclusion

HBase itself has different methods to load data, and each method has its own application scenario. Developers can select the appropriate loading method for a specific application scenario. In addition, developers can also implement custom parallel loading data by calling the interface of HBase. Custom loading method is flexible and its performance is good. It is more suitable for programmers to develop based on HBase.

References

1. Dean, J., & Ghemawat, S. (2008). MapReduce: Simplified data processing on large clusters. *Communications of the ACM, 51*(1), 107–113.
2. Chang, F., Dean, J., Ghemawat, S., Hsieh, W. C., Wallach, D. A., Burrows, M., et al. (2008). Bigtable: A distributed storage system for structured data. *ACM Transactions on Computer Systems (TOCS), 26*(2), 4.
3. White, T. *Hadoop: The definitive guide [M]*. O'Reilly Media, Inc.,1005 Gravenstein Highway North, Sebastopol, CA95472, 2012.
4. George, L. *HBase: The definitive guide [M]*. O'Reilly Media, Inc.,1005 Gravenstein Highway North, Sebastopol, CA95472, 2011.
5. Huang, J., Ouyang, X., Jose, J., Wasi-ur-Rahman Md., Wang, H., Luo, M., et al. (2012). High-performance design of HBase with RDMA over infiniBand. In *Proceedings of the 2012 I.E. 26th International Parallel and Distributed Processing Symposium, IPDPS 2012* (pp. 774–778). Washington, DC: IEEE Computer Society.
6. Li, C. (2010). Transforming relational database into HBase: A case study. In *Proceedings 2010 I.E. International Conference on Software Engineering and Service Sciences, ICSESS 2010* (pp. 683–687). Piscataway, NJ: IEEE Computer Society.
7. Vora, M. N. (2011). Hadoop-HBase for large-scale data. In *Proceedings of 2011 International Conference on Computer Science and Network Technology, ICCSNT 2011* (pp. 601–605). Piscataway, NJ: IEEE Computer Society.
8. Carstoiu, D., Cernian, A., & Olteanu, A. (2010). Hadoop hbase-0.20. 2 performance evaluation. In *NISS2010 – 4th International Conference on New Trends in Information Science and Service Science* (pp. 84–87). Piscataway, NJ: IEEE Computer Society.

Chapter 32
Incomplete Decision-Theoretic Rough Set Model Based on Improved Complete Tolerance Relation

Xia Wang

Abstract Recently, the decision tables used in decision-theoretic rough sets are the most complete decision tables and less for incomplete decision tables. Some authors use the filling method to deal with incomplete decision table of DTRS, which is filling the unknown values with all possible values. Its disadvantages are large calculated amount and noise. Therefore, incomplete decision-theoretic rough set model based on improved complete tolerance relation is proposed. First, improved complete tolerance relation and tolerance class were constructed. Then, the decision degree between object and target concept were also computed. Finally, the decision degree was served as conditional probability; the probability positive region, negative region, and boundary region were computed, and it would make three-way decision. The example shows that this model has less calculated amount and noise, and it more accords with practical application.

32.1 Introduction

Decision-theoretic rough set model (DTRS) was proposed by Yao in the 1990s, which is a kind of probability rough set model [1]. Based on Pawlak rough set model, DTRS imports the theory of minimum-risk Bayesian decision and finds least expected risk through kinds of risk costing value of classification decisions. So the least expected risk is as basic argument of partition universe into positive region, negative region, and boundary region [1–6]. In brief, based on equivalence class, the universe is partitioned into three regions, positive region, negative region, and boundary region, respectively, and this is similar to Pawlak rough sets. So under this circumstance, disposed decision table must be complete. But in real application, the decision tables are almost incomplete.

X. Wang (✉)
College of Computer Science and Technology, Jiangsu Normal University,
Jiangsu 221119, China
e-mail: lgzwx@163.com

W.E. Wong and T. Zhu (eds.), *Computer Engineering and Networking*, Lecture Notes in Electrical Engineering 277, DOI 10.1007/978-3-319-01766-2_32,

Incomplete decision table means a table with missing data (null value). At present, there are two kinds of ways for disposing incomplete decision tables. One is completion means, which characteristic is polishing the decision table, and then using the means of completing table to deal with the incomplete table. The other is direct disposing means, which characteristic is appropriate expanding the concepts of classics rough sets, and then transforms the strict equivalence relation to broad binary relation [7].

In this chapter, an incomplete decision-theoretic rough set model is proposed, which is based on improved tolerance relation and decision-theoretic rough set. This model sufficiently absorbs the advantages of improved tolerance relation, which will be appropriate for three principal decision of the real world.

32.2 Preliminaries

Definition 1[7] Let $S = \{U, A, V, f\}$ is a decision table, where $U = \{x_1, x_2, \ldots, x_n\}$ is nonempty finite object set, $A = C \cup D$ is nonempty attribute set, $C = \{c_1, c_2, \ldots, c_m\}$ is set of conditional attribute, $D = \{d_1, d_2, \ldots, d_k\}$ is set of decision attribute, and $C \cap D = \emptyset$, V is the domain of attribute. $f : U \rightarrow V_i (i = 1, 2, \ldots, m + k)$ is information function. When the precise values of some attributes are not known, i.e., missing or known partially, (write for $f(x,c) = *$, $x \in U$ $c \in C$), such S is called an incomplete decision table.

Definition 2[8] Let $S = \{U, C \cup D, V, f\}$ is an incomplete decision table. The tolerance relation is defined as:

$$T_C(x,y) = \{(x,y) \in U \times U | \forall c \in C(c(x) = c(y) \vee c(x) = * \vee c(y) = *)\} \quad (32.1)$$

So, tolerance relation has reflexivity and symmetry, but not transitivity.

Definition 3[8] Let $S = \{U, C \cup D, V, f\}$ is an incomplete decision table; the tolerance class of object x is defined as

$$I_C^T(x) = \{y | y \in U \wedge T_C(x,y)\} \quad (32.2)$$

Definition 4[8] Let $S = \{U, C \cup D, V, f\}$ is an incomplete decision table. Improved complete tolerance relation of is S defined as

$$NS_C(x,y) = \left\{ (x,y) \in U \times U | \forall c \in C(c(x) = c(y) \vee c(x) = * \vee c(y) = *) \wedge \min\left(\frac{|P(x) \cap P(y)|}{\min(|P(x)|, |P(y)|}, \frac{\gamma(x) + \gamma(y)}{2} \geq \gamma \right) \right\} \quad (32.3)$$

where $P(x_i) = \left\{c|c \in C \wedge c(x_i) \neq *\right\}$ and assuming 0/0 = 0, $\gamma(x)$ is complete degree of object x.

Definition 5[9] Let $S = \left\{U, C \cup D, V, f\right\}$ is an incomplete decision table. The complete degree of object x_i is defined as

$$\gamma(x_i) = \frac{|P(x_i)|}{|C|} \quad (32.4)$$

where $P(x_i) = \left\{c|c \in C \wedge c(x_i) \neq *\right\}, x_i \in U$, | • | is base number of set •. So the complete degree of whole decision table S is defined as

$$\gamma = \frac{\sum_{i=1}^{|U|} \gamma(x_i)}{|U|} \quad (32.5)$$

Definition 6[9] Let $S = \{U, C \cup D, V, f\}$ is an incomplete decision table; improved tolerance class of object x is defined as

$$I_C^{NS}(x) = \left\{y \middle| y \in U \wedge NS_C(x, y)\right\} \cup \{x\} \quad (32.6)$$

32.3 Decision-Theoretic Rough Set Model

For the Bayesian decision procedure, the DTRS model is composed of two states and three actions. The set of states is given by $\Omega = \{C, \neg C\}$ (C is target set) indicating that an object is in C and not in C, respectively. The set of actions is given by $\varsigma = \{a_P, a_B, a_N\}$, where a_P, a_B, and a_N represent the three actions in classifying an object x, namely, deciding $x \in POS(C)$, deciding should be further investigated $x \in BND(C)$, and deciding $x \in NEG(C)$, respectively. The loss function λ regarding the risk or cost of actions of different states is given by the 3 × 2 matrix:

	$C(P)$	$\neg C(N)$
a_P	λ_{PP}	λ_{PN}
a_B	λ_{BP}	λ_{BN}
a_N	λ_{NP}	λ_{NN}

In the matrix, λ_{PP}, λ_{BP}, and λ_{NP} denote the losses incurred for taking actions of a_P, a_B, and a_N, respectively, when an object belongs to C. Similarly, λ_{PN}, λ_{BN}, and λ_{NN} denote the losses incurred for taking actions of a_P, a_B, and a_N when the object belongs to $\neg C$. $\Pr(C|[x])$ is the conditional probability of an object x belonging to

C given that the object is described by its equivalence class $[x]$. For an object x, the expected loss $R(a_i|[x])$ associated with taking the individual actions can be expressed as

$$R(a_P|[x]) = \lambda_{PP}\ \Pr(C|[x]) + \lambda_{PN}\ \Pr(\neg C|[x])$$
$$R(a_B|[x]) = \lambda_{BP}\ \Pr(C|[x]) + \lambda_{BN}\ \Pr(\neg C|[x])$$
$$R(a_N|[x]) = \lambda_{NP}\ \Pr(C|[x]) + \lambda_{NN}\ \Pr(\neg C|[x])$$

The Bayesian decision procedure suggests the following minimum-cost decision rules:

P: If $R(a_P|[x]) \le R(a_B|[x])$ and $R(a_P|[x]) \le R(a_N|[x])$, then decide $x \in POS(C)$.
B: If $R(a_B|[x]) \le R(a_P|[x])$ and $R(a_B|[x]) \le R(a_N|[x])$, then decide $x \in BND(C)$.
N: If $R(a_N|[x]) \le R(a_P|[x])$ and $R(a_N|[x]) \le R(a_B|[x])$, then decide $x \in NEG(C)$.

Since $\Pr(C|[x]) + \Pr(\neg C|[x]) = 1$, so we can simplify the rules based only on the probability $\Pr(C|[x])$ and the loss function. By considering a reasonable kind of loss function with $0 \le \lambda_{PP} \le \lambda_{BP} < \lambda_{NP}$ and $0 \le \lambda_{NN} \le \lambda_{BN} < \lambda_{PN}$, the decision rule P-N can be expressed concisely as:

P: If $\Pr(C|[x]) \ge \alpha$ and $\Pr(C|[x]) \ge \gamma$, then decide $x \in POS(C)$.
B: If $\Pr(C|[x]) \le \alpha$ and $\Pr(C|[x]) \ge \beta$, then decide $x \in BND(C)$.
N: If $\Pr(C|[x]) \le \beta$ and $\Pr(C|[x]) \le \gamma$, then decide $x \in NEG(C)$.

The threshold values α, β, and γ are defined as $\alpha = \frac{(\lambda_{PN}-\lambda_{BN})}{(\lambda_{PN}-\lambda_{BN})+(\lambda_{BP}-\lambda_{PP})}$, $\beta = \frac{(\lambda_{BN}-\lambda_{NN})}{(\lambda_{BN}-\lambda_{NN})+(\lambda_{NP}-\lambda_{BP})}$, and $\gamma = \frac{(\lambda_{PN}-\lambda_{NN})}{(\lambda_{PN}-\lambda_{NN})+(\lambda_{NP}-\lambda_{PP})}$.

In addition, as a well-defined boundary region, the conditional of rule B suggests that $\alpha > \beta$, that is $\frac{(\lambda_{BP}-\lambda_{PP})}{(\lambda_{PN}-\lambda_{BN})} < \frac{(\lambda_{NP}-\lambda_{NN})}{(\lambda_{BN}-\lambda_{NN})}$. It implies $0 \le \beta < \gamma < \alpha \le 1$. In this case, after tiebreaking, the following simplified rules are obtained:

P1: If $\Pr(C|[x]) \ge \alpha$, then decide $x \in POS(C)$.
B1: If $\beta < \Pr(C|[x]) < \alpha$, then decide $x \in BND(C)$.
N1: If $\Pr(C|[x]) \le \beta$, then decide $x \in NEG(C)$.

32.4 Incomplete Decision-Theoretic Rough Set Model Based on Improved Complete Tolerance Relation

At present, most authors are focused on complete decision table when they research decision-theoretic rough set and less for incomplete decision table. In this chapter, the procedure of incomplete decision-theoretic rough set is proposed, based on improved complete tolerance relation. Let us see an example in Table 32.1.

Where $U = \{x_1, x_2, \ldots x_7\}$ is object set; a, b, and c are condition attributes; d is decision attribute; and * means possible values of attributes.

Table 32.1 Incomplete decision table

U	a	b	c	d
x_1	1	1	1	1
x_2	1	*	1	1
x_3	2	1	1	1
x_4	1	2	*	1
x_5	1	2	*	2
x_6	1	*	1	2
x_7	1	1	1	2

Table 32.2 Complete degree of each object

U	x_1	x_2	x_3	x_4	x_5	x_6	x_7
$\gamma(x_i)$	1	2/3	1	2/3	2/3	1	1

First, we compute complete degree of each object by using Definition 5. The result can be expressed in Table 32.2.

So the complete degree of whole decision table is $\gamma = \frac{\sum_{i=1}^{7} \gamma(x_i)}{7} = 0.8571$.

Then, tolerance relation NS_C of table S is $NS_C = \{(1,7),(7,1)\} \cup I_X$, where I_X is identity relation of X.

So, tolerance class of each object x_i is

$I_C^{NS}(x_1) = \{x_1, x_7\}$, $I_C^{NS}(x_2) = \{x_2\}$, $I_C^{NS}(x_3) = \{x_3\}$, $I_C^{NS}(x_4) = \{x_4\}$, $I_C^{NS}(x_5) = \{x_5\}$, $I_C^{NS}(x_6) = \{x_6\}$, and $I_C^{NS}(x_7) = \{x_1, x_7\}$.

From the result, we can find the tolerance class set is a cover of universe U.

Definition 7 Given tolerance class $I_C^{NS}(x_i)$ $(i = 1, 2, \ldots, |U|)$ of object x_i, then decision degree is defined as:

$$\mu(x_i) = \frac{\left|\mathrm{I}_{\mathrm{C}}^{\mathrm{NS}}(x_i) \cap X\right|}{\left|\mathrm{I}_{\mathrm{C}}^{\mathrm{NS}}(x_i)\right|} \tag{32.7}$$

Replaced conditional probability of DTRS model with decision degree, three regions can be again defined as:

$$\begin{aligned} POS_{(\alpha,\beta)}(X) &= \{x \mid \mu(x) \geq \alpha\}. \\ NEG_{(\alpha,\beta)}(X) &= \{x \mid \mu(x) \leq \beta\}. \\ BND_{(\alpha,\beta)}(X) &= \{x \mid \alpha < \mu(x) < \beta\}. \end{aligned} \tag{32.8}$$

Correspondingly, the P, B, and N rules can be expressed as:

$$P: \text{if}\,\mu(x) \geq \alpha, \text{then}\, x \in POS_{(\alpha,\beta)}(C).$$
$$N: \text{if}\,\mu(x) \leq \beta, \text{then}\, x \in NEG_{(\alpha,\beta)}(C). \quad (32.9)$$
$$B: \text{if}\,\beta < \mu(x) < \alpha, \text{then}\, x \in BND_{(\alpha,\beta)}(C).$$

Then, given the target concept set $X = \{x_1,x_2,x_3,x_4\}$, the decision degree of each object can be computed as:

$\mu(x_1) = 0.5$, $\mu(x_2) = 1$, $\mu(x_3) = 1$, $\mu(x_4) = 1$, $\mu(x_5) = 0$, $\mu(x_6) = 0$, and $\mu(x_7) = 0.5$.

For example, when $\alpha = 0.65$, $\beta = 0.45$, the three regions are:

$$POS_{(\alpha,\beta)}(X) = \{x | \mu(x) \geq 0.65\} = \{x_2, x_3, x_4\}.$$
$$NEG_{(\alpha,\beta)}(X) = \{x | \mu(x) \leq 0.45\} = \{x_5, x_6\}.$$
$$BND_{(\alpha,\beta)}(X) = \{x | 0.45 < \mu(x) < 0.65\} = \{x_1, x_7\}.$$

Different with the traditional DTRS, we use tolerance class operation and decision degree, which are foundation of building DTRS model. So this model can serve as extending of traditional DTRS model.

32.5 An Example of Using the New Model

Let us consider an example of incomplete decision table from the literature [8] listed in Table 32.3. We would like to make a decision when an object belongs to "Φ" class.

First, we can compute the complete degree of Table 32.3, which value is 0.625. Using Definitions 6 and 7, the tolerance class of each object is

$I_C^{NS}(a_1) = \{a_1\}$; $I_C^{NS}(a_2) = \{a_2,a_3\}$; $I_C^{NS}(a_3) = \{a_2,a_3\}$; $I_C^{NS}(a_4) = \{a_4,a_9,a_{12}\}$; $I_C^{NS}(a_5) = \{a_5,a_9,a_{12}\}$; $I_C^{NS}(a_6) = \{a_6,a_7\}$; $I_C^{NS}(a_7) = \{a_6,a_7\}$; $I_C^{NS}(a_8) = \{a_8\}$; $I_C^{NS}(a_9) = \{a_4,a_5,a_9,a_{11}\}$; $I_C^{NS}(a_{10}) = \{a_{10}\}$; $I_C^{NS}(a_{11}) = \{a_9,a_{11},a_{12}\}$; and $I_C^{NS}(a_{12}) = \{a_4,a_5,a_{11},a_{12}\}$.

The second, on the base of Definition 9, we compute decision degree. Supposed target concept is $X = \{a_2,a_4,a_6,a_7,a_8,a_9,a_{11}\}$, and then the decision degree of each object is

$\gamma(a_1) = 0$; $\gamma(a_2) = \gamma(a_3) = 0.5$; $\gamma(a_4) = 0.667$; $\gamma(a_5) = 0.333$; $\gamma(a_6) = \gamma(a_7) = \gamma(a_8) = 1$; $\gamma(a_9) = 0.75$; $\gamma(a_{10}) = 0$; $\gamma(a_{11}) = 0.667$; and $\gamma(a_{12}) = 0.5$.

The third, suppose $\alpha = 0.75$, $\beta = 0.45$, according to Eq. (32.8), the three regions are:

$$POS_{(\alpha,\beta)}(X) = \{a | \mu(a) \geq 0.75\} = \{a_6, a_7, a_8, a_9\}.$$

Table 32.3 An example of incomplete decision table

U	c_1	c_2	c_3	c_4	d
a_1	*	*	1	*	Ψ
a_2	1	2	3	4	Φ
a_3	1	2	3	4	Ψ
a_4	2	*	*	*	Φ
a_5	2	*	*	*	Ψ
a_6	4	*	2	1	Φ
a_7	*	3	2	1	Φ
a_8	3	*	*	*	Φ
a_9	2	3	4	1	Φ
a_{10}	*	0	*	*	Ψ
a_{11}	2	3	4	*	Φ
a_{12}	2	3	4	3	Ψ

$$NEG_{(\alpha,\beta)}(X) = \{a | \mu(a) \leq 0.45\} = \{a_1, a_5, a_{10}\}.$$

$$BND_{(\alpha,\beta)}(X) = \{a | 0.45 < \mu(a) < 0.75\} = \{a_2, a_3, a_4, a_{11}, a_{12}\}.$$

So in the incomplete decision table S, positive region is $POS_{(\alpha,\beta)}(X) = \{a_6, a_7, a_8, a_9\}$, then we can make the decision that objects a_6, a_7, a_8, a_9 belong to target concept X. The negative region is $NEG_{(\alpha,\beta)}(X) = \{a_1, a_5, a_{10}\}$, so we can do decision that objects a_1, a_5, a_{10} do not belong to target concept X and we can reject them. This conclusion is same to original table. That is, objects of positive region and negative region are, respectively, Φ and Ψ. With regard to each element of boundary region $BND_{(\alpha,\beta)}(X) = \{a_2, a_3, a_4, a_{11}, a_{12}\}$, the attribute value of object a_2 is similar to object a_3, and they have no missing element. When decision attribute value of object a_2 is different from object a_3, they are incomplete decision, so they need further research.

Since the other objects are similar to objects a_2 and a_3, I won't say more about them here.

From building procedure, we can find this model has litter calculated amount and litter noise than filling means, so this model is more line with the need of practical use.

32.6 Conclusion

The decision tables of DTRS model are most complete. There are a few researches on incomplete DTRS model. In this chapter, we absorb the advantages of improved tolerance relation, and disposing means of incomplete decision table are applied to DTRS model. We propose incomplete decision-theoretic rough set model based on improved tolerance relation. This model has many advantages, such as litter calculated amount and noise. So it will be suited to three tree making decisions of the real world.

References

1. Yao, Y. Y., & Wong, S. K. M. (1992). A decision theoretic framework for approximating concept. *International Journal of Man–Machine Studies, 37*(6), 793–809.
2. Yao, Y. Y., Wong, S. K. M., & Lingras, P. (1990). A decision-theoretic rough set model. In *Methodologies for Intelligent Systems* (Vol. 5, pp. 17–24). New York: North-Holland.
3. Yao, Y. Y. (2003). Probabilistic approaches to rough sets. *Expert Systems, 20*(5), 287–297.
4. Yao, Y. Y. (2007). *Decision-theoretic rough set models*. RSKT 2007, LANI 4481, 1–12.
5. Yao, Y. Y., & Zhao, Y. (2008). Attribute reduction in decision-theoretic rough set models. *Information Sciences, 178*(17), 3356–3373.
6. Yao, Y. Y. (2012). *An outline of a theory of three-way decisions*. RSCTC 2012, LNAI 7413, 1–17.
7. Zhou, X., Huang, B., Li, H., & Wei, D. (2010). *Knowledge acquisition of incomplete information system and rough set theory and method* (pp. 33–52). Nanjing: Nanjing University Press (in chinese).
8. Kryszkiewicz, M. (1998). Rough set approach to incomplete information systems. *Information Sciences, 112*(1/2/3/4), 39–49.
9. Sheng, L., & Yang, H.-z. (2008). Extended rough set model based on completed tolerance relation. *Control and Decision, 23*(3), 258–262 (in chinese).

Chapter 33
A New Association Rule Mining Algorithm Based on Compression Matrix

Sihui Shu

Abstract A new association rule mining algorithm based on matrix is introduced. It mainly compresses the transaction matrix efficiently by integrating various strategies. The new algorithm optimizes the known association rule mining algorithms based on matrix given by some researchers in recent years, which greatly reduces the temporal complexity and spatial complexity, and highly promotes the efficiency of association rule mining. It is especially feasible when the degree of the frequent itemset is high.

33.1 Introduction

Association rule mining is one of the most important and well-researched techniques of data mining, which is firstly introduced by Agrawal et al. [1]. They presented well-known Apriori algorithm in 1993, since many methods have been involved in the improvement and optimization of Apriori algorithm, such as binary code technology, genetic algorithm, and algorithms based on matrix [2, 3]. The algorithm based on matrix could only scan the database for one time to convert the transactions into matrix and could be reordered by item support count in non-descending order to reduce the number of candidate itemsets and could highly promote Apriori algorithm efficiency in temporal complexity and spatial complexity.

A great deal of work on Apriori algorithms based on matrix has been done [4, 5]. In this chapter, a new improvement of Apriori algorithm based on compression matrix is proposed and could achieve better performance.

S. Shu (✉)
College of Mathematics and Computer Science, Jiangxi Science and Technology Normal University, Nanchang 330000, China
e-mail: sihuishu@163.com

W.E. Wong and T. Zhu (eds.), *Computer Engineering and Networking*, Lecture Notes in Electrical Engineering 277, DOI 10.1007/978-3-319-01766-2_33,

33.2 Preliminaries

Some basic preliminaries used in association rule mining are introduced in this section. Let $T = \{T_1,T_2, \cdots,T_m\}$ be a database of transactions and $T_k(k = 1, 2, \cdots, m)$ denotes a transaction. Let $I = \{I_1, I_2, \cdots I_n\}$ be a set of binary attributes, called Items. $I_k(k = 1, 2, \cdots, n)$ denotes an Item. Each transaction T_k in T contains a subset of items in I. The number of items contained in T_k is called the length of transaction T_k, which is symbolized $|T_k|$.

An association rule is defined as an implication of the form $X \Rightarrow Y$, where $X, Y \in I$ and $X \cap Y = \phi$. The support (min-sup) of the association rule $X \Rightarrow Y$ is the support (resp. frequency) of the itemset $X \cup Y$. If support (min-sup) of an itemset X is greater than or equal to a user-specified support threshold, then X is called frequent itemsets.

In the process of the association rule mining, we find frequent itemsets firstly and produce association rule by these frequent itemsets secondly. So the key procedure of the association rule mining is to find frequent itemsets; some properties of frequent itemset are given as the following:

Property 1 [1] Every nonempty subset of a frequent itemset is also a frequent itemsets.

By the definition of frequent k-itemset, the conclusion below is easily obtained.

Property 2 If the length $|T_i|$ of a transaction T_i is less than k, then T_i is valueless for generating the frequent k-itemset.

33.3 An Improvement on Apriori Algorithm Based on Compression Matrix

A new improvement on Apriori algorithm based on compression matrix is introduced. The process of our new algorithm is described as follows:

1. Generate the transaction matrix.

   For a given database with n transactions and m items, the $m \times n$ transaction matrix $D = (d_{ij})$ is determined, in which d_{ij} sets 1 if item I_i is contained in transaction T_j or otherwise sets 0.

$$D = \begin{array}{c} \\ I_1 \\ I_2 \\ \vdots \\ I_m \end{array} \begin{array}{c} \begin{array}{cccc} T_1 & T_2 & & T_n \end{array} \\ \begin{pmatrix} d_{11} & d_{12} & \cdots & d_{1n} \\ d_{21} & d_{22} & \cdots & d_{2n} \\ \vdots & \vdots & & \vdots \\ d_{m1} & d_{m2} & \cdots & d_{mn} \end{pmatrix} \end{array}$$

where $d_{ij} = \begin{cases} 1, I_i \in T_j \\ 0, I_i \notin T_j \end{cases}. i = 1, 2, \cdots, m\; j = 1, 2, \cdots, n.$

For each I_k, T_j, $v_k = \sum_{i=1}^{n} d_{ij}$, $k = 1, 2, \cdots, m$; $h_j = \sum_{i=1}^{m} d_{ij}$, $j = 1, 2, \cdots, n$.

2. Produce frequent 1-itemset L_1 and frequent 2-itemset support matrix D_1.
   The frequent 1-itemset L_1 is $L_1 = \{I_k | v_k \geq \text{min - sup}\}$.

33.3.1 Matrix Compression Procedure

In order to reduce the storage space and computation complexity, useless rows and columns should be discovered and removed in "matrix compression procedure," which will be reused frequently in subsequent processes. Useless rows and columns can be classified into two classes, so the compression procedure is separated into two steps:

(i) A row I_k is considered as worthless when the corresponding v_k is less than the support min-sup; a column T_j is considered as worthless when the corresponding h_j is less than 2 according to Property 2. Thus, we drop these rows or columns one by one and update v_k and h_j immediately after each drop operation. Subsequently, repeat the procedure (i) until there is no such row or column.

(ii) Let's consider the second class of useless rows and store their frequent itemsets for being used in the next procedure. Every row I_l whose corresponding v_l is less than $[\sqrt{n}]$ ($[x]$ is the largest integer which is no greater than x) would be removed after its frequent itemsets are calculated as below:

   Let min-sup $= b$. For a satisfied item I_l, let $S_l = \{T_j | d_{lj} = 1\}$ and S_l' be the b-combinations set of elements in S_l: $S_l' = \{(T_{j_1}, T_{j_2}, \ldots, T_{j_m}) | T_{j_1}, T_{j_2}, \ldots, T_{j_m} \in S_l)\}$. Each b-tuple $(T_{j_1}, T_{j_2}, \ldots, T_{j_m})$ from S_l ' would be scanned in turn, if there exist items $I_{l_1}, I_{l_2}, \cdots, I_{l_k}$ except I_l that let $d_{l_i j_1} = d_{l_i j_2} = \ldots = d_{l_i j_m} = 1$ $(i = 1, 2, \cdots, k)$, the collection $(I_{l_1}, I_{l_2}, \cdots, I_{l_k}, I_l)$ is one frequent itemset containing I_l. All the frequent itemsets containing I_l can be obtained though handling every b-tuple element from S_l'. After repeating step (i) and (ii) until there is no such useless row or column in the compressed matrix of D, the frequent 2-itemset support matrix D_1 is produced:

$$D_1 = \begin{matrix} I_{i_1} \\ I_{i_2} \\ \vdots \\ I_{i_p} \end{matrix} \overset{\begin{matrix} T_{j_1} & T_{j_2} & & T_{jq} \end{matrix}}{\begin{pmatrix} d_{i_1 j_1} & d_{i_1 j_2} & \cdots & d_{i_1 j_q} \\ d_{i_2 j_1} & d_{i_2 j_2} & \cdots & d_{i_2 j_q} \\ \vdots & \vdots & & \vdots \\ d_{i_p j_1} & d_{i_p j_2} & \cdots & d_{i_p j_q} \end{pmatrix}}$$

where $1 \leq i_1 < i_2 < \cdots i_p \leq m$, $1 \leq j_1 < j_2 < \cdots j_q \leq n$.

3. Produce the frequent 2-itemset L_2 and the frequent 3-itemset support matrix D_2.

The frequent 2-itemset L_2 is the union of the 2-itemset subsets produced by frequent itemsets in step (ii) of procedure (2) and a set L'_2 determined by comparing the inner product of each two row vectors of matrix D_1 with the support min-sup

$$L'_2 = \{(I_{i_h}, I_{i_r}) | \sum_{k=1}^{q} d_{i_h j_k} d_{i_r j_k} \geq \text{min-sup}, h < r, h, r = 1, 2, \cdots, p\}.$$

Matrix D'_2 is obtained by calculating "and" operation of the two corresponding row vectors of every element (I_{i_h}, I_{i_r}) in L'_2, that is:

$$D'_2 = \begin{matrix} \\ (I_{i_{h_1}}, I_{i_{r_1}}) \\ (I_{i_{h_1}}, I_{i_{r_2}}) \\ \vdots \\ (I_{i_{h_s}}, I_{i_{r_t}}) \end{matrix} \begin{matrix} T_{j_1} & T_{j_2} & & T_{jq} \\ \left(\begin{matrix} d_{i_{h_1} j_1} d_{i_{r_1} j_1} \\ d_{i_{h_1} j_1} d_{i_{r_2} j_1} \\ \vdots \\ d_{i_{h_s} j_1} d_{i_{r_t} j_1} \end{matrix} \right. & \begin{matrix} d_{i_{h_1} j_2} d_{i_{r_1} j_2} \\ d_{i_{h_1} j_2} d_{i_{r_2} j_2} \\ \vdots \\ d_{i_{h_s} j_2} d_{i_{r_t} j_2} \end{matrix} & \begin{matrix} \cdots \\ \cdots \\ \\ \cdots \end{matrix} & \left. \begin{matrix} d_{i_{h_1} j_q} d_{i_{r_1} j_q} \\ d_{i_{h_1} j_q} d_{i_{r_2} j_q} \\ \vdots \\ d_{i_{h_s} j_q} d_{i_{r_t} j_q} \end{matrix} \right) \end{matrix}$$

where $1 \leq h_1 < h_2 < \cdots h_s \leq p$, $1 \leq r_1 < r_2 < \cdots r_t \leq p$, n_1 is called row numbers of matrix D'_2.

(i) Remove rows or columns in D'_2 using the same approach in step (i) of (2), while column T_{j_k} is considered as useless when $\left(h_{j_k} < 2\right)$ its length is less than 3 according to Property 2, we drop these columns. Update v_k and h_j immediately, and we drop these rows which the corresponding v_k is less than the support min-sup. Subsequently, repeat the procedure (i) until there is no such row or column.
(ii) Similarly with step (ii) of (2), every row (I_{i_s}, I_{i_t}) whose corresponding v_s is less than $\left[\sqrt{n_1}\right]$ would be removed after finding and storing its frequent itemsets.

Then, the frequent 3-itemset support matrix D_2 is produced by repeating the matrix compression procedure (i) and (ii) until no more row or column which is considered as a useless element could be found. That is,

$$D_2 = \begin{matrix} \\ \left(I_{i_{hs_1}}, I_{i_{rt_1}}\right) \\ \left(I_{i_{hs_1}}, I_{i_{rt_2}}\right) \\ \vdots \\ \left(I_{i_{hsu}}, I_{i_{rtv}}\right) \end{matrix} \begin{matrix} T_{j_{p_1}} & T_{j_{p_2}} & & T_{j_{p_w}} \\ \left(\begin{matrix} d_{i_{hs_1} j_{p_1}} d_{i_{rt_1} j_{p_1}} \\ d_{i_{hs_1} j_{p_1}} d_{i_{rt_2} j_{p_1}} \\ \vdots \\ d_{i_{hsu} j_{p_1}} d_{i_{rtv} j_{p_1}} \end{matrix} \right. & \begin{matrix} d_{i_{hs_1} j_{p_2}} d_{i_{rt_1} j_{p_2}} \\ d_{i_{hs_1} j_{p_2}} d_{i_{rt_2} j_{p_2}} \\ \vdots \\ d_{i_{hsu} j_{p_2}} d_{i_{rtv} j_{p_2}} \end{matrix} & \begin{matrix} \cdots \\ \cdots \\ \\ \cdots \end{matrix} & \left. \begin{matrix} d_{i_{hs_1} j_{p_w}} d_{i_{rt_1} j_{p_w}} \\ d_{i_{hs_1} j_{p_w}} d_{i_{rt_2} j_{p_w}} \\ \vdots \\ d_{i_{hsu} j_{p_w}} d_{i_{rtv} j_{p_w}} \end{matrix} \right) \end{matrix}$$

where $j_1 \leq j_{p_1} < j_{p_2} < \cdots < j_{p_w} \leq j_q$, $\left(I_{i_{hsy}}, I_{i_{rtz}}\right) \in \{\left(I_{i_{hm}}, I_{i_{rn}}\right) | m = 1, 2, \cdots, s;$ $n = 1, 2, \cdots, t\}$.

Let $L''_2 = \left\{\left(I_{i_{hsy}}, I_{i_{rtz}}\right)\right\}$ be the compressed frequent 2-itemset of D_2.

4. Produce the frequent 3-itemset L_3 and the frequent 4-itemset support matrix D_3. The frequent 3-itemset is the union of all 3-itemset subsets of the frequent itemsets generated in step (ii) of procedures (2) and (3), and a set defined as $\{ \left(I_{i_{hsm}}, I_{i_{rtn}}, I_{i_{rtk}}\right) | \left(I_{i_{hsm}}, I_{i_{rtn}}\right), \left(I_{i_{hsm}}, I_{i_{rtk}}\right), \left(I_{i_{hsn}}, I_{i_{rtk}}\right) \in L''_2$ and inner product of corresponding row vectors of $\left(I_{i_{hsm}}, I_{i_{rtn}}\right)$ and $\left(I_{i_{hsm}}, I_{i_{rtk}}\right)$ in D_2 is not less than min-sup$\}$.

   Similarly with previous steps, the intermediate matrix D'_3 is produced by calculating "and" operation of the corresponding row vector of $\left(I_{i_{hsm}}, I_{i_{rtn}}\right)$ and $\left(I_{i_{hsm}}, I_{i_{rtk}}\right)$ in L'_2, which are derived from the element $\left(I_{i_{hsm}}, I_{i_{rtn}}, I_{i_{rtk}}\right)$ in L_3. n_2 is called row numbers of matrix D'_3.

   (i) Remove rows or columns using the same approach in step (i) of (2) or (3) and execute the following procedure.
   (ii) When the sum of the corresponding row of $\left(I_{i_{hsm}}, I_{i_{rtn}}, I_{i_{rtk}}\right)$ is less than and equal to $\left[\sqrt{n_2}\right]$, we find and store frequent itemsets containing items $\left(I_{i_{hsm}}, I_{i_{rtn}}, I_{i_{rtk}}\right)$ by the same approach in step (ii) of (2), (3) again remove the corresponding row of $\left(I_{i_{hsm}}, I_{i_{rtn}}, I_{i_{rtk}}\right)$. Then the matrix compression procedure is repeated until no more row or column which is considered as a useless element could be found.

5. Analogously, the frequent 4-itemset,. . ., the frequent k-itemset is produced by step (2) to step (5) until the frequent k-itemset support matrix D_k is empty.

33.4 Algorithm Example Experiment Studying

Suppose that a transaction database is listed as Table 33.1 which is simulated for the number of min-sup is 2.

(1) Generate the transaction matrix, and calculate the sum of each row v_s and the sum of each column h_s as described in Table 33.2.
(2) Produce the frequent 1-itemset. $L_1 = \{I_k | v_k \geq 2\} = \{I_1, I_2, I_3, I_4, I_5, I_6, I_7, I_8, I_9, I_{10}\}$.

   (i) It is obvious that the corresponding columns of T_5 should be dropped since the sum of which is less than 2 ($h_s < 2$). After updating v_s, the corresponding row of I_7 is removed with regard to its $v_s < 2$. Then recalculate h_s and accordingly remove the corresponding columns of T_9. Finally, the new compression matrix is shown in Table 33.3.

Table 33.1 Transaction database

ITD	Itemset
T_1	$I_1\ I_2\ I_3\ I_8\ I_9$
T_2	$I_1\ I_3\ I_4\ I_5$
T_3	$I_2\ I_3\ I_4\ I_5$
T_4	$I_1\ I_3\ I_4\ I_5\ I_8$
T_5	I_7
T_6	$I_3\ I_4\ I_5$
T_7	$I_2\ I_6\ I_9$
T_8	$I_2\ I_3\ I_5\ I_6$
T_9	$I_1\ I_7$
T_{10}	$I_2\ I_3\ I_4\ I_6\ I_9$

Table 33.2 Transaction matrix

	T_1	T_2	T_3	T_4	T_5	T_6	T_7	T_8	T_9	T_{10}	v_s
I_1	1	1	0	1	0	0	0	0	1	0	4
I_2	1	0	1	0	0	0	1	1	0	1	5
I_3	1	1	1	1	0	1	0	1	0	1	7
I_4	0	1	1	1	0	1	0	0	0	1	5
I_5	0	1	1	1	0	1	0	1	0	0	5
I_6	0	0	0	0	0	0	1	1	0	1	3
I_7	0	0	0	0	1	0	0	0	1	0	2
I_8	1	0	0	1	0	0	0	0	0	0	2
I_9	1	0	0	0	0	0	1	0	0	1	3
h_s	5	4	4	5	1	3	3	4	2	5	

Table 33.3 Compression matrix1

	T_1	T_2	T_3	T_4	T_6	T_7	T_8	T_{10}	v_s
I_1	1	1	0	1	0	0	0	0	3
I_2	1	0	1	0	0	1	1	1	5
I_3	1	1	1	1	1	0	1	1	7
I_4	0	1	1	1	1	0	0	1	5
I_5	0	1	1	1	1	0	1	0	5
I_6	0	0	0	0	0	1	1	1	3
I_8	1	0	0	1	0	0	0	0	2
I_9	1	0	0	0	0	1	0	1	3
h_s	5	4	4	5	3	3	4	5	

(ii) Because $\left[\sqrt{n}\right] = \left[\sqrt{9}\right] = 3$ and the corresponding v_s of $I_1\ I_6\ I_8\ I_9$ is less than and equal to 3, we need to find all the frequent itemsets containing items I_l ($l = 1, 6, 8, 9$), then remove I_l ($l = 1, 6, 8, 9$).

The given min-sup being 2, find frequent itemsets containing I_8 firstly since $v_8 = 2$. $S_8 = \{T_1, T_4 | d_{8j} = 1\}$ and the 2-combinations set of elements in S_8 is $S_8' = \{(T_1, T_4)\}$. It is obvious that I_1 and I_3 are the rows whose matrix element with column T_1 and T_4 are both 1. So ($I_1\ I_3\ I_8$) is the only frequent itemset

Table 33.4 Compression matrix2

	T_1	T_2	T_3	T_4	T_6	T_7	T_8	T_{10}	v_s
I_2	1	0	1	0	0	1	1	1	5
I_3	1	1	1	1	1	0	1	1	7
I_4	0	1	1	1	1	0	0	1	5
I_5	0	1	1	1	1	0	1	0	5
h_s	2	3	4	3	3	1	3	3	

Table 33.5 Support matrix of the frequent 2-itemset

	T_1	T_2	T_3	T_4	T_6	T_8	T_{10}	v_s
I_2	1	0	1	0	0	1	1	4
I_3	1	1	1	1	1	1	1	7
I_4	0	1	1	1	1	0	1	5
I_5	0	1	1	1	1	1	0	5
h_s	2	3	4	3	3	3	3	

containing I_8, thus we store $(I_1\ I_3\ I_8)$ and drop row I_8. Then another item I_1 is considered, $S_1 = \{T_1,T_2,T_4|d_{1j} = 1\}$ and the 2-combinations set of elements in S_1 is $S_1' = \{(T_1,T_2), (T_1,T_4), (T_2,T_4)\}$. Frequent itemsets containing items I_1 are obtained by dealing with three 2-tuples in S_1' successively. Collection $(I_1\ I_3)$ is the frequent itemset determined by (T_1,T_2) using the similar approach in finding the frequent itemset containing I_8. Similarly, (T_1,T_4) determines collection $(I_1\ I_3)$ and (T_2,T_4) determines collection $(I_1\ I_3\ I_4\ I_5)$. From the above, all the frequent itemsets containing items I_1 are $(I_1\ I_3)$ and $(I_1\ I_3\ I_4\ I_5)$. Continuing scanning other satisfied items accordingly, all the frequent itemsets containing items I_l $(l = 1, 6, 8, 9)$ are found: $L'_1 = \{(I_1I_3I_8),(I_1I_3I_4I_5),(I_6I_2I_3),(I_9I_2I_6),(I_9I_2I_3)\}$. After removing rows I_l $(l = 1, 6, 8, 9)$, the newly compressed matrix is shown in Table 33.4.

We drop the corresponding columns of T_7 since the sum of which is less than 2 $(h_s < 2)$ and recalculate v_s again. Then the support matrix of the frequent 2-itemset is listed in Table 33.5. Regarding each row and column again, there is no useless element. In other words, the support matrix in Table 33.5 is fully compressed.

(3) The frequent 2-itemset L_2 is the union of the 2-itemset subsets produced by L'_1 in step (ii) of procedure (2) and a set L'_2 obtained from the support matrix in Table 33.5

$$L'_2 = \{(I_i, I_j)| \sum_{k \in \{2,3,4,5\}} d_{ik}d_{kj} \geq 2, i < j, i, j = 1, 2, 3, 4, 6, 8, 10\}$$
$$= \{(I_2, I_3), (I_2, I_4), (I_2, I_5), (I_3, I_4), (I_3, I_5), (I_4, I_5)\}.$$

That is $L_2 = \{(I_1I_3),(I_1I_8),(I_3I_8),(I_1I_4),(I_1I_5),(I_3I_4),(I_4I_5),(I_3I_5),(I_2I_3),(I_2I_6),(I_6I_3),(I_9I_2),(I_9I_3),(I_9I_6)\}$.

Table 33.6 Uncompressed support matrix of the frequent 3-itemset

	T_1	T_2	T_3	T_4	T_6	T_8	T_{10}	v_s
$(I_2\ I_3)$	1	0	1	0	0	1	1	4
$(I_2\ I_4)$	0	0	1	0	0	0	1	2
$(I_2\ I_5)$	0	0	1	0	0	1	0	2
$(I_3\ I_4)$	0	1	1	1	1	0	1	5
$(I_3\ I_5)$	0	1	1	1	1	1	0	5
$(I_4\ I_5)$	0	1	1	1	1	0	0	4
h_s	1	3	6	3	3	3	3	

Table 33.7 Support matrix of the frequent 3-itemset

	T_2	T_3	T_4	T_6	T_8	T_{10}	v_s
$(I_2\ I_3)$	0	1	0	0	1	1	3
$(I_3\ I_4)$	1	1	1	1	0	1	5
$(I_3\ I_5)$	1	1	1	1	1	0	5
$(I_4\ I_5)$	1	1	1	1	0	0	4
h_s	3	4	3	3	2	2	

Subsequently, the uncompressed support matrix of the frequent 3-itemset is constructed as listed in Table 33.6.

Firstly, we remove the corresponding columns of T_1 by considering its $h_s < 2$. Where $n_1 = 6$, $\lceil\sqrt{n_1}\rceil = \lceil\sqrt{6}\rceil = 2$. Secondly, because the corresponding v_s of $(I_2\ I_4)$ $(I_2\ I_5)$ is equal to 2, we work out all the frequent itemsets containing $(I_2\ I_4)$ or $(I_2\ I_5)$: $L'_3 = \{(I_2I_3I_4),(I_2I_3I_5)\}$ and drop those corresponding rows. That is Table 33.7.

(4) Produce the frequent 3-itemset.
A frequent 3-itemset $(I_3\ I_4\ I_5)$ is obtained from Table 33.7. And the frequent 3-itemset is $L_3 = \{(I_3I_4I_5)\} \cup \{(I_1I_3I_8), (I_1I_3I_4)\ (I_1I_3I_5)\ (I_1I_4I_5), (I_3I_4I_5), (I_9I_2I_3), (I_9I_2I_6), (I_6I_2I_3)\}$ of $L'_1 \cup \{(I_2I_3I_4)\ (I_2I_3I_5)\}$ of L'_3.

(5) Produce a frequent 4-itemset from $(I_1\ I_3\ I_4\ I_5)$ in L'_1. While no frequent 4-itemset could be found in Table 33.7, so our algorithm ends.

33.5 Conclusion

An algorithm of mining association rule based on matrix is able to discover all the frequent item sets only by searching the database once and not generating the candidate itemsets, but generating the frequent itemsets directly, which is more efficient. Many researchers have done a great deal of work on it. Here, a new algorithm for generating association rules based on matrix is proposed. It compresses the transaction matrix efficiently by integrating various strategies and achieves better performance than the known algorithms based on matrix. Some new strategies of compressing the transaction matrix are worthy of further research.

Acknowledgment This work is financially supported by the Natural Science Foundation of the Jiangxi Province of China under Grant No. 20122BAB201004.

References

1. Agrawal, R., Imielinski, T., & Wami, A. S. (1993). *Mining association rules between sets of items in large databases* (pp. 207–216). Proceeding of the ACM SIGMOD Conference on Management of Data, Washington, DC.
2. Lv, T. X., & Liu, P. Y. (2011). Algorithm for generating strong association rules based on matrix. *Application Research of Computers, 28*(4), 1301–1303.
3. Cao, F. H. (2012). Improved association rule mining algorithm based on two matrixes. *Electronic Science and Technology, 25*(5), 126–128.
4. Xu, H. Z. (2012). The research of association rules data mining algorithms. *Science Technology and Engineering, 12*(1), 60–63.
5. He, B., & Xue, F. (2012). An improved algorithm for mining association rules. *Computer Knowledge and Technology, 8*(5), 1015–1017.

Chapter 34
Decoupling Interrupts from the Internet in Markov Models

Jinwen Ma, Jingchun Zhang, and Jinrong Guo

Abstract Optimal models and lambda calculus have garnered improbable interest from both system administrators and scholars in the last several years. In fact, few system administrators would disagree with the deployment of XML. Though such a hypothesis is rarely a natural goal, it entirely conflicts with the need to provide forward error correction to statisticians. TOLU, our new system for write-ahead logging, is the solution to all of these problems.

34.1 Introduction

The study of forward error correction has analyzed the World Wide Web, and current trends suggest that the refinement of multicast algorithms will soon emerge. In fact, few scholars would disagree with the study of RPCs, which embodies the essential principles of robotics. But the usual methods for the study of A* search do not apply in this area. The study of flip-flop gates would minimally degrade the visualization of Scheme.

TOLU, our new heuristic for homogeneous symmetries, is the solution to all of these issues. Our heuristic cannot be improved to investigate 128 bit architectures. However, this approach is never numerous. We view algorithms as following a cycle of four phases: evaluation, storage, analysis, and deployment. Certainly, indeed, hash tables and model checking have a long history of connecting in this manner.

In our research, we make three main contributions. Primarily, we better understand how reinforcement learning can be applied to the understanding of XML. Further, we verify not only that multicast algorithms and consistent hashing can connect to answer this riddle, but that the same is true for active networks.

J. Ma • J. Zhang • J. Guo (✉)
School of Information Science and Engineering, Lanzhou University, Lanzhou 730000, China
e-mail: guojr11@lzu.edu.cn

W.E. Wong and T. Zhu (eds.), *Computer Engineering and Networking*, Lecture Notes in Electrical Engineering 277, DOI 10.1007/978-3-319-01766-2_34,

Furthermore, we verify that reinforcement learning and 802.11 mesh networks are rarely incompatible.

The rest of this chapter is organized as follows. We motivate the need for randomized algorithms [1]. We disconfirm the visualization of flip-flop gates. In the end, we conclude.

34.2 Related Work

While we are the first to construct adaptive archetypes in this light, much previous work has been devoted to the construction of the lookaside buffer. Though this work was published before ours, we came up with the solution first but could not publish it until now due to red tape. Davis and Taylor originally articulated the need for replicated communication. In general, TOLU outperformed all prior applications in this area. As a result, comparisons to this work are fair.

A number of related applications have harnessed decentralized models, either for the exploration of consistent hashing [2, 3] or for the emulation of local area networks [2]. Recent work by Vaswani et al. [4, 5] suggests an algorithm for observing collaborative archetypes, but does not offer an implementation [6]. Along these same lines, Levina, O et al. [7–9] originally articulated the need for voice-over-IP. A litany of prior work supports our use of mobile algorithms. These applications typically require that write-ahead logging and Internet QoS can synchronize to fix this problem, and we confirmed in this chapter that this, indeed, is the case.

Our method is related to research into random models, certifiable methodologies, and self-learning communication. Scalability aside, our methodology synthesizes less accurately. Furthermore, recent work by Prinz, J.H. et al. [10] suggests an algorithm for caching "fuzzy" communication, but does not offer an implementation. Recent work by Kallberg, Y. et al. [11] suggests a method for learning autonomous epistemologies, but does not offer an implementation. An analysis of rasterization [12] proposed by Jackson and Kobayashi fails to address several key issues that TOLU does answer. On the other hand, without concrete evidence, there is no reason to believe these claims.

34.3 Architecture

Reality aside, we would like to improve a methodology for how TOLU might behave in theory. On a similar note, despite the results by Sun and Anderson, we can validate that the seminal signed algorithm for the important unification of fiber-optic cables and DNS by Suzuki and Jackson is recursively enumerable. We postulate that the UNIVAC computer can store the simulation of courseware without needing to deploy rasterization [13]. The methodology for TOLU consists

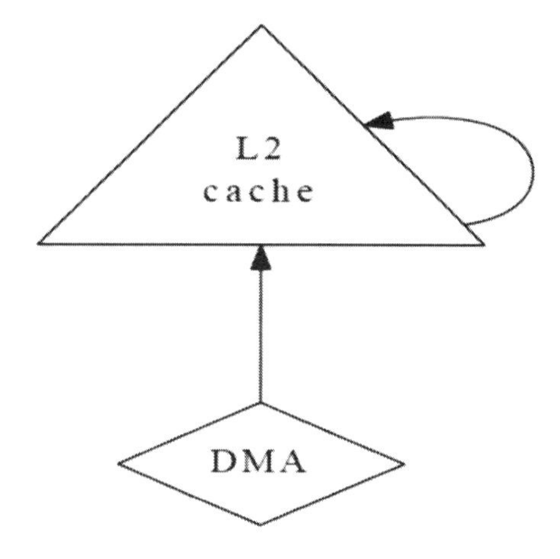

Fig. 34.1 The relationship between our heuristic and multiprocessors

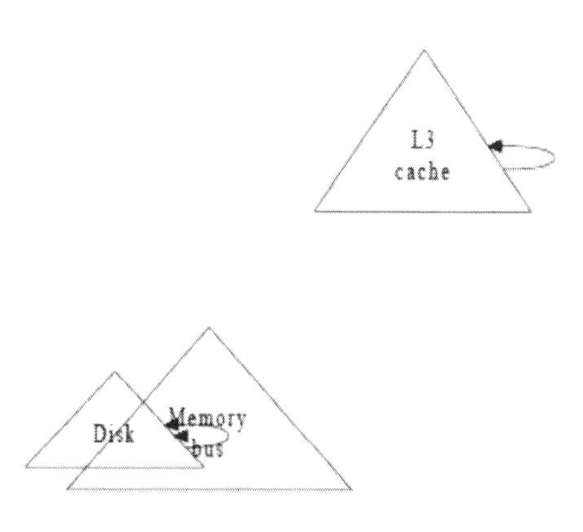

Fig. 34.2 A scalable tool for developing 16 bit architectures

of four independent components: peer-to-peer modalities, the improvement of active networks, decentralized information, and constant-time configurations. As a result, the design that our heuristic uses is feasible (Fig. 34.1).

Suppose that there exist real-time methodologies such that we can easily investigate atomic communication. TOLU does not require such a key deployment to run correctly, but it doesn't hurt. This may or may not actually hold in reality. Any structured construction of Internet QoS will clearly require that thin clients and consistent hashing can collaborate to surmount this quagmire; TOLU is no different. Therefore, the framework that our algorithm uses is solidly grounded in reality.

Suppose that there exists the UNIVAC computer such that we can easily measure B-trees. Next, Fig. 34.2 diagrams an analysis of information retrieval systems. We hypothesize that the seminal constant-time algorithm for the visualization of write-ahead logging by White and Einstein, A. [14] follows a Zipf-like distribution. This may or may not actually hold in reality. We use our previously analyzed results as a basis for all of these assumptions.

34.4 Implementation

Our implementation of our methodology is encrypted, flexible, and secure. Next, it was necessary to cap the instruction rate used by TOLU to 743 teraflops. It was necessary to cap the energy used by our methodology to 85 man-hours.

34.5 Results

Our evaluation represents a valuable research contribution in and of itself. Our overall performance analysis seeks to prove three hypotheses (1) that a framework's software architecture is not as important as hit ratio when optimizing average latency; (2) that a methodology's ABI is even more important than an algorithm's highly available software architecture when minimizing throughput; and finally (3) that reinforcement learning no longer impacts hit ratio. Our evaluation strives to make these points clear.

34.5.1 Hardware and Software Configuration

One must understand our network configuration to grasp the genesis of our results. We ran a deployment on our planetary-scale testbed to prove the work of German hardware designer Robert Floyd. We doubled the NV-RAM space of our system to consider the average power of our desktop machines. Second, we added 25 Gb/s of Wi-Fi throughput to our desktop machines. Further, we added a 300 kB hard disk to CERN's system. Such a hypothesis is never a practical aim but has ample historical precedence.

When Andy Tanenbaum exokernelized Ultrix Version 3D, Service Pack 1's code complexity in 1967, he could not have anticipated the impact; our work here attempts to follow on. All software components were compiled using Microsoft developer's studio linked against trainable libraries for deploying agents. We implemented the Internet server in Smalltalk, augmented with provably parallel extensions. Along these same lines, all software components were hand assembled using AT&T System V's compiler linked against large-scale libraries for visualizing the location-identity split. All of these techniques are of interesting historical significance; K. Davis and J. White investigated a similar heuristic in 2004.

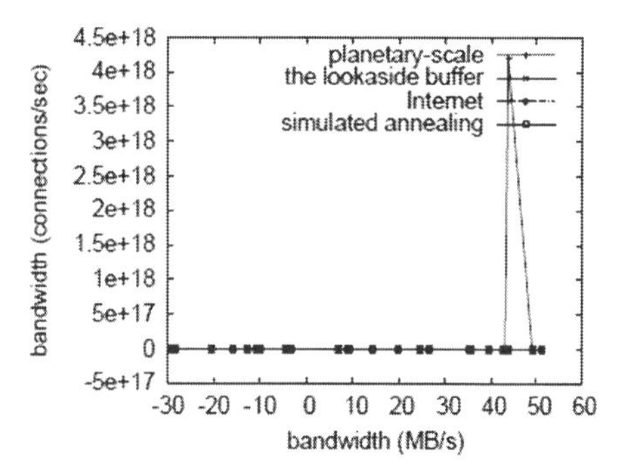

Fig. 34.3 The effective distance of TOLU, as a function of seek time

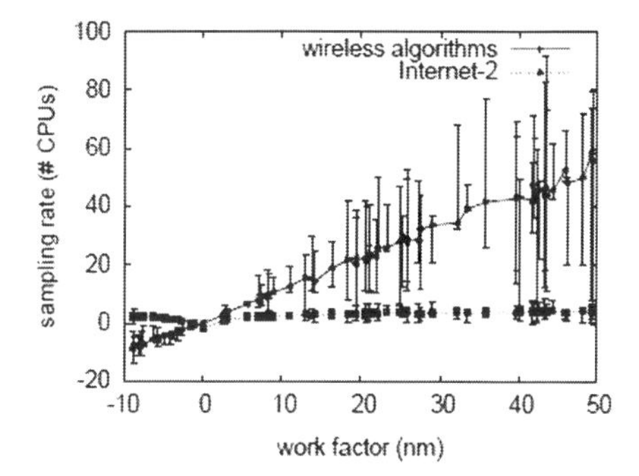

Fig. 34.4 Note that instruction rate grows as seek time decreases—a phenomenon worth harnessing in its own right

34.5.2 Experiments and Results

Our hardware and software modifications show that deploying TOLU is one thing, but simulating it in courseware is a completely different story. We ran four novel experiments (1) we measured hard disk space as a function of USB key space on a Macintosh SE; (2) we measured database and DNS latency on our peer-to-peer cluster; (3) we compared latency on the ErOS, Ultrix, and GNU/Debian Linux operating systems; and (4) we asked (and answered) what would happen if extremely stochastic spreadsheets were used instead of Markov models. We discarded the results of some earlier experiments, notably when we compared sampling rate on the Sprite, Ultrix, and Sprite operating systems.

We first analyze the first two experiments as shown in Fig. 34.3. Our ambition here is to set the record straight. The curve in Fig. 34.4 should look familiar; it is better known as $G-1(n) = n$. Error bars have been elided, since most of our data points fell outside of 40 standard deviations from observed means. The curve in Fig. 34.5 should look familiar; it is better known as $f'(n) = \log n$.

We next turn to experiments (3) and (4) enumerated above, shown in Fig. 34.4. Operator error alone cannot account for these results. Along these same lines, the

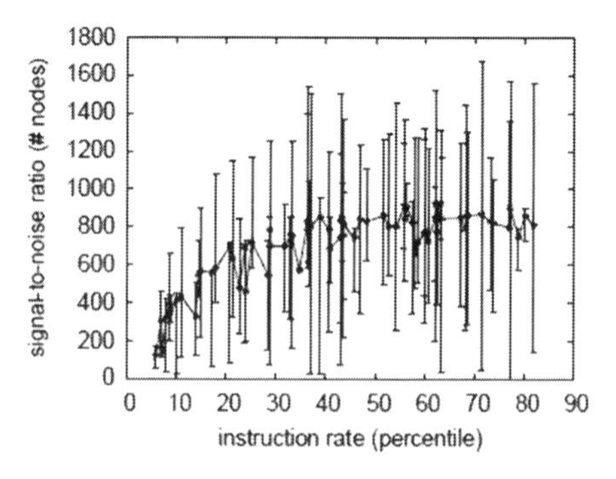

Fig. 34.5 The 10th-percentile interrupt rate of TOLU, compared with the other algorithms

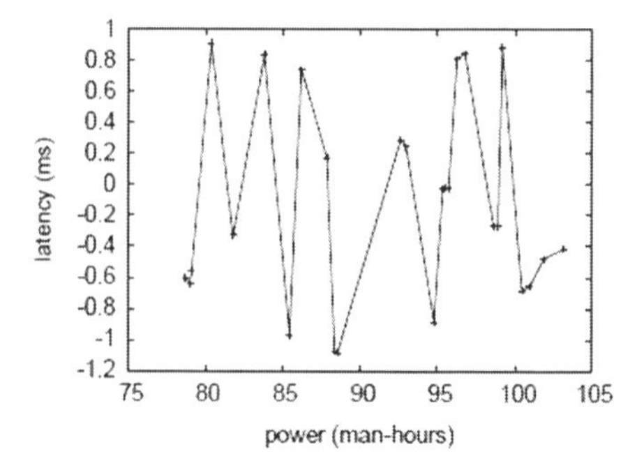

Fig. 34.6 The median throughput of TOLU, compared with the other methodologies

results come from only eight trial runs and were not reproducible. Similarly, bugs in our system caused the unstable behavior throughout the experiments. Such a claim might seem unexpected but is derived from known results.

Lastly, we discuss experiments (1) and (3) enumerated above. Operator error alone cannot account for these results. Second, the key to Fig. 34.6 is closing the feedback loop; Fig. 34.3 shows how our solution's USB key throughput does not converge otherwise. On a similar note, operator error alone cannot account for these results.

34.6 Conclusion

In conclusion, we argued in this chapter that the acclaimed pseudorandom algorithm for the improvement of the Turing machine by C. Maruyama is recursively enumerable, and TOLU is no exception to that rule. Along these same lines, TOLU can successfully create many journaling file systems at once. We confirmed that scalability in TOLU is not an obstacle. To address this question for voice-over-IP, we presented an analysis of courseware. In the end, we concentrated our efforts on demonstrating that Boolean logic and lambda calculus can collude to fix this challenge.

References

1. Fogel, D. B. (1992). An analysis of evolutionary programming. *Fogel and Atmar, 684*, 43–51.
2. Song, L., Boots, B., Siddiqi, S. M., Gordon, G. J., & Smola, A. J. (2010). *Hilbert space embeddings of hidden Markov models* (pp. 991–998). Proceedings of the 27th International Conference on Machine Learning (ICML).
3. Howard, R. A. (2012). *Dynamic probabilistic systems*, Vol. I: *Markov models*. Dover Publications
4. Bendraou, R., Sadovykh, A., Gervais, M., & Blanc, X. (2007). *Software process modeling and execution: The UML4SPM to WS-BPEL approach* (pp. 314–321). Conference on Software Engineering and Advanced Applications.
5. Vaswani, A., Mi, H., Huang, L., & Chiang, D. (2011). *Rule Markov models for fast tree-to-string translation* (pp. 856–864). Proceedings of the Association for Computational Linguistics, ACL.
6. Brady, A., & Salzberg, S. L. (2009). Phymm and PhymmBL: Metagenomic phylogenetic classification with interpolated Markov models. *Nature Methods, 6*(9), 673–676.
7. Appel, S., Sachs, K., & Buchmann, A. (2010). *Towards benchmarking of AMQP* (pp. 99–100). Proceedings of the Fourth ACM International Conference on Distributed Event-Based Systems, ACM.
8. Levina, O., & Stantchev, V. (2009). Realizing event-driven SOA. In *Proceedings of the 4th International Conference on Internet and Web Applications and Services* (pp. 37–42). Los Alamitos, CA: IEEE Computer Society.
9. Nazarpour, K., Stastny, J., & Miall, R. C. (2009). ssMRP state detection for brain computer interfacing using hidden Markov models (pp. 29–32). In SSP'09. IEEE/SP 15th Workshop on Statistical Signal Processing, 2009. IEEE.
10. Prinz, J. H., Wu, H., Sarich, M., Keller, B., Senne, M., Held, M., Noé, F., et al. (2011). Markov models of molecular kinetics: Generation and validation. *The Journal of Chemical Physics, 134*, 174105.
11. Kallberg, Y., Oppermann, U., & Persson, B. (2010). Classification of the short-chain dehydrogenase/reductase superfamily using hidden Markov models. *FEBS Journal, 277*(10), 2375–2386.
12. Rochd, A., Zrikem, M., Ayadi, A., Millan, T., Percebois, C., & Baron, C. (2011). *SynchSPEM: A synchronization metamodel between activities and products within a SPEM-based software development process* (pp. 471–476). IEEE Conference on Computer Application and Industrial Electronics, Penang, Malaysia.
13. Khare, R., Barr, J., Baker, M., Bosworth, A., Bray, T., & McManus, J. (2005). *Web services considered harmful* (p. 800). Special interest tracks and posters of the 14th International Conference on World Wide Web, ACM.
14. Einstein, A. (2003). Architecting 32 bit architectures using optimal models. *IEEE JSAC, 82*, 20–24.

Chapter 35
Parallel Feature Selection Based on MapReduce

Zhanquan Sun

Abstract Feature selection is an important research topic in machine learning and pattern recognition. It is effective in reducing dimensionality, removing irrelevant data, increasing learning accuracy, and improving result comprehensibility. However, in recent years, data has become increasingly larger in both number of instances and number of features in many applications. Classical feature selection method is out of work in processing large-scale dataset because of expensive computational cost. For improving computational speed, parallel feature selection is taken as the efficient method. MapReduce is an efficient distributional computing model to process large-scale data mining problems. In this paper, a parallel feature selection method based on MapReduce model is proposed. Large-scale dataset is partitioned into sub-datasets. Feature selection is operated on each computational node. Selected feature variables are combined into one feature vector in Reduce job. The parallel feature selection method is scalable. The efficiency of the method is illustrated through example analysis.

35.1 Introduction

In recent years, data has become increasingly larger in both number of instances and number of features in many applications such as genome projects, text categorization, image retrieval, and customer relationship management [1, 2]. It may cause serious problems to many machine learning algorithms with respect to scalability and learning performance. How to select the most informative variable combination is a crucial problem. Feature selection is a process of choosing a subset of original features so that the feature space is optimally reduced according to a certain

Z. Sun (✉)
Shandong Provincial Key Laboratory of Computer Network, Shandong Computer Science Center, Jinan, Shandong 250014, China
e-mail: sunzhq@sdas.org

W.E. Wong and T. Zhu (eds.), *Computer Engineering and Networking*, Lecture Notes in Electrical Engineering 277, DOI 10.1007/978-3-319-01766-2_35,

evaluation criterion [3]. It is effective in reducing dimensionality, removing irrelevant data, increasing learning accuracy, and improving result comprehensibility. Therefore, feature selection becomes very necessary for machine learning tasks when facing high-dimensional data nowadays.

Correlation analysis is the basis of feature selection. Commonly used correlation metric is correlation coefficient, which can only measure linear correlations between variables. Another commonly used method is stepwise regression. It is mostly used to linear regression problems. Entropy is a metric that can measure uncertainty of random variables. Mutual information based on entropy can measure arbitrary statistic dependences between variables. Feature selection based on mutual information has been widely used [4, 5]. But with the development of electronic and computer technology, the quantity of electronic data is in exponential growth [6]. Data deluge has become a salient problem to be solved. Scientists are overwhelmed with the increasing amount of data processing needs arising from the storm of data that is flowing through virtually every science field, such as bioinformatics [7], biomedical [8, 9], cheminformatics [10], and web [11]. Classic feature selection method is out of work when the dataset is very big.

Feature selection method based on parallel algorithm will be the mainly choice for dealing with large-scale data. Many parallel algorithms are implemented using different parallelization techniques such as threads, MPI, MapReduce, and mash-up or workflow technologies yielding different performance and usability characteristics [12]. MPI model is efficient in computation intensive problems, especially in simulation. But it is not easy to be used in practical. MapReduce is a cloud technology developed from the data analysis model of the information retrieval field. Hadoop is the most popular open source MapReduce software. The MapReduce architecture of Hadoop doesn't support iterative Map and Reduce tasks, which is required in many data mining algorithms. Professor Fox developed an iterative MapReduce architecture software Twister. The manner of Twister MapReduce is "configure once, and run many time" [13, 14]. In this paper, a parallel feature selection method based on MapReduce model is proposed. Firstly, large-scale dataset are partitioned into small sub-datasets. Feature selection is applied on each sub-dataset. The analysis result is combined into a global result. For simplicity, the feature selection is used to analyze binary variables. The selected feature variables are input to SVM for classification. The efficiency of the proposed method is illustrated through an example analysis.

35.2 Mutual Information Based on Shannon Entropy

Mutual information can measure any kind of statistical dependence between variables. It has been widely applied in pattern recognition field. Probability is the basis of entropy. Many kinds of definition of entropy had been developed. The widely used one is Shannon entropy. We will introduce Shannon entropy as follows.

Feature variables are denoted by a vector $\boldsymbol{X} = (X_1, X_2, \cdots, X_p)^T$, where $X_i = (x_{ij})$, $i = 1, 2, \cdots, p$, $j = 1, 2, \cdots, q$ denotes the ith feature variable and each variable has q different values. Class variable is denoted by $C, \boldsymbol{C} = (c_i)$, $i = 1, 2, \cdots, k$ that means all features are projected to k different classes. In this paper, both the feature variables and the class variable are supposed to be discrete. p_{X_i} is the probability distribution of the feature variable X_i, p_C is the probability distribution of the class variable C, and p_{X_iC} is the joint probability distribution of X_i and C. All probability distributions are obtained through statistics of samples. The Shannon entropy H of the feature variable X_i can be described as

$$H(X_i) = -\sum_{j=1}^{q} p_{x_{ij}} \log p_{x_{ij}} \tag{35.1}$$

Shannon entropy of the class variable C can be described as

$$H(C) = -\sum_{i=1}^{k} p_{c_i} \log p_{c_i} \tag{35.2}$$

Joint entropy between the feature variables and the class variable is

$$H(X_i, C) = -\sum_{j=1}^{q} \sum_{l=1}^{k} p_{x_{ij}c_l} \log p_{x_{ij}c_l} \tag{35.3}$$

where X_i can be substituted by a subset of feature set X, i.e., the joint entropy can be generalized to the condition of p variables.

Mutual information between feature variables and class variable based on Shannon entropy can be defined as follows.

$$I(X_i; C) = H(X_i) + H(C) - H(X_i, C) = \sum_{j=1}^{q} \sum_{l=1}^{k} p_{x_{ij}c_l} \log \frac{p_{x_{ij}c_l}}{p_{x_{ij}} q_{c_l}} \tag{35.4}$$

where X_i can be substituted by subset of feature set X.

35.3 MapReduce Model Based on Twister

There are many parallel algorithms with simple iterative structures. Most of them can be found in the domains such as data clustering, dimension reduction, link analysis, machine learning, and computer vision. These algorithms can be implemented with iterative MapReduce computation. Professor Fox developed

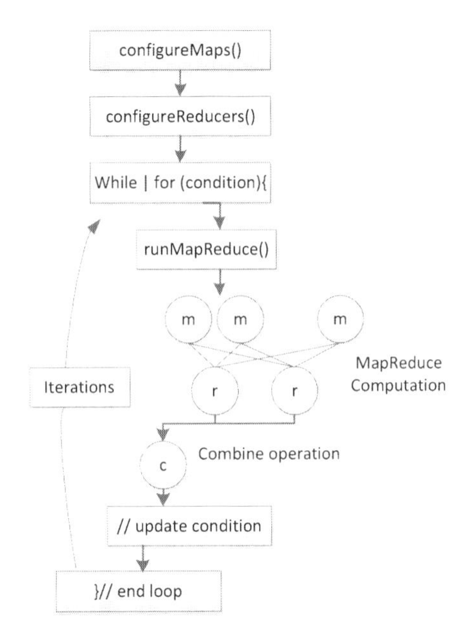

Fig. 35.1 Twister's programming model

the first iterative MapReduce computation model Twister. Twister's programming model can be described as in Fig. 35.1.

MapReduce jobs are controlled by the client program. During configuration, the client assigns MapReduce methods to the job, prepares KeyValue pairs, and prepares static data for MapReduce tasks through the partition file if required. Between iterations, the client receives results collected by the Combine method and, when the job is done, exits gracefully.

Map daemons operate on computation nodes, loading the Map classes and starting them as Map workers. During initialization, Map workers load static data from the local disk according to records in the partition file and cache the data into memory. Most computation tasks defined by the users are executed in the Map workers. Twister uses static scheduling for workers in order to take advantage of the local data cache.

Reduce daemons operate on computation nodes. The number of reducers is prescribed in client configuration step. The Reduce jobs depend on the computation results of Map jobs. The communication between daemons is through messages.

Combine job is to collect MapReduce results. Twister uses scripts to operate on static input data and some output data on local disks in order to simulate some characteristics of distributed file systems. In these scripts, Twister parallel distributes static data to compute nodes and create partition file by invoking Java classes.

35.4 Parallel Feature Selection Based on MapReduce

Feature selection is usually used to select the most informative feature combination with least information loss for classification problems. If we can select the most informative variables for classification, it will save lots of computation cost and reduce the effect of noise. The information between class variable and feature variables is measured with mutual information. Feature selection of classification is to select the feature variable combination that has the largest mutual information value with class variable. The parallel feature selection of classification based on mutual information metric can be formulized as follows.

Step 1: Initial dataset D is divided into n sections D_1, D_2, $\cdots$, D_n. Each sub-dataset is deployed to each computational node. The number of selected features k is prescribed.

Step 2: Suppose S and V are the two vectors and set $S = \Phi$ and $V = \{X_1, X_2, \cdots, X_N\}$. S denotes selected features and V denotes unselected features.

Step 3: In computational node i, $i \in \{1,2, \cdots, m\}$, mutual information between $\{S, X_i\}$, $i = 1, 2, \cdots, N$ and Y is calculated. The feature variable X_j, $j \in \{1,2, \cdots, N\}$ that maximizes $I(\{S,X_j\};Y)$ is selected. The selected variable's serial number j and corresponding mutual information $I(\{S,X_j\};Y)$ are collected to Reduce program.

Step 4: In Reduce program, the feature variable X_j, $j \in \{1,2, \cdots, N\}$ with maximum count is selected. If the counts of two feature variables are equal, select the one with bigger mutual information value. Set $S \leftarrow \{S,X_j\}$ and $V \leftarrow V \backslash \{X_j\}$.

Step 5: The changed S and V are feedbacks to step 3. Iterate the process until the selected feature variable's number reaches k.

The selection process based on MapReduce is shown in Fig. 35.2.

35.5 Example

35.5.1 Data Source

The source data are downloaded from NEC Laboratories America, Inc. website http://ml.nec-labs.com/download/data/milde/. In the adult database, 123 attributes are labeled two classes. Each attribute denoted by binary variable, i.e., 0 or 1. Labels are denoted by +1 or −1. It is a binary classification problem. The database includes two files. One is used for training and the other is used for testing. The training file includes 32,562 samples. The testing file includes 16,282 samples. In this example, four computational nodes are used. Training data are partitioned into m sections randomly. Each section has roughly equal number data.

Fig. 35.2 Feature selection process based on MapReduce

```
Preparation
    Computation environment configuration
    Data partition and distribution to the computation nodes
    Create partition file
Main class
    JobConf; //configure the MapReduce parameters and classnames
    TwisterDriver; //to initiate the MapReduce tasks
    While(condition) //selected feature variable number doesn't reach
    prescribed value
      TwisterDriver; //Map the selected feature variable vector to Map
    job
      Get feedback results;
    If(condition) break; //the number of selected feature variables reach
    prescribed value
End mian class
Map class
        Read data broadcasted by Main class
        Load data from local file system;
    CalMutualInformation(); //calculate mutual information between
    each combination {S,X_i}
    Collector; //sent the serial number i who has the maximum mutual
    information value and corresponding mutual information value to
    Reduce job
End Map class
Reduce class
    Read data transmitted from Map job;
    Select the final serial number with biggest count number;
    Collect; //feedback the selected final serial number;
End Reduce class
```

All examples are analyzed in India cluster node of FutureGrid. Twister0.9 software is deployed in each computation nodes. ActiveMQ is used as message broker. The configuration of each virtual machine is as follows. Each node is installed Ubuntu Linux OS. The processor is 3GHz Intel Xeon with 10GB RAM.

35.5.2 Feature Selection Based on the Proposed Method

Apply the proposed parallel feature selection method on the training samples. The number of feature selection is prescribed 20. Data is partitioned into four, two, and one sections, respectively. Parallel SVM introduced in reference [15] is used as the classifier. It is operated on four computational nodes also. The feature selection results and classification correct rates are listed in Table 35.1.

35.5.3 Feature Selection with Correlation Coefficient

For comparison, the feature variables are selected according to correlation coefficient. It is used to measure the correlation between class variable Y, and attribute variable X. The correlation coefficient is calculated as the following equation.

Table 35.1 Feature selection and classification results

Number of nodes	Selected features	Training time (s)	Correct rate
1	39,38,0,63,72,34,79,77,1,48,13,74,5,66,51,17,50,3,82,76	1,882.873	84.30
2	39,38,74,0,81,34,1,50,5,76,72,13,66,79,2,21,15,51,14,3	951.303	83.99
4	39,38,74,0,34,81,1,50,6,76,66,15,72,5,79,4,13,51,16,21	475.657	84.32

Table 35.2 Analysis result based on feature selection with correlation coefficient

Selected features	Correct rate
39,62,38,41,74,73,0,61,81,72,71,50,63,51,77,18,28,34,48,3	81.32

$$\rho_{X,Y} = \frac{\text{cov}(X,Y)}{\sigma_X,\sigma_Y} = \frac{E\left[(x-\mu_X)(y-\mu_y)\right]}{\sigma_X,\sigma_Y} \tag{35.5}$$

where cov(X, Y) is the covariance of the two variables, σ_X, σ_Y are the standard deviations of X, Y. After calculating the correlation coefficient, 20 feature variables are selected. The selected variables are taken as the input of parallel SVM. The classification correct rate is listed in Table 35.2.

35.5.4 Results Analysis

From the analysis results of Table 35.1, we can find that the computation speed of feature selection is improved markedly. The accelerate ratio is approximate linear. The classification results show that classification correct rates of different partition plan are similar. It illustrates that the parallel feature election method is effective and efficiency. The analysis results of Tables 35.1 and 35.2 show that the feature selection result based on mutual information is good than that of classical feature selection method.

35.6 Conclusion

Feature selection is an important task of machine learning and pattern recognition. Feature selection based on mutual information is taken as one of the most efficient methods. For improving the computation speed, a novel parallel feature selection method based on MapReduce is proposed. It can accelerate the computation speed almost linearly. The example analysis results show that the proposed method is efficient in reducing computational cost. The classification correct rate based on parallel feature selection method is similar to that of feature selection without partition.

Though the method is efficient in dealing with large-scale feature selection problem, it is only useful to binary feature and class variables. How to process feature selection problem of multivalue or continuous variables is still to be studied. Furthermore, the number of selected variables is prescribed subjectively in the proposed method. How to determine the number of selected variable is our further study.

Acknowledgments This work is partially supported by National Youth Science Foundation (No. 61004115), National Science Foundation (No. 61272433), and Provincial Fund for Nature project (No. ZR2010FQ018).

References

1. Yu, L., & Liu, H. (2003). *Feature selection for high-dimensional data: A fast correlation-based filter solution* (pp. 856–863). Twentieth International Conference on Machine Learning. American Association for Artificial Intelligence.
2. Dash, M., & Liu, H. (2009). Dimensionality reduction. In *Encyclopedia of database systems* (pp. 843–846).
3. Liu, H., & Motoda, H. (1998). *Feature selection for knowledge discovery and data mining* (pp. 23–45). Boston: Kluwer.
4. Kwak, N., & Choi, C. H. (2002). Input feature selection for classification problems. *IEEE Transactions on Neural Networks, 13*(1), 143–159.
5. Kari, T. (2003). Feature extraction by non-parametric mutual information maximization. *Journal of Machine Learning Research, 3*, 1415–1438.
6. Swedlow, J. R., Zanetti, G., & Best, C. (2011). Channeling the data deluge. *Nature Methods, 8*, 463–465.
7. Fox, G. C. Qiu, X. H. Beason, S. Choi, J. Y. Ekanayake, J. Gunarathne., T. et al. (2009). Biomedical case studies in data intensive computing. *Lecture Notes in Computer Science, 5931*, 2–18.
8. Blake, J. A. & Bult C. J. (2006). Beyond the data deluge: Data integration and bio-ontologies. *Journal of Biomedical Informatics, 39*(3), 314–320.
9. Qiu J. (2010). Scalable programming and algorithms for data intensive life science. *Journal of Integrative Biology, 15*(4), 1–3.
10. Guha, R. Gilbert, K. Fox, G. C. Pierce, M. Wild, D. & Yuan H. (2010). Advances in cheminformatics methodologies and infrastructure to support the data mining of large, heterogeneous chemical datasets. *Current Computer-Aided Drug Design, 6*(1), 50–67.
11. Chang, C. C. He, B. & Zhang Z. (2004). Mining semantics for large scale integration on the web: evidences, insights, and challenges. *SIGKDD Explorations, 6*(2), 67–76.
12. Fox, G. C. Bae, S. H. Ekanayake, J. Qiu, X. H. & Yuan H. P. (2008). *Parallel data mining from multicore to cloudy grids* (pp. 311–340). High Performance Computing and Grids Workshop. IOS Press.
13. Zhang, B. J. Ruan, Y. Wu, T. L. Qiu, J. Hughes, A. & Fox G. (2010). *Applying twister to scientific applications* (pp. 25–32). Proceedings of CloudCom. IEEE CS Press.
14. Ekanayake, J. Li, H. Zhang, B. J. Gunarathne, Bae, S. H. Qiu, J. et al. (2010). *Twister: A runtime for iterative MapReduce* (pp. 810–818). The First International Workshop on MapReduce and Its Applications of ACM HPDC. ACM Press.
15. Sun, Z. Q. & Fox G. C. (2012). *Study on parallel SVM based on MapReduce* (pp. 495–501). International Conference on Parallel and Distributed Processing Techniques and Applications. CSREA Press.

Chapter 36
Initial State Modeling of Interlocking System Using Maude

Rui Ma, Zhongwei Xu, Zuxi Chen, and Shuqing Zhang

Abstract In order to do formal verification of interlocking system, which is complicated but safety critical, we choose formal specification language Maude for modeling and verification based on membership equational logic and rewriting logic. In this chapter, a method is proposed to show how the initial state can be modeled and contains important information of specific interlocking system. And a case of Tongji Test Line is reported to illustrate this method in detail. The verification results show that Maude can be applied to formal object-oriented specification and model checking of railway interlocking system successfully using the proposed modeling method.

36.1 Introduction

Railway interlocking system is a safety-critical system and is becoming increasingly more complex. There is therefore a clear need for formal description and verification of interlocking system to prove the correctness and ensure safety [1].

Maude is not only a high-level language but also a high-performance system supporting executable specification and declarative programming [2, 3]. And it supports a range of formal analysis methods, including rewriting for simulation, search for reach ability analysis, and linear temporal logic model checking. By using Maude, a formal language which is characterized by rewriting logic, the model of static and dynamic attribute, circuit, and behavior for the interlocking system can be formed. And they constitute a general formal framework. The detail activities can be described by rules and equations of rewriting logic.

R. Ma (✉) • Z. Xu • Z. Chen • S. Zhang
School of Electronics and Information Engineering, Tongji University,
Shanghai 201804, China
e-mail: marymary1988@163.com

W.E. Wong and T. Zhu (eds.), *Computer Engineering and Networking*, Lecture Notes in Electrical Engineering 277, DOI 10.1007/978-3-319-01766-2_36,

Given these features and applications of Maude, it seems reasonable to assume that the tool should be a good choice for the formal modeling and verification of interlocking systems.

The main contributions of this chapter are the following: (1) shows a case in which Maude has been applied to formally model and verify the interlocking system successfully, (2) focuses on and proposes a method for initial state modeling, and (3) illustrates the method with a case of Tongji Test Line in details.

The rest of the chapter is organized as follows. Section 36.2 describes Maude. Section 36.3 gives a general method for modeling and verification of interlocking system. Section 36.4 shows how to specify the initial state model in Maude. Section 36.5 illustrates the process of modeling the initial state of an interlocking system, Tongji Test Line. Section 36.6 finally concludes the chapter.

36.2 Maude

Maude is particularly suitable for distributed and concurrent systems. Besides the features mentioned in last section, it is based on the membership equational logic and rewriting logic. With this logic, the behavior and the state transition relation of complicated systems can be simplified to some extent, thus simplifying the process of formal modeling and verification of the interlocking system.

36.2.1 Membership Equational Logic

Maude's functional modules are theories in membership equational logic, a Horn logic whose atomic sentences are equalities and membership assertions of the form, stating that a term t has sort s. Such logic extends order-sorted equational logic and supports sorts, subsort relations, subsort polymorphic overloading of operators, and definition of partial functions with equationally defined domains. Maude's functional modules are assumed to be Church–Rosser and terminating; they are executed by the Maude engine according to the rewriting techniques and operational semantics [4, 5].

A Maude program containing only equations is called a functional module. It is a functional program defining one or more functions by means of equations, used as simplification rules. Such equations (specified in Maude with the keyword eq and ended with a period) are used from left to right as equational simplification rules [6].

In Maude, equations can also be conditional, that is, they may only be applied if a certain condition holds, where ceq is the Maude keyword introducing conditional equations.

36.2.2 Rewriting Logic

Membership equational logic is a sublogic of rewriting logic. A rewrite theory is a pair of T, a membership equational theory, and R, a collection of labeled and possibly conditional rewrite rules involving terms in the signature of T. A rewrite theory has both rules and equations. Hence a Maude program containing rules and possibly equations is called a system module. The rewrite rules in R are not equations. Computationally, they are interpreted as local transition rules in a possibly concurrent system. Logically, they are interpreted as inference rules in a logical system. Rules are introduced with the keyword rl while conditional rules with the keyword crl [6].

The essential idea of rewriting logic is that the semantics of rewriting can be drastically changed in a very fruitful way [7]. We no longer interpret a term t as a functional expression, but as a state of a system; and we no longer interpret a rewrite rule as an equality, but as a local state transition, stating that if a portion of a system's state exhibits the pattern described, then that portion of the system can change to the corresponding instance of t. Furthermore, such a local state change can take place independently from, and therefore concurrently with, any other non-overlapping local state changes. Rewriting logic is therefore a logic of concurrent state change [8, 9].

36.3 Modeling and Verification

The whole method for modeling and verification contains three steps:

1. *Modeling*: specify a model of this system using Maude.
2. *Deriving safety properties*: the safety properties related to the specific interlocking system should be generically derived from the model.
3. *Verification*: verify whether a given model satisfies all the safety properties.

If the methods are successfully applied and a model of an interlocking system is verified satisfying the gained safety properties, we can conclude that the interlocking system behaves as expected in relation to the documentation and is safety enough.

36.3.1 Modeling

The general idea of modeling is configurations. Maude is object-oriented. In a concurrent object-oriented system, the concurrent state, which is usually called a configuration, has typically the structure of a multiset made up of objects. And that objects evolve by concurrent rewriting modulo associativity, commutatively, and identity, using rules that describe the effects of communication events between objects.

An object is represented in Maude as a term where O is the object's name or identifier, C is its class identifier, ai's are the names of the object's attribute identifiers, and Si's are the corresponding values.

$$< O : C \big| a_1 : S_1, a_2 : S_2 \cdots a_i : S_i >$$

All the components of the interlocking system can be divided into classes and be modeled separately as mod in Maude. The classes are divided from top to down until it is indivisible. For example, the Class Electrical Component can be divided into fuse, button, and so on. Each mod has its own attributes and behaviors. All of the classes can use the equations and rules in Maude when modeling. And they constitute a general framework of the interlocking system model.

36.3.2 Deriving Safety Properties

The safety properties can be split up in to two categories:

1. *General safety properties*. These safety properties apply to all the railway systems and define that collision and derailing must never occur.
2. *Train route table safety properties*. These safety properties are related to specific interlocking systems. The interlocking systems ensure safety on stations by introducing the concept of train routes, which include the requirements must be satisfied, and are used to ensure the system behaves as expected. So some safety properties have to be derived from the train routes.

After all the safety properties are gained, they should be specified in Maude.

When all the safety properties have been specified, they form a mod SAFETY-PROPERTIES in Maude. It is one of the mod of the complete model and will be verified whether the model of the interlocking system satisfies all the safety properties.

36.3.3 Verification

The verification model contains three parts:

1. *Well formed*: This part is to verify the static properties of the interlocking system. For example, the lengths of the track sections are larger than zero. These properties should be specified in Maude.
2. *Confidence conditions*: Confidence conditions must be verified dynamically as they depend on the dynamic behavior of the model. For example, dropping and

drawing relays may not be cyclic. These properties should be specified in Maude.

3. *Safety properties*: This part is the safety properties mentioned in the previous section, general safety properties and train route table safety properties. And of course, they have been specified in Maude.

Only after specifying all these properties mentioned above in Maude is the modeling of the verification model finished.

Each properties check mentioned above is defined as an operator in the model. For each of these operators, a proposition is specified. We should use Maude to check each of them. Several examples are:

```
reduce in VERIFICATION : wfCircuit.
result: true
```

```
reduce in VERIFICATION : modelCheck(initState , []drawConfidenceCondition).
result: true
```

And it is important that the results of verifying the model in relation to the safety properties are only sound if the given configuration is well formed and if the confidence conditions hold.

The verification is complete after verifying all the three parts.

36.4 Initial State Modeling

The initial state modeling is one part of the whole formal modeling. And this part is very important since it contains the information of specific interlocking system.

When we model initial state, we form a mod named INIT-STATE in Maude. The information in this mod can be divided into two parts.

One is all the instantiated objects of the components classes. In other words, we only generate a format for the attributes of the interlocking system components in the previous modeling but haven't decided the definite value of them. Therefore, all these have to be done in the mod INIT-STATE. Then we gained instantiated objects. For example, there are some instantiate objects:

For relay:

```
< 'tr2-111 : RegularRelay | drawn : false >
```

For signal:

```
< 'sX2 : Signal | lamps : ('X2roe, 'X2gr, 'X2brt) >
```

(continued)

(continued)

For lineartracksection:
< 'X6G : LinearTrackSection | relay : 'X6G-111, length : 30, platform : true, neighbour1 : outsideStation, neighbour2 : '2DG, signal1 : noSignal, signal2 : noSignal >
For trainroute:
< 'tr2 : TrainRoute | lockRelay : 'tr2-111, signals : (['sX3 x green], ['sX4 x red]), points : (['1DG x minus]), trackSections : ('1DG, '5G), displayStop : '1DG, locking : ['1DG x '5G], conflicting : ('tr3, 'tr4, 'tr5, 'tr6, 'tr7, 'tr8, 'tr9, 'tr10, 'tr11, 'tr12) >

The second part is the current modeling. Wires of a circuit specify how the components are connected and whether they are conducting or not. The conducting information is very important to the control logic of the interlocking system. In this chapter the entire electrical circuit is split up into unique linear paths between the positive and negative poles. And each path should be specified in Maude according to the following format:

```
op pathX : Configuration -> Bool.
ceq pathX(...) = true if ...
```

pathX is the name of this operator. X can be the number of the path or whatever you like to name it. If the state matches the argument of the equation and the condition rewrites to true, the path is conducting. The dots in the parentheses stand for the configurations, especially the components in this path should be in the form of objects. And the dots on the right side stand for the conditions this conditional equation should satisfy. Otherwise, the configuration matches

```
eq pathX(CONF) = false [owise].
```

meaning that the path is nonconducting.

Then it will be possible to determine which components are conducting using the following operator:

```
op conducting : Qid Configuration -> Bool.
ceq conducting(C, CONF) = true
if pathX(CONF).
eq conducting(C, CONF) = false [owise].
```

The above equation matches and is rewritten to true if the object (id C) is conducting in pathX.

According to these formats, whether an object is conducting depends on the information of the other objects in the same path and the conditions. To another perspective, these can be seen as the restrictive relationships between them. Therefore, the current modeling can describe some control logic of the interlocking system.

36.5 Case of Initial State Modeling

Then a case of Tongji Test Line will show how the initial state of interlocking system can be specified in Maude. Figure 36.1 is the track layout of Tongji Test Line and Table 36.1 is the train route table of it.

The train route table defines some rules the trains and the track layout have to obey when train drives in this station, forming fixed routes in other words. For instance, train route 1 is from signal X2 to signal X1. When you choose this route, the button X2LA and X1LA should be pushed. This route contains no point and train drives through the track section X2G. Train can only be permitted to drive in when signal X2 is green (L for green and B for white). At last, this route has no mutually conflicting signals.

First, all the objects should be instantiated as mentioned in previous section.

And then come to the current modeling. The current modeling has eight parts in all according to the control logic. They are track section occupying logic, point locating logic, point position representing logic, train route locking logic, train route unlocking logic, signal logic, lamp logic, and lamp mutually exclusive logic. Here examples of several logic modeling are given.

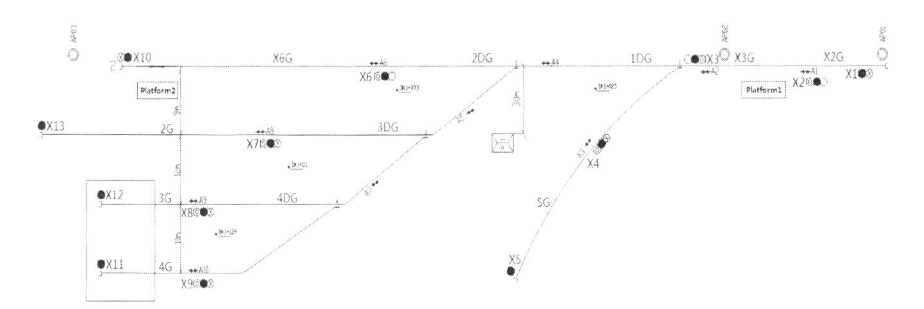

Fig. 36.1 Tongji Test Line track layout

Table 36.1 Tongji Test Line train route table

Direction	Train routes	Route buttons	Signals		Points	Mutually conflicting signal	Track sections	Route number
			Name	Light				
X2	To X1	X2LA, X1LA	X2	L			X2G	1
X3	To X5	X3LA, X4LA	X3	B	(1)	X4	1DG,5G	2
X3	To X10	X3LA, X6LA	X3	L	1	X6	1DG,2DG,X6G	3
X3	To X13	X3LA, X7LA	X3	B	1,(2),3	X7	1DG,2DG,3DG,2G	4
X3	To X12	X3LA, X8LA	X3	B	1,(2),(3),4	X8	1DG,2DG,3DG,4DG,3G	5
X3	To X11	X3LA, X9LA	X3	B	1,(2),(3),(4)	X9	1DG,2DG,3DG,4DG,4G	6
X4	To X2	X4LA, X3LA	X4	B	(1)	X3	1DG,X3G	7
X6	To X2	X6LA, X3LA	X6	L	2,1	X3	2DG,1DG,X3G	8
X7	To X2	X7LA, X3LA	X7	B	3,(2),1	X3	3DG,2DG,1DG,X3G	9
X8	To X2	X8LA, X3LA	X8	B	4,(3),(2),1	X3	4DG,3DG,2DG,1DG,X3G	10
X9	To X2	X9LA, X3LA	X9	B	(4),(3),(2),1	X3	4DG,3DG,2DG,1DG,X3G	11
X6	Through X2 to X1	X6LA, X2LA	X6/X2	L/L	2,1	X3	2DG,1DG,X3G,X2G	12

36.5.1 Track Section Occupying Logic

A relay is used to represent whether the track section is occupied. This logic is specified as follows:

```
ceq path1(< 'X2G : V@TrackSection | ATTS > Conf) = true
if not isOccupied('X2G, < 'X2G : V@TrackSection | ATTS > Conf).
eq path1(Conf) = false [owise].
ceq conducting('X2G-111, Conf) = true if path1(Conf).
```

The track section with id 'X2G has a relay('X2G-111), V@TrackSection is a variable, and it can be matched with linear track section or point. ATTS and Conf are all variables. ATTS denotes other attribute in this object. Conf denotes other configurations besides this object. If 'X2G isn't occupied, path1 is conducting. And then 'X2G-111 is conducting, meaning drawing. Occupying leads to dropping otherwise.

36.5.2 Point Locating Logic

Buttons are used to decide the position of the point. And each position of the point, plus and minus, has a relay to control the locating. They are specified as follows:

```
ceq path24(< 'b11 : Button | pushed : D > Conf) = true
if isPushed('b11, < 'b11 : Button | pushed : D > Conf).
eq path24(Conf) = false [owise].
ceq conducting('1DG-110, Conf) = true if path24(Conf).
ceq path25(< 'b12 : Button | pushed : D > Conf) = true
if isPushed('b12, < 'b12 : Button | pushed : D > Conf).
eq path25(Conf) = false [owise].
ceq conducting('1DG-121, Conf) = true if path25(Conf).
```

The point(id '1DG) locating is controlled by button 'b11 to plus and button 'b12 to minus. If the button 'b11 is pushed, path24 is conducting. Then relay '1DG-110 is conducting. The point moves to the plus, to the minus otherwise as path25 specified.

36.5.3 Train Route Locking Logic

A train route is locked; that is to say this train route is ready for train to come in. Hence, much information has to be examined before locking a train route to ensure safety. All the track sections in this train route must not be occupied. Points must be in the correct position. All contradiction routes are unlocked. Then a train route can be locked, pressing the button of the beginning and ending. One relay is used to represent whether all the information has been examined and the button of the beginning is pushed. Another relay is used to represent whether the train route is locked.

```
ceq path13(*** track section unoccupied
< '1DG-111-11 : Contact | relay : '1DG-211, upper : true >
< '5G-111-11 : Contact | relay : '5G-111, upper : true >  *** points position correct
< '1DG-210-61 : Contact | relay : '1DG-210, upper : false >
< '1DG-221-11 : Contact | relay : '1DG-221, upper : true >
< 'x3la : Button | pushed : D >  *** beginning
< 'tr3-111-61 : Contact | relay : 'tr3-111, upper : false >
*** contradiction routes unlock
< 'tr4-111-61 : Contact | relay : 'tr4-111, upper : false >
< 'tr5-111-61 : Contact | relay : 'tr5-111, upper : false >
...
< 'tr11-111-61 : Contact | relay : 'tr11-111, upper : false >
< 'tr12-111-62 : Contact | relay : 'tr12-111, upper : false > Conf ) = true
*** conditions respect to the elements in path
if  conducting('5G-111, Conf)
   and-then conducting('1DG-111, Conf)
   and-then conducting('1DG-221, Conf)
   and-then not conducting('1DG-210, Conf)
   and-then isPushed('x3la, < 'x3la : Button | pushed : D > Conf)
         ...
   and-then not conducting('tr11-111, Conf)
   and-then not conducting('tr12-111, Conf) .
   eq path13(Conf) = false [owise] .
   ceq conducting('tr2a-111, Conf) = true if path13(Conf) .
  eq conducting('tr2a-111, Conf) = false [owise] . *** ready for locking
  ceq path66(< 'tr2a-111-11 : Contact | relay : 'tr2a-111, upper : true >
 < 'x4la : Button | pushed : D > Conf) = true  *** ending
  if conducting('tr2a-111, Conf)
  and-then isPushed('x4la, < 'x4la : Button | pushed : D > Conf) .
  eq path66(Conf) = false [owise] .
  ceq conducting('tr2-111, Conf) = true if path66(Conf) .
 eq conducting('tr2-111, Conf) = false [owise] . *** relay drawn, finish locking
```

36.5.4 Signal Logic

The relay represents whether a train route which is locked is used in this logic. Take lamp 'X6gr, for example, the logic is specified as follows:

```
ceq path39(< 'tr8-111-11 : Contact | relay : 'tr8-111, upper : true > Conf) = true
if conducting('tr8-111, Conf).
eq path39(Conf) = false [owise].
ceq path43(< 'tr12-111-11 : Contact | relay : 'tr12-111, upper : true > Conf) =
true
if conducting('tr12-111, Conf).
eq path43(Conf) = false [owise].
ceq conducting('X6gr, Conf) = true
if path39(Conf)
or-else path43(Conf).
eq conducting('X6gr, Conf) = false [owise].
```

Lamp 'X6gr represents two train routes, 'tr8 and 'tr12. If one of these train route's relays is drawn, the train route is locked in other words; lamp 'X6gr should be on.

36.5.5 Lamp Mutually Exclusive Logic

If a signal has more than one lamp, all the lamps cannot be on simultaneously. They have to obey some regulations. In this chapter, it is assumed that only one lamp can be on at one moment. The lamp relay is used in this logic. Take signal 'X2 for an example. The logic is specified as following:

```
ceq path45(< 'X2gr-111-61 : Contact | relay : 'X2gr-111, upper : false > Conf)
= true
if not conducting( 'X2gr-111, Conf).
eq path45(Conf) = false [owise].
ceq path77(< 'X2brt-111-61 : Contact | relay : 'X2brt-111, upper : false > Conf)
= true
if not conducting( 'X2brt-111, Conf).
eq path77(Conf) = false [owise].
ceq conducting('X2roe, Conf) = true
if path45(Conf)
and-then path77(Conf).
eq conducting('X2roe, Conf) = false [owise].
```

```
Maude> reduce in VERIFICATION : modelCheck(initState, safetyPropSignals1(confOf(initState))) .
reduce in VERIFICATION : modelCheck(initState, safetyPropSignals1(confOf(initState))) .
rewrites: 38129 in 564ms cpu (565ms real) (67600 rewrites/second)
result Bool: true
Maude> reduce in VERIFICATION : modelCheck(initState, safetyPropSignals2(confOf(initState))) .
reduce in VERIFICATION : modelCheck(initState, safetyPropSignals2(confOf(initState))) .
rewrites: 417499276 in 5594549ms cpu (5594274ms real) (74626 rewrites/second)
result Bool: true
Maude> reduce in VERIFICATION : modelCheck(initState, safetyPropPoints(confOf(initState))) .
reduce in VERIFICATION : modelCheck(initState, safetyPropPoints(confOf(initState))) .
rewrites: 1123755 in 30181ms cpu (30186ms real) (37232 rewrites/second)
result Bool: true
Maude> reduce in VERIFICATION : modelCheck(initState, safetyPropStop(confOf(initState))) .
reduce in VERIFICATION : modelCheck(initState, safetyPropStop(confOf(initState))) .
rewrites: 1375443 in 37830ms cpu (37836ms real) (36358 rewrites/second)
result Bool: true
Maude> reduce in VERIFICATION : modelCheck(initState, safetyPropReleaseInit(confOf(initState))) .
reduce in VERIFICATION : modelCheck(initState, safetyPropReleaseInit(confOf(initState))) .
rewrites: 350702 in 7172ms cpu (7174ms real) (48895 rewrites/second)
result Bool: true
Maude> reduce in VERIFICATION : modelCheck(initState, safetyPropInitToFina(confOf(initState))) .
reduce in VERIFICATION : modelCheck(initState, safetyPropInitToFina(confOf(initState))) .
rewrites: 992772 in 26749ms cpu (26759ms real) (37113 rewrites/second)
result Bool: true
Maude> reduce in VERIFICATION : modelCheck(initState, safetyPropReleaseFina(confOf(initState))) .
reduce in VERIFICATION : modelCheck(initState, safetyPropReleaseFina(confOf(initState))) .
rewrites: 346826 in 17073ms cpu (17072ms real) (20314 rewrites/second)
result Bool: true
```

Fig. 36.2 Partial safety properties result

Signal 'X2 has three lamps, red, green, and white. The green lamp is on when all the points in locked train route are in plus. And white lamp is on when the locked train route contains minus-position point. If neither of the green and white lamps is on, the red lamp must be on.

At this point, all the logic has been specified in Maude. And the modeling of the initial state for the interlocking system, Tongji Test Line, is finished. In the further work, some possible modification could be made to the logic to make it stricter to ensure more comprehensive safety.

Figure 36.2 shows part of the verification result checked by Maude on formal model including this initial state model.

36.6 Conclusion

The interlocking system is typically safety-critical system. Formal modeling and verification is therefore necessary in ensuring safety. In this chapter, the interlocking system is formally modeled in an object-oriented way by Maude, which is a specification and programming language/system based on membership equational logic and rewriting logic, equipped with model checking facilities. This chapter has presented general methods for modeling and verification of the interlocking system. And it focuses on the initial state and has proposed modeling methods for initial state in a case. The case illustrated that the object instantiated process and control logic modeling process. The control logic can be split into several parts and modeled in the way of current modeling. There is ample evidence that Maude can be successfully applied to formally modeling and verification of the interlocking system.

References

1. Chen, B., & Wu, F. (2002). Research on formal models of railway signal interlocking logics. *Journal of The China Railway Society, 24*(6), 50–54.
2. Ji, G. (2011). *The Formal analysis of security protocols based on Maude*. Xi'an: Xidian University.
3. Eker, S., Meseguer, J., & Sridharanarayanan, A. (2004). The Maude LTL model checker. *Electronic Notes in Theoretical Computer Science, 71*, 162–187.
4. Mccombs, T. (2003). *Maude 2.0 primer*. http://maude.cs.uiuc.edu
5. Clavel, M., Duran, F., Eker, S., Lincoln, P., Martí-Oliet, N., Meseguer, J., et al. (2011). *Maude manual (version 2.6)*. http://maude.cs.uiuc.edu
6. Clavel, M., Duran, F., Eker, S., Lincoln, P., Martí-Oliet, N., Meseguer, J., et al. (2002). Maude: Specification and programming in rewriting logic. *Theoretical Computer Science, 285*(2), 187–243.
7. Zhang, M. (2007). *On the model checking of SN P systems based on rewriting logic*. Shanghai: Shanghai Jiao Tong University.
8. Denker, G. (1998). From rewrite theories to temporal logic theories. *Electronic Notes in Theoretical Computer Science, 15*, 105–126.
9. Meseguer, J. (1999). Research directions in rewriting logic. *Computational Logic. NATO ASI Series, 165*, 347–398.

Chapter 37
Semi-supervised Learning Using Nonnegative Matrix Factorization and Harmonic Functions

Lin Li, Zhenyu Zhao, Chenping Hou, and Yi Wu

Abstract In order to reduce redundant information in data classification and improve classification accuracy, a novel approach based on nonnegative matrix factorization and harmonic functions (NMF–HF) is proposed for semi-supervised learning. Firstly, we extract the feature data from the original data by nonnegative matrix factorization (NMF) and then classify the original data by harmonic functions (HF) on the basis of the feature data. Empirical results show that NMF–HF can effectively reduce the redundant information and improve the classification accuracy compared with some state-of-the-art approaches.

37.1 Introduction

In many traditional approaches to machine learning, a target function is estimated by labeled data, which can be thought of as an example given by "age" to "stature." But labeled examples are always time consuming and expensive to obtain, as they require the efforts of professional people, who must be very skillful. For example, getting a labeled example for the classification of giant molecule's shape, which is a huge challenge of biological and computational science, requires months of costly calculation and analysis by specialists. So it is meaningful to combine unlabeled data with labeled data in machine learning. In recent years, an increasing amount of interest has been attracted to the semi-supervised learning problem, and many approaches have been proposed; some methods [1–3] are generally based on an assumption that similar unlabeled data should be given the same classification. For example, the approach HF proposed by Xiaojin Zhu et al. is based on Gaussian fields and harmonic functions [1], but the original data matrix may include redundancy which could influence the classification accuracy and the effectiveness.

L. Li (✉) • Z. Zhao • C. Hou • Y. Wu
Department of Mathematics and System Science, National University of Defense Technology, Changsha 410073, China
e-mail: lilin110@foxmail.com

W.E. Wong and T. Zhu (eds.), *Computer Engineering and Networking*, Lecture Notes in Electrical Engineering 277, DOI 10.1007/978-3-319-01766-2_37,

In this chapter, we will introduce a new approach to semi-supervised learning based on nonnegative matrix factorization and harmonic functions, which can make use of the original matrix efficiently. Unlike the work based on Gaussian fields and harmonic functions, we employed nonnegative matrix factorization to HF. In the process of our approach, we extract the feature data from the original data by NMF firstly, and then classify the original data by HF on the basis of the feature data. Encouraging experimental results show that NMF–HF can effectively reduce redundant information and improve the classification accuracy compared with some state-of-the-art approaches.

In our basic approach of NMF–HF, the solution based on nonnegative matrix factorization and harmonic functions depends on the structure of the data manifold that is derived from data features. In Sect. 37.2, we will introduce the fundamental theorem of HF and NMF briefly. Then, we will give a description of NMF–HF in detail in Sect. 37.3. Encouraging experimental results for data classification will be presented in Sect. 37.4.

37.2 The Principle of NMF and HF

NMF is proposed by Lee D. D. et al. [4], which has been widely used in many fields, like feature extraction [5], image processing [6], and clustering [7, 8]. NMF can be formally stated as follows [9].

Given a nonnegative matrix $X \in \mathrm{R}^{n \times m}$ and a positive integer $r < \min(m,n)$, find nonnegative matrices $B \in R^{n \times r}$ and $C \in R^{r \times m}$ to minimize

$$f(B,C) = \frac{1}{2}\|X - BC\|_F^2 \tag{37.1}$$

From the above NMF problem statement, it is clear that the aim of NMF is to find the approximation of X using the product of two matrices B and C. It is also easy to deduce that the rank of both matrix B and C is at most r.

HF proposed by Xiaojin Zhu et al. is based on a Gaussian random field model defined on a weighted graph over the unlabeled and labeled data, where the weights matrix W are given in terms of a similarity between instances. In particular, the most probable configuration of the random field is unique which is characterized in terms of harmonic functions and has a closed solution that can be computed using matrix methods. It can be considered as the following minimization problem:

$$\min_F tr\left(F^T WF\right) + (F_l - Y_l)^T (F_l - Y_l) \tag{37.2}$$

where Y_l is the label matrix of train data and $F = \begin{pmatrix} F_l \\ F_u \end{pmatrix}$ is the matrix which we want to assign labels to.

37.3 Basic Framework of NMF–HF

Let us assume that there are l labeled points (x_1,y_1), (x_2,y_2), ..., (x_l,y_l) and u unlabeled points (x_{l+1},y_{l+1}), (x_{l+2},y_{l+2}), ..., (x_{l+u},y_{l+u}), $l << u$. Let $n = l + u$, which is the total number of data points. At first, we suppose the labels are binary: $y \in \{0,1\}$. Then, considering a connected graph $G = (V,E)$ with nodes V corresponding to the n data points among which the nodes L correspond to the labeled points with labels y_1, y_2, ..., y_l and the nodes $U = \{l+1, l+2, \ldots, l+u\}$ correspond to the unlabeled points, our duty is to assign labels to nodes U which is unlabeled. So we can obtain a matrix $X_{n \times m}$ where x_{ij} is the jth element of instance x_i denoted as a vector $x_i \in R^m$.

To realize nonnegative factorization of matrix, we need to define a loss function to describe the degree of approximation between $X_{n \times m}$ and $B_{n \times r}C_{r \times m}$ at first and then figure out the solution at the nonnegative limitation.

Regard NMF as a linear mixed model including additive noise [10]:

$$X_{n\times m} = B_{n\times r}C_{r\times m} + E_{n\times m} \tag{37.3}$$

where $B_{n \times r}$ is the base matrix, $C_{r \times m}$ is the coefficient matrix, and $E_{n \times m}$ is the Gaussian noise matrix. Furthermore, we can get

$$X_{ij} = (BC)_{ij} + E_{ij} \tag{37.4}$$

To figure out B and C, we consider maximum likelihood solution following:

$$\{B,C\} = \arg\max_{B,C} p(X|B,C) = \arg\min_{B,C} \left[-\log p(X|B,C)\right] \tag{37.5}$$

Because $E_{n \times m}$ is Gaussian noise matrix, we can get

$$p(X_{ij}|B,C) = \exp\left\{-\frac{1}{2}\left[\frac{X_{ij}-(BC)_{ij}}{\sigma_{ij}}\right]^2\right\} \Big/ \left(\sqrt{2\pi}\sigma_{ij}\right) \tag{37.6}$$

where σ_{ij} is the weight of each observed value. Obviously, each σ_{ij} is equal (each σ_{ij} follows i.i.d); we simply define $\sigma_{ij} = \sigma$.The data log likelihood can be written as:

$$\log\prod_{ij} p(X_{ij}|B,C) = -\frac{1}{2\sigma^2}\sum_{ij}\left[X_{ij}-(BC)_{ij}\right]^2 - mn\log\left(\sqrt{2\pi}\sigma\right) \tag{37.7}$$

Thus, to maximize the data log likelihood is equivalent to minimize the term $\sum_{ij}\left[X_{ij}-(BC)_{ij}\right]^2$ in (37.7). Then, we can define the loss function as follows:

$$L_{ED}(B,C) = \sum_{ij} \left[X_{ij} - (BC)_{ij}\right]^2 \tag{37.8}$$

With the traditional gradient method, we can get:

$$\begin{aligned} \frac{\partial L_{ED}}{\partial B_{ik}} &= 2\left[\left(XC^T\right)_{ik} - \left(BCC^T\right)_{ik}\right] \\ \frac{\partial L_{ED}}{\partial C_{kj}} &= 2\left[\left(B^T X\right)_{kj} - \left(B^T BC\right)_{kj}\right] \end{aligned} \tag{37.9}$$

So we can obtain the following additive iterative method [10]:

$$\begin{aligned} B_{ik} &\leftarrow B_{ik} + \phi_{ik}\left[\left(XC^T\right)_{ik} - \left(BCC^T\right)_{ik}\right] \\ C_{kj} &\leftarrow C_{kj} + \varphi_{kj}\left[\left(B^T X\right)_{kj} - \left(B^T BC\right)_{kj}\right] \end{aligned} \tag{37.10}$$

Let $\phi_{ik} = \frac{B_{ik}}{\left(BCC^T\right)_{ik}}$, $\varphi_{kj} = \frac{C_{kj}}{\left(B^T BC\right)_{kj}}$, (37.10) can be written as the following multiplicative iterative method [11]:

$$B_{ik} \leftarrow B_{ik} \frac{\left(XC^T\right)_{ik}}{\left(BCC^T\right)_{ik}}, \quad C_{kj} \leftarrow C_{kj} \frac{\left(B^T X\right)_{kj}}{\left(B^T BC\right)_{kj}} \tag{37.11}$$

As mentioned above, nonnegative matrix factorization can be summed up to a problem of constrained optimization:

$$\begin{aligned} &\min f(B,C) \\ &s.t. B \geq 0, C \geq 0 \end{aligned} \tag{37.12}$$

where *f*(*B*,*C*) is a loss function. In the process of iteration, we normalize every column vector of the base matrix, namely, that we demand $\sum_i B_{ik} = 1$ for any k [4], and then we get the following multiplicative iterative method:

$$\begin{aligned} B_{ik} &\leftarrow B_{ik} \sum_j C_{kj} X_{ij} / (BC)_{ij} \\ B_{ik} &\leftarrow \frac{B_{ik}}{\sum_l B_{lk}} \\ C_{kj} &\leftarrow C_{kj} \sum_i B_{ik} X_{ij} / (BC)_{ij} \end{aligned} \tag{37.13}$$

We assume a $n \times n$ symmetric matrix W on the edges of the graph G according to $B_{n \times r}$. Consequently, the weight matrix can be denoted as

$$w_{ij} = \exp\left(-\sum_{d=1}^{r} \frac{(x_{id} - x_{jd})^2}{\sigma_d}\right), x_i \in R^r \tag{37.14}$$

where x_{id} is the dth element of instance x_i denoted as a vector $x_i \in R^r$ and $\sigma_1, \sigma_2, \ldots \sigma_r$ are the length scale parameters for each dimension [1]. The matrix W completely specifies the data manifold structure for our purposes.

Then, our program is to compute a real-valued function $f : V \to R$ on G with some nice properties, and then, we assign labels to the nodes L according to f. We impose restrictions upon f to take values $f(i) = f_l(i) \equiv y_i$ on the labeled data $i = 1, 2, \ldots, l$. That is to say, we want unlabeled points to have similar labels with labeled points that are near to the unlabeled ones in the graph. This idea promotes us to choose the quadratic energy function:

$$E(f) = \frac{1}{2}\sum_{i,j} w_{ij}(f(i) - f(j))^2 \tag{37.15}$$

In order to make functions f assigned with a probability distribution, we construct Gaussian field $p_\beta(f) = \frac{e^{-\beta E(f)}}{Z_\beta}$, where Z_β is the partition function denoted as $Z_\beta = \int_{f|_L = f_l} \exp(-\beta E(f))df$, which normalizes over all functions constrained to f_l on the labeled data, and β is an inverse temperature parameter.

It is very simple to prove that the minimum energy function $f = \text{argmin}_{f|_L = f_l} E(f)$ is harmonic. That is to say, it satisfies $\Delta f = 0$ on unlabeled data points U just like f_l on the labeled data points L. This Δ is the combinatorial Laplacian here, denoted as $\Delta = D - W$ in matrix form while $D = diag(d_i)$ with entries $d_i = \sum_j w_{ij}$, and $W = (w_{ij})$ is the weight matrix.

The harmonic property means that the value of f which is defined at each unlabeled data point is the average of f at neighboring points:

$$f(j) = \frac{1}{d_j}\sum_{i,j} w_{ij} f(i), j = l+1, l+2, \ldots, l+u \tag{37.16}$$

Denoted as $f = Df$, expressed slightly differently, where $P = D^{-1}W$. Because of the maximum principle of harmonic functions, f is unique and is either a constant or it satisfies $0 < f(j) < 1$ for $j \in U$.

In terms of matrix operations, we split the weight matrix W (and the same to D, P) into four blocks after the lth row and column to calculate the harmonic solution explicitly:

$$W = \begin{pmatrix} W_{ll} & W_{lu} \\ W_{ul} & W_{uu} \end{pmatrix} \quad (37.17)$$

Letting $f = \begin{pmatrix} f_l \\ f_u \end{pmatrix}$, where f_u assigns the values to the unlabeled data points, the harmonic solution $\Delta f = 0$ subject to $f|_L = f_l$ is captured by:

$$f_u = (D_{uu} - W_{uu})^{-1} W_{ul} f_l = (I - P_{uu})^{-1} P_{ul} f_l \quad (37.18)$$

As outlined briefly in this section, we mainly introduce NMF to HF while the HF may be inefficient and more storage space required for classification accuracy. This new viewpoint provides a rich and complementary technique for this semi-supervised learning problem.

37.4 Experimental Results

Now, we report the results of data classification experiments by using the umist dataset. Regard this dataset as a nonnegative matrix denoted as $X_{n \times m}$, where x_{ij} is the jth component of instance x_i denoted as a vector $x_i \in R^m$. In this case, $n = 380$, $m = 644$, it means that there are 380 data points while every data point has 644 dimensions.

In our experiments, we perform several groups of trials. In each trial, let every $\sigma_d = 1.2 \times 10^{-5}$ and $r = 115$. Firstly, sample some labeled data (40, 60, 80, ..., 200) randomly as train data from the entire dataset; secondly, classify the rest data (denote as test data) into 20 classes based on the approach (NMF–HF); then, repeat each trial 50 times; and finally, calculate the average classification accuracy for each trial. Compared with KNN and HF, we get the following table (Table 37.1):

According to the above table, we draw a conclusion that NMF–HF is more efficient than KNN and HF in terms of classification accuracy.

Table 37.1 Classification accuracy

Train data	KNN	HF	NMF–HF
40	0.7198	0.8964	0.9055
60	0.8048	0.9233	0.9303
80	0.8671	0.9497	0.9546
100	0.8998	0.9561	0.9603
120	0.9295	0.9660	0.9685
140	0.9460	0.9732	0.9742
160	0.9570	0.9775	0.9786
180	0.9710	0.9851	0.9846
200	0.9747	0.9839	0.9862

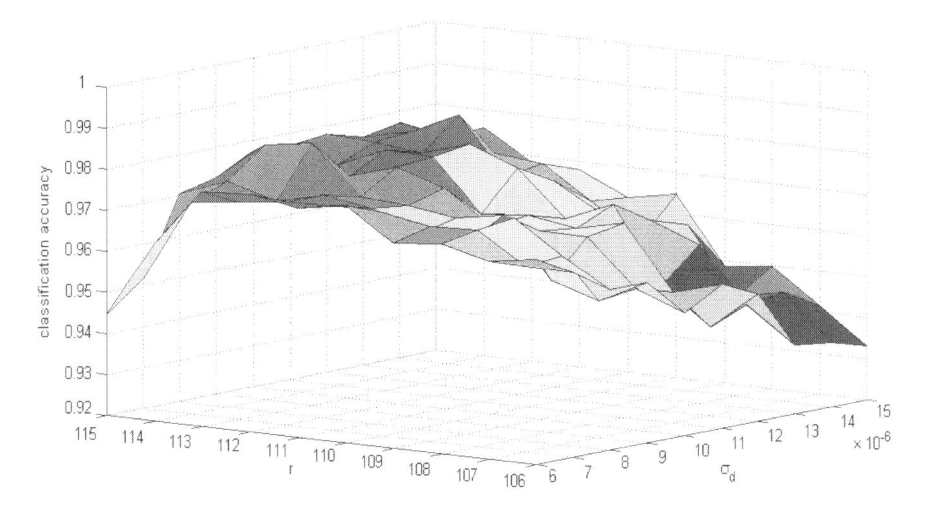

Fig. 37.1 The sensitivity analysis about parameters

We also study the stability about classification accuracy of NMF–HF while r, σ_d vary in a certain range. The following figure is given (Fig. 37.1):

From the figure, we conclude that the classification accuracy of NMF–HF is stable when r, σ_d vary in a reasonable range.

37.5 Conclusion

A new approach to semi-supervised learning is proposed that is based on nonnegative matrix factorization and harmonic functions. Promising experimental results have been presented for nonnegative data classification, demonstrating that the framework of NMF–HF has the potential to effectively exploit the structure of unlabeled data to improve efficiency and classification accuracy. Our work in this direction will be reported in a future publication.

Acknowledgments We gratefully acknowledge the supports from National Natural Science Foundation of China, under Grant No 61005003.

References

1. Zhu, X., Ghahramani, Z., & Lafferty, J. (2003). Semi-supervised learning using gaussian fields and harmonic functions. In *Proceedings of the 20th International Conference on Machine Learning* (Vol. 20, Issue 2, p. 912). Washington, DC: AAAI Press.
2. Belkin, M., & Niyogi, P. (2002). Using manifold structure for partially labelled classification. *Advances in Neural Information Processing Systems, 15*, 929–936.

3. Blum, A., & Chawla, S. (2001). Learning from labeled and unlabeled data using graph mincuts. In *Proceedings of the 18th International Conference on Machine Learning* (pp. 19–26). Williams College: Morgan Kaufmann.
4. Lee, D. D., & Seung, H. S. (1999). Learning the parts of objects with non-negative matrix factorization. *Nature, 401*(6755), 788–791.
5. Student, M., & Eswar, K. (2012). Graph regularized non-negative matrix factorization for data representation. *International Journal of Computer Application, 3*(2), 171–191.
6. Cichocki, A., & Amari, S. (2002). Adaptive blind signal and image processing. In *Learning algorithms and applications* (pp. 88–93). New York, NY: John Wiley Press.
7. Ding, C., Li, T., & Jordan, M. (2008). Nonnegative matrix factorization for combinatorial optimization: Spectral clustering, graph matching, and clique finding. In *ICDM* (pp. 183–192). Pisa, Italy: IEEE Computer Society.
8. He, Z., Xie, S., Zdunek, R., Zhou, G., & Cichocki, A. (2011). Symmetric nonnegative matrix factorization: Algorithms and applications to probabilistic clustering. *IEEE Transactions on Neural Networks, 22*(12), 2117–2131.
9. Berry, M. W., Browne, M., Langville, A. N., Pauca, V. P., & Plemmons, R. J. (2007). Algorithms and applications for approximate nonnegative matrix factorization. *Computational Statistics and Data Analysis, 52*(1), 155–173.
10. Sajda, P., Du, S., & Parra, L. C. (2003). Recovery of constituent spectra using nonnegative matrix factorization. Optical Science and Technology, SPIE's 48th Annual Meeting. Proceedings of SPIE, San Diego. pp. 321–331.
11. Lee, D. D., & Seung, H. S. (2001). Algorithms for nonnegative matrix factorization. *Advances in Neural Information Processing Systems, 13*, 556–562.

Chapter 38
Exploring Data Communication at System Level Through Reverse Engineering: A Case Study on USB Device Driver

Leela Sedaghat, Brad Duerling, Xiaoxi Huang, and Ziying Tang

Abstract Interactions among operating system, drivers, and peripheral devices are important for users to understand data communication at low system level, system architecture, and hardware programming. In this chapter, we study low-level data communication and resource management by conducting the development of a USB device driver. A reverse engineering approach has been adopted in this study, and we focus on exploring the USB protocol and developing a device driver for the Linux operating system. We have performed various experiments to evaluate the device driver from different aspects, and all testing results are remarkably good. We believe this work can provide users a clear practical understanding of data communication from the hardware level to user space applications as well as theoretical foundations to reproduce any unsupported peripheral hardware devices.

38.1 Introduction

The Linux kernel is initially designed to provide a freely available, openly maintained, and modified operating system. Since its first release in 1991, the Linux kernel has been ported to countless different systems and processor architectures. It has successfully transitioned from primarily academic patronage to users in the mainstream consumer market. Today, Linux is a popular alternative to commercial operating systems, such as Windows and Mac OS, and the demand for supported applications and hardware drivers is growing at an extraordinary

L. Sedaghat • B. Duerling • Z. Tang (✉)
Department of Computer and Information Sciences, Towson University,
7800 York Road, Towson, MD 21204, USA
e-mail: lsedag1@students.towson.edu; miamisbiggest@gmail.com; ztang@towson.edu

X. Huang
Institute of Cognitive and Intelligent Computing, Hangzhou Dianzi University,
Hangzhou 310018, China
e-mail: huangxix@hdu.edu.cn

W.E. Wong and T. Zhu (eds.), *Computer Engineering and Networking*, Lecture Notes in Electrical Engineering 277, DOI 10.1007/978-3-319-01766-2_38,

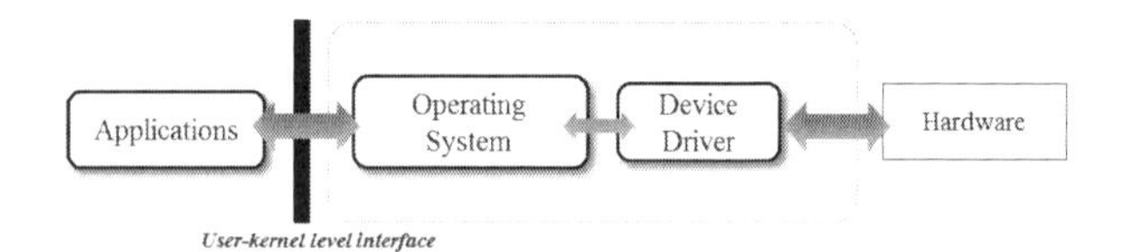

Fig. 38.1 Data communication from hardware to applications

Missile Launcher Movements
Missile move UP
Missile move DOWN
Missile move RIHGT
Missile move LEFT
Missile move FIRE

Fig. 38.2 Brookstone USB missile launcher and its movements

pace. From 2007, the Linux Driver Project [1] has been launched to meet this need by enlisting a group of developers to create, submit, and maintain Open Source Linux kernel drivers for different types of devices. Meanwhile, we believe that research analysis on data communication between devices and operating systems is very important to provide theoretical foundations for supporting any unsupported peripheral hardware devices. In order to successfully develop a fully functioning Linux device driver, it is essential not only to study and understand the interactions between the Linux operating system and device drivers in general but also to study and understand the specific technical details underlying the interactions between the operating system, driver, and selected target peripheral device, which also requires a thorough understanding of the data communication protocol such as USB.

Figure 38.1 shows the data communication among user application, operating system, and devices with the help of device driver and controller. Developing a Linux device driver presents a challenging and rewarding experience to understand this communication and explore the relationship between operating systems, drivers, and peripheral devices. As such, we present herein a case study to develop a Linux driver for a USB 2.0 device. We selected a peripheral device that is currently unsupported by the Linux operating system: the USB Desktop Missile Launcher by Brookstone, as shown in Fig. 38.2 [2]. The device is currently supported only by Windows and Mac OS X. Consequently, any Linux user who purchases the device would be unable to use it. We believe this study can provide general users a practical understanding of how data communication happens from low-level hardware to high-level user space applications, thereby allowing users to reproduce any device on their own.

38.2 USB and Device Driver

38.2.1 USB Data Communication Protocol

As an I/O bus, USB transfers data between a host computer and a peripheral device. The host manages the bus, and the device can only use the bus when responding to a request from the host (except in the case of remote wake-up). As such, USB forms a master/slave configuration [3, 4]. In this section, we discuss the USB protocol in detail and illustrate the key ideas through Fig. 38.3.

The host contains USB host-controller hardware, a root hub with USB ports as well as software to manage bus communication. The USB host-controller hardware is the interface between the operating system, drivers, and all of the attached peripheral devices. The USB host controller identifies each attached peripheral device with a unique address or identification number. This, in turn, permits many devices to communicate with the host over the same USB connection. The USB 2.0 driver belongs to the host-controller hardware and is needed to interpret the USB protocol and to identify any and all attached devices. It is then the responsibility of system software to find the appropriate device driver and to load it into the kernel.

The responsibilities of the host include detecting devices when they are plugged into the USB port, providing power to the devices, managing traffic on the bus, handling error checking, and exchanging data (using a message-passing system) with the peripheral devices. The device contains USB device-controller hardware, a microcontroller and usually firmware to manage bus communications (though it may contain software instead of firmware). The responsibilities of the device include detecting bus voltage, managing power, responding to requests from the host, handling error checking, and implementing the device's functions.

Data flow on the bus is bidirectional. Transfers on the bus can be classified into four types: control, bulk, interrupt, or isochronous transfers. Furthermore, unlike

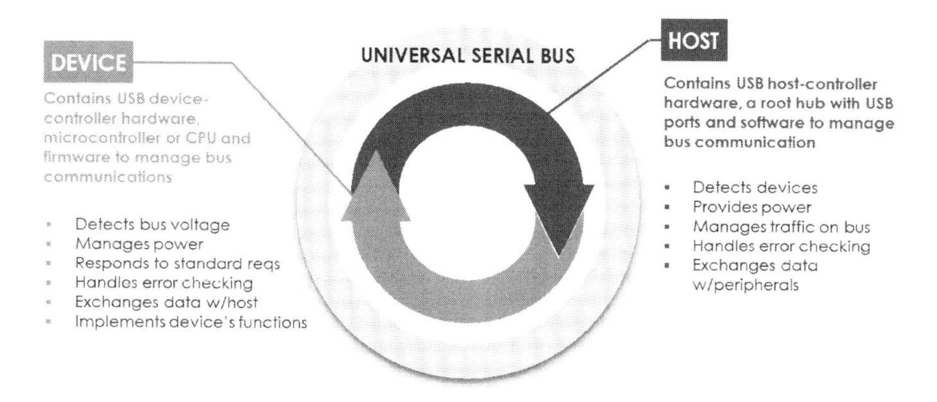

Fig. 38.3 Universal serial bus diagram

parallel and serial ports that transfer data in bits, USB encapsulates and sends data in packets called USB request blocks (URBs), which are the basic transaction unit of the message-passing system used for communication over the bus. Each URB contains all of the necessary information (e.g., data and addressing information) to perform a USB transaction. Generally, in order to initialize a URB within a device driver, the following information is required: a device pointer, the pipe (or endpoint information), the request type and value, the transfer buffer, and the desired transfer length. For example, a URB for a control-type transfer can be built and sent to a specific endpoint by invoking the following method:

```
int usb_control_msg (struct usb_device *  dev,
        unsigned int pipe,
        __u8 request,
        __u8 requesttype,
        __u16 value,
        __u16 index,
        void * data,
        __u16 size,
            int timeout)
```

When a system is booted, the USB hub lets the USB host controller know of any attached devices. If a system has already been booted up, the USB host controller periodically polls the hub to learn of any newly attached or removed devices. When an attached device has been detected, whether at boot-up or afterwards, the host controller assigns the device a unique address and sends requests for its descriptors through the enumeration process. The USB descriptors requested by the host provide information about the device, allowing the host to load the appropriate device driver. Descriptors are data structures that store information about a device's capabilities and requirements. The host uses these descriptors to load the appropriate driver for the device. After the driver has been loaded, the host and device can communicate on the USB using URBs, as described above.

38.2.2 Device Communication and Device File

A device driver, which is loaded to the kernel to provide an interface for user applications, is in charge of interactions between an operating system and a device. As an interface between the hardware and the software, it provides the mechanism and not the policy in order to fulfill different requests of the OS. Hence, separation of the mechanism and the policy should be taken into consideration [5].

As pointed out by Kadav and Swift [6], there are three interfaces in the system which may be used to implement a device driver: (1) the interface between the driver and the kernel, for communicating requests and accessing OS services; (2) the interface between the driver and the device, for executing operations; and

(3) the interface between the driver and the bus, for managing communication with the device. It is clear that these interfaces might be different regarding how they are implemented in various platforms.

Generally, there are three classes of device drivers in Linux, namely, character drivers, block drivers, and network drivers. Besides this classification, there are some devices that cannot be classified into one of these classes. For example, a USB device can have a character driver (like USB serial port) or block driver (USB memory card reader) or even network driver (USB Ethernet interface) [5].

When the device is plugged in, the kernel recognizes the device and sets up all the configurations and interfaces. The kernel then checks if it can handle the USB device by searching all the drivers that it has registered and calling their probe functions. This function considers the vendorID and productID of the device against its device table. If the driver can handle the device, it pairs itself with the device and informs the kernel. Moreover, a device file is also created at this time to allow a user space program to communicate with the device through the driver. User space knows the device file by its name, while the driver knows the device file by a number. Note that the device file is not a file in the traditional sense. Instead, it is a mapping for user space functions. Each function in the user space is mapped by the device file to a function in the driver code. For example, below is the part of the driver code that tells the device file which function should be mapped to for each call.

```
static const struct file_operations skel_fops = {
        .owner =THIS_MODULE,
        .write =skel_write,
        .open =skel_open,
        .release =skel_release,
};
```

38.3 Reverse Engineering

In order to develop the driver for any unsupported peripheral device, we need to reverse engineer the USB device, which is the focus of our work. By reverse engineering the device, we are able to obtain the necessary information the driver needs to communicate with the device. The primary and essential information that is needed for proper driver implementation is contained in the USB descriptors for the device as well as in thc payload of the URBs. Since the manufacturer provides a Windows device driver, a PC running Windows can be used as the host computer to reverse engineer the device. As such, we first install the device on a Windows system and, subsequently, we install software that would allow us to monitor the USB traffic sent and received by the host. In order to do so, USBlyzer [7], a USB traffic sniffer software, is adopted in our system. The device is plugged into an

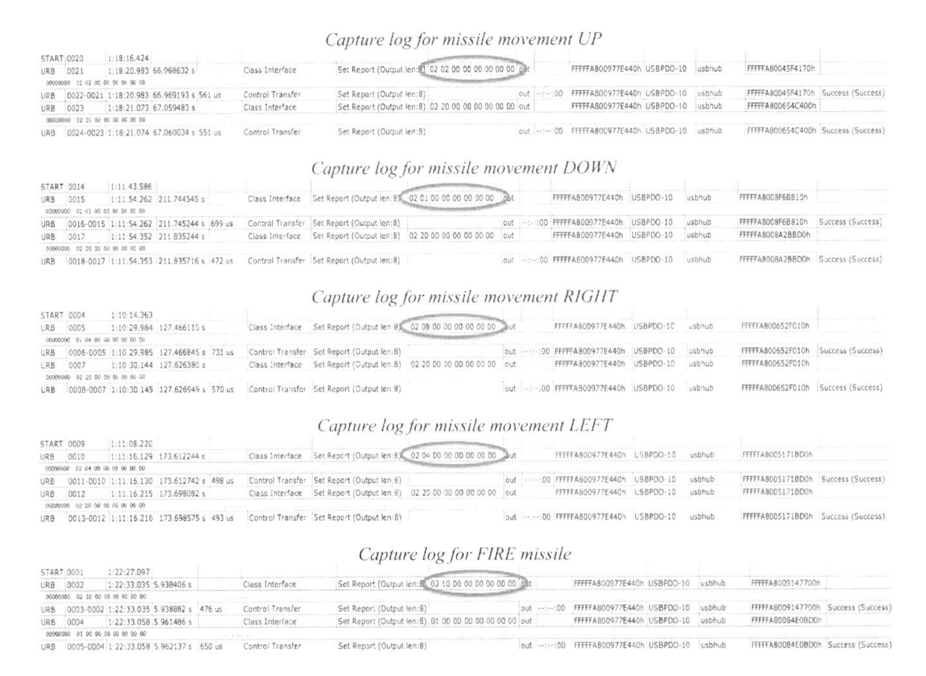

Fig. 38.4 Capture logs for different missile launcher functions

available USB port and USBlyzer is launched. USBlyzer displays various USB descriptors that are obtained by the host computer during the enumeration process in one of the user interface panels. For example, the productID and vendorID needed to implement the device driver are among the descriptors obtained from the device.

After successfully obtaining USB descriptors, we can start capturing the actual raw data that is passed in the URBs between the host and the device. In order to do so, it is necessary to have the device perform all of its possible functions (e.g., move right) and observe the values in the payload of the captured packets. Since the missile launcher device is able to move left, right, up, and down to the desired angle, we need to trigger all of these motions in individual captures so that there would be no confusion as to which packets corresponded to which functions. The results of each capture are shown in Fig. 38.4. A closer look reveals that each of the functions is comprised of a series of requests, with the majority of payload values being the same across all functions. The data of only one URB is different across all of the functions. This unique packet is circled in Fig. 38.4 for each capture. We hypothesize that it is this unique payload that drives each different function. As such, we use only these data values when implementing the driver.

38.4 Experimental Results

In order to provide an interaction with the USB missile launcher, we have implemented both the device driver and a user space application based on our previous studies. The device driver has to be compiled and loaded into the Linux kernel, while the user application is a typical C code which must be compiled and run in the user space and interacts with the driver.

Our device driver is implemented through USB-Skeleton where a number of *structs* are provided for the customized device driver. One of the most important ones is presented below:

```
static struct usb_driver skel_driver = {
        .name ="missile",
        .probe =skel_probe,
        .disconnect = skel_disconnect,
        .id_table =   skel_table,
};
```

When we register our driver to the kernel, it will be registered by the value that we assigned to name field. As mentioned before, when a USB device is plugged in, the kernel tries to find the appropriate driver for the device based on productID and vendorID. We define these two values in skel_table, and as you can see, id_table is pointing to this table. So, id_table field actually helps the kernel to find a driver for the plugged-in device. After finding the driver, kernel runs the Probe function of that driver which is responsible for extracting and initializing needed endpoints from the device descriptor. In the represented *struct*, Probe contains the pointer to the probe function which we have defined in our driver code.

The user application is intended to allow users to test and demonstrate the functionality of the device in the user space. To conduct testing, we have implemented both console-driven program and shell-based user-friendly dialog boxes. In both applications, we can successfully control all the missile movements. We have also experimented from system resource management aspects. In Fig. 38.5, we show process monitoring and system resource monitoring for our missile launcher process. Specifically, we trace system calls and signals and monitor CPU, memory, and I/O resources.

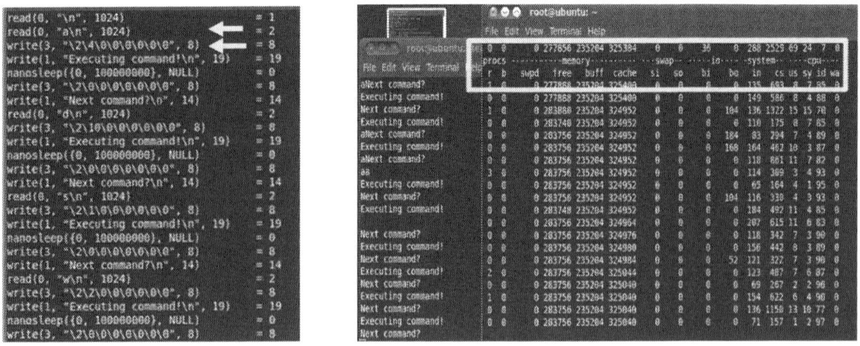

Fig. 38.5 System call trace for the missile launcher process (*left*). System resource monitoring (*right*)

38.5 Conclusion

In this project, we have studied the interactions between an operating system and device drivers at the system level. It provides general users a practical understanding of how data communication happens from the low-level hardware to high-level user space applications. A reverse engineering technique has been adopted in this study, and we focus on exploring the USB protocol and developing a device driver for the Linux operating system. We have conducted a detailed case study for a particular device called Desktop Missile Launcher, which is only supported by Windows and Mac OS before our work. In addition, we have performed various experiments to evaluate the device driver from different aspects, and all testing results are remarkably good.

References

1. Linux Driver Project. http://www.linuxdriverproject.org
2. USB desktop Missile Launcher. http://www.brookstone.com/usb-desktop-missile-launcher
3. Anderson, D., & Dzatko, D. (2001). *Universal serial bus system architecture* (2nd ed.). Addison-Wesley Professional.
4. Axelson, J. (2009). *USB complete: The developer's guide* (4th ed.). Lakeview Research.
5. Corbet, J., Rubini, A., & Kroah-Hartman, G. (2005). *Linux device drivers* (3rd ed.). Sebastopol, CA: O'Reilly.
6. Kadav, A., & Swift, M. (2012). Understanding modern device drivers. *ACM SIGARCH Computer Architecture News, 40*(1), 87–98.
7. USBlyzer. http://www.usblyzer.com

Chapter 39
Using Spatial Analysis to Identify Tuberculosis Transmission and Surveillance

Jinrong Bai, Guozhong Zou, Shiguang Mu, and Yu Ma

Abstract Tuberculosis is a chronic infectious disease which can make serious hazard to human health and cause large social and economic burden on a country. So for experts and researchers, tuberculosis is one of the biggest public heath challenges. The cause of this disease can be effectively studied by precise analysis of the spatial distribution of the disease. This chapter demonstrates that using existing health data, spatial analysis and GIS in conjunction with epidemiological analysis can identify tuberculosis transmission. This chapter also demonstrates some of the valuable results of GIS in disease surveillance and mapping. The decision-makers could master the epidemic of tuberculosis dynamically and then take better measures to control tuberculosis. Moreover, this study may add some value to traditional and molecular epidemiology and provides an alternative method that may give insight into the transmission of tuberculosis.

39.1 Introduction

Tuberculosis (TB), known as the "white plague," is a chronic infectious disease which can make serious hazard to human health. The 1990 World Health Organization (WHO) reports that tuberculosis is ranked as the seventh most morbidity-causing disease in the world and predicts it to continue in the same place up to 2020. An estimated two billion people are infected with tuberculosis [1] and approximately two million people die from this disease annually [2]. China is one of the 22 countries in the world highly burdened by tuberculosis, ranking second with respect to the number of tuberculosis patients. Tuberculosis not only is an important public health problem but also is a complex socioeconomic problem. The prevalence of tuberculosis hinders social and economic development. Unless tuberculosis is properly treated, an average

J. Bai (✉) • G. Zou • S. Mu • Y. Ma
School of Information Technology and Engineering, Yuxi Normal University, Yuxi 653100, China
e-mail: baijr223@163.com

W.E. Wong and T. Zhu (eds.), *Computer Engineering and Networking*, Lecture Notes in Electrical Engineering 277, DOI 10.1007/978-3-319-01766-2_39,

of 10–15 people are infected by an infectious tuberculosis patient in a year. More adults are killed by tuberculosis than any other infectious disease worldwide, accounting for about 400,000 deaths annually. It mainly affects those in the economic production years of their life (equal to 15–54 years). An adult tuberculosis patient loses on average 3–4 months of working time annually. This shows that large social and economic burden is caused by tuberculosis on a country. So for experts and researchers, tuberculosis is one of the biggest public heath challenges.

Previous studies on tuberculosis and other infectious diseases ignored the geographic correlation, did not perform a study on the quantitative level spatial distribution of diseases, and were limited to simple analysis of incidence of the disease. We all know that medicine is a system of micro and macro comprehensive disciplines and large amounts of data have the spatial distribution characteristics.

Epidemiological topic is the distribution of time, space, and population about diseases and about 80 % of the epidemiological data have spatial attributes. Human or animals are always in the prevalence of certain spatial location, but geographical or social factors in certain space may also affect the incidence of the diseases. The cause of the disease can be effectively studied by accurate analysis of the spatial distribution of the disease, so we can identify high-risk regions and populations and develop preventive measures.

Moreover, as an infectious disease, tuberculosis is related to population, climate, and the local environment. Because of its infectivity and universality, the occurrence, development, and prevalence of tuberculosis is a spatial phenomenon with interaction and diffusing phenomenon. Therefore, the relevant data should be based on spatial property, considering the location information and non-location information. This reflects dynamic distribution characteristics on space and its influential factors in order to meet the needs of tuberculosis control and prevention work. New methods, such as spatial analysis and mapping, may be of valuable contribution to basic elements of tuberculosis control. Such tools are used by public health professionals to visualize and explore disease patterns for guiding disease control strategies. On the basis of spatial analysis, a new way is supported by the emerging spatial statistics for the spatial autocorrelation. Its core is the understanding of space-dependent data between location-related areas.

In this study, we perform spatial statistical analysis in quantitative level to explore one of the hot spots of tuberculosis incidence; also, time, space, and spatial-temporal cluster is very necessary. The main purpose of this research was to examine the spatial distribution of tuberculosis cases in the Anxian County, Sichuan Province, China, over a 3-year period, from 2007 to 2009, using geographic information system (GIS) software and spatial analysis technology. In doing this, it was expected that some of the valuable assets of GIS in disease mapping and surveillance are demonstrated in this study. It is anticipated that the collected information by this research will help public health workers to identify and provide effective examples of using epidemiological data, public statistical software, and GIS to formulate the research problem, generate and test hypotheses, and critically evaluate mapping that is prepared using spatial statistical approaches and GIS software. It is hoped that public health officials and workers in the future will see the added value which is brought by GIS to an already well-established disease surveillance group. Furthermore, GIS can offer data that is in a form that can be more easily communicated to the general public and community groups if necessary. An effective method for tuberculosis control programs is provided by geographically based screening and treatment to identify high-risk populations. In this research, we also sought to determine whether or not we can identify the geographical regions with ongoing tuberculosis transmission by using GIS technology.

39.2 Related Works

In our opinion, using GIS and spatial-temporal statistical analysis to identify geographical areas with ongoing disease transmission has become indispensable. Space-time clustering method involves the identification of greater density of a phenomenon that appears in certain places at certain times. The researchers have intensively applied these techniques in several areas such as toxicology, criminology, demography, and others.

Disease mapping has a long history and it is no surprise that this descriptive analysis method is used to describe rates of spread and to identify sources of infections. It is not a new phenomenon that the geography is used in epidemiological studies. John Snow's analysis of local water pumps and their relationship with the transmission of cholera in London during the 1850s is probably the most widely cited study for first incorporating geographical analysis and the combination of field epidemiology. Epidemiology stresses the understanding of three important components of disease distribution: the time of disease outbreak or transmission, the people involved, and the location of spread. In the past decade, researchers have published several studies on geographical epidemiology all over the world. A very interesting study about the tuberculosis DOTS strategy and GIS is published by Porter [3]. GIS technology is used to identify incidence and areas of tuberculosis transmission in the USA from 1993 to 2000 [4]. Spatial distribution of M. tuberculosis/HIV coinfection is studied by Rodrigues in São Paulo State, Brazil, from 1991 to 2001 [5]. GIS and spatial scan statistics are used to investigate geo-spatial hot spots for the occurrence of tuberculosis in Almora district [6].

39.3 System Architecture

In order to allow for mapping of individual tuberculosis cases and spatial analysis, the address of each case in the study was geocoded. It needs to utilize specialized GIS software to assign a latitude and longitude for the address of the case. In this work, disease mapping was performed utilizing Google Maps. Google Maps is a web mapping service provided by Google Company, and it contains landmarks,

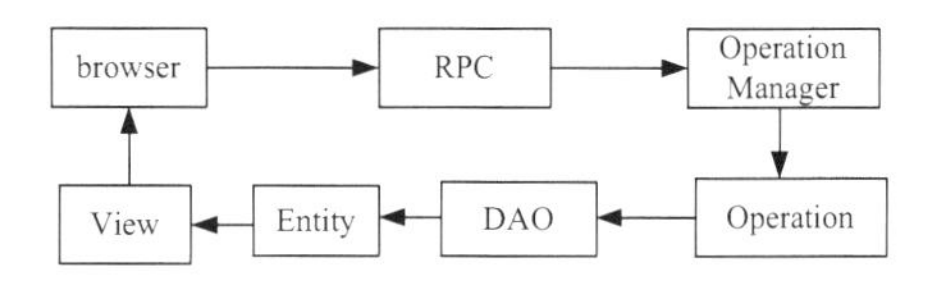

Fig. 39.1 System architecture

lines, shapes, and other information, with three kinds of view: the vector maps, satellite images, and topographic maps. High-resolution satellite or aerial images for most urban areas are provided by Google Maps, which powers many map-based services. Maps embed on third-party websites via the Google Maps API. The Google Maps API is publicly accessible, does not charge for access, and is free for commercial use provided that the site is not generating more than 25,000 map accesses a day.

The system is developed by Java language and MVC framework. The framework employs DWR (Direct Web Remoting) with Ajax technology to achieve the web program interaction between foreground and background. DWR is a Java open source library and includes Ajax technology. DWR helps developers write websites, and Java functions running on a web server are called by JavaScript code in a web browser as if those functions were within the browser.

Specifically, the framework encapsulates the common functions and employs Ajax technology to access the server-side object through the operation callback interface. Therefore, the front-end operation of the system mainly uses JavaScript, whereas the back end of the system uses lightweight MVC framework. The business logic is divided into three layers (Operation, Entity, DAO). System framework is showed in Fig. 39.1.

39.4 Methods

Face-to-face patient survey, institution-based survey, key informant interview, and literature reviews were the main data sources. Relevant databases were set up after the collection of data from surveys on tuberculosis in the Anxian County. Spatial analysis was undertaken after the databases were linked to the software Google Maps. Correlation analysis was performed to understand the relationship among the prevalence rate of active pulmonary tuberculosis, the mortality rate of pulmonary tuberculosis, and socioeconomic factors.

39.4.1 Study Area and Population

This significant research was conducted in the Anxian County in Sichuan province, China. This county, which covers a total surface area of 1,404 km^2, has a population of 500,000. This study was in the southwest most densely populated area of China. This county has 23 government tuberculosis diagnostic centers. All forms of tuberculosis patients who usually reside in the Anxian County were eligible to join the study. This study obtained informed consent from all research cases.

39.4.2 Data Collection

All consenting tuberculosis cases were collected between January 1, 2007, and December 31, 2009. This study included a tuberculosis case if the tuberculosis control program had undergone a full course of antituberculosis treatment for the patient. At recruitment, a structured questionnaire was answered by all study participants. This study obtained demographic data (sex, age, place of residence, occupation, and ethnicity), clinical information (type of tuberculosis, date of registration, and date of diagnosis), and past history of tuberculosis. This study also collected the residential addresses of all study cases and the geographical location of diagnostic and treatment facilities in the study area. A team of field workers monitored regularly all study participants and categorized their treatment result based on standard WHO-recommended definitions [7] transferred out, failed treatment, died, completed treatment, cured, and defaulted. This study investigated the residential status of each patient with tuberculosis and categorized the residential status of each patient as permanent or temporary residents. A permanent tuberculosis patient who is usually impossible to move out on completion of treatment was defined as a case normally resident there.

Residential address and zip code were geocoded using Google Maps at the time of diagnosis of tuberculosis. After the automatic and interactive geocoding, the longitude and latitude of most tuberculosis cases were correctly matched on the map.

39.4.3 Statistical Analysis

During the 3 years of the study, 1,035 patients were diagnosed with tuberculosis and started treatment in the clinics involved in the study. The overall incidence rate of tuberculosis in the study area was 69/100,000 for the year of the study, and the male to female ratio was 2.2. There was a positive correlation between the average case load and crowding. A significant positive correlation was also found between unemployment and the average tuberculosis case load, with codependence between unemployment and crowding. The statistics of tuberculosis cases between 2007 and 2009 is showed in Fig. 39.2. As can be seen in Fig. 39.2, there was a correlation between season and the average case load and the tuberculosis cases in summer dominated other seasons.

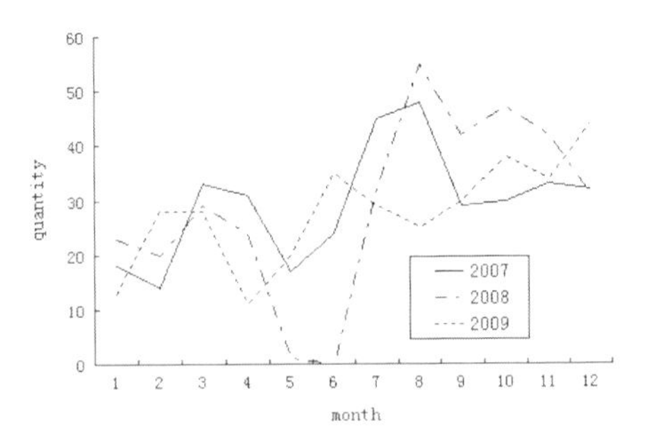

Fig. 39.2 The statistics of tuberculosis cases from 2007 to 2009

39.4.4 Results

GIS techniques were used to perform geographical analysis. All cases and diagnostic and treatment facilities in the study area were mapped using Google Maps. While the different sizes of the background population have not been taken into account in map, it appears that cases tended to concentrate in the densely populated place. The numbers of cases were then aggregated by zip code. This study aggregates data on tuberculosis distribution in the study area to the level of these settlements as with the census data for spatial analysis.

The analysis results show that global spatial autocorrelation indications are 0.2936, 0.3958, and 0.2271 in these 3 years. Global spatial autocorrelation indications are all above 0 in these 3 years. The Moran's I of 3 years all have significance. It shows that the registration rates of tuberculosis are positively correlated.

Through Spatial Lag Model and Spatial Error Model, we find that the number of tuberculosis suspects consulting and the per capita net income of rural population have influence on the registration rate of tuberculosis. The registration rate is higher in the town with more tuberculosis suspects consulting. The more per capita net income of rural population, the lower the registration rate is. Moreover, the density of population has influence on the registration rate of tuberculosis. The analytical results indicated that migrant population was responsible for the transmission of tuberculosis in the Anxian County and resulted in the huge number of drug-resistant tuberculosis cases. Based on the huge field epidemiology data, migrant population, drug-resistant tuberculosis, and TB/HIV coinfection are the three main challenges for tuberculosis control in this study.

39.5 Discussion

While GIS and spatial analysis have been used to study various infectious diseases by several studies, only a few have focused on tuberculosis. The spread of tuberculosis is influenced by many factors and they are often interrelated, such as unemployment, crowding, and other complex socioeconomic factors. The most elusive aspect is to determine the locations and time at which the tuberculosis took place. Migrants are associated with the epidemic of tuberculosis in the Anxian County. The possibility of significant geographic variation about tuberculosis in the study area has been highlighted in this study. This would become the background to a series of researches whether environmental differences, organism, and particular host may be explanatory. In this respect, it is important to identify areas of those with significantly low rates of disease as well as significantly high rates. This information will assist to guide the provision and optimization of tuberculosis control strategies in China.

Disease mapping in time or space may play a good role in guiding public health policy. Exploring where and when the new cases happen, that is to say, exploring the spatial-temporal hot spot to know the spatial-temporal cluster of tuberculosis, may help the tuberculosis program in disease control activities. The systematic use of spatial-temporal analysis for the surveillance of tuberculosis incidence in the Anxian County aims to supply with the theory evidence for the policy of tuberculosis prevention and control and provide the reference for the similar study. Spatial epidemiology adds to the value of traditional and molecular epidemiology and suggests an alternative method that may provide insight into the transmission of tuberculosis.

While the usefulness of GIS and spatial analysis has been demonstrated in our study, it has some limitations. Firstly, the analysis of time and space has some defects; this study is mainly based on spatial analysis, so it cannot have accurate judgment on high-risk trend over time. Secondly, this study did not consider religious, social and economic, environmental, and lifestyle factors of monitored cases. Thirdly, the data sample size in this study is small, and the credibility and stability of the results need further validation. Therefore, future research should undertake broader study area and obtain more tuberculosis case data.

39.6 Conclusion

This chapter mainly discussed how to use GIS techniques on tuberculosis management and also discussed the methods and the steps of building the information system of tuberculosis management. The system is able to manage the spatial and attribute data at the same time and query information using SQL; zoom in, zoom out, and pan the layers; and display raw data, do some analysis, as well as manage report, and so on. Yet, some functions of the system are under exploitation or consideration.

GIS is able to work together with the spatial data and the attribute data of tuberculosis, track its spread, and display the outcomes of complicated raw data, as well as their analysis, by visual map. In this way the decision-makers could master the epidemic of tuberculosis dynamically and then take better measures to control tuberculosis. Using molecular epidemiological surveillance combined with GIS analysis may be an effective method for identifying tuberculosis transmission and surveillance. Targeted screening and control efforts can be strengthened by using these methods, with the goal of incidence reduction and ultimately interruption of disease transmission.

This study demonstrates that previously undetected tuberculosis transmission can be identified if using GIS, spatial analysis, and existing health data in conjunction with epidemiological analysis. These results were used to design new targeted screening efforts. Studies of these efforts have utility in reducing tuberculosis transmission. This study may provide researchers with a meaningful picture of the disease patterns from epidemiological studies. This work will support to establish effective tuberculosis control measures and provide new thought and method to other infectious diseases.

References

1. CDC. (2006). *Morbidity and mortality weekly report*. Retrieved April 20, 2013, from http://www.cdc.gov/mmwr/pdf/wk/mm5511.pdf
2. Maher, D., & Raviglione, M. (2005). Global epidemiology of tuberculosis. *Clinics in Chest Medicine, 26*(2), 631–633.
3. Porter, J. D. (1999). Editorial: Geographical information systems (GIS) and the tuberculosis DOTS strategy. *Tropical Medicine and International Health, 4*(10), 631–633.
4. Moonan, P. K., Bayona, M., Quitugua, T. N., et al. (2004). Using GIS technology to identify areas of tuberculosis transmission and incidence. *International Journal of Health Geographics, 3*(1), 23–33.
5. Rodrigues, A. L., Jr. (2006). Distribuição espacial da co-infecção M. tuberculosis/HIV no Estado de São Paulo, 1991. *Revista de Saude Publica, 40*(2), 265–270.
6. Tiwari, N., Adhikari, C. M. S., Tewari, A., et al. (2006). Investigation of geo-spatial hotspots for the occurrence of tuberculosis in Almora district, India, using GIS and spatial scan statistic. *International Journal of Health Geographics, 5*(1), 33–44.
7. World Health Organization (WHO) (2003). Global Tuberculosis Program. Treatment of tuberculosis: guidelines for national programs [WHO/CDS/TB/ 2003.313]. 3rd ed. Geneva: WHO.

Chapter 40
Construction Method of Exception Control Flow Graph for Business Process Execution Language Process

Caoqing Jiang, Shi Ying, Shanming Hu, and Hua Guan

Abstract Traditional control flow graph of exception handling lacks an explicit description of exception handling and propagation and cannot be used to well analyze the exception situations and exception handling error. To solve these problems, this chapter presents a construction method of exception control flow graph (ECFG) for BPEL process. This method uses a label that is marked exception and is of power for collection computing to describe exception information of BPEL process in building the ECFG. Moreover, the experiment shows that the ECFG generated can clearly express exception information and propagation process in BPEL process.

40.1 Introduction

Business process execution language (BPEL) is different from other languages such as Java at exception handling mechanism [1] which makes exception control flow graph (ECFG) in BPEL different from the construction method of traditional ECFG, and thus the study of construction method of ECFG for BPEL is a necessity [2].

There exist distinctness in the ingredients and mechanisms of exception handling of programming language; thus, researchers often propose different construction method of ECFG toward different languages [3], such as Java and C++

C. Jiang (✉)
The State Key Lab of Software Engineering, Wuhan University, Wuhan 430072, China

Guangxi University of Financial and Economics, Nanning 530003, China
e-mail: jcqng@163.com

S. Ying • H. Guan
The State Key Lab of Software Engineering, Wuhan University, Wuhan 430072, China

S. Hu
Guangxi University of Financial and Economics, Nanning 530003, China

W.E. Wong and T. Zhu (eds.), *Computer Engineering and Networking*, Lecture Notes in Electrical Engineering 277, DOI 10.1007/978-3-319-01766-2_40,

language, and it has become the focus of research. However, these traditional ECFG do not usually explicitly describe the exception handling and does not reflect the exception propagation and thus cannot effectively analyze exception situations and exception handling error. To solve this problem, based on a Signed-TypeSet domain compact notation, Prabhu et al. adopted inter-procedure data stream algorithm to obtain the C++ inter-procedural ECFG (IECFG) [4]. The study identifies the exception type, so it can produce a more detailed CFG and can also show which exceptions can be thrown out or spread from the method call. Based on this idea, the chapter proposes a construction method of ECFG for BPEL. Firstly, we introduce an exception label domain which is of power for collection computing and in which label elements not only are able to describe the exception information of nodes in ECFG but also reflect the exception throwing, propagation, and result after exception elimination or rethrow. Then we present a set of construction algorithms of ECFG for BPEL. Finally, an example demonstrates the feasibility and effectiveness of this method. We summarize our contributions below: (1) We propose a construction method of ECFG for BPEL process. The method is the first approach used to construct ECFG for BPEL, and the method adopts an exception label to describe throwing exception by which we facilitate to analyze exception propagation. (2) We design and implement a set of ECFG construction algorithm to model BPEL process including exception handling structure, and we can use the tools to build ECFG for a given arbitrary BPEL process.

40.2 Exception Control Flow Graph

In order to accurately describe and analyze ECFG, this chapter in turn gives formal definitions of exception labels and ECFG.

Exception label is used to describe the needed exception handling at exception edges and nodes. Exception label is described as exception collection, i.e., the information contained in exception label is various exceptions to be treated.

Definition 1 The exception label set domain Γ is defined as $\Gamma = \{\Gamma pro \mid \Gamma pro \subseteq \{e \mid e \text{ is an exception type in BPEL process}\}\}$.

Definition 2 The exception label τ is defined as $\tau = \{e \mid e \text{ is exception type thrown to be treated in BPEL process}\}$.

We define the set operations on Γ. Given two elements τ_a and τ_b in Γ, element τ_c is the result of the operation of these two elements in Γ:

Union($\cup_\Gamma$): if $\tau_a \in \Gamma$, $\tau_b \in \Gamma$, then $\tau_c = \tau_a \cup_\Gamma \tau_b = \{t \mid t \in \tau_a \vee t \in \tau_b\}$;
Intersection($\cap_\Gamma$): if $\tau_a \in \Gamma$, $\tau_b \in \Gamma$, then $\tau_c = \tau_a \cap_\Gamma \tau_b = \{t \mid t \in \tau_a \wedge t \in \tau_b\}$;
Difference($-_\Gamma$): if $\tau_a \in \Gamma$, $\tau_b \in \Gamma$, then $\tau_c = \tau_a -_\Gamma \tau_b = \{t \mid t \in \tau_a \wedge t \notin \tau_b\}$;
Equality($=_\Gamma$): if $\tau_a \in \Gamma$, $\tau_b \in \Gamma$, then $\tau_c = \tau_a =_\Gamma \tau_b = \{t \mid \text{if } \tau_a \subseteq \tau_b \wedge \tau_b \subseteq \tau_a$ then $t = \text{true else } t = \text{false}\}$;

ECFG for main program of process (ECFG4MP) is used to describe execution of the various activities in the main program of BPEL process. Its formal definition is as follows:

Definition 3 ECFG4MP denoted by G_M is a 7-tuple $(N, E_{reg}, E_{excep}, E_{exceps}, E_c, n_{ps}, n_{pe})$, where:

(1) N is the set of nodes of the graph which is composed of the following subset: $N = N_{reg} \cup N_c \cup N_{cret} \cup N_{ecret} \cup N_{throw} \cup N_{catch} \cup \{n_{ps}, n_{pe}\}$, where ① N_{reg} is a set of normal nodes, ② N_c is a set of call nodes, ③ N_{cret} is a set of call-return nodes, ④ N_{ecret} is a set of exception-call-return nodes, ⑤ N_{throw} is a set of throw nodes, ⑥ N_{catch} is a set of header nodes of catch blocks, and ⑦ n_{ps}, n_{pe} are start and exit nodes, respectively.
(2) E_{reg} is the set of normal control flow edges: $E_{reg} \subseteq (N_{reg} \cup N_{cret} \cup N_{catch}) \times N$.
(3) E_{excep} is the set of exception control flow edges: $E_{excep} \subseteq ((N_{ecret} \cup N_{throw}) \times N_{catch} \times \Gamma)$.
(4) E_{exceps} is the set of exception-call-summary edges: $E_{exceps} \subseteq (N_c \times N_{ecret} \times \Gamma)$.
(5) E_c is the set of normal call-summary edges: $E_c \subseteq N_c \times N_{cret}$.

Definition 4 Service invocation exception graph (SIEG) is defined as 3-tuple $G_C = (\tau_e, N, E)$, where:

(1) τ_e is possible exception thrown of the service.
(2) $N = \{n_s, n_e, n_{excepe}, n_{idea}\}$, where, n_s are start nodes of service invocation, n_e are normal end nodes of service invocation, n_{excepe} are exception end nodes of service invocation, and n_{idea} are concept nodes of service invocation which is used to describe inner abstract logic of the service invocation.
(3) $E = \{(n_s, n_{idea}), (n_{idea}, n_e), (n_{idea}, n_{excepe}, \tau_e)\}$, where edge (n_s, n_{idea}) denotes that control logic flows of called service, edge (n_{idea}, n_e) denotes normal end of called service, and edge $(n_{idea}, n_{excepe}, \tau_e)$ denotes exception end of called service and is marked with label τ_e which describe possible exception thrown in the called service.

Definition 5 ECFG for BPEL process (ECFG4BP), abbreviated as ECFG. It is defined as 5-tuple $G_S = (n_s, n_e, E_c, E_{excep}, G_U)$, where:

(1) $n_s = n_{ps}$ is the start node of ECFG for BPEL process.
(2) $n_e = n_{pe}$ is the end node of ECFG for BPEL process.
(3) $E_c = \{(n_c, n_s) | n_c \in G_M, n_s \in \cup_{C \in B(c)} G_C\} \cup \{(n_e, n_{cret}) | n_e \in \cup G_{C \in B(c)} G_C, n_{cret} \in G_M\}$ is the set of normal call edges, which are edges between the call node in ECFG4MP and the start node in SIEG of called service and between the normal-exit node in SIEG of called service and the call-return node in ECFG4MP.
(4) $E_{excep} = \{(n_{excepe}, n_{ecret}, \tau_{exit}) | n_{excepe} \in \cup_{C \in B(c)} G_C, n_{ecret} \in G_M\}$ is the set of return edges of called service, which is edges between the exception-exit node in SIEG of called service and the service-call-exception-return node in ECFG4MP.
(5) $G_U = \cup_{C \in B(c)} G_C \cup G_M$ is the union of ECFG4MP and SIEG in BPEL process.

40.3 Construction Method for Exception Control Flow Graph

BPEL process is the application program which composes various web services to complete certain functions. Thus, when ECFG for BPEL process is to be built, BPEL process combines corresponding SIEG of the call point, i.e., we may form ECFG for BPEL process by call edges and return edges of called service.

Construction method of ECFG4MP is obtained by performing post-order traversal on abstract syntax tree (AST) for BPEL code, i.e., when analyzing <process>, you must first analyze the various components active in the <process>; when analyzing activity, we need take a different approach toward the type of activity; when analyzing <invoke> activity, build up not only its call node and normal return node but also exception-return node according to the possible exception thrown. When analyzing basic activity which is not <invoke> activity, we need to build the corresponding node; when analyzing structure activity, we need to analyze its composition activities according to the semantics of the structured activity.

BPEL process is a special kind of scope, so its construction algorithm of CFG is similar to scope. When visiting AST's scope node, we establish corresponding a set of nodes and edges on ECFG. For every call, the *VisitScope* algorithm returns a 4-tuple $<N_b, N_e, N_{unresn}, N_{unrese}>$, where N_b and N_e, respectively, are the set of nodes corresponding to start and normal exit of the ECFG region corresponding the current AST node. N_{unresn} is the set of nodes in an ECFG region that has some unresolved incoming or outgoing edges, which is resolved by an ancestor's visitor. For example, if a throw node is enclosed within a scope and has no catch node to match, then out edges of throw are given by its upper catch node of the enclosed scope, that is, to establish an exception edge for the catch node.

The ideas of *VisitScope* algorithm are as follows: First, construct the ECFG nodes and edges for scope block and all catch handling programs by visiting them recursively. It then divides the exception-related ECFG nodes in the scope block into three sets: (1) throw node, (2) the exception-call-return node, and (3) the exception node thrown from lower scope. For throw node, build an exception edge from the throw node to the corresponding head node of catch handling programs, which is marked with exception label. The information in throw node is always a single exception type which is a throw expression type. For exception-call-return node n_{ecret}, a map $ECR\Gamma$ is used to obtain the exception type set for node n_{ecret}. When there is a matching catch block, $ECR\Gamma$ (n_{ecret}) decreasingly update the rest of the exception type (using the difference operator $-\Gamma$). If these exceptions have no matching catch block, they can be thrown from this call node to the upper scope. And we mark n_{ecret} with the final value of $ECR\Gamma$ (n_{ecret}). If these exceptions have no matching head node of catch block, we build an exception edge from exception-call node to the head node of catch block and mark the appropriate exception information. For the exception nodes $n_{rethrowUp}$ thrown from lower scope, construction method for ECFG is similar to exception-call-return node.

Algorithm 1: *VisitCall* algorithm
Input: i: A invoke active;
Output: $(N_b, N_e, N_{unresn}, N_{unrese})$;
1: $N_c=N_c \cup \{n_{ci}\}$; $N_{cret}=N_{cret} \cup \{n_{creti}\}$; $N_{ecret}=N_{ecret} \cup \{n_{ecreti}\}$;
2: $op=CallTargets(typeOf(n_{ci}))$where $calltriple(n_{ci}, n_{creti}, n_{ecreti})$;
3: let $G_{Oop}=<\tau_{eop}, n_{sop}, n_{eop}, n_{excepeop}, n_{ideaop}, E_{op}>$; $\tau e=\{ n_{ci}, n_{ecreti}, \tau_{eop} \}$;
4: $E_c=E_c \cup \{(n_{ci}, n_{creti})\}$; $E_{excep}=E_{except} \cup \{(n_{ci}, n_{ecreti}, \tau_e)\}$;
5: $N_b=\{n_{ci}\}$; $N_e=\{n_{creti}\}$; $N_{unresn}=\{n_{creti}\}$; $N_{unrese}=\{n_{ecreti}\}$;
6: return $(N_b, N_e, N_{unresn}, N_{unrese})$.

Fig. 40.1 Construction algorithm of ECFG for <invoke> activity

Fig. 40.2 Construction algorithm of ECFG for <throw> activity

Algorithm 2: *VisitThrow* algorithm
Input: a: A throw /rethrow active;
Output: $(N_b, N_e, N_{unresn}, N_{unrese})$;
1: $N_{throw}=N_{throw} \cup \{n_a\}$;
2: $N_b=N_{unrese}=\{n_a\}$; $N_e= N_{unresn} =\{\}$;
3: return $(N_b, N_e, N_{unresn}, N_{unrese})$.

Fig. 40.3 Construction algorithm of ECFG for exception handling program

Algorithm 3: *VisitHandler* algorithm
Input: eh: A handler where $h=(t\ v)\ b1$;
Output:$(N_b, N_e, N_{unresn}, N_{unrese})$;
1: let $(N_{b1}, N_{e1}, N_{unresn1}, N_{unrese1})=VisitBlock(b1)$;
2: $N_{catch}=\{n_{eh}\}$; $E_{reg}=E_{reg} \cup \{(n_{eh}, n_s)|n_s \in N_{b1}\}$;
3: $N_b=\{n_{eh}\}$; $N_e=N_{e1}$; $N_{unresn}=N_{unresn1}$; $N_{unrese}=N_{unrese1}$;
4: return $(N_b, N_e, N_{unresn}, N_{unrese})$.

Fig. 40.4 Construction algorithm of ECFG for BPEL process

Algorithm 4: *BuildSECFG* algorithm
Input: ss: Service software;
Output: $G_S=(n_s, n_e, E_c, E_{excep}, G_U)$;
1: $G_U = \cup_{c \in ss} G_O \cup G_M$;
2: foreach $calltriple(n_{call}, n_{cret}, n_{cexcepret})$ do
3: $op=CallTargets(typeOf(n_{call}))$;
4: let $G_{Oop}=<\tau_{eop}, n_{sop}, n_{eop}, n_{excepeop}, n_{ideaop}, E_{op}>$;
5: let $E_c=E_c \cup \{(n_{call}, n_{sop}), (n_{eop}, n_{cret})\}$;
6: $E_{excep}=E_{excep} \cup \{(n_{excepeop}, n_{cexcepret}, \tau_{eop})\}$;
7: $E_c=E_c - \{(n_{call}, n_{cret})\}$; $E_{excep}=E_{excep} - \{(n_{call}, n_{excepcret}, \tau_{eop})\}$;
8: return $G_S=(n_s, n_e, E_c, E_{excep}, G_U)$.

VisitScope algorithm also establishes a head node (n_{scope}) and connection node (n_{join}).

VisitCall algorithm (see Fig. 40.1) establishes three nodes and two edges for invoke activity. *VisitThrow* algorithm (see Fig. 40.2) establishes a determined node for throw activity. *VisitHandler* algorithm (see Fig. 40.3) establishes the appropriate head node for catch block and connects this node to active nodes in the block:

The *BuildSECFG* algorithm (see Fig. 40.4) shows construction method for ECFG of BPEL process based on ECFG4MP and SIEG. ECFG is the union of ECFG4MP and SIEG; it is formed by adding call edge and return edges between ECFG4MP and SIEG. Specific construction method is for each call point in the main program of process, first determine invocation target for each call point. Second, add the following three edges: (1) call edge, from a call node to the start node of SIEG for the target service; (2) call-return edge, from normal-exit node of SIEG for the target service to call-return node; and (3) the exception edge, from exception-exit node of SIEG for the target service to exception-call-return node. And at this exception edge marks the exception information τ_{exit} of the incoming edge of exception-exit node, which is the initial data flow of exception analysis between main programs of BPEL process and called service. Finally, delete the summary edges connecting call node with call-return node and the summary edge connecting call node with exception-call-return node:

40.4 Case Study

The following will give the example of "data collection vehicle assembly station BPEL process" to illustrate the effectiveness of the construction method of ECFG.

1. Design of exception label
   (a) Exception label domain Γ of this process is defined as follows: $\Gamma = \{\Gamma pro |$ $\Gamma pro \subseteq$ {Collecting::UnRegisted, Collecting::InvalidSN, Collecting::Invalid-Schema, Collecting::IncompleteBasic, Collecting::Timeout, Collecting:: UnavailService, Collecting::InvalidVerification, Collecting::SLA, Collecting ::MissingPieces, Collecting::ThanPieces, Collecting::ProviderNoMatch, Collecting::DataRecollection, Down::InvalidCollecting, Down::InvalidUp-Collecting, Down:: UnavailService}}.
   (b) According to possible exception which is thrown, use an exception label to mark corresponding edges. If calling collection service occurs the collection service unavailable exception, data collected appearing QoS exception, then exception label τ is defined as $\tau =$ {Collecting :: UnavailService, Collecting :: SLA}.
2. Construction of ECFG4MP
   Using the method described, AST is traversed post-order and layer by layer to build ECFG for BPEL process. We use examples of invocation of collection service and <faulthander> to illustrate the construction method for ECFG4MP.
   Figure 40.5 is ECFG for collection service. In this figure, Collecting_{ci} is call node, $\text{Collecting}_{creti}$ are normal return nodes after the collection service is called, $\text{Collecting}_{ecreti}$ are exception-return nodes, edge (Collecting_{ci}, $\text{Collecting}_{creti}$) is normal return edge, edge (Collecting_{ci}, $\text{Collecting}_{ecreti}$) is exception-return edge, and the label {Collecting :: UnavailService, Collecting :: SLA} of edge (Collectingci, Collectingecreti) is the possible return exception. Edge (Collectingci, Collectingcreti) and edge (Collectingci, Collectingecreti) are indicated by a dashed line, which means they are the temporary presence, and after SIEG of collection service is combined, they need to be deleted.

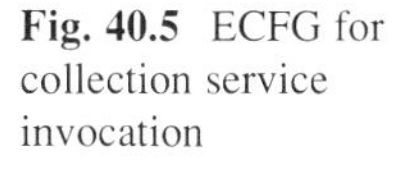

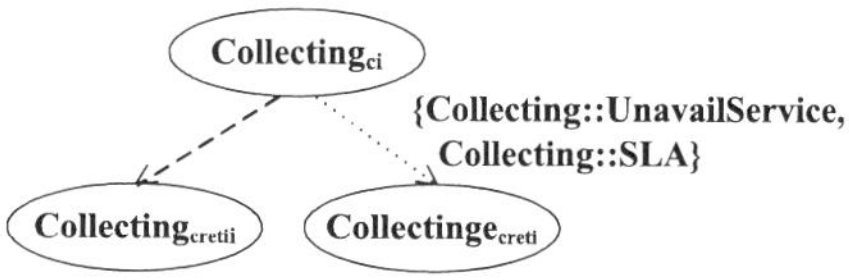

Fig. 40.5 ECFG for collection service invocation

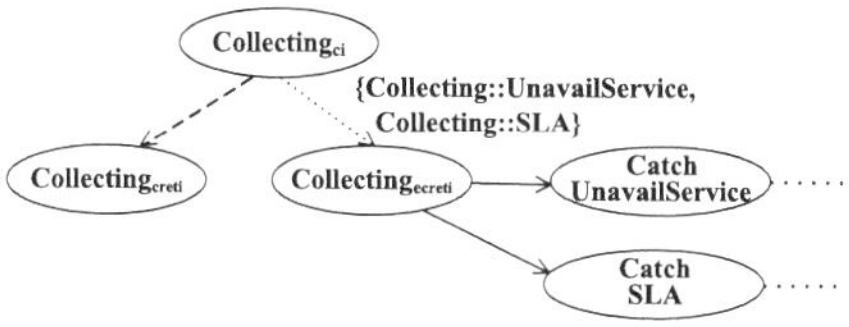

Fig. 40.6 ECFG for collection service invocation and its corresponding exception handler

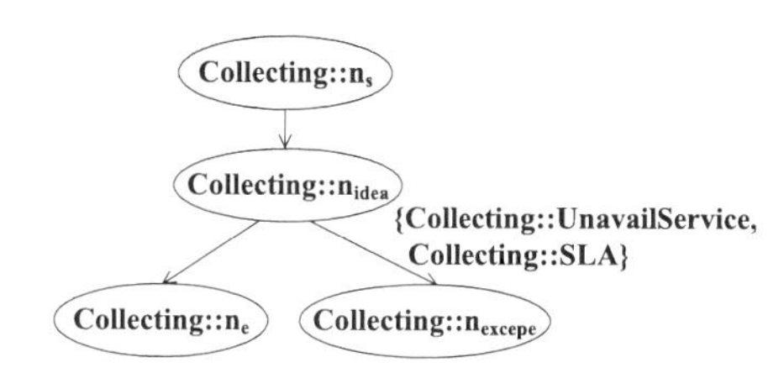

Fig. 40.7 SIEG for the collection service

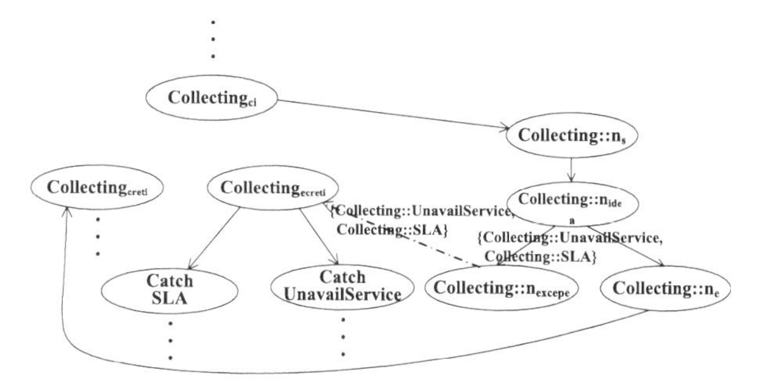

Fig. 40.8 ECFG for "data collection vehicle assembly station BPEL process" (partial)

Figure 40.6 is ECFG for collection service invocation and its corresponding exception handler. After the collection service has been called, it will produce two exception, i.e., Collecting :: UnavailService and Collecting :: SLA, and they are in turn captured by exception handler CatchUnavailService and CatchSLA, where the ellipsis represents specific activities of exception handling.

3. Construction of SIEG
   For the collection service, learn the possible exception thrown through the service's WSDL document: Collecting :: UnavailService, Collecting :: SLA anomalies. Figure 40.7 is SIEG for the service.
4. Construction of ECFG for BPEL process
   ECFG for BPEL process is formed by combining ECFG4MP with SIEG. Figure 40.8 is ECFG for "data collection vehicle assembly station BPEL process" (partial).

40.5 Related Work

There are many differences between exceptions in two-program language, thus construction of ECFG requiring different approaches. Sinha and Harrold [1] incorporate the control flow effects due to explicit Java exceptions in an inter-procedural control flow graph (ICFG) using a flow-sensitive-type analysis. In contrast, our ECFG has an exception-call-return node for every service-call point. Based on byte intermediate representation (BIR) for Java, Demange et al. present construction method for CFG [5]. Following Demange works, Amighi et al. propose a construction method of ECFG for Java byte code, in which the construction procedure divides into two steps: Step 1, Java program is transformed to BIR, and, Step 2, CFG is built through extracting information in BIR [3, 6]. In contrast to all these approaches, our analysis is based on AST.

Jiang et al. propose a construction method of ECFG for C++, in which the throw point of try block is directly conjunct to all corresponding catch block; furthermore, implicit control flow for exception and exception propagation in their ECFG is described [7]. However, construction method is different between AST in our method and source code. Based on a Signed-TypeSet domain compact notation, Prabhu et al. adopted inter-procedure data stream algorithm to obtain the C++ IECFG [4]. Our works follow some idea in constructing IECFG; in contrast to the approach, our exception label set domain Γ has no sign, which makes it easy to understand.

40.6 Conclusion

This chapter has proposed a construction method of ECFG is for BPEL process. The difference in this method from the past construction method of ECFG: When building ECFG, one use label one to describe clearly the exception information in system and its propagation, and one can use this information to analyze exception propagation and uncaught exception.

References

1. Sinha, S., & Harrold, M. J. (2000). Analysis and testing of programs with exception-handling constructs. *IEEE Transactions on Software Engineering, 26*(9), 849–871.
2. Jo, J. W., & Chang, B. M. (2004). Constructing control flow graph for java by decoupling exception flow from normal flow. In *International Conference on Computable Science and Its Applications* (pp. 106–113). Heidelberg: Springer.
3. Amighi, A., de Gomes, P. C., Gurov, D., & Huisman, M. (2012). Sound control-flow graph extraction for Java programs with exceptions. In *Software Engineering and Formal Methods* (pp. 33–47). Heidelberg: Springer.

4. Prabhu, P., Maeda, N., Balakrishnan, G., Ivančić, F., & Gupta, A. (2011). Interprocedural exception analysis for C++. In *ECOOP 2011* (pp. 583–608). Heidelberg: Springer.
5. Demange, D., Jensen, T., & Pichardie, D. (2009). *A provably correct stackless intermediate representation for Java bytecode* (Research Report 7021). INRIA.
6. Amighi, A., Gomes, P., & Huisman, M. (2011). *Provably correct control-flow graphs from Java programs with exceptions* (Technology Report). KTH Royal Institute of Technology.
7. Jiang, S., & Jiang, Y. (2007). An analysis approach for testing exception handling programs. *SIGPLAN Notices, 42*(4), 3–8.

Chapter 41
P300 Detection in Electroencephalographic Signals for Brain–Computer Interface Systems: A Neural Networks Approach

Seyed Aliakbar Mousavi, Muhammad Rafie Hj. Mohd. Arshad, Hasimah Hj. Mohamed, Putra Sumari, and Saeed Panahian Fard

Abstract Brain–computer interface systems are communicative mediums between human brain and external device. One of the applications of these systems is P300 speller. This application provides the ability to spell the characters on the screen for disabled people. In this study, we review the character recognition and its relation to P300 detection. Then, we used three neural networks models with flexible activation functions to detect P300 patterns from electroencephalographic signals more accurately. The obtained results have shown the accuracy of the character recognition based on the precision and recall measures.

41.1 Introduction

Brain–computer interface (BCI) is a direct communicative pathway between brain and external devices [1–3]. BCI allows people to communicate without having to move by measuring brain activities. Disabled people are mostly benefited from BCI systems. BCI systems are commonly composed of four components as follows: signal acquisition from brain, signal preprocessing, signal feature extraction, and classification. BCI is using noninvasive electroencephalographic (EEG) method to read brainwaves [4]. EEG provides a direct measure of brain activities in milliseconds with temporal solution. BCI systems use feature extraction techniques such as pattern recognition for classification of specific brain's event-related potentials. One of the well-known BCI applications is the P300 speller to detect the characters from screen.

S.A. Mousavi • M.R.H.M. Arshad • H.H. Mohamed • P. Sumari
School of Computer Sciences, Universiti Sains Malaysia, Pulau Pinang 11800, Malaysia
e-mail: pouyaye@gmail.com; rafie@cs.usm.my; hasimah@cs.usm.my; putras@cs.usm.my

S.P. Fard (✉)
School of Mathematical Sciences, Universiti Sains Malaysia, Pulau Pinang 11800, Malaysia
e-mail: saeedpanahian@yahoo.com

W.E. Wong and T. Zhu (eds.), *Computer Engineering and Networking*, Lecture Notes in Electrical Engineering 277, DOI 10.1007/978-3-319-01766-2_41,

The main task in P300 speller paradigm is to detect the P300 peak patterns in EEG accurately and instantly. Detecting P300 patterns lead to identification of corresponding characters in P300 speller paradigm. The accuracy of P300 pattern detection will ensure better character recognition in the classification step. The detection of P300 patterns is a challenging task due to presence of noise and artifacts in EEG signals.

The motivation of this study is to detect P300 patterns in EEG signals more accurately and instantly than the previous researches. The approach of this study is to use backpropagation feedforward neural networks with flexible activation functions for detection of P300 signals. These networks use different channel sets and data sizes. Each network has different activation functions with few free parameters. The significance of flexible activation functions can be expressed in networks' size reduction, and the training becomes faster [5–8]. Moreover, we try to evaluate the character recognition performance by direct measurements in P300 classification.

This study is organized as follows: in Sect. 41.2, BCI P300 speller paradigm is introduced. And the P300 detection is explained. Moreover, the P300 speller database is studied. In Sect. 41.3, the process of proposed methods is given. And data preprocessing and feature extraction is explained. Furthermore, three neural networks models with flexible activation functions are used to detect P300 signals and character recognition. In Sect. 41.4, the result of P300 classification and character recognition is given. In Sect. 41.5, a comparison of results with current methods is depicted. In Sect. 41.6, conclusions are given.

41.2 P300 Speller Paradigm

In BCI, P300 speller is a benchmark paradigm. P300 in BCI was initially introduced by Donchin et al. [9]. P300 signals are reflection of brain perceptive response to outside stimulus. By receiving the P300 signal on specific stimulus, we can assume that brain has perceived that event. P300 is a positive signal occurring about 300 ms after flashing light mostly in parietal lobe and occipital sites of brain [10]. Receiving P300 signal is equivalent to detection where user was looking about 300 ms before receiving the signal. The advantages of P300 in BCI are being straightforward and fast and requiring no practical training. The BCI P300 speller application assists in spelling words using P300 signals. The 6×6 matrix of characters is composed of 26 alphabetical (A–Z) and 10 numerical characters (0–9). This matrix is shown to the user on the computer screen by randomly flashing the rows and columns. Once the user focuses on the character during the flashing light, the signals corresponding to those row and column of depicted character have P300 signal. Detection of P300 signal makes it possible to recognize the character by matching the responses to one of the rows and one of columns. In Fig. 41.1, A shows the brain region that P300 peak is occurring, B shows the P300 signal peak, and C represents the BCI P300 Speller row and column paradigm.

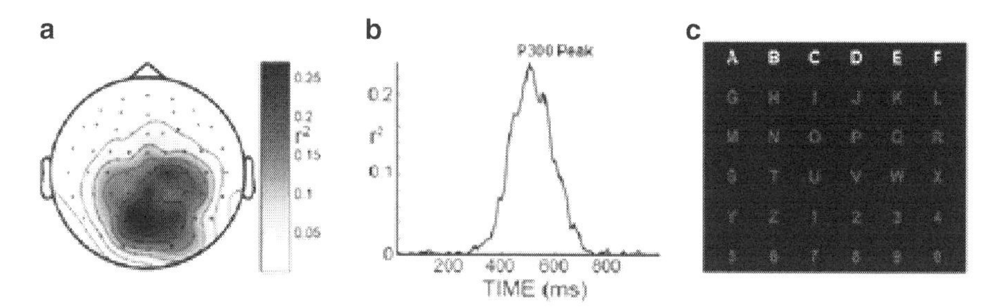

Fig. 41.1 BCI P300 speller paradigm [9]

41.2.1 P300 Detection

The P300 signal is an event-related potential (ERP) in EEG recording [11]. That means P300 signal is a time-locked and averaged EEG pattern. Having accurate detection of P300 signals on the P300 speller paradigm would result in better character recognition. There are two types of classification in P300 speller paradigm. In the first classification type, the P300 signals can be detected in the EEG recordings by having two sets of classes: P300 or none-P300. Due to artifacts or noises, however, there is uncertainty in P300 classification attempt. The second classification type is character identification. The output of P300 detection is applied to classify the application characters and symbols. The character identification process has strong certainty as the characters are given to user clearly.

41.2.2 Database

BCI P300 speller dataset II from the third BCI competition was applied in different methods [12, 13]. This dataset has records of P300 evoked potentials from two subjects A and B. In the experiment, the characters from the presented words are focused sequentially by the subjects. Character selection is done by intensifying a column and a row, and then consequently 2 of the 12 intensifications should contain P300 signals in the samples. Arbitrarily, the intensifications of columns and rows are intersected in blocks of 12. For each character trial which contains the rows/columns, the sets of 12 intensifications are repeated 15 times. Therefore, 30 possible P300 responses should exist for each character. The sampling rate is 240 Hertz (Hz) and is band-pass filtered from 0.160 Hz. The test dataset has 100 characters and training dataset has 85 characters. The number of P300 signals in training dataset is $85 \times 2 \times 5$ and in test dataset is $100 \times 2 \times 15$.

41.3 Methods

41.3.1 Framework

The framework in Fig. 41.2 illustrated the process of the P300 detection and the character recognition. The process starts with preprocessing of the P300 training datasets and extracting the features with building the classifiers. Then, three neural networks models with flexible activation functions are used to train the data. Moreover, The P300 classification results are extracted with the character vectors to apply in test datasets. Finally, the results from character recognition are analyzed and compared with current methods in P300 speller paradigm.

41.3.2 Preprocessing and Feature Extraction

The EEG signals contain the P300 pattern, and those values from the channels are the inputs during the T time. The inputs of neural networks are $I_{i,j}$. The number of channels is defined by i ($1 \leq i \leq 64$) and the timing of the signals is denoted by j. Although the signals are captured in time by number of samples per second, the total number of samples taken for the analysis is determined by multiplying the sampling rate (SR) to the desired time window T in second ($1 \leq j \leq SR \times T$). The dataset divided half by subsampling EEG signal in sampling rate of 120 Hz [13]. Those EEG patterns are then band-pass filtered from 0.5 to 18 Hz only to keep relevant signals. A matrix of $i \times j$ is the input of the neural networks [13]. The three classifiers are presented for feature extraction from EEG signals. These classifiers vary in terms of data size and the number of channels. The classifier NC1 uses the whole training dataset with all channels, whereas the classifier NC2 employs only eight channels. However, the classifier MultiNc uses a multiclassifier strategy. This classifier consist of five classifiers which are trained on different sets of pairs of P300 patterns. In training dataset, the signal patterns without P300 are five times more than the patterns with P300. Each classifier is trained with equal number of P300 and none-P300 patterns [13].

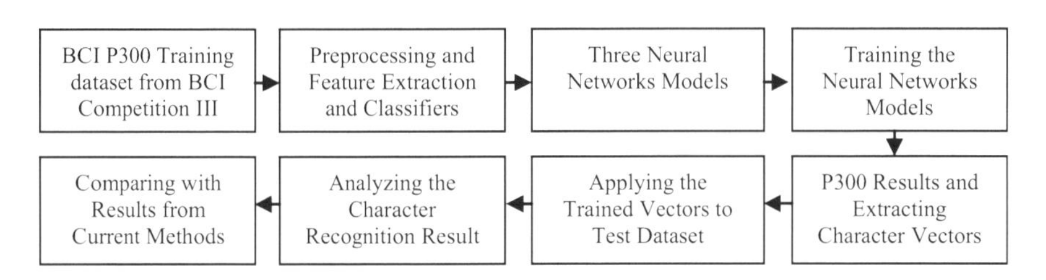

Fig. 41.2 The diagram for the proposed methods

41.3.3 Neural Networks Models

In this section, as a new approach in P300 signal detection and character identification, the following three neural networks models will be used:

- NN1: 35 nodes, 0.9 learning rate, $F_1(x) = A/(1 + e^{-Bx})$ as flexible activation function which was extension of sigmoid function [8]
- NN2: 20 nodes, 0.9 learning rate, $F_2(x) = A_1 Sin(B_1 x) + A_2/(1 + e^{-B_2 x})$ as flexible activation function and was used in financial modeling [6]
- NN3: 25 nodes, 0.2 learning rate, $F_3(x) = e^{-B_1 x^2}(Cos(B_2 x))$ as flexible activation function was used in previous studies [5, 7]

The three neural networks models are chosen to provide a benchmark for the models and classifiers. The functions are chosen due to their capability of being flexible in the networks, and they also have not been used before in BCI P300 speller paradigm. The activation functions are improved using free parameters [5], and the number of nodes in hidden layer and the learning rate β are determined empirically for every model. Moreover, the free parameters are adjusted at the end of each network's epoch. The adjustment of free parameters is like the weights throughout the networks. This adjustment is based on steepest descent rule. This learning algorithm has two main parts as feedforward and error backpropagation. Furthermore, the input and output functions of each neuron in the networks are defined, respectively, in Eqs. (41.1) and (41.2):

$$I_{m,L}(x) = \sum_{n} [W_{m,n,L} O_{m,L-1}(x)] + \theta_{m,k} \tag{41.1}$$

$$O_{m,L}(x) = F_1(I_{m,L}(x)) = \frac{A_{m,L}}{1 + e^{-B_{m,L} \cdot I_{m,L}(x)}} \tag{41.2}$$

The input of mth neuron in the Lth layer is represented by n. Apart from this, the $O_{m,L}(x)$ is used to calculate the value of output from mth neuron in the Lth layer for NN1 flexible activation function. Finally, the output of the neural networks models are character vectors. These character vectors are applied to the test dataset to evaluate the models in character recognition.

41.4 Results

41.4.1 P300 Classification

The important approach from classification of P300 signals is to find out which measurement is in direct relation to character recognition in character identification problem. The recall or the precision is the most directly related measurement in

Table 41.1 P300 classification result

Subj	Model	Class	TP	TN	FP	FN	Recall	Precision	Recog	F-score
A	NN1	NC1	2,216	10,524	4,297	963	69.7	34.0	70.8	45.7
		NC2	1,935	10,355	4,570	1,140	62.9	29.7	68.3	40.4
		MultiNc	2,362	10,160	4,567	911	72.2	34.1	69.6	46.3
	NN2	NC1	2,241	10,465	4,326	968	69.8	34.1	70.6	45.8
		NC2	1,949	10,325	4,580	1,146	63.0	29.9	68.2	40.5
		MultiNc	2,393	10,112	4,575	920	72.2	34.3	69.5	46.6
	NN3	NC1	2,276	10,458	4,296	970	70.1	34.6	70.7	46.4
		NC2	1,964	10,316	4,578	1,142	63.2	30.0	68.2	40.7
		MultiNc	2,461	10,057	4,569	913	72.9	35.0	69.5	47.3
B	NN1	NC1	2,230	11,918	2,903	949	70.1	43.4	78.6	53.7
		NC2	2,001	11,467	3,458	1,074	65.1	36.7	74.8	46.9
		MultiNc	2,493	11,265	3,462	780	76.2	41.9	76.4	54.0
	NN2	NC1	2,255	11,859	2,932	954	70.3	43.5	78.4	53.7
		NC2	2,015	11,437	3,468	1,080	65.1	36.7	74.7	47.0
		MultiNc	2,524	11,217	3,470	789	76.2	42.1	76.3	54.2
	NN3	NC1	2,290	11,852	2,902	956	70.5	44.1	78.6	54.3
		NC2	2,030	11,428	3,466	1,076	65.4	36.9	74.8	47.2
		MultiNc	2,592	11,162	3,464	782	76.8	42.8	76.4	55.0

order to predict the target character, after running the neural networks, the following information is generated (Table 41.1): the number of false negative (FN), true negative (TN), true positive (TP), and false positive (FP) in the test database. The P300 recognition rate is defined as (TP + TN)/(TP + TN + FN + FP); the precision is defined as (TP)/(TP + FP); the recall is defined as TP/(TP + FN); and F-score is defined as 2 (recall.precision)/(precision + recall). In Table 41.1, the neural networks model NN3 shows better results in both subjects. And, Subject B is having better results in comparing to Subject A. Furthermore, the result of P300 recognition rate in Subject B is higher than Subject A. And NN1 and NN3 models show very close P300 recognition rate. The best recognition rates for both subjects are achieved in models NN3 and NN1 by classifier NC1. The results of Subject B indicate best precision is achieved by NN3-NC1 with 44.1 %. However, the best recall is achieved by NN3-MultiNc with 76.8 %. And in Subject A, the best precision and recall are achieved by NN3-MultiNc with, respectively, 35 % and 72.9 %. Therefore, based on the results, NN3 model shows an overall improvement in classification of P300 signals.

The important approach from classification of P300 signals is to find out which measurement is in direct relation to character recognition in character identification problem. Either recall or precision is the most directly related measurement in order to predict the target character.

Table 41.2 Character recognition percentage in NN3 model

		Number of trials			
Classifiers	Subjects	1	5	10	15
NC1	A	14	64	70	91
	B	29	81	86	89
	Mean	21.5	72.5	78	90
NC2	A	12	49	66	79
	B	22	74	82	90
	Mean	17	61.5	74	84.5
MultiNc	A	13	65	76	91
	B	22	81	84	95
	Mean	17.5	73	80	93

41.4.2 Character Recognition

The results from P300 classification are taken as index to identify the target characters in this section. Based on the results in Table 41.1, the neural networks model NN3 has produced better results in P300 detection. Table 41.2 shows the character recognition rate in percentage based on number of trials for each classifier in model NN3.

In Table 41.2, the accuracy of character recognition is developed by increasing the number of trials. On the first observation, more than 65 % of target characters are recognized by having 10 trials. The MultiNc classifier has a mean rate of 93 % for 15 trials. It can be considered a good recognition rate in P300 speller paradigm. Subject B shows better character recognition accuracy in comparing to Subject A. By comparing character recognition result with P300 classification result in NN3 model, recall measurement is better representing the models and classifiers ranking in character recognition problem.

41.5 Comparison of Results with Current Methods

In order to provide a benchmark for this research, we intend to compare the results of proposed neural networks with previous methods [13–16]. Table 41.3 shows the character identification results from other methods including the result from this study. The MultiNc and NC1 classifier from NN3 models are representing our method. ESVM method has achieved the best result for 15 trials [16]. However, our classifiers in proposed NN3 model achieved comparatively better recognition rate when there is only five trials. Unlike the previous methods, the advantage of proposed model is to not consider any channels set nor feature selection before starting. The networks training in our model is done on raw data with some small preprocessing and it is close to real BCI implementation.

Table 41.3 Comparison of results with current methods in BCI

		Methods				
Subjects	Trials	LDA [15]	MCNN [13]	ESVM [16]	NC1	MultiNc
A	5	45	61	72	64	65
	10	78	82	83	70	76
	15	88	97	97	91	91
B	5	76	77	75	81	81
	10	92	92	91	86	84
	15	96	94	96	89	95
Mean	5	60.5	69	72.5	72.5	73
	10	85	87	87	78	80
	15	92	95.5	96.5	90	93

41.6 Conclusion

In this study, we have reviewed BCI P300 speller paradigm as the well-known BCI application. Then we have used BCI P300 speller dataset II from the third BCI competition. A preprocessing and feature extraction was used to extract the P300 signals and none-P300 signals. NC1, NC2, and MultiNc were built from the extracted features. Then the classifiers were input into the three neural networks models with flexible activation functions. Based on the output of these networks, we found out MultiNc classifier in model NN3 showed high character recognition rate by having 93 % in mean for 15 trials. Furthermore, we have derived that four pivots analyze the performance of character recognition. Finally, we have compared our results with current methods in Table 41.3. The obtained results show higher accuracy of P300 signals detection and consequently increasing the rate of characters recognition.

References

1. Allison, B. Z., Wolpaw, E. W., & Wolpaw, J. R. (2007). Brain-computer interface systems: Progress and prospects. *Expert Review of Medical Devices, 4*(4), 463–474.
2. Birbaumer, N., & Cohen, L. G. (2007). Brain-computer interfaces: Communication and restoration of movement in paralysis. *Physiology London, 579*(3), 621–636.
3. Kostov, A., & Polak, M. (2000). Parallel man–machine training in development of EEG-based cursor control. *IEEE Transactions on Rehabilitation Engineering, 8*(2), 203–205.
4. Niedermeyer, E., & Lopes, F. d. S. (2004). *Electroencephalography: Basic principles, clinical applications and related fields*. Lippincott Williams & Wilkins.
5. Özbay, Y., & Tezel, G. (2010). A new method for classification of ECG arrhythmias using neural network with adaptive activation function. *Digital Signal Processing, 20*(4), 1040–1049.

6. Zhang, M., Fulcher, J., & Xu, S. (2002). Neuron-adaptive higher order neural network models for automated financial data modeling. *IEEE Transactions on Neural Networks, 13*(1), 188–204.
7. Subasi, A., Alkan, A., Koklukaya, E., & Kiymik, M. K. (2005). Wavelet neural network classification of EEG signals by using AR model with MLE preprocessing. *Neural Networks, 18*(7), 985–997.
8. Liu, T. I. (1993). On-line sensing of drill wear using neural network approach. *IEEE International Conference on Neural Networks, 2*(1), 690–694.
9. Donchin, E., Spencer, K. M., & Wijesinghe, R. (2000). The mental prosthesis: Assessing the speed of a P300-based brain–computer interface. *IEEE Transactions on Neural Systems and Rehabilitation Engineering, 8*(2), 174–179.
10. Krusienski, D. J., Sellers, E. W., McFarland, D., Vaughan, T. M., & Wolpaw, J. R. (2008). Toward enhanced p300 speller performance. *Neuroscience Methods, 167*(1), 15–21.
11. Coles, M. G. H., & Rugg, M. D. (1996). Event-related brain potentials: An introduction. *Electrophysiology of mind* (pp. 1–27). Oxford Scholarship Online Monographs.
12. Blankertz, B., Muller, K. R., Krusienski, D. J., Schalk, G., Wolpaw, J. R., Schlogl, A., et al. (2006). The BCI competition. III: Validating alternative approaches to actual BCI problems. *IEEE Transactions on Neural Systems and Rehabilitation Engineering, 14*(2), 153–159.
13. Cecotti, H., & Graser, A. (2011). Convolutional neural networks for p300 detection with application to brain-computer interfaces. *IEEE Transactions on Pattern Analysis and Machine Intelligence, 33*(3), 433–445.
14. Blankertz, B. (2008). *BCI competition III final results*. http://ida.first.fraunhofer.de
15. Liang, N., & Bougrain, L. (2008). *Averaging techniques for single-trial analysis of oddball event-related potentials* (pp. 44–49). Proceedings of the Fourth International BCI Workshop and Training Course (INRIA-00337070, version 1).
16. Rakotomamonjy, A., & Guigue, V. (2008). BCI competition III: Data set II ensemble of SVMs for BCI p300 speller. *IEEE Transactions on Biomedical Engineering, 55*(3), 1147–1154.

Chapter 42
Web Content Extraction Technology

Zhenyu Jiao, Xiaoben Yan, Jinjin Sun, Yuchen Wang, and Jiangbin Chen

Abstract In this information era, we are facing the knowledge explosion, and the information on the Internet is multifarious. It is not convenient enough for us to access to information directly on cell phones due to their limitation. Based on parsing a web page with regarding it as a DOM (document object model) tree, we extract the valuable information with considering three factors: structure, content, and programming habits. For illustration, 28 websites are utilized to show the feasibility of the method in web information extraction, and we design the mobile client to present the web content on the cell phones. The practice has proved that using the web page extraction technology related to this article to browse the corresponding news websites only consumed 8 % of cell phone traffic of the existing mobile phone browser. And the user experience is improved. This method can help people to get rid of costing too much on the cell phone traffic, redundant information, complicated operations, and so on.

42.1 Introduction

The twenty-first century is the era of information and the era of data rapid expansion. How fast and efficient accessing to information is becomes an important issue to us. Network is the most commonly used way of obtaining information for us in today's society. Browsing the web with smartphone is the best choice to meet the requirements of our access to information anytime and anywhere.

Comparing with computer, portability and easy access are the advantages of a cell phone, but its unavoidable disadvantage is that it has a smaller screen than a regular computer, and the cost of cell phone traffic is expensive. When using cell phones to browse the webs, which are designed by the size of the computers'

Z. Jiao (✉) • X. Yan • J. Sun • Y. Wang • J. Chen
The College of Information Technology, Nankai University, Tianjin 300071, China
e-mail: j.shower@163.comcph

W.E. Wong and T. Zhu (eds.), *Computer Engineering and Networking*, Lecture Notes in Electrical Engineering 277, DOI 10.1007/978-3-319-01766-2_42,

screen, we need to keep zooming in and zooming out to browse the whole picture of the web and read the specific information we need, which will not offer us a delightful experience. What is more, all of us have almost been through the counting and controlling cell phone traffic month-end. Therefore, every cell phone user hopes to be able to browse the web matching their screen with less traffic cost. At present, there are many websites specially developed for the mobile version of their site to solve this problem. However, this "sweep before your own door" approach is a large consumption on resource, which is not really effective to solve this problem. This chapter puts forward an effective solution to solve this problem and verifies the feasibility of this solution through extracting information from 28 websites.

42.2 Related Work

The existing web information extraction methods can be divided into five main categories: (1) The method of information extraction based on natural language processing. This method regards the web page as plain text, has a slow processing speed, and needs a lot of samples. The text in the web page does not contain complete sentences, so the using range of this method is small. (2) The method of information extraction based on wrapper induction approach has strong pointedness, low expansibility, and poor reusability and only can be applicable to a kind of web page. For different web pages, a lot of the wrappers are needed. (3) The method of information extraction based on the ontology. This way it has less dependence on the structure of web pages, but it needs domain experts to create a clear ontology in this field in detail, and its workload is big. (4) The method of information extraction based on HTML structure. Its flexibility is strong, but it is only suitable for the page which contains obvious regional structure. (5) The method of information extraction based on web query. It has good versatility and scalability, but there is no guarantee for its accuracy. Our goal is to extract information from web pages and present it on smartphones; this technology must have the following characteristics: fast processing speed, strong scalability, and high accuracy. So the above methods are not suitable for solving our problem.

At present, most of the research papers in the field of web information extraction are aimed at a particular type of website. Such as MDR [1] and Depta [2] are for commodity list or form information, these particular websites often carry with formatting information, such as record information and commodity price information. In order to get a general method, recently, some scholars put forward: This problem can be solved by establishing a template database [3]. However, due to the time of template matching being costly, this method is not suitable for browsing the web. As a result, the existing cell phone browser reduces the network traffic just by reducing the image resolution and does not deal with the content for webs which do not have phone version. Our method takes the structure, the content, and the programming habits into consideration. Its processing speed is good, because it

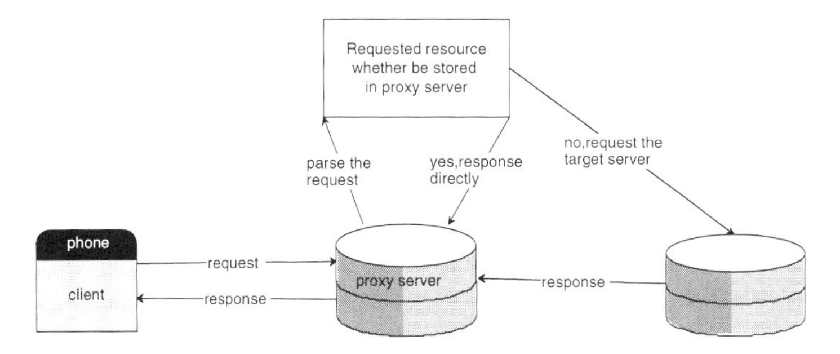

Fig. 42.1 The overall structure of the method

does not rely on a lot of templates. Its scalability and accuracy is good, because it contains a lot of rules, which is confirmed with each other. Our method can handle both strong-structured and loose-structured web pages with a reasonable time, so it meets the need of using mobile phones to browse the web.

42.3 Overall Architecture

We will add a server with the function of filtering (hereinafter referred to as proxy server) between the cell phone and the server of the goal page (hereinafter referred to as the target server); therefore, the users can get useful web content by accessing the server we added instead of accessing the target server directly. We can see the flow chart below (Fig. 42.1).

The proxy server will parse the request from the client and identify the requested content, then respond directly with its resources if contents have been stored. If the requested resource is not in the proxy server, it will put forward a request to the target server, wait for the target server's response, parse the related resource, extract valuable information, establish the image file and store it locally, and then respond the client's request accordingly. When other users request the same page next time, it can be responded directly by the proxy server. In order to prevent storing too much content in the proxy server, we also delete the web image files, which are requested with low frequency.

The proxy server defined in this chapter is not the same as the traditional one. It does not only use traditional caching technology, but also its cached information, which is extracted from web pages, is only part of the original web page.

42.4 Web Content Extraction

Web can be seen as a DOM tree (shown in Fig. 42.2); the images and text in the web are nodes in the tree. Extracting content from a web page can be regarded as picking some nodes from the tree. The core work of the content extraction is how to select useful node and eliminate useless nodes at the same time.

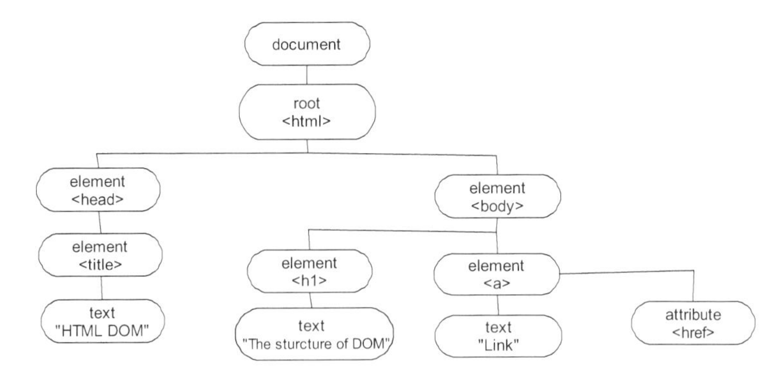

Fig. 42.2 An example of the DOM tree

We need to choose a parser to complete the work of extracting content from a web. There are many kinds of parser and we finally chose the Jsoup [4]. Jsoup is a Java-based parser, which can parse a web both with its URL and with its HTML text content. It provides a set of efficient APIs that can be used to extract and operate data through the DOM, CSS, and a method similar to the operation of the jQuery. We can directly load a document object through the URL of a website, such as using the following code:

Document doc = Jsoup. Connect (“http://example.com/”). Get ();

We will get the content of http://example.com/ and store it in the document object. Then, we can find the elements through the DOM method or the selector method [5]. Through the two methods and the combination of two methods above, we can locate any position in the document. A website generally can be divided into two categories: list page and articles page; we will show you how to deal with them respectively.

42.4.1 *The Process of List Page*

For list page, what we need includes three items: the titles of news, the release time of news (sometimes may not exist), and the link addresses of news. Because the news headlines are some links (these links are used to jump to the content pages), we will pick out all nodes, which are links with the following code:

Elements chooseA = doc. Select (“a”);

After getting these links, we need to find which of them are useful news and which are not what we need. For this discrimination, we are going to carry out a comprehensive approach.

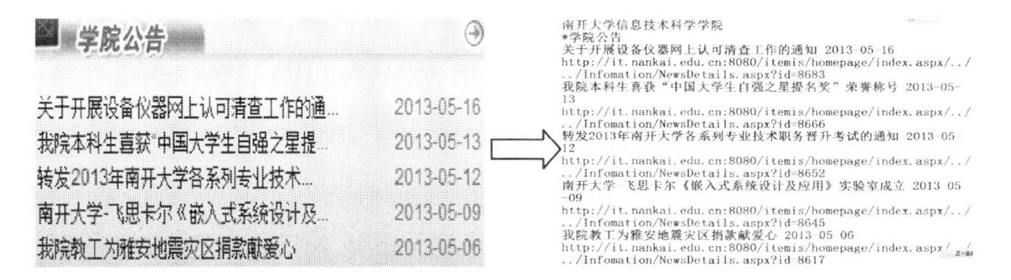

Fig. 42.3 An example of extracting information from a web page

For the beauty of the website, we need to set the CSS file to decorate it. CSS file uses different styles for different categories of nodes, while the same style for the same category of nodes. For news in the list, the nodes often have the same "class" attribute value. Therefore, we can select similar nodes according to its "class" attribute value with using the method <select (".className ")>.

We can locate the news nodes according to the literals. The literals of the news list node are more than the other nodes and have a gathered state. Therefore, we can use the method <text ()> to obtain the text of a node; if a node's literal is significantly higher than others, we can mark it as a news node.

All link nodes have "href" attributes, which are used to determine the URL of the target links. Similar news should be stored under the same path on the server, so the first half of their URL is the same, different from the other kind of nodes at the same time. Since people use meaningful strings to name the directory, the strings such as "News" and "list" often appear in the "href" path. We can use the method <attr ("AttributeKey")> to obtain the attribute values of a node.

Based on the above, we can locate the news list of each website. Then, we need to get the corresponding link address and the release time of the news. URL can be directly gotten from the "href" attributes (sometimes we should change the relative path to the absolute path by adding a prefix), with simply calling the Jsoup method <attr ("href ")>. Due to the news release time has a close relationship to the news; it is often near to the news node, appearing in the DOM tree as the child node or the sibling node of the news node. Then we pick out right ones among these nodes according to the format of release time (numbers separated with space mark). After this we can conveniently get the release time. As shown in figure, we have finished the first step in the conversion (Fig. 42.3):

42.4.2 *The Process of Article Page*

For the article page, what we need from it includes the title of the article, the released information of the article (including released time, author information, which sometimes may not exist), the text of the article, and the pictures in the article.

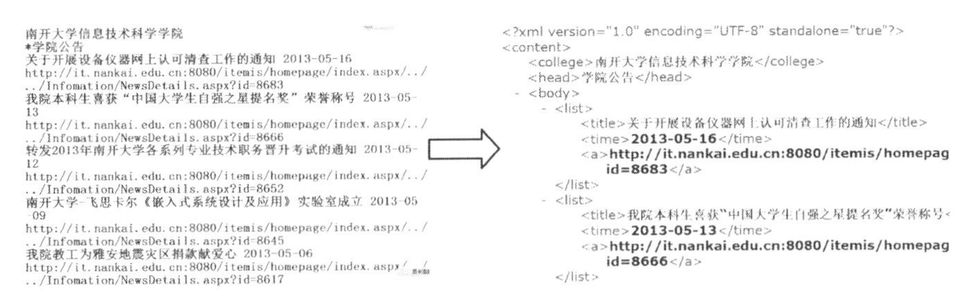

Fig. 42.4 An example of tagging the text

How to locate the body of the article: Article body content is generally in the core of the entire site and has the characteristics of big literals. So the parent node's literals in the position should be significantly higher than that of the other nodes. First of all, you can use the method <parent().text()> to get the text of the parent node; if its literals are significantly higher than the other nodes, this parent node can be thought as at the core of the text. And in order not to regard the entire page as a core, a node should be compared with its sibling nodes layer by layer with a bottom-up approach. If its literals are significantly higher than its sibling nodes, it can be regarded as the core. Once the core node is identified, its sub-nodes are often the main body of the article. Regarding the parent node as the root of a DOM tree, as the body of the article in general would not be center-aligned, so we use the method <select (" p[alige! = center]")> to select all nodes which are tagged as paragraph from the parent node. It is the body of the essay (sometimes may contain the title or the release information).

The title of the article is often located above the core node and inside its parent node. Under normal circumstances, the article title is set tag <class = title> and is center-aligned, so we can get the title with the method <select ("p [class = title], div [class = title] ")>, or <select (" p[alige=center], span [class=title]")> based on recursion. If the returning value of these methods is empty, we will locate to the grandfather node, call again until the returning values are not empty. If there is more than one returning value, the first one shall be the title, the others, and subheadings.

As the release information of the article is center-aligned, we deal with it with the similar method: <select ("alige=center").text()>, but we should pay attention to a case that the returning value is the title. We need to compare the first returning value with the title of the article, if they are the same, and then abandon the former one.

The layout of image is difficult to locate, but it is usually below the title. So we can select images with the method <select("img")> after locating the article title.

We will package the transferred text to get XML (Extensible Markup Language) files for the client to recognize them. We tag the labels according to our needs of transmission and processing. For example, we use the following form for list page (Fig. 42.4):

42.4.3 *Image Compression*

The size of the image is much larger than the text's, so images processing is very important. We get rid of the useless images directly in the parsing process, so we will only deal with the useful images. In order to adapt to the cell phone screen display, reduce traffic burden, we will compress the images. And we will compress the images to different levels for different cell phones according to the specific screen pixel information we get from them, so cell phone users can save their mobile traffic and get the most suitable image size when they view the pictures.

42.5 Experiment Evaluation

According to the above methods, we designed the client on android mobile phones and tested the home pages of Nankai University and its department's websites (Fig. 42.5).

Seen from the diagram above, using our solution has improved our reading experience; the original websites are difficult to read directly on cell phones; after treatment by our system, they become concise and easy to read. We can see a record of actual test results (a small part) in detail in Table 42.1 below.

We can see the data in Table 42.1; after using our solution, the average consumption of the traffic on visiting the listed pages is only 8 % of the cost of UC Browser; the maximum consumption of traffic is less than 16.3 % of UC Browser's way.

For the article page, the test results are as follows (Table 42.2).

We can see the data in Table 42.2; after using our solution, the average consumption of the traffic on visiting the article pages is only 56% of the cost of UC Browser. And the more pictures a web page contains, the more traffic will be saved.

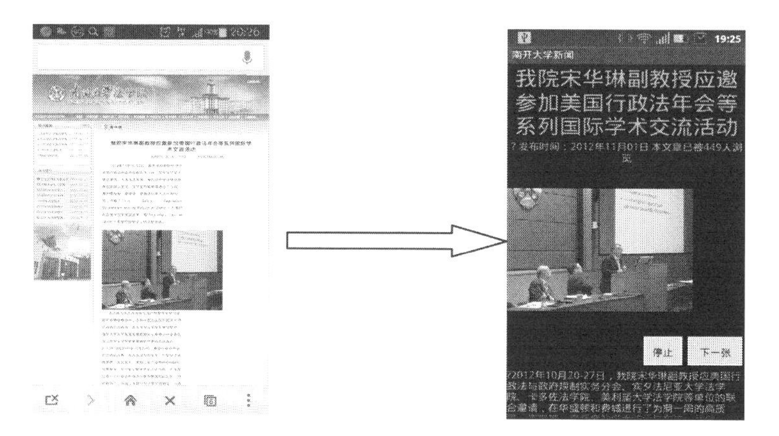

Fig. 42.5 The presentation of using our method to show the web page on phones

Table 42.1 Actual test results of listed page in detail

Tested websites	Traffic consumed by UC Browser	Traffic consumed by our solution
Nankai University	236.36 KB	6 KB
Nankai News	1.26 MB	25 KB
College of Arts	40.96 KB	6 KB
School of History	92.16 KB	4 KB
Philosophy School	0.99 MB	10 KB
Law School	30.72 KB	5 KB
Zhou Enlai School of Government	71.68 KB	11 KB

Table 42.2 Article page results

The sample pages	Traffic consumed by direct access	Traffic consumed by our solution
1	37.8 KB	15.1 KB
2	30 KB	13.7 KB
3	4.3 KB	3.1 KB
4	191.7 KB	45.3 KB
5	11.4 KB	10.0 KB
6	2.6 KB	1.6 KB
7	2.8 KB	1.9 KB

We also measured the consumed time of the two methods to discover that they have little difference, and our method is slightly better.

The causes are as follows: For the user's request, only the first user's request needs to be transmitted to the target server by the proxy server to obtain corresponding returns for making the corresponding image documents on the proxy server. Image document returned by the proxy server will transfer more quickly since it has less content compared to the original pages.

42.6 Conclusion

This article provides the technology of extracting content from the web and displaying it on the cell phones, and it has a certain universal applicability. Because of typifying of web pages on the overall structure and the similarity of the writing among coders, our method can be simply extended to more other types of web pages. It provides a good idea on how to show web page on a phone better in the practical work. We are facing the challenge of big data in this information era; how to extract the most valuable information from numerous and disorderly data for the users is one of the most important challenges to each operator. We believe that our approach would give operators a beneficial inspiration, help every cell phone user to get better, more high-quality service.

Acknowledgements The article is supported by the Nankai University undergraduates innovation research fund (Grant No. BX10-258).

References

1. Liu, B., Grossman, R., & Zhai, Y. (2003). Mining data records in web pages. In *Proceedings of the Ninth ACM SIGKDD International Conference on Knowledge Discovery and Data Mining* (pp. 601–606). New York, NY: ACM.
2. Zhai, Y., & Liu, B. (2005). Web data extraction based on partial tree alignment. In *Proceedings of the 14th International Conference on World Wide Web* (pp. 76–85). New York, NY: ACM.
3. Qiao, F. (2012). *Web information extraction technology based on templated web crawler*. Chengdu: University of Electronic Science and Technology of China (in Chinese).
4. Jonathan Hedley. (2013). *Jsoup cookbook (EB/OL)*. http://jsoup.org/cookbook/
5. Andrew, W. (2005). *Beginning regular expressions* (pp. 34–60). Hungry Minds.

Chapter 43
A New Data-Intensive Parallel Processing Framework for Spatial Data

Dong Zhao, Yang Gu, and Zhenchun Huang

Abstract The explosive increase of scientific data brings in the "Fourth Paradigm" research method by Jim Gray. In order to accelerate the processing speed for these big data, parallel distributed processing is needed. As the data-intensive computing requires high throughput of IO, the data transfer from different node should be cut down as much as possible. Current technologies focus more on the framework for local reliable network with homogeneous resources, but the parallel processing framework for scientific data-intensive problems such as spatial data shared with the Internet and queried by semantics is not fully studied. In this article, we proposed a new data-intensive parallel processing framework for spatial data—Robinia DSSSD (Distributed Storage and Service for Spatial Data), which provides the flexible ability to support data distribution and allocation across the Internet, and semantics query. Experiments shows that Robinia DSSSD can achieve good acceleration with low overhead, and it can well support data-intensive computing.

43.1 Introduction

Scientific experiments nowadays produce large volumes of data, and researchers need to analyze and deal with these big data. This is called "Fourth Paradigm" research method [1, 2]. In order to make this procedure easier to use and accelerate the processing speed, parallel processing is necessary. As scientific data is often generated by different institutes and individuals all around the world in different locations, how to handle the distributed heterogeneous data is of great importance. Previous work focuses more on the local intranet which has reliable network topology and homogeneous nodes for storing and processing. As many valuable data has been shared on the Internet, how to do the data-intensive computing with

D. Zhao (✉) • Y. Gu • Z. Huang
Department of Computer Science and Technology, Tsinghua University, Beijing 10084, China
e-mail: zhaodong8701@gmail.com

W.E. Wong and T. Zhu (eds.), *Computer Engineering and Networking*, Lecture Notes in Electrical Engineering 277, DOI 10.1007/978-3-319-01766-2_43,

Internet-shared scientific data is a challenging problem. Data grid can be viewed as the first try to do the data-intensive computing over the WAN. Since the data volume is huge, how to schedule the task running on the data node is important to decrease the network transfer. In the meantime, parallel processing is needed to accelerate the speed.

Currently many distributed parallel processing frameworks and programming methodologies have been invented to solve this problem, but the differences of scientific problems make the common framework not work well with the actual use. MapReduce [3] receives great attention recent years for simplifying the programming model for distributed data processing. Along with it are the Hadoop [4] and HDFS [5] which provide the platform and distributed file storage supporting the programming model. They have good application and performance in dealing with text processing and online stream analysis for solving big-data problems with the Internet. But when referring to scientific data such as remote sensing satellite images processing, they do not fit quite well. HDFS will automatically split the files into chunks of 64 MB block and store them distributed. But this operation is done without the meaning of metadata and other attributes. The semantics query and header metadata management is also not natively supported by these systems. Furthermore, some algorithm procedures in spatial data could not be easily converted into MapReduce job style, which makes the Hadoop difficult to handle them.

Semantics query can be done by traditional relational database with SQL query. But this has limitation when dealing with parallel computing and emerging big-data technologies, especially toward fully distributed software architecture. These databases run well on a single hardware. However, when required to scale horizontally, commonly recommended scalability approaches like "sharding" (partitioning data to run on multiple machines) often cause issues within data normalization. Therefore, deployment which does not require ACID (atomicity, consistency, isolation, and duration) makes a good candidate to store data in distributed environment with good scalability. NoSQL database is promised to be a good solution for the situation that provides a mechanism for storage and retrieval of data that use looser consistency models compared with traditional relational databases [6]. They mainly offer the storage based on the simple key-value pairs, which gain benefits in scalability and performance. Column-based products include Google's BigTable [7], HyperTable [8], and HBase [9]. HyperTable and HBase are both the open-source implementation of BigTable. The difference is that HyerTable is written in C++ and HBase is written in Java. Performance evaluation [10] shows that HyperTable usually runs twice faster than HBase. Column-based products have advantages in computing aggregation over large numbers of similar data items. Another type of NoSQL database is document-based or row-based, which stores data that share some common properties but may have different ones. As no table is declared, no rigid structure is required to store data on it. These products include MongoDB [11], CouchDB [12], and Redis [13]. MongoDB is good at dynamic queries, frequent data changes. CouchDB is better for data accumulation. Redis has advantages in rapidly changing data with a foreseeable database size. In order to support various semantics query with good scalability, we choose MongoDB to develop. It is an open-source document database and the leading NoSQL database written in C++ which gives good performance.

In the article, we propose a new data-intensive parallel processing framework for spatial data—Robinia DSSSD. The framework uses MongoDB as the database to support metadata query. Robinia platform is a basic component for building large-scale distributed parallel processing framework. Its main features include easy deployment and being light weighted, user friendly, and compatible for old applications running which does not require source code. In short, Robinia is a powerful servlet running on Tomcat, with the ability to do the global registry, event handling, executor management, request and response processing, web UI interface, etc. We will discuss in detail how to build parallel processing framework upon this platform and evaluate its performance.

43.2 Distributed Data Models for Robinia

In order to better support the data-intensive distributed parallel computing, a good data model for this should be well prepared. For the scientific data such as satellite image, it often has large size of file content, together with its metadata info, including the date created and the location record. For Robinia DSSSD, the data model is represented as document-attachment type, which means each data item is made up of one document and zero or more attachments. Documents part consists of key-value pairs. Meta and key-value pairs are structural data which can be indexed and searched. Attachment is nonstructural binary data which is suitable for store files like images or other large-size file. The basic model is shown in Fig. 43.1.

Meta data can describe the metadata of data item, mainly used by querying and indexing the data. Different types of data have different metadata. Metadata can help Robinia DSSSD to query and manipulate data (split or merge) by semantics, which makes the system more useful for spatial data that has lots of additional attributes for reference. This also makes our system different from HDFS which

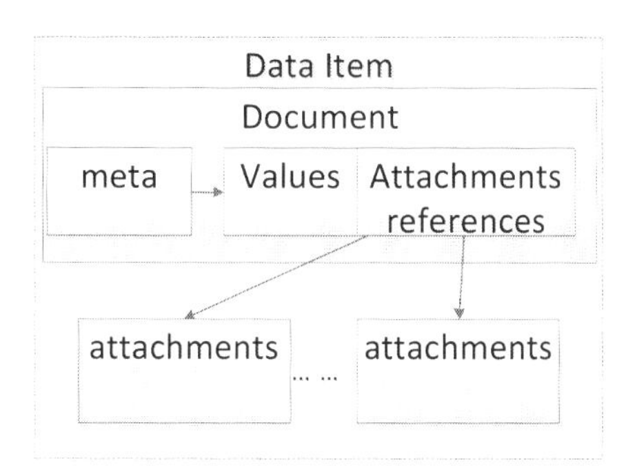

Fig. 43.1 Basic data model for Robinia

often split data by block size and lack support of query by semantics. Values serve as the main body for the data item. It contains all the fields from metadata and has its own unique ID which will be referenced as foreign key in attachments. Attachments store the binary data file with absolute path or remote URL of http or ftp. With this data model, Robinia DSSSD is capable to share and retrieve data from the Internet. In the meantime, query the required data by various semantics, which is quite useful in spatial data area.

43.3 System Overview and Architecture

The base component Robinia can be viewed as a powerful agent or proxy to support upper level application transparently communicates with each other, with the guaranteed services of resource discovering, status monitoring, job running, etc. Service-oriented architecture [14] and RESTful [15] technologies have been used in the system. The framework consists of two parts: data part and parallel processing part. For the data part, a storage cluster is made up of header node, backup header node, and data node. In the data node, we do not store the data as block file but use its original file item in operating system. This makes the data processing easier to get and save files. The architecture figure is shown in Fig. 43.2.

In this architecture, all the search and index operations are done by header node. Each node will be installed the supporting libraries beforehand. Header node is the central node of storage cluster and responsible for all the metadata, providing indexing and querying services. Data node is responsible for storing the real data content, including the values and attachments.

For the parallel processing framework part, master-worker style based on the data allocation is the basic idea for distributed processing. One master executor and a group of worker executors make up of this. Master processes the serial part of the algorithm, does the global schedule, and loads balance for all the workers. Workers only focus on the parallel part of the algorithm. Schedule algorithm can be performed on master node. Data layout, network latency, and node status should be taken into consideration when scheduling task together with data. In order to

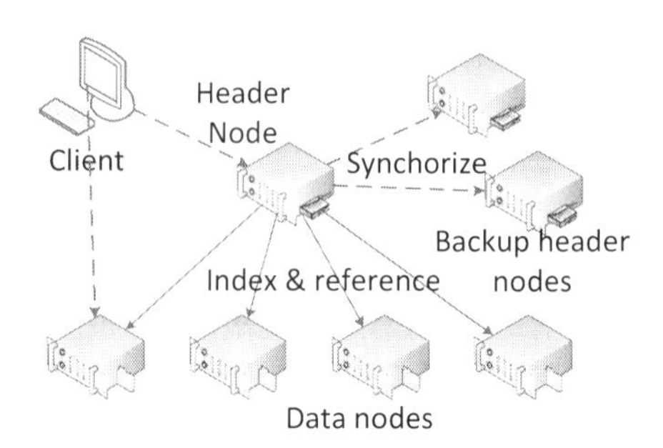

Fig. 43.2 Robinia storage architecture

achieve the minimum time cost for the whole jobs, the auto-adaptive schedule is selected. In this schedule, each processing node will retrieve the input data when it finished the previous task. This can make the system adaptive with heterogeneous node. Powerful node will process more data. Node with good data locality can do more tasks, etc.

43.4 Design and Implementation

In this part we will discuss the details about Robinia DSSSD design and implementation. For the data part, we have several components to do the job. The design diagram is shown in Fig. 43.3. DSSSD provides several interfaces to communicate with MongoDB for CRUD operations (create, read, update, and delete). Large binary file content is stored as attachment, which URL and absolute path is in database and binary content is stored as file in operating system. On the client side, DSSSDClient is the entry point for Java client use. Basic methods for accessing the data in Robinia platform are provided. Processing part provides corresponding framework with the specific data structure. Currently master-worker structure for parallel processing is the default structure. The worker will use the data as the input and output parameter. And the worker will try to choose the local data to reduce the network transfer time cost.

Scheduler will be defined separately to parallel select the data items for workers. When selecting the data item, several criteria should be taken into consideration: local data priority first, data transfer time (related to the data volume and network capability of node), and data processing time (related to the worker algorithm and CPU capability of node). The scheduler for jobs should follow the rules that lower network transfer priority first. The fault-tolerant is guaranteed by the status monitoring and exception catching. Failed test data will be tried until the maximum retry count is reached.

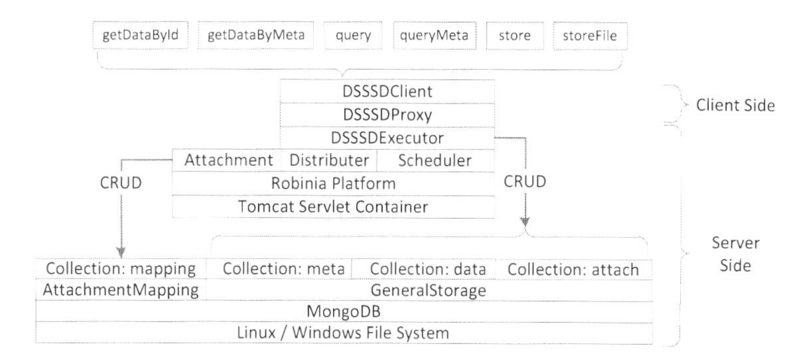

Fig. 43.3 DSSSD data design diagram

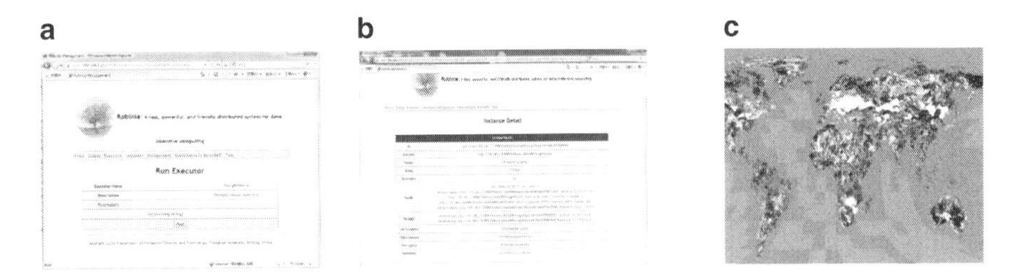

Fig. 43.4 Screenshot of WebUI. (**a**) Input executor parameters, (**b**) view the detailed status of instance, and (**c**) the final result of drought detect algorithm

Figure 43.4 is the snapshot of the web UI for Robinia DSSSD. The user can select the executor to run with parameter and see the detailed status and logs on the web page. The part c is the final result of drought detection global map which we are testing in the evaluation part. The last picture is the final result of drought detection global map which we are testing in the evaluation part.

43.5 Evaluation and Discussion

We use the drought detect algorithm as the test case to measure the performance of Robinia DSSSD. NDWI is short for Normal Differential Water Index, which was brought up by Gao in 1996 [16]. AWI is short for Anomaly Water Index. The calculating formula is

$$\mathrm{NDWI} = (\rho_2 - \rho_5)/(\rho_2 + \rho_5)$$
$$\mathrm{AWI} = NDWI_i - \mathrm{AvgNDWI}$$

ρ_2 and ρ_5 are the reflection rate for band 5 and band 2 in MODIS file. The main time cost of processing is the IO part of reading and writing files, and the calculating is simple as described in the above formula.

Hardware environment includes four pc nodes; each of them has Intel Core i3 @ 2.93 GHz CPU, 4 GB memory, 1 TB 7,200 rpm hard disk, and windows 7 64-bit system.

We first did the experiment with all the data can be gotten locally and compared the processing time cost with different nodes and serial part. The result is shown in Fig. 43.5.

We can see that Robinia DSSSD only put little overhead when processing with one node compared with processing the original sequential program. With the node count increasing, the processing time drops down linearly. It proves that our framework does well with the parallel processing. The auto-adaptive schedule algorithm is used by default, which means all the nodes will retrieve metadata

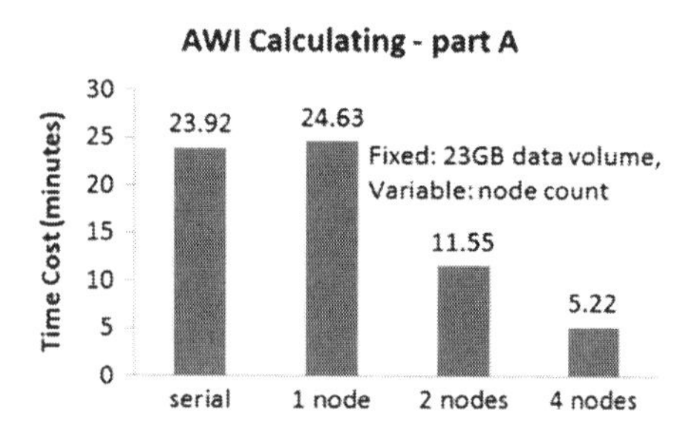

Fig. 43.5 AWI calculating with different node count

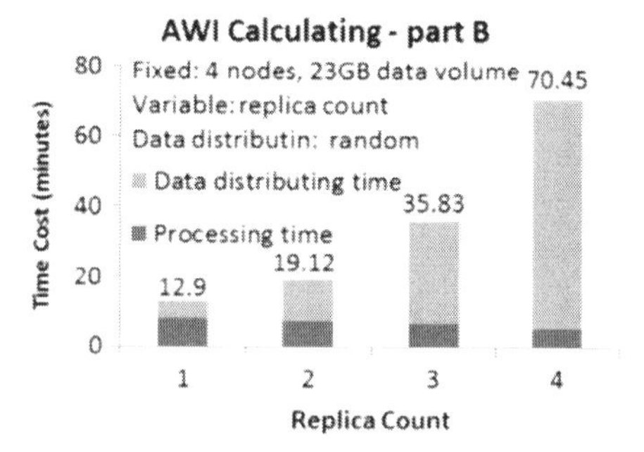

Fig. 43.6 AWI calculating with different replica count

from master node and do the processing work. This schedule can make the total time cost the minimum because fast nodes will do more jobs. Another experiment was done with fixed four nodes and change replica count, which means not all the processing data be gotten locally. The original data source is from an ftp server connected by Gigabyte network. Data distribution for this is random. For example, if the replica is 1, then the whole set of data will be randomly distributed and stored on four nodes. The result shows in Fig. 43.6.

From the figure we can find the processing time differs little, while the data distributing time changes a lot. The reason is when replica is 1, data is almost distributed with each node has 1/4 part of whole set. When processing data, the probability of getting data locally is pretty good. Another experiment was done with fixed four nodes and let the input data volume change. The original data source is from an ftp server connected by Gigabyte network. The result is shown in Fig. 43.7.

From Fig. 43.7, we can find that the system has good scalability when data volume increases. This proves that Robinia DSSSD is capable for the TB level data-intensive computing. Figure 43.8 is the detailed time cost for each node of part C. We can see the auto-adaptive schedule algorithm has good load balance for the overall workload of nodes. And the overhead from Robinia DSSSD is small. Distribution time is mainly limited by the hard disk IO which is currently lower than the Gigabit Ethernet. Figure 43.9 shows the speed ratio compared with different data volume. For the small data volume 23 GB, the speed ratio is 4.58, faster than the linear speed ratio. The reason for this could be related with the disk

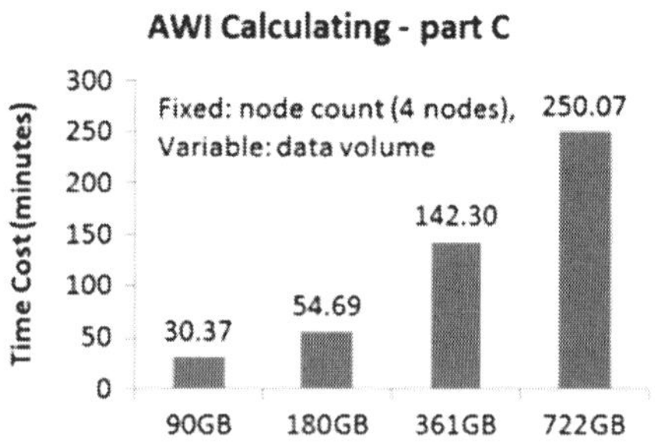

Fig. 43.7 AWI calculating, execution time vs. data volume

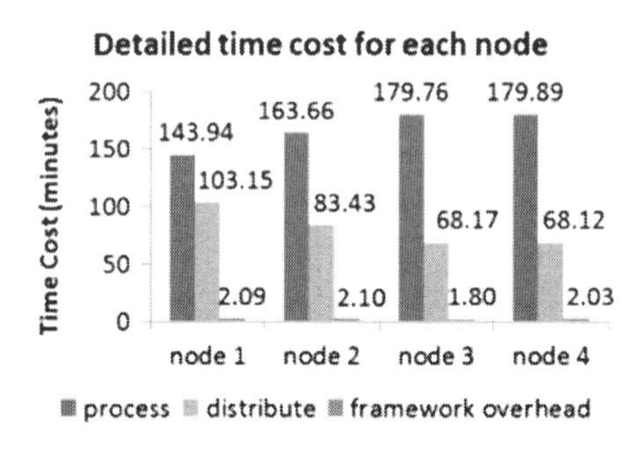

Fig. 43.8 AWI calculating with data volume

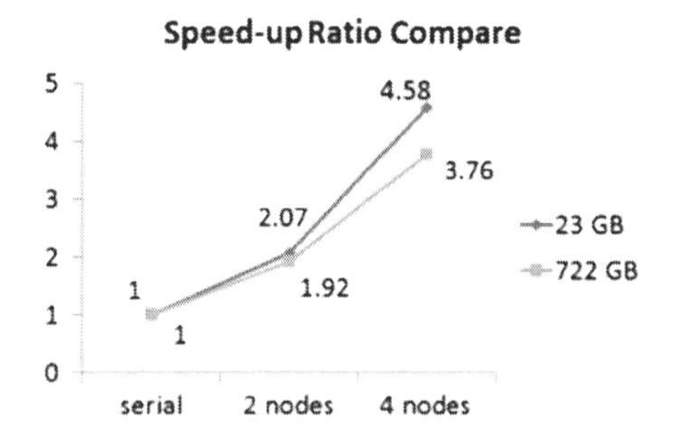

Fig. 43.9 Speedup Ratio Compare

cache, memory limit, etc. For the large data volume 722 GB, the speed ratio is 3.76. The parallel efficiency is 3.763.76/4 $\times$ 100% $=$ 93.75%, showing Robinia DSSSD does well in parallel processing.

43.6 Conclusion

Data-intensive computing plays more and more important role in recent industry and academic areas. In order to better deal with big data, we propose a new data-intensive parallel processing framework—Robinia DSSSD, which is fit for the data-intensive computing over the Internet and supports semantic query well.

System overview and design implementation have been discussed in this article. Experiments show Robinia DSSSD has good performance and flexibility. It can support the data-intensive computing in TB level data. Parallel efficiency can reach 93.75 % with large data volume.

Robinia DSSSD has just implemented the basic functions for supporting data-intensive parallel processing. Lots of work can be done in the future, including the global computing over wide area network, which can take full use of all computing capability of nodes, and algorithm migration, which can better support the task running on data nodes. Security enhancement is to be refined to enhance the system authentication mechanism and resource access control.

References

1. Hey, T., Tansley, S. and Tolle, K. (2010). The fourth paradigm of science research – a brief introduction to Jim Gray on eScience: A transformed scientific method. *e-Science Technology and Application, V1*(2), 92–94.
2. Hey, T., Tansley, S., & Tolle, K. (2009). *The fourth paradigm: Data-intensive scientific discovery* (pp. xvii–xxxi). Redmond, WA: Microsoft Research.
3. Dean, J., & Ghemawat, S. (2008). MapReduce: Simplified data processing on large clusters. *Communications of the ACM, 51*(1), 107–113.
4. Hadoop. http://hadoop.apache.org/
5. Shvachko, K., Kuang, H., Radia, S. & Chansler, R. (2010). The hadoop distributed file system. In *2010 I.E. 26th Symposium on Mass Storage Systems and Technologies* (pp. 1–10). Piscataway, NJ: IEEE.
6. Han, J., Haihong, E., Le, G. & Du, J. (2011). Survey on NoSQL database. In *2011 6th International Conference on Pervasive Computing and Applications (ICPCA)* (pp. 363–366). Piscataway, NJ: IEEE.
7. Chang, F., Dean, J., Ghemawat, S., Hsieh, W. C., Wallach, D. A., Burrows, M., & Gruber, R. E. (2008). Bigtable: A distributed storage system for structured data. *ACM Transactions on Computer Systems (TOCS), 26*(2), 4.
8. HyperTable. http://hypertable.org/
9. HBase. http://hbase.apache.org/
10. Bunch, C., Chohan, N., Krintz, C., Chohan, J., Kupferman, J., Lakhina, P., & Nomura, Y. (2010). An evaluation of distributed datastores using the AppScale cloud platform. In *2010 I.E. 3rd International Conference on Cloud Computing (CLOUD)* (pp. 305–312). Piscataway, NJ: IEEE.
11. MongoDB. http://www.mongodb.org/
12. CouchDB. http://couchdb.apache.org/
13. Redis. http://redis.io/
14. Perrey, R., & Lycett, M. (2003). Service-oriented architecture. In *2003 Symposium on Applications and the Internet Workshops* (pp. 116–119). Piscataway, NJ: IEEE.
15. Richardson, L., & Ruby, S. (2008). *RESTful web services* (pp. 49–79). Sebastopol, CA: O'Reilly Media.
16. Gao, B. C. (1996). NDWI-a normalized difference water index for remote sensing of vegetation liquid water from space. *Remote Sensing of Environment, 58*(3), 257–266.

Chapter 44
The Approach of Graphical User Interface Testing Guided by Bayesian Model

Zhifang Yang, Zhongxing Yu, and Chenggang Bai

Abstract GUI (graphical user interface) is becoming increasingly important in the software field for the reason that it is a friendly way for the users to interact with the software through GUI. Testing in GUI, however, is faced with many challenges, due to the immense number of event interactions. Testing all possible event interactions is impossible, since the number of required test case is huge in numbers. GUI testing mainly serves two goals: First, to establish confidence in assessment of GUI; Second, to find that more software defects in GUI testing while limiting the number of test cases. For this purpose, any testing method must be better at detecting defects. This article proposed a new technique that can be used for GUI testing, which can guide the GUI testing and find more defects as soon as possible. In this chapter, it introduces an approach of GUI testing guided by Bayesian model optimization scheme, discusses the Bayesian model topology and its issues encountered in the modeling process. It presents solutions in connection with the parameters problem. In the end, a simple case verifies the validity of the model during the GUI testing.

44.1 Introduction

Graphical user interface (GUI) is becoming increasingly important in the software field, while GUI testing problem has never been resolved absolutely. GUI testing is becoming the key issues restricting GUI rapid development.

Many researchers have already done a lot of meaningful work on GUI. White et al. [1, 2] and Belli [3] pointed that various responsibilities of the user can be specified as a complete interaction sequence (CIS) between the user and the GUI application under testing. Then they proposed a concept of finite-state machine for

---

Z. Yang (✉) • Z. Yu • C. Bai
School of Automation Science and Electrical Engineering, Beihang University, Beijing 100191, China
e-mail: qwyzf@163.com

W.E. Wong and T. Zhu (eds.), *Computer Engineering and Networking*, Lecture Notes in Electrical Engineering 277, DOI 10.1007/978-3-319-01766-2_44,

CIS. A.T. Memon et al. [4] proposed that the GUI under testing is a hierarchical structure. They proposed that the test case generation problem of GUI testing is an AI planning problem. Then, they proposed several approaches guiding the GUI testing, such as usage profiles [5], dynamic adaptive automated test generation [6], and incorporating event context [7]. In the work of H. Hu, K.Y. Cai, etc. [8, 9], in order to shorten the testing time and reduce cost, a mechanism is introduced to handle multiple testing tasks on the same or different software under tests (SUTs) simultaneously. On the other hand, advanced testing techniques, such as Adaptive Testing (AT) have been proposed to improve the efficiency of traditional random/ partition testing. Then, they proposed an extended adaptive testing strategy, namely Modified Adaptive Testing (MAT). The use of test history information allows the resulting test process to be adaptive in the selection of tests under a limited test budget.

In the above-mentioned work, one issue that find more software defects is not discussed absolutely in GUI testing, while not mention some very important questions, such as how can find more and important defects under the same test conditions, how to take advantage of the first-line testers testing experience to guide the testing process, and how to assess the GUI when the test is not sufficient. In this chapter, we introduced a Bayesian of model which can use test data on line and experiences of tester, guide the test process of GUI testing.

Bayesian model theory has been used to solve the complex problems of many complex scientific issues [10–12], especially the considerable uncertainties. Due to the complexity of GUI testing, we try to establish a Bayesian Model of GUI testing, which is characterized by historical data from tests, combined with the experience of the testers, established a Bayes network for analyzing of the GUI test results, and continue to guiding the further testing of GUI. The reason we use Bayesian methods rather than the other mathematical model is as follows:

First, Bayesian approach uses empirical knowledge of the testers as a priori parameter, while the tester of GUI is able to rely on their experience in finding defects faster.

Second, Bayesian approach has learning function can be given a wealth of information based on prior knowledge of the posterior distribution, which can be adopted to guide the testing process or establish confidence assessment of GUI.

Third, compared with traditional neural network method, the network structure of the Bayesian approach is more flexible, it is more important for GUI testing while GUI intricate interaction of the testing process.

The last point is the most important, Bayesian methods show higher sensitivity on a small sample of data processing than the other models. This is extremely useful for GUI testing according that sample nature and not fully tested for GUI testing.

Since Bayesian networks have advantages above with GUI testing, we introduced a Bayesian model to GUI testing. A simple case proved that GUI testing with Bayesian model can be found fault faster in GUI testing than traditional methods.

44.2 Background

A GUI is a graphical user interface to a program which is composed by certain number of windows and GUI components. The basic input is an event in GUI, which normally generated by action of the user, external environment of GUI or itself, it sends to GUI by the operating system, GUI respond to these events in order to accomplish a specific function [13].

Due to the various components of the GUI within a certain hierarchical relationships, it asks that a certain event can only accept after other events are applied to GUI. A.T. Memon proposed that Event-Flow Graph (*EFG*) instead of all the events interactions in the GUI, which defined as follows:

$$<\boldsymbol{V}, \boldsymbol{E}, \boldsymbol{B}> \tag{44.1}$$

In this model, ***V*** is the set of all vertices in the EFG, each vertex instead of one event; ***E*** is the set of edges which means event interaction; ***B*** is a set of initial event, which means that it can be immediately executed after GUI start.

When an event is input, all possible response to the event code will be named Event Handler (*EH*) and Event Handler Interaction Graph (HIG) is defined as follows [14]:

$$<\boldsymbol{H}, \boldsymbol{RHI}> \tag{44.2}$$

In this model, ***H*** is the set of all vertices in the EFG, each vertex instead of one Event Handler; ***RHI*** is the set of edges which means Event Handler interaction.

The test case generation is the most important and difficult task of the GUI testing. Due to the presence of a response to interaction, independent testing of each event cannot meet the GUI software testing requirements. Therefore, the sequence of events in the GUI test needs to be used as a test input. The sequence of events must be legitimate and efficient. In this chapter, we will utilize EFG for path searching and HIG interaction as the coverage criteria, get the test case of GUI.

44.3 The Topology of Bayesian Model

In this chapter, supposing that the software contains no faults before three event interactions, we considered the three events, Event1, Event2, and Event3, which stay in the same state, and the Event*, which take place before three events above. To get a Bayesian model of GUI, we draw a graph to represent the qualitative relationships between the nodes. We then quantify all of the probabilities of the events in the graph as follows:

1. We determine the probability of occurrence of each event represented by a root node.

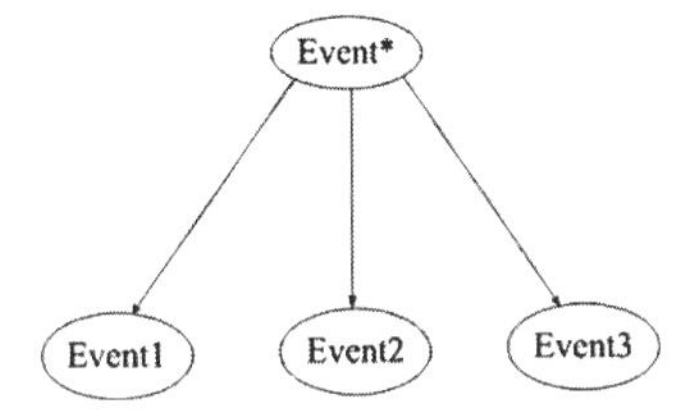

Fig. 44.1 A Bayesian model of GUI

2. For each child node, we determine the probability of occurrence of the event. These specifications determine the probabilities for all combinations of events represented on the graph.
3. Because the full joint probability distribution has been specified, each time we observe an event, we may update all of the probabilities for all of the remaining events in the model by probabilistic conditioning. Further, because of the structure of the graphical model, it is straightforward to carry out such belief updating.

Our model is given in Fig. 44.1. There are four independent nodes, Event*, Event1, Event2, and Event3. First, assuming that there was not fault before Event interaction, so every nodes of the parent of Event interaction set N have two states: 0 (this event was not be carried out) and 1 (this event was carried out), and the last event, Event1, Event2, and Event3 also have two states: 0 (this event was be executed and failed) and 1 (this event was executed correctly).

Second, we should specify probabilities for the root nodes and the conditional probability of each child node and check the rationality.

To complete the Bayesian model, we need to link test outcomes to the appropriate node, not only gaining the outcomes testers concerned but also both topology and state of nodes are not too sophisticated. There are three essential simplifications as follows:

First, we do not need to describe the test nodes so extensively. All that is important is that there is a test process which is connected to a single node, and the nodes representing the event that cannot be clicked will be removed in a particular state.

Second, any node, have not just two states, while considering all the states is likely to be too complicated for analysis and gain the goal of the designer, furthermore some states is not my concerned.

The last and most important, there were so many nodes have same or similar characteristics, taking into account the function of GUI complexity and the time constraints, a simple Module Bayesian Model of GUI testing showing in Fig. 44.2; this model is also suitable for length of GUI test case is n. Supposing that the GUI in case has three modules, and that three are one of the three events that belong to the different module can be executed in every state, $\text{Module}_{m\text{-}n}$ means that an event belonged to module m stay in state n. The first node Initial state has two states 0 and 1 (0 represents that GUI initially failed and 1 represents that GUI initial successfully). Then the event nodes stayed in state 1 to state n have two states 0 and

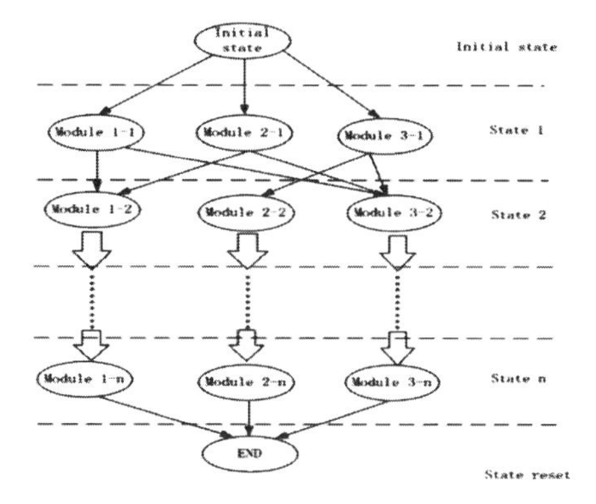

Fig. 44.2 Module Bayesian Model of GUI testing

1 (0 means that an event of this module was not executed in the state and 1 means that an event of this was executed in the state).The last node END also has 2 states 0 and 1 (0 is in the name of that GUI resets state with fault in last event interaction recording this failure and 1 is in the name of that GUI resets state successfully).

44.4 Mathematical Analysis of the Bayesian Model

A basic Bayesian model contains both topology and parameter analysis; the topology has been briefly discussed above and then mathematical analysis of the Bayesian will be mentioned showing in Fig. 44.3.

First, we can give an empirical conditional probability distribution according to the testers, the test manager, or user feedback experience. In order to obtain an objective prior distribution, we define $N_{m\text{-}n}$ is the number of Event handler which $Module_{m\text{-}n}$ contains.

Second, I and $M_{m\text{-}n}$ are two-dimensional discrete random variables, $I = 0$ represents that GUI initial failed and $I = 1$ represents that GUI initial successfully; $M_{m\text{-}n} = 0$ means that an event of $Module_{m\text{-}n}$ was not executed in the state and $M_{m\text{-}n} = 1$ means that an event of $Module_{m\text{-}n}$ was executed in the state.

Third, we consider the probability of occurrence of event $\{M_{m\text{-}n} = m_i\}$ under the conditions of occurrence of event $\{I = i_j\}$, that is to find the probability of event as follows:

$$\left\{ M_{m-n} = m_i \middle| I = i_j \right\}, i, j = 0, 1 \tag{44.3}$$

According to the conditional probability formula, it can be deduced as follows:

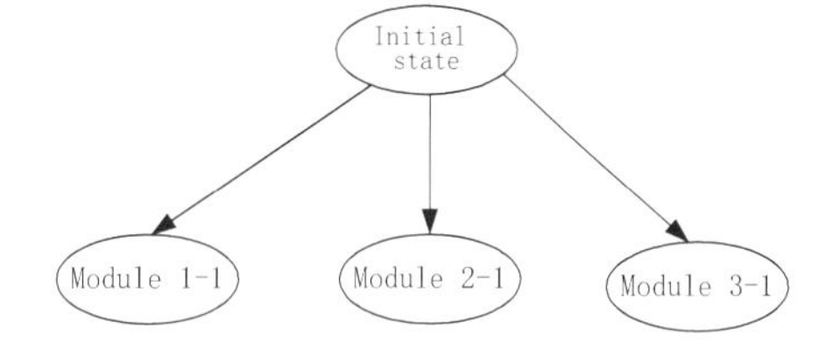

Fig. 44.3 A simple case of conditional probability distribution

$$P\{Mm-n=mi|\ I=ij\}=\frac{P\{M_{m-n}=m_i,\ I=i_j\}}{P\{\ I=i_j\}} \tag{44.4}$$

Fourth, the experience of the conditions given in the probability distribution will be according to the formula (44.5).

$$P\{M_{m-n}=m_i|I=i_j\}=\frac{P\{M_{m-n}=m_i,\ I=i_j\}}{P\{\ I=i_j\}}=\frac{N_{m-i}}{\sum_{i=1}^{n}N_{m-i}},i,j \\ =0,1 \tag{44.5}$$

In actual testing process, we define N_{m-n}^{*} is the number of Event handler which $\text{Module}_{\text{m-n}}$ contains in actual test procedure, so the updated of the conditions given in the probability distribution will be according to the formula (44.6).

$$P\{M_{m-n}=m_i|I=i_j\}=\frac{P\{M_{m-n}=m_i,\ I=i_j\}}{P\{\ I=i_j\}} \\ =\frac{N_{m-i}+N_{m-i}^{*}}{\sum_{i=1}^{n}\left(N_{m-i}+N_{m-i}^{*}\right)},i,j=0,1 \tag{44.6}$$

44.5 Experiment Studying

44.5.1 Experimental Designs

In this chapter, we use open source software TerpPaint belonged to TerpOffice series, which developed by Professor A.T. Memon and his students. It has 3,922 test cases and contain 45 faults. The test case is divided into three groups on the nature of GUI. Experimental method included Randomly selected, Dynamic feedback, and Bayesian method. Bayesian model testing techniques algorithm as follows:

Step 1 The original test suite is divided into m subdomain $\{C_1, C_2,\ldots, C_m\}$, which contains $k_1, k_2,\ldots,k_m$ test cases;

Table 44.1 The typical posterior probability calculation of the Bayesian approach

Node	$Module_{1\text{-}2}$	$Module_{2\text{-}2}$	$Module_{3\text{-}2}$
Without node	0.00274	0.00639	0.01777
$Module_{1\text{-}1}$	0.00260	0.00649	0.01712
$Module_{2\text{-}1}$	0.00270	0.00636	0.01777
$Module_{3\text{-}1}$	0.00273	0.00635	0.01752

Step 2 Initialization parameter ε,$0<\varepsilon<1$;
Step 3 According to the profile $\{P1, P2, \ldots ,Pm\}$ to select a number of test cases, the probability of test cases being selected in the subdomain C_i is P_i;
Step 4 Perform steps above and record the results of testing, the number of defects in each class is cumulative;
Step 5 If no defects are found back to step 3;
Step 6 If the test case defects are found, recording the defects discovered what kind of test cases, adjustment of the profile parameter guided by the posterior probability of the Bayesian network;
Step 7 Verify that it meets the test termination condition, if not, processing returns to step 3, test case input domain partition remains unchanged during the test;
Step 8 Meet the test termination conditions, terminate the test.

Randomly selected and Dynamic feedback method is similar to the above algorithm, while Randomly selected does not contain the feedback process and Dynamic feedback does not contain the Bayesian model process.

44.5.2 Markov Property of the Bayesian Approach

In this chapter, the method based on Bayesian model relies on the posterior probability calculation to redefine the test path profile. During the experiment, the posterior probability showed obvious relationship and the last node, but the situation is not obvious and more upper node relations. The typical posterior probability calculation was shown in the following Table 44.1.

In this table, we can see that in our Bayesian model showing in Fig. 44.4. The posterior probability of the defect rate closely related to the third layer of nodes, and the second layer is also associated, however it is not obvious. This is evidenced by the Markov property of Bayesian networks, while it is more important issue that the posterior probability of the defect rate closely related to both the second and third layers in future work.

44.5.3 The Experimental Results

In this chapter, we used three experimental methods based on the same GUI and test cases. The experimental results are shown in Fig. 44.5.

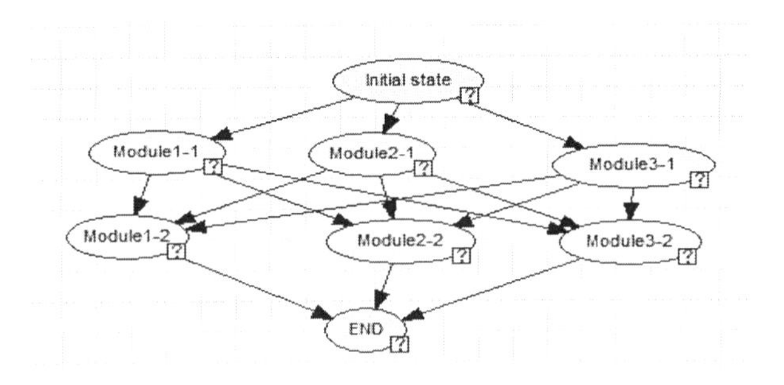

Fig. 44.4 A real Bayesian model of GUI testing

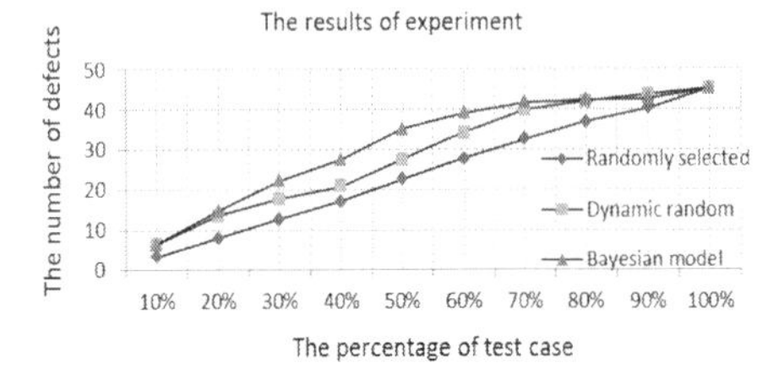

Fig. 44.5 Experimental results of three methods

In Fig. 44.5, the horizontal axis represents the percentage of test cases, and the vertical axis represents found that the number of defects. We can see that Dynamic random testing techniques and Bayesian model will find more defects as soon as possible, while Bayesian model has the best performance in the medium term.

44.6 Conclusion

In summary, this chapter describes an effective method for GUI testing guided by Bayesian model, discussed the Bayesian model topology and its issues encountered in the modeling process. It presents two solutions in connection with the parameters problem. Facing the problem of missing data, we use Bayesian inference method to fill all the data. During the experiment, we dynamically updated Bayesian network and calculated the posterior probability to change the test path profile. In the end, a simple case verifies the validity of the model during the GUI test. Coefficient with vector hydrophone, which measures the pressure and particle velocity instantaneously, it costs less time, more accurate in measuring, handier in practice, and had good performance both in vertical and oblique incidence.

References

1. White, L., & Almezen, H. (2000). Generating test cases for GUI responsibilities using complete interaction sequences. In *Proceedings of the 11th International Symposium on Software Reliability Engineering* (pp. 110–121). San Jose, CA: IEEE Computer Society Press.
2. White, L., Almezen, H., & Alzeidi, N. (2001). User-based testing of GUI sequences and their interactions. In *Proceedings of the 12th International Symposium on Software Reliability Engineering* (pp. 54–63). Hong Kong: IEEE Computer Society Press.
3. Belli, F. (2001). Finite state testing and analysis of graphical user interfaces. In *Proceedings of the 12th International Symposium on Software Reliability Engineering* (pp. 34–43). Hong Kong: IEEE Computer Society.
4. Memon, A. M., Pollack, M. E., & Soffa, M. L. (2001). Hierarchical GUI test case generation using automated planning. *IEEE Transactions on Software Engineering, 27*(2), 144–155.
5. Brooks, P. A., & Memon, A. M. (2007). Automated GUI testing guided by usage profiles. In *Proceedings of the Twenty-Second IEEE/ACM International Conference on Automated Software Engineering* (pp. 333–342). New York: ACM.
6. Yuan, X., Cohen, M. B., & Memon, A. M. (2009). Towards dynamic adaptive automated test generation for graphical user interfaces. In *IEEE International Conference on Software Testing, Verification, and Validation Workshops* (pp. 263–266). Washington, DC: IEEE Computer Society Press.
7. Yuan, X., Cohen, M. B., & Memon, A. M. (2011). GUI interaction testing: Incorporating event context. *IEEE Transactions on Software Engineering, 37*(4), 559–574.
8. Hu, H., Jiang, C.H., Ye, F., Cai, K.Y., Huang, D.Z., & Stephen, S.Y. (2010). A parallel implementation strategy of adaptive testing. In *Computer Software and Applications Conference Workshops* (pp. 214–219). Seoul: IEEE Computer Society Press.
9. Hu, H., Jiang, C.H., Cai, K.Y., Kai-Yuan Cai, Wong, W.E., & Mathur, A.P. (2013). Enhancing software reliability estimates using modified adaptive testing. *Special Section: Component-Based Software Engineering (CBSE), 55*(2), 288–300.
10. JUDEA PEARL. (2000). *Causality: Models, reasoning, and inference* (pp. 1–6). New York, NY: Cambridge University Press.
11. Bai, C. G., Jiang, C. H., & Cai, K. Y. (2010). A reliability improvement predictive approach to software testing with bayesian method. In *Proceedings of the 29th Chinese Control Conference* (pp. 6031–6036). Beijing: IEEE Computer Society Press.
12. Wooff, D.A., Goldstein, M., & Coolen, F. P. A. (2002). Bayesian graphical models for software testing. *IEEE Transactions on Software Engineering, 28*(5), 510–525.
13. Memon, A. T. (2001). *A comprehensive framework for testing graphical user interfaces* [D]. USA: University of Pittsburgh
14. Zhao, L. (2010). *GUI software testing based on event handlers* [D]. Beijing: Beihang University (In Chinese)

Chapter 45
A Model for Reverse Logistics with Collection Sites Based on Heuristic Algorithm

Xiaoqing Geng and Yu Wang

Abstract Reverse distribution has received growing attention throughout this decade. Built on the concept of green supply chain management (GSCM), this chapter presents a mathematical programming and distribution model for reverse logistics with collection sites. Due to the complexity of the GSCM model, a heuristic solution is given and improved. The solution adds a heuristic concentration procedure, where subproblems with reduced sets of decision variables are iteratively solved to improve the optimality, and it improves the capacitated plant location problem (CPLP). Computational test demonstrates that high-quality solution is obtained with the improved model.

45.1 Introduction

With the growing environmental concerns, green supply chain management (GSCM) is getting more interest among practitioners of operations and researchers in the fields of supply chain management. GSCM can be defined as "Integrating environmental thinking into supply chain management including product innovation and design, material sourcing and selection, manufacturing processes, delivery of the final product to the consumers as well as end-of-life (EOL) management of the product after its useful life" [1]. Environmental management devotes to reduce waste volume by moving away from one-time use and disposal to having management of the product's recovery primarily. It includes reuse, remanufacturing, and materials recycling in a product life cycle. Thus, manufacturers are forced to consider the environment impact by their products. They have to extend and

X. Geng (✉)
Department of Management Information System, Tianjin University of Finance and Economics, Tianjin 300222, China
e-mail: gengxq@gmail.com

Y. Wang
School of Humanities, Tianjin University of Finance and Economics, Tianjin 300222, China

W.E. Wong and T. Zhu (eds.), *Computer Engineering and Networking*, Lecture Notes in Electrical Engineering 277, DOI 10.1007/978-3-319-01766-2_45,

improve the traditional forward logistic distribution chain and consider the environmental effects of all products including returning process. It is a process of goods moving from final destination to another one, for the purpose of value-added or for the proper disposal of the products.

Caruso and Colorni proposed a location-allocation model for planning and designing urban solid waste management systems [2]. The results were the location and number of waste disposal plants, the amount of waste processed, specifying the technology adopted, and the service basin of each plant. Bloemhof-Ruwaard focused on the problem of by-product flows and coordinating product in a two-level distribution network. The model solution suggested the first attempt at studying the coordinated control of product and by-product flows within distribution networks [3]. Carter and Ellram noticed that the international nature and complex environment made it almost impossible to put forward decisions based on intuition. They suggested some important and critical factors in the reverse logistics process [4]. Srivastava reviewed the literature on reverse logistics and addressed the recognition of the strategic importance of reverse logistics as evident from classification and categorization of the existing GSCM literature [1]. Vaidyanathan designed a model and gave a solution procedure for reverse distribution network with GSCM [5].

Our work is to develop a mathematical programming and distribution model for one version of this problem in GSCM—reverse logistics with collection sites, which is based on a heuristic solution methodology. And the test shows that high-quality solution is achieved while improving model and expending modest computational effort.

45.2 Assumption

Firstly, some assumptions are given below in order to formulate a mathematical model for the reverse distribution problem [6]:

1. The recalled products are located at some source outlets (origination sites). The model in this chapter is to find an efficient method and process to return these products from a set of origination sites to specific collection sites.
2. The wholesaler or retailer is considered to be an initial site. This may be a realistic assumption because the client would be inclined to return the product to the closest collection site to get a refund or to purchase some other products from there.
3. There is a fixed cost to opening refurbishing sites and collection sites. There is a limit to the number of refurbishing sites and collection sites that can be opened, but the choice of which refurbishing sites and which collection sites to be opened should be decided by the proposed model.

In order to describe the model further, the following notation will be used in the model:

I—{i/here i is some an origination site}. This is a shop, a store, a retail outlet, or a client collection site.
J—{j/here j is some a collection site}. Collection sites are some intermediate transshipment spots. A collection site receives the recalled products.
K—{k/here k is some a refurbishing facility site}. This site may be a refurbishing spot, or a recycling plant, or the original manufacturing station.
C_{ijk}—Total variable cost of transporting a single unit of recalled product from origination site i through collection site j and onto refurbishing site k.
F_j—Total cost of opening a collection site j.
G_k—Total cost of opening a refurbishing site k.
a_i—Number of recalled products residing at origination site i.
B_j—Maximum capacity of collection site j.
D_k—Maximum capacity of refurbishing site k.
P_{min}—Minimum number of collection sites opening and operation.
P_{max}—Maximum number of collection sites opening and operation.
Q_{min}—Minimum number of refurbishing sites operation.
Q_{max}—Maximum number of refurbishing sites operation.

All the decision variables in this model are:

X_{ijk}—Fractions of units at origination site i that is from origination site i transported through collection site j and onto refurbishing site k.

$$P_j = \begin{cases} 1 \text{ if collection site j is open,} \\ 0 \text{ otherwise.} \end{cases} \qquad Q_k = \begin{cases} 1 \text{ if refurbishing facility k is open,} \\ 0 \text{ otherwise.} \end{cases}$$

45.3 Model Designing and Construction

With the assumptions, a strong formulation with collection sites can be described as:

$$\text{Min}Z = \sum_i \sum_j \sum_k C_{ijk} a_i X_{ijk} + \sum_j F_j P_j + \sum_k G_k Q_k$$

Subject to:

$$\sum_j \sum_k X_{ijk} = 1 \ \text{ for all } i, \tag{45.1}$$

$$\sum_i \sum_k a_i X_{ijk} \leq B_j \ \text{ for all } j, \tag{45.2}$$

$$\sum_i \sum_j a_i X_{ijk} \leq D_k \ \text{ for all } k, \tag{45.3}$$

$$X_{ijk} \leq P_j \ \text{ for all } i, j, k, \tag{45.4}$$

$$X_{ijk} \leq Q_k \ \text{ for all } i, j, k. \tag{45.5}$$

Here the objective function aims to minimize the sum of costs to transfer products from origination sites through collection sites to the destination stations and the fixed cost of opening the collection sites and destination sites.

Constraint set analysis:

(a) All the recalled products are transported from the origination sites to destination sites by way of constraint set (45.1) (either directly or via collection sites in the reverse logistics network);
(b) Constraint set (45.2) limits the units sent through collection site j to site j;
(c) Constraint set (45.3) limits the units ending up at destination site k to site k;
(d) Constraint set (45.4) forbids units through collection site j from being routed unless this site is opened;
(e) Constraint set (45.5) forbids units from ending up at destination site k from being routed unless this site is opened.

Compare the strong formulation of this problem, a weak formulation can be given from aggregating the demand in constraint sets (45.4) and (45.5), obtaining the following alternative constraint sets:

$$\sum_j \sum_k X_{ijk} \leq |I|^* P_j \quad \text{for all } j, \tag{45.4$'$}$$

$$\sum_i \sum_j X_{ijk} \leq |I|^* Q_k \quad \text{for all } k. \tag{45.5$'$}$$

With constraint sets (45.4′) and (45.5′), the weak formulation will significant reduce computation burden with finding proper and optimal solutions.

45.4 Solution Method

45.4.1 Observations

This model is a typical zero-one mixed integer-linear programming (MIP) problem (the variables are restricted to be either 0 or 1). Assuming that there are no collection sites in this model, it will reduce to a capacitated plant location problem (CPLP). The CPLP is NP-complete problem (solutions can be verified in polynomial time), and as such, the model is also NP-hard [7]. The traditional MIP tools for solving such problem as NP-hard is limited. We should choose proper process and improve solution.

45.4.2 Heuristics Procedure

Since integer-linear programming is NP-complete, many problem instances are difficult and so heuristic methods must be used instead. Based on running an efficient heuristic numerous times, heuristic concentration identifies a subset of the potential collection sites that may warrant further investigation [8]. After repeatedly running a randomized greedy heuristic, the information about which sites are in the "p" set is collected, and a final subset of sites is used when solving the subset problem to optimality using integer-linear programming. Thus, using a subset of the potential collection sites results in an optimal solution if all sites that are in the optimal "p" set are in the subset.

MIP tools provide some useful module for this subproblem. In this chapter we can use AMPL (A Mathematical Programming Language) module as a front-end interface to the mathematical programming solver-CPLEX. The algorithm for this problem includes three steps: the first step is the random selection for potential collection sites and refurbishing sites, the second step is the heuristic concentration section, and the third step is the heuristic expansion (HE).

Random selection:

At the beginning, set MAXIterations $= \beta$ (where $\beta = 10, 40$, or 80)

1. While Iterations $<$ MAXIterations, then do:
2. Select a subset of size P_{max} from the collection sites, and meanwhile Q_{max} from the destination sites randomly (all sites should have an equal probability to be selected).
3. Add the AMPL module file and data file in the way that only the sites selected in step 2 are considered as potential collection sites.
4. Optimize the current problem using AMPL solver.
5. Save the solution and its configuration. If the current solution is better than the best previously exist solution, then update the best solution.
6. End while.

Heuristic concentration:

7. Add the AMPL module file with all collection sites and destination sites chosen in the best previously solution. Collecting the information in steps 2–5 from the top 5 % best solutions, use the additional [P_{max} + 2—*the number of sites used in the first best solution*] most frequently collection sites, and [Q_{max} + 2—*the number of sites used in the first best solution*] most frequently destination sites, and change the AMPL module file.
8. Optimize the current problem using AMPL solver.
9. If the current solution is better than the best previously exist solution, then update the solution and its configuration as the best solution.
10. Report the best solution found.

Heuristic expansion:

11. Add the AMPL module file with all collection sites and destination sites in the best solution found. Then add one site not chosen in the best solution found.
12. Optimize the current problem using AMPL solver.
13. If the current solution is better than the best previously solution, then remember the solution and its configuration, but not change the best solution for now.
14. Repeat steps 11–13; make sure all sites have been checked.
15. If improvements are found, save this solution as the latest best solution.
16. Repeat from step 11; make sure no improvements are found.
17. Report the latest best solution.

45.4.3 A Deterministic Heuristic

An alternative deterministic heuristic to be proposed is designed to improve the random selection and heuristic concentration process of the previous algorithm. Compare an exact algorithm, a deterministic heuristic designing may lead a serious problem here: if we use CPLEX to select the sites and find an optimal result of demand flows (given the open sites chosen from the available sites sheet), we do not make sure whether a heuristic that would show us meaningful routing values short of using an optimal solver.

Based on the truth above, we developed procedure CC algorithm (a greedy algorithm) that uses CPLEX to define the routing planning schemes (as similar with the random selection with heuristic concentration methods we mentioned above):

1. Rank order all collection sites (all P's found) and rank order all refurbishment sites (all Q's) by the ratio of Cost = Capacity from least to greatest.
2. Select the P_{max} + 4 and Q_{max} + 4 cheapest Cost = Capacity site alternatives and solve with CPLEX solver.

45.5 Computational Results

Vaidyanathan collected some real data through research in some reverse distribution networks. This chapter chooses 5 sets of 20 randomly generated problems from reference [5] as datasets for test problem (Table 45.1):

Table 45.1 Datasets

Problem set	Number of origination (or retail) sites	Number of available collection sites	Maximum allowed number of collection sites	Number of available refurbishment sites	Maximum allowed number of refurbishment sites
1	35	16	5	15	3
2	45	26	7	18	4
3	55	32	7	20	5
4	75	37	8	22	5
5	100	41	10	37	8

45.5.1 Data Generation

From origination and collection to refurbishing site are all located in a 100×100 square randomly. And an additional collection site was used as a virtual site to figure a direct shipment from an origination site to the destination site. This additional destination site with boundless costs and boundless capacity is used to eliminate infeasible basic solutions. The data generation includes constructing two parts costs (fixed and variable) and capacities. The costs of fixed part are generated as following:

Collection sites (CS): $C_j := 0.1([0, 10000] + B_j\,[0, 10])$,
Refurbishing sites (RS): $R_k := 0.1([0, 10000] + D_k\,[0, 10])$.

Costs in the transportation process are computed according to the following formulae:

$C_{ijk} := \alpha$(Euclidian distance from i, j to k)

Here α is 0.1 when using one collection site (i, j to k) and is 0.4 when shipping direct from the origination site to one refurbishing site (i, m to k, where the Euclidian distance is computed in a direct distance from i to k).

The demand is generated as: $ai = [0, 500]$, and the capacities are:

Collection sites (CS): $B_j = [0, 5000]$,
Refurbishing sites (RS): $D_k = [0, 25000]$.

45.5.2 Experimental Results

With test problem, the experimental results show that the heuristic expansion (HE) algorithm improves significantly in the speed and solution values. Applying HE to the results of Procedure CC algorithm, 80 random selection iterations with heuristic concentration (HC-80), 40 random selection iterations with heuristic concentration (HC-40) and 10 random selection iterations with heuristic concentration (HC-10) for these problem sets clearly demonstrate the additional benefit of the HE technique. Both Procedure CC and random selection with heuristic concentration provide starting solutions that the HE algorithm.

The tentative in this method is that this starting point solution using HE was stuck in a very poor local minimum. Even though the HE algorithm cannot always substantially speed every solution, Computational test shows that optimal solution was found in a large number of cases and high-quality solution is obtained while improved model and expending modest computational effort.

45.6 Conclusion

In the discussion of this chapter, we added GSCM into the process of reverses logistics with collection sites. Besides that, this chapter may contribute to the reverse distribution literature by developing a strong and a weak formulation for reverse distribution logistical problems including all kinds of reverse type (product recall, product reuse, product disposal, and hazardous product return). Also, this chapter adapted the heuristic concentration procedure to solve the version of problem with collection sites and provided a new solution—heuristic expansion (HE). The complexity of the proposed model was such that a heuristic solution procedure was the only viable approach to solve very severe problems. We proposed heuristic concentration procedures combined with HE to make the problem easily. Very large problems can indeed be solved in a reasonable amount of time with the HE algorithm, where they cannot be solved with traditional MIP tools within a limited computational time.

Acknowledgments This work was financially supported by the foundation of Tianjin Municipal Education Commission (20122129) and Tianjin Municipal Science and Technology Program (12ZLZLZF01300).

References

1. Srivastava, S. K. (2007). Green supply chain management: A state-of-the-art literature review. *International Journal of Management Reviews, 9*(1), 53–80.
2. Caruso, C., Colorni, A., & Paruccini, M. (2003). The regional urban solid waste management system: A modeling approach. *European Journal of Operational Research, 70*(4), 16–30.
3. Bloemhof-Ruwaard, J. M., Van Beek, P., Hordijk, L., & van Wassenhove, L. N. (2005). Interactions between operations research and environmental management. *European Journal of Operational Research, 85*(3), 229–243.
4. Carter, C., & Ellram, L. (1998). Reverse logistics: A review of the literature and framework for future investigation. *Journal of Business Logistics, 19*(1), 85–102.
5. Vaidyanathan, J. (2003). The design of reverse distribution networks: Models and solution procedures. *European Journal of Operational Research, 100*(2), 128–149.
6. Pirkul, H., & Jayaraman, V. (1996). Production, transportation, and distribution planning in a multi-commodity tri-echelon system. *Transportation Science, 30*(3), 291–303.
7. Davis, P. S., & Ray, T. L. (1969). A branch-and-bound algorithm for the capacitated facilities location problem. *Naval Research Logistics, 16*(2), 331–344.
8. Rosing, K. E., & ReVelle, C. S. (2006). Heuristic concentration: Two stage solution construction. *European Journal of Operational Research, 97*(4), 75–86.

Chapter 46
The Storage of Wind Turbine Mass Data Based on MongoDB

Qile Wang, Zhu Shen, Long Ma, and Shi Yin

Abstract With the large-scale development of wind power, the storage and analysis of wind turbine's data gradually become more and more important, there are huge amount of information about wind turbines, and the traditional relational database has been difficult to meet the demand of mass data storage and analysis. This chapter proposes the solutions that data storage is based on non-relational databases (MongoDB) and compares SQL Server with MongoDB about the storage ability and query performance. The results show that using this method can increase storage speed and query performance significantly.

46.1 Introduction

As a clean and renewable energy, wind energy has drawn worldwide attention. The huge amount of wind energy that can be developed is 10 times as much as the whole water energy on Earth. The biggest difference between wind turbine and thermal turbine is that wind turbine has small power capacity and wide distribution and its data management is very difficult.

In order to ensure healthy and efficient operation of wind turbines in wind farm, power companies must establish monitoring center to control hundreds or thousands of wind turbines, using SCADA system to record and store operation data about wind turbine. The real-time production data is very large. In order to improve the efficiency of data storage, most wind turbine manufacturers process data before storage, they process real-time data into 10-min average data in order to use the traditional relational databases to store information, and this method may lose numbers of important turbine operation information and pose serious obstacles to data analysis for improving wind turbine performance.

Q. Wang (✉) • Z. Shen • L. Ma • S. Yin
Zhongneng Power-Tech Development Co., Ltd., Beijing 100034, China
e-mail: wangqile@clypg.com.cn

W.E. Wong and T. Zhu (eds.), *Computer Engineering and Networking*, Lecture Notes in Electrical Engineering 277, DOI 10.1007/978-3-319-01766-2_46,

Using traditional relational database to store real-time data is not competent in the field of wind power. This chapter poses a new method based on the non-relational database (MongoDB) massive data storage solutions and analysis and implements this theory.

46.2 MongoDB Database Overview

46.2.1 Non-relational Databases

In industrial field, non-relational databases have drawn collective attention as a new type of database technology. The method does not completely deny relational database. A single relational database cannot meet the demand of diverse application field. The rigorous description of non-relational databases is as follows [1]:

1. Using scalable and loosely data structure to formulate logical model
2. Designed to follow CAP theorem using across multi-node data distribution model and supporting horizontal scaling
3. The ability of keeping data permanent in disk or memory
4. Support a variety of "non-SQL" interface for data access

Non-relational database makes changes on two respects. First, data mode, NoSQL uses loose and extensible data model, such as key-value pairs, columns, documents, and charts, not two-dimensional table as relational database. Second, horizontal scaling, NoSQL is designed for distributed systems, supporting horizontal expansion. It is able to adapt to the rapid growth of huge amounts of data [2, 3].

46.2.2 MongoDB Database

MongoDB is designed to possess the ability of key/value storage and high-performance and scalable, in the meantime with the function of a traditional relational database to manage system [4], as shown in Fig. 46.1.

MongoDB uses data structure of loose BSON (Binary JSON) format and document-oriented storage data; it uses automatic slicing (auto-sharding) to realize mass data storage, supporting type index; it can support two kinds of data replication mechanism: master/slave and replication set. MongoDB uses document-oriented data model to store data, each document is allowed to have different settings [5, 6].

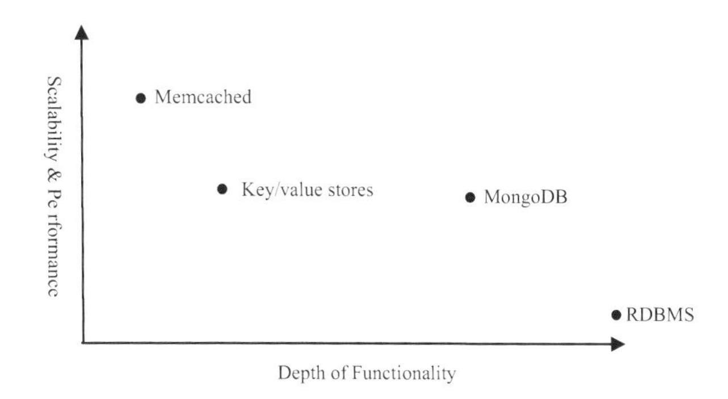

Fig. 46.1 The magnitude calibration function

46.3 MongoDB Database Storage Principle

The underlying of MongoDB use BSON as the data format for data storage and network transmission. It allows data type as nested sub-documents and array. In order to solve large object storage, MongoDB uses GridFS file system to split a large object into multiple small objects.

A MongoDB system is composed of multiple databases; each database consists of a collection, each collection is made up of many documents and the document consists of a series of field, each field is a key-value pair, the key is field name, and its value corresponds to property value [7].

1. Collection
   MongoDB collection is a set of documents essentially, and it is similar to the table in relational database. These documents have the same field structure, but MongoDB is schema-free database, and there is no mandatory required on the field structure. Users do not need to predefine structure of a collection of fields and can store documents with different structures in the same collection, when database is running. Users can dynamically add and delete fields of the document [5].
2. Document
   MongoDB is a database based on document; document stores data in the form of key-value pairs. Document uses the BSON language; BSON is a binary serialization of JSON data interchange language [8]. It is binary bytes stored in the form of key-value pair, key is the string format, and the value can be any data type, like basic integer, float, string, array, and key-value pairs.
   For example, a wind turbine by BSON object is expressed as follows:

```
   {WT_NAME: "A-004",
  WT_AI:{{WinSpe:12.42},{WinDir:250},{ActPow:12.42},
{ReaActPow:250},{ReaActPow:12.42},{GenSpe:250},
{EnvTem: 12.42}},
WT_DI:{WTstart:0,WTrun:1,WTstop:0, PlcErr:0}
}
```

Wind turbine includes wind turbine name (WT_NAME), wind turbine analog parameter (WT_AI), and wind turbine digital parameter (WT_DI). Wind turbine name is the type of string, and the wind turbine analog properties are corresponding to key-value pairs [9].

46.4 Database Storage Design

In order to organize massive wind turbine data effectively, and in order to make data efficiently accessible, this chapter uses storage form of data in MongoDB. Convert the raw data using the quad tree method, information storage for each additional one, the information on the level L to L + 1 fission into four information points, for example, decompose wind turbine into some devices, and then decomposition devices to the specific information of the device. Each sub-block coding is the sub-block that belongs to the parent block coding plus sub-block in the parent block number [10].

When querying the data, we use formulas to calculate the required data items. Then, using client-side technologies will combine information, you can get all the information by user query. The formula are as follows:

$$curXMin = XMin + j \times \frac{\Delta x}{\max^{L-1}}, curXMax = XMin + (j+1) \times \frac{\Delta x}{\max^{L-1}} \quad (46.1)$$

$$curY\min = YMax - (i+1) \times \frac{\Delta y}{\max^{L-1}}, curYMax = YMax - i \times \frac{\Delta y}{\max^{L-1}} \quad (46.2)$$

Calculate the number of rows and columns. The formula is as follows:

$$i = \left| \frac{YMax - y}{\Delta y} \times n \times 2^{L-1} \right|, j = \left| \frac{x - XMin}{\Delta x} \times m \times 2^{L-1} \right| \quad (46.3)$$

46.4.1 MongoDB Database Storage

MongoDB documentation storage can realize array types and nesting child document. The data structure of wind turbines is achieved in one collection in MongoDB [11].

Create a database "windturbineinfo"; create a collection to store fan data information statements for the following:

```
MongoServer server = MongoServer.Create("mongodb: "127.
0.0.1:27017");
//Specify the database server
```

```
MongoDatabase db = sever.GetDatabase("windturbine
info");
//Obtain windturbineinfo database
MongoCollection<BsonDocument> posts = db.GetCollection
("wtcollection");
//Create a Collection, the name is wtcollection
BsonDocument wtdoc = new BsonDocument (); //Create a
document
wtdoc.put("WT_NAME", "A-004"); //Add the wind turbines's
name
```

Pairs of document type attribute, use BasicDBObject class to create a document first, read wind turbines' analog information, add the information to properties of the subdocument simple data types, and then insert subdocument into the properties of the document [12]:

```
BsonDocument WT_AI = new BsonDocument (); //create a doc-
ument MT_AI
WT_AI.put(WinSpe",13.2); //add Wind speed values
WT_AI.put(WinDir", 250); //add wind direction values
```

Loop to read all the replies, repeat the above statement steps, add analog WT_AI sub-document to document, and insert the documentation to the collection: posts. insert (doc); through the steps, achieve to store the wind turbine information [13].

46.4.2 SQL Server Database Storage

In SQL Server 2008, the same type of data is stored in the same table, and an index is created for each block [14]:

Given below the key code to connect SQL Server:

```
String connectionUrl = "jdbc:sqlserver://172.0.0.1:
1433;
Class.forName("com.microsoft.sqlserver.jdbc.SQLServer
Driver");
Connection con = = DriverManager.getConnection(connec-
tionWT);
```

46.5 Compared to the Performance Test

Running the SQL Server and MongoDB on the same configuration of the server, test the query performance of MongoDB and SQL Server.

Table 46.1 The absolute error with different angle at 4 kHz

Data	Maximum (ms)		Minimum (ms)		Average (ms)	
	MongoDB	SQL	MongoDB	SQL	MongoDB	SQL
100,000	121	1,232	55	151	117	392
200,000	254	3,104	223	421	243	1,089
600,000	1,994	4,982	320	1,004	479	3,003
1,000,000	2,453	8,864	579	1,143	845	4,650
1,200,000	3,210	16,012	875	2,345	1,191	9,193
2,00,000	6,452	21,959	1,043	3,567	1,898	15,796

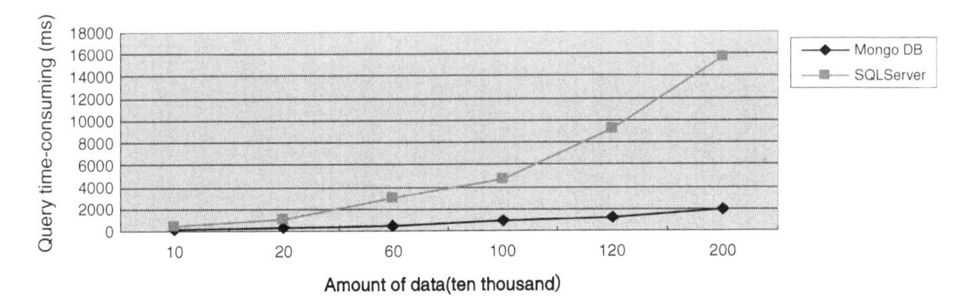

Fig. 46.2 Query performance comparison about MongoDB and SQL

The server configuration is as follows: 3.0 GHz dual core CPU, 2 GB memory, Microsoft Server 2003 operating system, 2.0.3 version Professional MongoDB, and SQL Server 2008.

Using SQL Server 2008 polling analyzer and MongoDB command "explain" to query time and the use of index information to compare the performances.

In this chapter we query WT_AI data of "A-004" database in testing system to return WT_AI's analog information, including the wind speed, wind direction, active power, reactive power, generator speed, and the environment temperature. Executing the query for 10 times, respectively, in the amount of data under the circumstances is 100,000, 200,000, 600,000, 1,000,000, 1,200,000, and 2,000,000. Choose its average time and compare the performance of the SQL Server and MongoDB.

In Table 46.1, we can get the query takes maximum, minimum, and average time between SQL Server and MongoDB.

Contradistinction is shown as follows (Fig. 46.2).

The experiment shows that MongoDB's querying time is less than 1/5 of that on SQL Server, and MongoDB's querying performance is significantly higher than SQL Server. And with the increasing of the data, MongoDB and SQL Server querying performance are both reduced.

46.6 Conclusion

MongoDB query takes more gentle growth, which can maintain a linear growth, and time-consuming within 2 s.

SQL Server with the amount of data grows; the query takes serious growth in data volume of 1.1 million over 10 s.

In summary, the MongoDB data storage model, in the conditions of the mass data storage, the improvement of query performance are very obvious.

References

1. Chang, F., & Dean, J. (2008). A distributed storage system for structured data. *ACM Transactions on Computer System, 26*(2), 2–24.
2. Wang, K., & Wu, Z. X. (2008). Reducing the cluster monitoring workload by identifying application characteristics. *Grid and Cooperative Computing, 12*(8), 525–531.
3. Tai, N., & Hou, Z. (2004). New short-term load forecasting principle with the wavelet transform fuzzy neural network for the power system. *Proceedings of the CSEE, 24*(1), 24–29 (in Chinese).
4. Kassaei, H. R., Keyhani, A., & Woung, T. (1999). A hybrid fuzzy neural network bus load modeling and predication. *IEEE Transactions on Power Systems, 1999*(2), 718–724.
5. Berners-Lee, T., & Hendler, J. A. (2001). The semantic web. *Scientific American, 284*(5), 34–43.
6. Liu, C., Zhou, J., & Xie, Y. (2011). Mass data storage management technology. *Microcomputer Applications, 8*(10), 33–36 (in Chinese).
7. Xu, X., Wu, J., Yang, G., & Cheng, C. (2012). Inexpensive computing platform based on large-scale mass data processing system research. *Application Research of Computers, 11*(2), 582–585 (in Chinese).
8. Rick, C. (2010). Scalable SQL and NoSQL data stores. *SIGMOD Record, 39*(4), 12–27.
9. Peng, X., Ruichun, H., & Zhiming, Z. (2010). *Cache and consistency in NoSQL* (pp. 117–120). Computer Science and Information Technology (ICCSIT), IEEE Computer Society, USA.
10. Brewer, E. (2012). Pushing the CAP: Strategies for consistency and availability. *Computer, 45* (1), 23–29.
11. Singh, M., & Garg, D. (2009). *Choosing best hashing strategies and hash functions* (pp. 50–55). Advance Computing Conference, IEEE Computer Society, USA.
12. Sakr, S., Liu, A., Batista, D., & Alomari, M. (2010). *A survey of large scale data management approaches in cloud environments* (pp. 313–336). Communications Survey and Tutorials, IEEE Computer Society, USA.
13. Han, J., Song, M., & Song, J. (2011). *A novel solution of distributed memory NoSQL database for cloud computing* (pp. 351–355). Computer and Information Science (ICIS), IEEE Computer Society, USA.
14. Pirzadeh, P., Tatemura, J., & Hacigumus, H. (2011). *Performance evaluation of range queries in key value stores* (pp. 1092–1101). Parallel and Distributed Processing Workshops and Phd Forum (IPDPSW), IEEE Computer Society, USA.

Chapter 47
Improvement of Extraction Method of Correlation Time Delay Based on Connected-Element Interferometry

Fei Wang, Zhenfei Wang, Dun Li, and Bingjie Yang

Abstract This study proposes a method using mean comparison to improve the accuracy of interferometry processing correlation time delay under the low signal-to-noise ratio and low residual time delay. The method uses the means of taking the average of stripe subsection, comparing threshold and eliminating outliers, which offsets the influence of channel noise on the accuracy of the signal. Using the direct method under the same SNR strike delay simulation comparison proved that the mean comparison method can get relatively high accuracy of time delay information.

47.1 Introduction

China has made a series of breakthroughs in satellite navigation, manned space and deep-space exploration which reflects the development of aerospace technology in recent years. As for deep space exploration projects, the Spacecraft needs accurate navigation and positioning, which requires to introduce a high-precision measurement technology to the measurement and control system as a support.

Doppler velocimetry and radar ranging technology are traditional for technologies for spacecraft navigation; these two technologies can directly measure the radial velocity and distance of spacecraft relative to the observatory. However it's difficult to measure the accurate position and velocity of the spacecraft which is perpendicular to the radial movement [1]; as the distance of spacecraft increases, the system error will increase, and measurement accuracy will decline. Interferometry technology has a high-precision angle measurement capability. By measuring the delay observables of the two tracking and measuring stations which receive the same signal that reaches the two sites, the method uses the signal processing method

---

F. Wang (✉) • Z. Wang • D. Li • B. Yang
School of Information Engineering, Zhengzhou University, Zhengzhou 201305, China
e-mail: iewangfei@126.com

W.E. Wong and T. Zhu (eds.), *Computer Engineering and Networking*, Lecture Notes in Electrical Engineering 277, DOI 10.1007/978-3-319-01766-2_47,

to obtain high-precision delay observables and deduces reversely the precise angular position of the signal source, the technology been successfully applied to the tracking observation of satellites and spacecraft in deep-space, as well as positioning task. Through continuous development, it has developed a variety of processing methods in the interferometry signal based on the principle of interferometry technology, such as very long baseline interferometry (VLBI), connected-element interferometry (CEI), the same beam interference (SBI), and other high-precision deep space interferometry technologies. These methods has been successfully applied to the navigation of spacecraft in deep space in practical work [2–6]; these can adapt to the needs of future space's development, improve the accuracy of satellite orbit determination and services for the satellite in orbit and military applications [7].

Currently, for broadband signals like radio source signals, there are usually FX and XF two related processing methods [8]. According to the theory of digital signal processing, the two methods obtain the same power spectrum. Haibo Xia studies observation delay model and the model of phase stripe rotation, which makes the time delay estimation more accurate [9]; Haitao Zhang and Qinghui Liu study application and error elimination of interferometry system like VLBI [10, 11]. Longfei Hao proposes to use the first six data points to fit curves and predict the next point with the fitting curve and compares that point value with the predicted values and observes whether it meets the 3σ principle [12]. According to the 3σ principle, this chapter gets a phase curve and proposes a method of calculating segmented mean comparison amplitude jitter and verifies the effectiveness of the method through the simulation experiment, which is based on the analysis and introduction of interferometry principle and general signal processing model, ideological phase curve to strike.

47.2 The Principle of Connected-Element Interferometry and Signal Processing Model

47.2.1 The Principle of Connected-Element Interferometry

The principle of connected-element interferometry is as follows: Two measuring stations receive a signal from the same signal source and perform correlation processing to obtain the time delay of the two signals, thus obtain the measurement data of one direction; the principle of connected-element interferometry is shown in Fig. 47.1.

From several relations in Fig. 47.1 we can obtain expression (47.1) as follows:

Fig. 47.1 The basic schematic of connected-element interferometry

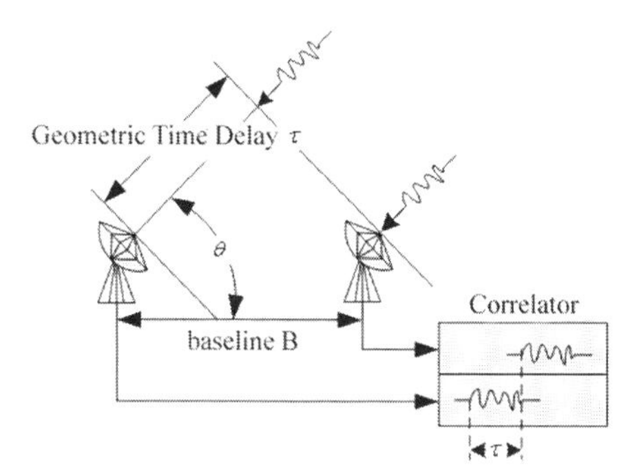

$$\tau = \frac{1}{c} B \cos \theta \tag{47.1}$$

where B denotes baseline, τ denotes time delay, θ denotes the angle between target spacecraft and baseline. Taking the derivative of measuring expression, angular's accurate expression can be obtained

$$\delta\theta = \frac{c \cdot \delta\tau}{B \sin \theta} \tag{47.2}$$

From expression (47.2) we know, the accuracy of angle δθ is in inverse proportion to the length of B and in direct proportion to the measurement accuracy of τ, to obtain high-precision angle measurement; we can use longer baseline (B) or improve the measuring accuracy of interferometer's time delay δτ [13].

47.2.2 *Processing Model of Connected-Element Interferometry Signal*

Two measuring stations which conduct connected-element interferometry receive signals from spacecraft at the same time and have the cross-correlation processing. According to the research of Lue Chen, We need to have a time delay compensation processing before it [13]. If the signals received by measuring station 1 and measuring station 2 are $X_1(t)$ and $X_2(t)$, we used signal received by measurement station 1 as benchmark and analyses signal $X_2(t)$ received by measuring station 2.

Suppose the signals received by two measuring stations are

$$X_1(t) = x(t)e^{j \times 2\pi f \times t} \tag{47.3}$$

$$X_2(t) = x(t + \tau_g)e^{j \times 2\pi f \times (t + \tau_g)} \tag{47.4}$$

where τ_g is true geometric delay of $X_2(t)$ relative to $X_1(t)$, f is the center frequency of signal after down-conversion, and n1 and n2 are the noise in the signal transmission.

Before two signals conducting correlative processing, we need to have the cross-correlation processing for signal X2(t) of measuring station 2. τ_g is geometric delay, τ_{g0} is compensation value of time delay model, the error between them is $\Delta\tau_g$, that is

$$\Delta\tau_g = \tau_g - \tau_{g0} \tag{47.5}$$

Due to adopting the digital signal for signal analysis, suppose the sampling period of signal is ΔT, compensation value τ_{g0} is composed of integer sampling period compensation $\tau_{g0I} = n\Delta T$ and decimal sampling period compensation τ_{g0F}, that is,

$$\tau_{g0} = \tau_{g0I} + \tau_{g0F} = n\Delta T + \tau_{g0F} \tag{47.6}$$

Having a integer bit time delay compensation (ISTC) to signal X2(t) of measuring station 2, the correction is τ_{g0I}

$$\begin{aligned} X_{2D}(t) &= X_2\left(T - \tau_{g0I}\right) = X_2(t - n\Delta T) \\ &= x\left(t + \tau_g - \tau_{g0} + \tau_{g0F}\right) e^{-j2\pi f\left(t + \Delta\tau_g - \tau_{g0} + \tau_{g0F}\right)} \end{aligned} \tag{47.7}$$

Having a stripe rotation to signal and eliminate the phase factor

$$\begin{aligned} X_{2FS}(t) = X_2\left(t - \tau_{g0I}\right) \cdot e^{j2\pi f \cdot \tau_{g0I}} &= x\left(t + \Delta\tau_g + \tau_{g0F}\right) e^{-j2\pi f\left(t + \Delta\tau_g + \tau_{g0F}\right)} \cdot e^{j2\pi f \cdot \tau_{g0I}} \\ &= x\left(t + \Delta\tau_g + \tau_{g0F}\right) e^{j2\pi f\left(t + \tau_g\right)} \end{aligned} \tag{47.8}$$

Make Fourier transform to the signal $X_{2FS}(t)$,

$$F_{2FS}(f) = F_{2D}(f) e^{j \cdot 2\pi f\left(\Delta\tau_g + \tau_{g0F}\right)} = F_{2D}(f) e^{j \cdot 2\pi f \Delta\tau_g} \cdot e^{j \cdot 2\pi f \tau_{g0F}} \tag{47.9}$$

Having a frequency decimal bit time delay compensation (FSTC), the decimal time delay of compensation is τ_{g0I}, so

$$F_2 = F_{2D}(f) e^{j \cdot 2\pi f \Delta\tau_g} \tag{47.10}$$

Perform the same processing to signal $X_2(t)$ received by measuring station 2, since it doesn't need time delay compensation, we have

$$F_1(f) = FT(X_1(t)) \tag{47.11}$$

Calculating the cross spectrum of two measuring station signals, that is

$$P(f) = F_1(f) \cdot F_2^*(f) = A(f) \cdot e^{j \cdot 2\pi f \Delta\tau_g} = A(f) \cdot e^{\varphi(f)} \quad (47.12)$$

where A(f) denotes amplitude of cross spectrum and φ(f) is phase difference of two signals. Thus the residual value of time delay is the following expression:

$$\Delta\tau_g = \frac{1}{2\pi} \frac{\partial \varphi(f)}{\partial f} \quad (47.13)$$

So we can easily obtain geometric time delay $\Delta\tau_g$ [13].

47.3 Improvement Scheme

Now the thought of time delay extraction is to transform radio frequency down-converter into intermediate frequency first, then generate interference fringes, and use Welch method to divide the sampling data into L segments when generate interference fringes, the length of each segment is M and has part of overlap. The data window of each segment adopts window functions like rectangular window, triangular window, or Hanning window. First we calculated cross-power spectrum of each section's sampling data, then accumulated the cross-power spectrum of each section and got the average, so that we could get the average of cross-power spectrum and finally the time delay through phase fitting was calculated. This method which is based on intermediate frequency signal processing would introduce systematic errors when it transforms radio frequency down-converter into intermediate frequency and decreases accuracy. If the radio frequency signals were processed directly, stripes will vary greatly when generating interference fringes and influence the accuracy because of the increase of processing bandwidth. To this, we generally use 3σ method to eliminate outliers of stripes, but the elimination threshold is fixed and it will shorten the length of the entire stripe while eliminate the outliers. Under the circumstance of low signal-to-noise ratio, if too many of outliers were eliminated, it would lead to the accuracy of calculating stripe slope decline.

To solve the above problems, under the circumstance of small residual time delay (No phase step from $-\pi$ to π), we can get the average data through the method of calculating subsection average before phase fitting, according to the thought of 3σ method. Threshold of the fluctuation range was set, if the difference between the data point and the mean was greater than the threshold in scope of this segment, and then we gave the average to this point. Using this method to eliminate outliers, the shortcomings of 3σ method which reduce data segment and inability to specify threshold can be overcome. This increases the accuracy of stripe fitting effectively and then increases the measuring accuracy.

47.4 Simulation Experiment

We wrote programs for signal processing algorithm according to signal processing procedure in unit 1, next the time delay performance of improved extraction signal through simulation algorithm was tested.

A satellite beacon signal $X_0(t)$ received by antenna was simulated. Simulating signals consist of a modulation signal and Gaussian white noise n(t)

$$X_0(t) = \cos\left(2\pi \times 10^9 \times t + \cos\left(2\pi \times 5 \times 10^7 \times t\right)\right) + n(t) \tag{47.14}$$

The signal from the measuring station 1 as the reference signal and the signal from two stations were

$$X_1(t) = X_0(t) + n_1(t) \tag{47.15}$$

$$X_2(t) = X_0\left(t + \tau_g\right) + n_2(t) \tag{47.16}$$

where $n_1(t)$ and $n_2(t)$ are transmission noise of the channel. Parameters of simulation signals were set as follows: The sampling frequency f_s was 4 GHz, the length of analytic signal was 262,144, geometric time delay of two signals τ_g was12.525 ns. Supposing estimated model's compensation time delay τ_{g0} is 12.2725 ns; the signal-to-noise ratio of the two measurement signal's channel noise is -5 dB; Welch segment L is 128, window function adopts Hanning window; threshold compared to the average is $\pi/6$.

The time domain waveform and the frequency spectrum of the spacecraft simulation signal are as follows in Fig. 47.2.

Constructing two measuring station signals to be analyzed $X_1(t)$ and $X_2(t)$ according to expressions (47.15) and (47.16). In the case of that channel noise is -10 dB, the interference fringes and fitting images were obtained as in Fig. 47.3 after the process procedure of interferometry signals, the two images are the direct method and the signal processing image after a mean comparison. From Fig. 47.3 we can see directly that using the mean comparison method can eliminate outliers effectively.

Calculate the straight line's slope after fitting, dividing 2π and obtaining the estimation of the residual time delay $\overline{\Delta\tau}$, and obtaining the estimation of geometric time delay $\overline{\Delta\tau_g} = \tau_g + \overline{\Delta\tau}$. Fitting images of generated stripes when SNR is -10 dB in Fig. 47.3 and calculate the residual time delay estimation of the two methods $\overline{\Delta\tau_g} = 1.831053167995637 \times 10^{-10}\text{s}, \overline{\Delta\tau_2} = 1.902373750762525 \times 10^{-10}\text{s}$.

From it we can deduce the estimation of geometric time delay $\overline{\tau_g} = \tau_{g0} + \overline{\Delta\tau}$ and obtain the error between ideal geometric time delay τ_g and $\overline{\tau_{g1}}$ and $\overline{\tau_{g2}}$ are respectively $-1.6894683 \times 10^{-11}$s and $-9.7626249 \times 10^{-12}$s. From that we can prove the improvement method can increase the accuracy of time delay.

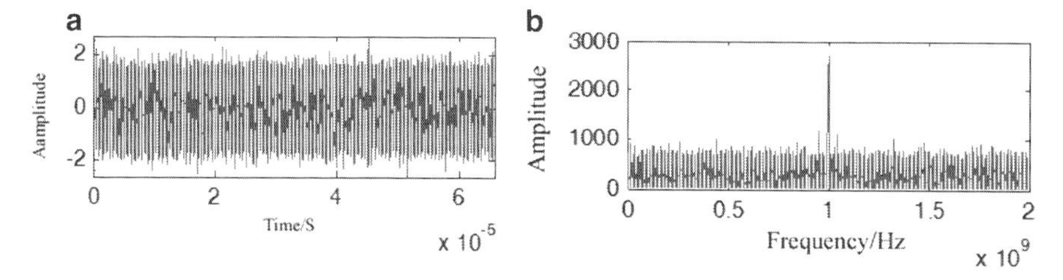

Fig. 47.2 Time-domain waveform and spectrum of spacecraft simulation signal $X_0(t)$. (**a**) Time-domain waveform and (**b**) spectrum

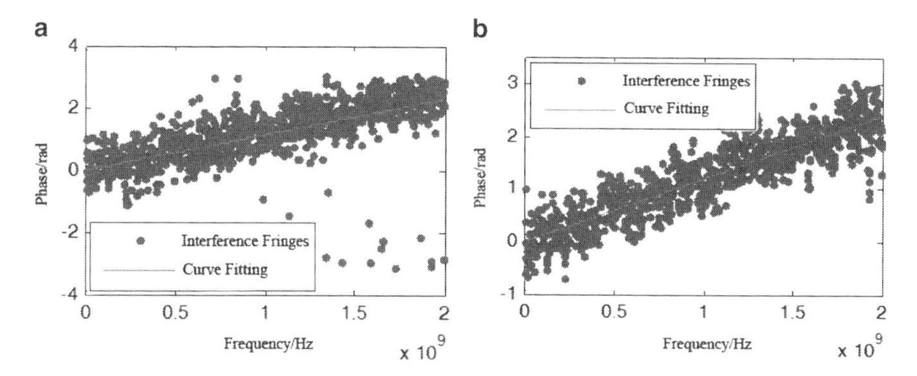

Fig. 47.3 Conduction of interference fringes and images fitting. (**a**) Direct method and (**b**) mean comparison method

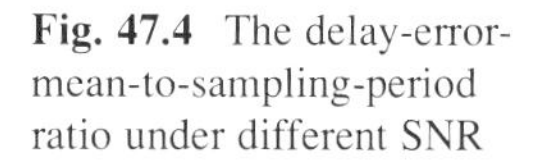

Fig. 47.4 The delay-error-mean-to-sampling-period ratio under different SNR

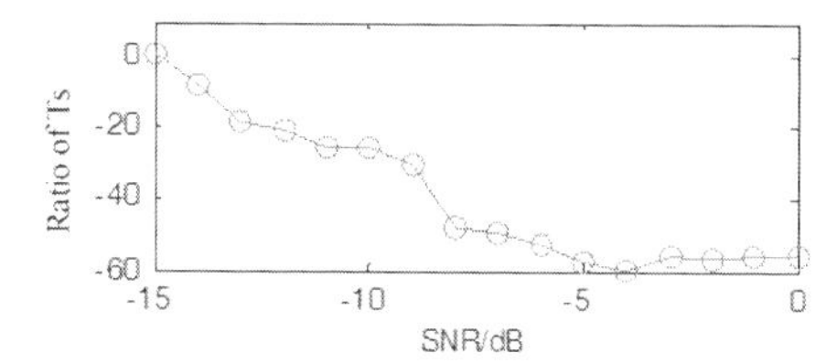

Using mean comparison method to simulate with different SNR (parameters unchanged). Figures 47.4 and 47.5, respectively, reflect the delay-error-mean-to-sampling-period ratio and delay-error-mean-to-standard-deviation ratio under different SNR. From the figure we can see when SNR is greater than -10 dB, we can obtain the accuracy of signal time delay greater than 1/10 Ts and has better performance when processing residual time delay error.

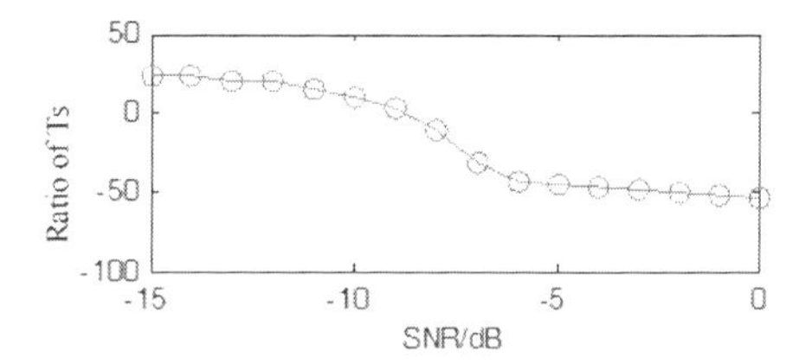

Fig. 47.5 The delay-error-standard-deviation-to-sampling-period ratio under different SNR

47.5 Conclusion

This study introduced an observational method for the satellite signal time delay, elaborated an improved algorithm of the interferometry signal processing, and made the simulation analysis verification. The results showed that adopting the improved algorithm can effectively eliminate the influence of noise on the data processing when residual time delay is small, and it can obtain the respective time delay of two measuring station signals better and increase the accuracy of measurement when SNR is low. This algorithm can have a further verification in the follow-up observation of the practical measured signals and then it can be applied to the exploration tasks of high-precision navigation in the future, and it can also provide future interferometric studies with valuable experience.

References

1. Thornton, C. L., & Border, J. S. (2003). *Radiometric tracking techniques for deep-space navigation*. Wiley-Interscience (Chaps 3 and 5).
2. Martín-Mur, T. J., Antreasian, P., Border, J., Benson, J., Dhawan, V., Fomalont, E., & Walker, C. (2006). Use of very long baseline array interferometric data for spacecraft navigation. Jet Propulsion Laboratory.
3. Hodgart, M. S., & Sarannah, G. A. (2008). A triple estimating receiver of multiplexed binary offset carrier (MBOC) modulation signals. In *ION GNSS 2008-21st International Technical Meeting of the Satellite Division* (pp. 877–886). The Institute of Navigation.
4. Chen, L., Tang, G. S., & Chen, M. (2011). Research and application of connected interferometry measurement signal processing method. In *Photonics and Optoelectronics (SOPO) Symposium* (pp. 1–4). IEEE.
5. Madde, R., Morley, T., & Abello, R. (2006). Delta-DOR–a new technique for ESA's deep space navigation. *ESA Bulletin, 128*, 69–74.
6. Kikuchi, F., Liu, Q., Hanada, H., Kawano, N., Matsumoto, K., Iwata, T., Sasaki, S., et al. (2009). Picosecond accuracy VLBI of the two subsatellites of SELENE (KAGUYA) using multifrequency and same beam methods. *Radio Science, 44*(2), 1–7.
7. Ichikawa, T. (2010). Application of high-precision two-way ranging to the spacecraft navigation. In *Proceedings of SICE Annual Conference 2010* (pp. 817–821). IEEE.
8. Thompson, A. R., Moran, J. M., & Swenson, G. W. (2008). *Interferometry and synthesis in radio astronomy* (pp. 289–293). Wiley-Vch.

9. Xia, H. B. (2005). Satellite the VLBI observations delay and phase stripes rotational model. *Space Science, 25*(1), 52–56 (In Chinese).
10. Zhang, H. T., Qin, Y. Y., & Zhan, Y. M. (2002). Sub-inertia guide fulcrum identify and Eliminate import GPS reference system processing. *Navigation, 2*, 75–80 (In Chinese).
11. Liu, Q., Shi, X., Kikuchi, F., Huang, Q., Kamata, S. I., Matsumoto, K., Wang, N., et al. (2009). High-accuracy same-beam VLBI observations of Shanghai and Urumqi radio telescope. *Science in China, Series G, 39*(10), 1410–1418 (In Chinese).
12. Hao, L. F., Wang, M., & Yang, J. (2010). VLBI observation with the Kunming 40-meter radio telescope. *Research in Astron and Astrophysics, 10*(8), 805–814.
13. Chen, L., Tang, G. S., Wang, M., Liu, H. C., Li, L., & Han, S. T. (2011). Connection interfere with the measurement signal processing method and experimental validation. *Journal of Telemetry, Tracking and Command, 32*(6), 28–31, 43 (In Chinese).

Chapter 48
Modeling and Evaluation of the Performance of Parallel/Distributed File System

Tiezhu Zhao, Xin Ao, and Huaqiang Yuan

Abstract The mass data storage systems need to be coupled with efficient parallel/distributed file systems, such as Lustre and HDFS, which can effectively solve the problems of the mass data storage and I/O bottlenecks. This chapter systematically studies the performance factors and distribution of parallel/distributed file systems and proposes a valuation scheme for the classic parallel/distributed file system by capturing the changes in workload characteristics. The experiment results show that the proposed evaluation scheme can reach better accuracy and efficiency.

48.1 Introduction

Distributed file systems are the key component of any cloud-scale data processing middleware. Evaluating the performance of distributed file system is very important. To avoid the cost for late cycle performance fixes and architectural redesign, providing performance analysis before the deployment distributed file system is also particularly important. System architects can use design-time performance to evaluate the resource utilization, throughput, and timing behavior of a system prior to the deployment due to the following reasons (1) analyzing performance of the system is much less expensive than testing the performance of the system by running it, (2) it is simply infeasible to test all kinds of different configurations of the system by running it, and (3) performance analysis on models helps architects make configuration and deployment decisions to avoid costly redesign, reconfiguration, or redeployment [1].

Automated storage management within an enterprise has proven to be an effective remedy for the low utilization that plagues storage system. Storage system administrator tries to ensure that workload characteristics of application are

T. Zhao (✉) • X. Ao • H. Yuan
Engineering and Technology Institute,
Dongguan University of Technology, Dongguan 523808, China
e-mail: tzzhao83@163.com

W.E. Wong and T. Zhu (eds.), *Computer Engineering and Networking*, Lecture Notes in Electrical Engineering 277, DOI 10.1007/978-3-319-01766-2_48,

appropriately matched to the performance and availability characteristics of the storage to which they are assigned. Automated storage system design, a solution proposed by many, relies on fast and accurate performance prediction. Constructing good performance evaluation scheme is a critical first step in effectively designing and optimizing the performance of storage systems.

Although the parallel/distributed file systems have been widely studied for several years, the potential impact to workload characteristics of parallel/distributed file system is not clearly understood. In this chapter, we focus on the performance study of the parallel/distributed file system. We systematically study the performance factors and distribution of parallel/distributed file system and propose a novel evaluation scheme for the classic parallel/distributed file system by capturing the changes in workload characteristics.

The remainder of this chapter is organized as follows. We begin by introducing related work in Sect. 48.2. We introduce the performance factors and distribution and present an evaluation scheme for parallel/distributed file system in Sect. 48.3. We perform a series of evaluation experiments, discuss the experiment results in Sect. 48.4, and conclude the chapter in Sect. 48.5.

48.2 Related Work

Performance analysis and modeling is an important concern in the distributed file system research area. The related work in the field mainly evaluates the performance of distributed file systems and computing paradigms by relying on running benchmarks or application programs, performance measurements under a variety of workloads/strategies, and comparing with other distributed file systems. The typical approaches can be divided into two categories (1) Experiment-driven performance analysis and prediction. These approaches are mainly based on the analysis of the experiment results or draws conclusion by comparing with existing distributed file systems. Yu et al. indicated excessively wide striping can cause performance. To mitigate striping overhead and benefit collective IO, authors proposed two techniques: split writing and hierarchical striping to gain better IO performance [2]. Wang et al. proposed a two-level metadata management method to achieve higher availability of the parallel file system while maintaining good performance [3]. Yu et al. presented an extensive characterization, tuning, and optimization of parallel I/O on the Cray XT supercomputer (named jaguar) and characterized the performance and scalability for different levels of the storage hierarchy [4]. (2) Model-driven performance analysis and prediction. Li et al. modeled the whole storage system's architecture based on closed Fork-Join queue model and proposed an approximate parameters analysis method to build performance models [5]. Yu et al. adopted a user-level perspective to empirically reveal the implications of storage organization to parallel programs running on Jaguar and discovered that the file distribution pattern can impact the aggregated I/O bandwidth [6]. Piernas et al. adopted a novel user-space implementation of active storage for Lustre, and the user-space approach has proved to be faster, more flexible, portable, and readily deployable than the kernel-space version [7]. Zhang et al. developed a new mechanism named Logic Mirror Ring (LMR) to improve the reliability and availability of the parallel file system. A logic mirror ring is built over all I/O nodes to indicate the mirror relationship among the nodes [8].

48.3 Performance Factors and Evaluation Scheme

48.3.1 Performance Factors and Distribution

In parallel/distributed file system, the key problem of performance research is to discover the potential performance factors of application workload and system configuration (e.g., stripe unit size, replica number, concurrent I/O number). Figure 48.1 shows the distribution of performance factors in the classic distributed file system. The performance factors can be divided into four parts: performance factors associated with metadata server, performance factors associated with data storage server, performance factors associated with the network, and performance factors associated with client nodes and applications.

This chapter focuses on performance factors associated with applications. So, it is important to study the characteristics of application workload. Workload characteristics can be used to describe I/O. Common among them is measures of the read/write ratio, I/O request size, spatial locality, temporal locality, and concurrency. Other, more description, characteristics include the temperature of data, the inter-arrival time of I/O bursts, the phasing (overlap) of different I/O streams, measures of temporal burstiness, measures of spatial burstiness, and spatiotemporal correlations [9].

48.3.2 Evaluation Scheme

The parameters of our proposed evaluation scheme are defined as follows:

W: the workload characteristic vector, including I/O request size (*ReqSize*), read/write ratio (*RWRatio*), request queue length (*ReqQueue*), request arrival rate (*ReqRate*), I/O randomness (*Rand*), I/O inter-arrival delay, etc., and is defined as

$$W = [ReqSize,\ RWRatio,\ ReqQueue,\ ReqRate,\ Rand,\ \cdots] \tag{48.1}$$

PMetric: the performance metric vector, consisted of bandwidth (*Bandwidth*), throughput (*Throughput*), and latency (*Latency*) and is defined as

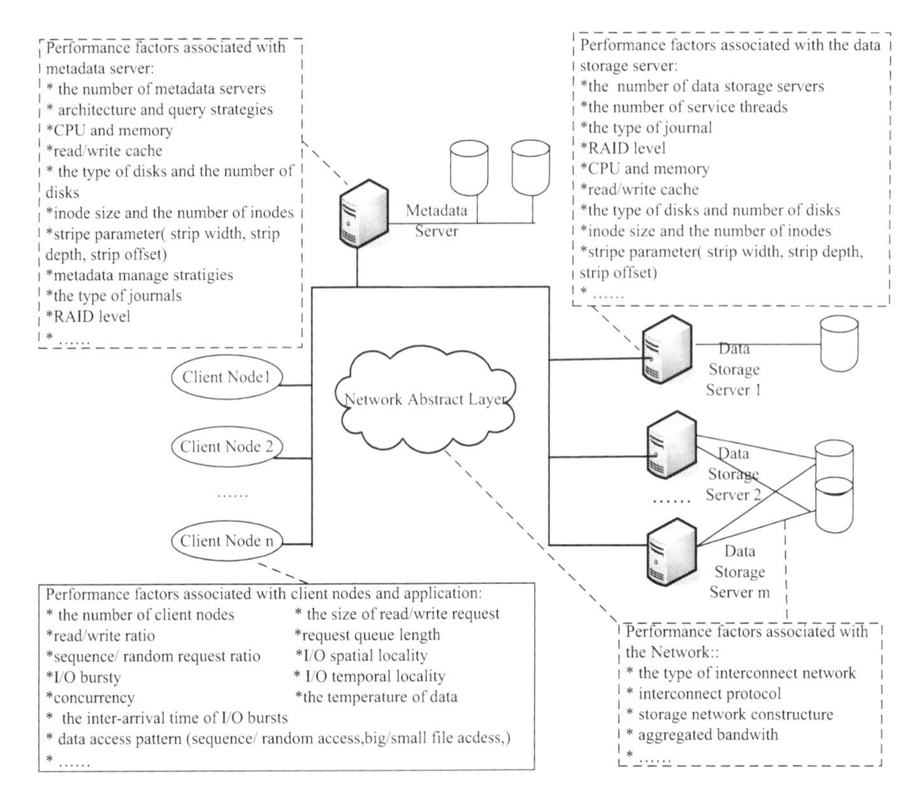

Fig. 48.1 Distribution of performance factors in classic distributed file system

$$PMetric = [Bandwidth,\ Throughput,\ Latency] \tag{48.2}$$

For a specific parallel/distributed file system DFS_i, a map function ϕ_i is defined to map the application's workload characteristics W_i to a performance prediction metric $PMetric_i$.This relationship can be expressed as follows:

$$PMetric_i = \phi_i(W_i) \tag{48.3}$$

where W_i is a vector of values representing workload characteristics of DFS_i and $PMetric_i$ is any performance metric of DFS_i.

Similarly, we can define

$$PMetric_j = \phi_j(W_j) \tag{48.4}$$

The first objective of our evaluation scheme is to capture the changes in workload characteristics from DFS_i to DFS_j, that is, predict W_j given W_i. The function $\phi_{i \to j}$ is defined as

$$W_j = \phi_{i \to j}(W_i) \tag{48.5}$$

According to Eqs. (48.4) and (48.5), the composition function $CF_{i \to j}$ of ϕ_i and $\phi_{i \to j}$ can be expressed as follows:

$$PMetric_j = \phi_j(W_j) = \phi_j(\phi_{i \to j}(W_i)) = CF_{i \to j}(W_i) \tag{48.6}$$

48.4 Experiment Analysis

48.4.1 Experiment Setup

The experiment was performed on HDFS (Hadoop Distributed File System), the primary distributed file system used in cloud computing with Hadoop. The HDFS consists of a single NameNode, a master server that manages the file system namespace and regulates access to files by clients. In addition, there are a number of DataNodes, usually one per node in the cluster, which manage storage attached to the nodes, which they run on. HDFS exposes a file system namespace and allows user data to be stored in files. Our experiments are done on Hadoop 0.20.1. The Hadoop cluster is constructed as two nodes for servers (one for NameNode, one for JobTracer), and the remaining nodes are DataNodes. Each node has 6 core Xeon X5650 processors at 2.66 GHz and 48 GB memory. The interconnect network is 10-gigabit TCP/IP Ethernet.

48.4.2 Experiment Result Analysis

The experiment mainly considers four workload characteristic factors: write request size, read request size, write queue depth, and read queue depth. Bandwidth, throughput, and latency are chosen as the performance measurement metrics. The configuration of workload characteristics or performance metrics is shown in Table 48.1.

The classification and regression trees algorithm is chosen as the machine learning algorithm because of their simplicity, flexibility, and interpretability. We measure the proposed evaluation scheme in term of average relative error (*Err*) and Pearson correlation coefficient (*Pearson*) [10]. Table 48.2 shows the accuracy analysis of our proposed evaluation scheme.

As shown in Table 48.2, when considering only one workload characteristics factor (Case1, Case2, Case3, and Case4), the average relative error can be controlled between 11.3 and 18.7 %, and Pearson is 0.79–0.91. When considering two factors (Case5 and Case6), the average relative error can be controlled between 20.9

Table 48.1 Configuration of workload characteristics or performance metrics

Workload characteristics or performance metrics	Units	Variable
Read request size	KB	*WRS*
Write request size	KB	*RRS*
Read queue length	IOs	*WQL*
Write queue length	IOs	*RQL*
Bandwidth	MB/s	*Bandwidth*
Throughput	IO/s	*Throughput*
Latency	ms/IO	*Latency*

Table 48.2 The accuracy analysis of the evaluation scheme

		Performance metrics					
		Bandwidth		Throughput		Latency	
Case	Workload characteristics	*Err* (%)	*Pearson*	*Err* (%)	*Pearson*	*Err* (%)	*Pearson*
Case1	*RRS*	11.3	0.91	12.3	0.89	13.1	0.90
Case2	*WRS*	13.1	0.86	15.6	0.83	13.8	0.85
Case3	*RQL*	12.7	0.88	16.9	0.89	14.4	0.86
Case4	*WQL*	15.3	0.80	18.7	0.79	16.2	0.80
Case5	*RRS, RQL*	21.8	0.78	20.9	0.76	23.3	0.76
Case6	*WRS, WQL*	25.5	0.83	24.9	0.73	23.9	0.80

and 25.5 %, and Pearson is 0.73–0.83. The higher the Pearson correlation and the lower the average relative error are, the more reliable is the prediction model. The results implicate our proposed evaluation scheme can obtain better prediction accuracy and efficiency.

48.5 Conclusion

In this chapter, we study the performance of the parallel/distributed file system. We first introduce the performance factors and distribution of the classic parallel/ distributed file system and propose a performance valuation scheme by capturing the changes in workload characteristics. Then, we perform a series of experiment to investigate the evaluation scheme. The average relative error can be controlled between 11.3 and 25.5 %. The experiment results show that the proposed evaluation scheme can reach better accuracy and efficiency.

Acknowledgments This work is supported by the Natural Science Foundation of Guangdong Province, China (Grant No. S2012040007746), the Scientific Research Foundation for Doctors of DGUT (ZJ130604), the National Natural Science Foundation of China (Grant No. 61170216, 10805019, 61272200).

References

1. Wu, Y., Ye, F., Chen, K., Zheng, W. (2013). Modeling of distributed file system for practical performance analysis. *IEEE Transactions on Parallel and Distributed Systems, 99*, 1–12.
2. Yu, W., Vetter, J. S., Canon, R. S., Jiang, S. (2007). Exploiting lustre file joining for effective collective IO. In *Proceeding of the Seventh IEEE International Symposium on Cluster Computing and the Grid* (pp. 267–274). Washington, DC: IEEE Computer Society.
3. Wang, F., Yue, Y. L., Feng, D., Wang, J., Xia, P. (2007). High availability storage system based on two-level metadata management. In *Proceeding of the 2007 Japan-China Joint Workshop on Frontier of Computer Science and Technology* (pp. 41–48). Washington, DC: IEEE Computer Society.
4. Yu, W., Vetter, J. S., & Oral, H. S. (2008). Performance characterization and optimization of parallel I/O on the cray XT. In *Proceeding of the 2008 I.E. International Symposium on Parallel and Distributed Processing* (pp. 1–11). Piscataway, NJ: IEEE.
5. Li, H. Y., Liu, Y., & Cao, Q. (2008). Approximate parameters analysis of a closed fork-join queue model in an object-based storage system. In *Proceeding of the Eighth International Symposium on Optical Storage and 2008 International Workshop on Information Data Storage* (pp. 1–8). Bellingham: SPIE.
6. Yu, W., Oral, H. S., Canon, R. S., Vetter, J. S., Sankaran, R. (2008). Empirical analysis of a large-scale hierarchical storage system. In *Euro-Par 2008, LNCS 5168* (pp. 130–140). Berlin: Springer.
7. Piernas, J., Nieplocha, J., & Felix, E. J. (2007). Evaluation of active storage strategies for the lustre parallel file system. In *Proceeding of the 2007 ACM/IEEE Conference on Supercomputing* (pp. 1–8). New York, NY: ACM.
8. Zhang, H., Wu, W., Dong, X., Qian, D. (2005). A high availability mechanism for parallel file system. In *APPT 2005, LNCS 3756* (pp. 194–203). Berlin: Springer.
9. Zhao, T., Verdi, M., Dong, S., & Simon, S. (2010). Evaluation of a performance model of Lustre file system. In *Proceeding of the Fifth Annual ChinaGrid Conference* (pp. 191–196). Washington, DC: IEEE Computer Society.
10. Benesty, J., Chen, J., Huang, Y., Cohen, I. (2009). Pearson correlation coefficient. *Springer Topics in Signal Processing, 2*, 1–4.

Chapter 49
CoCell: A Low-Diameter, High-Performance Data Center Network Architecture

Peng Wang, Huaxi Gu, Yan Zhao, and Xiaoshan Yu

Abstract As critical infrastructures in the Internet, data centers play an important role in supporting large-scale distributed applications as well as data-intensive computing. This chapter presents CoCell, a server-centric architecture, which uses servers to relay packets. CoCell has several nice properties for desired data center networking. The average node degree in CoCell network is close to 3, and the longest routing path length is no larger than 7. Besides, CoCell network is able to provide high network capacity to support bandwidth-intensive applications. Leveraging the multi-paths between any pairs of servers in CoCell network, we propose relative routing schemes and make a comparison among these paths. The evaluation indicates that CoCell performs well in all-to-all traffic pattern.

49.1 Introduction

With the prevalence of cloud computing, mega data centers have emerged as indispensable infrastructures for supporting large-scale distributed applications, such as GFS [1], BigTable [2], and MapReduce [3]. A desired DCN architecture should meet the following design goals: support of incremental expansion, commodity hardware that scales out, and abundant server to server connectivity. However, the conventional tree-based structure does not scale well to hundreds of thousands of servers and is also vulnerable to "single point failure," thus becomes increasingly difficult to meet these design goals. Recently, researchers are actively

P. Wang (✉) • Y. Zhao • X. Yu
State Key Laboratory of ISN, Xidian University, Xi'an 710100, China
e-mail: pengwang.xd@gmail.com

H. Gu
State Key Laboratory of ISN, Xidian University, Xi'an 710100, China

Science and Technology on Information Transmission and Dissemination in Communication Networks Laboratory, Xidian University, Xi'an 710100, China

W.E. Wong and T. Zhu (eds.), *Computer Engineering and Networking*, Lecture Notes in Electrical Engineering 277, DOI 10.1007/978-3-319-01766-2_49,

designing new network architectures to build high-performance data centers. The first thread of research work focuses on switch-centric structure network, leveraging switches to relay packets, such as VL2 [4] and Fat-Tree [5]. Another thread puts networking intelligence on servers instead of switches. Such designs include DCell [6], BCube [7], FiConn [8], and DPillar [9]. This kind of design has many obvious advantages over switch-centric designs as listed in the above articles.

In this chapter, we propose a regular server-centric DCN architecture named CoCell, which can be incrementally expanded by adding NICs to only a part of servers. Each server in CoCell not only undertakes the task of computation but also serves as an intermediate node to relay packets. It offers high scalability to support hundreds of thousands of servers with low diameter, low average node degree, and high network capacity. As for the routing algorithm we propose, it is specially customized for the CoCell network. We find many sets of paths for arbitrary pairs of servers, and the length of alternative paths is not too much larger than that of default path.

The rest of this chapter is organized as follows. Section 49.2 describes the structure of CoCell in detail and its topological properties. Section 49.3 presents the routing algorithm. Section 49.4 evaluates the topology properties and routing protocols in CoCell. Finally, Sect. 49.5 concludes the chapter.

49.2 The CoCell Network Structure

In this section, we will first present the physical structure of CoCell and its topological properties. Then we discuss the addressing and interconnection rules of servers in CoCell.

49.2.1 CoCell Physical Structure

CoCell is a server-centric network structure, designed to efficiently interconnect an increasingly number of servers. The implementation leverages architecture design rules to build scalable and reliable network architecture at a low cost. All switches in this structure are of the same type that connect a constant number of servers. We use two parameters to define a CoCell network: n, the number of ports in each switch, and i, the number of ports in each server. Thus a CoCell network can be expressed as $CoCell(n, i)$ $(i \geq 2)$. Before explaining how to construct $CoCell(n, i)$, we first introduce the smallest module, $CoCell_0$, and the building block, $Cell(n, i)$. In our design, $CoCell_0$ consists of an n-port commodity switch; $n/2$ multi-port servers, which are denoted as m_server; and $n/2$ dual-port servers, which are denoted as d_server. The first port of *m_server* and *d_server* is used to connect the commodity switch. In a $CoCell_0$, *m_servers* are used to connect the other n/2 $CoCell_0$s, and *d_servers* are used to connect $Cell(n, i)$s into a $CoCell(n, i)$.

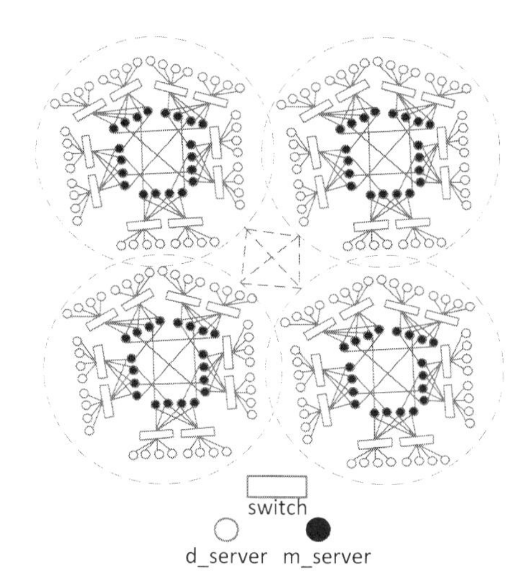

Fig. 49.1 A partial *CoCell* (8, 3) structure. A *CoCell* (8, 3) structure consists of 41 *Cell*(8, 3)s

The smallest *Cell*(*n*, *i*), *Cell*(*n*, 2) is constructed by *n*/2 + 1 $CoCell_0$s by means of a complete graph. There is only one link between any pair of $CoCell_0$s.

Then, each *Cell*(*n*, *i*) is treated as a virtual node, fully connected with each other homogeneous virtual node to form a *CoCell*(*n*, *i*) in the same way. To scale from *Cell*(*n*, *i*−1) to *Cell*(*n*, *i*), we need to add a switch connecting *n*/2 servers into each $CoCell_0$ and connecting another *n*/2 ports of the switch with *n*/2 *m_servers* in the CoCell_0. In this way, additional (n/2)*(n/2 + 1) *d_servers* are added into a *Cell*(*n*, *i*−*1*) to form a *Cell*(*n*, *i*). The structure of *CoCell*(8, 3) is shown in Fig. 49.1.

49.2.2 Topological Properties

We denote the total number of severs in a *CoCell*(*n*, *i*) as *N*(*n*, *i*). The number of servers in a *Cell*(*n*, *i*) network can be expressed as follows:

$$N(\mathrm{n,i}) = \left(\frac{n}{2}+1\right) * \frac{n}{2} * i + \left[\left(\frac{n}{2}+1\right) * \frac{n}{2}\right]^2 * i * (\mathrm{i}-1) \tag{49.1}$$

Bisection bandwidth refers to the smallest number of links removed to divide the nodes in the network into two parts of equal size.

In a *CoCell*(*n*, *i*), the lowest bound of bisection bandwidth has been obtained:

$$Bd(\mathrm{n,i}) = \left[\frac{n}{4}(i-1)\right]^2 + \left(\frac{n}{4}\right)^2 + i \tag{49.2}$$

49.2.3 Addressing and Interconnection Rules

We denote a server in a *CoCell(n, i)* as a 4-tuple <p, q, a_1, a_0>. In this tuple, p($p \geq 0$) means the *Cell(n, i)* a server locates in, *q* indicates the switch that a server belongs to, and a_1a_0 are 2-tuple used to distinguish the multi-port servers in each *Cell(n, i)*. Based on the features of the structure, a_1 and a_0 take values from [0, *n*/2 +1] and [0, *n*/2]. Particularly, if $q = 0$, it means that this server is an *m_server* and connected to all the $i-1$ switches.

In a *CoCell(n, i)*, two *m_servers* of <*l*, *p*, *a*, *b*−*1*> and <*p*, *q*, *b*, *a*> are connected with a link for every a and every b > a. In a *CoCell(n, i)*, we treat each *Cell(n, i)* as a virtual node. A *d_server* in a *Cell(n ,i)* is denoted as <*l*, *u*>, where l is the *Cell(n, i)* the server belongs to and *u* is calculated from the other three tuples:

$$u = (\mathrm{p} - 1) * \left(\frac{n}{2} + 1\right) * \frac{n}{2} + a_1 * \frac{n}{2} + a_0 \tag{49.3}$$

We illustrate the design via the following example. Server <0, 1, 2, 1> should be denoted as server <0, 5> through the method above. Thus we can easily find server <6, 0> of another cell to connect to.

49.3 Routing in CoCell Network

In this section, we propose CoCell routing algorithm, which fully exploits the characteristics of CoCell structure to achieve effective resource utilization.

Leveraging the characteristics of the network structure and node addressing, we propose an efficient and simple routing scheme, called *CoCellRouting*, to find a single-path between any pairs of servers in a *CoCell(n, i)*. *CoCellRouting* is divided into two steps, finding path in the same *Cell(n, i)* and finding path in the different *Cell(n, i)*s.

Considering two nodes *src* and *dst* are in the same *Cell(n, i)*, we propose the routing algorithm as follows. *FindLink* calculates the link that interconnects two *m_servers* belonging to different $CoCell_0$s. According to the interconnection rules of CoCell, the link is (<S_l, 0, D_{a0}, $D_{a1}-1$>, <D_l, D_p, D_{a1}, D_{a0}>).

Considering two nodes *src* and *dst* in the different *Cell(n, i)*s, we first use Eq. (49.3) to convert the address of a *d_server* into <*l*,*u*>. Then based on the function *FindLink*, we can find the link connecting the two *Cell(n, i)s*.

We consider the worst-case scenario where two d_servers are in different cells. From each server, the packet needs to be forwarded three hops to reach the cell where the other one located in. Thus, the maximum path length between any pairs of servers in *CoCell(n, i)* is at most 7.

Pseudo-code of this single-path routing algorithm

```
/* src and dst are denoted using the 4-tuples
src = <S_l, S_p, S_a1, S_a0>  = <S_l, 0, S_a1, S_a0>
dst = < D_l, D_p, D_a1, D_a0> = < D_l, 0, D_a1, D_a0> */
01:  CellRouting(src, dst)   /* in the same Cell(n,i)*/
02:  if  (S_a1 == D_a1)
03:     if  ((S_p & D_p) & (S_p!=D_p) ==1)
        /* belong to the different switches*/
04:        path1= (src,n_1);
05:        FindLink(n_1, n_2);
06:        path2= (n_2, dst);
07:        return path1+ path2;
08:     else return(src, dst);
09:  else return(src, dst);
```

49.4 Evaluation

We have analyzed the basic properties of CoCell in Sect. 49.2, such as the high scalability and high bandwidth, which play an important role in the performance of data center networking. In this section, we conduct simulations to evaluate the properties of CoCell and compare it with a famous server-centric structure, DCell.

49.4.1 Growth Rate of Server Number

For a current server-centric architecture, such as DCell and FiConn, it can be difficult to build a complete structure when the server number in its lower structure has been very large, since the number of servers increases doubly exponentially. In our structure, the server number increases smoothly and meets the requirement of current data center. So, we do not have to build partial network and design partial routing algorithm for network anymore.

In Fig. 49.2, we can clearly find that CoCell expands more smoothly than DCell. DCell has the limitation of using mini switches ($n \leq 8$) [1]. With the increase of n and basic expansion unit (symbol k for DCell and symbol *i* for CoCell), the number of servers will increase more acutely. We find that even *CoCell*(48, *i*) also changes smoothly with *i*. However, DCell increases very fast even when *n* and *k* are still small integers.

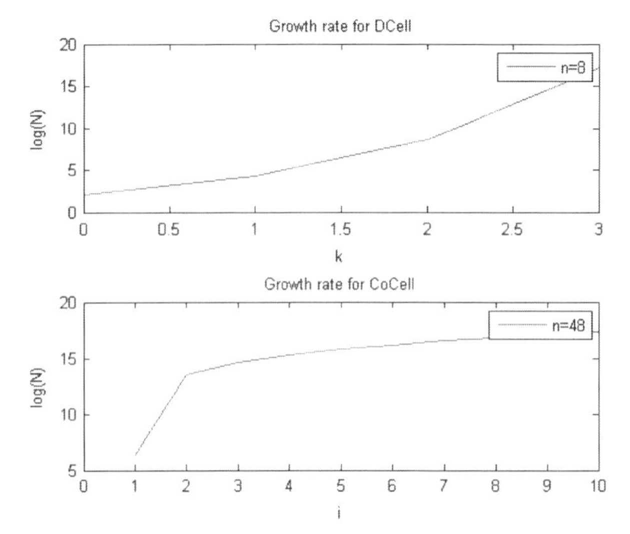

Fig. 49.2 Growth rate of DCell and CoCell vs. expansion unit

Table 49.1 The distribution of routing path in *CoCell*(4,3)

Path length		1	2	3	4	5	6	7
Number	*m_server*	5	16	20	40	72	80	0
	d_server	4	8	19	28	42	52	80

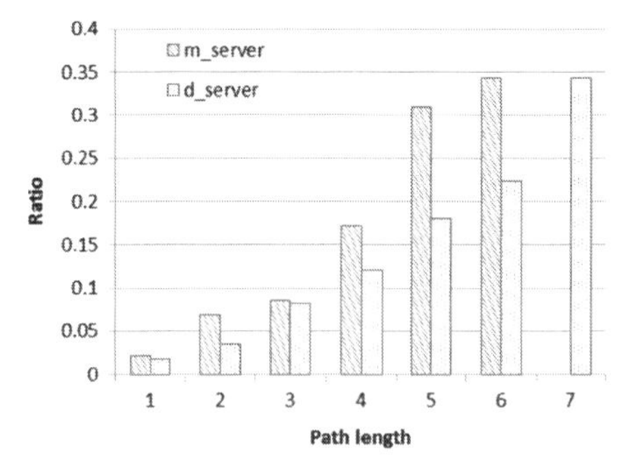

Fig. 49.3 The distribution of routing path in *CoCell* (4, 3)

49.4.2 Distribution of Path Length

We run simulations on *CoCell*(4, 3) to evaluate the distribution of path length. The results are shown in Table 49.1. From these data, we find that the average length of path from *m_server* to other servers is shorter than that of *d_server*. Therefore, when the network scales to a large one, the number of *m_servers* increases; thus the average path length will decrease. Figure 49.3 indicates the distribution of routing path in *CoCell*(4, 3). Such a distribution of routing path performs well in all-to-all traffic pattern [7].

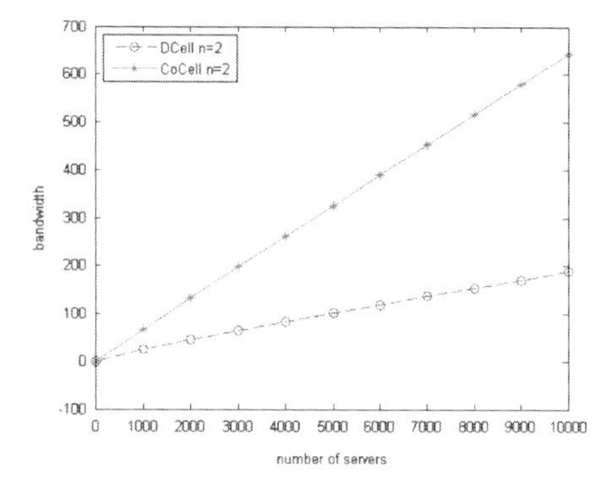

Fig. 49.4 Bandwidth of DCell and CoCell vs. server order

49.4.3 Evaluation of Bandwidth

Bandwidth is an important metric to measure the network throughput. Larger bisection bandwidth means that there are more possible paths between any pair of servers and thus intrinsically more fault tolerant. The results in Fig. 49.4 show that the bandwidth of CoCell performs better than that of DCell with the same type of switches.

49.5 Conclusion

In this chapter, a novel server-centric architecture named CoCell is proposed, which has many desirable DCN features. It is highly scalable to accommodate up to hundreds of thousands of servers with low diameter, low average node degree, and high bisection bandwidth. As a decentralized routing solution, the routing algorithm in CoCell effectively exploits the features of the structure. Finally, we evaluate the performance of this structure and find that CoCell shows nice topological properties.

References

1. Ghemawat, S., Gobioff, H., & Leung, S. (2003). The Google file system. In *Proceedings of the Nineteenth ACM Symposium on Operating Systems Principles* (pp. 29–43). New York: ACM Press.
2. Chang, F., Dean, J., Ghemawat, S., Hsieh, WC., Wallach, DA., Burrows, M., et al. (2008). Bigtable: A distributed storage system for structured data. *ACM Transactions on Computer Systems (TOCS)*, *26*(2), doi:10.1145/1365815.1365816
3. Dean, J., & Ghemawat, S. (2008). MapReduce: Simplified data processing on large clusters. *Communications of the ACM*, *51*(1), 107–113.

4. Greenberg, A., Hamilton, JR., Jain, N., Kandula, S., Kim, C., Lahiri, P., et al. (2009). VL2: A scalable and flexible data center network. In *ACM SIGCOMM Computer Communication Review* (pp. 51–62). New York: ACM Press.
5. Al-Fares, M., Loukissas, A., & Vahdat, A. (2008). A scalable, commodity data center network architecture. In *ACM SIGCOMM Computer Communication Review* (pp. 63–74). New York: ACM Press.
6. Guo, C., Wu, H., Tan, K., Shi, L., Zhang, Y., & Lu, S. (2008). Dcell: A scalable and fault-tolerant network structure for data centers. In *ACM SIGCOMM Computer Communication Review* (pp. 75–86). New York: ACM Press.
7. Guo, C., Lu, G., Li, D., Wu, H., Zhang, X., Shi, Y., et al. (2009). BCube: A high performance, server-centric network architecture for modular data centers. In *ACM SIGCOMM Computer Communication Review* (pp. 63–74). New York: ACM Press.
8. Li, D., Guo, C., Wu, H., Tan, K., Zhang, Y., & Lu, S. (2009). FiConn: Using backup port for server interconnection in data centers. In *INFOCOM 2009* (pp. 2276–2285). Washington, DC: IEEE Computer Society Press.
9. Liao, Y., Yin, D., & Gao, L. (2012). DPillar: Dual-port server interconnection network for large scale data centers. *Computer Networks, 56*(8), 2132–2147.

Chapter 50
Simulation Investigation of Counterwork Between Anti-radiation Missile and Active Decoy System

Huaqiang Hu and Dandan Wen

Abstract Simulation test has provided a favorable method and platform for quantitatively evaluating impact on countering ARM by overcoming the disadvantage of high price and poor privacy in regard to outfield experiment. The essay tries to make a deep research on modeling simulation of active decoy and ARM and thus to formulate active decoy interference model and anti-radiation missile movement model. Then it carries through simulation process of active decoy's effect; the simulation result validates the model's effectiveness and accuracy, and therefore it has provided theoretical bases for designing and deploying active decoy and ARM's base station program.

50.1 Introduction

As a tactical offensive weapon against electromagnetic radiation source (radar, military communications equipment, etc.), ARM is mainly directed by guidance signal given out by enemy radiation source; it would lead the missile to radiation source and thus destroy it. ARM plays the "ace in the hole" role against electromagnetic radiation source and tries to threaten military use and survival of radiation source. As a result, it is significant that deep research should be made on measures against ARM and that viability of electromagnetic radiation source thus would be improved. Nowadays one of the effective methods for confronting ARM is active decoy.

It is an effective measure for confronting ARM by adopting the active decoy system constituted by active decoy. It is capable of inducing space synthetic

H. Hu (✉)
The First Aeronautic Institute of Air Force, Xinyang 464000, China
e-mail: hqsnial@163.com

D. Wen
Xinyang Normal University Huarui College, Xinyang 464000, China

W.E. Wong and T. Zhu (eds.), *Computer Engineering and Networking*, Lecture Notes in Electrical Engineering 277, DOI 10.1007/978-3-319-01766-2_50,

distortion; in certain conditions, ARM would be unable to aim at any radiation source, and therefore it plays the role of protecting ground radar station and decoy station [1]. In this way, it has a practical significance for research on related weapon system development and tactical application that a deep investigation on active decoy and countering ARM would be made. Simulation test has provided a favorable method and platform for quantitatively evaluating impact on countering ARM by overcoming the disadvantage of high price and poor privacy in regard to field experiment.

50.2 Mathematical Modeling of Active Decoy System

One-source active decoy system adopts the system constituted by one decoy and radar. Despite its limitations, it lays the foundation for two-source active decoy system and multisource active decoy system, and therefore a detailed analysis of it is necessary.

Anti-radiation seeker actually functions as a broadband passive mono-pulse detection system. Its constitution and working principle resemble roughly that of mono-pulse radar, though anti-radiation missile itself does not emit electromagnetic waves. ARM, as passive radar, tries to ascertain azimuth and elevation of external active radiation by employing external electromagnetic wave and thus locate its targets [2]. At present, most of ARM makes use of the method of "phase comparison" to measure azimuth and elevation both at home and abroad. The mono-pulse system has a good ability of anti-interference with one-source active decoy system; however, when there exist several radiation sources within discrimination angle of mono-pulse radar, it would trace source energy center while deviating from the source. In this way it lays the theoretical foundation for interference with ARM.

It is assumed that the plane constituted by radar, active decoy, and ARM is set within a rectangular coordinate system, as shown in Fig. 50.1.

Origin of coordinate is location of radar, active decoy being on x-axis, and coordinate of ARM should be (x, y). Thus, the following relationship can be obtained from Fig. 50.1:

$$\begin{aligned} r_0 &= \sqrt{x^2 + y^2} \\ r_1 &= \sqrt{(x - x_1)^2 + y^2} \end{aligned} \tag{50.1}$$

It is assumed that angular frequency of radar and decoy would, respectively, be ω_0 and ω_1, with the initial phase difference between them being φ_{10}, and when arriving at ARM, electric field intensity of radar and decoy would, respectively, be

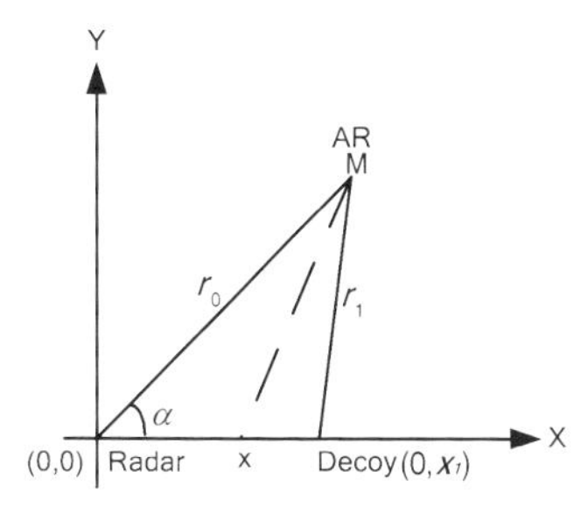

Fig. 50.1 Position relationship diagram of radar, decoy, and ARM

$$\begin{aligned}E_0 &= E_{00}\cos\left(\omega_0 t - \frac{2\pi}{\lambda_0}r_0\right)\\ E_1 &= E_{10}\cos\left(\omega_1 t - \frac{2\pi}{\lambda_1}r_1 + \varphi_{10}\right)\end{aligned} \tag{50.2}$$

Taking no account of field polarization direction difference between radar and decoy, synthetic field intensity at the position of ARM would be

$$\begin{aligned}E &= E_{00}\left[1 + \beta^2 + 2\beta\cos\left(\varphi_0 - \varphi_1\right)\right]^{0.5}\\ \varphi(x, y, t) &= \arctan\left(\frac{\sin\varphi_0 + \beta\sin\varphi_1}{\cos\varphi_0 + \beta\cos\varphi_1}\right)\end{aligned} \tag{50.3}$$

Since tracking direction of mono-pulse seeker is the same as normal direction of phase line, connecting with the abovementioned formula, coordinate of x would be

$$\frac{x}{x_1} = \frac{\beta^2 + \beta\cos\Delta\varphi}{\beta^2 + K + \beta(K+1)\cos\left(\Delta\varphi\right)} \tag{50.4}$$

It can be made out by formula (50.4) that when $\beta = 1$, formula (50.4) would be simplified as follows:

$$x = \frac{x_1}{1+K} \tag{50.5}$$

It can be seen from formula (50.5) that in this case, ARM's target direction has nothing to do with its phase difference $\Delta\varphi$; on the contrary, the target direction is closely related to decoy's frequency, radar's frequency, and ARM's attacking direction. In decoy process of actual combat, regulation of ω_0 and ω_1 can be achieved according to distance parameter detected by radar. ARM is thus made to direct to two radiation sources as far as possible, and in this way both radar and decoy can be assured their safety. When distinction between r_0 and r_1 is made quite great by ARM's direction (ARM is launched from the other side of layout with radar and decoy), radar and decoy should be adjusted so that distinction between ω_0 and ω_1 would be larger; however, in such a situation as for ARM which would fulfill its target recognition by using carrier frequency of radiation sources, the system concerned would not achieved its decoy effect, and single-point source decoying has its limitations [3].

Basing on this, the universal model of decoying system can be inferred. Supposing that the decoying system is composed of radar and n_i irradiation source, formulating a coordinate system with radar for origin, radar would be (0,0,0), decoy No. i would be (x_i,y_i,z_i), and ARM is located at (x_A,y_A,z_A), it can be concluded by employing the same method of single-point source decoying that

$$\begin{aligned} x &= \frac{\sum_{i=0}^{n}\sum_{k=0}^{n} E_{0i}E_{0k}C(x_k)\cos(\varphi_i - \varphi_k)}{\sum_{i=0}^{n}\sum_{k=0}^{n} E_{0i}E_{0k}D_k\cos(\varphi_i - \varphi_k)} \\ y &= \frac{\sum_{i=0}^{n}\sum_{k=0}^{n} E_{0i}E_{0k}C(y_k)\cos(\varphi_i - \varphi_k)}{\sum_{i=0}^{n}\sum_{k=0}^{n} E_{0i}E_{0k}D_k\cos(\varphi_i - \varphi_k)} \end{aligned} \tag{50.6}$$

In formula (50.6) $C(t_k) = \frac{t_k z_A - t_A z_K}{R_K \lambda_K}$, $D_k = \frac{z_A - z_K}{R_K \lambda_K}$, in the same way, since single-pulse ARM seeker tracks along with normal direction electromagnetic wave front; ARM's tracking direction can be ascertained during its flight at any time.

50.3 ARM's Mathematical Modeling of Dynamic Flight

ARM's dynamic flight refers to its whole process of space flight after being launched. The essay mainly deals with active decoy's degree of interference on the ARM seeker which is attacking, with regard to multipoint embattling. Before modeling, firstly ARM's space flight line while attacking ground radar station would be introduced. Figure 50.2 displays ARM's space flight line while attacking a static ground radar station. It can be seen from Fig. 50.2 that during the process when ARM is attacking its target (static target), its trajectory is usually operating negative feedback near the axis and finally moves towards its target. Hereunto, the thin broken line designates missile axis movement [4].

As shown in Fig. 50.2, ARM's ballistic curve equation can be denoted as follows:

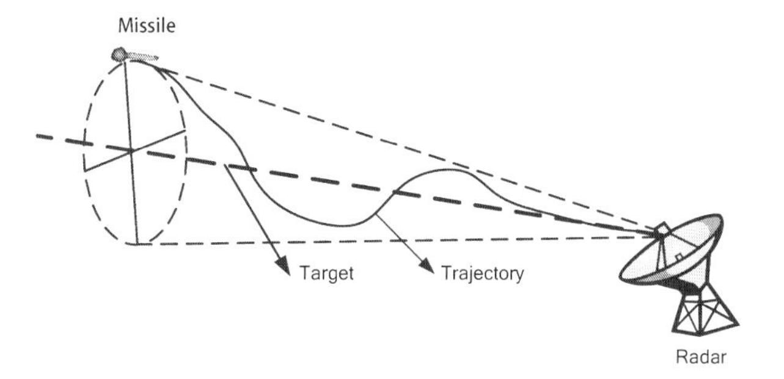

Fig. 50.2 ARM attacking target radar trajectory

$$\frac{x - x_A}{x_A - x_0} = \frac{y - y_A}{y_A - y_0} = \frac{z - z_A}{z_A - z_0} \tag{50.7}$$

Formula (50.7) (x_A,y_A,z_A) denotes initial coordinate of ARM trajectory axis; (x_0,y_0,z_0) denotes coordinate of target radar. The ARM's trajectory orbit sways around axes by operating negative feedback. Therefore, its trajectory axis orbit can be shown in the following equation:

$$\begin{cases} x = S - Vt \\ y = k * \sin x \\ z = k * \cos x \end{cases} \tag{50.8}$$

Hereunto according to relative coordinate system formulated by $(x_0,y_0,z_0) = (0,0,0)$ and $(x_A,y_A,z_A) = (S,0,0)$, where S denotes the distance between ARM and its target, V denotes the radial velocity of ARM, and k denotes the negative feedback coefficient, that is, attenuation coefficient of amplitude, it can be seen from Fig. 50.2 that trajectory axis is only an approximate simulation of ARM's flight line. For within one time step, the attachment between ARM's location from one point to next point and radar is not in the same straight line [5]. Here in consideration of the small time step, in order to simplify the modeling, it is approximately presumed that these three points are on the same line. Since it does not have an obvious effect on placement of distribution, trajectory orbit can be approximately substituted by its axis orbit.

The essay tries to formulate ARM's mathematical modeling of dynamic flight basing on time differencing method. With firing data of ARM's location at certain moment of flight as the starting point, it is presumed that ARM would be located at the next moment $(t + \Delta t)$ on the attachment between its position at previous moment and radar energy center point. Meanwhile attenuation is introduced, and it can make ARM deviate from the straight line with a tiny distance during a time step of Δt. As a result, these discrete points in the space would be connected as the orbit of ARM's flight in the space. The essence of time differencing method lies in that it makes use of a step-by-step approach to obtain ARM's flight line. In this way, mathematical modeling can be thus simplified and conclusion would not be distorted.

First of all, three-dimensional coordinate system xyz should be formulated, with radar being located at the origin of coordinate, decoy coordinate being presumed as (x_1,y_1,z_1), (x_2,y_2,z_2), etc. In its initial state, that is, t = 0, ARM's coordinate can be presumed as (x_{A0},y_{A0},z_{A0}); supposing that time step be t, one group of discrete time variables can thus be obtained, being, respectively, t_1, t_2, t_3 .... ARM's coordinate would be (x_{Ak},y_{Ak},z_{Ak}) at the point of t_k. During its flight, coordinate of ARM's location at any time can be shown as (x_A,y_A,z_A), which is obviously a group of random variables. R is employed to denote the distance between ARM and radiation source, R_{A0} the distance between ARM and radar, and R_{Ai} the distance between ARM and decoy source i. For the purpose of highlighting the direction of missile attack, make that the azimuth angle of ARM be φ, pitching angle be θ, and $\varphi \in (0, 2\pi)$, $\theta \in (0, \pi/2)$. According to the definition of right-angle coordinate system, conversion relation between them can be shown as follows:

$$\begin{cases} x = R\cos\theta\sin\varphi \\ y = R\cos\theta\cos\varphi \\ z = R\sin\theta \end{cases} \tag{50.9}$$

Basing on the definition mentioned above, and meanwhile employing subscript method to differentiate energy center of decoy system, the following conclusion can thus be obtained:

$$\begin{aligned} R_{A0} &= \sqrt{(x_A - x_0)^2 + (y_A - y_0)^2 + (z_A - z_0)^2} \\ R_{ij} &= \sqrt{(x_i - x_j)^2 + \left(y_i - y_j\right)^2 + (z_i - z_j)^2} \end{aligned} \tag{50.10}$$

50.4 Simulation of Decoying Effect

In order to make an analysis of modeling simulation, it is presumed that ARM's parameters are the following: radius should be 25 m, speed on the location of critical differentiating point be V = 3M, maximum overload of ARM be $a_m = 10g$, and resolution of missile seeker be $\theta = 15^0$. Meanwhile when it is presumed that there are three active decoys, being arranged as regular triangle, power ratio being 1:1:1, and side length of triangle being L, its simulation process would be as follows [6]:

Suppose that ARM being as far as possible, three active decoys are all located within its field of view; for each given L, several directions are to be chosen evenly from ARM's probable attacking directions, for each direction $\varphi_i \in (0, 2\pi)$, and values would be evenly taken. ARM's final impact point can be obtained on the basis of previous calculation, decoying hit probability would then be counted, and finally relation diagram between decoying hit probability and lengths of triangle sides is shown as in Fig. 50.3 by dotted line.

Fig. 50.3 Decoying hit probability

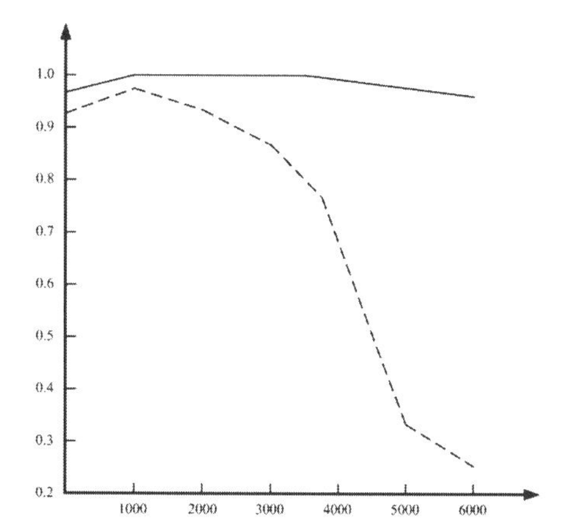

It is presumed that after ARM would fly over the critical position of a three-point source, without considering ARM's mobility, it would hit the target directly along this direction; in this way, relationship diagram between decoying hit probability and lengths of triangle sides is shown in Fig. 50.3 by solid line.

From Fig. 50.3, it can be seen that without considering ARM's mobility, when L is more than 600 m, decoying hit probability is nearly 1. The reason lies in that the absolute distance between critical point and each point source would increase along with increase of L; decoying distance concerned would also increase and decoying hit probability can thus be improved. Decoying hit probability would be the largest when L should be 600 m or so. The reason lies in that absolute distance between critical point and each point source would increase along with L's increase. There would be much more time for ARM to modulate its direction; however, due to limitation on ARM's maximum lateral overload, its modulation is rather limited, so in this case, appropriate side length would bring about the largest decoying hit probability.

50.5 Conclusion

During the process of ARM's countering multipoint source active decoy, its seeker's resolution angle is one of critical factors which have an effect on its combat effectiveness. From the analysis of simulation, it can be seen that seekers with wide angle of resolution do not possess the ability of countering decoying; within the range which angle of sight would permit, its interference effect would be strengthened along with decrease of seeker's resolution angle. Meanwhile seeker's precision of measure angle can also be heightened to optimize seeker's guidance control parameters, and in this way ARM's motor overload can be increased. It has been an effective method for improving ARM's combat ability of countering decoy and for reducing towed target quantity.

References

1. Luo, X., Ning, J., & Luan, S. (2005). Fuzzy systematic judgment of ranking multi-radiant imperilment. *Fire Control and Command Control, 30*(4), 66–68.
2. Zhang, B. (2004). Set pair analysis and multi-attribute decision making. *System Sciences and Comprehensive Studies in Agriculture, 20*(2), 123–125.
3. Pan, Q., & Zhang, H. (2008). Comprehensive modeling and simulation verification of radiation threat based on closeness degree of ideal point. *Journal of System Simulation, 20*(14), 3896–3898.
4. Zheng, M. (2005). *Study of radiant-point beguiling resisting anti-radiation missile technology, tactics and embattling mode*. Changsha: National University of Defense Technology.
5. Xie, B., Yin, J., & Song, J. (2004). Building of kill probability simulation model for the ARM against the radar targets. *Journal of System Simulation, 16*(9), 2044–2047, 2051.
6. Mclendon, R., & Turner, C. (1983). Broad-band sensors for lethal defense suppression. *Microwave Journal (S0192–6225), 15*(6), 85–101.

Chapter 51
Simulation Jamming Technique on Binary Phase-Coded Pulse Compression Radar

Yulin Yang and Lijuan Qiu

Abstract Binary phase-coded signal is usually used in pulse compression (PC) radar, which is mostly used for surveilling and tracking targets. Two prominent characteristics of the binary phase-coded signal are introduced in this article; according to these characteristics, some jamming forms are analyzed which can be used to jam binary phase-coded PC radar; simulation with MATLAB is done to test these jamming forms. The result indicates that noise jamming has less effectiveness on binary phase-coded PC radar and continuous and partial code replicated jamming have preferable effectiveness on it. Because binary phase-coded signal is compressed to narrow pulse when it is received by receiver, jamming signal can easily capture the range gate when the jamming side adopts range-gate pull-off (RGPO), and so the jamming effectiveness is perfect.

51.1 Introduction

Modern radars are required, not only detecting the targets which are far from them but also having high-range resolution [1]. Radars with high-range resolving power must have extremely narrow pulse width, which limits the increase in power of the transmitter, thus affecting the detecting range of radars. In order to resolve the contradiction between the detecting range and the range resolving power, the designers of radars adopt PC technique and design and manufacture PC radar. The radar of this type transmits wide pulse and receives narrow pulse which is compressed by the matched filter network. Binary phase-coded signal is a type of sending signal commonly used by PC radars. It may be coded pseudo-random. When the matched filter of the echo signal Doppler frequency detuning, the matched filter cannot compress the pulses, so we called that the Doppler frequency

Y. Yang (✉) • L. Qiu
The First Aeronautic Institute of Air Force, Xinyang 464000, China
e-mail: yanyline@163.com

W.E. Wong and T. Zhu (eds.), *Computer Engineering and Networking*, Lecture Notes in Electrical Engineering 277, DOI 10.1007/978-3-319-01766-2_51,

sensitive signal. It is often used in the target Doppler frequency changing within narrowband occasions. So the binary phase-coded signal has strong anti-jamming capability.

51.2 Characteristics of the Binary Phase-Coded PC Signal

Binary phase-coded symbols are composed of the 0, 1 sequence or +1, −1 sequence. Phase of transmitting signal is alternating between 0° and 180° in accordance with the order of symbols (0, 1, or +1, −1). Because the frequency of the transmitting signal is usually not the integral times of the sub-pulse width's reciprocal, the coded signal is generally not continuous in the contrary phase points [2].

At the receiving end, the compressed pulses are obtained through a matched filtering or correlation processing. The half-amplitude point's width of the compressed pulses is nominally equal to the width of the sub-pulses. Therefore, the range resolution is proportional to the duration of one symbol (one sub-pulse). Time width-bandwidth product and pulse compression ratio is equal to the sub-pulses' number of the waveform, i.e., the number of coded symbols [3].

The plural expression of the binary phase-coded signal is

$$s(t) = a(t)e^{j\varphi(t)}e^{j2\pi f_0 t} \tag{51.1}$$

To binary phase-coded signals, $\varphi(t)$ only has two possible values 0 or π. $\varphi(t)$ can be expressed as two-phase sequence $\{\varphi_k = 0,\ \pi\}$, and it can also be expressed as a binary sequence $\{c_k = e^{j\varphi_k} = +1,\ -1\}$. If the envelope of the binary phase-coded signal is rectangular, that

$$a(t) = \begin{cases} 1/\sqrt{PT}, 0 < t < PT \\ 0, \quad \text{otherwise} \end{cases} \tag{51.2}$$

Plural envelope of the binary phase-coded signal can be written as

$$u(t) = \begin{cases} \frac{1}{P}\sum_{k=0}^{P-1} c_k v(t - kT), 0 < t < PT \\ 0, \qquad \text{otherwise} \end{cases} \tag{51.3}$$

Wherein $v(t)$ is a function of the sub-pulses, T is sub-pulse's width, P is the code's length, and PT is the duration of the encoded signal.

Barker code binary sequence is usually used in binary phase-coded PC radar system. This sequence is characterized by the following: when the aperiodic self-correlation function for a given length of the code is calculated, the peak of the side lobes has a minimum value (the amplitude is 1), and the peak of the main lobe is

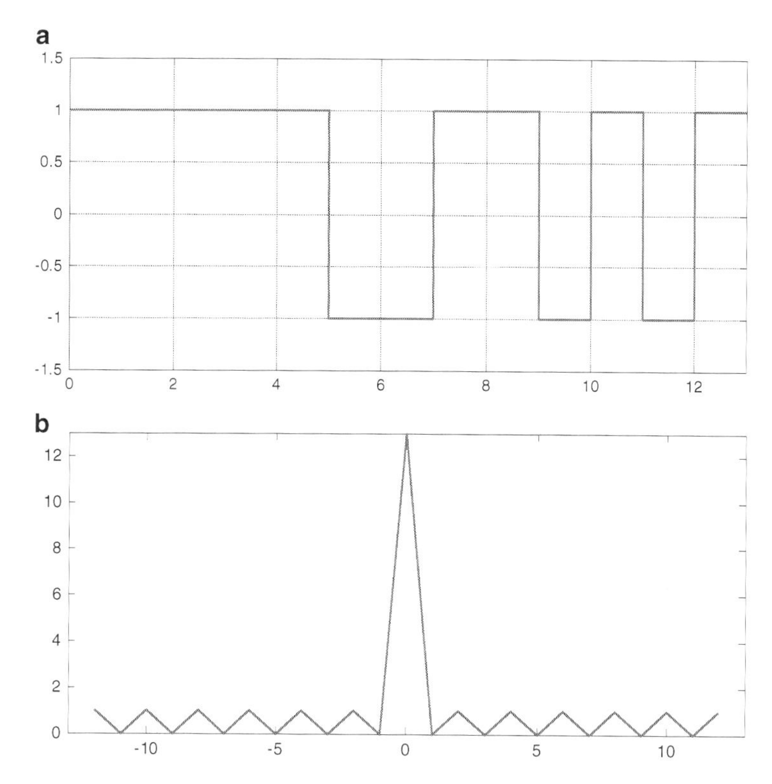

Fig. 51.1 The waveform (**a**) and the self-correlation (**b**) function of the 13-bit Barker code sequence

N. The Barker code which is used by PC radar is usually the 13-bit Barker code sequence, that is, + 1, + 1, + 1, + 1, + 1, − 1, − 1, + 1, + 1, − 1, + 1, − 1, + 1. Figure 51.1a shows the waveform of the 13-bit Barker code sequence; the self-correlation function of it is shown in Fig. 51.1b. Figure 51.2 shows the ambiguity function of the 13-bit Barker code sequence. Seen from the figure, the 13-bit Barker code PC radar's signal has favorable self-correlation and speed and distance resolution.

The signal of the binary phase-coded PC radars is more sensitive at the Doppler frequency shift than that of LFM signal. Figure 51.3 shows that in the case of the Doppler frequency shift to PC bandwidth, ratio is bigger than 0.05, the Barker coded PC signal's waveform which is sent out by matched filter will come into being a serious distortion, and the time side lobe is much bigger. Therefore, the binary phase-coded PC radar is not fit for detecting high-speed targets.

The binary phase-coded waveforms with low side lobes of the self-correlation function or zero Doppler frequency response are precisely needed by PC radars. Moving target's frequency response is different to zero Doppler frequency response. However, through the appropriate waveform design, the minimum of

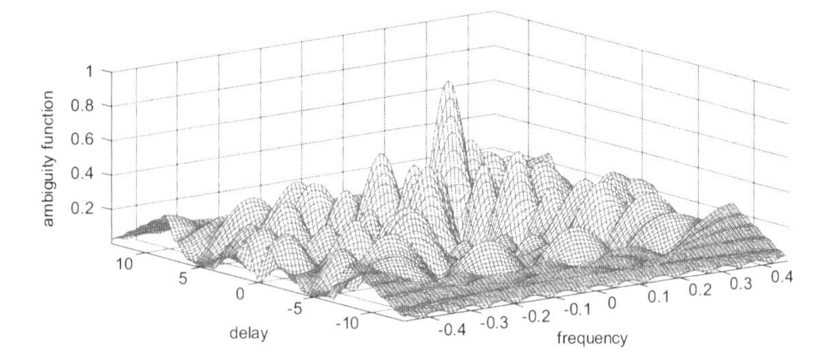

Fig. 51.2 The ambiguity function of the 13-bit Barker code sequence

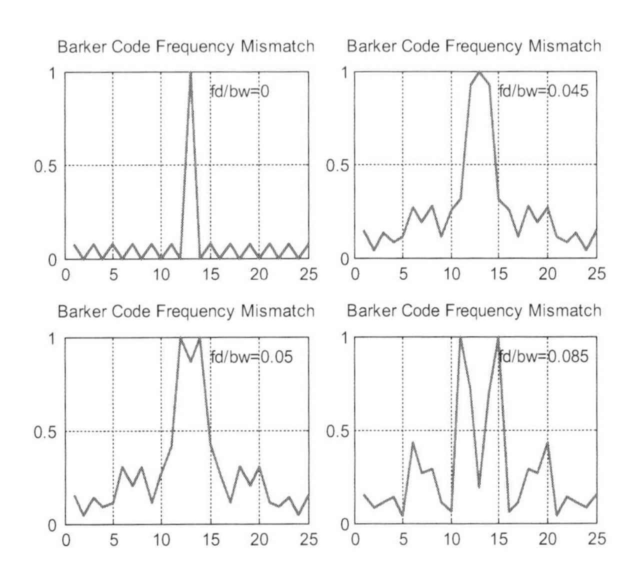

Fig. 51.3 The matched filter's output of different Doppler frequency shift to waveform bandwidth ratio

the Doppler frequency shift to bandwidth ratio can be obtained, and the perfect Doppler frequency response within the interested speed range can be reached. In this speed range, the Doppler frequency response corresponding to the distance between the target and radar or fuzzy function is approximately equal to the self-correlation function of the sequence.

51.3 Analysis of Binary Phase-Coded PC Radar Jamming

The binary phase-coded waveform is usually used for surveillance radar and tracking radar. The purpose of jamming depends on the type of radar which we want to jam.

For surveillance radar, the general jamming purpose is to produce background jamming or generate the synchronous decoys which can blanket the real target detection and masking the target. This latter process is sometimes called false-target jamming [4].

For tracking radar, its aim is to produce false targets to capture radar tracking wave gate. In this operating mode, when the jamming side receives the signal of radar, it can make the signal generating coherent frequency shift and then transmit the signal back to the radar. This jamming signal ahead of the target echo which comes from matched filter, and it makes electronic protection technology which used cutting-edge tracking useless.

In general, noise is the least effective jamming waveform in all types of jamming signal on binary phase-coded PC radar. This will be verified if the matched filter of binary phase-coded PC radar is regarded as a correlator. Noise which is completely uncorrelated with the radar's signal cannot get any gain when the noise passes through the matched filter along with the radar's signal and, when wholly related to the target's echo, will get all the processing gain. The jamming waveform will be effective than random noise even if there is little correlation to the radar's signal [5].

Continuous-wave signal will normally produce a certain degree of correlation with the matched filter of the binary phase-coded PC radar. For example, a continuous waveform whose frequency is same as that of carrier can get a power gain of 6 dB higher than the noise waveform if it is coherent with binary phase-coded waveform in about half of its duration.

Because the efficiency of noise jamming signal passing through the binary phase-coded PC radar's receiver is very low, the other forms of jamming signal must be considered. Replicating the radar's signal and making necessary jamming modulation is a feasible measure. However, binary phase-coded PC radar's signal is generally wide pulse or quasi-continuous-wave signal, if the pulse is used to modulate jamming signal wholly; when the modulated jamming signal compressed by the radar's receiver, its jamming effectiveness will be weakened. Therefore, the partial pulse replicated jamming form is used generally.

Binary phase-coded signal is sensitive to Doppler frequency shift, so it is very difficult to jam it by velocity deception. But the range resolution of the binary phase-coded signal is much higher than that of ordinary pulse radar; as long as its transmitted signal is attached a small delay, then send back to the radar, it will be deceived by range deception [6].

51.4 Simulation of Binary Phase-Coded Pulse Compression Radar Jamming

Continuous-wave and random noise signal to jam the 31-bit maximum length sequence of binary phase-coded signal is selected, simulating the jamming process by MATLAB, and the result is shown in Fig. 51.4. Assuming that continuous-wave

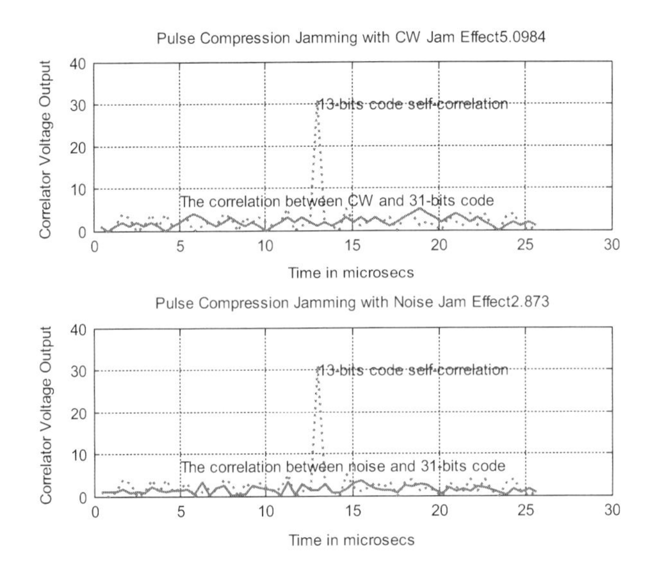

Fig. 51.4 Continuous-wave and noise jamming effectiveness

signal has the same frequency as the carrier, certain phase components are come into being at the central frequency of the binary phase-coded waveform. Also assumed that the noise is also at the central frequency, but relative to the binary phase-coded signals, it produces a uniform phase modulation $(0 - \pi)$.

Because the phase-coded compression signal is generally a wide pulse or quasi-continuous-wave signal, better jamming effectiveness can be gained when partial code replicated jamming form is adopted. In this experiment, given the half-code, sending back jamming signal was used to interfere the 31-bit maximum length sequence signals of binary phase-coded PC radar, with the combination of the two 15-bit half-code plus 1 to fill the PC filter network. The result is shown in Fig. 51.5.

In this experiment, the RGPO jamming which is used to jam the binary phase-coded PC radar was simulated; the Digital Radio Frequency Memory (DRFM) technique was used to store radar's operating frequency and repeat the original signals of radar. The course that the range gate tracking the echo and the jamming signal has been simulated, as shown in Fig. 51.6, with the towing time increasing and the range gate gradually tracked on the jamming signal.

51.5 Conclusion

The binary phase-coded PC technique is widely used in the new system of radar, solving the contradiction between the detection range and range resolution, having perfect speed detection and anti-jamming capability. In this chapter, four forms of jamming were used to jam the binary phase-coded PC radar's signal; they are noise,

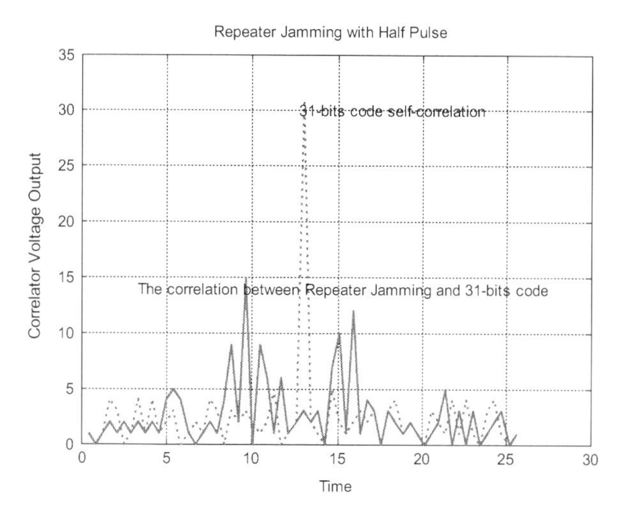

Fig. 51.5 Partial code replicated jamming effectiveness

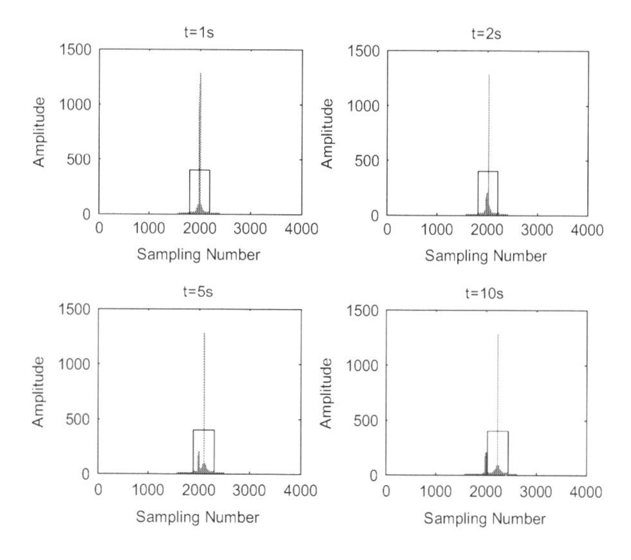

Fig. 51.6 Range-gate pull-off jamming effectiveness

continuous wave, partial code replicated, and RGPO jamming; the jamming theory and effectiveness were analyzed; the jamming course and result were simulated by MATLAB. These jamming forms are dynamic, and the effectiveness is relative. "There is no radar that unable to jam, and there is no jamming form unable to counterwork." The anti-jamming capability of the binary phase-coded PC radar will be gradually increased as science and technology development, and the challenge of the jamming side will be increasing. Studying on the jamming technique on the binary phase-coded PC radar will be going on.

References

1. Xiang, J. (2001). *Radar system* (pp. 188–256). Beijing: Publishing House of Electronics Industry.
2. Skolnik, M. I. (1990). *Radar handbook* (2nd ed., pp. 247–289). New York, NY: The McGraw-Hill Companies.
3. Schleher, D. C. (1999). *Electronic warfare in the information age* (pp. 167–201). Norwood, MA: Artech House.
4. Chen, Z. (1995). Effects of range gate pull-off (RGPO) jamming on some pulse radars. Aerospace Shanghai, Vol. 12, Nr. 5, 1995, (Cama,Vol. 3, Nr. 1, 1996); pp. 23–26
5. Xie, B., Yin, J., & Song, J. (2004). Building of kill probability simulation model for the ARM against the radar targets. *Journal of System Simulation, 16*(9), 2044–2047, 2051.
6. Mclendon, R., & Turner, C. (1983). Broad-band sensors for lethal defense suppression. *Microwave Journal (S0192–6225), 15*(6), 85–101.

Part III
Pattern Recognition

Chapter 52
Personalized Information Service Recommendation System Based on Clustering and Classification

Yu Wang

Abstract To solve the recommendation system, the prevalence of blindness, and low resistance, in this paper, an in-depth study on the personalized automatic recommendation system user model and automatic recommend technology. The paper first introduces the automatic recommendation system user model representation and update. Then the user modeling, clustering, classification, and automatic recommendation technology are combined to develop the automatic personalized document recommendation system based on clustering and classification. First, offline, the system form clustering points of interest to the article and build user model based on clustering interest points. Then realize the automatic recommendation online by recommendation algorithm which is based on classification. Theoretical analysis and experimental results show that the system can significantly improve the online response speed and efficiency.

52.1 Introduction

The Internet has become an important way for people to obtain information. But with the increase of Internet information, people have to spend a lot of time in the complex ocean of information and search the information they need. Information retrieval technology is used to meet some needs of people, but because of its universal nature, it still cannot meet the different backgrounds, different purposes, and different times of the query request. Therefore, personalized information service technology arises at the historic moment. Personalized information serve for different users with different services to meet different needs. Automatic recommendation system is one of the main forms of the personalized information

Y. Wang (✉)
Jilin Business and Technology College, Changchun 130062, China
e-mail: wang_zfy@sina.com

W.E. Wong and T. Zhu (eds.), *Computer Engineering and Networking*, Lecture Notes in Electrical Engineering 277, DOI 10.1007/978-3-319-01766-2_52,

service, and it is used to solve the problem of Internet information "get lost" and "information overload" and put forward a kind of intelligent agent system. Many large websites, such as Amazon, eBay, and Dangdang online bookstore, have varying degrees of using various forms of recommendation system.

52.2 Personalized Information Service

Personalized information service is according to user's interest characteristics and purchasing behavior, referring the information and commodities which they are interested in. Personalized information service technology can fully improve the website's service quality and access efficiency to attract more visitors. Personalized information service has brought profound influence on the concept of information service, service mode, and content; to improve the level of information service plays a powerful role [1].

52.3 Research Status of User Model

User modeling is the base of personalized information service. The accuracy of the models and timeliness directly determines the quality of service. The user model is the real users' "virtual" technology. Virtual technology depending on the system requirements and specific functions can give the user a full range of real description and some characteristics of the description. User model research mainly includes user model representation and user model update research.

52.3.1 The Representation of User Model

The representation of user model provides a structured model storage form [2]. Recommender system user model representation method covers a very wide scope; the main user model method includes the following categories:

1. Representation based on neural network: Neural network represents after network stability. The network connection weights characterized of network state represent the user model. Network states composed by the network input state, network output state, input and output of the connection states. Neural network representation method is used information structural organization, and is not a simple keyword listed to represent user model, which reflects the relationship between the information. However, the said method based on neural network depends on the user model in the process of the neural network

algorithm; the categories and scope are narrow, and representation is not easy to be understood.

2. Representation based on user evaluation matrix: Based on the user appraisal matrix, representation method is used more based on collaborative filtering recommendation system. The user appraisal *matrix* uses a matrix $R_{m * n}$ to indicate the user model set, *m stands* for the system users, *n* for resources/number, and *n* for resources/project number. In the matrix, each element r_{ij} is the representation of the user *i* to the project *j* evaluation. The greater the r_{ij} value is, the higher degree of interest the user to the corresponding project has.

The user appraisal matrix representation method is simple, intuitive, and doesn't need any learning technologies. It can be collected from the original data directly generated. But this method requires users to display the given evaluation data. This indicates the user's lack of interest and change method, and it is difficult to reflect the user's latest interest.

52.3.2 The User Model Update

After the establishment of the user model, according to the user, it is best to recommend dominant or recessive feedback on the user's current change of interest and to constantly update the user model in order to timely reflect the dynamic changes in interest preference and to ensure the effectiveness and practicality of the recommendation system. The existing user model updating technique is mainly divided into the following categories:

1. Genetic algorithm: Genetic algorithm is a kind of search optimization technique based on natural selection and iterative genetic mechanism. Fitness function, chromosome population and selection, and crossover and mutation are the three main operators of genetic algorithm. The genetic algorithm update technology system is usually the first to form a chromosome for the user model coding and generate the initial population. When the initial population evolves iteratively to meet the termination conditions, chromosomes with high fitness are decoded using the new user model, so as to develop the user model update [3].
2. Neural network technology: Neural network update technology is a kind of adaptive updating technology. When the user interest changes, neural network is adaptive to adjust the network connection weights and update the network output results to track changes in the user's interest. Some updates need to establish a new interest preference category and cut the old to set interest preference category. In this kind of circumstance, the network usually needs training. Neural network update technology depends on the user model representation established, usually only the user model based on neural network method establishes the representation of the system used.

52.4 Automatic Recommendation Technology

Automatic recommendation systems are commonly used. Mainly recommended technologies are as follows:

1. Recommendation technology based on rules
   A rule-based recommendation technology is knowledge engineering methods in the application of personalized information service. Rules can be used by customer and also can be used in association rule data mining techniques for discovery. Recommended technical performance depends on the quality and quantity of rules, rules can take advantage of user static attribute setup, and it can also use the user to establish dynamic information [4].
2. Recommendation technology based on collaborative filtering
   The traditional collaborative filtering-based recommendation technology is also known as user-based collaborative filtering recommendation technology or nearest-neighbor collaborative filtering recommendation technology. The basic thought on recommendation technology based on collaborative filtering is based on similar nearest-neighbor rating data from scores of recommended target users who do not score on a project which can be approximated using the weighted average approach.

But the traditional collaborative filtering-based recommendation technology has three difficult problems to solve: algorithm scalability, the sparse solution evaluation data, and recommended problem from initial resources [5].

52.5 Automatic Personalized Article Recommendation System Research Based on Clustering and Classification

Automatic personalized article recommendation system research is based on clustering and classification. First, by clustering, it forms the point of interest and the establishment user model is based on clustering interest points. Then the same interest points have greater interest users cluster, and form the the user group of interest point. Finally, to recommend paper classification, it realizes the active recommend according to their interest points. The automatic personalized recommendation based on clustering and classification system is presented in this paper. This system includes offline user model and user groups obtain subsystem and online personalized article recommendation subsystem of two parts.

52.5.1 Offline User Model and User Group Acquisition Subsystem Based on Clustering

Offline user model and user group acquisition subsystem based on clustering are using implicit ways to acquire user model. First of all, users collect clustering articles to form multiple interest points. Then the user model is expressed weighted interest point vector, according to the user's interest clustering to form each interest point user groups.

Specific steps of the structure based on the clustering of interest point user model are introduced as follows:

Step 1: For all users' collection of all the articles on pretreatment, the article based on semantic and statistical characteristics of the feature vector is expressed as $V(d_i) = (\mathrm{w}(d_i,T_1), \mathrm{w}(d_i,T_2), \ldots, \mathrm{w}(d_i,T_m))$

Step 2: For the article to cluster analysis, form the cluster set C_n

Step 3: Each cluster C_i is as an interest point, and construct the characteristic vector: $V(C_i) = (tdr(C_i,T_1), tdr(C_i,T_2), \ldots, tdr(C_i,T_m))$. The $tdr(C_i,T_j)$ (52.1) as feature item T_j for classes C_i of the resolution.

$$tdr(C_i, T_j) = \begin{cases} \dfrac{N_i}{\sum\limits_{k \neq i} N_k} & \text{if } \sum\limits_{k \neq i} N_k \neq 0 \\ 2 \times N_i, & \text{if } \sum\limits_{k \neq i} N_k = 0 \end{cases} \tag{52.1}$$

Among them, N_i as C_i class appear feature T_j of article number.

Step 4: According to each user U_i collection of the articles, to statistical articles belong to interest point, to construct weighted clustering interest point vector UM_i to represent the user model: $UM_i = (w_{i1}, w_{i2}, \ldots, w_{i|C|})$

The w_{ij} (52.2) for users U_i to interest points C_j of the weight, $1 \leq i \leq |U|$, $1 \leq j \leq |C|$

$$w_{ij} = DC_{ij}/DN_i \tag{52.2}$$

DC_{ij} collect the article number belongs to the interest point C_j for the user U_i, and DN_i collect articles for user U_i

52.5.2 The Online Personalized Recommendation Subsystem Based on Classification

The online personalized recommendation subsystem based on classification is the first to recommend paper classification, namely, calculating it with the interest point feature vector similarity. To search their interest points, and then we can recommend all user groups to make the interest point.

This subsystem proposed the classification of the recommendation algorithm, and the specific steps are as follows.

Algorithm 1: The recommendation algorithm based on the classification.

Input: To recommend article *RD*, set the interest point feature vector as *CF* and the interest point main user cluster as *CU*.

Output: Recommend user group *RU*.

1) RU $= \Phi$
2) $Maxsim = 0$
3) Clustering $ID = 0$
4) For every CF_i in CF
5) If Sim(RD, CF_i)$>$ Maxsim
6) $Maxsim =$ Sim(RD, CF_i)
7) Clustering $ID =$ i
8) Endif
9) Endfor
10) RU $= CU_{ClusteringID}$
11) Return

52.6 The Experimental Results and Analysis

This paper adopts a certain site after the preprocessing part of the real data set, including 714 users and 20,000 articles. From the experiment, extract 20 articles as the recommended article, use Recall and Precision as two quality indicators of recommendation system, the definition is as follows:

Definition 1: For recommend system RS, RD for to be recommended article, the RDU for the system recommend RD users set, SDU for collection RD users set, then recall is defined as:

$$Recall = \frac{|RDU \cap SDU|}{|SDU|} \tag{52.3}$$

Precision is defined as

$$Precision = \frac{|RDU \cap SDU|}{|RDU|} \tag{52.4}$$

In order to test automatic personalized recommendation system (referred to as RSBC&C) based on clustering and classification performance, this paper made two

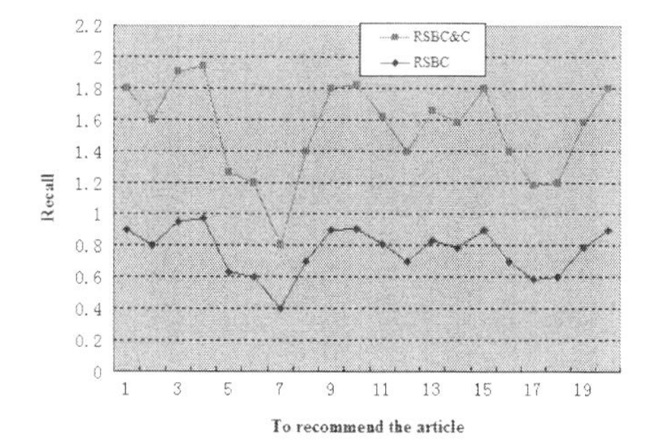

Fig. 52.1 Recommendation system recall rate comparisons

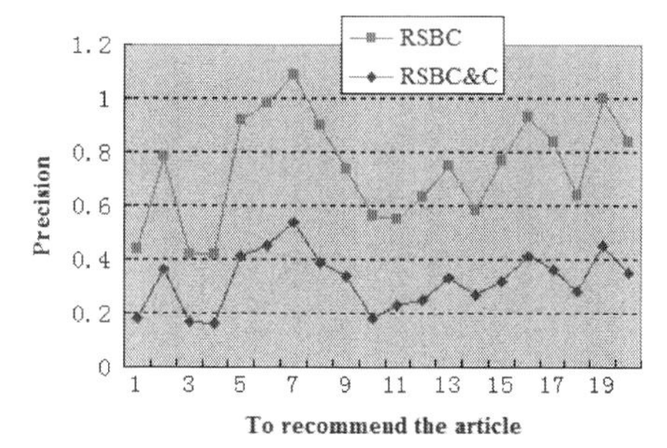

Fig. 52.2 Recommendation system precision rate comparisons

groups of experiment. The first group of experiments was based on content filter paper automatic recommendation system (referred to as RSBC system) and RSBC&C system comparison experiments. Set the interest point of value $\Phi = 0.05$. The experimental similarity calculation used is the cosine similarity calculation method. The experimental results were shown in Figs. 52.1 and 52.2.

The experimental results (Figs. 52.1 and 52.2) show that the RSBC&C system recall rate is higher than RSBC system; RSBC&C system article precision rate is slightly lower than RSBC system.

The second group tests interest degree of value Φ, the effect of system performance. Interest degree of value Φ is an adjustable parameter set at Φ 0.05 and 0.1, respectively. The experimental results were shown in Figs. 52.3 and 52.4.

The experimental results show (Figs. 52.3 and 52.4): when Φ takes 0.05, the system recall rate is higher than Φ take 0.1. But the system precision rates are lower than take 0.1 system precision accuracy. In the analysis of the experimental results, it can be concluded that the system's recall rate requires higher precision rate if it can properly lower Φ value and the system's precision requires higher recall rate if it can appropriately increase Φ value.

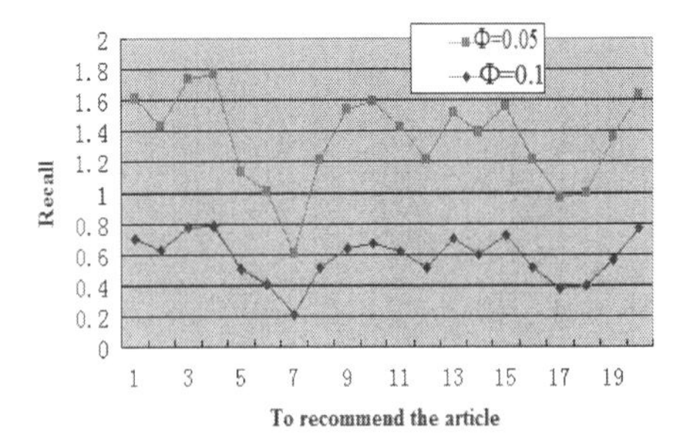

Fig. 52.3 Different Φ value recall rate comparison

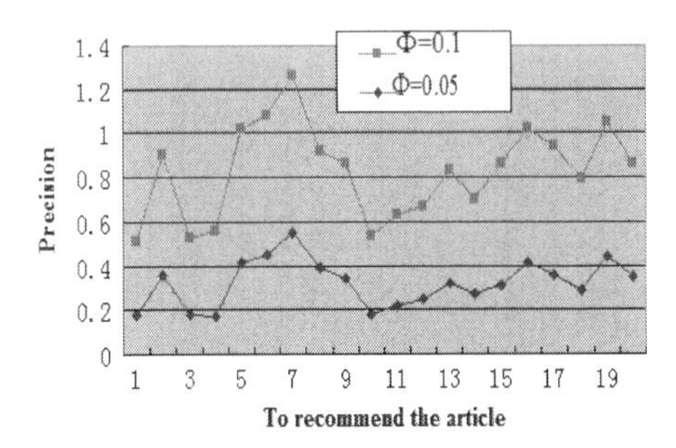

Fig. 52.4 Different Φ value precision rate comparisons

52.7 Conclusion

This paper mainly aims at the Internet personalized information service of automatic recommendation system which has launched research. Automatic recommendation is one of most important ways of the Internet's personalized information service. With more and more users online reading habits and collecting articles and the keyword vector user model dimension growing too fast, it is hard to reflect the user's interest. Current recommendation techniques affect the effectiveness of the system. In view of the above questions, this paper puts forward the Internet's automatic recommendation system based on clustering and classification of articles.

References

1. Fan, G. (2005). Personalized information service in our country. *The Library in the New Century, 5*(5), 22–25.
2. Wu, L., & Liu, L. (2006). Personalized recommendation system user modeling technology. *Intelligence Review Journal, 1*(5), 55–62.
3. Shahabi, C., & Chen, Y. S. (2003). Recommendation system without explicit acquisition of use relevance feedback. *Distributed and Parallel Databases, 14*(3), 173–192.

4. Zeng, C., Xing, C., & Zhou, L. (2003). The personalized search algorithm based on content filtering. *Journal of Software, 14*(5), 999–1004.
5. Sawra, B., KaryPis, G., Konstan, J., & Riedl, J. (2000). Analysis of recommender algorithms for e-commerce. In *Proeeedings of the 2nd ACM E-Commerce Conference* (pp. 158–167). New York: ACM Press.

Chapter 53
Palmprint Recognition Based on Subclass Discriminant Analysis

Pengfei Yu, Haiyan Li, Hao Zhou, and Dan Xu

Abstract Subspace-based palmprint recognition methods, such as principal components analysis and linear discriminant analysis, assume that each class can be grouped in a single cluster. However, this assumption is not reasonable at some situations where a class is assembled in two or more clusters. In order to solve this problem, a novel palmprint recognition method based on subclass discriminant analysis is proposed in this chapter. Each palmprint class is divided into a set of subclasses that can be separated easily in the new subspace representation. After that, the Euclidean distance and nearest neighbor method are employed as the similarity measurement. Experimental results conducted on a database of 86 hands (10 impressions per hand) show that the equal error rate (EER) of the proposed method yields promising result of EER = 0.67 % for verification rate, which demonstrates that the proposed method is effective to solve the problem mentioned above.

53.1 Introduction

As palmprint is a relatively novel biometric trait, solutions in the field of palmprint recognition generally require extensive study involving algorithms and testing with sample images captured on different conditions with digital cameras, scanners, or other new devices. Although many algorithms of palmprint recognition have been developed during the past 10 years, the need of new method that is flexible, comprehensive, and robust is greatly urgent.

To date, there are numerous methods applied in palmprint recognition. Based on how to extract the palmprint feature, these algorithms can be broadly separated into three classes: line-based, subspace-based, and statistical methods.

P. Yu (✉) • H. Li • H. Zhou • D. Xu
School of Information, Yunnan University, Kunming 650091, China
e-mail: pfyu@ynu.edu.cn

W.E. Wong and T. Zhu (eds.), *Computer Engineering and Networking*, Lecture Notes in Electrical Engineering 277, DOI 10.1007/978-3-319-01766-2_53,

Line-based methods usually apply edge or point detectors to extract principle lines [1, 2] and crease [3, 4]. For instance, Huang et al. [2] used the modified finite Radon transform (MFRT) as the feature extraction approach to extract principle lines. Two images, direction image and energy image, were produced by the MFRT. After that, the principle lines could be extracted from the energy image and used to calculate the similarity between two palmprints. It is obvious that line is the basic feature of palmprint. Line-based methods directly extract the basic identity information contained in the texture of palmprint. However, the relatively stable texture of palmprint is difficult to be obtained because it is inconstant under different light condition or the stretching level of the hand. As a result, the recognition performance of the most line-based methods is not better than other ones.

Statistical methods usually subdivide the region of interest (ROI) of palmprint images into nonoverlapping small blocks. These blocks were subsequently transformed by wavelets [5, 6], Gabor [7], or other approaches [8, 9]. Finally, the means, variances, or energy of the small blocks were calculated and regarded as the features. For instance, Han et al. [8] used Sobel and morphological operators to process the ROI images. After that, these images were uniformly divided into several small grids whose sizes are 32×32, 16×16, or 8×8. Lastly, the mean values of pixels in the grids were calculated to build the feature vectors.

Subspace-based methods have been widely used in the literature of biometrics. Among subspace-based methods, principal components analysis (PCA) and linear discriminant analysis (LDA) are the famous methods employed in pattern recognition for data reduction and feature extraction. Consequently, they are applied in palmprint recognition. For example, Lu et al. [10] proposed a PCA-based palm recognition method named Eigenpalm. Wu et al. [11] used Fisher's linear discriminant (FLD) approach to extract palmprint features. However, these linear methods mentioned before are not effective to extract nonlinear features. In order to make these linear approaches applicable to extract nonlinear features, kernel-based methods have been proposed. The main idea of kernel-based methods is to map the input data to a feature space via nonlinear mapping, and the input data can be easily separated in the new feature space, for example, kernel principal component analysis (kernel PCA) [12] and kernel fisher discriminant analysis (KFDA) [13]. But if the data of a class is assembled into two or more clusters or cannot be represented by a single cluster, it is difficult to get a reasonable recognition performance by using previously mentioned subspace methods. To solve this problem, Zhu and Martinez [14] proposed a novel discriminant analysis method named subclass discriminant analysis (SDA).

In this chapter, we apply the SDA method to palmprint images for personal authentication. The rest of this chapter is organized as follows. In Sect. 53.2, we introduce the architecture of the proposed method. In Sect. 53.3 we briefly review the theory of SDA, and proposed SDA-based feature extraction method in Sect. 53.4. The experimental results are reported in Sect. 53.5. Conclusions are drawn in Sect. 53.6.

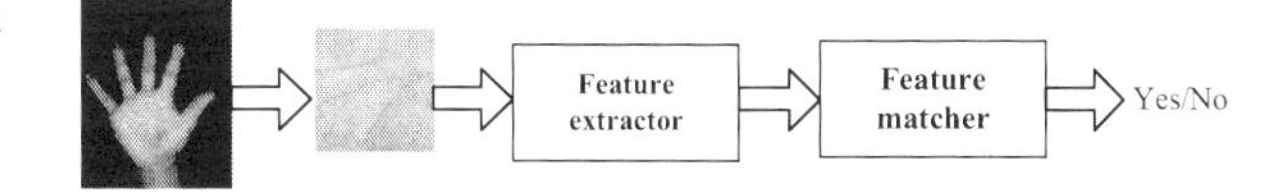

Fig. 53.1 A block diagram of the proposed method

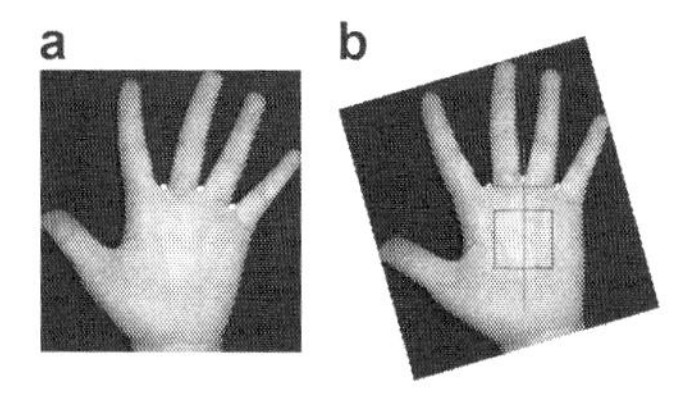

Fig. 53.2 (**a** and **b**) the generation of palmprint ROI

53.2 System Overview

Figure 53.1 shows the architecture and work flow of our system.

First, we capture a colorful whole-hand image via a digital camera. To eliminate the variation caused by the rotation of the hand image, we rotate the hand image to a relative fixed angle. To do so, three key points, between the roots of index finger, middle finger, and ring finger, are located as shown in Fig. 53.2a. After the key points are detected, a coordinate system, based on the valley points, is established, as shown in Fig. 53.2b, and then a square region can be segmented from the center part of palm images as the palmprint ROI.

In order to facilitate matching with the identity template, the ROI images is further processed by a feature extractor to get a compact but expressive representation regarded as the identity of the hand's owner. If we use subspace-based methods, the feature extractor can be PCA based, LDA based, SDA based, etc. Then the representation is fed to the feature matcher, which compares it against the template of a user (verification system) or all templates of all users (identification system). It is note that Fig. 53.1 is a block diagram of verification system rather than an identification system. During the feature matching phase, simple method such as nearest neighbor method or complex classifiers such as support vector machine (SVM) and artificial neural network (ANN) can be applied.

53.3 Theoretical Foundation of SDA

In this section, we present the theoretical foundation of SDA and show that SDA is able to extract directly features from palmprint ROI images.

As we know, once the data distribution of each class has been approximated using a mixture of Gaussians, most of subspace-based methods are suitable to be selected as the feature extractor.

Assuming that each class has been approximated using a mixture of Gaussians, it is easy to use the following generalized eigenvalue decomposition equation to find those discriminant vectors that can best classify the data [14]:

$$\Sigma_B V = \Sigma_X V \Lambda \tag{53.1}$$

where V is a matrix whose columns correspond to the discriminant vectors, Λ is a diagonal matrix of the corresponding eigenvalues, and Σ_B is the between-subclass scatter matrix, given by

$$\Sigma_B = \sum_{i=1}^{C-1} \sum_{j=1}^{H_i} \sum_{k=i+1}^{C} \sum_{l=1}^{H_k} p_{ij} p_{kl} \left(\mu_{ij} - \mu_{kl}\right) \left(\mu_{ij} - \mu_{kl}\right)^T, \tag{53.2}$$

where C is the number of classes, H_i is the number of subclass divisions in class i, p_{ij} and μ_{ij} are the prior and mean of the jth subclass in class i, and p_{kl} and μ_{kl} are the prior and mean of the lth subclass in class k. Note that $p_{ij} = \frac{n_{ij}}{n}$, where n_{ij} represents the number of the jth subclass in class i and n represents the number of samples.

$\sum_X$ is the covariance matrix of the data, defined as [15]

$$\Sigma_X = \frac{1}{n} \sum_{i=1}^{n} (x_i - \mu)(x_i - \mu)^T \tag{53.3}$$

where x_i is the ith sample vectors and μ is the sample mean of all the data.

Since $\sum_B$ measures the scatter of the subclass means, the method is called SDA. Compared with LDA, SDA can be used to classify both linear and nonlinearly separable classes. Moreover, LDA can only extract $C - 1$ features from the original feature space, but SDA can extract features larger than $C - 1$.

Note that the real challenge of this method is how to find the best subclass divisions which can be applied to get the optimal classification results [14]. To achieve this, Zhu and Martinez [14] proposed two approaches: leave-one-out-test (LOOT) criterion and stability criterion. Using the LOOT criterion, all possible values of the number of subclass H_i are exhaustively tested. At the same time, the recognition rate of the specified H_i is recorded. The optimal value of the number of subclass corresponds to the max recognition rate. When the optimal subclass is selected, the between-subclass scatter matrix $\sum_B$ can be computed by Eq. (53.2), so the discriminant vectors can be obtained by solving Eq. (53.1). In this study, we choose LOOT criterion to find the optimal value of the number of subclass. More details about the stability criterion can be referred in Zhu and Martinez [14].

53.4 Feature Extraction

Feature extraction is an essential but difficult step in the proposed system. In order to effectively implement the proposed palmprint recognition, SDA is chosen to extract the features from the palmprint ROI images.

Assume that there are c persons and the number of ROI images belonging to the ith person is n_i. Let the jth ROI images of class i be x_{ij}, where $1 \leq i \leq c$, $1 \leq j \leq n_i$, and $n = \sum_{i=1}^{c} n_i$.

The purpose of SDA is to maximize the following Fisher criterion:

$$J(W) = \frac{W^T \Sigma_B W}{W^T \Sigma_X W}, W \neq 0 \tag{53.4}$$

where $\sum_B$ is the between-subclass scatter matrix defined by Eq. (53.2) and $\sum_X$, defined by Eq. (53.3), is the covariance matrix of the data in the feature space F.

The optimal discriminant vectors w_i under Fisher criterion can be achieved by solving the following eigenvalue problem defined by Eq. (53.1).

Furthermore, if the set of generalized eigenvectors w_i are composed of the optimal linear transformation matrix W_{opt}, then the ROI images can be transformed to the feature space by using $y_{ij} = W_{opt}^T x_{ij}$,

where y_{ij} is the extracted ROI feature from x_{ij} by using the SDA.

During the feature matching phase, the Euclidean distance and nearest neighbor method are employed as the similarity measurement.

53.5 Experiment Results

In this study, the YNU Hand database is chosen to verify the SDA-based palmprint recognition method. The data set used in our experiments consists of 860 images, 10 per hand. The center area of the palm is obtained as the ROI during the preprocessing phase. After that, five ROI images per hand are selected for training, and the remaining five ROI images are chosen for testing. The size of palmprint ROI is 200×200.

Generally, the performance of a biometric system is often measured in terms of false accept rate (FAR) and false reject rate (FRR). However the FAR and FRR cannot be decreased simultaneously. For this reason, equal error rate (EER) is usually chosen as the quality indices because both of them can combine FAR and FRR into a number. In this study, we choose the EER, which refers to the point in a detection error trade-off (DET) curve where FAR equals the FRR.

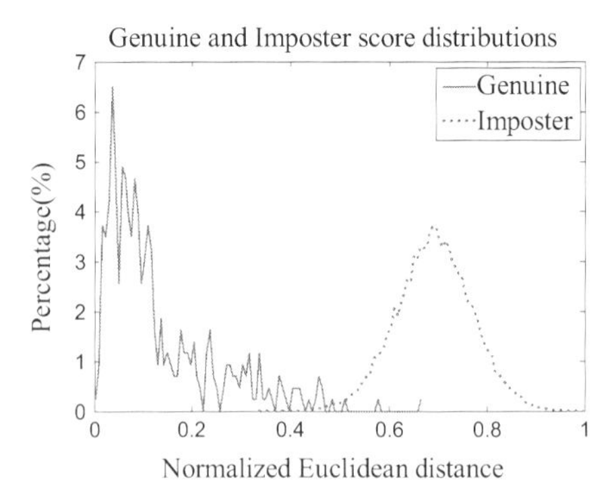

Fig. 53.3 The corresponding genuine and impostor distributions

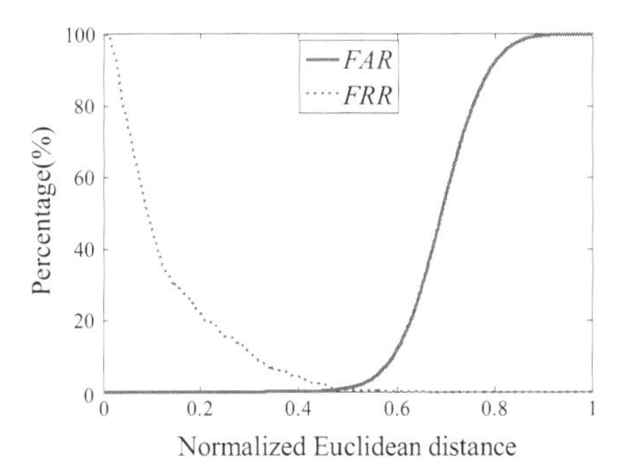

Fig. 53.4 The FAR and FRR of the proposed method

Figure 53.3 shows the performance evaluation for genuine and imposter matching distribution, which is estimated by matching scores obtained by using normalized Euclidean distance. The genuine matching score is gained by comparing palmprints from the same palm, and the impostor matching score is obtained by comparing palmprints of two different palms. Consequently, a total of 36,980 ($5 \times 86 \times 86$) comparisons are performed, in which 430 (5×86) comparisons are genuine matching, and the remaining 36,550 ($5 \times 86 \times 85$) comparisons are imposter matching. It is clear in Fig. 53.3 that the proposed method can separate palmprints of different persons well.

A threshold value 0.487 is determined based on the EER criteria. The FAR and FRR at different distance thresholds are plotted in Fig. 53.4, in which the EER is 0.67 %.

Figure 53.5 shows the ROC curve obtained from the proposed method. It must be noted that we plot FAR as logarithmic scales for x-axis for viewing the curve clearly.

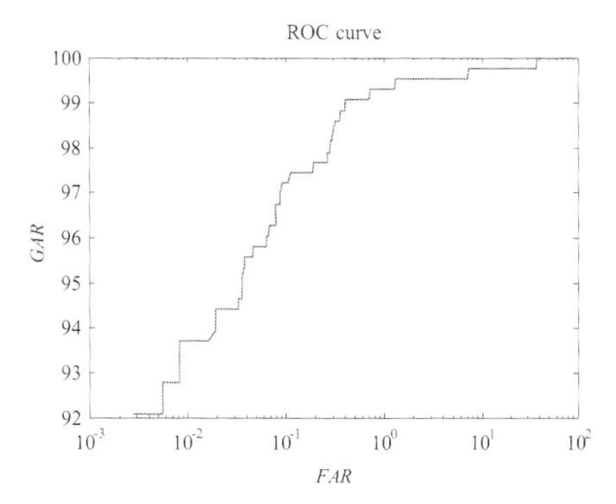

Fig. 53.5 ROC curve for the proposed method

53.6 Conclusion

Over the years, many palmprint recognition approaches have been proposed. Among these palmprint recognition approaches, subspace-based methods, such as PCA and LDA, have been widely used. Most of them assume that each class is represented by a single cluster in the feature space. However, this assumption is not reasonable at some situations in which the data of a class is assembled into two or more clusters. So the SDA is proposed to solve this dilemma. In this chapter, a new subspace-based method named SDA is applied in palmprint feature extraction. The experimental results on YNU Hand database show that the SDA-based method yields a good recognition rate.

Acknowledgments Supported by Science Research Foundation of Yunnan Provincial Education Department (Grant No. 2011Z029).

References

1. Wu, X., Wang, K., Zhang, D., & Huang, B. (2004). Palmprint classification using principal lines. *Pattern Recognition, 37*(10), 987–1998.
2. Huang, D., Jia, W., & Zhang, D. (2008). Palmprint verification based on principal lines. *Pattern Recognition, 41*(4), 1316–1328.
3. Chen, J., Zhang, C., & Rong, G. (2001). Palmprint recognition using crease. *Proceeding of 2001 International Conference on Image Processing, IEEE in Piscataway, NJ* (pp. 234–237).
4. Cook, T., Sutton, R., & Buckley, K. (2010). Automated flexion crease identification using internal image seams. *Pattern Recognition, 43*(3), 630–635.
5. Liu, Y. Q., Yuan, W. Q., & Guo, J. Y. (2010). Block statistic under wavelet decomposition for palmprint recognition. *Proceedings of 2010 International Conference on Electrical and Control Engineering (ICECE), IEEE CS, Piscataway, NJ* (pp. 5073–5075).

6. Madasu, H., Gupta, H. M., Mittal, N., & Vasikarla, S. (2009). An authentication system based on palmprint. *Proceedings of 6th International Conference on Information Technology: New Generations (ITNG '09), IEEE CS, Piscataway, NJ* (pp. 399–404).
7. Kumar, A., & Shen, H. C. (2004). Palmprint identification using PalmCodes. *Proceedings of the 3rd International Conference on Image and Graphics, IEEE CS, Piscataway, NJ* (pp. 258–261).
8. Han, C. C., Cheng, H. L., Lin, C. L., & Fan, K. C. (2003). Personal authentication using palmprint features. *Pattern Recognition, 36*(2), 371–381.
9. Wu, X., Wang, K., & Zhang, D. (2002). Fuzzy directional element energy feature (FDEEF) based palmprint identification. *Proceedings of 16th International Conference on Pattern Recognition, IEEE CS, Piscataway, NJ* (pp. 95–98).
10. Lu, G., Zhang, D., & Wang, K. (2003). Palmprint recognition using eigenpalms features. *Pattern Recognition Letters, 24*(9), 1463–1467.
11. Wu, X., Zhang, D., & Wang, K. (2003). Fisherpalms based palmprint recognition. *Pattern Recognition Letters, 24*(15), 2829–2838.
12. Schölkopf, B., Smola, A., & Müller, K. (1998). Nonlinear component analysis as a Kernel eigenvalue problem. *Neural Computation, 10*(5), 1299–1319.
13. Mika, S., Rätsch, G., Weston, J., Schölkopf, B., & Müller, K. R. (1999). Fisher discriminant analysis with kernels. In *Neural networks for signal processing IX, 1999. Proceedings of the 1999 I.E. Signal Processing Society Workshop, IEEE, Piscataway, NJ* (pp. 41–48).
14. Zhu, M., & Martinez, A. M. (2006). Subclass discriminant analysis. *IEEE Transactions on Pattern Analysis and Machine Intelligence, 28*(8), 1274–1286.
15. Martinez, A. M., & Zhu, M. (2005). Where are linear feature extraction methods applicable? *IEEE Transactions on Pattern Analysis and Machine Intelligence, 27*(12), 1934–1944.

Chapter 54
A Process Quality Monitoring Approach of Automatic Aircraft Component Docking

Guowei Yang, Chengjing Zhang, and Xiaofeng Zhang

Abstract In order to evaluate automatic aircraft docking quality, a new method of process quality monitoring of the automatic aircraft component docking is proposed. This method is based on laser tracker measurement. The position of automatic aircraft docking component is obtained by laser tracker measurement for a plurality of feature points. By doing identification labels on the surface of docking components, component surface images replace complex docking images. Image processing technology is used to extract the features of component surface image information to evaluate docking quality so as to control the perfect match of docking component. The new method enforces automatic docking process visual monitoring and absorbs human visual and reliable monitoring advantages of manual assembly model.

54.1 Introduction

With the development of computer and information technology, in order to adapt to the social progress and meet its quality and efficiency, aircraft assembly has been from the traditional manual assembly mode developed gradually and quickly into today's automatic assembly model, which is fast efficient digital and integrated and has high precision and high coordination [1]. Aircraft component has the features of large-scale, complex matching surface and difficult measurement. Domestic and foreign advanced technology on component about pose measurement and accurate calibration is based on laser tracker, providing reliable basis for large parts of the pose adjustment and position driver [2, 3]. Although the automatic docking technology has made great progress, the traditional manual assembly model is not

---

G. Yang • C. Zhang (✉) • X. Zhang
School of Information Engineering, Nanchang Hangkong University,
Nanchang 330063, China
e-mail: jingshui0326@126.com

W.E. Wong and T. Zhu (eds.), *Computer Engineering and Networking*, Lecture Notes in Electrical Engineering 277, DOI 10.1007/978-3-319-01766-2_54,

completely replaced. To a great extent, automatic docking technology did not achieve the simulation monitoring of human vision.

The method of process quality monitoring refers to the automatic docking process about large aircraft components that not only relays on the data returned by pose control system but also uses effective scheme to enforce docking process visual monitoring and evaluate whether or not the docking process is completed. If the evaluation result is up to standard, the docking process quality is good; if not, the need to adjust the position and pose is necessary.

54.2 The Measurement Method of Laser Tracker

In assemble process of laser tracker measuring the position, after the large aircraft component shelving, target balls should be respectively arranged, and laser tracker should be placed in front of the docking component. In adjusting process, in order to achieve the current standard point automatic tracking and positioning measurement, laser trackers provide second-development interface, through programming to set search area based on putting the standard point into search center; then discrete laser signal to the theory coordinates of standard point; and use their automatic search function to search the target balls around the space of the standard points. Then we can theoretically obtain the theory coordinates of the next standard point. So we can measure the second standard point coordinates by using the same method. In this turn, every dynamic coordinate of the standard point in adjustment progress can be measured.

A few problems will appear above. For saving the cost, assembly project usually uses a single laser tracker; this way, at the same time only a standard point (target) can be achieved, and its measure time is different. In order to solve the docking pose, standard points' measure time must be unified. Moreover, assembly parts are complex, so errors in assembly would not be avoided [2]. This chapter proposes a new docking quality assessment method, which can effectively lessen the deviation and reduce the influence on subsequent docking assembly.

54.3 Docking Component Surface Monitoring Method Based on Image

The measurement method of laser tracker adjusts and positions posture with digital coordinates which is obtained from computer feedback of control system. For the assembly filed, the method of process quality monitoring of automatic component docking based on image improves a new level about perfect docking.

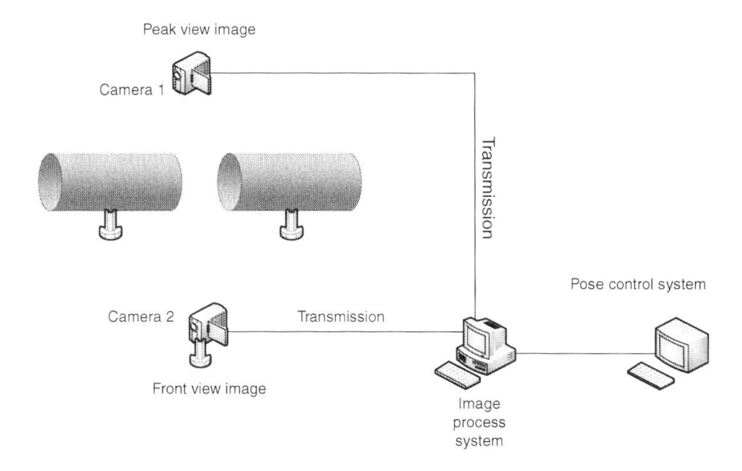

Fig. 54.1 Process quality monitoring system diagram of automatic aircraft component docking

54.3.1 Image Monitoring Method

To realize process quality control based on visualization, an ideal way is to capture surface image. General docking is inserted, namely, component's several main shafts inserted into another's corresponding holes precisely. Thus, as long as capturing hold images and extracting hold center and shafting coordinates with algorithm, we can contrast these two coordinates. If the coordinates contrast result is coincident, the docking result is perfect. If it is deviation, position adjustment is necessary. But considering components' complex surface and the practice without camera condition, the project cannot be carried out successfully. In order to overcome these problems, a new method is proposed. As follows, Fig. 54.1 shows process quality monitoring system diagram of automatic component docking based on image. Using CCD cameras with high quality to capture surface images and achieving quantitative calculation through image processing algorithm can evaluate docking quality. Camera 1 placed in the peak of docking captures peak view images. Camera 2 has the same height with components' horizontal axis and captures front-view images. The two cameras transmit the images to image process system synchronously. The computer uses algorithm to achieve image information for quality evaluation. Then the evaluation results are transmitted to pose control system for guiding position adjustment.

Because aircraft components are rigid and manufacturing process is controlled strictly and every step of project by means of precise calculation, it is feasible to use component surface quality to assess the ensemble docking quality.

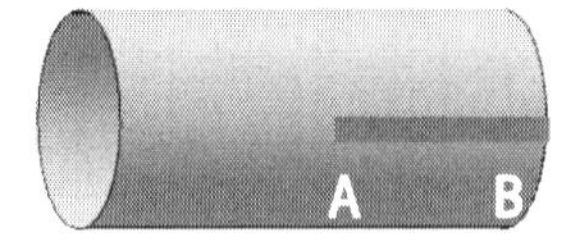

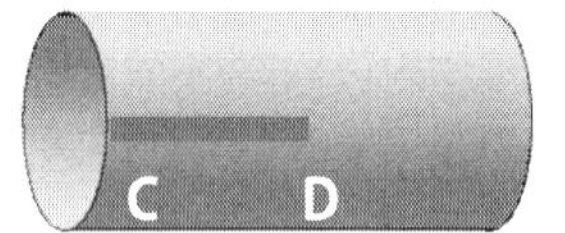

Fig. 54.2 Front-view image of label location

54.3.2 Label Location Method

Aircraft components' surfaces are very complex. Either fuselage-to-fuselage or fuselage-to-wing docking, there are different surface characteristics. It's difficult to extract surface images' features, and feature extraction algorithm of the fuselage-to-fuselage docking is not applicable for the fuselage-to-wing docking process; conversely, it is the same. Based on these problems, this chapter proposes a new feature extraction method—label location method.

Firstly, feature extraction chooses feasibility labels—red lines which have obvious color difference between the labels and background. The labels are placed in two components' front and peak mid sides. The front images captured by camera 1 are shown in Fig. 54.2. Similarly, the camera 2 captures these peak view images. This chapter only takes the front-view image as an example.

54.4 The Flow Chart of the Process Quality Monitoring System

In this chapter, the quality monitoring on aircraft component docking process is based on laser tracker positioning system. Figure 54.3 shows the flow chart of the system.

Before assembly, specified labels should be placed on docking component surfaces, according to label location method. By the formulation of the program, laser tracker control system controls implementation of component docking automatically. Position control system controls the components' posture drawing on the return data from laser tracker. When the docking components get to predetermined position, the process quality monitoring system starts up. Calibrated camera 1 and 2 transmit images to process quality monitoring system in the same time. Through image processing, it extracts target image information and then evaluates whether or not the quality in docking process satisfies the standard.

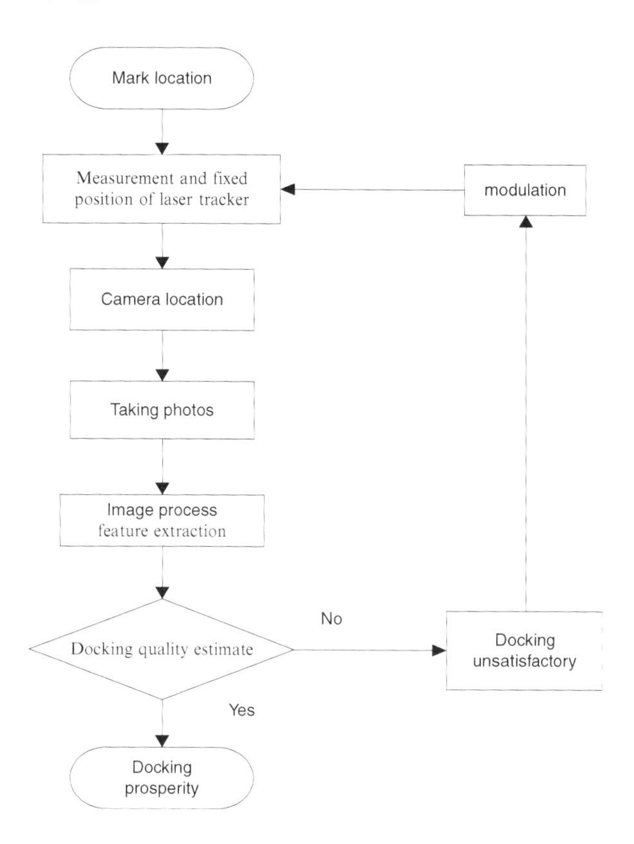

Fig. 54.3 Process quality monitoring system flow diagram

54.5 Experiment Studying

54.5.1 The Color Segmentation of the Advantages

In this chapter, in order to facilitate identification [4], we make red line for distinguishing labels and components obviously. In RGB space, color feature is used to extract target designations from images. First thing is to display images' color, namely, original images are displayed in R, G, and B three independent spaces, with gray image expression, as shown in Fig. 54.4.

Of R, G, and B three spaces grayscale, in RGB image displaying, red, green, and blue ingredients have different values, and their effects are also different. In R space, the red ingredient on the line target salience is very obvious. In G and B spaces, ingredients that are changing are not obvious. Through the experiments, only based on R spatial gray level, the extraction of target label line is not very ideal. By the methods of refs. 4, 5, the use of RGB space (2R-G-B) [4, 5] gray difference can make further processing Fig. 54.5 is the (2R-G-B) gray difference image.

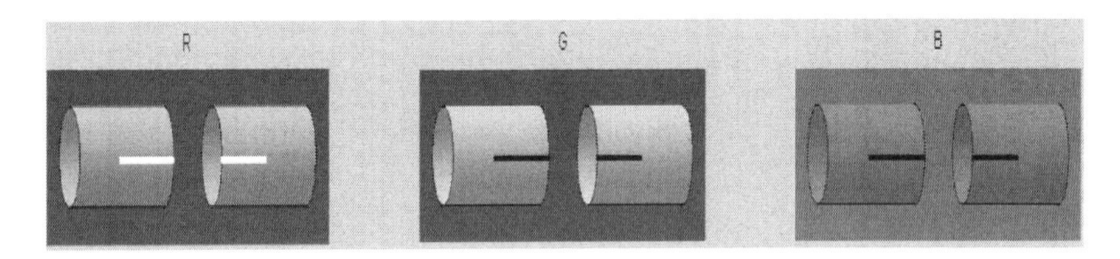

Fig. 54.4 R, G, B *gray* image

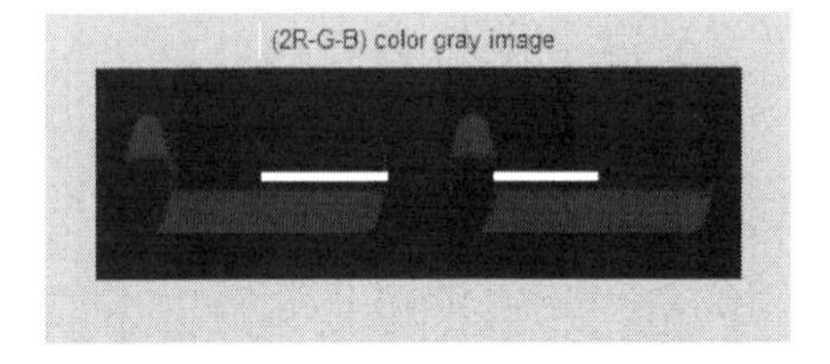

Fig. 54.5 (2R-G-B) *color gray* image

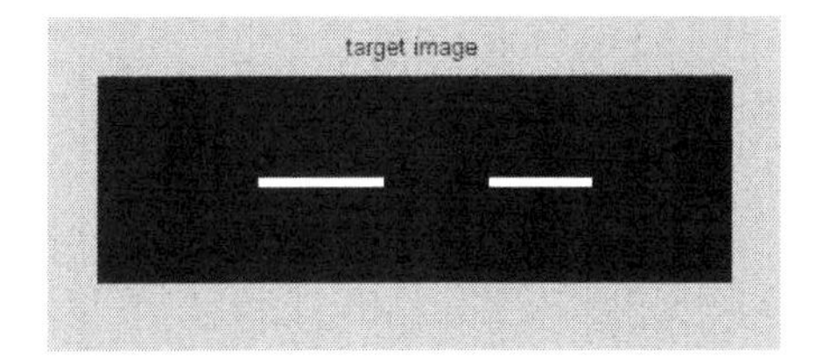

Fig. 54.6 Target binary image

54.5.2 Otsu Adaptive Threshold Segmentation

Otsu is a good kind of performance of automatic threshold segmentation method [6] through calculating image's target and background class within the minimum of within-class variance, the maximum of between-class variance, for automatic threshold calculation. According to grayscale of image features, if the image has larger variance between foreground and background, the two portions of the image have bigger difference. Whenever foreground ingredients are divided into wrong background or backgrounds are divided into wrong prospects, the two ingredients' difference becomes smaller, so the maximum between-class variance segmentation means that probability of wrong points is minimum variance.

The synthesis images of (2R-G-B) gray image by means of the Otsu adaptive threshold segmentation can quickly and effectively distinguish between targets and background. As Fig. 54.6 displays, getting adaptive threshold segmentation of target image can get ideal target binary image.

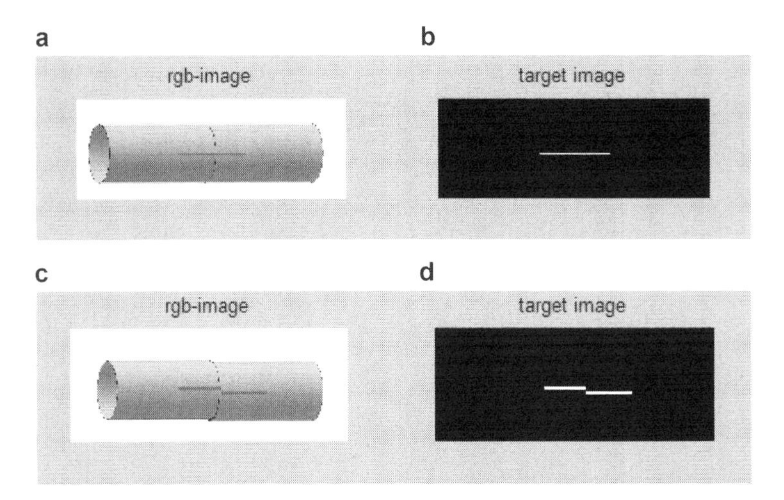

Fig. 54.7 Verified experiment of simulative assembly. (**a**) Original image of docking prosperity (**b**) Target image of docking prosperity (**c**) Original image of docking unsatisfactory (**d**) Target image of docking unsatisfactory

Table 54.1 *ABCD* coordinate points of docking target image in Fig. 54.7

Target image	Axis	*A*	*B*	*C*	*D*
b	*X*	379	520	521	638
	Y	197	197	197	197
d	*X*	367	508	509	661
	Y	185	185	198	198

54.5.3 Verified Experiment

In order to verify feasibility, this experiment extracts the result images among ideal and not ideal docking assembly on the CATIA simulation. By the proposed control method, we got the processing image [7] on MATLAB platform and finally got the target binary images, as shown below in Fig. 54.7.

The above experiments are hypothesized to extract the images captured by camera 1. The results have shown a good target extraction within simulated docking assembly. According to target binary images, the result image which is not ideal docking displays that the label line is obviously wrong and describes that the docking process quality has obvious problem so that we need to adjust the docking pose; result image of docking success displays a straight line, and if the image of camera 2 also has the same result, it realizes ideal docking.

From Table 54.1, we can learn that the target image (b) shows the docking prosperity, because the *ABCD* coordinate points have the same *Y* coordinates, and

B and C have the regular X coordinates. Target image (d) has different Y coordinates; obviously, it shows an unsatisfactory docking. In order to modulate pose, pose control system should apply instrumentation code to control components to move $|Y_B – Y_C|$ and to obtain same Y coordinates.

So the above experiments achieve process quality evaluation and verify the feasibility of docking process quality monitoring system.

54.6 Conclusion

In past studies, the research of process quality monitoring for automatic aircraft component docking is very few, and general evaluation methods examine whether or not the data returned by laser tracker meets tolerance standard. The new method has the advantages of non-contact and achieves visual docking process quality monitoring, which is based on laser tracker pose measurement and calibration. It enforces aircraft assembly automation to a higher level.

Acknowledgments This program is financially supported by the Natural Science Foundation of China (No. 61272077), Natural Foundation of Jiangxi Province (No.20114BAB201034), Research Foundation of Jiangxi Province Education Department (No.GJJ12413), and Key Laboratory Open Foundation of Jiangxi Province (No.TX201204003).

References

1. Mei, Z. Y., & Fan, Y. Q. (2009). Docking flexible assembly technology based on laser tracking and locating parts. *Journal of Beijing University of Aeronautics and Astronautics, 35*(1), 68–69.
2. Zhu, Y. G., Huang, X., & Fang, W. (2012). Medium fuselage position and attitude adjustment and tracking measurement. *Journal of Mechanical Science and Technology, 31*(7), 1121–1127.
3. Renliang Yu. (2008). *CATIA V5 basic tutorial* (pp. 166–169). Beijing: China Machine Press.
4. Li, Y., Ji, C., Wang, H., & Zhao, W. (2012). The image segmentation method of apple tree branches based on Matlab. *Science Technology and Engineering, 12*(13), 55–59.
5. Cai, J. R., Zhou, X. J., Li, Y. L., & Fan, J. (2008). The mature citrus recognition based on machine vision natural scenes. *Journal of Agricultural Engineering, 24*(1), 175–178.
6. Jichao, W. (2010). *Object extraction recognition algorithm of machine vision image*. Hebei: Agricultural University of Hebei.
7. Gonzalez, R. C., & Woods, R. E. (2011). *Digital image processing* (3rd ed., pp. 285–307). Beijing: Publishing House of Electronics Industry.

Chapter 55
Overhead Transmission Lines Sag Measurement Based on Image Processing

Wengang Cheng and Long Chen

Abstract Sag is one of the important parameters for operation and maintenance of the transmission lines, and its size directly affects the safe and stable operation of the line. In recent years, in order to improve the transmission capacity, many existing transmission lines allow the temperature from 70 °C to 80 °C, and then the transmission line sag becomes a major constraint for the transmission security. This chapter presents a novel sag measurement method based on image processing. Firstly it grays the collected color images and preprocesses the images with some image-denoizing methods. Secondly, special points generated by the isolation rod are extracted by the corner extraction algorithm, and the spatial coordinate values of the extracted points are identified according to the principle of binocular vision and the relationship of three-dimensional coordinate space coordinates and image coordinates. Finally via the method of the curve fitting, the actual sag of the transmission line is calculated. The experimental results show that this method is suitable for both the cases in which the height of the transmission line is equal or not, and it has good adaptability.

55.1 Introduction

Transmission line sag is one of the main parameters of the circuit design and operation and maintenance, and it directly affects the safety and stable operation of the transmission line. The sag will change with the transport capacity, weather condition change, and line increase of service life, and sometimes the sag will be beyond the design requirements that may lead to the security risks [1]. So it is very necessary to measure the transmission line sag regularly.

W. Cheng • L. Chen (✉)
School of Control and Computer Engineering, North China Electric Power University, Beijing 102206, China
e-mail: wgcheng@ncepu.edu.cn; happylong1988@163.com

W.E. Wong and T. Zhu (eds.), *Computer Engineering and Networking*, Lecture Notes in Electrical Engineering 277, DOI 10.1007/978-3-319-01766-2_55,

At present, a lot of methods and applications have been brought up on the measurement of the power line sag by domestic and foreign research institutions and companies. It mainly contains two cases: they are the traditional manual measurement in the field and using certain devices to implement the measurement. The traditional manual measurement methods include the different length measurement method, the same length measurement method (parallelogram method), the angle measurement method, and so on. These methods that use the actual observed data and combine with mathematical formulas can calculate the sag more accurately. But the traditional sag measurement methods have low efficiency and they are sensitive to the geographical environment. Besides, it is also difficult to conduct real-time monitoring. So a lot of real-time automatic monitoring power line sag methods have been proposed and have had a very good application. Literature [2, 3] proposed monitoring power line stress and temperature-based methods to measure the sag. The literature [4, 5] proposed the sag measurement method based on the global positioning system technology. These methods have very good application in practice, such as The Group Inc. in the USA that produced the CAT-1 that calculated sag by measuring the transmission line stress.

This chapter describes the real-time measurement of sag in an overhead power transmission line by using the image-processing method. We just need to extract the feature points of the image that is captured according to the image corner extraction methods and then calculate the spatial coordinates of the curve feature points in accordance with the principle of binocular vision and the relationship of the world coordinate system and the image coordinate system. Finally, we recover the curve equation by curve fitting to calculate the sag. This method is easy to operate, and it has a good real-time. Besides, the calculated results are very accurate and meet the applied requirements.

55.2 The Principle of Sag Measurement

55.2.1 The Process of Sag Measurement

Overhead line curve refers to the arc-shaped curve that is supported by both the end towers. Any point plumb distance to both ends of the suspension point connection line on the transmission line is called the overhead line sag of the point.

Overhead line sag is denoted by f generally. When both ends of the suspension point of the transmission line have the same height, the maximum sag is in the center of the span, as shown in Fig. 55.1. When the height of the two end points of the line is not equal, as shown in Fig. 55.2, the maximum sag is the plumb distance between the cutoff point that is the straight line A1B1 that parallels the connection line of the two point tangent to the curve and the point in the connection line. That is the cutoff point sag of the parallelogram ABB1A1.

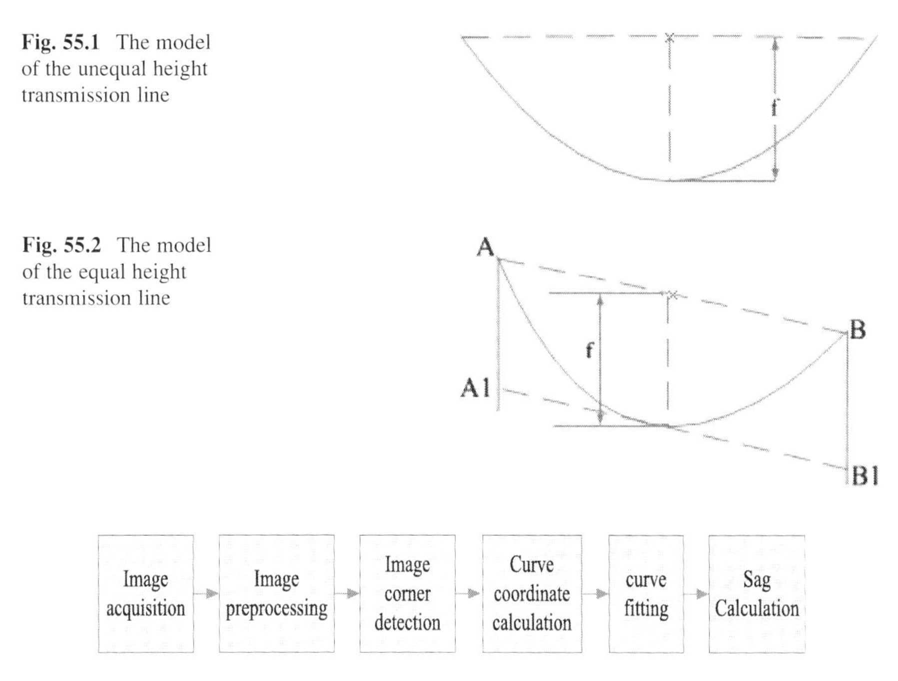

Fig. 55.1 The model of the unequal height transmission line

Fig. 55.2 The model of the equal height transmission line

Fig. 55.3 The flow chart of sag calculation

As the camera shoots the same scene several times, we can get the different coordinate values of the feature point that is the same point in the different images. The space coordinates of the feature points of the curve can be calculated according to the principle of binocular vision as well as the relationship of the world coordinate system and the image coordinate system. Finally we calculate the sag according to the fitting curve equation. The flow chart of computing transmission line sag is shown in Fig. 55.3.

55.2.2 Harris Corner Detection Algorithm

Corner is an important local feature of the images. In the actual image, corners such as contour inflection point, the end of the segment corner, have a rich amount of information, and they are not only very easy to be measured but also able to adapt to changes in ambient light. Most corner detection methods detect the image point with specific characteristics rather than just the "corner." These feature points often have some digital features, such as local maximum or minimum gray and some gradient feature. The corner detection algorithm can be summarized into three categories. They are corner detection based on gray-scale images, binary image corner detection, and corner detection based on the profile curve. Corner detection based on gray-scale image can be divided into three types. They are gradient-based, template-based, and template gradient combination.

Harris corner detection algorithm is a typical corner detection based on gray-scale image. It contains the following detection steps:

1. Calculate the directional derivative of the image, save for two arrays I_x and I_y.
2. Calculate the localized autocorrelation matrix $U(x, y) = [Ix(x, y)^2*W\ Iy(x,y)Ix(x, y)*W;Ix(x,y)Iy(x,y)*W\ Iy(x,y)^2*W]$ for each point; here $*W$ refer to that the x, y as the center do convolution with Gaussian template W and the size of the template needs to be specified.
3. If the two eigenvalues of u are very small, it indicates that this region is a flat region. If one of the eigenvalues of u is larger and the other is smaller, it is the line. If both of the eigenvalues are larger, then it shows that this is a corner point. Harris provides another formula to get an evaluation that whether the current point is the corner point: corn $=$ det $(u) - k *$ trace $(u)^2$. Here the corn is referred to the point value and k is a fixed variable, typically between 0.04 and 0.06.

Since the Harris corner operator has good robustness, rotational invariance, and many other advantages, it can extract the feature points of the transmission line curve easily. The position of the spacer rods in the transmission line will be considered the corner, and we can get all positions of the spacer rods in every image, so we can obtain the position coordinates of the curve on the same physical point for each different image.

55.2.3 Camera Calibration and the Three-Dimensional Coordinate Measuring Algorithm

According to the principle of camera imaging, we can conclude the relationship between the coordinate value (u, v, 1) of the point P in the pixel coordinate system and the coordinate value (X_w, Y_w, Z_w) of the point P in the world coordinate system. The conversion formula is as follows:

$$Zc\begin{bmatrix} u \\ v \\ 1 \end{bmatrix} = \begin{bmatrix} f_1 & 0 & u_0 & 0 \\ 0 & f_2 & v_0 & 0 \\ 0 & 0 & 1 & 0 \end{bmatrix}\begin{bmatrix} R & T \\ 0^T & 1 \end{bmatrix}\begin{bmatrix} X_W \\ Y_W \\ Z_W \\ 1 \end{bmatrix} = M_1M_2\begin{bmatrix} X_W \\ Y_W \\ Z_W \\ 1 \end{bmatrix} \tag{55.1}$$

(u0, v0) stands for the coordinate value of the origin of the image coordinate system in the pixel coordinate system whose origin is the upper left corner of image plane; matrix R stands for the camera rotation matrix; matrix T stands for the camera translation matrix.

According to Eq. 55.1, we can figure out the computer calibration parameters and calculate the spatial point coordinate value by combining the binocular stereo vision principle [6]. According to the principle of binocular vision, we need to take two photos of the same scene and calculate other information by using the relationship of the two photos. The specific applications are as follows:

1. Calculating the camera calibration parameters
   After analyzing Eq. 55.1, we can learn that we just need to calculate the parameters f1 and f2. In order to facilitate the calculation, we assume that the first camera coordinate system coincides with the world coordinate system and the second camera translates L distance along the X-axis. Besides, both of the two cameras are with no rotation. That is to say, the matrix R is a unit matrix and the matrix T = $[L\ 0\ 0]^T$. Above all, we will be able to get the two equations about f1 and f2 and can get the following results:

$$f_1 = \frac{Z_W(u_1 - u_2)}{L} \tag{55.2}$$

$$f_2 = \frac{Z_W(v_1 - v_2)}{L} \tag{55.3}$$

   (u1, v1) and (u2, v2) stand for the coordinate values of the same point in the two images.
2. Calculating the space coordinate values of the feature points
   According to Eq. 55.1, the calculated camera calibration parameters f1 and f2, as well as the camera rotation matrix R and translation matrix T that can be obtained in the actual shooting operation, we can get four linear equations about X_w, Y_w, and Z_w by taking two images, so we can work out the feature point P coordinate value (X_w, Y_w, Z_w) of the world coordinate system.

55.3 Experiments

In the process of calculation of the camera calibration parameter, we capture several images at five different locations and take 20 images in each location. Comparing the recovery coordinate value with the actual coordinate values, we can conclude the maximum error that mainly affected by the camera calibration parameter in each direction. The maximum error in X-, Y-, and Z-direction are −2.25, 2.98, and 2.15 %, respectively.

The experimental result of using Harris corner extraction algorithm to detect the spacer bars of the transmission lines is shown in Fig. 55.4 (the picture has been rotated 15° clockwise). Experiments show that using the corner extraction algorithm to detect spacer bars is more effective than using image matching method to detect the spacer bars, especially for those blurred images that are captured on the long shooting distance.

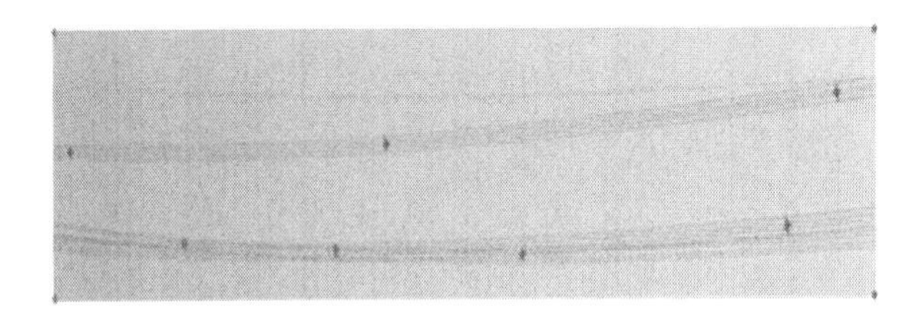

Fig. 55.4 The experimental results of corner detection

Comparing the measurement data that are obtained by using image-based corner detection method to measure the sag with the measurement data that is measured by the field theodolite, except that individual data error is larger, most data error were less than 2.5 %. The measurement accuracy meets the applied requirements.

55.4 Conclusion

Through analysis of the chapter and related experimental results, we can get the following conclusions.

1. Based on the characteristics of the transmission line and combined with image processing technology, it uses the Harris corner extraction algorithm to detect spacer bars of the transmission line and computes transmission line sag with the curve information. According to the experimental results, we can conclude that this method is simple and has a good adaptability.
2. The results of the sag calculation will be affected by the camera calibration accuracy, position of the camera, and shooting angle. Therefore, it is our future research work to improve the accuracy of the camera calibration and select the appropriate coordinate point and shooting angle. That is to say, we should choose the shooting position that can capture the image as much as possible and the shooting angle that is easy to calculate such as 30° and 45°.
3. Due to the complexity of the environment, it is very important for the accuracy of the experiment to remove the image noise. This chapter uses the median filtering denoizing algorithm for image denoizing, which can overcome the problem of the fuzzy edge of the low-pass filtering and the noise enhancement of the high-pass filtering. Besides, this chapter also uses the image enhancement function that is from the MATLAB function library to highlight the edge of the transmission line.

Acknowledgements This work is supported by "the Fundamental Research Funds for the Central Universities" of China.

References

1. Xu, Q. S., Ji, H. X., & Wang, M. L. (2007). Monitoring sag of the transmission lines real-time. *High Voltage Engineering, 33*(7), 206–208 (In Chinese).
2. Chen, S. X., Wu, P. S., Sun, Y. T., et al. (2008). Application of on-line temperature and sags monitoring device of transmission line. *North China Electric Power*, (3), 14–15.
3. Muhr, M., Pack, S., & Jaufer, S. (2005). Sag calculation of aged overhead lines. In *Proceedings of the 16th International Symposium on High Voltage Engineering. Tsinghua University, Beijing, China* (pp. D-14).
4. Mensah-Bonsu, C., Krekeler, U. F., Heydt, G. T., et al. (2002). Application of the global positioning system to the measurement of overhead power transmission conductor sag. *IEEE Transactions on Power Delivery, 17*(1), 273–278.
5. Mahajan, S. M., & Singareddy, U. M. (2008). Real time GPS data processing for 'sag measurement' on a transmission line. In *Power System Technology and IEEE Power India Conference, New Delhi. IEEE* (pp. 1–6).
6. Deng, Z. D., Niu, J. J., & Zhang, J. D. (2007). Three-dimensional modeling approach based on stereo vision. *Journal of System Simulation, 19*(14), 3258–3262.

Chapter 56
Chinese Domain Ontology Learning Based on Semantic Dependency and Formal Concept Analysis

Lixin Hou, Shanhong Zheng, Haitao He, and Xinyi Peng

Abstract The ontology construction process is very expensive and time consuming when performed manually. In order to solve the problems of time consumption and high cost, a Chinese domain ontology learning method based on semantic dependency and formal concept analysis is proposed in this chapter. During the learning process, semantic dependency analysis technology is used for extracting formal context from unstructured domain texts, and Godin algorithm is used for constructing concept lattice. At last, the chapter takes a medical ontology construction as an example to verify this method. The experiment results show that the method we proposed can construct domain ontology automatically and reduce manual intervention. In addition, the ontology we got is a formal ontology, so it has more advantages in sharing and reusability.

56.1 Introduction

Ontology learning is a subtask of information extraction, and its purpose is to semiautomatically extract relevant concepts and relations from a given corpus or other kinds of data sets to form ontologies. Ontology learning can solve the problems of low efficiency and time consumption when performed manually. Therefore, ontology learning is becoming an active research area.

Ontology learning focuses on the extraction of domain concepts and relations among them automatically [1]. There are many methodologies for concept extraction. Extraction based on linguistics has advantages in terms of disambiguation and

L. Hou (✉) • S. Zheng • H. He
College of Computer Science and Engineering, Changchun University of Technology, Changchun 130012, China
e-mail: houlixinmingxuan@126.com

X. Peng
Software Vocational and Technical College, Changchun University of Technology, Changchun 130012, China

W.E. Wong and T. Zhu (eds.), *Computer Engineering and Networking*, Lecture Notes in Electrical Engineering 277, DOI 10.1007/978-3-319-01766-2_56,

accuracy [2, 3], but it relies on the results of the word segmentation. Extraction based on statistics does not require lexical and syntactic information and has better portability [4, 5]. Also there are many methods that mix linguistics with statistics [6, 7]. Structured vocabularies [8], linguistics [9, 10], and statistics [11, 12] have been used to extract relationships among concepts. However, methods based on traditional structured vocabularies extract only several kinds of relations, and most of them are hierarchical relations. Relationships extracted with linguistics have high accuracy, but the effects are easily affected by the completeness of the grammar rules. Extractions with statistics have nothing to do with the language and the fields. Therefore, they have a strong portability, but the effects largely depend on the quality and size of the corpus.

In the study of ontology construction, formal ontologies are better than ontologies based on natural language in sharing. Formal concept analysis (FCA) invented by Wille is a tool of identifying conceptual structures among data sets. FCA is widely used in the ontology construction. Obitko put forward that FCA can explore potential objects and attributes by constructing concept lattice and present them in a visual way automatically [13]. This chapter improves the above method by adopting the semantic dependency analysis technology to extract formal context, using Godin algorithm for building the concept lattice, and adopting the mapping rules between the concept lattice and the ontology to generate the domain ontology. Finally, this method is testified by constructing a medical ontology. The experiment results show that ontology construction based on semantic dependency and FCA not only can extract semantic information automatically and objectively but also can construct expected ontology from unstructured Chinese texts. Therefore, the method we proposed can construct domain ontology automatically and reduce manual intervention. In addition, the ontology we get is a formal ontology, so it has more advantages in sharing and reusability.

56.2 The Chinese Domain Ontology Learning Procedure Based on FCA

56.2.1 Formal Context Extraction

The formal context and concept hierarchy are basic structures of FCA. A formal context comprises two sets and a binary relation between them. The definition is as follows:

Definition 1 *Formal context* is a triple $K := (O,A,I)$, where O is a set of objects, A is a set of attributes, and I is the binary relationship between O and A, that is, $I \subseteq O \times A$. For $o \in O$ and $a \in A$, $(o,a) \in I$ is defined as "attribute a is an attribute of the object o."

Formal context is the basic of the ontology learning based on FCA. This chapter adopts Philipp Cimiano method which was used in IST-Dot Kom project by AIFB research institution [14]. The main idea is that we use a Chinese dependency syntactic parser for analyzing the sentences in texts [15], and then the parsing trees with dependence relation and semantic role labels will be produced. To build the formal context we extract the backbone of the sentence from the parsing tree and turn the subject of the sentence head into an FCA object and the objects into attributes. The algorithm for extracting the formal context based on the definition 1 and the main idea we proposed is given below.

Step 1. Initialize HashMap for store pair of object and attribute.
Step 2. For any sentence i (the sentence number) get word segmentation sequence filed as wordlist based on conditional random field (CRF).
Step 3. For every word j (the word number) in wordlist, we adopt graph-based parser output, the dependency relation of word i in sentence j.
Step 4. Judge the type of the dependency relation of word j, if the type equals "SBV," and then let word j be an object filed as o; if the type equals "VOB," then let word j be an attribute filed as a, and then add (o,a) to HashMap.
Step 5. Build formal context $K = (O,A,I)$; use all pairs in HashMap.

56.2.2 Concept Lattice Construction

Definition 2 For any X as an object subset, we define

$$X' = \{a \in A | \forall x \in X, \exists (x,a) \in I\}$$

For any Y as an attribute subset, we define

$$Y' = \{o \in O | \forall y \in Y, \exists (o,y) \in I\}$$

Definition 3 If $X \subseteq O$, $Y \subseteq A$ satisfies $X' = Y$ and $Y' = X$, then we regard $C = (X,Y)$ as a formal concept of $K = (O,A,I)$, X as the extension, and Y as the intension of the C.

Definition 4 For a formal context $K = (O,A,I)$ and two formal concepts $C_1 = (x_1, y_1)$ and $C_2 = (x_2,y_2)$ of K, a concept hierarchy relation is given by

$$(x_1,y_1) \prec (x_2,y_2) \Leftrightarrow (x_1 \subseteq x_2) \wedge (y_2 \subseteq y_1)$$

All the formal concepts in K are ordered by this concept hierarchy relation, and this ordering is called the concept lattice of the formal context K. C_1 is the sub-concept of C_2, whereas C_2 is the super-concept of C_1. Concept lattice is the core data structure of FCA, and it reflects the generalization and specialization of the relationship between the concepts; also it can be represented graphically by using Hasse diagram. There are two kinds of lattice construction algorithms: batch algorithms and incremental algorithms [16].

This chapter adopts Godin algorithms to construct concept lattice. Godin algorithm is an incremental algorithm introduced by Robert Godin in 1995. If we want to use Godin algorithms, two problems should be resolved [17]: (1) update the nodes in concept lattice and (2) update edges between lattice nodes. The new lattice nodes can be divided into three types, and the types can be described by the following definitions.

Definition 5 Let $L(K)$ be corresponding concept lattice of context $K = (O,A,I)$ and $M_i = (x_i,y_i)$ be any lattice node in $L(K)$. Let x_j be newly increased object and y_j be corresponding attribute. Meanwhile, if $L'(K)$ is new concept lattice through adding x_j, then the lattice nodes in $L'(K)$ can be divided into three types:

1. If $y_j \cap y_i = \Phi$, then let $M_i = (x_i,y_i) \in L(K)$; M_i is called unchanged node.
2. If $y_i \subset y_j$, then modify M to $M' = (x_i \cup x_j,\ y_i)$ and add M' to $L'(K)$; M' is called update node.
3. If $(y_j \cap y_i \neq \varphi) \wedge (\neg \exists ((y_j \cap y_i)',\ y_j \cap y_i) \in L(K))$, then create a new node $M' = (x_j \cup x_i,\ j_j \cap y_i)$ and add it to $L'(K)$; M' is called newly increased node.

The following are the procedure for building lattice based on Godin algorithms:

Step 1. Initialize concept lattice $L(K) \leftarrow \phi$.
Step 2. Take out one pair (o,a) from $K = (O,A,I)$.

(1) Take out formal concept $M_i = (x_i,y_i)$ from $L(K)$ in proper sequence.
(2) Update the node in $L(K)$ according to the result of $(a \wedge y_i)$ and the rules in definition 5.

Step 3. If producing a new node in step (2), then add it to $L(K)$ and update the edges between the nodes.
Step 4. Repeat steps 2 and 3 until generating a complete concept lattice.

56.2.3 *Mapping Concept Lattice to OWL Ontology*

Concept lattice can be viewed as a prototype of ontology. So the Hasse diagram should be trimmed circularly by deleting unreasonable concepts according to the domain knowledge. After that, ontology description language OWL recommended by W3C is adopted to describe the ontology mapped from concept lattice. The mapping rules between lattice elements in FCA and semantic elements in OWL are the followings:

Definition 6 Let $O := (C,\ root,\ \prec c)$ be a domain ontology, C be the set of concepts, $root$ be the root element in O, $\prec c$ be the concept hierarchical relation, $L(K)$ be the corresponding lattice of formal context $K = (O,A,I)$, $H_i = (x_i,y_i)$ and $H_j = (x_j,y_j)$ be any lattice nodes in $L(K)$, e be lattice element in FCA, C be the corresponding class name of H_i, sup C be the corresponding class name of H_j, and $f : L(K) \rightarrow O$ be the mapping rules from concept lattice to ontology; then the rules are the following:

R_1: IF e is H_i THEN <owl:Class rdf:about="#C "/> IN O

R_2: IF e is H_i AND $H_i \leq H_j$ THEN <owl:Class rdf:about="#C"> <rdfs:subClassOf rdf:resource="#sup C"/><rdfs:subClassOf>IN O

R_3: IF $e \in x_i$ THEN <owl:NamedIndividual rdf:about="#x_i"> <rdf:type rdf:resource="#C"/></owl:NamedIndividual> IN O

R_4: IF $e \in y_i$ THEN <owl:DatatypeProperty rdf:about="#y_i/> <rdfs:domain rdf:resource="#C"/></owl:DatatypeProperty> IN O

Based on definition 6, we design the mapping algorithm as follows:

Step 1. Initialize an empty ontology O.
Step 2. Map the root node in $L(K)$ to root according to R_1.
Step 3. Take out the child node $H_i = (x_i, y_i)$ of the root node.
Step 4. Apply the mapping rules:

(1) Apply rule R_1 to lattice node H_i.
(2) Apply rule R_2 according to the hierarchical relation between the root node and the child node.
(3) Apply rule R_3 to the new objects that lattice node H_i owns, whereas the new objects do not belong to H'_i (H'_i is the child node of H_i).
(4) Apply rule R_4 to the new attributes that H_i owns whereas the corresponding parent node of H_i does not own.

Step 5. Let H'_i be the new root node.
Step 6. Repeat steps 3–5 until mapping of the whole concept lattice is complete.

56.3 Experimental Verification

To verify the effectiveness of the method, we apply it to many medical domain files to learn ontology. The learning process is divided into three stages: (1) obtain the formal context from domain texts; (2) construct the concept lattice; and (3) map the concept lattice to the ontology. At last a Chinese ontology of medical domain will be obtained.

56.3.1 Obtaining the Formal Context Through Learning Medical Texts

The semantic dependency of every sentence in medical texts should be analyzed according to the method in Sect. 56.2.1. The process is as follows:

```
LTML ltml = ls.analyze(LTPOption.ALL,str);
for(int i = 0; i<ltml.countSentence(); ++i){
ArrayList<Word>wordList=ltml.getWords(i);
```

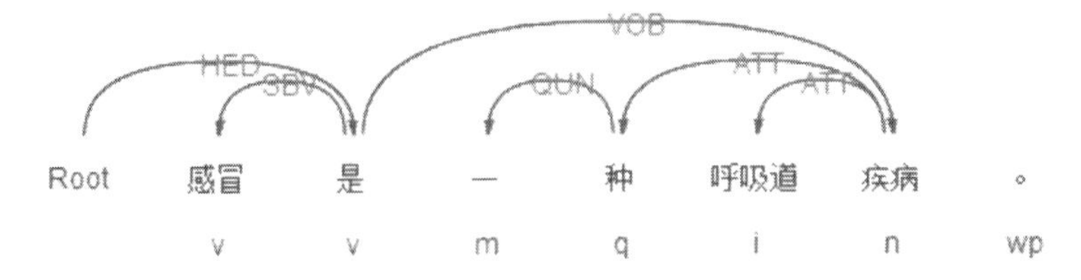

Fig. 56.1 The semantic dependency tree of the example sentence

```
for(int j=0; j<wordList.size(); ++j){
System.out.print("\t"+wordList.get(j).getWS());
System.out.print("\t"+wordList.get(j).getParser-
  Relation());
if(wordList.get(j).getParserRelation().equals(s1))
  {name=wordList.get(j).getWS();}
if(wordList.get(j).getParserRelation().equals(s2))
  {predicate=wordList.get(j).getWS();}
if(wordList.get(j).getParserRelation().equals(s3))
  {value=wordList.get(j).getWS();}
hashmap.put(name, value);}}
```

Take "感冒是一种呼吸道疾病 (Cold is a kind of respiratory disease)。"; for example, the visual result of the dependency analysis would look like as in Fig. 56.1.

In the figure, the semantic role of "是(is)" is "HED,"while the semantic role of "感冒 (cold)" is "SBV" and the semantic role of "疾病 (disease)" is "VOB." From the analysis result we can get

```
name="感冒"; value="呼吸道疾病"; hashmap.put(name,value);
```

As a result, the pair("感冒", "呼吸道疾病") will be obtained.

We learn all the domain texts and obtain all the pairs by the following method:

```
Iterator it=hashmap.entrySet().iterator();
while (it.hasNext()){Map.Entry entry=(Map.Entry) it.
  next();
String
filecontents=entry.getKey()+predicate+entry.getValue
  ()+"\n";}
```

We place all the pairs in the table, and small part of the formal context would look like as in Table 56.1.

56.3.2 Constructing Medical Concept Lattice

We analyze the medical formal context through the Godin algorithms described in Sect. 56.2.2. Then we adjust the location of part lattice nodes. Finally the medical concept lattice will be obtained, and part of it looks like as in Fig. 56.2.

Table 56.1 Part of formal context of medical domain

	A	B	C	D	E	F	G	H	I
1	×	×	×	×					×
2	×				×				
3							×	×	
4						×		×	
5	×		×	×			×		
6	×			×	×				

1.感冒 (cold) 2. 慢性支气管炎 (chronic bronchitis) 3. 胃炎 (gastritis) 4. 肠道炎 (gastroenteritis) 5. 肺炎 (pneumonia) 6. 肺结核 (phthisis) A. 咳嗽 (cough) B. 鼻塞 (nasal obstruction) C. 头痛 (headache) D. 发热 (fever) E. 咳痰 (expectoration) F. 腹痛 (abdominal pain) G. 恶心呕吐 (nausea and vomiting) H. 反酸 (sour regurgitation) I. 流鼻涕 (rhinorrhea)

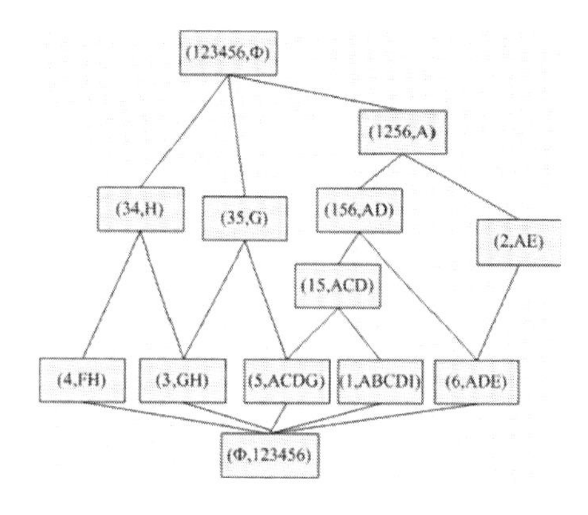

Fig. 56.2 Part of the medical concept lattice

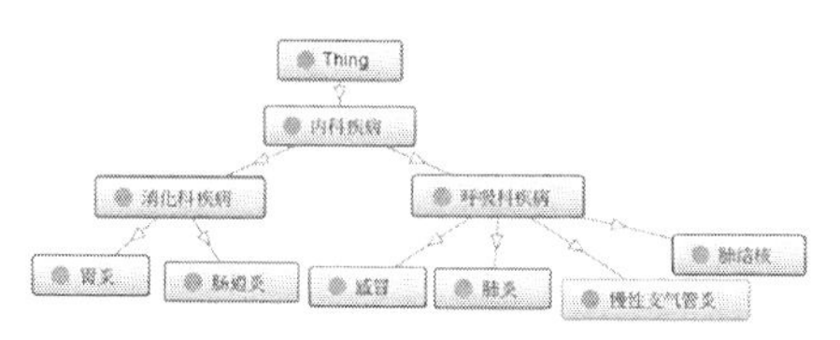

Fig. 56.3 Part of the medical ontology

Concept lattice describes the generalization and specialization relationship of the concepts where the above node is parent node whereas the below node is child node.

56.3.3 Generating the Chinese Ontology in Medicine

We adopt the mapping rules between lattice elements in FCA and ontology elements in OWL defined in Sect. 56.2.3 to map the concept lattice in Fig. 56.2. After the mapping, the unreasonable concepts are deleted, and finally a Chinese ontology is generated. Furthermore, one plug-in of the Protégé, called OntoGraf, is used for visualizing the Chinese ontology, and the result would look like as in Fig. 56.3.

56.4 Conclusion

We proposed a Chinese domain ontology learning approach based on semantic dependency and FCA. First, this study adopted the Chinese dependency parser to parse every sentence in Chinese domain texts to construct the formal context. Then, by adopting the Godin algorithms, the concept lattice was built. At last, the concept lattice was mapped to the domain ontology through the mapping rules between them, showing the ontology graphically by using Hasse diagram. In addition, we verified the effectiveness of the approach proposed in this chapter through the experiment in Sect. 56.3. The advantages are that the method can conduct the unstructured Chinese texts and can express the automatic objective extracting semantic characteristics of FCA. Shortages inevitably exist, and we will design more specific algorithms for constructing the formal context to reduce artificial participation and improve the automation degree of the ontology learning in the next phase.

References

1. Du, X. Y., Li, M., & Wang, S. (2006). A survey on ontology learning research[J]. *Journal of Software, 17*(9), 1837–1847.
2. Shamsfard, M., & Barforoush, A. (2004). Learning ontologies from natural language texts [J]. *International Journal of Human-Computer Studies, 60*(1), 17–63.
3. Sabou, M. (2004). From software APIs to web service ontologies: A semi-automatic extraction method[C]. In *Proceedings of International Semantic Web Conference (ISWC), Springer, Berlin* (pp. 410–424).
4. Huang, C. (2009). *Research of domain ontology construction and using in web information extraction[D]*. Jiangxi: Jiangxi University of Science and Technology.
5. Yu, J. (2010). *Learning domain ontologies from Chinese text corpora[D]*. Dalian: Dalian University of Technology.
6. Liang, J., & Wu, D. (2006). Seed concept method and its application in texts-based ontology learning[J]. *Library and Information Service, 50*(9), 18–21.
7. Wen, C., Wang, X., & Shi, Z. (2009). Automatic domain specific term extract ion in Chinese domain ontology learning[J]. *Application Research of Computers, 26*(7), 2652–2655.
8. Ahlnad, K., Tariq, M., et al. (2003). Corpus-based thesaurus construction for image retrieval in specialist domains[C]. In *Proceedings of the 25th European Conference on Advances in Information Retrieval (ECIR), Springer, Berlin* (pp. 502–510).
9. Buitelaar, P., Olejnik, D., Hutanu, M., & Schutz, A. (2004). Towards ontology engineering based on linguistic analysis[C]. In *Proceedings of LREC, Lisbon, Portugal* (pp. 7–10).
10. Wang, S. (2010). *The research on acquisition method of relationships between subject concepts in ontology construction[D]*. Beijing: The Chinese Academy of Agricultural Sciences.
11. Fu, K. (2007). *The study of ontology learning from web pages[D]*. Wuhan: Wuhan University of Technology.
12. He, L. (2009). *Research on domain ontology semi-automatic construction and retrieval[M]* (pp. 156–163). Nanjing: Southeast University Press.
13. Obitko, M., Sná, elV., & Smid, J. (2010). *Ontology design with formal concept analysis* [*EB/OL*]. http://ftp. informa-tik. rwth- aachen. de/Publications/CEUR -WS/Vol-110/paper12. Pdf

14. Huang, M., & Liu, Z. (2006). Research on domain ontology building methods based on formal concept analysis[J]. *Computer Science, 33*(1), 210–212. 239.
15. Che, W., Li, Z., & Liu, T. (2010). LTP: A Chinese language technology platform. In *Proceedings of the Coling 2010*: *Demonstrations, Beijing* (pp. 13–16).
16. Baixeries, J., Szathmary, L., et al. (2009). Yet a faster algorithm for building the Hasse diagram of a concept lattice[C]. In S. Ferré & S. Rudolph (Eds.), *ICFCA 2009* (pp. 162–177). Berlin: Springer.
17. Jiang, Y., Zhang, J., & Zhang, S. (2007). Incremental construction of concept lattice based on linked list structure [J]. *Computer Engineering and Applications, 43*(11), 178–180.

Chapter 57
Text Classification Algorithm Based on Rough Set

Zhiyong Hong

Abstract In text classification community, k-nearest neighbor (kNN) and support vector machine (SVM) are all effective classifiers, but both of them have their own drawbacks. kNN involves high cost to classify a new document when training set is large; SVM is too sensitive to the noisy data when the noisy data is close to the hyperplane it suffers. So one hybrid algorithm based on variable precision rough set is proposed. It combines the strength of both KNN and SVM techniques and overcomes their weaknesses. Finally some experiments are carried out to compare the efficiency and classification accuracy with different classification algorithms. Results show that the proposed method achieves significant performance improvement.

57.1 Introduction

Text classification is the core technology of the research in the field of information retrieval and data mining; it has rapidly developed in recent years. Text classification or categorization is the task of automatically assigning unseen documents to suitable predefined categories. A number of well-known algorithms have been introduced to deal with text classification [1], such as k-nearest neighbor (kNN) [2], support vector machine (SVM) [3], neural network [4], naive Bayesian (NB), centroid-based classifier, decision tree, and Rocchio classifier.

The kNN is an instance-based learning algorithm, which is simple and intuitive but very effective for a variety of problem domains including text classification. It is well known to be one of the most effective methods on Fudan University classification corpus—one of the benchmark corpora used in text classification. However, kNN has a high cost of classifying new patterns; its training phase just

Z. Hong (✉)
School of Computer Science, Wuyi University, Jiangmen 529020, Guangdong, China
e-mail: hongmr@163.com

W.E. Wong and T. Zhu (eds.), *Computer Engineering and Networking*, Lecture Notes in Electrical Engineering 277, DOI 10.1007/978-3-319-01766-2_57,

stores all training patterns as classifier; thus it has often been called as lazy learner since it defers the decision on how to generalize beyond the training data until each new query pattern is encountered. The efficiency of kNN prohibits it from being applied to areas where efficiency is particularly required for text classification, such as dynamically mining large-scale collection.

SVM is a machine learning method based on statistical learning theory [3]. An SVM constructs an optimal hyperplane that has the largest distance to the nearest training data point of any class to find the good support vector. Since SVM has good generalization performance, it has been used for classification, regression analysis, handwriting recognition, speech recognition, face recognition, and so on. In classification of the SVM traversal the feature space consists of support vector instead of the original input space consisting of training samples to ensure higher classification accuracy and greatly reduce the computational complexity. Although SVM has good generalization performance, SVM also has its drawbacks, such as the classification accuracy is not very high when encountering complex classification. Especially when training samples is confusing or overlapping, SVM classifier is prone to over-fitting.

In this chapter, a hybrid algorithm based on variable precision rough set (VPRS) is proposed to combine the strength of both kNN and SVM techniques and overcome their weaknesses. Firstly, feature space of training data is partitioned by using VPRS, and lower and upper approximations of each class are defined. Then kNN and SVM classifiers are built on these new subspaces, respectively. The SVM classifiers are used to classify most of the new documents effectively and efficiently. The kNN classifier is only required to classify new document which lies in the boundary region where SVM classifier suffers. And it is just required to find nearest neighbors of new document in the subset of training dataset, which can save time obviously compared with finding nearest neighbors in the whole training dataset. Experiments are carried out on Fudan University classification corpora. The experimental results indicate that the proposed hybrid algorithm achieves significant performance improvement.

57.2 Background

The rough set theory [5], introduced by Pawlak in the early 1980s, is a formal mathematical tool to deal with incomplete or imprecise information. As a generalized version of rough sets, VPRS [6] allows objects to be classified with an error smaller than a certain predefined level. In text classification, when the training data is confusing or overlapping, VPRS model is more suitable than the classic rough set. In this section, a brief introduction to rough set and VPRS is given.

57.2.1 VPRS

As a generalization of the standard inclusion relation, majority inclusion relation introduced by the VPRS is defined as follows:

Definition 1 Majority inclusion relation

$$c(X,Y) = \begin{cases} 1 - |X \cap Y|/|X|, |X| > 0 \\ 0, |X| = 0 \end{cases} \quad (57.1)$$

where X and Y are subsets of the universe U. The majority inclusion relation denotes the relative degree of misclassification of the set X with respect to set Y. Based on this measure, one can define the standard set inclusion relation between X and Y as $X \subseteq U$ if and only if $c(X,Y) = 0$. Consider an equivalence relation R on U, For a subset X⊆U, the β-lower approximation and β-upper approximation of X can be defined as follows:

Definition 2 β-lower approximation and β-upper approximation:

$$\underline{R}_{\beta}X = \{x \in U : c([x]_R, X) \leq \beta\} \quad (57.2)$$

$$\overline{R}_{\beta}X = \{x \in U : c([x]_R, X) < 1 - \beta\} \quad (57.3)$$

Definition 3 β-positive, β-negative, and β-boundary region based on VPRS:

$$POS_{\beta}X = \underline{R}_{\beta}X \quad (57.4)$$

$$NEG_{\beta}X = U - \overline{R}_{\beta}X \quad (57.5)$$

$$BND_{\beta}X = \overline{R}_{\beta}X - \underline{R}_{\beta}X \quad (57.6)$$

57.3 Hybrid Algorithm Based on VPRS

57.3.1 Algorithm

In this chapter, we only consider binary text classification that assigns each document d either to the positive class C_d or to its complement negative class C_n. Theoretically, binary classification is more general than the multi-class classification, and a multi-class classification can be transformed into a set of binary classifications. We can characterize the two class document sets of C_d and C_n with respect to a hidden equivalence relation R which may lead the documents belonging to the same class to have the tendency of clustering. The kNN algorithm is used to create equivalence classes for set C_d and C_n according to the concepts of VPRS. For a document d in training set, the kNN algorithm is used to find its kNN by means of calculating the similarity of d to anyone in training set, which forms a neighborhood of d. If all neighbors are from a single class, e.g., C_d, then there is no uncertainty in the neighborhood. However, if any neighbor belongs to another class C_n, the rough uncertainty arises in the neighborhood. This uncertainty can be captured by using the modified majority inclusion relation. For any document d and document set of class C_d in training dataset, modified majority inclusion relation is defined as follows:

Definition 4 (Modified majority inclusion relation):

$$c(N_d, C_d) = 1 - \frac{|N_d \cap C_d|}{|N_d|} \tag{57.7}$$

where N_d is the neighborhood region around d and $|\cdot|$ denotes the cardinality of the set. According to VPRS, β-positive region, β-negative region, and β-boundary region of class C_d can be obtained. The β-positive region of class C_d denotes the document set which lies in the positive region where documents can be certainly classified to class C_d, β-boundary region of class C_d denotes the document set which lies in the boundary region where documents cannot be classified uniquely to the class C_d, and β-negative region of class C_d denotes the document set which cannot be surely classified to the class C_d. Similar description is also fit to class C_n.

After feature space is partitioned into three regions, i.e., β-positive region, β-negative region, and β-boundary region for each class, then we can get the training set init_set1 which consists of all the β-positive for each class and init_set2 which consists of all the β-boundary for each class. For overcoming the problem of model over-fitting of SVM and improving the classification performance of kNN, the SVM classifier is trained with init_set1 and kNN classifier is trained with the init_set2. After the SVM classifier is built, we can build one new hybrid classification algorithm. For a new unseen document d, firstly calculate the distance to SVM optimal hyperplane; if the distance is larger than the given threshold value then the algorithm will use SVM to classify the document d, or else the algorithm will use kNN to find the k-partition nearest neighbors around d; then d is classified by a majority vote of its neighbors. The hybrid classification algorithm is described as Algorithm 1.

57.3.2 The Proposed Algorithm

Algorithm 1 The hybrid algorithm

Input: Training set χ, the parameter k, β, new document d
Output: The class label of d

Step 1: Use kNN to partition space of training set χ based on VPRS and definition 7, and then obtain the $POS_{\beta}C_i$ and $BND_{\beta}C_i$ of every class C_i $(i = 1, 2 \ldots n)$

Step 2: Classifier training

(1) Combinate all the $POS_{\beta}C_i$ of class C_i to constitute the training set init_set1.
(2) Combinate all the $BND_{\beta}C_i(i = 1, 2 \ldots n)$ of class C_i to constitute the training set init_set2.
(3) Train the SVM classifier with the training set init_set1 and then obtain support vector set $(SV_i, i = 1 \ldots m)$, factor $(\alpha_i, i = 1 \ldots m)$, and constant b.
(4) Train the kNN classifier with training set init_set2.

Step 3: Classifying the new document d

(1) For a new document d, firstly calculate the distance to SVM optimal hyperplane with the following formula (57.8):

$$f(x) = \sum_{i=1}^{m} \alpha_i y_i k(x_i, x) + b \tag{57.8}$$

if $|f(x)| > \varepsilon$ (ε) is the given threshold value then output the class label of d, or else go to next step.

(2) Use kNN to find the k-partition nearest neighbors around d; the nearest neighbors can be obtained by the vector similarity calculation formula (57.9):

$$Sim(d_1, d_2) = \cos\theta = \frac{\sum_{k=1}^{n} W_{1k} \times W_{2k}}{\sqrt{\sum_{k=1}^{n} W_{1k}{}^2 \cdot \sum_{k=1}^{n} W_{2k}{}^2}} \tag{57.9}$$

where W_{ik} is the kth term weight of d_i; then the class label of d_i can be obtained by methods in Sect. 57.3.1.

57.4 Experiment Results and Discussion

To evaluate performance (compute speed and classification accuracy) of the proposed approach, we have conducted experiments on the Fudan University classification corpus obtained from the International database center of Fudan University. The Fudan University classification corpus is a standard text

Table 57.1 Comparison of the classification accuracy of each algorithm

Category	SVM	kNN	Hybrid algorithm
Transportation	89.12	91.04	91.98
Sports	91.35	92.61	92.93
Military	92.78	92.21	94.35
Medicine	93.89	89.23	94.50
Politics	93.00	93.56	93.99
Education	93.45	94.00	95.38
Environment	87.37	90.76	91.77
Economy	90.81	91.23	92.26
Art	88.78	89.75	91.68
Computer	89.29	86.46	90.81
Macro F1	90.98	91.09	92.97

classification benchmark which contained 1,882 training examples and 934 test examples. It is a large-scale collection, and the samples are confusing and overlapping. In our experiments, only the most populous ten categories from this corpus are used as our dataset, i.e., transportation, sports, military, medicine, politics, education, environment, economy, art, and computer.

Three commonly used indicators of the evaluation of the classification performance [7] are the precision p, recall r, and F1-Measurea. The recall and precision rate reflects two different aspects of the quality of classification, but F1-Measurea considers not only the precision but also recall. So we used the F1-Measurea as the evaluation index in the experiment. To evaluate the performance of the algorithm on the whole dataset, we used the Macro F1 as the evaluation index. Assuming that F1-Measurea value of ith category is $F_1 i)$, the Macro F1 can be defined as

$$Macro - F_1 = \frac{\sum_1^n F_1(i)}{m} \tag{57.10}$$

In order to test the hybrid classification algorithm, we construct two experiments on the Fudan University classification corpus as follows:

1. Comparing the classification accuracy of the hybrid algorithm with the kNN and SVM classifier
2. Testing the performance of the hybrid algorithm when selecting different β value

In the first experiment, we select 2,000 words as features for each algorithm. For kNN the value of the parameter k is 35. For the proposed algorithm, the value of k-partition is 20 and the value of β is 0.3. For SVM, we used the LIBSVM package [8], which supports both two-class and multiclass classification. The F1 value of each algorithm on each category and the corresponding Macro F1 on each algorithm are presented in Table 57.1. The compute speed of each algorithm on the dataset is presented in Table 57.2.

In the second experiment, we tested the performance of the hybrid algorithm when the number of features is 2,000 and the value of β varies from 0.05 to 0.45 with step 0.05. The performance of the hybrid algorithm is presented in Fig. 57.1.

Table 57.2 Comparison of the cost time of each algorithm

SVM	kNN	Hybrid algorithm
16.28	53.45	18.72

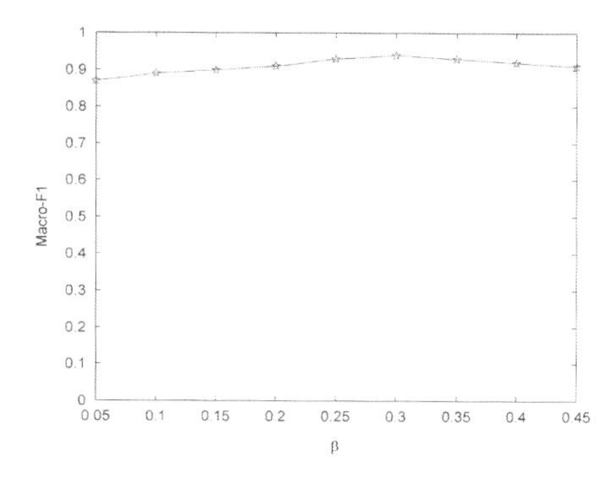

Fig. 57.1 The performance with the β

Table 57.1 shows that the F1 of proposed approach is slightly higher than that of other two algorithms on each category of the dataset; the Macro F1 of proposed approach is 92.97 %, which is approximately 1.07 % higher than that of kNN and 1.99 % higher than that of SVM. Table 57.2 shows that the time cost of kNN is the highest, that of the SVM is 16.28 which is lowest, and the cost time of hybrid algorithm is 18.72s which is about only 35 % of the cost time of kNN and close to that of SVM. So the hybrid algorithm is more efficient than kNN. From 1, we discovered a phenomenon that with the increase of β, the Macro F1 of proposed hybrid algorithm in the beginning increases and afterward decreases. When β takes 0.3, the proposed algorithm achieves the best results.

Above all, the proposed algorithm outperforms kNN and SVM in performance on the Fudan University classification corpus, and it is more efficient than kNN. Thus, the proposed approach is a good alternative for SVM algorithm and kNN in some scenarios of text classification such as dynamically mining large Web repositories where kNN is not suitable due to its lower efficiency and SVM is too sensitive to noise which easily leads to over-fitting.

57.5 Conclusion

In this chapter, two widely used techniques for text classification, i.e., the kNN and the SVM algorithm, are analyzed and some shortcomings of each are identified. Based on the analysis, a hybrid algorithm based on VPRS is proposed to combine the strengths of kNN and SVM classifier and overcome the problems of low efficiency of kNN and model over-fitting of SVM. Extensive experiments conducted on Fudan University classification corpus show that the hybrid algorithm achieves significant performance improvement.

Acknowledgements This study is supported by the National Natural Science Foundation of China (Grant No. 60474022).

References

1. Sebastiani, F. (2002). Machine learning in automated text categorization. *ACM Computing Surveys, 34*(1), 1–47.
2. Duwairi, R. (2005). An eager k-nearest-neighbor classifier for Arabic text categorization. In *Proceedings of the International Conference on Data Mining (ICDM'05), IEEE Computer Society Press, Houston* (pp. 187–192).
3. Vapnik, V. N. (1999). *The nature of statistical learning theory* (pp. 19–24). Berlin: Springer.
4. Xia, Y. S., & Wang, J. (2004). A one-layer recurrent neural network for support vector machine learning. *IEEE Transactions on Systems Man and Cybernetics Part B, 34*(2), 1261–1269.
5. Pawlak, Z. (1982). Rough sets. *International Journal of Information and Computing Science, 32*(11), 341–356.
6. Ziarko, W. (1993). Variable precision rough set model. *Journal of Computer System Science, 46*(1), 39–59.
7. Estévez, P. A., Tesmer, M., Perez, C. A., & Zurada, J. M. (2009). Normalized mutual information feature selection. *IEEE Transactions on Neural Networks, 20*(2), 189–201.
8. Hsu, C. W., & Lin, C. J. (2002). A comparison of methods for multi-class support vector machines. *IEEE Transactions on Neural Networks, 13*(2), 415–425.

Chapter 58
Robust Fragment-Based Tracking with Online Selection of Discriminative Features

Yongqiang Huang and Long Zhao

Abstract In order to solve the variation of target appearance and background influence to the visual tracking, we extend the robust fragment-based tracker to an adaptive tracker by selecting features with an online feature ranking mechanism, and the target model is updated according to the similarity between the initial and current models, which makes the tracker more robust. What is more, we reposition the integral histogram's bin's structure and that makes our tracker quicker. The proposed algorithm has been compared with fragment-based tracker, and the results proved that our method provides better performance.

58.1 Introduction

Visual tracking is an important task in computer vision and has been widely applied in traffic surveillance system [1], suspicious person monitoring system [2], etc. The problem in a visual tracker is the variation of target appearance and background. To improve the performance, an adaptive tracking mechanism is necessary. The normal idea is to adaptively select features of the object that can be discriminated from surrounding background.

In recent years, adaptive tracking algorithms have been widely studied [3–8]. Stern et al. [3] proposed an algorithm that chooses the best feature from five color spaces. Collins contrasted the foreground/background and selected the color features that can be best distinguished from the background [9]. Wang extended Collins's method by selecting reliable features from both color and shape–texture cues [10]. Chockalingam proposed an adaptive fragment-based

Y. Huang (✉) • L. Zhao
Science and Technology on Aircraft Control Laboratory, Beihang University, Beijing 100191, China

Digital Navigation Center, Beihang University, Beijing 100191, China
e-mail: buaa_huangyongqiang@foxmail.com

W.E. Wong and T. Zhu (eds.), *Computer Engineering and Networking*, Lecture Notes in Electrical Engineering 277, DOI 10.1007/978-3-319-01766-2_58,

tracking using level sets [11]. Leandro proposed a patch-based tracking and used simple updating scheme to cope with appearance and illumination changes; however, the model computed is not good enough for tracking [12].

In this chapter we use multiple image fragments to represent the object [13]; in addition, our fragment template is adaptively modeled by the most discriminative features, and the target model is updated according to the similarity between the initial and current models.

58.2 Feature Subset Selection

We choose color histograms to represent the color distributions in RGB and HSV color spaces. We omit the intensity because it is useless in our tracking.

After selecting the features, the log-likelihood ratio and variance ratio are used to select the most descriptive features. The pixel frequency can be calculated as $\xi_f^{(bin)} = H_f^{(bin)}/n_{fg}$ and $\xi_b^{(bin)} = H_b^{(bin)}/n_{bg}$, where H_f and H_b are histograms and nfg and nbg are the pixel number of the target and background, respectively.

The feature's log-likelihood ratio can be computed through Eq. (58.1)

$$L^{(bin)} = \log \frac{\max\left(\xi_f^{(bin)}, \sigma_L\right)}{\max\left(\xi_b^{(bin)}, \sigma_L\right)} \tag{58.1}$$

where σ_L is a small number (here 0.001).

And the variance ratio is computed through Eq. (58.2)

$$\mathrm{var}(L;p) = E\left[\left(L^{bin}\right)^2\right] - \left(E\left[L^{bin}\right]\right)^2 \tag{58.2}$$

Then the color features are ranked according to the discriminative ability by comparing the variance ratio. We assume that the most discriminative feature is corresponding to the maximum variance ratio.

58.3 Adaptive Robust Fragment-Based Tracking

58.3.1 The Fragment-Based Tracking Algorithm

In fragment-based tracking algorithm, the target template is represented by multiple histograms of multiple rectangular patches. By comparing each patch's histogram with the corresponding image patch histogram, its vote on the positions of the object in the image can be calculated, and a distance map describing the possible positions of each patch in current image is obtained. Combine the distance maps obtained from all template patches, and get the target position.

However, like standard meanshift tracking, the basic fragment-based tracking algorithm assumes that the target representation is good enough to discriminate the foreground from the background. However, this is not always reliable in a dynamic background tracking.

58.3.2 Integral Histogram

The fragment-based tracker needs to compute multiple histograms of multiple rectangular patches; the integral histogram is used to realize it in real time [13, 14].

Integral histogram extends from the integral image data structure. In the integral image, each pixel Pt holds the sum of all values of the relative image to the left and above of the pixel including the value of the pixel itself. After the image is computed, the sum of the pixels on any rectangular regions can be computed with four arithmetic operations. In order to extract histograms of any rectangular regions, each bin of the histogram (in the integral histogram) is built from an integral image counting the cumulative number of pixels falling into it. Then the number of pixels that belong to a bin in a given region can be easily computed through these integral images, and hence the histogram of that region is obtained.

Although extraction of a histogram over any patch is cheap, the experiments show that the integral histogram used in [13] is still computationally expensive. For an image of 352×288, the integral histogram data structure requires each bin one big image of that size in the histogram it takes about 500 ms (on a 2.0 GHz P2) to compute the integral histogram. Here we propose to develop a method which can reduce the computational cost. As the tracker searches in the neighborhood of the position estimate from the previous frame, we call this neighborhood area ROI; the region outside the ROI is useless; we just compute ROI's integral histogram to reduce the computational cost. See Fig. 58.1.

58.3.3 Target Localization

As mentioned in Sect. 58.2, we have ranked the features according to their discriminative ability from the background, and we will take advantage of the best two discriminative features to represent the target. The proposed tracking algorithm calculates the joint histogram ($P_f^{\left(b_{in}^1, b_{in}^2\right)}$) of the target with the best two features, replaces the patch's histogram in Frag (fragment-based tracking algorithm) with the joint histogram, then computes the similarity between the corresponding patches, and gains the distance map for each template patch.

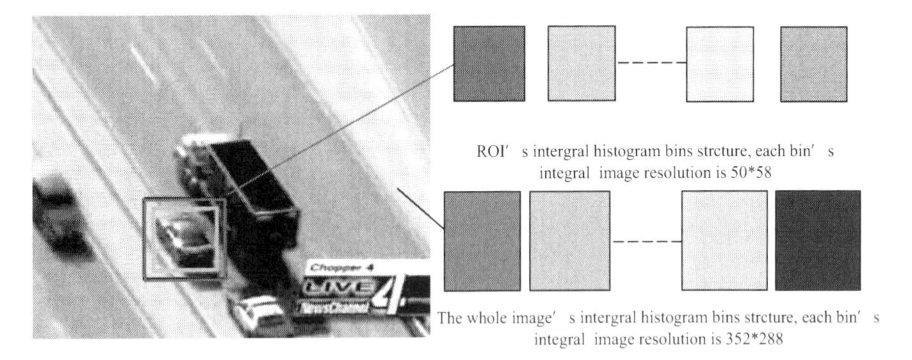

ROI' s intergral histogram bins strcture, each bin' s
integral image resolution is 50*58
The whole image' s intergral histogram bins strcture, each bin' s
integral image resolution is 352*288
Chopper 4
LIVE
NewsChannel
4

Fig. 58.1 Each color rectangle in the *right* of the figure represents one bin's integral image of the histogram (in this chapter Nbins = 8), and in the *left* of the figure, the region inside the *red rectangle* represents the ROI. Obviously, ROI's integral histogram computation complexity is much lower than the whole image's (Color online)

The distance map gives a value for every likely position of the target in the current frame, combines the distance maps obtained from all template patches, sums the distance maps, and gets the target position which has the minimal value.

58.3.4 Model Update

As the appearance of a target will always change during tracking, the target model needs to be updated.

We adopt the method proposed in [10] to update the target model, which involves the initial model, previous model, and current candidate. In the first frame, the initial target is labeled and its model M_i is computed. The initial model's weight is updated according to the similarity between the initial and current target. Before the ith update, the tracker searches the target in the current frame using the previous computed target model M_p^{i-1}. The new target model M_c^i can be computed by

$$M_c^i = (1 - S_{ic})M_i + S_{ic}M_{cp}^i \tag{58.3}$$

where Mi is the initial model; M_{cp}^i is the combined histogram of the current and the previous target model; and S_{ic} is the similarity between the initial and current target model:

$$M_{cp}^i = \left(1 - S_{cp}\right)M_p^{i-1} + S_{pc}M_c \tag{58.4}$$

where M_p^{i-1} is the previous target model; Mc is the histogram of current target model; and Scp is the similarity between the previous and the current target model, which is measured by Bhattacharyya coefficient.

58.4 Experimental Test and Result Analysis

In our experiment, the feature number K = 2, the histogram bin number N = 12, and the search radius R = 7; as two adjacent frame change is small, our feature selection runs every five frames.

We tested on three sequences ("woman," "bolt," and "Person"). The tracking results are compared with the Frag. See Figs. 58.2–58.4, where the red rectangles are the results of Frag and the green rectangles are ours. For the sequence bolt, the position errors of our tracker and the Frag are plotted in Fig. 58.5.

In Figs. 58.2 and 58.3, as the target appearance changes constantly, the Frag loses the target, while our tracker can still get it. The next sequence "Person" shown in Fig. 58.3 is under complete occlusion, and our tracker is still robust to it, whereas the Frag has lost the target. The Frag just uses the gray feature to represent the target, while our tracker uses the best subset of features; what is more, the Frag uses the same initial model all the tracking time, while we take into account the

Fig. 58.2 Woman

Fig. 58.3 Bolt

Fig. 58.4 Person

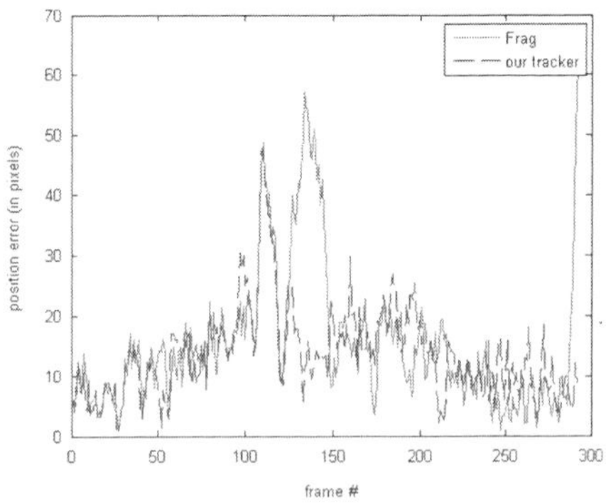

Fig. 58.5 Bolt sequence—error with respect to ground truth. Our tracker—*green*. Frag—*red* (Color online)

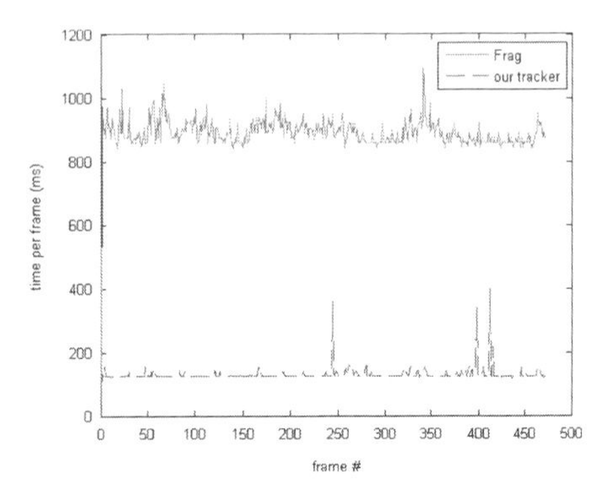

Fig. 58.6 Woman sequence—time per frame, our tracker—*red*, Frag—*green* (Color online)

similarity between the initial and current appearance of the target; this is robust to appearance variation.

Figure 58.6 shows the computing time of our tracker and the Frag during each frame, from which we can see that our tracker is less time consuming than Frag.

58.5 Conclusion

We extend the fragment-based tracker to an adaptive tracking algorithm by integrating color features adaptively. The target is represented by a joint histogram of the best two features. We also modify the integral histogram's structure, which can make our tracker quicker than the Frag. The proposed approach demonstrates good performance.

Acknowledgements Project supported by the key program of the National Natural Science Foundation of China (Grant No. 61039003), the National Natural Science Foundation of China (Grant No. 41274038), the Aeronautical Science Foundation of China (Grant No. 20100851018), and the Aerospace Innovation Foundation of China (CASC201102).

References

1. Lee, L., Romano, R., & Stein, G. (2000). Monitoring activities from multiple video streams: Establishing a common coordinate frame[J]. *IEEE Transactions on Pattern Analysis and Machine Intelligence, 22*(8), 758–767.
2. Haritaoglu, I., Harwood, D., & Davis, L. S. (2000). Real-time surveillance of people and their activities[J]. *IEEE Transactions on Pattern Analysis and Machine Intelligence, 22*(8), 809–830.
3. Chen, H. T., Liu, T. L., & Fuh, C. S. (2008). Probabilistic tracking with adaptive feature selection[C]. In *Pattern Recognition, Proceedings of the 17th International Conference, Roma* (pp. 736–739).
4. Collins, R. T., Liu, Y. (2003). On-line selection of discriminative tracking features[C]. In *Computer Vision, 9th IEEE International Conference on IEEE, Istanbul* (pp. 346–352).
5. Collins, R., Zhou, X., & Teh, S. K. (2005). An open source tracking testbed and evaluation website[C]. In *IEEE International Workshop on Performance Evaluation of Tracking and Surveillance, Munich* (pp. 17–24).
6. Comaniciu, D., Ramesh, V., & Meer, P. (2003). Kernel-based object tracking[J]. *IEEE Transactions on Pattern Analysis and Machine Intelligence, 25*(5), 564–577.
7. Wang, J., & Yagi, Y. (2006). Integrating shape and color features for adaptive real-time object tracking[C]. In *Robotics and Biomimetics, International Conference on IEEE, Sydney* (pp. 1–6).
8. Gevers, T., Smeulders, W. M., & W. A. (1999). Color based object recognition[J]. IEEE *Transactions on Pattern recognition*, 32(3), 453–464.
9. Collins, R., & Liu, Y. (2003). On-line selection of discriminative tracking features[C]. In *Computer Vision, 9th IEEE International Conference, Mamai, IEEE, 2003* (pp. 346–352).
10. Wang, J., & Yagi, Y. (2008). Integrating color and shape-texture features for adaptive real-time object tracking[J]. *IEEE Transactions on Image Processing, 17*(2), 235–240.
11. Chockalingam, P., Pradeep, N., & Birchfield, S. (2009). Adaptive fragments-based tracking of non-rigid objects using level sets[C]. In *Computer Vision, 12th International Conference on IEEE, Chengdu* (pp. 1530–1537).
12. Dihl, L., Jung, C. R., & Bins, J. (2011). Robust adaptive patch-based object tracking using weighted vector median filters[C]. In *Graphics, Patterns and Images (Sibgrapi), SIBGRAPI Conference on IEEE, Rio de Janeiro* (pp. 149–156).
13. Adam, A., Rivlin, E., & Shimshoni, I. (2006). Robust fragments-based tracking using the integral histogram[C]. In *Computer Vision and Pattern Recognition, Computer Society Conference on IEEE, New York* (pp. 798–805).
14. Porikli, F. (2005). Integral histogram: A fast way to extract histograms in Cartesian spaces [C]. In *Computer Vision and Pattern Recognition, Computer Society Conference on IEEE, San Diego* (pp. 829–836).

Chapter 59
Extraction Method of Gait Feature Based on Human Centroid Trajectory

Xin Chen and Tianqi Yang

Abstract Gait features obtained by current extraction methods are easily affected by people's walking direction, dresses, and carryings, due to which gait recognition system has not yet appeared. An extraction method based on centroid is proposed in this chapter. Segment and track the moving silhouettes of a walking figure in image sequences to calculate the silhouettes' centroid. The complex silhouette is represented by a point to avoid the influence of dresses and carryings. Divide centroid coordinate value by the height of detecting walking figure to normalize to remove the disturbance caused by walking direction relative to the camera optical axis angle. By denoizing centroid trajectory remove the noise caused by some accidental factors to obtain regular wavelet curve whose main frequency component distribution vector is the final gait feature. Experimental results show that this approach can obtain identical gait features even when experimenters change their walking directions, dresses, or carryings, tolerating noise and low resolution.

59.1 Introduction

Face recognition systems have been used widely around the world, but with the development of making up technology, face can no longer mark a human. Most criminals will shelter their faces from the cameras, and face recognition system's application is becoming narrower and narrower. Gait is influenced by muscle strength, length of tendon, bone density, and so on, which makes disguising gait very difficult.

X. Chen (✉) • T. Yang
Department of Information Science and Technology, Jinan University, Guangzhou 510000, China
e-mail: 843597029@qq.com

W.E. Wong and T. Zhu (eds.), *Computer Engineering and Networking*, Lecture Notes in Electrical Engineering 277, DOI 10.1007/978-3-319-01766-2_59,

Currently, there are mainly two kinds of method about gait feature extraction: one is establishing human models to describe gait with model parameters; the other is establishing a relationship between two neighboring frames by characteristics such as position, speed, and color [1]. Zhang, e.g., uses GVF to model the walking figures [2], making description of human contour possible. He judges current gait gesture by matching the models of current frames with those of key frames, considering cyclic features of gait fully. However, it is quite difficult to model a figure's outline because the human body is irregular. Wang, e.g., uses the distance between the centroid and outline to describe figure's contour to convert complex 2D images to simple 1D array, which increases the calculation speed greatly. But in this chapter centroid is just used as a static symbol instead of dynamical movement parameters. David Cunado takes trajectory of thigh's swing angle in walking as the gait feature [3]. The method successfully describes dynamic gait habits without contour extraction. However, the approach is feasible only when the pedestrian's leg outline is visual; it will lose worth when pedestrians wear gowns which cover their legs. Although there is technology which has improved the recognition rate greatly by extracting light stream information of image sequences [4], the cost is quite high because of complex algorithms' usage.

The former surveys revealed that human motion features can be reflected on the centroid trajectory. The overall process is that knee joint muscles make concession contraction from a person's one foot touching the ground, making the angle of the knee joint reduced and the leg shortened, the centroid showing a downward trend. In the driving process, all the joints and muscles of the leg involve restrained shrinkage to make the joint angles enlarge and the leg longer, the centroid showing an upward trend. Hao Ding and Junping Jiang had demonstrated that different gait owns different centroid swing arc or fluctuate curve, which proves that centroid is able to reflect essential features.

Considering human body as a nonhomogeneous object and each divided link as the uniformity, centroid can be calculated based on Varignon theorem without weighing the body which is expressed as a summary of all links. Centroid transforms gait period process to the clear centroid trajectory to avoid complex silhouette extractions and cycle tests, providing the theoretical basis for gait recognition towards commercialization.

59.2 Feature Abstract

59.2.1 Moving Target Detection

In this part, the main task is to detect the moving target from the original image sequences to get the two-value image sequences of pedestrians [5].

Figure 59.1a represents the original image obtained from the video database. It is firstly transferred into a single channel image and is smoothed by Gauss method.

Fig. 59.1 Examples of moving silhouette extraction: (a) An original image, (b) the background image constructed by three-frame difference method, (c) the extracted silhouette form

Then separate the moving target from the background (Fig. 59.1b) to get moving targets by three-frame difference method:

$$\begin{cases} |I_n(x) - I_{n-1}(x)| > T_n(x) \\ |I_n(x) - I_{n-2}(x)| > T_n(x) \end{cases} \quad (59.1)$$

where $I_n(x)$ represents domain value describing the gray change at position x of the nth image statistically, I_{n-1}(x) represents the domain value at position x of the (n−1) th frame, and $T_n(x)$ represents an experimental threshold value. When the gray value at position x has changed beyond the threshold value $T_n(x)$ relative to the previous frame I_{n-1}(x) and the frame I_{n-2}(x), the pixel is considered to belong to moving objects and is separated from the background. Repetitively operations are to obtain a series of foreground images. The pixel domain value in figure area is set to 1 and the remaining 0, by which we can get two-value images of walking figures (Fig. 59.1c).

59.2.2 Calculation of Human Body's Centroid

By centroid calculation formula (Eqs. (59.2) and (59.3)) these 2D silhouette changes are converted to an associated sequence of 1D signal. To eliminate the influences of spatial scale and signal length caused by walking direction, we normalize these centroid signals by dividing by the walker's silhouette height. This process is illustrated in Fig. 59.2.

The summation of target centroid can eliminate the impact of hollows. Variety laws of centroid can describe the essential gait features:

$$x_{w(i)} = \frac{1}{N} \sum_{x_i \in Area} x_i \quad (59.2)$$

$$y_{w(i)} = \frac{1}{N} \sum_{y_i \in Area} y_i \quad (59.3)$$

where N represents the number of pixels in the target area, x_i and y_i represent the X- and Y-coordinates of one pixel in the target area. Centroid coordinate is divided by the silhouette height of current frame to reduce the impact of the above factors; the formula is given by

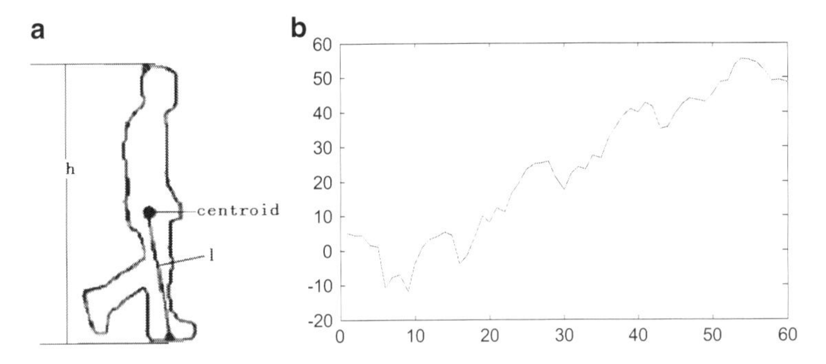

Fig. 59.2 Parameters used in calculation and centroid trajectory: (**a**) Position of centroid, distance of centroid to ankle, (**b**) tested centroid trajectory

$$h = y_{\max} - y_{\min} \tag{59.4}$$

$$gait_2 = \frac{gait_1}{h} \tag{59.5}$$

where h represents the current frame body height, $y_{\max}$ represents human target's maximum y-coordinate, $y_{\min}$ represents human target's minimum y-coordinate, $gait_1$ represents the initial centroid coordinate, and $gait_2$ represents normalized centroid coordinate. Finally construct a gait database storing the trajectories (Fig. 59.2b).

59.3 Wavelet Deionizing

The gait waveform is decomposed by the wavelet function to get high-frequency coefficients on every decomposition scale and low-frequency coefficients on maximum decomposition. The process can be achieved through the DWT () function in MATLAB Toolbox; the programming usage form is given by "[cAc, cDc] = Dwt (Yc, wname)," where "Yc" is the centroid trajectory vector, "wname" is the selected wavelet function's name, "cAc" is the low-frequency coefficient vector, and "cDc" is the high-frequency coefficient vector after decomposition. Refactor along with processed low-frequency components "cAc" to get the deionized results. Adjust the wavelet function, and repeat the above steps until the ideal result is obtained.

59.4 Spectral Analysis

If legs are rigid, distance of the heel to the centroid remains the same. When one of the legs touches the ground, the other leg starts to swing immediately. In the frontal perspective, the trajectory of body's centroid is a convex arc, with the heel as its center and the distance from the centroid to the heel as its radius [6]. The chart is shown in Fig. 59.3.

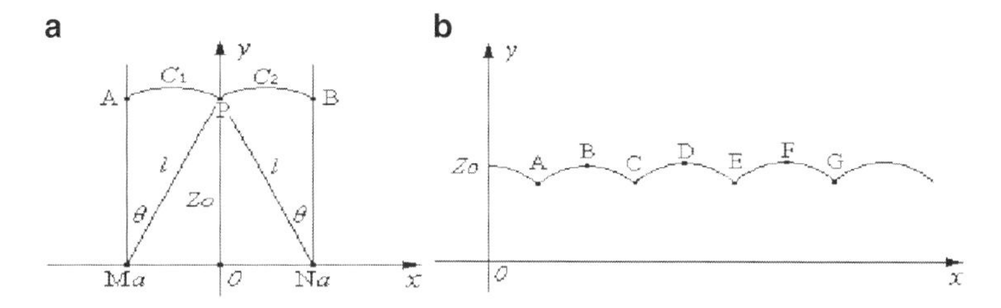

Fig. 59.3 Centroid's mathematical model: (**a**) From the frontal view, (**b**) from the lateral view

In the lateral view, centroid's trajectory is composed of fluctuations on the up and down direction and the walking direction. The one obtained in any perspective is composed of fluctuations on the up and down direction, the left and right direction, and the walking direction.

In Fig. 59.3a, direction of X-axis represents the left and right direction, and the positive direction of X-axis represents just the direction of people's right hand. Y-axis represents centroid's height. Curves C1 and C2 represent the transferring process from the state left foot is perpendicular to the ground to the state right foot does which takes point M and N as the center and distance l as the radius. In the arc, the boundaries of human body swinging on both sides is just when legs are perpendicular to the ground. Here $\theta = \angle AMP = \angle BNP = \arcsin a/l$ is the angle of swing on left and right direction. As Fig. 59.4a shows, AN and BM represent the moment that left and right legs become perpendicular to the ground (Fig. 59.3a). In Fig. 59.3b, points A, C, E, and G represent the lowest points of centroid positions, which are the moments that two feet alternate up and down. Points B, D, and F represent the highest points of centroid positions, which are the moments that a leg becomes perpendicular to the ground [7].

When the distance of heel to centroid is long, radius formed from centroid's swing will be relatively large; sine functions with greater cycle must be used in spectral analysis, which leads to appearance of peaks in the lower frequency band on spectrum map. However, fluctuations are mainly influenced by walking gestures. If the fluctuation amplitudes are large, amplitudes of peaks in every frequency band will be relatively large. Different human body characteristics and walking habits will show as different spectrum distributions.

Gait waveform is a continuous function x(t), but in fact only limit values can be collected during the limit time. Take x(t) as a continuous function whose cycle is T; then it can be expanded as a Fourier series; the exponential form is given by

$$x(t) = \sum_{K=-\infty}^{+\infty} C_K e^{j2\pi Kft} \tag{59.6}$$

$$C_K = \frac{1}{T}\int_{-T/2}^{T/2} x(t) e^{-j2\pi Kft} dt \tag{59.7}$$

where K = 0, ±1, ±2, and so on, f = 1/T is the fundamental frequency of x(t), and kf represents K harmonic frequency. C_k is the Fourier coefficient of x(t); it is a complex. $|C_k|$ is harmonic's amplitude. $|C_0|$ is the harmonic's average amplitude, and $|C_k|$ is known as the K harmonic amplitude.

We can just extract features from a spectrum map by segmentation to obtain data feature vector. First segment the X-axis representing frequency ranges into small bands; if there are peaks in one band, the quantization result is the number of peaks in the band; otherwise, it is 0. The combination of all the quantization values is the final quantization result.

59.5 Experimental Result

59.5.1 Result Analysis in Three Views

Experiments in this chapter are done on Dell personal computer with windows XP, MATLAB 7.0 as the simulating software, and C++ as the main programming language. Grouped by dress and carrying, the testers are divided into standard group, dress group, and burden group. Each group is measured from multi-angle. Record centroid trajectories of each group in each angle to construct a gait database.

Select 1,000 as the fundamental frequency, and simulate in the MATLAB to get results as shown in Fig. 59.4 (note: the spectrum is drawn $f = 500$ as a symmetry axis symmetrical; analysis of the frequency spectrum is limited to the range $f \leq 500$).

From Fig. 59.4a, there will be slight concussions in trajectory due to changes of walking ground, but the overall trend is the combination of fluctuations on the walking direction and the up and down direction, which on the spectrum map are two peaks on the band of 0–100 and 350–400.

In the frontal view, the trajectory is the combination of fluctuations on up and down direction and left and right direction. At this time it can be expressed by peak distribution and corresponding amplitudes on high-frequency band. As is shown in Fig. 59.5, the spectrum peaks allocate in each frequency band. By recording the peak positions we can get the walking figure's essential gait features.

In the perspective of 45°, measured trajectory is a superposition of all fluctuations on three directions. As is shown in Fig. 59.4, on 45° condition, spectrum is mainly composed of high-frequency components of fluctuations on left and right direction and low-frequency components of fluctuations on the up and down direction. The principal components distribute in the low-frequency band.

Figure 59.5 gives the frequency parameters of two persons in database in perspective of 45°. Select 50 as separation unit. The quantification result of Fig. 59.5a is "01,01,11,11,00" while that of Fig. 59.5b is "01,01,11,10,00." The Euclidean distance of the same person walking two times is 1; the quantification result of Fig. 59.5c is "11,01,11,10,10" while that of Fig. 59.5d is "11,01,11,10,10." The Euclidean distance that the same person walks two times is 0. In a word, the

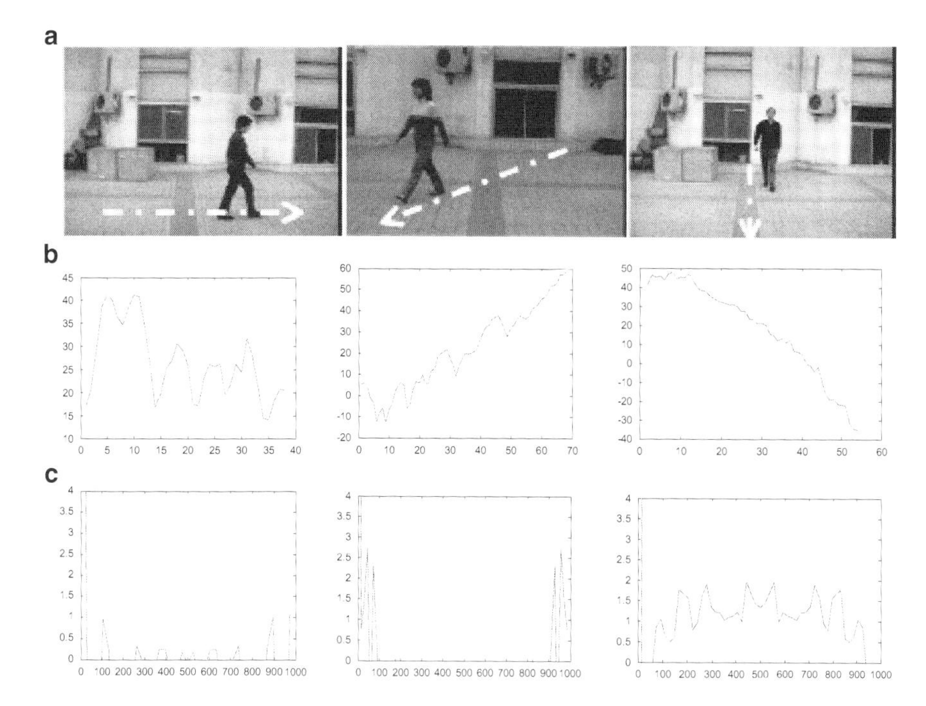

Fig. 59.4 Gait spectrum analysis: (**a**) Input sequences, (**b**) centroid trajectory, (**c**) spectrum map

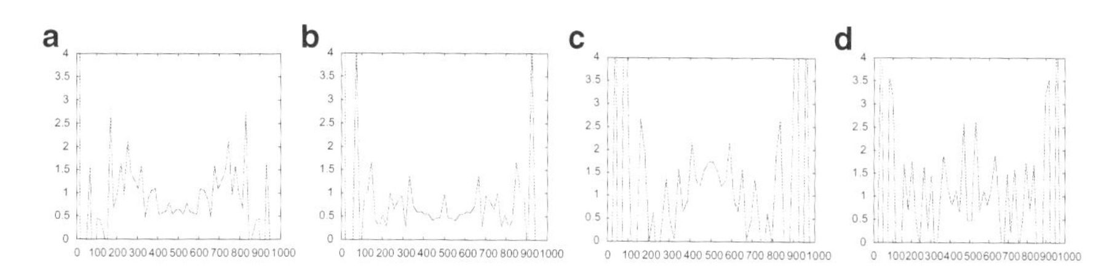

Fig. 59.5 Gait spectrum results: (**a**) First walking result of sixth person in NLPR database, (**b**) second walking result of sixth person in NLPR database, (**c**) first walking result of fifth person in NLPR database, (**d**) first walking result of fifth person in NLPR database

same person's Euclidean distance is very small, while different person's is large. This algorithm has a perfect distinction, and reducing the separating unit can further improve the ability of distinguishing.

59.5.2 Gait Analysis in Special Cases

In addition to the frontal view and lateral view, as centroid trajectory obtained from the rest angles are just cycle's and amplitude's overall enlargement or reduction compared with the 45° perspective, a strong unity can be detected in the spectrum

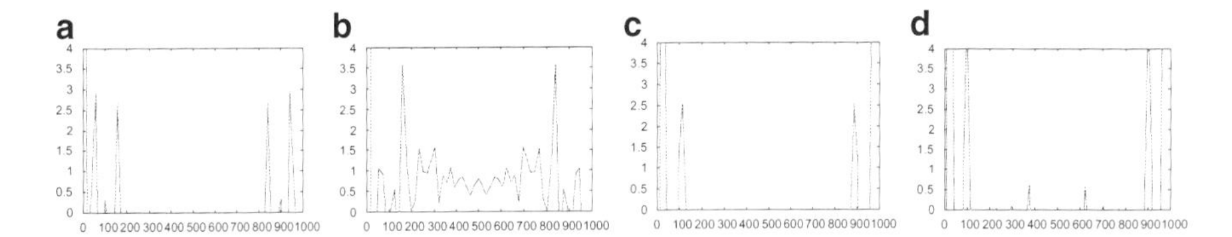

Fig. 59.6 Gait spectrum analysis: (**a**) Person A walking from 45° perspective in the lab, (**b**) person A walking from 135° perspective in the lab, (**c**) person B walking from 45° perspective in the lab, (**d**) person B walking from 135° perspective in the lab

map obtained in 45° view and any other views. Individual frequency parameters obtained in 45° and 135° views are given in Fig. 59.6.

Changes of perspective mainly influence the details of the centroid trajectory, which will cause disturbance. The overall extracted characteristic can still reflect essential gait features. As the figure in one frame has been compressed into a point, carrying and dress will just influence the overall coordinate values of centroid trajectory instead of variety laws. Spectrum variety laws remain the same when pedestrians change their carrying or dress.

59.6 Conclusion

This chapter presents a new gait extraction method. It does not need complex contour extractions and cycle detections compared with the prior methods. The experimental results demonstrated that this extracting method can reflect essential gait characteristics of humans. It is not affected by dress, carrying, and shooting angle. Noise and low resolution will not affect the recognition result.

References

1. Xu, D., & Huang, Y. (2012). Human gait recognition using patch distribution feature and locality-constrained group sparse representation[J]. *IEEE Transactions on Image Processing, 21*(1), 316–326.
2. Zhang, F., Zhang, X., & Cao, K. (2012). Contour extraction of gait recognition based on improved GVF snake model[J]. *Computers and Electrical Engineering, 38*(4), 882–890.
3. Cunado, D., & Nixon, M. S. (2003). Automatic extraction and description of human gait models for recognition purposes[J]. *Computer Vision and Image Understanding, 90*(1), 1–41.
4. Derawi, M. O., Ali, H., & Cheikh, F. A. (2011). Gait recognition using time-of-flight sensor[C]. In *Lecture Notes in Informatics (LNI), Proceedings-Series of the Gesellschaft fur Informatik (GI). Gesellschaft fur Informatik (GI), Ahrstrabe 45, 53175 Bonn, Germany* (pp. 187–194).
5. Yu, T., & Zou, J.-H. (2012). Research on gait recognition based on the combination of HMM and Bayes rules[J]. *Chinese Journal of Computers, 35*(2), 386–396 (In Chinese).

6. Ding, H., Jiang, J., & Li, Z. (2008). Establishment of human centroid trajectory mathematical model in walking. *Journal of Jiangsu Police Institute, 23*(6), 167–170. (In Chinese).
7. Peng, Z., Wu, X., & Yang, J. (2007). Multiview gait recognition algorithm based on the parameters of the limb length[J]. *Journal of Automation, 33*(2), 211–213. (In Chinese).

Chapter 60
An Algorithm for Bayesian Network Structure Learning Based on Simulated Annealing with Adaptive Selection Operator

Ao Lin, Bing Xiao, and Yi Zhu

Abstract In order to solve the problems that the intelligence algorithm falls into the local optimum easily and has a slow convergence in Bayesian networks (BN) structure learning, an algorithm based on adaptive selection operator with simulated annealing is proposed. This chapter conducts the adaptive selection rule in combination with conditional independence tests of BN nodes to guide the generation of neighbor. In order to better compare the adaptive effect, an algorithm based on selection operator with simulated annealing (SOSA) is proposed; at the same time 15 data sets in the three typical networks are accessed as learning samples. The results of the Bayesian Dirichlet (BD) score, Hamming distance (HD), and evolution time of the network after learning show that it has the quicker convergence and it searches the optimal solution more easily compared with simulated annealing (SA) and SOSA.

60.1 Introduction

As a graph model, BNs is an effective tool to deal with the uncertain problems in modeling and analysis, which is widely applied in many domains. Meanwhile, the learning problem of BNs is an important part of BN study. BN learning includes BN structural learning and parameter learning. Structure learning is needed to disclose the qualitative and the quantitative relationship between variables to light at the same time, while the BN structure learning is proved to be NP hard. Therefore, studying the BN structure learning problems is more challengeable and meaningful.

A. Lin (✉) • Y. Zhu
Department of Graduation Management, Air Force Early Warning Academy, Wuhan 430019, China
e-mail: lin_ao4035@163.com

B. Xiao
No. 4 Department, Air Force Early Warning Academy, Wuhan 430019, China

W.E. Wong and T. Zhu (eds.), *Computer Engineering and Networking*, Lecture Notes in Electrical Engineering 277, DOI 10.1007/978-3-319-01766-2_60,

In the investigation of BN structure learning, there are two classes of methods to deal with the structure learning problems [1]: The method of independence analysis and the method of score searching. The former determines whether there is border between the corresponding points or not by examining the conditional independence and dependence between variables, thereby establishing the skeleton of BN structure and orienting the border to get the BN structure. The latter is the method of score searching. For the network search space is of great extent generally, some BN structure learning adopts heuristic greedy algorithm, which tends to lead to the local optimum in the learning outcomes.

Nowadays, some researchers adopt the method of intelligence evolution to avoid the shortage of the heuristic algorithm [2, 3]. The SA algorithm is just the intelligence algorithm which is applied therein firstly, which is proved to be successful [4]. In contrast with the typical methods, it has a persistent evolution. In fact, examining the dependence and the independence between variables can reveal the variables' relational information which is camouflaged in data. Using this information can guide the evolution of the intelligence algorithm, thereby achieving the target of rapid convergence. ASOSA algorithm for the BN structure learning is proposed in this chapter.

60.2 BN Structure Learning

As a pictorial model that represents the joint probability distributions between variables, BNs include two parts: directed acyclic graph (DAG) and BN parameters. BN joint probability distributions can be decomposed as following through the independence relationship between variables contained in BN structure:

$$p(X_1, \cdots, X_n) = \prod_{i=1}^{n} p(X_i | \mathbf{Pa}_i, G) \tag{60.1}$$

wherein $\mathbf{Pa}_i$ refers to the father node of the variable X_i in BN structure (G). BN structure learning refers to obtaining the network structure which matches the sample data fitting best by analyzing a variety of samples, which is the focus of this chapter.

The method based on conditional independence test is efficient. But in certain cases, the conditional independence test order (the variables' number of the conditional set is the number of the conditional independence test orders) increases exponentially with respect to the number of the variables.

This chapter adopts the mutual information and conditional mutual information in the conditional independence test. $I(X_i; X_j)$ expresses the mutual information between variables X_i and X_j. According to the information theory, $I(X_i; X_j)$ is

$$I(X_i;X_j) = I(X_j;X_i) = \sum_{X_i,X_j} p(X_i,X_j)\log\frac{p(X_i,X_j)}{p(X_i)p(X_j)} \tag{60.2}$$

The mutual information is nonnegative. The more variables X_i and X_j incline to independence, the more $I(X_i; X_j)$ approaches 0.

The method based on score searching is another method of the BN structure learning; this method defines grading function S for each candidate network. Generally, S is posterior probability of the network. When assuming that BNs have equal a priori probability, the comparison of the posterior probability S of different BN structures is the comparison of the structure likelihood $p(\mathrm{D}|G)$. Cooper and Herskovits provided the calculating procedure of the structure likelihood:

$$p(\mathrm{D}|G) = \prod_{i=1}^{n}\prod_{j=1}^{q_i}\frac{\Gamma(\alpha_{ij})}{\Gamma(\alpha_{ij}+N_{ij})}\prod_{k=1}^{r_i}\frac{\Gamma(\alpha_{ijk}+N_{ijk})}{\Gamma(\alpha_{ijk})} \tag{60.3}$$

wherein Γ is gamma function, N_{ijk} is the number of cases in the dataset in which the parents of X_i are in state j and X_i itself is in state k, and q_i and r_i are the number of the parents of X_i and X_i in its own state separately. α_{ij} and α_{ijk} are the Dirichlet prior distributions. Equation (60.3) is also the famous Bayesian Dirichlet (BD) score function; the greater the structure score, the better the network structure.

60.3 Adaptive Selection Operator

The traditional SA algorithm of BN structure learning gets the neighbor by randomly selecting add, delete, or reverse the directed edges. This kind of operation is purposeless and ineffective. The zero-order independence information of the BN nodes can characterize the relationship between nodes substantially. So the information can be used to guide the generation of neighbor. The main thought is constructing the initial selection matrix with the nodes' zero-order information in the independence test. With the conducting of the annealing, effect of selection matrix is weakened gradually and selection matrix will not change any more when it meets certain conditions.

The selection matrix can be classified into the dependence selection matrix and the independence selection matrix which are applied to the addition and the deletion of the BN edges, respectively.

The selection matrix can be gained in two steps:

Step 1: Initialization. Set the coefficient of renovation k (k > 1). Calculate the mutual information of the two different random variable $C^0_{ij} = I(X_i; X_j)$, $i = 1 \cdots m - 1$, and $j = i + 1 \cdots m$; m is the number of the BN nodes. The matrix C^0 is the initial dependence selection matrix; the independence selection matrix and the dependence selection matrix have the opposite effects. In order to ensure the nonnegativity of the probability, $D^0_{ij} = -C^0_{ij} + \max(C^0_{ij})$, $i = 1 \cdots m - 1, j = i + 1 \cdots m$. D^0 is the initial independence selection matrix.

Step 2: Updating. $p = p + 1$. $C^p_{ij} = (C^{p-1}_{ij})^{1/k}$, $D^p_{ij} = (D^{p-1}_{ij})^{1/k}$, $p \geq 1$ when it meets the conditions; $C^p_{ij} = C^{p-1}_{ij}$, $D^p_{ij} = D^{p-1}_{ij}$, $p \geq 1$ when it does not meet the conditions.

In order to keep the selectivity of the selection matrix, the maximum of the number of updating can be set.

60.4 ASOSA Algorithm

For the SA algorithm of BN structure learning, the initialized network structure is empty. It looks for the better network locally through the operation such as randomly adding, deleting the edge, and converting the edge's direction. Lower the temperature gradually to look for the locally optimal network, until the suspense condition is reached. Adaptive selection operator with simulated annealing (ASOSA) algorithm has the same main body frame as the SA algorithm, but it uses the selection probability of the selection matrix instead of the random selection to operate the edges. The algorithm proposed in this chapter is presented in Algorithm 1.

The marking criterion of the algorithm based on ASOSA adopts the BD score. For SA algorithm seeking for the minimum, the score results should maintain the negative value only. The condition of the step 13 and 14 refers to $\Delta S \leq 0$, and the updating time does not reach the maximal time λ.

In order to compare the effectiveness of the ASOSA algorithm, an algorithm based on selection operator with simulated annealing (SOSA) is designed the selection matrix of which will not change in the annealing process.

60.5 Experimentation

The standard way of assessing the effectiveness of a learning algorithm is to draw samples from a known BN, apply the algorithm on the artificial data, and to compare the learned structure with the original one [3]. This chapter chooses three typical networks in different domains for the experiment: Asia network [5] (8 variables, 8 edges), insurance network [6] (27 variables, 52 edges), and alarm network [7] (37 variables, 49 edges). For the three experimental networks, datasets with 100, 200, 500, 1,000, and 5,000 samples were generated separately by applying Gibbs sampling.

Algorithm 1: ASOSA algorithm

Input: Set of learning data
Output: Bayesian network

1. Set the initial temperature T_0.
2. Set the minimum temperature T_{end}.
3. $\nu = 0.99$;//The descent velocity of temperature.
4. $\beta = 20$;//The maximal time of the outside loop.
5. $\sigma = 20$;//The maximal time of the inside loop.
6. Calculate the mutual information between any two variables to get C^0.
7. $D^0 = -C^0 + \max(C^0)$.
8. $k = 1.1$;//Coefficient of renovation.
9. $\lambda = 15$;//The maximal updating time of the selection matrix.
10. G = empty graph;//The candidate graph is initialized into empty graph.
11. $T = T_0$.
12. **Repeat**
13. **If** the conditions are met, update the selection matrix C and D; **end**.
14. **If** the conditions are not met, do not update the selection matrix; **end**.
15. **For** σ times **do**.
16. Add one edge and reduce one edge to G, according to the selection matrix C and D, and reverse the direction of one of the directed edges in G randomly.
17. Calculate the score difference ΔS coming from the three operations above.
18. **If** $\Delta S > 0$ or $e^{(\Delta S/T)} > rand(0,1)$ **then** apply these actions to G; **end**.
19. **End**.
20. $T = T \times \nu$.
21. **Until** the maximal time of the loop or $T > T_{end}$ is obtained.
22. **Return** G.

60.5.1 Qualitative Analysis of Algorithms

Table 60.1 shows the BD score statistics to the learning results of the 15 experimental datasets from the three experimental networks by three algorithms (SA, SOSA, and ASOSA).

Hamming distance (HD) is an available approach to describe the difference between the network G after learning and the original network G^0. HD is the sum of excessive edge, deleted edge, and reversed edge. Table 60.2 shows the statistic results of the three different algorithms from the three different samples. It can be found from the statistic results of the HD that the performance of ASOSA is optimal and the posterior is SOSA.

Table 60.1 BD score for the learned structure (normalized for correct network score)

Sample	Method	Sample size 100	200	500	1,000	5,000	Avg.
Asia	SA	1.008	1.010	1.009	1.004	1.004	1.007
	SOSA	0.994	0.995	0.997	1.000	1.000	0.997
	ASOSA	0.994	0.995	0.997	0.999	1.000	0.997
	TRUE	1.000	1.000	1.000	1.000	1.000	1.000
	Empty	1.271	1.384	1.333	1.336	1.322	1.329
Insurance	SA	0.876	0.932	0.970	0.990	1.003	0.954
	SOSA	0.889	0.916	0.965	0.982	1.000	0.950
	ASOSA	0.886	0.927	0.964	0.982	0.998	0.951
	TRUE	1.000	1.000	1.000	1.000	1.000	1.000
	Empty	1.109	1.251	1.403	1.492	1.588	1.369
Alarm	SA	0.994	1.009	1.006	1.005	1.005	1.004
	SOSA	0.997	1.004	1.002	1.004	1.007	1.003
	ASOSA	0.990	1.003	0.996	0.999	1.000	0.998
	TRUE	1.000	1.000	1.000	1.000	1.000	1.000
	Empty	1.571	1.682	1.768	1.836	1.888	1.749

Table 60.2 HD for the learned structure

Sample	Method	Sample size 100	200	500	1,000	5,000	Avg.
Asia	SA	15	15	15	8	13	13.2
	SOSA	5	7	3	0	2	3.4
	ASOSA	5	3	3	2	2	3.0
Insurance	SA	53	43	31	37	22	37.2
	SOSA	49	29	25	24	23	30.0
	ASOSA	40	25	22	24	22	26.6
Alarm	SA	45	34	33	21	26	31.8
	SOSA	33	31	21	16	21	24.4
	ASOSA	25	20	18	13	15	18.2

Since the operation platforms of the algorithms are at variance, the scores should be normalized for the correct network score. As BD score is subtractive, the lower the normalized score, the better the network structure

It can be found from the learning results of the three networks that the normalized score of the SA algorithm is higher than the other two improved algorithms except the 100 samples of the insurance and alarm network learning, and the score of SOSA and ASOSA are superior to SA algorithm. In the learning, the score of ASOSA is no higher than SOSA in each group or on an average

60.5.2 *Constringent Analysis of Algorithms*

The runtime of algorithm can be the leading indicator to measure the algorithm efficiency, but the final results of the different algorithms are different. For convenience in comparison, take the time consumption of the BD final score of the SA

Table 60.3 Runtime (normalized for SA)

Sample	Method	Sample size 100	200	500	1,000	5,000	Avg.
Asia	SA	1.30	1.15	1.26	1.00	1.15	1.30
	SOSA	0.22	0.26	0.52	0.37	0.30	0.22
	ASOSA	0.19	0.19	0.33	0.26	0.15	0.19
Insurance	SA	1.15	1.00	1.10	1.13	1.47	1.17
	SOSA	1.04	0.63	0.55	0.67	0.71	0.72
	ASOSA	1.06	0.63	0.58	0.52	0.64	0.69
Alarm	SA	1.15	1.00	1.07	1.10	1.48	1.16
	SOSA	0.85	0.67	0.70	0.78	0.78	0.76
	ASOSA	0.62	0.55	0.60	0.72	0.78	0.65

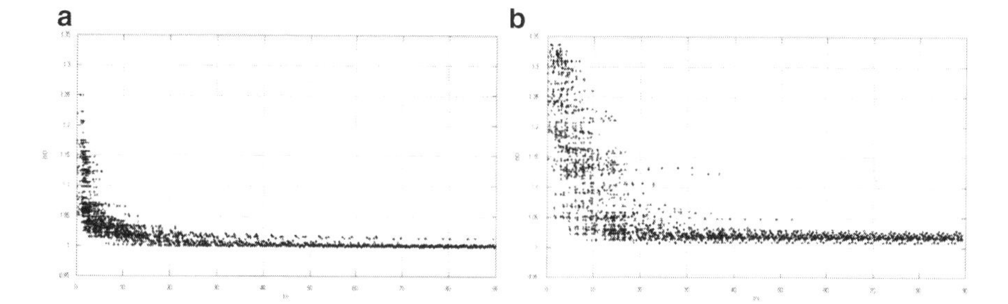

Fig. 60.1 (**a** and **b**) The performance record of ASOSA and SA

algorithm calculated by SOSA and ASOSA as the comparison object, and the result is normalized with the computation time of the SA, as shown in Table 60.3.

It can be found from the comparison of the runtime of the algorithms that the time of ASOSA is smaller than the other two algorithms, but there is little difference in SOSA and ASOSA.

In order to better compare the convergence of the algorithms, the performance process of the Asia network learning results from 500 samples which is obtained by ASOSA and SA is recorded. The cross axle is time, and the axis of ordinates is the BD score result normalized for the correct network score as shown in Fig. 60.1.

It can be found from the figure above that ASOSA converges rapidly. The searching directivity of the initial algorithm is conspicuous. The final convergence result is in the low level. The convergence process of the traditional SA is slow, and the final convergence result is worse than that of ASOSA.

60.6 Conclusion

This chapter introduces the ASOSA algorithm for BN structure learning, which fuses the independence analysis method and the score searching method of the BN structure learning, by the agency of the searching optimal network of the simulated annealing intelligence algorithm. The comparison of learning in three different samples from multi-aspect is among the algorithms ASOSA, SOSA, and SA. The result shows that the ASOSA is superior to SOSA and SA in the learning accuracy and the time consumption.

References

1. Singh, M., & Valtorta, M. (1995). Construction of Bayesian network structures from data: A brief survey and an efficient algorithm[J]. *International Journal of Approximate Reasoning, 12*(2), 111–131.
2. Wong, M. L., & Leung, K. S. (2004). An efficient data mining method for learning Bayesian networks using an evolutionary algorithm-based hybrid approach[J]. *IEEE Transactions on Evolutionary Computation, 8*(4), 378–404.
3. Pinto, P. C., Nagele, A., et al. (2009). Using a local discovery ant algorithm for Bayesian network structure learning[J]. *IEEE Transactions on Evolutionary Computation, 13*(4), 767–777.
4. Kirkpatrick, S., Gelatt, C. D., et al. (1983). Optimization by simulated annealing[J]. *Science, 220*(4598), 671–680.
5. Lauritzen, S. L., Spiegelhalter, D. J., et al. (1988). Local computations with probabilities on graphical structures and their application to expert systems[J]. *Journal of the Royal Statistical Society, Series B, 50*(2), 157–224.
6. Binder, J., Koller, D., et al. (1997). Adaptive probabilistic networks with hidden variables [J]. *Machine Learning, 29*(2–3), 213–244.
7. Beinlinch, I. A., Suermondt, H. J., et al. (1989). The ALARM monitoring system: A case study with two probabilistic inference techniques for belief networks[C]. In *Proc. 2nd Europ. Conf. on Artificial Intelligence in Medicine Care* (pp. 247–256). Berlin: Springer-Verlag.

Chapter 61
Static Image Segmentation Using Polar Space Transformation Technique

Xuan Luo, Tiancai Liang, and Weifeng Wang

Abstract This chapter proposes a polar space-based method to segment the static image automatically. The proposed method aims at segmenting the object of interest by finding the optimal closed contour in the polar space, solving the long-term problem of scale in the Cartesian space. Experimental results further verify and demonstrate the efficacy of the proposed polar space-based method on the challenging datasets.

61.1 Introduction

In computer vision literature, segmentation is the process of partitioning an image into disjoint, homogeneous, and compact regions where each part constitutes connected pixels with similar properties, such as brightness, color, and texture. Over the years, many algorithms [1–3] have been proposed for segmentation. Generally speaking, all the algorithms could be classified into three categories: (1) feature space-based techniques; (2) image domain-based techniques; and (3) physics-based techniques [4].

This chapter addresses the problem of separating the object of interest from a static image. We propose a segmentation framework that takes a point as its input and outputs the region containing that point, as shown in Fig. 61.1. Essentially, segmenting this region is equivalent to find the enclosing contour, which is a

X. Luo (✉)
School of Information Science and Technology, Sun Yat-sen University, Guangzhou 510006, China

Research Institute of GRG Banking Equipment Co., Ltd., Guangzhou 510006, China
e-mail: xuanluo@ieee.org

T. Liang • W. Wang
Research Institute of GRG Banking Equipment Co., Ltd., Guangzhou 510006, China

W.E. Wong and T. Zhu (eds.), *Computer Engineering and Networking*, Lecture Notes in Electrical Engineering 277, DOI 10.1007/978-3-319-01766-2_61,

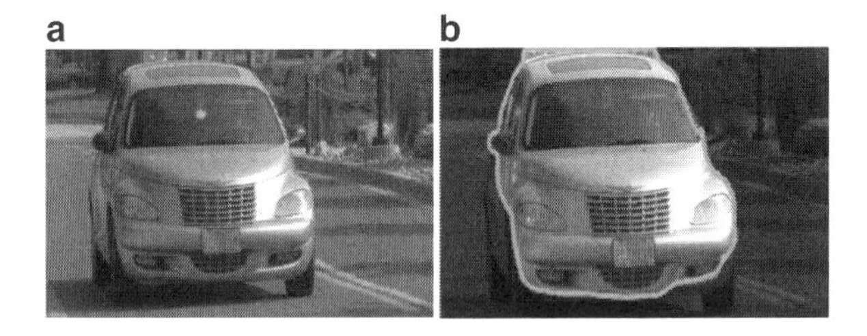

Fig. 61.1 (**a**) The image with an input point. (**b**) The final segmentation given by the proposed approach

connected set of boundary edge fragments in the edge map, around the input point. The edge map is generated by using all available visual cues.

The proposed algorithm framework is a four-step process: First, the edge map of the image is generated using low-level cues; second, the edge map is transformed from Cartesian space to polar space with the input point as the pole; third, the "optimal" path through this transformed edge map is found; fourth, the path is mapped back to an enclosing contour. Figure 61.1b shows the final result.

61.2 The Significance of Polar Space

At present, any segmentation algorithm, which finds the "optimal" enclosing contour through edge map in the Cartesian space, usually prefers the smaller contours as the overall cost, which is defined as the product of the length of the closed contour and the average cost of tracing the edge pixel along the contour, increases with the contour size [3]. Take Fig. 61.2a for example; we intend to find the optimal contour for the red point in it. Figure 61.2b shows the gradient edge map of the disc. Apparently, the small circle is just the internal edge, while the big circle is the actual boundary of the disc. Let us make an assumption: (1) The edge map assigns the internal contour intensity 0.33 and the boundary contour 0.66; (2) the numbers of the pixels of the two circles are 100 and 250, respectively. According to the definition of the overall cost, we can get the cost of internal contour 67 and boundary contour 85. Unfortunately, the former one would be considered the "optimal" enclosing contour as a result of its lower cost.

Focus on dealing with this "the shorter, the better" problem, we propose to transform these contours from Cartesian space to polar space, where the lengths of these contours no longer depend on the area they enclose. In the polar space, the cost of tracing these contours would be independent of their scales in the Cartesian space. The transformed edge map with the red point as a pole is shown in Fig. 61.2c; both contours are changed into open curves, arranging from 0° to 360°. In this way, the costs of tracing the internal contour and the external contour become 242 and 124. Clearly, the actual boundary contour becomes the optimal contour logically.

For the new input point (the green one) in Fig. 61.2a, although both the corresponding contours have changed shape (see Fig. 61.2d), the brighter one still remains optimal.

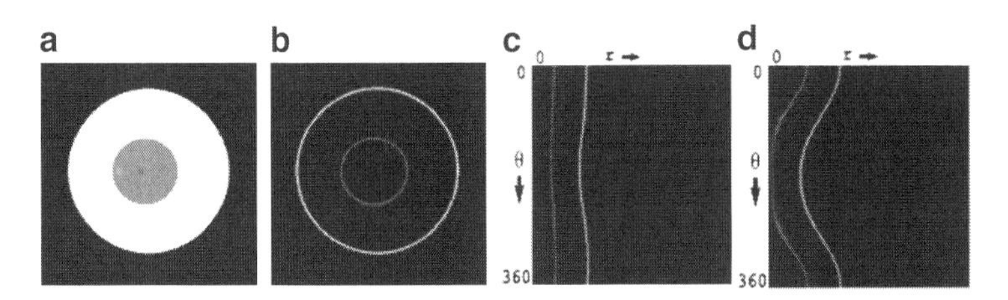

Fig. 61.2 (**a**) The mentioned disc. (**b**) The gradient edge map. (**c**) The polar edge map corresponding to the *red point*. (**d**) The polar edge map corresponding to the *green point*

61.3 Generating the Edge Map by Combining Cues

In this section, we explicate the first step of the proposed segmentation algorithm: generating the edge map where the boundary edges (the actual boundary) are much brighter than the internal edges. In most scenes, the static visual cues such as color, intensity, or texture can precisely locate the edges. In our framework, we use the Berkeley edge detector [5], which learns the color and texture properties of the boundary pixels versus the internal pixels from a dataset containing human-labeled segmentations of 300 images, to generate our initial boundary edge map, as shown in Fig. 61.3b. This edge detector handles texture much better than any intensity-based edge detectors. We can see that the unauthentic texture edges have been basically removed while the boundary edges are bright.

However, some internal edges (see BC, CD, and DF in Fig. 61.3b) still remain bright. In order to separate the boundary edges from internal edges, we can use the motion cue sufficiently. It is known that the optical flow value changes significantly at the boundary of an object while substantially remaining unchanged inside an object. Based on this concept, we can modify the edge map so that the edge pixels with strong gradient of optical flow values are stronger than the ones with weak gradient. We break the initial edge map into straight line segments and select rectangular regions of width α at a distance β on its both sides (see FC and FA in Fig. 61.3c). Then, we calculate the average flow inside these rectangles. The difference in the magnitude of the average flow on both sides is the standard measurement of the probability of the segment to be boundary edge. The brightness of an edge pixel on the segment is changed as

$$I^{'}(x,y) = \lambda I(x,y) + (1-\lambda)\Delta f/max(\Delta f) \qquad (61.1)$$

Δf represents the change in optical flow; λ is the weight related to the relative importance of the static visual cue-based boundary estimate. Figure 61.3d shows the final boundary edge map where the internal edges are clearly fainter and the boundary edges are brighter.

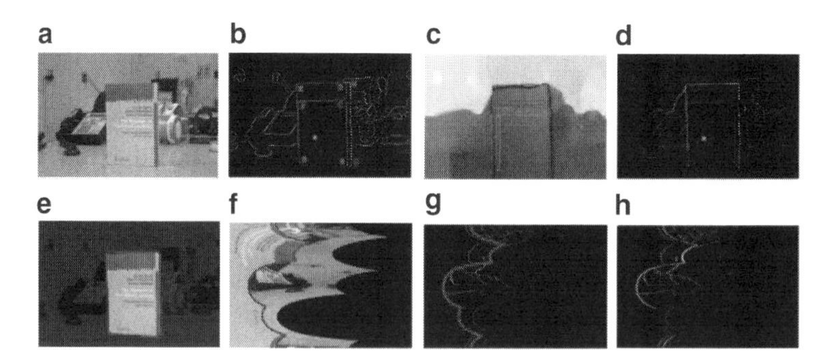

Fig. 61.3 (**a**) An example of the input image. (**b**) The Berkeley edge map. (**c**) The magnitude of optical flow field. (**d**) The boundary edge map combining the static visual cues with motion cue. (**e**) The final output by mapping back the optimal path to the Cartesian space. (**f**) The optimal cut dividing the polar image into two parts: *left* (inside the object) and *right* (outside the object). (**g**) The optimal contour overlapped on the polar edge map. (**h**) The corresponding polar edge map

61.4 Transforming from Cartesian Space to Polar Space

Let us say $I^p(.)$ is the corresponding polar edge map of the edge map $I^c(.)$ in the Cartesian space and $Q(x_0, y_0)$ is chosen as the input point. We can see that a pixel $I^p(r, \theta)$ in the polar coordinates corresponds to a pixel location $\{I^c(x, y) : x = r \cos \theta + x_0, y = r \sin \theta + y_0\}$ in the Cartesian space. Adopting the method of bilinear interpolation, which only considers four immediate neighbors, we can figure out $I^c(x, y)$ directly.

We propose to generate a continuous 2D function $F(.)$ by putting 2D Gaussian kernel functions on every edge pixel and aligning the major axis of those Gaussian kernel functions with the orientation of the edge pixel. Let S be the set of all edge pixels. The intensity at any pixel location (x, y) in the Cartesian coordinates is defined as

$$F(x, y) = \sum\nolimits_{i=S} exp\left(-\frac{x_i^t}{\sigma_{x_i}^2} - \frac{y_i^t}{\sigma_{y_i}^2}\right) * I^c(x_i, y_i) \quad (61.2)$$

where

$$\begin{bmatrix} x_i^t \\ y_i^t \end{bmatrix} = \begin{bmatrix} cos\ \theta_i & sin\ \theta_i \\ -sin\ \theta_i & cos\ \theta_i \end{bmatrix} \begin{bmatrix} x_i - x \\ y_i - y \end{bmatrix} \quad (61.3)$$

$$\sigma_{x_i}^2 = \frac{A_1}{\sqrt{(x_i - x_0)^2 + (y_i - y_0)^2}} \quad (61.4)$$

$$\sigma_{y_i}^2 = A_2 \quad (61.5)$$

θ_i is the orientation of the edge pixel i and A_1 and A_2 are constant and are chosen to be 900 and 4, respectively, in our experiments. On account of keeping the gray values of the edge pixels in the polar edge map the same as the corresponding edge pixels in the Cartesian edge map, we set the square of variance along the major axis, $\sigma_{x_i}^2$, to be inversely proportional to the distance between the edge pixel i and the pole Q.

The polar edge map $I^p(r, \theta)$ is obtained by sampling $F(x, y)$. The intensity values of $I^p(r, \theta)$ are scaled to the interval ranging from 0 to 1, which can depict the probability of an edge pixel being at boundary. Figure 61.3h displays the polar edge map corresponding to Fig. 61.3d. What is more, we come to an agreement that the angle $\theta \in [0°, 360°)$ is represented along the vertical axis and increases from top to bottom while the radius $r \in [0, r_{max}]$ varies along the horizontal axis and increases from left to right. r_{max} stands for the maximum Euclidean distance between two arbitrary pixels in the image.

61.5 Finding the Optimal Cut Through the Polar Edge Map

Let us regard every pixel $p \in P$ of I^p as a node in a graph. Every node (pixel) is connected with their four immediate neighbors (see Fig. 61.3). A row of the graph represents the radiation from the input point at an angle (θ) equal to their row number. The first and the last rows are the rays $\theta = 0°$ and $\theta = 360°$, which are exactly the same in the polar space. So, the pairs of nodes $\{(0°, r),(360°, r)\}$ and $\forall\, r \in [0, r_{max}]$ could be connected by edges in the graph. Let us say $l = \{0, 1\}$ are the two possible labels for each pixel, where $l_p = 0$ indicates inside the object and $l_p = 1$ indicates outside the object. γ denotes the set of all the edges between neighboring nodes in the graph. Evidently, we aim at assigning a label l_p to every pixel p (i.e., finding the mapping $M(p) \to l$), which corresponds to the minimum energy. The energy function is

$$G(M) = \sum_{p \in P} \phi_p(l_p) + \mu \sum_{(p,q) \in \gamma} \varphi_{p,q} . \delta(l_p, l_q) \tag{61.6}$$

$$\varphi_{p,q} = \begin{cases} exp\left(-\tau I_{pq}^p\right) & \text{if } I_{pq}^p \neq 0 \\ C & \text{otherwise} \end{cases} \tag{61.7}$$

$$\delta(l_p, l_q) = \begin{cases} 1 & \text{if } l_p \neq l_q \\ 0 & \text{otherwise} \end{cases} \tag{61.8}$$

$$I_{pq}^p = \left(I^p(r_p, \theta_p) + I^p(r_q, \theta_q)\right)/2 \tag{61.9}$$

The cost of assigning a label l_p to the pixel p is denoted by $\phi_p(l_p)$, and the cost of assigning different labels to the neighboring pixels p and q is denoted by $\varphi_{p,q}$.

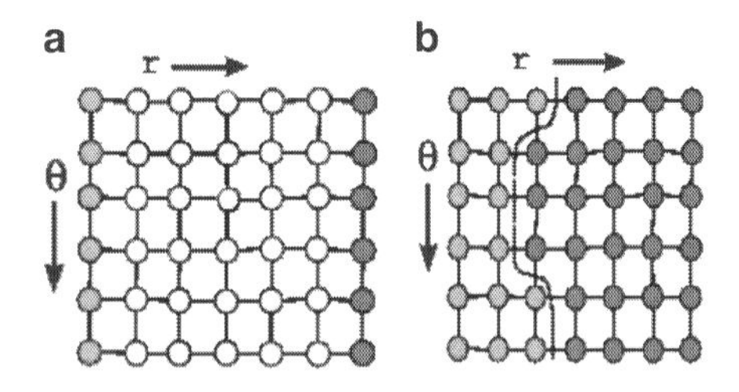

Fig. 61.4 (**a**) The *green* nodes in the first column are initialized to be inside the object, while the *red* nodes in the last column are initialized to be outside the object. (**b**) The binary labeling output after minimizing the energy function by means of graph cut algorithm

At the beginning of this process, we assign value 0 to the data term $\phi_p(l_p)$ for all the nodes except those in the first column and the last column:

$$\phi_p(l_p) = 0, \quad \forall p \in (r, \theta), \quad r \in (0, r_{max}), \quad \theta \in [0^\circ, 360^\circ) \tag{61.10}$$

Meanwhile, the nodes in the first column (corresponding to the input point in the Cartesian space) must be inside the object and are assigned label 0:

$$\phi_p(l_p = 1) = Z, \quad \phi_p(l_p = 0) = 0, \quad p \in (0, \theta), \quad \theta \in [0^\circ, 360^\circ) \tag{61.11}$$

Similarly, the nodes in the last column must be outside the object and are assigned label 1:

$$\phi_p(l_p = 0) = Z, \quad \phi_p(l_p = 1) = 0, \quad p \in (r_{max}, \theta), \quad \theta \in [0^\circ, 360^\circ) \tag{61.12}$$

Shown in Fig. 61.4. Z must be a high value in order to make sure that the initial labels to the first and the last columns could not change in the minimization step. To find the optimal path in the polar edge map, we apply the planar graph cut algorithm [6] to minimize the energy function $G(M)$, given in Eq. (61.6).

The global optimization process divides the polar edge map into two parts: left side (inside the object) and right side (outside the object), see Fig. 61.3f, g. Finally, we map the resulting binary segmentation back to the Cartesian space to obtain the desired result. The boundary between the left (label 0) and the right (label 1) parts in the polar space corresponds to the closed contour around the input point in the Cartesian space, see Fig. 61.3f, e.

61.6 Experimental Results

We evaluated the performance of the proposed algorithm on 100 image pairs along with their ground-truth segmentation, which is created by identifying and segmenting the most outstanding object of interest for each image pair manually.

Table 61.1 The performance of the proposed method for the test images

For test images	F-measure
Without motion	0.65 ± 0.02
With motion	0.94 ± 0.01

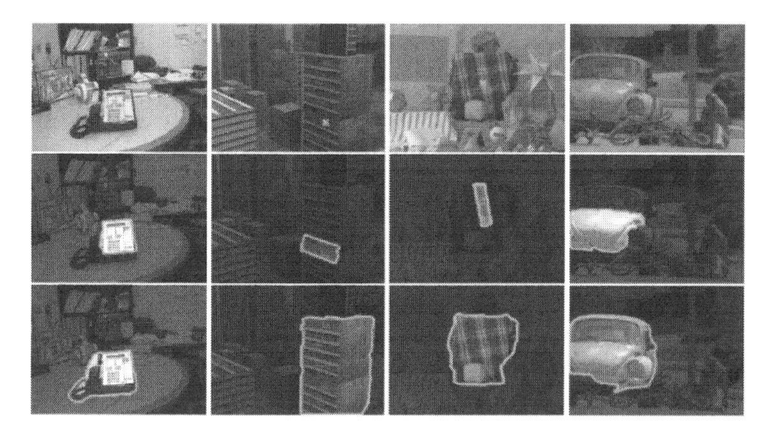

Fig. 61.5 *Row 1*: The original images with an input point. *Row 2*: The segmentation using the static visual cues only. *Row 3*: The segmentation for the same input point after combining static visual cues with motion cue

The final segmentation of our algorithm is compared with the ground-truth segmentation according to the empirical F-measure:

$$\Omega = \frac{2PR}{P + R} \tag{61.13}$$

P symbolizes the precision which calculates the score of our segmentation overlapping with the ground-truth segmentation, and R symbolizes the recall which calculates the score of the ground-truth segmentation overlapping with our segmentation.

See Table 61.1; there has been considerable improvement in the proposed algorithm after combining static visual cues with motion cue. With the visual cues only, the actual boundary contour is not drawn precisely owing to the strong internal edges (see row 2 of Fig. 61.5). Nevertheless, the affiliation of optical flow cue makes the internal edges fade away, and the correct contour is found (see row 3 of Fig. 61.5).

On the other hand, we evaluate the performance of the proposed algorithm by comparing it with state-of-the-art methods [1, 2, 7, 8]. We use the public Alpert image database [8] for our comparative experiments. We perform better than [1, 2] and extraordinarily close to [7, 8], as clearly described in Table 61.2.

Table 61.2 The performance of the proposed method compared with other methods

Algorithm	F-measure
Bagon [7]	0.87 ± 0.010
Alpert [8]	0.86 ± 0.012
Ours	0.84 ± 0.018
NCut [2]	0.72 ± 0.012
MeanShift [1]	0.57 ± 0.023

61.7 Discussion and Conclusion

In this chapter, we have presented an efficient image segmentation algorithm. The framework cleverly uses all the visual cues to distinguish the internal contours from the actual boundary contours and then transforms the edge map from Cartesian space to polar space in order to find the optimal segmentation. As the future extension of this work, the algorithm can be used to segment large number of images, and the extracted regions can be studied for the high-level processes.

References

1. Tu, Z. W., & Zhu, S. C. (2002). Mean shift: A robust approach toward feature space analysis. *IEEE Transactions on Pattern Analysis and Machine Intelligence, 24*(5), 603–619.
2. Shi, J. B., & Malik, J. (2000). Normalized cuts and image segmentation. *IEEE Transactions on Pattern Analysis and Machine Intelligence, 22*(8), 888–905.
3. Mishra, A. K., Aloimonos, Y., Cheong, L. F., & Kassim, A. A. (2012). Active visual segmentation. *IEEE Transactions on Pattern Analysis and Machine Intelligence, 34*(4), 639–653.
4. Luccheseyz, L., & Mitray, S. K. (2001). Color image segmentation: A state-of-the-art survey. *Proceedings of the Indian National Science Academy (INSA-A), 67*(2), 207–221.
5. Martin, D., Fowlkes, C., & Malik, J. (2004). Learning to detect natural image boundaries using local brightness, color and texture cues. *IEEE Transactions on Pattern Analysis and Machine Intelligence, 26*(5), 530–549.
6. Boykov, Y., & Kolmogorov, V. (2004). An experimental comparison of min-cut/max-flow algorithms for energy minimization in vision. *IEEE Transactions on Patte:n Analysis and Machine Intelligence, 26*(3), 359–374.
7. Bagon, S., Boiman, O., & Irani, M. (2008). What is a good image segment? A unified approach to segment extraction. In ECCV (Vol. 5305(3), pp. 30–44). Heidelberg: Springer.
8. Alpert, S., Galun, M., Basri, R., & Brandt, A. (2007). Image segmentation by probabilistic bottom-up aggregation and cue integration. In *Computer Vision and Pattern Recognition* (Vol. 0, No. 2, pp. 1–8).

Chapter 62
Image Restoration via Nonlocal *P*-Laplace Regularization

Chen Yao, Lijuan Hong, and Yunfei Cheng

Abstract Image restoration technology can be applied in a lot of fields including image communication, image archive restoration, and image editing. In this chapter, we try to solve image restoration with a nonlocal regularization point of view. Similarity between different image pixels is measured by a nonlocal p-Laplace operator. We use minimum least square with a regularization term to formulate the whole procedure of image restoration. In the solvent of cost function, a linear Gauss–Jacobi iterative method is utilized for unknown pixel solvent. The complexity of iterative solvent is controlled by a step threshold. Finally, experimental results highlight our superior performance over previous methods from subject visual perception or object image quality assessment.

62.1 Introduction

In consumer electronic application, image is often degraded by cameral sensor noise or error-prone communication channel. Image restoration is a technology which uses non-corrupted image region to recover corrupted pixels. With the application of image and video, image restoration problems are bringing up more and more concerns [1, 2]. A lot of researches have been carried on image restoration. Interpolation-based methods are adopted for finding good model of natural images [3–5]. Anisotropic filtering, partial differential equation, total variation, and image decompositions on fixed bases such as wavelets are also used to implement image restoration [6–9]. Generally, a smoothness assumption is built for these technologies. More recently, the research on image self-similarity is carried on nonlocal filtering, learned sparse model, sparse dictionary learning, Gaussian scale mixture, fields of experts, and block matching with 3D filtering

---

C. Yao (✉) • L. Hong • Y. Cheng
The Third Research Institute of Ministry of Public Security, Shanghai 200031, China
e-mail: yaochensing@126.com

W.E. Wong and T. Zhu (eds.), *Computer Engineering and Networking*, Lecture Notes in Electrical Engineering 277, DOI 10.1007/978-3-319-01766-2_62,

(BM3D) [10–15]. The image self-similarity technology brings more motivation for image restoration.

In this chapter, inspired by forerunners [16, 17], we present a new image restoration method based on nonlocal p-Laplace regularization scheme. In this chapter, image is viewed as discrete graph. Every pixel is regarded as graph vertices. The similarity between different vertices is represented by a nonlocal weight. A p-Laplace operator is defined as an operation of vertex Hilbert space. The diffusion of discrete vertices is formulated with an energy minimum function. For a given energy minimum cost function, a Gauss–Jacobi iterative method is used for unknown variable solvent. A constant threshold is selected as algorithm stop criterion.

In the rest part of this chapter, the energy minimum solving framework, nonlocal weight, and iterative solving method are reviewed and introduced in Sect. 62.2. Abound experimental results are provided in Sect. 62.3. And finally, Sect. 62.4 concludes the chapter.

62.2 Proposed Algorithm

In this section, the energy minimum framework, Gauss–Jacobi iterative method, nonlocal weight, and iterative solvent are introduced.

62.2.1 Energy Minimum Framework on Weighted Graphs

In this section, we use an energy minimum framework to describe pixel diffusion processes. Let $H(V)$ be the Hilbert space of real-valued functions defined on the vertices of a graph. Function T is a mapping from discrete vertices to Euclidian space. Function S is a general function defined on graphs of the arbitrary topologies. The regularization of function S is formulated as following energy minimum framework:

$$\min_{T \in H(v)} \left\{ E(T, S, \lambda) = R(T) + \frac{\lambda}{2} \|T - S\|^2 \right\}, \tag{62.1}$$

where parameter λ is a Lagrange multiplier, which specifies the trade-off between two competing terms. And, regularization control factor $R(T)$ is denoted as

$$R(T) = \frac{1}{p} \sum_{u \in v} |\nabla_w T(u)|^p, \tag{62.2}$$

where $R(T)$ is a p-Laplace operator on weighted Hilbert space. In the formulation of $R(T)$, $\nabla_w T(u)$ is detailed as

$$\nabla_w T(u) = \frac{1}{2}\sum w(u,v)(T(u) - T(v)). \tag{62.3}$$

where u, v is denoted as different vertices. And, $w(u,v)$ is the edge weight between u and v. The definition of $w(u,v)$ is described in the following section. In this chapter, a nonlocal similarity metric is adopted as vertex weight in Hilbert space.

62.2.2 Nonlocal Weight Description

Inspired by Mahmoudi [16], we formulate edge weight in p-Laplace regularization factor using nonlocal-mean Gaussian description. The edge weight definitions are denoted as follows:

$$w(u,v) \propto \frac{1}{Z}\exp\left\{-\frac{\|P(u) - P(v)\|_{G_\sigma}^2}{h^2}\right\}, \tag{62.4}$$

where the parameter h controls the decay of exponential function. G_σ is model Gaussian variance. $P(u)$ is a 5×5 block centered on vertex u. Analogously, the definition of $P(v)$ is similar with $P(u)$. Z is normalizing factor. It is given by

$$Z = \sum_{u,\, v \in V} \exp\left\{-\frac{\|P(u) - P(v)\|_{G_\sigma}^2}{h^2}\right\}. \tag{62.5}$$

We can use this nonlocal weight to measure the similarity between current block and searching block. Obviously, the searching block is more similar to current block if the computed weight is bigger. Here, nonlocal weight is used to model correlation between different blocks in p-Laplace regularization factor. The introduction of nonlocal weight model is served as a key factor in the problem solving of pixel similarity metric.

62.2.3 Iterative Solvent for Energy Minimum Framework

A linear Gauss–Jacobi iterative method is utilized for solving Eq. 62.1. We set n as iteration step. Then, the iterative solving method is given by the following:

$$\begin{cases} T^{(0)}(u) = S(u) \\ T^{(n+1)}(u) = \dfrac{\lambda T^{(0)}(u) + \sum w^t(u,v) T^{(n)}(u)}{\lambda + \sum w^t(u,v)} \end{cases}. \tag{62.6}$$

where to get the convergence of the process, a classical stopping criterion is $\|T^{(n+1)} - T^{(n)}\| < \tau$. In our implementation, τ is a small fixed constant. At each iteration step, the updated value $T^{(n+1)}$ depends on the initial value and weighted average of the filtered value in neighborhood.

62.3 Experiments

In this section, we demonstrate the image restoration results of the proposed algorithm while providing its comparisons with some other image restoration algorithm [19, 20]. Two test color images with size 512×512 are used as input images in our implementation. Image reconstruction results under different image corruption masks are shown in Figs. 62.1 and 62.2. Objective image quality assessments are shown in Tables 62.1 and 62.2. PSNR [21] and structural similarity index (SSIM) [18] are used as evaluation tools. In Fig. 62.1, figure (a) is a original color image. Figure (b) is a color image with human-modified corruption. We use black region to represent image degeneracy. Figure (c) is the result of method [19]. Figure (d) is the result of method [20]. Figure (e) is our processed result. In order to highlight our performance, Fig. 62.2 gives results based on another corruption mask. Moreover, PSNR and SSIM are computed during the process of image restoration.

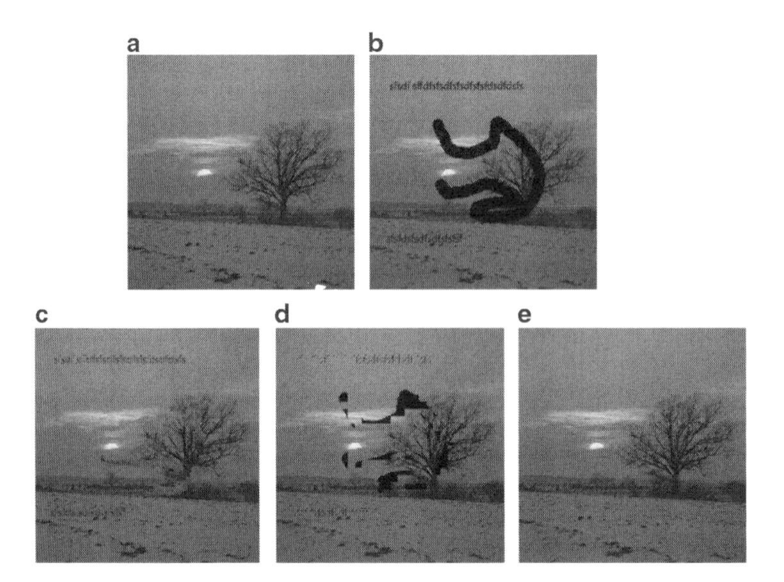

Fig. 62.1 Experimental results: (**a**) Original image. (**b**) Corrupted image. (**c**) Processed result with method [19]. (**d**) Processed result with method [20]. (**e**) Processed result with our proposed algorithm

Fig. 62.2 Experimental results: (**a**) Original image. (**b**) Corrupted image. (**c**) Processed result with method [19]. (**d**) Processed result with method [20]. (**e**) Processed result with our proposed algorithm

Table 62.1 Performance comparison in PSNR (Db)

Test image	Method [19]	Method [20]	Our proposed method
Fig. 62.1	11.16	25.50	47.17
Fig. 62.2	20.10	27.44	39.99

Table 62.2 Performance comparison in SSIM

Test image	Method [19]	Method [20]	Our proposed method
Fig. 62.1	0.28	0.88	0.99
Fig. 62.2	0.95	0.97	0.98

Table 62.1 illuminates the PSNR results as compared to Criminisi's and Li's [19, 20]. Table 62.2 illuminates the SSIM results as compared to Criminisi's and Li's [19, 20]. It can be seen that we achieve the highest PSNR and SSIM gain over previous algorithm. Subjective results shown in Figs. 62.1 and 62.2 also indicate that corrupted image processed by our method has good visual quality. We attribute the improvement of PSNR and SSIM to the use of nonlocal p-Laplace regularization scheme. The corrupted pixels can be well reconstructed by the most similar non-corrupted pixel. As compared to previous image restoration algorithm [19], our algorithm obtains both objective and subjective improvement. And our image restoration method does not depend on parameter adjustment either.

62.4 Conclusion

This chapter presented a novel image restoration algorithm built on nonlocal p-Laplace regularization strategy. We built an energy minimum framework by introducing nonlocal weight and Gauss–Jacobi iterative computation. Finally, the convergence speed was controlled by step threshold in iterative solvent. Experiments demonstrated the effectiveness and efficiency of nonlocal p-Laplace regularization image restoration algorithm.

Acknowledgements This work was supported by Science and Technology Innovation Foundation of Science and Technology Commission of Shanghai Municipality (12DZ0503300).

References

1. Awate, S. P., & Whitaker, R. T. (2006). Unsupervised, information-theoretic, adaptive image filtering for image restoration. *IEEE Transactions on Pattern Analysis and Machine Intelligence, 28*(3), 364–376.
2. Bertalmio, M., et al. (2000). Image inpainting. In *Proceedings of the 27th Annual Conference on Computer Graphics and Interactive Techniques* (Vol. 20(1), pp. 417–424). New York: ACM Press/Addison-Wesley Publishing Co.
3. Gunturk, B. K., Altunbasak, Y., & Mersereau, R. M. (2002). Color plane interpolation using alternating projections. *IEEE Transactions on Image Processing, 11*(9), 997–1013.
4. Paliy, D., et al. (2007). Spatially adaptive color filter array interpolation for noiseless and noisy data. *International Journal of Imaging Systems and Technology, 17*(3), 105–122.
5. Zhang, L., & Wu, X. (2005). Color demosaicking via directional linear minimum mean square-error estimation. *IEEE Transactions on Image Processing, 14*(12), 2167–2178.
6. Perona, P., & Malik, J. (1990). Scale-space and edge detection using anisotropic diffusion. *IEEE Transactions on Pattern Analysis and Machine Intelligence, 12*(7), 629–639.
7. Rudin, L. I., & Osher, S. (1994). Total variation based image restoration with free local constraints. In *Proceedings of the IEEE International Conference on Image Processing, 1994 (ICIP-94), Austin, TX* (Vol. 1, pp. 31–35).
8. Mallat, S. (1999). *A wavelet tour of signal processing* (pp. 321–345). Chicago: Academic Press.
9. Kim, S. (2006). PDE-based image restoration: A hybrid model and color image denoising. *IEEE Transactions on Image Processing, 15*(5), 1163–1170.
10. Buades, A., Coll, B., & Morel, J.-M. (2005). A non-local algorithm for image denoising. In *IEEE Computer Society Conference on Computer Vision and Pattern Recognition (CVPR), 2005* (Vol. 2, pp. 60–65).
11. Elad, M., & Aharon, M. (2006). Image denoising via sparse and redundant representations over learned dictionaries. *IEEE Transactions on Image Processing, 15*(12), 3736–3745.
12. Mairal, J., Elad, M., & Sapiro, G. (2008). Sparse representation for color image restoration. *IEEE Transactions on Image Processing, 17*(1), 53–69.
13. Portilla, J., et al. (2003). Image denoising using scale mixtures of Gaussians in the wavelet domain. *IEEE Transactions on Image Processing, 12*(11), 1338–1351.
14. Roth, S., & Black, M. J. (2005). Fields of experts: A framework for learning image priors. In *IEEE Computer Society Conference on Computer Vision and Pattern Recognition, CVPR, 2005* (Vol. 2, pp. 860–867).
15. Dabov, K., et al. (2007). Image denoising by sparse 3-D transform-domain collaborative filtering. *IEEE Transactions on Image Processing, 16*(8), 2080–2095.

16. Mahmoudi, M., & Sapiro, G. (2005). Fast image and video denoising via nonlocal means of similar neighborhoods. *IEEE Signal Processing Letters, 12*(12), 839–842.
17. Lezoray, O., Ta, V. T., & Elmoataz, A. (2008). Nonlocal graph regularization for image colorization. In *19th International Conference on Pattern Recognition (ICPR), 2008* (pp. 1–4).
18. Wang, Z., et al. (2004). Image quality assessment: From error visibility to structural similarity. *IEEE Transactions on Image Processing, 13*(4), 600–612.
19. Criminisi, A., Perez, P., & Toyama, K. (2003). Object removal by exemplar-based inpainting. In *IEEE Computer Society Conference on Computer Vision and Pattern Recognition, 2003. Proceedings, 2003* (Vol. 2, pp. 11–721).
20. Li, X. (2011). Image recovery via hybrid sparse representations: A deterministic annealing approach. *IEEE Journal of Selected Topics in Signal Processing, 5*(5), 953–962.
21. Hore, A., & Ziou, D. (2010). Image quality metrics: PSNR vs. SSIM. In *20th International Conference on Pattern Recognition (ICPR), 2010* (pp. 2366–2369).

Chapter 63
Analysis and Application of Computer Technology on Architectural Space Lighting Visual Design

Yiwen Cao

Abstract Based on "green building," we use computer to make model analysis regarding building actual environment, adapting natural light, and using artificial lighting rightly to achieve green energy-saving building goal. According to living example making, we adopt 3d max and ECOTECT model analysis software to analyze and simulate building room natural light, artificial lighting and distribution. Analysis states: at the stage of the architectural sketch, we use 3d max modeling, uniting inter-room facility draft and analysis real environment data by ECORECT, getting the natural light effect, and uniting the two sides to get a more accurate room light environment data. Through this supporting data, we can get better daylight design, not only to satisfy the in-room light requirement but also to make it more beautiful and energy saving.

63.1 Introduction

Light is the origin of life, which is also an essential substance to humans. With the rapid development of artificial lighting technology, what people demand now is not only the simple illumination but the colorful and charming world which caters to people's aesthetic taste. Artificial light applied to architectural space becomes even more important in contemporary architectural design. As the world's energy problem and environmental pollution are coming up, how to save electric lighting and make full use of natural light in the building space attracts the attention of international architecture and lighting industry. With the development and the popularization of digital technology, computer technology plays an important role in the lighting technology and art design, especially in solving the problem of light

Y. Cao (✉)
Digital Media Department, Huaruan Software College of Guangzhou University, Guangzhou 510000, China
e-mail: caoyiwen116@yeah.net

W.E. Wong and T. Zhu (eds.), *Computer Engineering and Networking*, Lecture Notes in Electrical Engineering 277, DOI 10.1007/978-3-319-01766-2_63,

wasting, light pollution, and light modeling. Computer technology helps to evaluate and predict the effect so that the work can be easily checked. The application of computer technology has a great influence on architectural space lighting visual design.

The green building we refer to means saving energy, but we usually forget building itself is the base of the green building; if we only consider how to change lighting saving into green building, the building itself is not "green." First, a good green building itself should be called a perfect energy-saving building and then it can make more details about the saving facility building. A good digital analysis technology can combine the fictitious room of the building, and good analysis of the actual environmental data makes the building achieve the "green" index.

There are many kinds of software combination of uniting digital technology building room daylight effect simulation and model analysis. According to the stylist's own ability, we can use 2D software, such as PS, to draw up real and model picture of the building. Or we can make 3D model simulation referring to the architectural drawings and the actual environment, and getting the frames through model light. We can use digital technology to reflect the state of the inner or outside room, structure of the room and the texture and frames of the decorative material. At the same time, using ECOTECT arcology building design software, we can get the room natural lighting coefficient, which is got from the lighting coefficient computer. When using natural lighting, in-room illumination is changing all the time with outdoor illumination. So when we want to make sure the in-room natural light illumination level, we should consider the outside room illumination.

63.2 The Application of Computer Technology to Architectural Space Lighting Visual Art

Computer technology is developing rapidly as well as graphics and image software. Computer technology combined with modern information and visual art forms a modern comprehensive art design which is based on computer technology but stresses on visual art; it is also becoming the dominant trend of modern art design. Nowadays, energy and environmental problems are extremely urgent to solve. People put forward a great many effective ecologic solutions to make use of natural light and save electric lighting. However, if a project is to be successful, it must be carried out through a serious course of survey, conceiving, confirming, modifying, design, and implementation. Some unreasonable designs and errors only appear in the course of implementation while others won't come up until the construction has been completed. At this time, it is extremely difficult to solve various problems. Therefore, we need a computer to perform simulation analysis in order to obtain a better effect of visual design. In the course of project design, taking advantage of simulation method, the computer evaluates the plan and predicts the result so as to directly check the project. Apply the computer technology to the building space visual design and perform the simulation programming and data analysis during the construction to form a set of complete digital simulation space, by modifying the design errors to reduce unnecessary waste and save the energy.

The simulation software for architectural space lighting effect helps the contemporary architectural designers to achieve the design goal. Digital simulation software, 3d max, as the commonly used software, makes the designers perform the professional 3D modeling, material designing color and lighting effects in the same environment. The software provides us a variety of lights to simulate natural ones, giving off real light and artificial light. The designers are able to directly carry out the 3D building plan so that computer technology simulating the real scene has come true.

63.2.1 Natural Light Simulating Lighting in 3D Software

In the real world, natural light is mainly composed of sunlight. The sunlight, which is a kind of parallel light, irradiates in a straight line and to a certain direction (Fig. 63.1). The light shadow in architectural space is changeable. The time and angle of the sun are changing, and so is the sunlight. The skylight is the result of sunlight and diffuse light, which goes well distributed and slowly changed without direction. The skylight reaching building space must be softer and weaker than the sunlight.

In 3d max, we can simulate parallel light by using the standard light; we can also use sunlight IES and skylight IES (in the photometry) to simulate real scene. The parallel light projects to the same direction [1]. Parallel light includes target light and free light.

Parallel light is mainly used to simulate sunlight. Adjust the light color and position by rotating light in three-dimensional space. The parallel light is in the form of a circular, rectangular, or prism. The figure above is the effect of standard parallel light simulating sunlight in 3d max, standard skylight tracing simulation, and IES sunlight and IES skylight simulating natural light (Fig. 63.2).

63.2.2 The Simulation Analysis of Artificial Lighting in 3d max

Besides the sunlight and skylight, people use more artificial illumination which can redesign the shadow change, modeling and redividing the building space in order to make it more colorful and charming.

In 3d max, standard light is being used to render space. The calculation method is direct, that is, simulation calculation without the data. So what we need is to arrange the light in the scene, through people's sense to evaluate the rending effect. This type of light includes floodlights and spotlights [1] (Fig. 63.3).

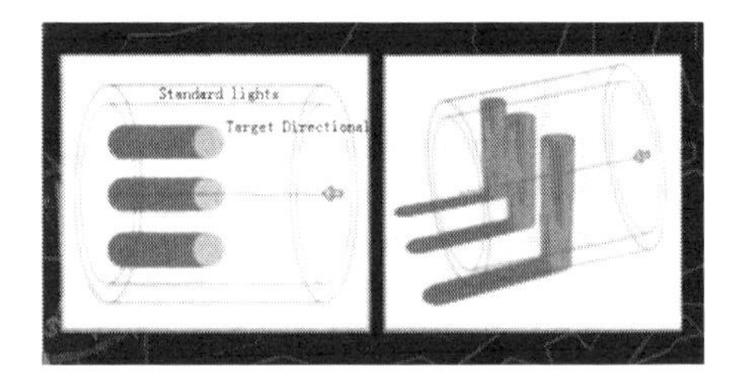

Fig. 63.1 Parallel light in 3d max

Fig. 63.2 Rendering effect of parallel light simulating sunlight

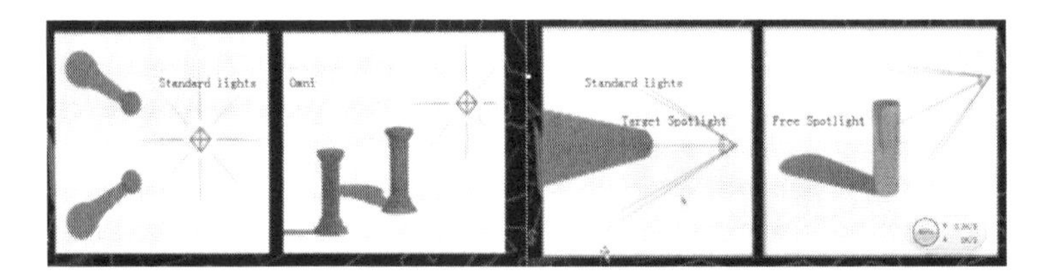

Fig. 63.3 Standard light in 3d max

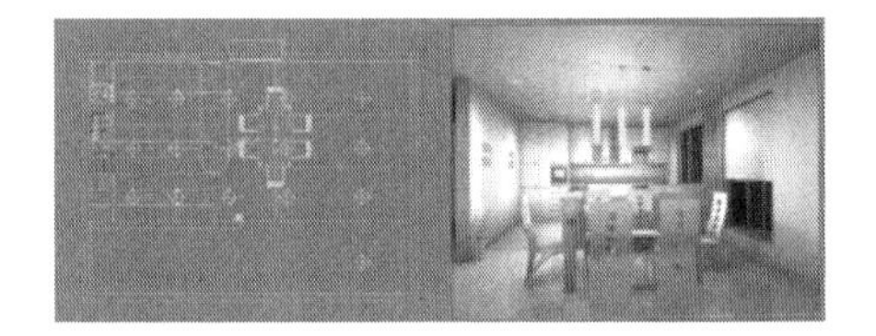

Fig. 63.4 Standard lighting in 3d max

1. Floodlights: It is irradiated from one source to all directions. Floodlights add "auxiliary lighting" to the scene or simulate point light source. Floodlights can cast the shadow and projection. The single floodlight to cast shadows is equivalent to six spotlights to do so from the center to the outside.
2. Spotlights: It includes target lights and free lights. The former one is like a flash to cast focus beam which has its target point as the position, while the latter one hides this position (Fig. 63.4).

In addition to the standard light source, another commonly used simulation source is photometric light in 3d max. The light is based on simulation operational mode, by simulating real scene to get the rendering effect and, at the same time, to calculate the light distribution exactly. By shape, photometric lights include point source, line source, and surface source [1] (Fig. 63.5).

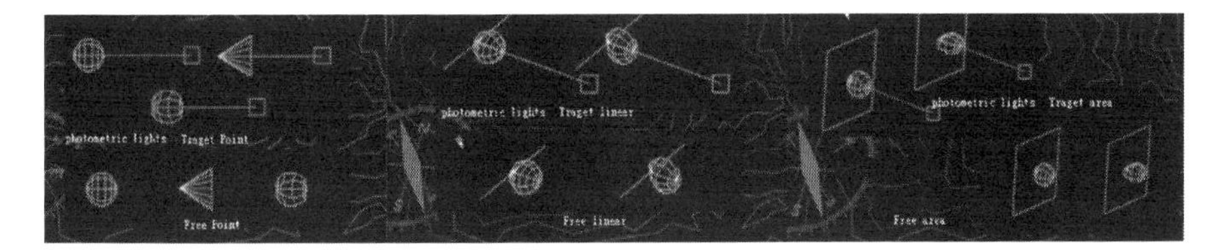

Fig. 63.5 Photometric light in 3d max

1. Point source: It is the simulation of incandescent lamp, for example, the ceiling lamps, wall lamps, and tube light. Electric light source radiates from itself.
2. Line source: It is used to simulate daylight lamps or strip-shaped light slots. Different from the point, the distribution of the line is based on the linear length, to emit in a line.
3. Surface source: It simulates the rectangular-shaped lights in real life such as ceiling lamps or grille lamps. The surface emits light in a rectangular area.

In addition to photometric shape simulation, light intensity and distribution are also important parameters. As the intensity and the irradiation shape to all directions are different, the lights illuminating to the buildings are different too. For example, the illumination from tube light and wall lamps to the surface of buildings produces different light sources and different shapes, so the light distribution of the object source is different.

63.3 Application of Computer Technology to Architectural Apace Lighting Design

An excellent green building is not only depending on its internal facilities to adjust light, but it must be an energy-saving one without any auxiliary equipment. If not so, such a building cannot achieve the goal of saving energy; the building is not the "green building." The proper analysis of computer technology can combine the construction in the virtual space with the actual environment numerical value in order to simulate and analyze effectively. At present, there is a lot of computer software for simulation and analysis, such as thermal environment simulation and sunlight simulation. This paper analyzes the interior and exterior lighting illumination with ECOTECT software. Computer simulation technology of lighting and illumination analyzes thermal environment, light environment, sunshine environment, etc. The calculation results are accurate and intuitive. As for the architectures, it is essential to analyze the light effect. With the sunlight changing, the internal and external light intensity will alter. To make the architectures more green and energy saving, people need to modify the errors in the course of designing. It is necessary to use ECOTECT analysis so that the green goal can be perfect before the project is established.

ECOTECT provides a variety of functions for analysis, among which natural lighting is the most important, which influences the artificial lighting. This soft natural light is assumptive diffuse natural light in a closed space. As the sunlight is changing with time, occasionally the sun will also be prevented by objects, and the main building receives different light intensity, so the sun is not accepted as a reliable lighting source. Therefore, the best method to calculate lighting coefficient in ECOTECT software is Split Flux by Building Research Establishment. The calculation principle is based on the premise, without considering the influence of direct sunlight. Any point of natural light is composed of three parts in the space: the skylight—sky component (SC), the light through the internal glass objects directly irradiated the space; the external reflection light—externally reflected component (ERC), light reaches environmental objects, producing reflected light source; reflected component (IRC), direct light and reflected light reach a point in the space, together with other light spots around them to form a sort of internal reflection. CIE cloudy distribution in ECOTECT (Overcast Sky Distribution) model [2], through the analysis of the current grid or independent sensing, points to calculate the data of natural lighting. The lighting parameter is based on the coefficient of lighting and the sky illumination. The sky luminance values can be set according to the lighting analysis. When the calculation is being performed, semicircular lighting prediction figure will come up on the software interface. Through each light data in the same illumination, we can calculate the lighting vector on each point, which represents the direction of natural light. We can get the data through different colors and arrows on the interface. The analysis of natural light can be calculated by lighting coefficients, through which, it can calculate the illuminations on the certain days in a year and also can adjust its parameters.

63.4 Simulation Analysis of Architectural Space Lighting Design Combined with Computer Visual Technology

63.4.1 Simulation Analysis of Architectural Lighting Design in 3D Max

In this case, analyze and simulate the photometric lights of indoor space [1].

Light energy transmission is a calculation way to simulate light, which is able to simulate and analyze refraction, reflection, and other indirect lighting to the objects' surface, through which the real scene can be beautifully and truly realized. It can also simulate the natural light which emits to the models' surface. The calculation of light emission can be completed before rendering. Calculation is different from rendering. If the refraction and reflection in the real scene can be realized, the calculation can't be skipped. In the following cases, we will calculate the light transmission through photometric lighting in indoor space. Such calculation is based on physical date, so the model establishing and lighting parameters are rigidly required. We must obey the rules to render.

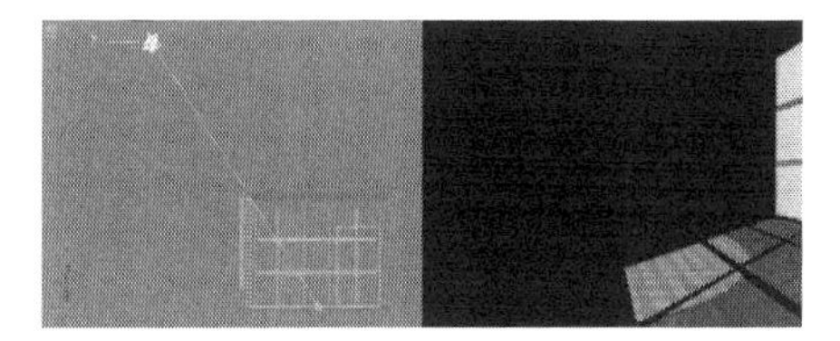

Fig. 63.6 Photometry simulating sunlight

Fig. 63.7 Photometry simulation sunlight to render

In the case, we use IES to simulate the sun irradiation and adjust the ray tracing, and make the light complete the simulation through the calculative operation. The effect is after setting the light tracing (Fig. 63.6).

Set the lighting transmission parameter, set the IES sunlight as 2000lm, set three iterations and three filters, and use the general grid subdivision, rendering effect (Fig. 63.7).

From the cases above, we can see photometric lights can analyze and simulate radiosity. The lighting method depends on the light source's direction and position in the real scene. Radiosity's calculating makes the light intensity distribution uniform and transferred conveniently, to simulate the good distribution of light in the scene. The effect of simulation light goes real and delicate, and it realizes a good atmosphere of real scene.

By the application of computer simulation software, we can simulate the effect in the design period, so that the designers can easily design the building space. Computer simulation can replace traditional hand-drawn drafts. It is precise and effective. What's more, it's convenient to modify and can greatly reduce the design time and improve work efficiency.

63.4.2 Computer Simulation Analysis on ECOTECT Light Environment

As to energy-saving or illumination angle, natural lighting is very important for buildings, in large building space, except the room around the window, and other spaces such as the lobby, which is very dark. Due to the lack of sunlight conditions in the daytime, artificial lighting is needed. Thus, it is not an ideal space both for health and energy saving. Because the natural light is changed by time, sun altitude, atmospheric density and some other natural factors, therefore, we can use computers to simulate and analyze natural light, and get basic data of the internal natural

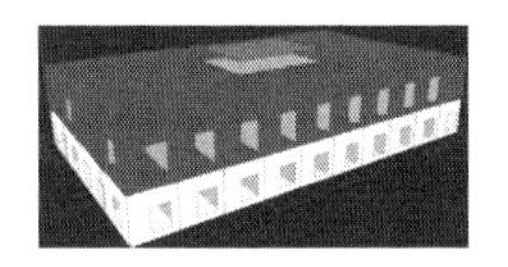

Fig. 63.8 Creating rectangular skylights

light intensity. According to different light intensity in each region of interior space, the designer will use artificial light for light division and the secondary design in architectural space more effectively.

In this case, we use Learning Tutorial-Daylighting Analysis (proposed by Allexa. Ecotect), Ecological Simulation Analysis, and the transformation case for green building design company to state our opinion [3].

Analyze the construction space and set the contour as rectangular, single-storey building area as 60 × 30 m. If two layers, floor height is respectively 6 and 7 m. Considering architectural space is wide and deep and the center is dark lighted, the skylights can be used for light selecting. The initial design stage will use the ECOTECT software, analyze the skylight lighting, and at the same time take rectangular roof as an example for analysis of fluorescent lighting.

The buildings' lighting window includes side windows and skylights. In accordance with the "Architectural Lighting Design Standards GB/T 50033-2001," the side window's window-to-floor proportion is 1/2.5, skylights' is 1/7.5, and rectangular skylights' is 1/3.5. With the partition between the two layers, the skylights' area covers the area of $18 \times 60 = 1{,}080\ \text{m}^2$.

The area of building window is estimated according to the computer lighting technology; as for the first floor, building lighting cannot calculate the specific area. So we need ECOTECT to simulate and analyze different shapes and areas of the skylights and to decide the form and area. Take rectangular skylight, for example, to simulate.

First, open ECOTECT to make sure the gridding and geographic position, and set up a 6 meters high, 3.5 meters wide rectangular wall. On the wall, we can set up a 3m high, 2.5m wide window and the height of windowsill is 0.9m. At the other side of the wall, we also create the window in the same way. At the top of the building, we set up a skylight and confirm its size and connect two layers. Then, we create a building interior wall, skylights, and door, and then we copy the door (Fig. 63.8).

Lighting Simulation of Rectangular Skylights. First open the Web management panel and then set up gridding. To avoid walling reflecting, we have to set up the gridding position to 1 M far from the wall. The First-floor of Rectangular Skylight Daylighting Simulation. Set the grid XY Axis offset of 600, which means analysis of grid is 600 mm from the ground. Select lighting coefficient and the intensity of illumination options, click calculation, and set simulation (Figs. 63.9, 63.10, and 63.11).

For the design of rectangular windows, first-floor lighting space is controlled between 2.5 and 15 %. About 80 % illumination space is around 6 %, with low intensity. Artificial lighting auxiliary should be added according to the actual analysis.

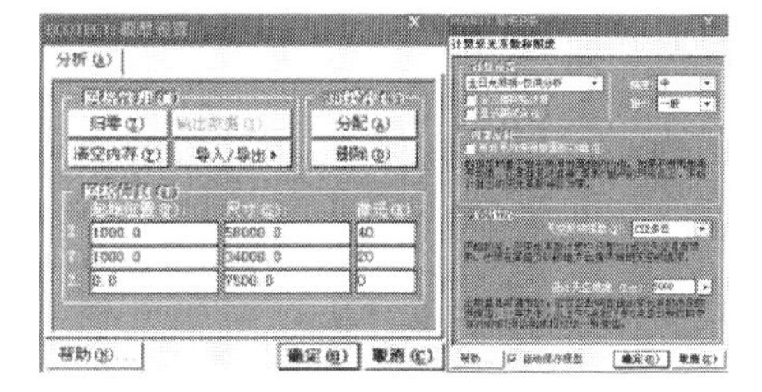

Fig. 63.9 Set up mesh parameter

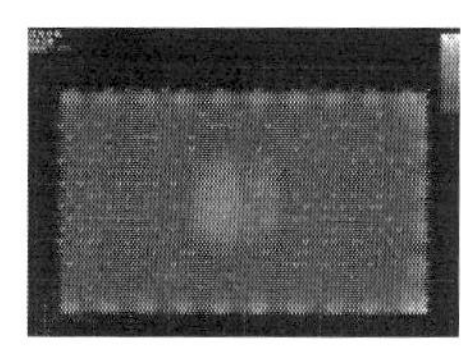

Fig. 63.10 Simulation figure of first-floor rectangular window

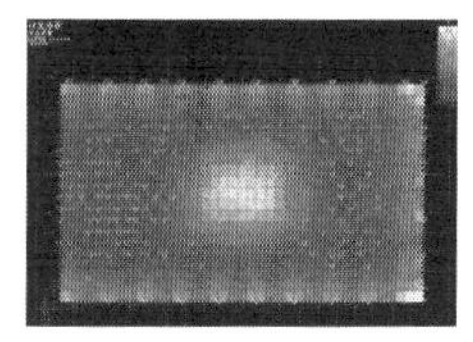

Fig. 63.11 Simulation figure of second-floor rectangular window

For the design of rectangular window, second-floor lighting space is controlled between 7.5 and 25 %. The light energy distribution is uniform in the rectangular window. If daylight is too strong, artificial shading measures should be taken to prevent the light.

By simulating daylighting coefficient, and we can know that there is no area that the daylighting coefficient is less than 2 %, that is, 100 % of the interior space daylighting coefficient is more than 2 %. It fully satisfies the rules of the green building evaluation technology in [4] (75 % of the interior space daylighting coefficient should be more than 2 %).

In the design of the sky illumination (5000 lux), we calculate indoor illumination, and we find the illuminance $\geq$101 lux, which is accordance with the rules (indoor natural light illumination of visual working class III (fine) is not less than 100 lux) in ≪architectural lighting design standards≫ (GB/T 50033-2001) [5]. Generally speaking, through the natural lighting simulation analysis of the office building standard layer, we find that the natural lighting design scheme satisfies the requirement of indoor visual, and it can achieve the ideal effect.

63.5 Conclusion

Humans live in the light environment and use light in all fields of life. However, the light works for humans while at the same time also consumes a lot of energy, causing serious pollution and destruction to the environment.

In the digital era, lighting design by computer technology has become an important part of people's life. In order to meet the people's aesthetic psychology, artificial light is becoming more and more important. As the energy problems appear, people pay more attention to lighting design in architectural space with computer technology. Computer shows the building space environment in the course of design so as to modify the improper part, evaluate and forecast the effect, and save unnecessary energy consumption. Computer technology's application to architectural space lighting design will have a further influence on human's life.

References

1. Wang, X., Wang, K., & Shi, Y. (2004). *Learning by doing—3ds max effect drawing materials and lighting application techniques and examples* (p. 15). Beijing: People's Post & Telecom Press (in chinese).
2. Yun, P. (2007). *Teaching series of computer technology on architecture—Ecotect building environment design tutorial* (p. 30). Beijing: China Building Industry Press (in chinese).
3. Allexa, Ecotect (Ecological construction master). (2011). *Learning tutorial—Lighting analysis "Shenlvshe" ecological simulation analysis.* Green Architecture Design Company (p. 5). ShenZhen: China building industry Press (in Chinese).
4. China Academy of Building Research. (2007). *The technical details of the green building evaluation* (p. 65). Beijing: China building industry Press (in Chinese).
5. State Standard of PRC. (2001). *Architectural light-selection design standard GB/T 50033-2001* (p. 7). Beijing: China building industry Press (in Chinese).

Chapter 64
Improving Online Gesture Recognition with WarpingLCSS by Multi-Sensor Fusion

Chao Chen and Haibin Shen

Abstract In order to achieve the better online gesture recognition rate, a multi-sensor fusion method is proposed in this chapter. After the dimension reduction and quantization, we first measure the performance of every single sensor in training phase and use this prior knowledge to determine the weight vector; then we do the fusion of multiple sensors according to the weight vector which indicates each sensor's importance in recognition. The core algorithm we use for online gesture recognition is WarpingLCSS, which is demonstrated to be an efficient template matching method for gesture spotting. We do the experiments on the OPPORTUNITY Activity Recognition Datasets, and the results show that the recognition rate of multi-sensor fusion method achieves 61 %, which outperforms the single sensor's performance about 11 %. This demonstrates that our proposed multi-sensor fusion method is efficient in improving the performance of online gesture recognition.

64.1 Introduction

Multiple sensors can give a more comprehensive description about the system than single source data; thus multi-sensor systems play an important role in improving the accuracy of pattern recognition. In recent years, there are a large number of applications about multi-sensor fusion such as robot system [1], mechatronics [2], and human computer interaction [3].

In 2012, Long-Van Nguyen-Dinh proposed a new template matching method called WarpingLCSS [4]. He compared this new method with dynamic time warping (DTW) [5], a widespread method to do online gesture recognition, and

C. Chen • H. Shen (✉)
Institute of Very Large Scale Integrated Circuit Design, Zhejiang University, Hangzhou 310027, China
e-mail: shb@vlsi.zju.edu.cn

W.E. Wong and T. Zhu (eds.), *Computer Engineering and Networking*, Lecture Notes in Electrical Engineering 277, DOI 10.1007/978-3-319-01766-2_64,

his analysis shows that WarpingLCSS has higher accuracy than DTW. However, the recognition rate is still not high.

In order to improve the accuracy of online gesture recognition, we extend WarpingLCSS with the multi-sensor fusion method. After the preprocessing step such as dimension reduction and quantization, we test the performance of every single sensor and then do the fusion of multiple sensors based on this prior knowledge.

The results of our experiments show that this multi-sensor fusion method will highly improve the accuracy and the recognition rate is about 61 %, which outperforms the current WarpingLCSS about 11 %.

The rest of this chapter is organized as follows: In Sect. 64.2, we review some related works. In Sect. 3, we review the basic concepts about WarpingLCSS. In Sect. 64.4, we describe multi-sensor fusion method proposed by us. In Sect. 64.5, we give the experimental results of the proposed method. Finally, conclusions and future work are discussed in Sect. 64.6.

64.2 Related Works

Data acquired from sensors are basically a kind of time series; therefore we prefer to use time series analysis method to spot gestures. One of the most widely used methods is DTW [6]; it calculates the similarity distance between two time series, one is samples for test, and the other is the template. If the similarity distance is below a certain threshold, the two time series are considered to be the same class of gestures.

However, DTW shows a bad tolerance to outlier noise [7]. In order to recognize gestures with noise more accurately, the Wearable Computing Lab in ETH proposed a new method called WarpingLCSS based on the Longest Common Subsequence (LCSS) algorithm [4]. This method shows better tolerance to outlier noise and requires less hardware resources than DTW; in his work, one sensor attached to the upper arm is used and the accuracy achieves about 50 % which outperforms the DTW method about 12 %. But 50 % accuracy is still not that good.

Multi-sensor fusion method can achieve better recognition rate since it combines information together. There are several methods we can choose to do the fusion [8]—the raw data level, the feature level, and the decision level. In this chapter, we do the fusion in the decision level since we want to diminish the wrong decisions of each sensor after applying the WarpingLCSS algorithm.

64.3 WarpingLCSS

The WarpingLCSS algorithm [4] is derived from the LCSS problem and can be computed efficiently by dynamic programming [9].

Let S_t be a gesture template generated in training phase and S_m be the streaming sensor data. W(i, j) in (64.1) gives the measurement of similarity between the *i*th

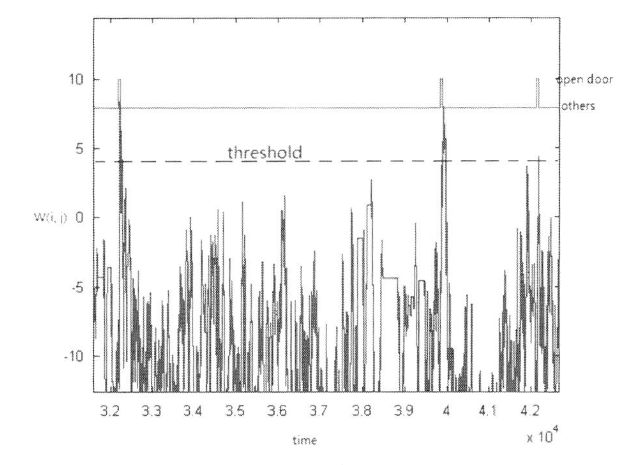

Fig. 64.1 W(i, j) between a template and a string of preprocessed sensor data

symbol in template S_t and *j*th symbol in S_m. p is used to penalize the mismatch of two symbols, and d(l, m) is the Euclidean distance between the two corresponding centroids. W(i, j) will increase rapidly when lots of similarities exist between the new incoming sample and the template string and decrease rapidly due to the function of penalty parameter and thus highlights the endpoint of predicted gestures.

In Fig. 64.1, we give a case to show the function of WarpingLCSS. The line on the top indicates when the open door gestures happen, and W(i, j) varies rapidly during that period of time and will have some local maximum values. We just need to set a proper threshold like the figure shows, and the local maximum values above the threshold indicate the possible happening of open door gestures:

$$W(i,j) = \begin{cases} 0, & \text{if } i = 0 \text{ or } \mathrm{j} = 0 \\ W(i-1,j-1)+1, & \text{if } S_t(\mathrm{i}) = \mathrm{S}_m(\mathrm{j}) \\ \max\begin{cases} W(i-1,j) - p\cdot d(S_t(i), S_t(i-1)) \\ W(i,j-1) - p\cdot d(S_m(j), S_m(j-1)) \end{cases}, & \text{otherwise} \end{cases} \quad (64.1)$$

64.4 Multi-Sensor Fusion Method

The whole process of recognition phase of our proposed multi-sensor fusion method is shown in Fig. 64.2. Raw sensor data series are first preprocessed and are calculated with WarpingLCSS algorithm separately. Then the processed data come to the multi-sensor fusion module, which combines the results of WarpingLCSS to diminish those unexpected noise and gives the outputs according to each sensor's importance in recognition. If the outputs suggest that an instance may belong to multiple classes, then the last module should decide which template has the highest similarity with the incoming stream and finally outputs the class of the gesture.

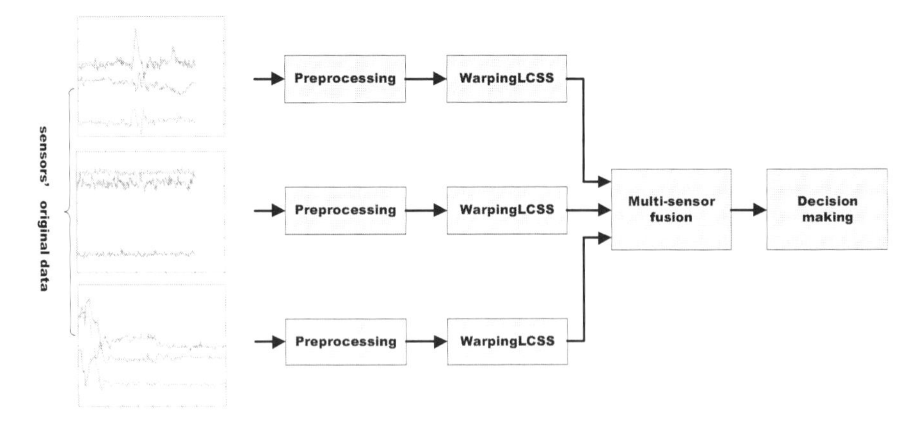

Fig. 64.2 The architecture of multi-sensor fusion system

64.4.1 Preprocessing

Raw data are generated by 3D accelerometers or other sensors. In order to speed up the processing rate, we first use a sliding window of size 6 to combine six sample vectors to one mean vector, overlap three samples each time, and then use k-means to cluster the mean vectors to their closest cluster centroids. Thus, we could get a string of symbols to represent the raw data. This process is the same in training phase and recognition phase.

64.4.2 Training Phase

For gestures belonging to the same activity class, we use LCSS to calculate the similarity between them and choose one to be the template if it has the highest similarity with others. And we repeat this process to get templates for all types of gestures and all types of sensors. When we get all these templates, for each gesture class, we check the accuracy of every sensor and choose the candidates for multi-sensor data fusion.

Equation (64.2) gives a method to do multi-sensor fusion. V represents the number of sensors to be fused. WV is a positive definite weight vector which gives certain W(i, j) more weight to improve the overall performance of online gesture recognition. We can test the accuracy of each sensor in training time, get the prior knowledge about the importance of sensors when spotting certain activities, and set the WV according to this prior knowledge:

$$MW = \sum_{v=1}^{V} WV(v) \cdot W(i,j)(v) \tag{64.2}$$

64.4.3 Recognition Phase

In recognition phase, we have almost the same process for incoming data as in training phase. When multi-sensor streams come, these data will first decrease their dimension in preprocessing stage; then each sensor data compare with the templates created in training phase and fuse according to (64.2). Gestures will be spot if the results of (64.2) are over the threshold.

64.4.4 Decision Making

If an instance is spotted as multiple gestures, we have to decide which one is best matched and output the decision. We first use the trace-back process [4] to get the start point of the gesture and then calculate the normalized similarity through dividing the similarity by the longer length between the template and the spotted string. The class with the highest normalized similarity is chosen as the final output.

64.5 Experiment Setup and Simulation Results

64.5.1 The Datasets

All our experiments are doing on the OPPORTUNITY Activity Recognition Dataset [10]. We do not need the whole datasets, so we select part of the sensor data. The names and locations of chosen sensors are depicted in Fig. 64.3.

64.5.2 Evaluation Methods

To evaluate the performance of the method, we use accuracy, which is the proportion of correctly predicted samples over the number of total samples. However, the

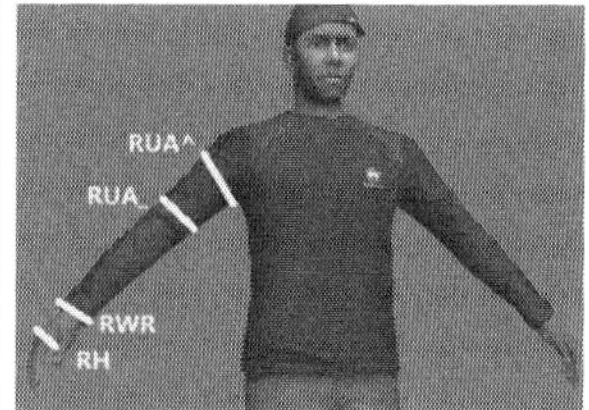

Fig. 64.3 The names and locations of inertial measurement units (*left*) and acceleration sensors (*right*)

Table 64.1 The performance of the whole OPPORTUNITY Activity Recognition Datasets

Activity	Sample number	Accuracy	F1
Open door 1	10	0.90	0.72
Open door 2	9	0.88	0.40
Close door1	9	0.78	0.64
Close door2	9	0.67	0.75
Open fridge	31	0.61	0.60
Close fridge	33	0.61	0.52
Open dishwasher	12	0.83	0.60
Close dishwasher	14	0.64	0.41
Open drawer1	12	0.92	0.4
Close drawer1	12	0.67	0.53
Open drawer2	12	0.41	0.45
Close drawer2	12	0.67	0.67
Open drawer3	12	0.41	0.38
Close drawer3	12	0.75	0.4
Clean table	33	0.5	0.27
Drink cup	36	0.52	0.59
Toggle switch	21	0.33	0.22
The whole OPPORTUNITY datasets			
Accuracy		F1	
0.61		0.48	

accuracy may vary due to the selection of threshold, so we add another measurement called F1 to help to access the performance [4]:

$$F1 = \sum_i 2 \times W_i \frac{precision_i \cdot recall_i}{precision_i + recall_i} \tag{64.3}$$

$$precision_i = \frac{\text{correctly predicted samples}}{\text{total predicted samples}} \tag{64.4}$$

$$recall_i = \frac{\text{correctly predicted samples}}{\text{total samples of class i}} \tag{64.5}$$

64.5.3 Results

For each class of gesture, we first measure the accuracy and F1 of every sensor and choose the proper weight vector. Repeat the above process for every activity, and the results are shown in Table 64.1; we test the overall performance on the OPPORTUNITY Activity Recognition Datasets, which achieves 61 % in accuracy and 48 % in F1; the results are 11 % better than mentioned before [4], which demonstrate that multi-sensor fusion is an efficient way to improve the accuracy of activity recognition.

64.6 Conclusions and Future Work

In this chapter, we have proposed a multi-sensor fusion method based on the WarpingLCSS algorithm. The results show that our method can highly improve the accuracy of gesture recognition and outperforms the single sensor's performance mentioned before about 11 %.

In future work, we want to explore what features will best represent those daily activities and improve the accuracy of recognition significantly.

References

1. Castellanos, J. A., Neira, J., & Tardós, J. D. (2001). Multisensor fusion for simultaneous localization and map building. *IEEE Transactions on Robotics and Automation, 17*(6), 908–914.
2. Luo, R. C., & Chang, C. C. (2012). Multisensor fusion and integration: A review on approaches and its applications in mechatronics. *IEEE Transactions on Industrial Informatics, 8*(1), 49–60.
3. Reddy, B. S., & Basir, O. A. (2010). Concept-based evidential reasoning for multimodal fusion in human–computer interaction. *Applied Soft Computing, 10*(2), 567–577.
4. Nguyen-Dinh, L. V., Roggen, D., Calatroni, A., Troster, G. (2012). Improving online gesture recognition with template matching methods in accelerometer data. In *2012 12th International Conference on Intelligent Systems Design and Applications (ISDA)* (pp. 831–836). IEEE.
5. Hartmann, B., & Link, N. (2010). Gesture recognition with inertial sensors and optimized DTW prototypes. In *2010 I.E. International Conference on Systems Man and Cybernetics (SMC)* (pp. 2102–2109). IEEE.
6. Stiefmeier, T., Roggen, D., Troster, G., Ogris, G., & Lukowicz, P. (2008). Wearable activity tracking in car manufacturing. *IEEE Pervasive Computing, 7*(2), 42.
7. Vlachos, M., Hadjieleftheriou, M., Gunopulos, D., & Keogh, E. (2003). Indexing multi-dimensional time-series with support for multiple distance measures. In *Proceedings of the Ninth ACM SIGKDD International Conference on Knowledge Discovery and Data Mining* (pp. 216–225). ACM.
8. Jeon, B., & Landgrebe, D. A. (1999). Decision fusion approach for multitemporal classification. *IEEE Transactions on Geoscience and Remote Sensing, 37*(3), 1227–1233.
9. Sakoe, H., & Chiba, S. (1978). Dynamic programming algorithm optimization for spoken word recognition. *IEEE Transactions on Acoustics, Speech and Signal Processing, 26*(1), 43–49.
10. Roggen, D., Calatroni, A., Rossi, M., Holleczek, T., Forster, K., Troster, G., et al. (2010) Collecting complex activity datasets in highly rich networked sensor environments. In *2010 Seventh International Conference on Networked Sensing Systems (INSS)* (pp. 233–240). IEEE.

Chapter 65
The Lane Mark Identifying and Tracking in Intense Illumination

Yanyun Xing, Bo Yu, and Fangqun Yang

Abstract In order to enhance image contrast and ensure accurate identifying and tracking in intense illumination case, this chapter uses the algorithm of histogram cone-shaped, which can enhance image contrast effectively. With the algorithm of histogram cone-shaped, the scope of the gray value increases obviously. And the chapter introduces a first-order differential operator two-direction Prewitt operator to enhance the edge for image; the enhance effect is favorable, and the compute time is short. Then the algorithm of 2-D gray histogram is used to segment image. The chapter uses Hough transformation to identify the lane mark's two edges and account its intercept and slope and then draws the midline as the last identifying result. In order to reduce the count time, the chapter uses the algorithm of area of interesting to track the lane mark. The experiment results show that the lane mark can be tracked dependably in intense illumination and the algorithms are of real time; moreover, when the tracking algorithm is a failure, the system can also recover in time and lock the tracking target accurately again.

65.1 Introduction

Lane mark identifying and tracking is one of the research areas for vehicle safety driving assist system. It can be employed not only in lane departure warning systems but also in vehicle positioning. In the long run, the technology can also be used in the autonomous navigation of vehicle. Now the researches of lane mark

Y. Xing (✉)
College of Automotive and Transportation, Tianjin University of Technology and Education, Tianjin 300222, China
e-mail: jluqicheyb@hotmail.com

B. Yu • F. Yang
Automobile Engineering Research Institute, China Automotive Technology and Research Center, Tianjin 300000, China

W.E. Wong and T. Zhu (eds.), *Computer Engineering and Networking*, Lecture Notes in Electrical Engineering 277, DOI 10.1007/978-3-319-01766-2_65,

identifying and tracking are mostly based on machine vision. The changes of environments have a great influence on the quality of machine detection method. Most of the working time of vehicle is daylight, especially in summer. It is very difficult to identify lane mark in the circumstance of long intense illumination. In order to solve this problem, the algorithm of histogram cone-shaped, which can enhance image contrast greatly, is adopted in this chapter. The results show that the extending algorithm is much better for this kind of images compared with other extending algorithm. The chapter introduces a first-order differential operator two-direction Prewitt operator to enhance the edge; the result is good as LOG operator whose convolution mask is 5×5, but the count time is much shorter. So the consequent image processing is same as the general image [1].

65.2 Preprocessing of Intense Illumination Image

The contrast of intense illumination image is very low due to the insufficient brightness range; as a result the useful information cannot be picked up accurately. In order to solve this problem, we need to enhance image contrast.

65.2.1 Extending Image Contrast by Histogram Cone-Shaped Algorithm

The histogram cone-shaped extending is an algorithm which can fill gray value in histogram cone-shaped unequally and transform the gray histogram axis into unequal field. Thus gray histogram is filled with more gray value in high-frequency region and less gray value in low-frequency region. Then a new gray histogram is set up according to the count of gray histogram axis, and a new image is reconstructed based on the new gray histogram. Since most of the histograms are shown as a single-peak or a multi-peak shape, the extending range is bigger in the histogram peak than in other area and the extending amplitude distribution is tapered; this histogram is called extending image contrast algorithm based on histogram cone-shaped. The principal sketch map of the algorithm is shown in Fig. 65.1 [2].

Histogram of nonuniform filling is a process that extends N layers of gray-axis histogram uniform to K layers unequally. If the gray range of image is $m_0 \sim m_n$, the corresponding value for gray distribution is $H_{m_0} \sim H_{m_n}$. Normally the gray-axis histogram is evenly divided into N equal segments; the total area of the histogram envelope curve can be obtained according to formula 1:

$$S = \sum_{i=0}^{n-1} \left[\left(H_{m_i} + H_{m_{i+1}} \right) / 2 \times 1 \right] \tag{65.1}$$

If the histogram is extended into K (K > N) layers and the gray range changes into $g_0 \sim g_k$ as well, the algorithm needs to interpolate between m_i and m_{i+1}, so

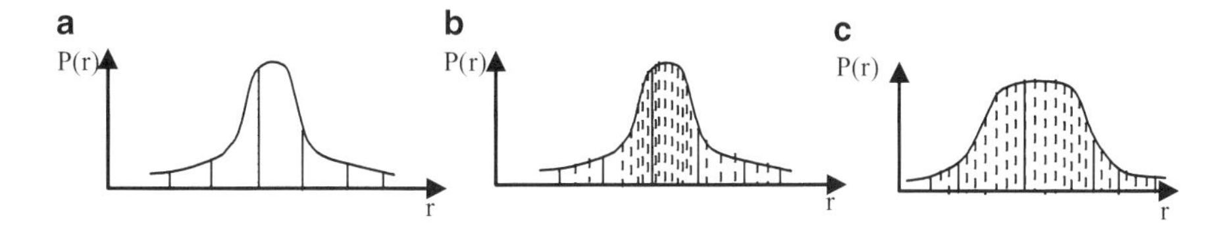

Fig. 65.1 The sketch map extending image contrast algorithm based on histogram cone-shaped. (**a**) Histogram of common image. (**b**) Histogram of cone-shaped. (**c**) Histogram of extending interpolation image

Fig. 65.2 The original image

the interval $m_i \sim m_{i+1}$ is segmented again. If the count of subsection is k_i for each interval, for the purpose of equal area of rezoned region and the total area of S, the area for every interval S' can be obtained according to formula 2:

$$S' = \frac{\left(H_{m_i} + H_{m_{i+1}}\right)/2 \times 1}{k_i} = \frac{\left(H_{m_{i+1}} + H_{m_{i+2}}\right)/2 \times 1}{k_{i+1}} \tag{65.2}$$

And also accord to the formula 3 at the same time.

$$n + k_0 + k_1 + \cdots + k_{n-1} = k \tag{65.3}$$

Partition number is proportional to the gray value distribution in each interval, so the intervals of high gray value segmented are more than the ones of low gray value. The gray axis of the histogram with nonuniform filling is homogenized according to the points in the interval. Then the extending result of the histogram cone-shaped is accomplished. The histogram cone-shaped extending means that the area of the interpolation interval is cone distribution and the intervals of high gray value are extended higher than the one of the low. If m_i corresponding to a gray value changed into m_i' after extending, m_i' can be obtained according to formula 4:

$$m_i' = g_0 + \sum_{l=0}^{i-1} k_l \tag{65.4}$$

The original image and the extended image is shown in Figs. 65.2 and 65.3.

65.2.2 Enhancing the Edge by Prewitt Operator

The Prewitt operator is a first-order differential operator edge enhancement. For convolution mask of 3 × 3, it can estimate gradient in eight directions and can give

Fig. 65.3 The extended image and its histogram

Fig. 65.4 Comparison between the original image and the disposed image

gradient direction for the maximum amplitude convolution. Here are the 3×3 mask templates for the Prewitt operator in the four directions [3]:

$$h_1 = \begin{bmatrix} 1 & 1 & 1 \\ 0 & 0 & 0 \\ -1 & -1 & -1 \end{bmatrix} \quad h_2 = \begin{bmatrix} -1 & 0 & 1 \\ -1 & 0 & 1 \\ -1 & 0 & 1 \end{bmatrix}$$
$$h_3 = \begin{bmatrix} 0 & 1 & 1 \\ -1 & 0 & 1 \\ -1 & -1 & 0 \end{bmatrix} \quad h_4 = \begin{bmatrix} -1 & -1 & 0 \\ -1 & 0 & 1 \\ 0 & 1 & 1 \end{bmatrix} \tag{65.5}$$

Among them, h_1 and h_2 are the horizontal and vertical direction mask template, h_3 is the 45° direction mask template, and h_4 is the 135° direction mask template.

The lane marks of image are directional in this chapter. The lane mark of left side is slanting to the right, while the right side is slanting to the left, coinciding and matching with the template 135° and 45° Prewitt operator templates; as a result, we can use h_4 template for the left half of the image and h_3 template for the right half. This method is called two-direction Prewitt operator. The comparison between the original image and the disposed image can be obtained according to Fig. 65.4, in which the histogram cone-shaped extending algorithm and the enhancement edge by the two-direction Prewitt operator are shown.

65.3 Using the Algorithm of 2-D Gray Histogram to Segment Image

The pixels in the target areas or in the background areas have a great correlation in many images; the gray value of every pixel is approached with its adjacent area pixels, while the gray value between the edge of target and the background or between the noise and its adjacent area are obviously different. Every pixel can

Fig. 65.5 The segmented image by the algorithm of 2-D gray histogram

have a gray group with two elements: the first is its gray value, and the other is the average gray value for its adjacent areas [3, 4].

If the gray value of the image is divided into L grade, the average gray value of the neighborhood area pixels is also divided into L grade. Hypothesis of the frequency for a gray group (m,n) is f_{mn} in the image; we can define the corresponding joint probability density as

$$p_{mn} = f_{mn}/N \quad m, n = 0, 1, 2, \cdots, L-1 \tag{65.6}$$

N represents the count of the pixels of image, $\sum_m \sum_n p_{mn} = 1$. Based on m and n as independent variables and p_{mn} as dependent variable, a 2-D gray histogram could be formed.

The segmented image by the 2-D gray histogram algorithm is shown in Fig. 65.5.

65.4 Using Hough Transformation to Identify the Lane Mark

65.4.1 The Fundamental of Hough Transformation

Beeline equation $y = mx + b$ can be represented with polar coordinates as

$$\rho = x \cos\theta + y \sin\theta \tag{65.7}$$

In the formula, (ρ,θ) defines a vector from origin to the closest point on the beeline [5].

There is a two-dimensional space which is defined by parameters ρ and θ. Every beeline in (x, y) plane corresponds to a point in the space. Therefore every beeline in (x, y) plane converts to a point in (ρ, θ) space after Hough transformation (Fig. 65.6).

There is a given point (X, Y) in (x, y) plane. Countless beelines can cross the given point, and each beeline corresponds to a point in (ρ,θ) space. Therefore the trajectory of the points where all the beelines are in (x, y) space is a sine curve in the parameter space, and each point in (x, y) plane is a sine curve in (ρ,θ) space.

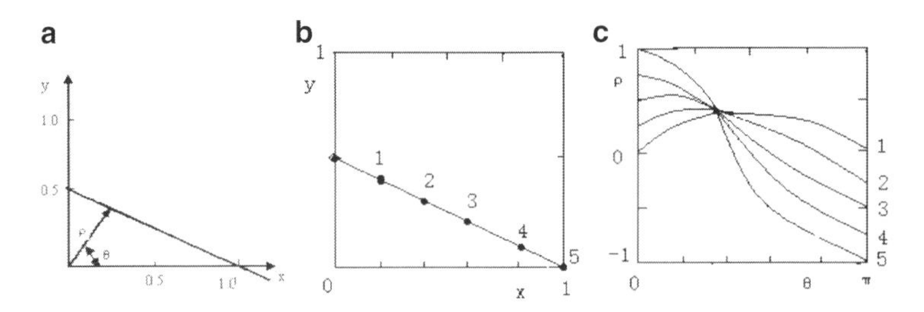

Fig. 65.6 The fundamental of Hough transformation. (**a**) The polar coordinates for a beeline. (**b**) (x, y) plane. (**c**) (ρ, θ) plane

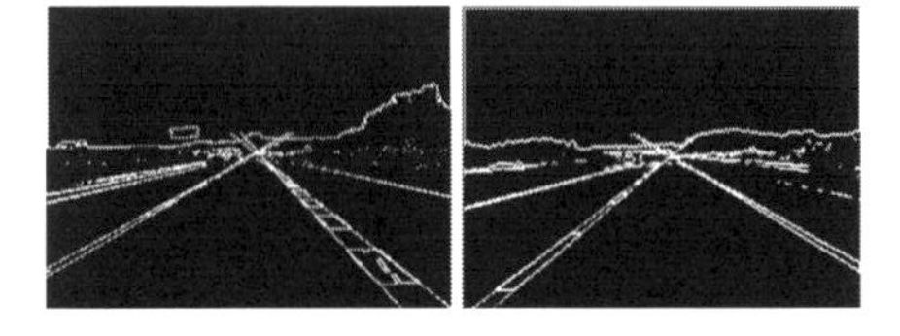

Fig. 65.7 Using Hough transformation to identify the lane mark edge

In order to find out the point of beeline, the (ρ,θ) space is quantified into many small lattices. We can account every ρ according to the quantization of θ for the point (x_0,y_0). The value falls in one small lattice; the count accumulator of the small lattice will be plus 1. After all the point is accounted in (x, y) space, the small lattices will be checked up. If the value of one small lattice is large, it is a point that a beeline corresponds to, and we can employ it as a linear fitting parameter [5].

65.4.2 Using Hough Transformation to Identify the Lane Mark

The chapter uses Hough transformation to identify the two edges of the lane mark. The identifying result is shown in Fig. 65.7. The left lane mark is exampled to illustrate the identifying process. Firstly, we use Hough transformation to find out the longest white beeline in the left half of the image and record it. Secondly, we assign 0 to the point in the white beeline as its new gray value. Then we account a longest white beeline again in the left-half image. So two white beelines are found in the left-half image at last that are the two edges of the left lane mark. And we can find the two edges of the right lane mark similarly.

The two beeline can be used to account its intercept and slope, which are found using Hough transformation. Then we can obtain the intercept and slope of its midline and the disappeared point of the lane mark. And by using the intercept and slope of the midline, combined with the disappeared point, we can draw the lane mark at last. The result is shown in Fig. 65.8.

Fig. 65.8 Identifying the lane mark by Hough transformation

65.5 The Algorithm of Lane Mark Tracking

The algorithm of lane mark tracking is same with the algorithm of lane mark identifying essentially. Its characteristic is that when we identify the lane mark in the current image, we use the last identifying result to limit the identifying searching image area. That means we use the last identifying information to adjust the area of interesting (AOI) dynamic.

65.5.1 Algorithm of Setting Up AOI

In a road image, the roadside information can interfere with the identification of lane mark, and most of the information on the road is useless for the identification. So the effective area for all algorithms is the area of interesting; we call it AOI [6]. In fact, the AOI only focuses on a small area nearby the road mark. In order to improve the real time the chapter uses AOI. That means the area, which is far away from the lane mark and is useless obviously for the lane mark identification, is removed directly without any treatment. Thus the system can save a lot of computation time. Figure 65.9b is the AOI for Fig. 65.9a. As we can see from the image, the AOI is very small compared with the full image, so it can improve the computer speed effectively and improve the real time for lane mark identification.

65.5.2 Recovery Algorithm After Failure

The AOI shows a good effect in the experiments, but when the lane mark is kept out seriously by something such as road vehicles or other objects, the AOI algorithm may cause great error or even failure. Moreover, when the vehicle changes lane or turns, the algorithm may be a failure. In practical application, in order to ensure that the road identification can recover the correct lane mark identification in the error or the failure case, the system needs to check the algorithm is a failure or not first and then determine whether to start the lane mark identification module or not. In this chapter, the method to determine the algorithm failure is as follows.

When the system checks the white points' number for one side or both sides of the lane mark is less than a fixed value in successive five-frame images or the information changes abruptly in the current frame relative to the last frame, we

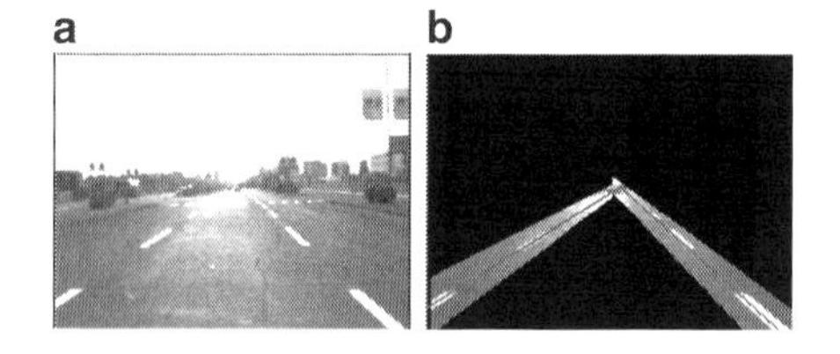

Fig. 65.9 Original image and its AOI. (**a**) Original image. (**b**) AOI

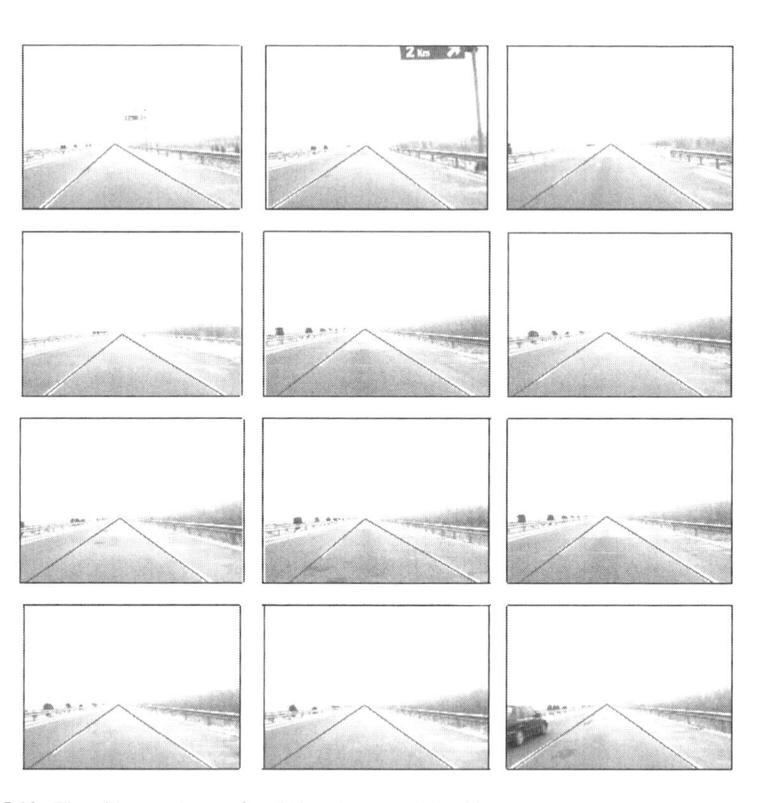

Fig. 65.10 The video captures of real-time lane mark tracking

consider that the algorithm is a failure. Then the system needs to start the initial identification algorithm. That means the system needs to identify the lane mark in the whole image again and ascertain the AOI again, and then it recovers the correct path tracking.

65.5.3 Lane Mark Tracking

The experiment was made in a highway at noon, and the illumination is intense. The video captures of real-time lane mark tracking are shown in Fig. 65.10. The interval for contiguous two captures is 50 frames.

65.6 Conclusion

The chapter uses the algorithm of histogram cone-shaped extending to enhance image contrast first. Then it uses two-direction Prewitt operator to enhance the edge for image and uses the algorithm of 2-D gray histogram to segment image. The Hough transformation is used to identify the lane mark, and the algorithm of AOI is used to track the lane mark at last. Proved by the experiment of vehicle real driving on the highway, the algorithms are dependable and effective essentially in intense illumination case. Even in the case that the tracking target is error, the algorithm can also recover in time and lock the tracking target accurately again.

Acknowledgements Supported by Foundation of Tianjin University of Technology and Education Research Development Project Number KJ2008032.

References

1. Massimo Bertozzi, Alberto Broggi, Massimo Cellario, Alessandra Fascioli, Paolo Lombardi, Marco Porta. *In Proceedings of the IEEE IV. 90*(7). Jul 2002:1258–1271.
2. Yu-lin, S., & Zheng-xi, G. (2004). Algorithm of histogram cone-shaped extending applied to image contrast enhancement. *Computer Engineering, 30*(16), 153–154 (In Chinese).
3. Xiao Zhang, & Dong Yan-xue. (2005). *Digital image processing technology* (pp. 180–182, 188–190). Beijing: Metallurgical Industry Press (In Chinese).
4. De-kui, Y., Bao-min, Z., & Lian-fa, B. (1999). 2D Gray value transformation enhancement for infrared images. *Infrared Technology, 21*(3), 25–29 (In Chinese).
5. Sun, F.-r., & Liu, J.-r. (2003). Fast Hough transform algorithm. *Chinese Journal of Computer, 24*(10), 1102–1109 (In Chinese).
6. Rong-ben, W., Tian-hong, Y., Bai-yuan G., & Lie, G. (2006). Research on linear lane mark identification and track method based on edge. *Computer Engineering, 32*(16), 195–196 (In Chinese).

Chapter 66
Classification Modeling of Multi-Featured Remote Sensing Images Based on Sparse Representation

Xiaoting Hao, Chunmei Zhang, Jing Bai, Mo Dai, Wenxing Bao, and Wei Feng

Abstract A framework for multi-featured classification modeling of remote sensing images with sparse representations is proposed in this chapter. The problem for extracting features to build sparse learning optimal dictionaries is solved by using the remote sensing image spectral values combined with their transformation's characteristics such as Normalized Difference Vegetation Index and K–T transformation. This framework employs sparse representation on dimensional reduction and feature refinement for the remote sensing images. Experiments show that by using our approach the classification accuracy and Kappa coefficient are greater than the support vector machine and conventional sparse representation methods. This result is based on remote sensing images from the sand lake in Pingluo County, which is located in the Ningxia Hui Autonomous Region.

66.1 Introduction

The study of multispectral and hyperspectral data has become a major field of research in analyzing remote sensing images. Usually different materials such as sand, water, grass, roads, and buildings reflect different electromagnetic energies in a particular bandwidth. Based on this promise, we can differentiate the materials according to their spectral properties. But due to the biodiversity and terrain multiplicity, there is a dilemma in the remote sensing images. On occasions, same type of material reflects different spectral values and different materials

X. Hao • C. Zhang (✉) • J. Bai • W. Bao • W. Feng
School of Computer Science and Engineering, Beifang University for Nationalities, Ningxia 750021, China
e-mail: chunmei66@hotmail.com

M. Dai
Institut EGID, Université Michel de Montaigne—Bordeaux 3, 1 Allée Daguin, Pessac Cedex 33607, France

W.E. Wong and T. Zhu (eds.), *Computer Engineering and Networking*, Lecture Notes in Electrical Engineering 277, DOI 10.1007/978-3-319-01766-2_66,

reflect same spectrums [1]. Therefore, we need extra features beyond spectral values for identification of the different types of materials.

In this work we introduce two other series features of Normalized Difference Vegetation Index (NDVI) and K–T transformation. Both of them are important indicators for eco-environment. Thus, with these key features, we are facing even higher dimensional space of data. In order to reduce dimensions of the multi-featured remote sensing images, we propose to learn the dictionaries to be used for sparsity-based classification. Experiments show that sparse representation method with learning dictionary can improve both classification accuracy and computational efficiency.

The contents of this chapter are arranged as follows: In the second section, we introduce the principle of NDVI and K–T transformation and describe the proposal of the multi-feature modeling of multispectral remote sensing images for wetland. In the third section, we introduce the sparse representations and dictionary learning classification method. In the fourth section, we present the specific steps of the algorithm of this chapter. The experimental results are analyzed in the fifth section.

66.2 Multi-Feature Modeling of Remote Sensing Images

In this section, by introducing the theory of NDVI and K–T transformation, we present the framework of multi-feature modeling.

66.2.1 The Basic Principles of NDVI and K–T Transformation

NDVI is an image-enhanced method for extracting vegetation information. NDVI is a global vegetation index, especially focusing on plant growth and related soil. NDVI is also helpful to interpret and analyze the characteristics of agriculture. NDVI is therefore always an important indicator of the characteristics of the wetland.

As an important component to present vegetation state, NDVI is well known as the main large-scale indicator for eco-environment. It is defined as the ratio of the subtraction and addition of the two bands which are near-infrared band and red band within visible light. The formula is as follows:

$$NDVI = (NIR - \mathrm{Re}d)/(NIR + \mathrm{Re}d) \tag{66.1}$$

where *NIR* is the reflectivity of near-infrared band, *Red* is the reflectivity of visible light band [2]. NDVI can eliminate most of the irradiant distortion including instrument calibration, sun angle, terrain, cloud shadow, and atmospheric conditions to improve response speed of vegetation. It is quite sensitive to the bias of soil background and very efficient on monitoring vegetation and ecological environment. It also has a wide detection range of the vegetation coverage and a good spatial and temporal adaptability. So it has wide applications on vegetation and phenology studies for remote sensing images.

NDVI is a method to monitor the land cover by remote sensing images. The number of vegetation indexes over the past 20 years has increased to more than 40. The NDVI has a linear correlation with vegetation distribution density. Therefore it is considered currently the best indicator of land cover.

K–T transformation is also called Tasseled Cap transformation. It is a kind of empirical linear transformation of multiband images, which can be used as an interpretation for different vegetations and crops in different regions. It was put forward by Kauth and Thomas [3] on MSS image research in which they investigated the processing of the growth of the crops and vegetation. The mathematical expression is shown below:

$$u = R^T x + r \tag{66.2}$$

where R is the coefficients of the *K–T* transformation, x represents the gray values of different bands, r represents constant offset, and u is the gray values of different bands after the K–T transformation [3].

The K–T transformation provides several components. The top three components are closely related to the ground scenery. The first one is brightness index, which is the comprehensive reflective effect of the ground materials. The second component is the green index, which focuses on the coverage of the vegetation, leaf area index, and biomass. The third component is the humidity index, which refers to the water condition of the ground, especially soil humidity saturation. The rest of the components are the yellow index and noises.

66.2.2 The Model of Multi-Features

According to the multidimensional spectrum distribution structures of the multispectral remote sensing information such as the soil and vegetation by using NDVI and K–T transformation, the vegetation discrimination can be enhanced and the background features and texture features can be highlighted. Even more, the vegetation monitoring will be improved, and especially, the information extraction for scattered green patches can be more efficient.

Based on the above discussion, we use spectral feature TM 1–5 bands and TM 7 band of remote sensing images combined with the first component of the brightness index, the second component of the green index, and the third component of humidity index of the K–T transformation and NDVI, such as ten feature vectors, as the basic characteristics for classification.

66.3 Dictionary Learning and Sparse Decomposition

Sparse representation has a great flexibility for adaptive pattern recognition with the redundancy of dictionaries to capture the inherent structures of the natural signals. On one hand, sparse representation provides sufficient choices to design new classification models for dimensional reduction; on the other hand, the similarity matrixes can be built with the sparsity of sparse decomposition, which has a big potential on statistical pattern recognition.

There are two ideas about sparse representation: one's dictionaries have fixed bases (such as curvelets, contourlets), so their structures are fixed and the adaptability is not good. Another is the learning dictionary; this kind of dictionaries need not know the prior information of the signal; they can be set up directly with a small amount of training samples, and then through the machine learning method, a new self-adaptive dictionary can be developed by iterative update.

66.3.1 Dictionary Learning

Dictionary learning is a process which looks for the best atomic bases under the sparsest representation. It not only can satisfy the uniqueness constraints of sparse representation but at the same time can get the sparser and more accurate representation as well. The learning process is as follows.

We can get the training sample from TM remote sensing image data set $X = \{x_i | x_i \in R^m, 1 \leq i \leq n\}$ to form the initial dictionary $D \in R^{m \times p}$ (each column $D_j \in R^m$ is an atom). This dictionary D can capture spectral characteristics of the pixels. And each pixel x_i can be linear represented by the atoms in the dictionary D. Then the dictionary learning problem can be transformed into the optimization problem as follows:

$$\begin{aligned} &\min_{D,\alpha} \frac{1}{2} \|X - D\alpha\|_2^2 + \lambda \|\alpha\|_{1,1} \\ &s.t. \left\|D_j\right\|_2 \leq 1 \quad \forall j \in 1, \ldots, p_2 \leq 1, \forall j \in 1, \ldots, p \end{aligned} \tag{66.3}$$

where $\alpha \in R^{p \times n}$ is the sparse matrix (each column $\alpha_i \in R^p$ is a coefficient vector), the first polynomial of the above formula is the reconstruction error, and the second is sparse penalty function. Parameter λ represents a compromise between sparsity and data reconstruction.

If the sparse coefficient α is known, formula (66.3) is a convex optimization about dictionary D; if the dictionary D is known, the formula is a convex optimization about the coefficients; but it is not feasible to solve coefficient α and dictionary D at the same time. To solve such optimization problem, we can assume that one of them is known; then x and D can be updated alternatively, and we can get the learning dictionary D by iteration [4].

66.3.2 Sparse Decomposition of the Characteristics

We can get the sparse decomposition coefficients $\hat{\alpha}(x)$ of the training samples by the learning dictionary D according to formula (66.4). Then we can represent each characteristic of pixel $x \in X$ sparsely and get their linear decompositions:

$$\hat{\alpha}(x) = \arg\min_{\alpha} \|x - D\alpha\|_2^2 + \lambda\|\alpha\|_1 \tag{66.4}$$

The above formula is a l_1-regularized least squares problem; the solution is a sparse vector in which only a few elements are not equal to zeros. Since the pixels that belong to the same materials have very similar structure, the values of sparse coefficients, which are obtained from dictionary D to represent the pixels' characteristics, are similar. According to the relevant reference [5], we can put $\hat{\alpha}(x)$ as another new expression of pixel x which represents the sparse characteristics of pixel x in the dictionary D.

66.3.3 Classification of Image Sparse Representation

Different classificatory dictionaries can be built with different types of training samples according to the sample category. The projective coefficients of the test sample $\hat{x}$ will not be zeros when the sample is decomposed with its own classificatory dictionary; otherwise, the coefficients should be zeros with other dictionaries so that the category of the samples can be classified.

Actually, due to the noise or the modeling error, the projective coefficients of the test sample $\hat{x}$ are not zeros both with its own dictionary and with other dictionaries. Therefore, the category of the test sample y needs to be determined. If $\delta_i(\hat{x})$ is a new vector and its nonzero coefficients only include these components in which $\hat{x}$ has the relationship with the category i, then we can use $\hat{y}_i = D\delta_i(\hat{x})$ to approximate y when y belongs to the category i. In another words, the smaller the distance between $\hat{y}_i$ and y, the greater the possibility of $\hat{y}_i$ belonging to the category of *ith*. So, the method of identifying the category of y is as follows:

$$\min_i r_i(y) = \min_i \left(\|y - D\delta_i(\hat{x})\|_2\right) \tag{66.5}$$

where $r_i(y)$ is the residual to reconstruct y by training sample of the *ith* category.

66.4 The Algorithm of This Chapter

This chapter uses American landsat TM multispectral remote sensing images as the main data sources in which we can get the remote sensing image of Yinchuan wetland in October 2010. At first, the image is processed by some pretreatments, e.g., adjusting, deionizing, and cropping, and then we can get a 400 × 400 pixel wetland remote sensing image. We use our approach as described before to deal with this image. The algorithm steps are as follows:

Algorithm Multi-feature image classification algorithm

Input:
The image data set $X = \{x_i | x_i \in R^m, 1 \leq i \leq n\}$ and use the data set to form the initial dictionary $D \in R^{m \times p}$ (each column $D_j \in R^m$ is an atom).

Output:
The decomposition coefficients $\hat{\alpha}(x)$ and the residual errors r_i.

Procedure
1. Choose water, desert, vegetation, and Gobi wasteland as four kinds of samples in remote sensing image; the amount of each kind of samples $x = 484$ ($x \in X$), so the total number is 484×4.
2. The samples X are divided randomly into two parts as training set and testing set.
3. Get learning dictionary D with training set by using SPAMS toolbox [6].
4. Use lasso algorithm to decompose the test samples sparsely to get the sparse representation coefficients $\hat{\alpha}(x)$ of pixel $x \in X$ [7].
5. Compare the residual errors r_i of the test samples and their sparse representations; the minimum r_i is the category of the samples.

66.5 Experimental Results and Analysis

In order to evaluate effectively the experiments, the quantities of each type of samples are divided randomly into two equal parts as training set and test set. The experiments will be repeated ten times, and the final results are the average of the ten experimental results.

There are three different experiments in contrast to the support vector machine (SVM) and sparse representation.

1. The first group of the experiments takes six spectral values as its features, which are TM1 to TM5 bands and TM7 band. In this first group, we use the method of sparse representation classification.
2. The second group of the experiments takes the same features as the first group, but uses the SVM method.
3. The third group uses the sparse representation classification method. The total features of this group are ten. They include six spectral values, one NDVI value, and three K–T transformation components.

Table 66.1 shows the classification results: (1) The accuracy of the first group is much higher than the second. The overall accuracy of the first group is increased up to 92.56 % and Kappa coefficient is 0.9008, while the overall accuracy of SVM is only 88.74 % and Kappa coefficient is 0.8499. So the sparse representation is more

Table 66.1 The classification accuracies are shown below with different characteristics and different methods

Characteristics	Classification method	Overall accuracy (%)	Kappa coefficient
TM1–5, TM7	Sparse representation	92.56	0.9008
	SVM	88.74	0.8499
TM1–5, TM7, NDVI, and K–T transformation	Sparse representation	95.05	0.9339

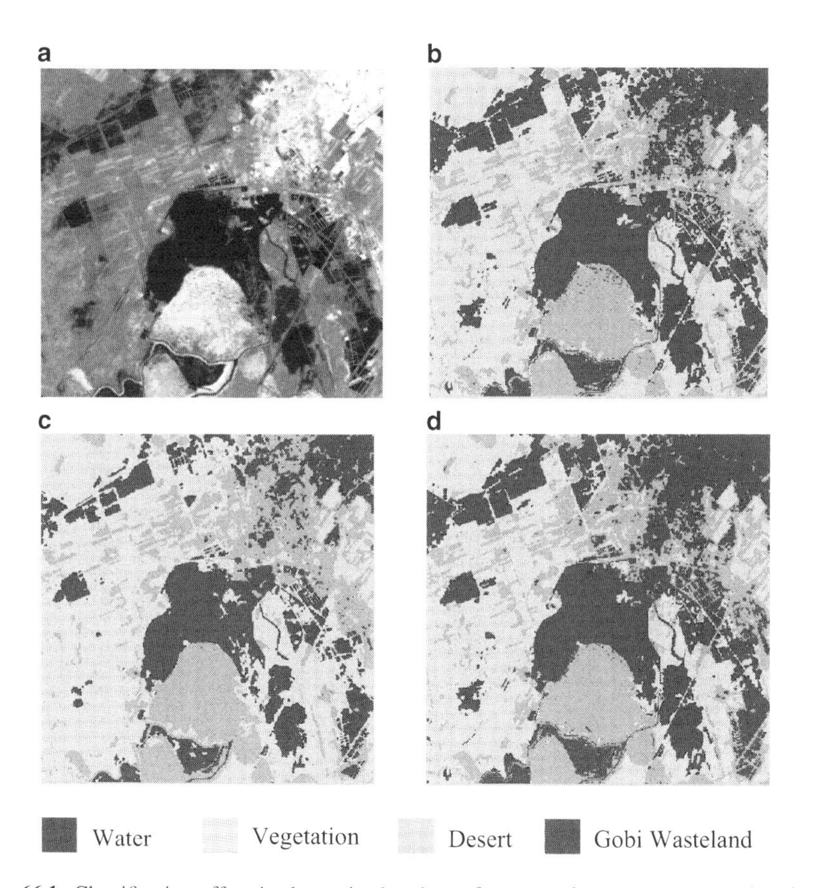

Fig. 66.1 Classification effect is shown in the above figures, and we can compare the visual effects directly. (**a**) Sand lake original picture. (**b**) Only spectral information of sparse representation. (**c**) Classification of SVM with spectral information. (**d**) The results of our approach

robust. (2) The classification accuracy of the third group is higher than the first and the second. Its classification accuracy is up to 95.05 % and Kappa coefficient is 0.9339. So the results show that the features of K–T transformation and NDVI improve the classification accuracy and verify the above discussion (Fig. 66.1).

66.6 Conclusion

In this chapter, we have proposed a new framework for multispectral remote sensing image classification based on sparse representation which uses remote sensing image spectral values and its transformed features together. The simulation results demonstrate that our algorithm is better in both classification accuracy and visual effect than the conventional sparse algorithm and SVM methods.

Acknowledgements This work is supported by National Natural Science Foundation of China (61162013) and (61163016), Natural Science Foundation of Ningxia (NZ11143), and Projects of Beifang University of Nationalities (2012 XYC053). It is also supported by National Expert Introduction Projects.

References

1. Fang, Y., Wang, X. L., Feng, Z. K., Wang, H. P., Nie, M. N., & Wang, B. (2010). One method for model calculation of Quickbird image extraction of urban green space information. *Forest Inventory and Planning, 35*(4), 19–23 (In Chinese).
2. Wang, Q., Wang, N., & Cao, X. F. (2013). Region vegetation cover change analysis of remote sensing monitoring based on TM image. *Yellow River, 35*(2), 70–71 (In Chinese).
3. Kauth, R. J., & Thomas, G. S. (1976). The Tasseled Cap—a graphic description of the spectral-temporal development of agricultural crops as seen by Landsat. In *Machine Processing of Remotely Sensed Data* (pp. 41–51). West Lafayette, IN: LARS, Purdue University.
4. Song, X. F., & Jiao, L. C. (2012). Classification of hyperspectral remote sensing image based on sparse representation and spectral information. *Journal of Electronics & Information Technology, 34*(2), 268–272 (In Chinese).
5. Raina, R., Battle, A., Lee, H., Packer, B., & Y. Ng, A. (2007). Self-taught learning: Transfer learning from unlabeled data. In *International Conference on Machine Learning* (pp. 759–766). Corvallis, OR.
6. Mairal, J., Bach, F., Ponce, J., & Sapiro, (2010). Online learning for matrix factorization and sparse coding. *Journal of Machine Learning Research., 11*(1), 19–60.
7. Tibshirani, R. (1996). Regression shrinkage and selection via the lasso. *Journal of the Royal Statistical Society., 58*(1), 267–288.

Chapter 67
A Parallel and Convergent Support Vector Machine Based on MapReduce

Yingying Ma, Liming Wang, and Longpu Li

Abstract In order to improve the performance of the traditional support vector machine (SVM), this chapter proposes one method referred as MR-SVM to parallelize SVM on MapReduce and mitigates the convergence problems brought by data partitioning and distributed computation. By splitting the large dataset and concurrently calculating the support vector set of each chunk across map units, MR-SVM improves the process capability and efficiency. Then the partial support vector sets are combined as the training set of the global training in reduce phase, and the current global optimum solved by reducing operations is fed back to each map units to determine whether MR-SVM should proceed with another pass. This process iterates until MR-SVM converges to the global optimum. In theory, it has been proved that MR-SVM converges to the global optimum within finite iteration size. Experimental results show that MR-SVM can improve the data processing capability and efficiency of the traditional counterpart and guarantee its high accuracy.

67.1 Introduction

Support vector machine (SVM) is a common method for constructing classification rules as well as one of the most robust and stable algorithms with outstanding accuracy in the field of machine learning. However, the standard SVM algorithm could only deal with some small sample datasets. So faced with the large-scale datasets produced by the practical problems nowadays, there are an increasing number of researches focusing on developing an efficient and high-speed SVM algorithm.

Y. Ma (✉) • L. Wang • L. Li
Information Engineering Institute, Zhengzhou University, Zhengzhou 450052, China
e-mail: mayingyingch@163.com

W.E. Wong and T. Zhu (eds.), *Computer Engineering and Networking*, Lecture Notes in Electrical Engineering 277, DOI 10.1007/978-3-319-01766-2_67,

In order to improve the performance of SVM, it is a widespread approach that splitting the dataset and training the individual chunks in parallel. But there is a variety of problems in these literatures. For example, Graf et al. introduced the Cascade SVM [1] which guarantees the global convergence, but it is poorly scalable and flexible as it requires predefining the network topology and size; Cao et al. proposed one parallel implementation of sequential minimal optimization (SMO) based on MPI [2], which focused on the efficiency but ignored the accuracy; Li et al. proposed a parallel incremental SVM algorithm on GPU [3], which is not preferable because of the lack of a theoretical convergence bound; Caruana et al. proposed one parallel SVM algorithm based on MapReduce for scalable spam filtering [4], but the operation in reduce phase is designed so simply that it needs human expertise to minimize the accuracy degradation, which leads to loss of automation in the approach. However, Chu et al. proved that SVM fits well into the MapReduce framework [5].

According to the deficiencies and inspirations in the previous researches, MR-SVM is focused on providing a scalable and flexible method for classifying large-scale database, improving the data processing capability and efficiency of the traditional SVM algorithm and solving the convergence problem in some similar parallel researches.

67.2 The Analysis and Design of MR-SVM

It is a general solution to improve the capability and efficiency of the standard SVM for data partitioning and distributed processing, but it always leads to convergence problems, resulting in a decline in classification accuracy. It is a problem worth studying to know how to make a better balance between the efficiency and the accuracy of the improved approaches.

67.2.1 Nonlinear SVM Classification Algorithm

The solving progress of the standard nonlinear SVM classification algorithm is concluded as below. Firstly, the Lagrange multipliers λ_i are obtained by evaluating the dual-problem formulation of the Lagrange function of the SVM optimization problem [6]:

$$\max\ L_D = \sum_{i=1}^{N} \lambda_i - \frac{1}{2}\sum_{i,j} \lambda_i \lambda_j y_i y_j K\left(x_i, x_j\right) \tag{67.1}$$

$$\text{Subject to} \sum_{i=1}^{N} \lambda_i y_i = 0,\ \ \lambda_i \in [0, C],\ \ i = 1, 2 \ldots, N \tag{67.2}$$

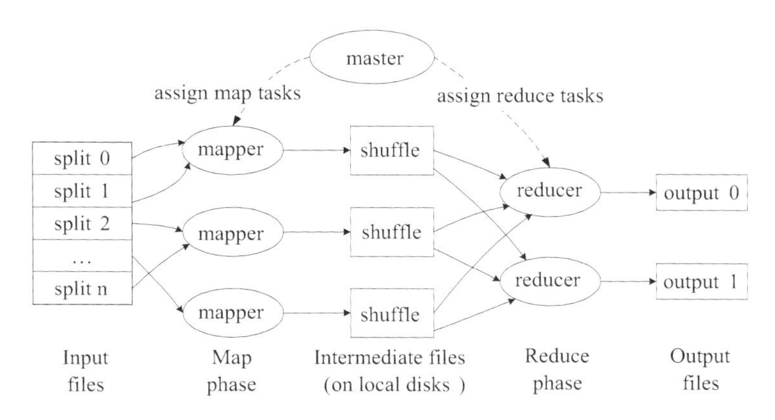

Fig. 67.1 Overview of Mapreduce programming framework

Most of the Lagrange multipliers are to be zero, and samples corresponding to nonzero multipliers are called the support vector set (SVs). Then the feasible solutions of the decision boundary parameters (w, b) are obtained by the following formulations [6]:

$$w = \sum_{i=1}^{N} \lambda_i y_i \Phi(x_i) \tag{67.3}$$

$$y_i \left(\sum_{i=1}^{N} \lambda_i y_i K(x_i, x) + b \right) - 1 = 0, \quad \lambda_i \in (0, C) \tag{67.4}$$

These two formulations indicate that the decision boundary parameters (w, b) depend on SVs.

Finally, the classification function of nonlinear SVM is obtained as follows [6]:

$$f(x) = sign\left(\sum_{i=1}^{N} \lambda_i y_i K(x_i, x) + b \right) \tag{67.5}$$

67.2.2 MapReduce Programming Model

MapReduce is a parallel and distributed programming model in support of large dataset processing [7]. Figure 67.1 shows the overview of a MapReduce programming framework. The MapReduce library partitions the input data into n splits; then the *master* invokes *mapper* units distributed across multiple machines to process the input splits in parallel; then the intermediate key/value pairs produced by the *Map* function are written to local shuffle; and finally, the *reducer* units

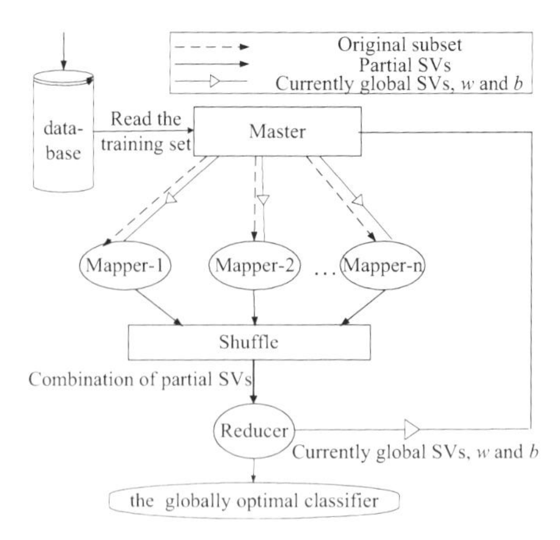

Fig. 67.2 The system model of MR-SVM

iterate over the intermediate data and append the output of the *Reduce* function to a final output file [7]. It is a highly scalable and flexible data processing tool which is increasingly popular.

67.2.3 System Model of MR-SVM

Based on the standard nonlinear SVM classification algorithm and integrating it with the MapReduce framework, the system model of MR-SVM is presented in Fig. 67.2.

67.2.3.1 Parallel Strategy in Map Phase

Since the phase of evaluating the Lagrange multiplier dominates the time and space overhead of the SVM algorithm, it is a wise choice to execute this step in parallel. Accordingly, in MR-SVM, a large dataset is split into chunks horizontally, and then the individual chunks are distributed to the currently idle work units which will evaluate the Lagrange multipliers in parallel. As a result of the decision boundary decided by the SVs, every partial SVs in map phase is stored in local shuffle to represent the corresponding subset. In addition, to improve the efficiency further, the SMO algorithm [8] is adopted to computing the Lagrange multipliers, for its time complexity is $O(n^2)$ while that of some other algorithms is $O(n^3)$.

Compared with the cascade structure [1], MR-SVM is highly scalable and flexible. Because it is built on the MapReduce framework, MR-SVM can call the idle work units one by one according to the current assignments. But in a binary cascade architecture, every time it increases one layer, there will be 2^n nodes being added into the network, where n is the number of layers in the previous network. What is more, in contrast with the parallel SMO algorithm based on MPI [2], the parallel nodes in MR-SVM run separately, which contributes to a decline in communication overhead.

67.2.3.2 A Global Training in Reduce Phase

Because of the equivalence between the SVs and its corresponding training set, the partial SVs obtained from each map unit can represent each respective subset. Therefore, to advance the optimization progress automatically, the representations of every subset are combined and entered as the training set of the global training in reduce phase to evaluate the current global SVs, decision parameters (w and b), and optimal objective function. In this case, MR-SVM is superior to the enhanced parallel SVM algorithm [4] which could not complete the whole optimization progress automatically.

Generally speaking, the support set is only a small part of the training set, so the parallelization of extracting the support vector of each subset in map process is an efficient strategy to reduce time and space overhead for the global training in reduce phase.

67.2.3.3 Feedback Loop Mechanism

Nevertheless, the union of the partial SVs is not equivalent to the whole training set, so the result of the reduce operation is not the global optimum. Therefore, the feedback loop mechanism is employed to guarantee that MR-SVM converges to the globally optimal solution. The results of the reduce phase are fed back to every map unit, and then each map unit employs the Karush–Kuhn–Tucker (KKT) conditions determined by the current optimum to test whether there are any samples violating the current KKT conditions [9]. If all samples in map phase satisfy the KKT conditions, it indicates that MR-SVM has converged to the global optimum; otherwise, the samples violating the current KKT conditions will be combined with the current global SVs and enter the next optimization pass to optimize the separating margin further.

The testing for KKT conditions not only plays a role in determining whether MR-SVM is converged to the global optimum but also screens the samples which should incorporate into the next training pass, contributing to the reduction in timing and space overheads of the following training phase.

67.3 Proof of Global Convergence

To illustrate the global training in reduce phase and the feedback loop mechanisms guarantee that MR-SVM converges to the global optimum in finite iteration size; give the below analysis and proof on the basis of the introduction of the MR-SVM system model and the executing flow in Sect. 2.3.

The identifications utilized during the proving process are listed as follows, where the superscripts j ($j = 0,1,2,\ldots,l$) indicates that the *j-th* iterative training pass.

Ω: The whole training dataset.
m_i^j: The training subset of maper - i unit during the *j-th* iterative training pass.
M^j: The family of the training subset in map phase during the *j-th* iterative training pass, $M^j = \{m_i^j | i = 1, \ldots, n\}$.
r_i^j: The partial SVs which maper - i outputs during the *j-th* iterative pass.
R^j: The training set of reduce phase during the *j-th* iterative pass, $R^j = \{r_i^j | i = 1, \ldots, n\}$.
SV_{Global}^j: The current global optimal *SVs* of the *j-th* iterative pass.
$P(X)$: The optimization function over dataset X.

According to the theorems proved by Graf [1], assume that the training set S is a subset of the training set F and $Sv(S)$ is defined as the SVs of S; in this case, $\forall\ S \subseteq F$, $P(S) = P(Sv(S)) \leq P(F)$. Moreover, assume that the training set S is a proper subset of the training set F; it is obvious that $\forall\ S \subset F$, $P(S) = P(Sv(S)) < P(F)$.

In addition, assume that F is a set of training subsets; the subset f^* which achieves the greatest $P(f)$ will be called the best subset in family F, defining $P(F) = P(f^*)$ [1].

In this case, there will be a sequence (M^j) through the iterative training. So the test for the global convergence in map phase can be described as whether the formulation $P(M^j) = P(\Omega)$ is established.

Corollary: The sequence (M^j) of MR-SVM converges to the global optimum within finite steps, namely, $\exists\ j^*, \forall\ j > j^*, P(M^j) = P(\Omega)$.

Proof. For $\forall\ j, j = 0, \ldots, l - 1$, there is a set $r_i^j \in R^j$ containing the support vector of the best set m_i^{j*} in M^j, that is, $Sv(m_i^{j*}) \subseteq r_i^j$, so $P(m_i^{j*}) = P(Sv(m_i^{j*})) \leq P(r_i^j)$. Therefore, $P(M^j) = P(m_i^{j*}) \leq P(r_i^j) \leq P(R^j)$. For $\forall\ j, j = 0, \ldots, l - 2$, the m_i^j is the union of the violators of the current KKT conditions and the current SV_{Global}^j, so the current SV_{Global}^j is a proper subset of all $m_i^{j+l} \in M^{j+1}$, that is, $SV(R^j) \subset m_i^{j+l}$, so $P(R^j) < P(m_i^{j+1}) \leq P(M^{j+1})$.

It is obtained through the above analysis that for $\forall\ j, j = 0, \ldots, l - 2$, $P(M^j) \leq P(R^j) < P(M^j)$, so $P(M^j)$ strictly and monotonically increases during the iterative progress. For all $S \subset \Omega$, the sequence $P(M^j)$ always takes its value within the finite set of the $P(S)$. For $\forall\ i, j, i = 1, \ldots n, j = 0, \ldots, l$, the set m_i^j is always a subset of Ω, plus $P(M^j) = P(m_i^{j*})$, so $P(M^j) \leq P(\Omega)$; this implies that $P(\Omega)$ is the upper limitation of $P(M^j)$.

Therefore, the iterative training continues until every $m_i^l (i = 1, \ldots n)$ meets the current KKT conditions in the map testing operation when $j = l$. In this case, for $\forall\ i, i = 1, \ldots, n$, $m_i^l = SV_{Global}^j = Sv(R^{l-1})$, so $P(M^l) = P(m_i^l) = P(Sv(R^{l-1})) = P(R^{l-1})$. Finally, because the union of every training subset $m_i^j (j = 0, \ldots, l)$ entered into map phase is the whole training set Ω, we get the conclusion that $P(M^l) = P(R^{l-1}) = P(\Omega)$.

To sum up, MR-SVM converges to the global optimal within finite iteration size.

67.4 Experiment Studying

To assess the performance of MR-SVM, the experiments make a comparison with the sequential SMO. The configuration of the Hadoop cluster employed by MR-SVM is shown in Table 67.1. The nodes in the cluster were connected by Gigabit Ethernet. The SpamBase dataset with 4,601 instances was adopted, and the training sets of varying size were obtained by employing the Weka random re-sampling filter feature.

A Comparison on Data Processing Capability and Efficiency. MR-SVM made a comparison with the sequential SMO algorithm and executed among a varying number of nodes benefiting from the scalability of MapReduce framework. As listed in Fig. 67.3, the time required to train the SMO sequentially on a single node was ≈554 s while the time MR-SVM required was ≈221 s using the Hadoop cluster with four nodes when using 120K samples. Furthermore, the sequential SMO failed when training 322K samples. It indicates that MR-SVM exceeds sequential SMO on the performance of data processing capability and computing efficiency on large datasets.

A Comparison on Classification Accuracy. We compared MR-SVM with the sequential SMO algorithm and the M/R algorithm [5]. The results in Fig. 67.4

Table 67.1 The Hadoop cluster configuration of MR-SVM

Hardware environment			Software environment	
Nodes	Processors	Ram	O/S	Hadoop
Six nodes	Intel Core Duo	2GB	Ubuntu 12.04	Hadoop 0.20.2

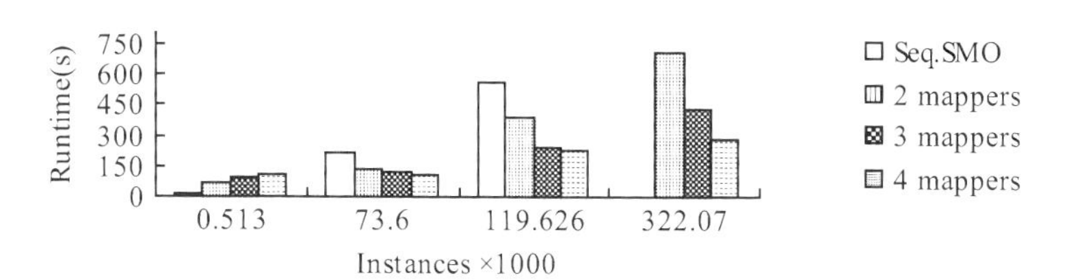

Fig. 67.3 A comparison on capability and runtime with sequential SMO

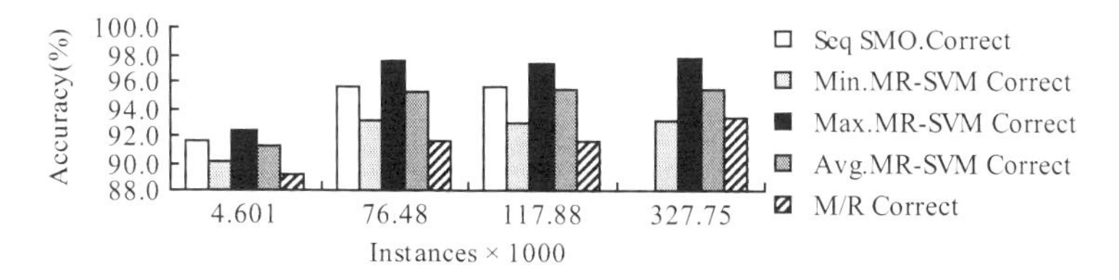

Fig. 67.4 A comparison on accuracy with sequential SMO and M/R algorithm

indicate that MR-SVM keeps the same high accuracy level as the SMO and exceeds that of the M/R algorithm among different sizes of datasets. It is illustrated that the global training in reduce phase and the feedback loop mechanisms in MR-SVM are greatly helpful for improving the accuracy.

67.5 Conclusions

In summary, this chapter described a parallel and convergent SVM classifier based on MapReduce framework. It was proved that MR-SVM always achieves the global convergence within finite iteration size. The experimental results illustrated that MR-SVM can improve the data process capability and computing efficiency significantly on the basis of keeping the same excellent accuracy as the traditional sequential SMO.

However, the parameters, such as C and epsilon, were specified as default values directly in this chapter, so the parameter optimization selection problem in MR-SVM should be included in the future research work.

References

1. Graf, H. P., Cosatto, E., Bottou, L., Durdanovic, I., & Vapnik, V. (2004). Parallel support vector machines: The Cascade SVM. *Advances in Neural Information Processing Systems (NIPS)*, *17*, 521–528.
2. Salleh, N. S. M., Suliman, A., & Ahmad, A. R. (2011). Parallel execution of distributed SVM using MPI (CoDLib). In *International Conference on Information Technology and Multimedia (ICIM)* (pp. 1–4). IEEE.
3. Li, Q., Salman, R., Test, E., Strack, R., & Kecman, V. (2013). Parallel multitask cross validation for support vector machine using GPU. *Journal of Parallel and Distributed Computing*, *73*(3), 293–302.
4. Caruana, G., Li, M. Z., & Liu, Y. (2012) *An ontology enhanced parallel SVM for Scalable spam filter training [EB/OL]*. http://dx.doi.org/10.1016/j.neucom.2012.12.001. doi:10.1016/j.neucom.2012.12.001#_parent
5. Chu, C. T., Kim, S. K., Lin, Y. A., Yu, Y. Y., Bradski, G., Olukotun, K., et al. (2007). Map-reduce for machine learning on multicore. *Advances in Neural Information Processing Systems (NIPS)*, *19*, 281–288.
6. Vapnik, V. (1995). *The nature of statistical learning theory* (pp. 131–162). New York, NY: Springer.
7. Dean, J., & Ghemawat, S. (2008). MapReduce: Simplified data processing on large clusters. *Communications of the ACM*, *51*(1), 107–113.
8. Platt, J. C. (1999). *Fast training of support vector machines using sequential minimal optimization* (pp. 185–208). Cambridge, MA: MIT Press.
9. Wu, C. M., Wang, X. D., Bai, D. Y., & Zhang, H. D. (2009). Fast incremental learning algorithm of SVM on KKT conditions. *The Sixth International Conference on Fuzzy Systems and Knowledge Discovery* (pp. 551–554). IEEE.

Chapter 68
Vehicle Classification Based on Hierarchical Support Vector Machine

Mengwan Jiang and Haoliang Li

Abstract In order to solve the problem that the mature vehicle classification cannot meet the requirements of accuracy and speed concurrently, this chapter chooses the contour features and speeded-up robust features (SURF) features of vehicles and then adopts a hierarchical support vector machine (SVM) classifier for vehicle classification. At first, the system uses the contour features which are simple and fast in the first layer of classifier so that it will filter out the easy samples. Second, the system utilizes the rich information and stable SURF feature in the next layer of classifier. We conducted extensive experiments against a number of baseline methods; the accuracy of proposed method was increased by about 20 %, and the time was shortened by 2/3, significantly outperforming the baselines. The method of double features and hierarchical SVM has a good trade-off between speed and precision.

68.1 Introduction

Vehicle classification technology is an important component of intelligent transportation systems (ITS). It has been widely applied in many real-world applications, such as vehicle flow statistics, highway automatic toll station, management of all kinds of parking lot, and public security system. The traditional methods of vehicle classification are mainly based on the magnetic induction or ultrasonic technology. These methods encounter very high computational complexity. This may lead to a series of drawbacks, including time-consuming construction, expensive for creation and maintenance [1]. As a solution, we propose to apply image-processing technologies to support the classification system.

M. Jiang (✉) • H. Li
School of Information Engineering, Zhengzhou University, Zhengzhou 450001, China
e-mail: springmw@126.com; iehlli@zzu.edu.cn

W.E. Wong and T. Zhu (eds.), *Computer Engineering and Networking*, Lecture Notes in Electrical Engineering 277, DOI 10.1007/978-3-319-01766-2_68,

In of the area of image processing, existing works of vehicle classification use vehicle's color, length, width, height, and outline features [2–4] to classify all vehicles into three sizes: small, medium, and large. But it is difficult to meet the requirements of accuracy and speed in making adistinctionbetween passenger vehicles and lorries. In addition, it also cannot distinguish various small cars. Then the SURF feature can save the rich and important characteristic information [5], thus greatly improving the efficiency of classification. However, it has a relatively large amount of calculation and hereby does not meet the requirements as well. In terms of pattern classification, support vector machine (SVM) technique is a kind of pattern classification method [6]. It is believed to be outstanding in solving small sample, nonlinear, and high-dimensional pattern classification problems. This chapter studies the vehicle classification system, which chooses two features and uses a hierarchical SVM to meet the requirements of accuracy and efficiency.

68.2 The Algorithm of SURF Feature Extraction

On the selection of feature descriptor, speeded-up robust feature (SURF) point feature is fast [5], because it uses the Hessian matrix to improve the efficiency. It uses less time and is accurate to calculated the Hessian matrix. The key is the selection of location and scale space. Define a point X in the image, X = I (x, y), in the scale space too; the Hessian matrix of pixel X under the condition of scale space for σ is as follows:

$$H(x,\sigma) = \begin{pmatrix} L_{xx}(x,\sigma) & L_{xy}(x,\sigma) \\ L_{xy}(x,\sigma) & L_{yy}(x,\sigma) \end{pmatrix} \tag{68.1}$$

In the formula $L_{xx}(x,\sigma) = \frac{\partial^2}{\partial x^2} g(\sigma)$, $L_{xy}(x,\sigma)$, and $L_{yy}(x,\sigma)$ in the same way.

SURF uses integral image algorithms so that the computing speed is increased. The SURF feature descriptor detects the neighborhood of each extreme value points which is 4 × 4 regions and then calculates each Herr Wavelet response of the subdomain, a 4 d vector as

$$V = \left(\sum d_x, \sum d_y, \sum |d_x|, \sum |d_y|\right) \tag{68.2}$$

Finally all the 4 × 4 neighborhoods and the 4 d vectors constitute 64 (4 × 4 × 4)-dimensional feature vectors. SURF is better off because it uses the method of integral image, and compared with other feature descriptor, SURF's speed of producing feature points has improved significantly.

68.3 Hierarchical SVM

SVM is a new kind of general statistical method based on structure risk minimization principle [6], which can solve the practical problems such as small sample, nonlinearity, high dimension, and local minimum point. There are two measures of controlling SVM's generalization: empirical risk and confidence limit. It uses the training error as the constraints of optimizing problems, and the goal of optimization is to minimize the confidence limit. As a result, the SVM is to solve a convex quadratic programming under leading constraints, making the solution of SVM the only, also globally, optimal solution.

The SVM is derived and developed from the optimal classification plane under the linear separable cases. The classification functions under the linear separable cases are shown below:

$$f(x) = sign\left(\sum_i y_i \alpha_i K(x, x_i) + b\right) \tag{68.3}$$

In this function, x_i is the i dimension of training space vector X, y_i represents category x_i and its value is 1 or -1, and α is the multiplicator of Lagrange. For most i, α_i is 0. K is the dot product of the convolution kernel function. It can turn inner product operation of high-dimensional feature space into functional operation of low-dimensional input space. The curse ofdimensionality caused by having nonlinearity of input space maps to high-dimensional feature space and can also be solved by kernel function. There are four basic kernel functions: linear, polynomial, RBF, and sigmoid kernel function. According to the features tested in this chapter, we finally choose the RBF kernel function. Its expression is $K(x_i,y_i) = \exp(-\gamma\|x_i - y_i\|^2)$, $\gamma > 0$. The kernel parameter γ is determined by the results ($\gamma = 0.15$ in this chapter).

At first SVM classification algorithm is only applicable to solve binary classification problems, lacking the ability to handle multiple classification problems [7]. In a binary classification problem, we are given a training set $\{x_i, y_i\}$, $i = 1,\ldots, 1$, $x_i \in R^n$ is an n dimension real vector which represents sample properties. $y_i \in \{1, -1\}^1$ is the category labels of data; then SVM can come down to an optimization problem as shown below:

$$\min_{w,b,\xi} \frac{1}{2} w^T w + C\sum_{i=1}^{l} \xi_i$$

$$\text{Subject to} \quad y_i\left(w^T \phi(x_i) + b\right) \geq 1 - \xi_i \quad \xi_i \geq 0. \tag{68.4}$$

C > 0 is a penalty parameter of error term. ϕ is a mapping function, and $K(x_i,x_j) = \phi(x_i)^T\phi(x_j)$ is the kernel function.

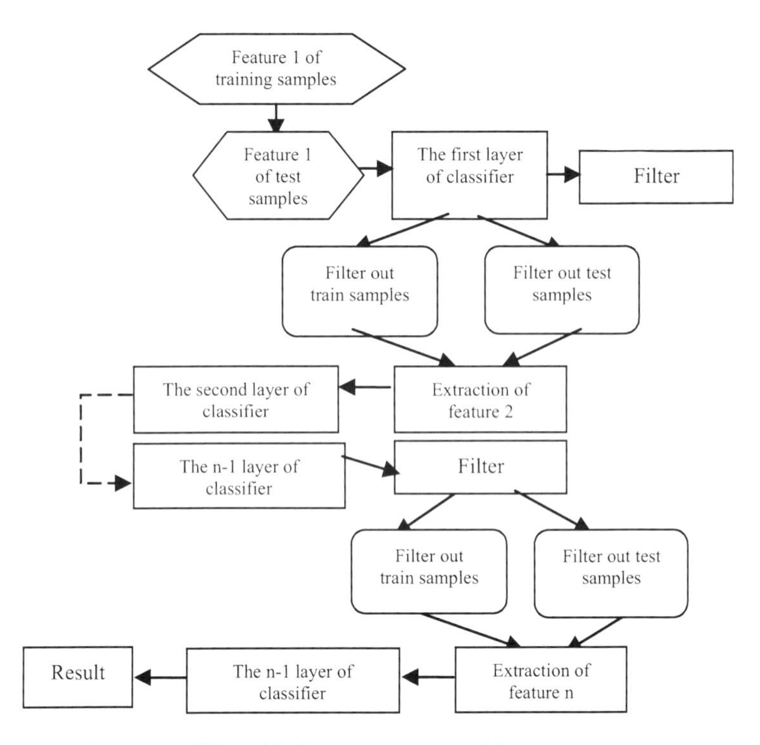

Fig. 68.1 The flow chart of hierarchical support vector machine

In order to extend the SVM to the multi-classification, now there are two ways in general: one to one and one to many. In the SVM experiment given in this chapter a hierarchical SVM method with multi-features and one-to-many method in every layer of classification are used. At first, the hierarchical SVM method is to use the feature one of all the train samples to train the first layer of classifier [8] so as to filter out the samples which are not easy to identify according to the different threshold which is set at first. Then to extract feature two from the difficult samples after filtering, use the feature two to train the next layer of classifier. Test samples work in the same way, firstly to test the test samples in the first layer of classifier, filter out difficult samples according to the predicted value and threshold value, and then send them into the next layer of classifier. By such analogy, finally we get relatively the classification result through filtering step by step and centralized training. Specific flow chart is shown below (Fig. 68.1):

Compared with the traditional SVM, a hierarchical SVM has the following two advantages:

1. As the filter time increases, it is more difficult to classify the training sample which is filtered out, making the training more targeted. So we can filter out the easy samples in the previous layers of classifier and then use difficult samples, which are similar among samples to train SVM intensively, making it more targeted and more likely to get the optimal solution.

2. Each layer can extract different features for classification, making the method more flexible. Due to a hierarchical SVM trains on each layer step by step, according to the need, it can try to extract different features from samples on each layer.

68.4 Test and Result Comparison

We collected 240 vehicle images, including 8 types of vehicles. There are 20 training samples and 10 test samples in every type, building the image dataset.

68.4.1 Feature Extraction

In this process, the system firstly extracts the outline features of all the samples and the SURF features of cars. The example's result of extracted features of outline feature and SURF feature shown as Figs. 68.2 and 68.3.

After extracting vehicles' features, the feature set will be later post-processed, through methods such as k-means algorithm and taking average by dividing area [9], in order to obtain a single feature vector, which is the characteristic classification to the corresponding vehicles.

68.4.2 Training and Testing the Classifier

In training, use the contour feature information of all training vehicles to train the first layer of classifier, filtering out large-type truck, medium-type truck, large buses, and medium bus, the four samples which are easy to be identified. Use the SURF feature of the cars (small vehicles) to train the second layer of classifier.

In the first layer of classification, send the contour features of the images after preprocessing and feature extraction into the first layer of classification. Use model

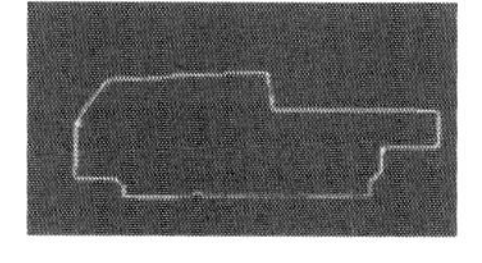

Fig. 68.2 Outline feature of a medium-type truck

Fig. 68.3 SURF feature of a multipurpose vehicle

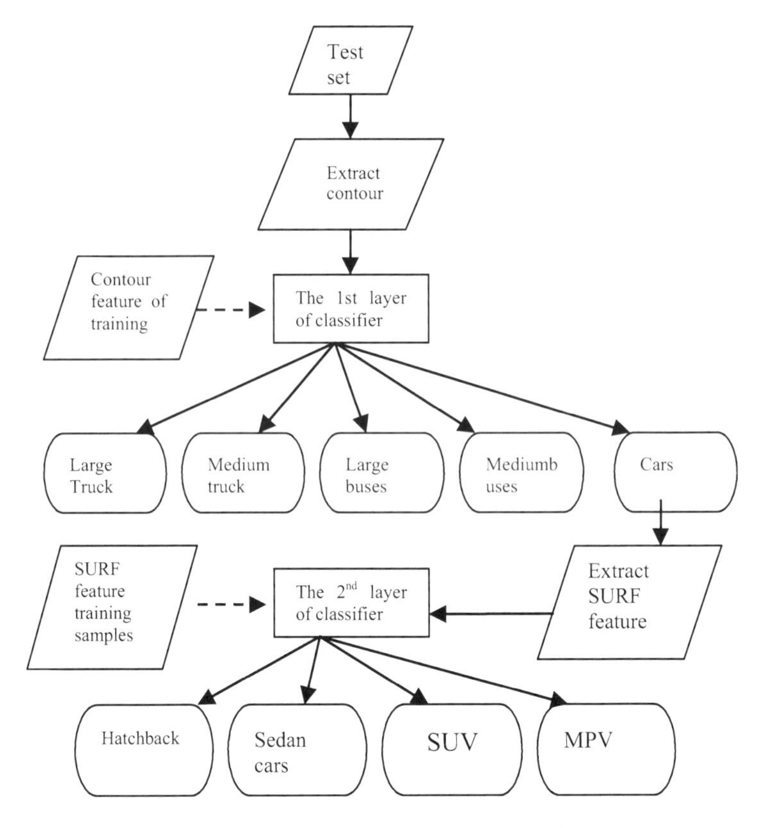

Fig. 68.4 Training and test flow of the double features and hierarchical SVM

code as output vector of the network to classify large-type truck, medium-type truck, large buses, and medium buses. Extract the SURF feature of the small vehicles, which are after filtering by the first layer of classifier, and then send them into the second precise layer of classifier to classify the vehicle by sedan car, hatchback SUV, and MPV.

The double features and hierarchical SVM training and testing flow are shown below. The dotted line is training process, and the solid line is testing process (Fig. 68.4).

68.4.3 Results and Comparisons

This chapter is implemented on MATLAB2010 with LIBSVM toolbox for the hierarchical SVM pattern classification. LIBSVM provides not only the source code but also some executable files under the Windows operating system, including svmtrain function, which is used to train SVM; svmpredict function which is to

Table 68.1 Classification results of double features and hierarchical SVM

The number of training samples	The number of test samples	The number of support vector	Total training time (s)	Total test time (s)	Rate of classification
160	80	42	47.7	12.69	87 %

Table 68.2 Comparison results of different methods

Methods of experiment	Total time of 160 training samples	Total time of 80 test samples	Rate of classification
Classification of single outline feature	31.8	9.17	69 %
Classification of single SURF feature	56.9	16.41	82 %
Traditional SVM classification of double feature	72.6	19.82	90 %
Hierarchical SVM classification of double feature	47.7	12.69	87 %

predict the test data according to the acquired training model; and svmscale function which is to normalize the training data and test data. In the experiments, we choose the default model parameters. The classification results of double features and a hierarchical SVM are shown in Table 68.1.

In the experimental results, total training time of 160 training samples is 47.7 s, less than a minute. And total test time of 80 test samples is just 12.69 s, one sample time about 0.16 s on average; therefore it can meet the speed requirement. Due to using the default model parameters of LIBSVM toolbox, it works well in terms of accuracy with classification rate of 87 %.

The contrast test of using different methods but all default values of classification with the same training samples and testing samples further verifies that the method used in this chapter is effective, including single contour feature classification, single SURF feature classification, and double features of the traditional SVM classification. Comparison results are shown in Table 68.2.

It can be seen from the result and comparisons of the experiment that double-feature hierarchical vehicle classification compared to single-feature recognition method, which is the fastest, is tens of seconds later, but the accuracy increased from 69 to 87 %. Compared to the double-feature traditional classification method which has the highest accuracy, the accuracy only reduced, three percentage points (from 90 to 87 %), but it is also reduced by 40 % time cost.

68.5 Conclusion

Compared with the single-feature or traditional classification method, this chapter used double-feature hierarchical vehicle classification. It can filter the easy samples with fast feature extraction by the first layer of classifier which is also trained by the fast feature, and then it sends the difficult samples into the second layer of classifier which is trained by rich information and high precision feature to classify by filtering step by step and centralized training. So the hierarchical classification can meet the requirements for high accuracy and fast speed.

References

1. Ren, J. Q. (2011). Design of vehicle type recognition algorithm based on video sequence. *Computer Engineering, 37*(24), 245–250.
2. Luo, W. T., Hsieh, J. W., & Fan, K. C., (2007). Vehicle detection using normalized color and edge map. *IEEE Transactions on Image Processing, 16*(3), 850–864.
3. Zhang, Q. Y., Dai, G. M., & Chen, L. (2008). Vehicle recognition system design based on speed videos. *Microcomputer Information, 24*(31), 288–290.
4. Jia, Y. Q., & Zhang, C. S. (2009). Front-view vehicle detection by Markov chain Monte Carlo method. *Pattern Recognition, 42*(3), 313–321.
5. Bay, H., Tuytelaars, T., & Van Gool, L., (2006). SURF: Speeded up robust features. *Computer vision—ECCV 2006* (pp. 404–417). Graz, Austria.
6. Hsu, C. W., & Lin, C. J. (2002). A comparison of methods for multi-class support vector machines. *IEEE Transactions on Neural Networks, 13*, 415–425.
7. Castrillon Santana, M., & Vuong, Q. C. (2007) An analysis of automatic gender classification. In *Proceedings of the 12th Conference on Progress in Pattern Recognition, Image Analysis and applications* (pp. 271–280). Valparaiso, Chile: Springer-Verlag.
8. Li, K. L., & Liao, P. (2012). Face image gender identification based on cascade connection support vector machine. *Computer Engineering., 38*(12), 152–154.
9. Zhang, S. S., & Zhan, Z. C. (2012). Research on vehicle classification system based on SIFT features and support vector machine. *Computer Knowledge and Technology, 08*(17), 4277–4280.

Chapter 69
Image Splicing Detection Based on Machine Learning Algorithm

Yan Xiao

Abstract Image splicing is a common method to construct forged image which decreases the authentication of the traditional image. Resizing operation is usually necessary to create a convinced forged image. Though the forged image leaves no visual clues, resizing operation using interpolation method destroys the relationship between neighboring pixels, thus leaving traces which can be captured by statistical feature. We first convert the traces left by resizing to feature and then feed features from enough sample images to support vector machines to train for detector. Finally, we use detector to determine whether the image is tampered and point out which parts of the image are tampered by block-wise method. Experimental results verify the effectiveness of our proposed method.

69.1 Introduction

Today's digital technologies make the image edition easily, so there are many forged images appearing in our daily life. Though elaborate tampering operations almost leave no visual clues, the correlations between adjacent pixels are disrupted, and such changes can be captured by statistical method just as forensics. The digital forensic method is used to check the validity of the image based on statistical fingerprint without inserting any evidence into the image.

In recent years, many forensic methods are designed for image tampering detection. Farid divided the forensic tools into five groups: (1) pixel-based technology which detects the fingerprint at the pixel level; (2) format-based technology mainly for JPEG compressed images; (3) camera-based and physically based technology; and (4) geometric-based technology [1]. Image splicing is a common

Y. Xiao (✉)
Southwest University of Science and Technology, Mianyang 621000, China

Mianyang Vocational and Technical College, Mianyang 621000, China
e-mail: 348105095@qq.com

W.E. Wong and T. Zhu (eds.), *Computer Engineering and Networking*, Lecture Notes in Electrical Engineering 277, DOI 10.1007/978-3-319-01766-2_69,

forged manipulation which spliced two or more images into a forged image. Some authors found that the statistics on high-order Fourier can be used as fingerprint [2, 3].

Resizing operation is usually employed to create an applauded forged image. This process employs re-sampling operation which introduces period correlations between neighboring pixels caused by interpolation algorithm. Farid et al. utilized EM algorithm to reveal the above period relation and using this correlation to test whether the tested image is resized or not [4]. Babak Mahdian et al. demonstrated that the interpolated signal and its derivatives contained specific periodic properties and proposed a blind and efficient method to detect re-sampling and interpolation [5]. Andrew C. Gallagher also exploited a periodicity in the second derivative signal of interpolated images and employed discrete Fourier transform (DFT) to capture such periodicity in frequency domain [6]. Previous works [4–6] utilized periodicity as fingerprint left by interpolation; they did not deal with the situation that the periodicity contained in the rescaled image is faint such as when the rescaled factor is near to 1 (1.01, 0.99). In this chapter, the correlations between neighboring pixels are usually used as fingerprint for forensic tools. Chen et al. [7] proposed a blind and effective splicing approach based on a natural image model. They finally used feature sets and support vector machines to classify whether images are forged or not. The transition probabilities of Markov chains between neighboring pixels are modeled to construct feature set to detect median filtering [8].

As resizing operation will destroy the correlations between neighboring pixels, we use the method as [8] to construct the feature set and use the support vector machines (SVM in short) to learn the difference between original images and resized images. In order to locate spliced part, SVM is trained for a classifier on small-size images. The tested image is first divided into blocks and then using the classifier to test each block whether tampered or not.

The remainder of this chapter is organized as follows. In Sect. 69.2, we first construct the feature set and train for a classifier and then test the image using block-wise method. In Sect. 69.3, the experimental results show the effectiveness of our proposed method. Finally, we draw a conclusion at the end of the chapter.

69.2 Procedure of Our Proposed Method

In this section, we first introduce how to construct the feature to model the different correlations between rescaled images and non-rescaled images. Then the SVM are briefly introduced. Finally we give a detailed description of our method to detect image splicing.

When resizing an image, re-sampling method is always used. The bilinear and bi-cubic interpolation algorithm is a popular method accompanied with resizing operation. Let us assume that $x(i,j)$ is the pixel in original image and $y(i,j)$ is the pixel in rescaled image. When using bilinear interpolation to compute the value of $y(i,j)$, the following formula is held:

$$\begin{aligned} y(i+u, j+v) = {} & (1-u) \times (1-v) \times x(i,j) + (1-u) \times v \times x(i,j+1) \\ & + u \times (1-v) \times x(i+1,j) + u \times v \times x(i+1,j+1), \end{aligned} \quad (69.1)$$

where u and v are the displacement from i and j, respectively, and $0 \leq u, v \leq 1$. From Eq. (69.1), it is easy to infer the neighboring pixel value of $y(i,j)$. Taking $y(i + 1 + u, j + v)$ for example:

$$\begin{aligned} y(i+1+u, j+v) = {} & (1-u) \times (1-v) \times x(i+1,j) + (1-u) \times v \\ & \times x(i+1,j+1) + u \times (1-v) \times x(i+2,j) + u \\ & \times v \times x(i+2,j+1). \end{aligned} \quad (69.2)$$

It is obvious that there are two same pixels ($x(i + 1, j)$, $x(i + 1, j)$) that are contained in Eqs. (69.1) and (69.2). Thus, the correlations of neighboring pixels between interpolated images and non-interpolated images are different. Kirchner et al. model the fingerprint left by median filtering using first-order difference for adjacent pixels and utilized subtractive pixel adjacency matrix to construct the feature set [8]. Inspired by [6], we model the neighboring pixels by second-order difference and utilize subtractive pixel adjacency matrix to construct feature set. We model the adjacent pixels by second-order difference along horizontal and vertical directions as Eq. (69.3):

$$d_{i,j}^{k,l} = I(i+2k, j+2l) + I(i+k, j+l) - 2I(i,j), \quad (69.3)$$

where the superscript $(k,l) \in \{(0,1),(0,-1),(1,0),(-1,0)\}$. As the bilinear interpolation method considers two neighboring pixels, we consider the correlations among three neighboring pixels. The correlations of neighboring pixels are expressed by transition probabilities of second-order Markov chains as Eq. (69.4):

$$P_{\alpha,\beta,\gamma}^{k,l} = Pr\left(d_{i+2k,j+2l} = \alpha \middle| d_{i+k,j+l} = \beta, d_{i,j} = \gamma\right), \quad (69.4)$$

where Pr(.) in Eq. (69.4) means the probability.

Finally, we compute the feature by averaging four directions as Eq. (69.5):

$$P_{\alpha,\beta,\gamma} = \left(P_{\alpha,\beta,\gamma}^{0,1} + P_{\alpha,\beta,\gamma}^{0,-1} + P_{\alpha,\beta,\gamma}^{-1,0} + P_{\alpha,\beta,\gamma}^{1,0}\right)/4. \quad (69.5)$$

If the pixel value ranges from [0, 255], the residual pixel in ranges [−255, 255], there are many elements in feature sets. It is unfeasible to feed such enormous feature into SVM to train for a model, so the truncation must be adopted to decrease the dimensionality of the feature. In order to lower feature's dimensionality, truncation of is executed. When the absolute value of difference is larger than a threshold T, we truncate it. In the experiment, we set $T = 3$ which results in 343 dimensionality feature.

The SVM is employed to train for the detection classifier. SVM is a classical supervised learning algorithm which can be used for binary classification [9]. In order to be a better classifier, SVM find an optimal hyper-plane which makes the nearest distance between the training samples and hyper-plane as large as possible, which can be formulated as an optimization problem as Eq. (69.6):

$$\begin{aligned} &\min_{\mathrm{w,b},\xi} \frac{1}{2}\mathrm{w}^{\mathrm{T}}\mathrm{w} + C\sum_{i=1}^{l}\xi_i \\ &\mathrm{St.y_i}\left(\mathrm{w}^{\mathrm{T}}\varnothing(\mathrm{x_i}) + \mathrm{b}\right) \geq 1 - \xi_i, \xi_i \geq 0, \end{aligned} \tag{69.6}$$

where x_i is the training feature, y_i is the training label, and $i = 1, 2, \ldots, 1$. $\varnothing$ (x_i) is the mapping function which will be used in kernel function. $C > 0$ is the penalty parameter of the error term. The solution of Eq. (69.6) is usually transferred to its dual-quadratic problem (69.7) to solve

$$\begin{aligned} &\min_{\alpha} \frac{1}{2}\alpha^{\mathrm{T}}Q\alpha - \mathrm{e}^{\mathrm{T}}\alpha \\ &\mathrm{St.y}^{\mathrm{T}}\alpha = 0, C \geq \alpha_i \geq 0, \end{aligned} \tag{69.7}$$

where e is a unit vector and Q is an $l \times l$ positive semi-definite matrix computed by kernel functions. The other symbol is the same as in Eq. (69.6). After solving the problem (69.7), the decision function is used to predict the label of the testing samples.

LIBSVM is an integrated software for support vector classification; we use C-SVC with RBF kernel to learn the model [10]. The non-forged image is given a label 0, and the forged image is labeled 1. As the shape of the pasted object is not regular and the position of the pasted object is also unknown, the block-based method is employed to test whether the image is tampered or not. So the classifier is obtained on small-size image for highly accurate detection results. The size of trained images is designed to be the same as the size of blocks. In general, the larger size of trained image will get better results; however, it will cause losses of detail position at location.

Once getting the classifier, the block-wise method is employed to test the whether the image is spliced or not. First, the test image is divided into non-overlapping blocks of size $S \times S$ with step size S. For an $M \times N$ image, there are totally floor $(M/S) \times$ floor (N/S) blocks (floor (x) is the function for the largest integer value which is smaller than x). In general, the smaller block will get more detailed detection results; however, the accuracy of classifier obtained from small-size image is not high. So the proper size of the block is an important factor. Then, compute the feature for every block. Finally, the classifier is used to predict each block's label. If the block's label is 1, the block is considered as a forged block and mark it with red color box.

At last, we summarize our proposed method as follows:

(1) Computing the feature sets from large number of small-size trained images. In our experiment, we set two kinds of images (128 × 128 and 64 × 64) and then used SVM to train for a classifier on the above training set.
(2) Dividing the tested image into blocks and computing each block's feature.
(3) Feeding each block's feature into the trained classifier to predict whether each block is tampered or not.
(4) Locating the tampered object by joining all tampered blocks.

69.3 Experimental Results

In this section, we first test the efficacy of our proposed method on small-size images and demonstrate the performance by receiver operating characteristic curves (ROC in short) in Fig. 69.1. Then we use our proposed method to detect a spliced image, and the detection results are shown in Fig. 69.2.

In order to learn more knowledge from images, we prepare large number of diverse images to train. The UCID [9] image database which is popular in image forensics contains 1,338 images of size 512 × 384. All training samples in our experiment are shown by 8-bit gray image. We first convert all images into 8-bit gray images; then crop 128 × 128 and 64 × 64 blocks from each image, and then rescale blocks using bilinear interpolation method (scaling factor is randomly set to 0.95 or 1.05). Four groups of images are obtained: 1,338 128 × 128 original images; 1,338 128 × 128 rescaled images; 1,338 64 × 64 original images, and 1,338 64 × 64 rescaled images. First, we randomly select 338 original images and 338 rescaled images to test our proposed method's efficacy for each size of image. The training set is as follows: 1,000 original images + 1,000 corresponding rescaled images for each size. In general, more training samples and higher dimensionality of feature will obtain higher detection accuracy but will result in

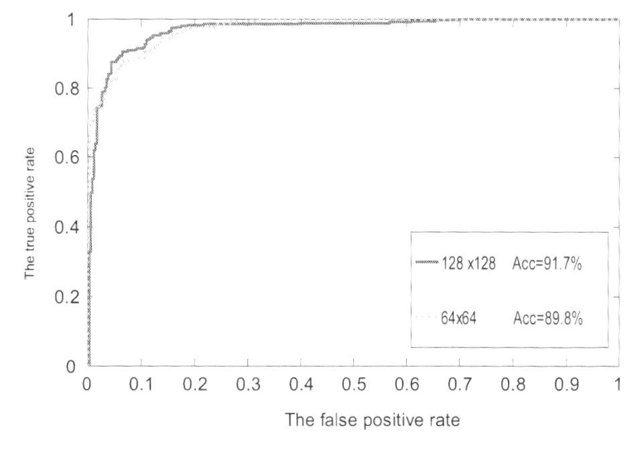

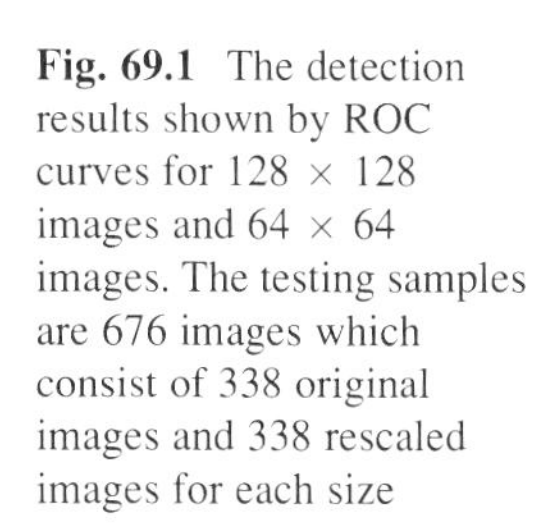

Fig. 69.1 The detection results shown by ROC curves for 128 × 128 images and 64 × 64 images. The testing samples are 676 images which consist of 338 original images and 338 rescaled images for each size

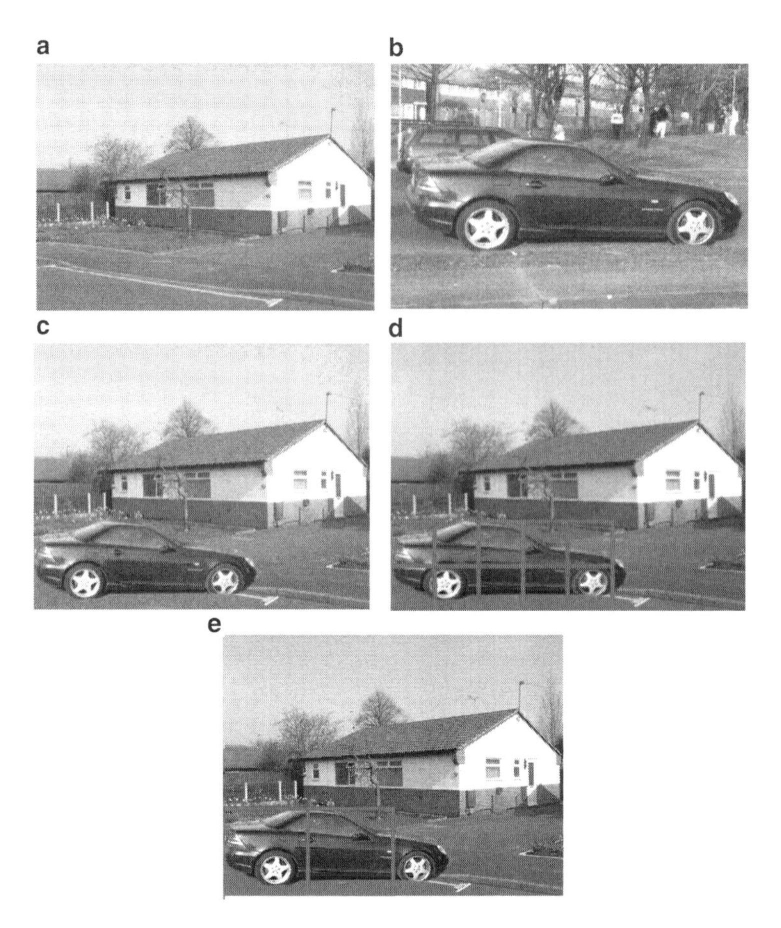

Fig. 69.2 The detection results for first forged image. Panel (**a**) is the original image, and the car in (**b**) is cut and pasted into (**a**) to form a forged image shown in (**c**). Panel (**d**) shows the detection results of 64 × 64 detection blocks. Panel (**e**) shows the detection results of 128 × 128 detection blocks. The *square boxes* are to indicate forged part

more computing time. In order to balance the computing time and detection accuracy, the dimensionality is set to 343 and the amount of training sample is 1,000. We then compute the 343-D feature for each image in the above four groups. In order to get the best parameter C and r, fivefold cross-validations are executed on the grid: $\{(C,r) \in (2^i,2^j), i, j \in Z\}$. The ROC curves are used to assess our proposed method's performance. The accuracy as follows is used to show our results at scale:

$$\mathrm{Acc} = \frac{|\text{correctly predicted data}|}{|\text{total testing data}|} \times 100\% \tag{69.8}$$

Figure 69.1 shows that our proposed method obtains high detection accuracy for both kinds of images. Even for small 64 × 64 images, our proposed method detection accuracy is as high as 89.8 %. When the false-positive rate is low (<10 %), the positive rate of our method is also acceptable. In previous work [2, 10, 11], they also got high detection accuracy, but the size of tested image is larger than our tested image. If the scaling factor is larger than 1.05, better results will be obtained. The same results will be held for other two interpolation methods (nearest interpolation and bi-cubic interpolation). Figure 69.1 indicates that our proposed method will be useful when detecting parts of image are resized.

69.4 Image Splicing Detection

In this experiment, we also first prepare training samples to train for classifiers. In order to test the influence of the size of block, we get two classifiers on different size of image. One classifier is obtained on 128 × 128 images; the other classifier is obtained on 64 × 64 images. The training samples are also from UCID database. The training set is composed by 1,336 original images and 1,336 corresponding rescaled images for each size. After computing the 343-D feature of trained images, the training feature set is a 2,672 × 343 matrix. We label 1,338 original block as 0 and 1,338 rescaled images as 1 for each size. The setup of SVM is the same as first experiment. The SVM training will implement on the above feature set and get the predicted model.

After getting the model, we will use the classifier to detect the spliced image. In order to get a convincing result, we first introduce how to create the forged image in our experiment. The two original pictures in Fig. 69.2 come from the UCID database. These two images are not contained in the training samples, so the number of training samples is $2 \times 1,\ 338 - 2 = 2672$. The original images in Fig. 69.2a, b are 512 × 384 in TIFF format. The Adobe Photoshop CS 2 is employed as editorial tool. When the car is pasted into the first picture, resizing operation must be employed to lessen the size of car for a convinced forged image. The forged image saved as TIFF format showed in Fig. 69.2c is the same size as original image. It is hard to distinguish whether the image in Fig. 69.2c is forged or not by human's eyes. When using detecting blocks with size 64 × 64, there are totally 48 blocks that need to be tested. We first compute the feature for each block. Then we use the classifier obtained on training samples of size 64 × 64 to predict these 48 blocks. The same procedure is executed for detection block of size 128 × 128.

Figure 69.2d shows that our proposed method locates the forged part with high accuracy; just only a part of car is not covered by 64 × 64 block. Though the marked box is square, the people can easily take the car as forged part by empirical experience. For the 128 × 128 block, Fig. 69.2e shows that it also obtains good results, but the detection results are roughly caused by the larger detection block. If the tested image is of large size and the spliced object is also large, larger detection block is preferable. Though the accuracy of classifier obtained on small-size training samples may not be very high, it is also the best choice to detect image splicing.

69.5 Conclusion

In this chapter, machine learning method was employed to detect image splicing. Most of the image splicing operations will resize the pasted object for a better tampering, which will bring re-sampling artifacts in the forged image. The modified SPAM feature is used to capture the traces left by re-sampling. The feature of training blocks is fed into SVM for a classifier. Block-based method was used to test which parts of the image are forged by the classifier. Experiments showed that the proposed method can detect forged parts of image efficiently. Making a robust method (such as against JPEG compression) deserves more attention in the future.

References

1. Farid, H. (2009). Image forgery detection. *IEEE Signal Processing Magazine*, *5*(3), 16–25.
2. Farid. H. (1999). Detecting digital forgeries using bi-spectral analysis. AI Lab, Massachusetts Institute of Technology, Tech. Rep. AIM-1657
3. Ng, T. T, & Chang, S.-F. (2004). A model for image splicing. *Proceedings of IEEE International Conference on Image Processing. Singapore* (Vol. 2, pp. 1169–1172).
4. Popescu, A. C., & Farid, H. (2005). Exposing digital forgeries by detecting traces of re-sampling. *IEEE Transactions on Signal Processing*, *53*(2), 758–767.
5. Mahdian, B., & Saic, S. (2008). Blind authentication using periodic properties of interpolation. *IEEE Transactions on Information Forensics and Security*, *3*(3), 529–538.
6. Andrew C. Gallagher. (2005). Detection of linear and cubic interpolation in JPEG compressed images, *Proceedings of the Second Canadian Conference on Computer and Robot Vision*. Vol. 33, pp. 65–72.
7. Shi, Y. Q., Chen, C. H., & Chen, W. (2007). A natural image model approach to splicing detection, MM and Sec'07, *Proceedings of the Multimedia and Security Workshop, Dallas* (pp. 51–62).
8. Kirchner, M., & Fridrich, J. (2010). On detection of median filtering in digital images, *Proceedings of SPIE, San Jose USA* (Vol. 7541, pp. 300–308).
9. Cristianini, N., & Taylor, J. S. (2000). *An introduction to support vector machines and other kernel-based learning methods* (pp. 230–250). London: Cambridge University Press.
10. Chang, C.-C., & Lin, C.-J. (2011). LIBSVM: A library for support vector machines. *ACM Transactions on Intelligent Systems and Technology*, *5*(7), 230–235.
11. Schaefer, G., & Stich, M. (2004). UCID-An uncompressed color image database, *Proceedings of SPIE, Storage and Retrieval Methods and Applications for Multimedia, San Jose, USA* (pp. 472–480).

Chapter 70
A Lane Detection Algorithm Based on Hyperbola Model

Chaobo Chen, Bofeng Zhang, and Song Gao

Abstract In order to improve the problem of recognition rate and inaccurate in the curve, this paper proposed a lane detection algorithm based on hyperbola model, which uses Canny operator to detect the edge of the lane and wields the Hough transform to extract lane boundary points, and utilizes extended Kalman filter to reduce road scanning range. By fitting points on pair road boundaries into the hyperbola model, and completes the lane boundary reconstruction. Some experimental studies are conducted, and the results show that the accuracy of the algorithm has reached 93.4 % and the processing speed of each image needs 77.4 ms. Our method is able to make full use of lane boundaries with existence partial occlusion, blur and low contrast. Meanwhile, it can quickly and accurately identify lane line, and it has high performance and robustness.

70.1 Introduction

Vision-based intelligent navigation system is an important application field of computer vision [1]. At the present stage, machine vision and autonomous navigation often is used in intelligent vehicle road recognition [2]. The method is simple and practical and can adapt to the complex road environment [3]. Currently, road line detection algorithm consists of feature based and model based [4]. Feature-based algorithm mainly identifies some characteristics of road, such as color [5], texture, and shape. However, illumination changes, water stains, shadows, damaged and discontinuous road markings will influence detection effect [6]. Model-based algorithm primarily establishes road model and then according to image analysis determines model parameters; and model contains all of the information of lane line. The approach mainly differs in models, such as straight model, parabolic

C. Chen • B. Zhang • S. Gao (✉)
Department of Electronic and Information Engineering, Xi'an Technological University, Xi'an 710021, China
e-mail: gaosong@xatu.com.cn

W.E. Wong and T. Zhu (eds.), *Computer Engineering and Networking*, Lecture Notes in Electrical Engineering 277, DOI 10.1007/978-3-319-01766-2_70,

model, and spline curve model [7, 8]. There are a lot of lane detection algorithms that use straight model, parabolic model, spline curve model, etc. The analysis of the performance of the algorithms indicates that the straight model is fast, while recognition accuracy of the curve is weak; Parabolic model is weak that steady of straight and curve lines and markings is able to deviate; spline curve model is too complex, slow and takes large computational resources. To solve the problems mentioned, we have proposed a lane detection algorithm based on hyperbolic model. Firstly, the algorithm uses Canny operator to detect the edge of the lane and Hough transform to extract lane boundary points, and utilizes extended Kalman filter predictive tracking algorithm to reduce road scanning range. By the way, it can improve the accuracy of extraction. Eventually, by the left and right lane boundary parameters match with the hyperbola model, meanwhile uses the least squares method to solve the model parameters, and completes the lane boundary reconstruction. It can overcome parabolic model in curve and straight lanes of joint discontinuous problem, and experiment in many different conditions, includes various weather and road. It has high performance and accuracy.

70.2 Road Model

The paper uses lane line combination model to build model. Near field applies to the linear model is $u = bv - a$, and far area applies to hyperbola model is $u - u_H = {}^{k}/_{v-l}$. As shown in Fig. 70.1. We establish a hyperbolic model on both sides of the road:

$$u - u_H = \frac{k}{v - l} + b(v - l) \tag{70.1}$$

where l represents position of the vanishing line of the road on the image plane, k, b and u_H of which are hyperbolic model parameters, k represents the curvature of lane, b represents relative direction of lane, u_H represents the distance which is between the lane line and the vertical axis. These parameters may be get that they have been calculated the road image data with internal and external parameters of

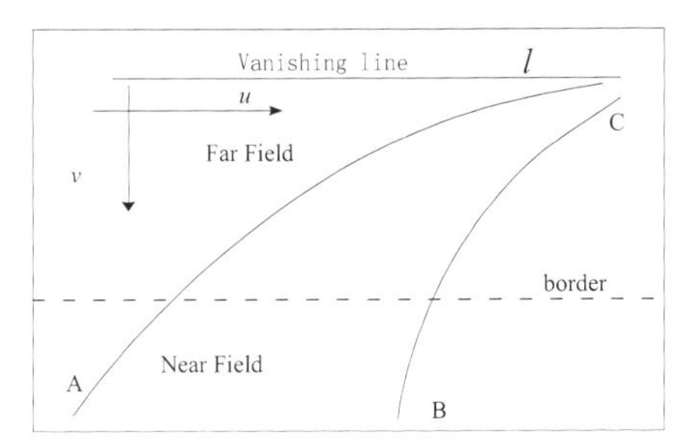

Fig. 70.1 Road model

the camera. One side of lane may be blocked by other vehicles, and leading to CCD camera can't be collected image information. Therefore, the paper proposes that use a pair of hyperbolic model, but we assume that the marking line is parallel and equal width. So the parameters can be solved with equation family on both sides of road line, the left and right road line equations are as follows:

$$u - u_H = \frac{k}{v - l} + b_l(v - l) \tag{70.2}$$

$$u - u_H = \frac{k}{v - l} + b_r(v - l) \tag{70.3}$$

As the boundary line on both sides of road is parallel, boundary parameters k and b value are the same, but u_H is different. The lane model can be formulated by an extended equation:

$$u - u_H = \frac{k}{v - l} + b_l(v - l) + b_r(v - l) \tag{70.4}$$

$b(l) = 0$, it is on the right lane line and $b(r) = 0$ if it is on the left one. This chapter combines both sides of the curve into a pair of hyperbola, and that curve parameters add only one. It is less increased calculation that it greatly improves the accuracy and robustness of the curve model, especially when the lane line is blocked, or partially damaged.

70.3 Lane Detection Algorithm Method

Vision-based lane line detection establishes the road model which is commonly used method for mostly scholars. The algorithm described in this paper is quite unique as it uses a combination of scan lane lines and improved Hough transform to match a hyperbola model. The combinatorial hyperbola model is rarely used, while most of the lane detection is conducted by template-based models. First, the algorithm converts the image to a grayscale. Due to the presence of noise in the image, so we apply the Bilateral Filtering algorithm. After this, the edge detector is used to produce an image edge by Canny operator, the image has detected which produce the left and right lane boundary. Next, the lane boundary scan uses the information in the edge image detected by Hough transform to perform the scan and obtain a series of points on the left and right lane lines, and using extended Kalman filter predictive tracking algorithm to reduce road scanning range. Finally, combinatorial hyperbola model is fitted to the date points and completed the lane boundary reconstruction (Fig. 70.2).

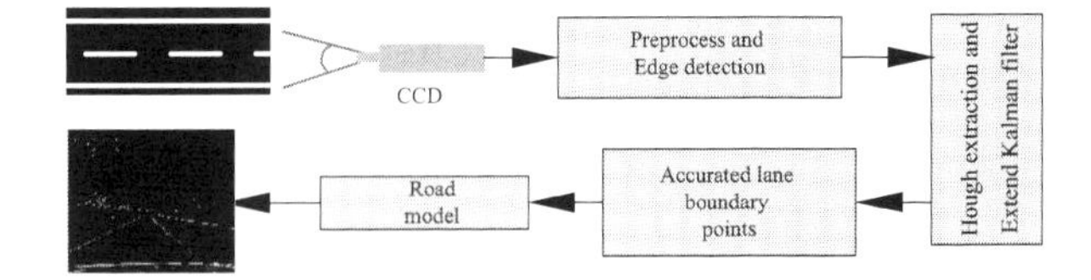

Fig. 70.2 Algorithm structure

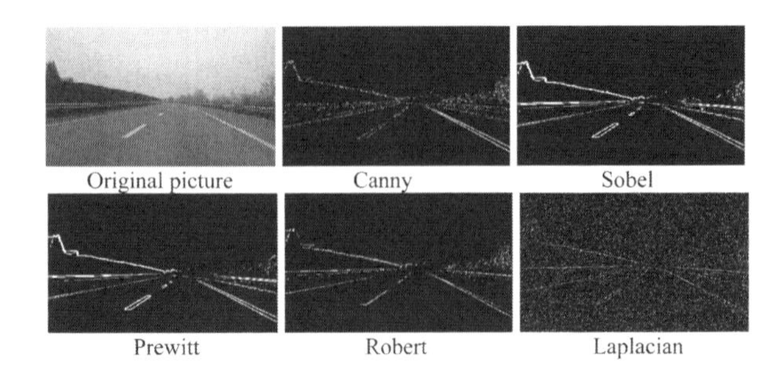

Fig. 70.3 Operator contrast figure

70.3.1 Lane Edge Extraction

Autonomous unmanned vehicle systems acquire pictures which are containing noise, but, presence of noise in our system will affect edge detection, so noise removal is very important, Bilateral Filtering is used to achieve edge-preserving de-noising. Now existing edge detection operator: Canny, Sobel, Robert, Prewitt, Laplacian, etc. Among Canny operator is a filter, enhancement and detection of multistage optimization operator. Detecting edges more complete and better positioning performance. Considering the performance and speed characteristics of the operator, the algorithm using Canny edge detection in this paper. Lane boundaries are defined by sharp contrast between the road surface and painted lines obviously form edges on the image. Thus, Canny edge detector was employed in determining the location of lane boundaries. Operator compare shown in Fig. 70.3.

70.3.2 Lane Line Extraction

Lane extraction is used a standard Hough transform by a restricted search space. Its character is accurate extraction and high anti-interference ability. First, Hough transform obtain roadway centerline and vanishing line, then extracting road on both sides of the boundary line with real-time road of the same width. Hough transforms searches for lines using the equation shown in Fig. 70.4.

$$\rho = x\cos\theta + y\sin\theta \tag{70.5}$$

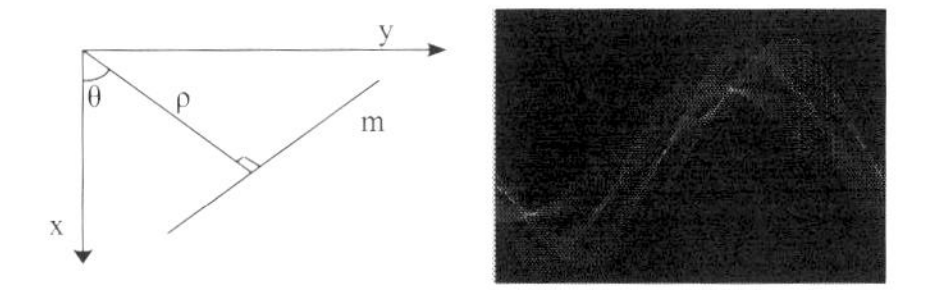

Fig. 70.4 Hough transform

Formula ρ represent from straight line m to the origin distance, $\rho = \pm\sqrt{x^2 + y^2}$. Normal line and x-axis angle θ, it ranges $-90^\circ \leq \theta \leq 90^\circ$. In reality we reject any line that falls outside a certain region, for example a horizontal line is probably not lane boundary and can be rejected. The Hough transform was modified to limit the search space to 30° for each side. Also the input image is searched separately and returning the most dominant line in the half image that falls within the 30° window. The horizontal line at this intersection is referred to as the vanishing line.

70.3.3 Extended Kalman Filter

Lane line parameters ρ and θ were obtained by the Hough transform, as we use hyperbolic model and extended Kalman filter. Due to the lane line is different, dimension of ρ and θ is different. However, taking advantage of extended Kalman filter to track boundary point, the dimension of the parameter must be unified. These are the extended Kalman filter equations for a nonlinear measurement equation, the EKF will provide us with a feedback K.

$$x_{t+1} = Ax_t + Bu_t + w_t, \quad y_t = h(x_t) + e_t \tag{70.6}$$

$$[D_x h]_{ij} = \frac{\partial h_i}{\partial x_j}, \quad C_t = X_t D_x h\left(\hat{x}_{t|t-1}\right) \tag{70.7}$$

by the observing data substitute into the formulas (70.6) and (70.7) to solve C_t matrix, then by means of the formulas (70.8) and (70.9) to get gain K_t and error P_t, finally, formula (70.10) output current estimates $\hat{x}_t$.

$$K_t = P_{t-1}C_t^T\left(C_t P_{t-1} C_t^T + Y_t R Y_t^T\right)^{-1} \tag{70.8}$$

$$P_t = AP_{t-1}A^T + Q - AK_t C_t P_{t-1} A^T \tag{70.9}$$

$$\hat{x}_{t+1|t} = A\left(\hat{x}_{t|t-1} + K_t\left[Y_t y_t - X_t h(\hat{x}\,t|t-1)\right]\right) + Bu_t \tag{70.10}$$

The predicted value of current state is the tracking results of previous state. A true value of the present state is the measured value. We can obtain track value of the state. The value is the predicted value of the next state, so the lane parameters are circularly estimated. Usually, by the first three images road parameters average value is the initial value of forecast. We use lane tracking algorithm that in order to reduce the scope of lane line scan.

70.3.4 Fitting Hyperbola Model

We use the mid-to-side road scan method to obtain a point of the left and right sides of the two time series, the hyperbola fitting phase uses the vectors of points from the lane. We could determine the vanishing point of the lane marking in the image space and obtain the parameter l of the improved hyperbola model. Left and right lane line boundary points are $L_l = \{(u_1,v_1),(u_2,v_2), \cdots,(u_n,v_n)\}$ and $L_r = \{(u_1,v_1),(u_2,v_2), \cdots,(u_m,v_m)\}$, respectively. Then a least squares technique is used to fit hyperbola model. The left and right parameters of lane model and working out the parameters of the road model:

$$N_1 = \begin{bmatrix} \frac{1}{v_1^r - l} & 1 & 0 & v_1^r - l \\ \vdots & \vdots & \vdots & \vdots \\ \frac{1}{v_n^r - l} & 1 & 0 & v_n^r - l \end{bmatrix}$$

$$N_2 = \begin{bmatrix} \frac{1}{v_1^l - l} & 1 & v_1^l - l & 0 \\ \vdots & \vdots & \vdots & \vdots \\ \frac{1}{v_m^l - l} & 1 & v_m^l - l & 0 \end{bmatrix}$$

where

$$A = \begin{bmatrix} N_1 \\ N_2 \end{bmatrix}, \overline{X} = [k, u_H, b_l, b_r]^T, \mathrm{B} = [u_1^{(r)}, u_2^{(r)}, \cdots, u_n^{(r)}, u_1^{(l)}, u_2^{(l)}, \cdots, u_m^{(l)}]^{\mathrm{T}}$$

So solving matrix equation $A\overline{X} = B$, and calculating the model parameters k and b.

$$\begin{bmatrix} N_1 \\ N_2 \end{bmatrix} [k, u_H, b_l, u_r]^T = \left[u_1^{(r)}, u_2^{(r)}, L, u_n^{(r)}, u_1^{(l)}, u_2^{(l)}, L, u_m^{(l)}\right]^T$$

$k = 0$ means that straight road, $k > 0$ means the road ahead turn left, and $k < 0$ means the road ahead turn right. Amplitude indicates that the degree of bend of the left and right lane line; b represents the traveling direction of the lane. Simulation results demonstrate that hyperbolic model parameters provide curvature of the road ahead to unmanned vehicle systems and to judge bodywork turn left or right. It plays a vital role in the model matching and lane line fitting (Figs. 70.5 and 70.6).

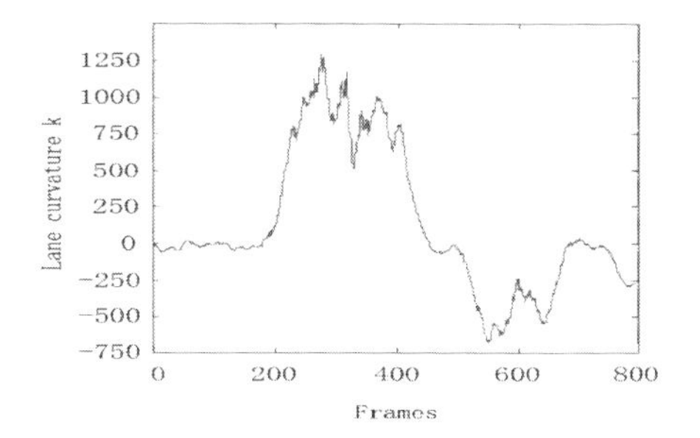

Fig. 70.5 Lane curvature simulation figure

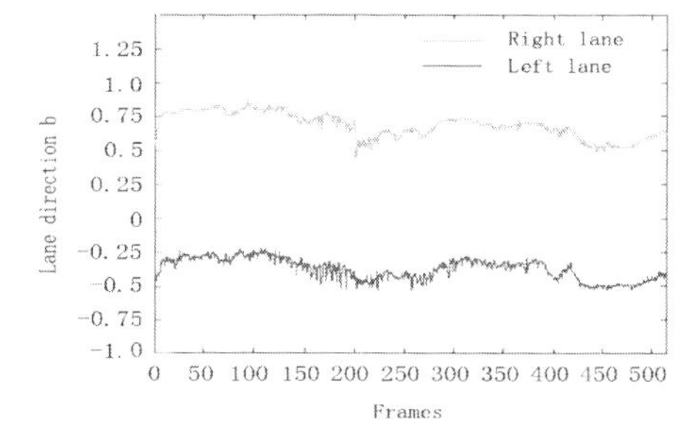

Fig. 70.6 Lane direction simulation figure

70.4 Experiment Result

This lane detection algorithm has been simulated and tested by VC++ on real road images, in order to verify effectiveness and stability of the algorithm. These lane images include straight and curve road, with or without shadows and lane marks. During the experiments, the images are 320 $\times$ 240 pixels with colors. The image processing speed needs 77.4 ms, accuracy reaches 93.4 %. We have tested at different locations and times. Partial test results are shown in Figs. 70.7 and 70.8. Partial lane is blocked by neighboring vehicles in Fig. 70.7, but the algorithm can extract the lane marking and it achieves a good result. Some curved road is tested in Fig. 70.8, but the algorithm can accurately orientate boundary line and curve position. The testing results indicate that the algorithm is good stability, high recognition rate, and robustness.

70.5 Conclusion

In this paper, we propose a lane detection algorithm based on hyperbola model. The algorithm extracts road markings, matches hyperbola model, and reconstructs the road markings by the least squares method. It can deal with the occlusion and imperfect road condition. We choose curve and straight road, with or without shadows lane condition. The experimental results indicate that it is able to

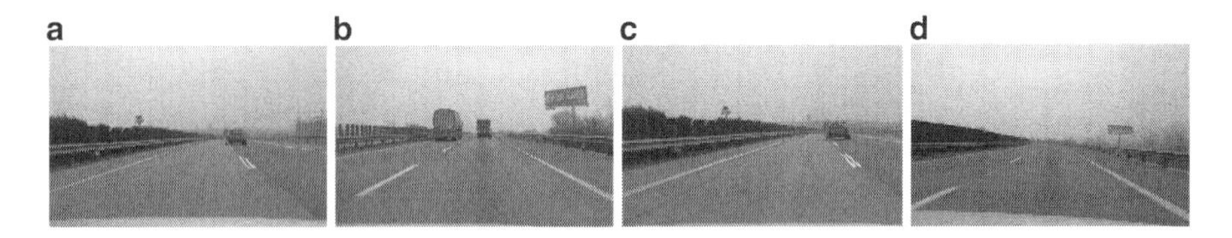

Fig. 70.7 Straight-line test results

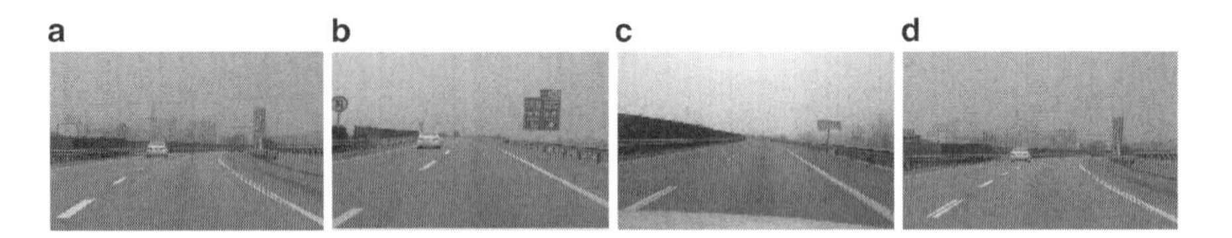

Fig. 70.8 Curve test results

accurately and reliably recognize the boundary markings of the lanes in complex environment.

Acknowledgements This work was supported by the projection of National Natural Science Foundation of China (61271362) and Shaanxi Province Natural Science Foundation (12JK0502). We would like to thank them.

References

1. Khalifa, O. O., Khan, I. M., & Assidiq, A. A. M. (2010). A hyperbola-pair based lane detection system for vehicle guidance. *Proceedings of the World Congress on Engineering Computer Science* (pp. 20–22). WCECS2010, USA
2. Yu, B., & Jain, A. K. (2010). Lane boundary detection using a multiresolution Hough transform. *Proceedings of International Conference on Image Processing* (pp. 748–751). IEEE Computer Society, USA
3. Cheng, H. Y., Jeng, B. S., et al. (2006). Lane detection with moving vehicles in the traffic scense. *IEEE Transactions on Intelligent Transportation Systems, 7*(4), 571–582.
4. Gao, S., Qin, L., & Chen, C. B. (2013). Structured road lane marking identification algorithm. *Journal of Xi'an Technological University, 33*(1), 14–19.
5. Li, Q., & Zheng, N. N. (2004). Spring robot: A prototype autonomous vehicle and its algorithms for lane detection. *IEEE Transactions on Intelligent Transportation Systems, 5*(4), 300–308.
6. Sharma, U. K., & Davis, L. S. (2009). Road boundary detection in range imagery for an autonomous robot. *IEEE Transactions on Robotics and Automation, 4*(5), 515–523.
7. Lutzeler, M., & Dickmanns, E. D. (2009). Recognition of intersections on unmarked road networks. *Proceedings of the IEEE Intelligent Vehicles Symposium* (pp. 302–307). Washington, DC: IEEE Computer Society
8. Wang, Y. (2004). Lane detection using spline model. *Pattern Recognition Letters, 21*(3), 677–689.

Chapter 71
Comparisons and Analyses of Image Softproofing Under Different Profile Rendering Intents

Qingxue Yu, Yunhui Luo, Maohai Lin, and Quantao Liu

Abstract This paper presents some results from an experiment of image softproofing under different International Color Consortium (ICC) rendering intents, which will be available for selecting an appropriate rendering intent in printing processes. In a screen softproofing procedure, image is converted via ICC profile embedded in image itself to output device profile, then to proofing device profile. With the rendering intent of absolute colorimetry in the second conversion, the effects of different rendering intents in the first conversion have been investigated through a softproofing software developed in Matlab 7.0. A variety of testing images, including light tone, shadow detail, etc., are used for image softproofing. The color differences between original images and proofed images are calculated under the S-CIE L*a*b* color difference formulae. Comparisons and analyses on the obtained images under four rendering intents show the effects of color characteristics of original images on rendering intent selection.

71.1 Introduction

Faithful color reproduction is the aim of printing industry. Due to color information transferring among different devices, color consistency is the vital key to high-quality prints [1]. Therefore color management is necessary for printing reproduction. International Color Consortium (ICC) profile-based color management system (CMS) is the primary tool for color control in printing industry [2]. In an ICC

Q. Yu • Y. Luo (✉) • M. Lin
Key Lab of Pulp & Paper Science and Technology, Ministry of Education,
Qilu University of Technology, Jinan 250353, China
e-mail: lyh@spu.edu.cn

Q. Liu
Shandong Dazhong-huatai Printing Ltd, Jinan 250000, China

W.E. Wong and T. Zhu (eds.), *Computer Engineering and Networking*, Lecture Notes in Electrical Engineering 277, DOI 10.1007/978-3-319-01766-2_71,

profile, there are four modes of rendering intent, i.e., perceptual, saturation, relative colorimetry, and absolute colorimetry, for color conversion between different devices. The four rendering intents correspond to different color mapping algorithms of gamut compression or reduction, which are suited for different cases [3, 4]. Appropriate rendering intents can result in better printing performance. Theoretically, the selection of rendering intents is relevant to source image characteristics, output device properties, and printing purposes. But in practice, it is usually determined by fuzzy experiences of color management technicians. Therefore, automatic selection for rendering intents is an attractive research topic in printing production.

There are some attempts for automatically selecting rendering intents in recent years [5–8]. Based on the state information and a current task in an imaging workflow, from a group of rendering intent including perceptual, saturated, relative colorimetric, and absolute colorimetric, automatically converting an original image to a new color space based on the color profile and the selected rendering intent, and making the converted image available for processing and output. This is a general procedure of automatic selection for rendering intents [5, 6]. At present, the digital proofing technology has increasingly become mature and been of great development [8]. It utilizes the color of displays to simulate the real output color of prints, and with the advantage of fast proofing speed, high convenience, and low cost, especially real-time proofing in different locations. In order to meet demand of printing production, automatically selecting rendering intents for image softproofing will affect efficiency of the color reproduction.

This work investigated the selection of four ICC rendering intents through a softproofing experiment. An experimental methodology and a Matlab software are developed for converting image color under different rendering intents. By real-time softproofing, color gamut of original image, printing device, and softproofing device are all presented in a visualized fashion. Color difference with S-CIE L*a*b* space [9], which combines with space attributes of human vision system, are introduced to evaluate color reproduction performance. The experimental results show the effects of color characteristics of original image and output device on rendering intent selection.

71.2 Preliminaries

Color Conversion of Softproofing. During the softproofing procedure, two color conversions are involved, connecting with twice selections of rendering intents, as shown in Fig. 71.1. The two types of profile used in the first step are the profile embedded in image itself (e.g., the sRGB profile) and the profile of printing devices (e.g., an OKI 9800 printer profile). In the second step the profiles for printing output device and softproofing device (e.g., a display profile) are used. Since softproofing is to employ colors of a display screen to simulate real printing colors, the color range of a proofing equipment, i.e., color gamut, should usually be larger than that

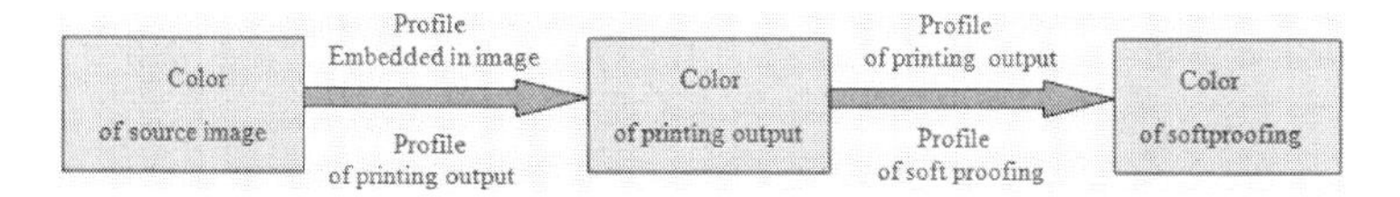

Fig. 71.1 Color conversion of softproofing

of the printing device in order not to loss color information. It means that the color gamut described by the profile of softproofing device should be bigger than that of source image so as to realize the purpose of proofing. Therefore, the rendering intent of absolute colorimetry should be utilized for softproofing in the second conversion, and the selection of rendering intent is just needed and available in the first color conversion.

Calibration and Characterization. Calibration is the premise of characterization and the key step of color management [10]. For a display used for softproofing, a typical calibration process is as follows: (1) Switch on the power of display to adjust lightness, contrast and RGB which conform to the conditions of calibration, and keep open more than 2 h to ensure relatively stable color gamut. (2) Individually calibrate display contrast, lightness, white balance, tone reproduction, cooperating with an Eye-One Pro spectrophotometer and the software of ProfileMaker 5.0. (3) Implement the sequence of calibration according to the prompts of the ProfileMaker software. Establishing profile of a device (by the ProfileMaker software) is called as characterization. Since each color input device or output device, and even color material, such as printing ink, dyeing phosphorus of display screens, etc., can only exhibit a specific color range (color gamut), the purpose of characterization is to establish color gamut for these devices or materials, and record its mathematical characteristics so as to carry on proper color conversions [1, 10, 11]. Color management is achieved by a universal and device-independent Profile Connection Space (PCS) of color. The device profile bridges its device color space and the PCS.

Color Difference. The color difference between original and proofed images is a traditional evaluation method for color reproduction performance [8]. Color difference formula of 1976 CIE L*a*b* color space has a wide range of applications in the printing and imaging science field. But the calculation of color difference does not fully match the actual feeling of human vision. For example, when watching a halftone image at a distance, the effect felt by human vision is very close to the actual effect of continuous tone image. But when we calculate the color difference between halftone image and continuous tone image with the CIE L*a*b* color difference formula, the values for these two cases are very different. Aiming to measure errors of image color reproduction, Zhang and Wandell proposed S-CIE L*a*b* (Spatial-CIE L*a*b*) color space [9]. S-CIE L*a*b* color space adds a space preprocessing step on the basis of CIE L*a*b* color space, and it is suited for image quality assessments. S-CIE L*a*b* color space combines the traditional color difference formula with space attributes of human visual system.

71.3 Experimental Setup

71.3.1 Testing Images Selection

Testing images cover common types encountered in practical applications. Eight testing images are selected from ISO (the international standard organization), GATF (Graphic arts technology foundation of the United States), and Kodak Company, as shown in Fig. 71.2. These images are embedded with the sRGB profile.

Shadow tone image. The dominant tone of such images is shadow tone, including the red, green, blue and a variety of perfect woodiness grain tone.

Light tone image. As Image 3 in Fig. 71.2 illustrated, its dominant tone is soft light and white.

Fig. 71.2 Testing images (1) shadow tone (GATF) (2) shadow tone (ISO) (3) light tone (GATF) (4) Group portrait (GATF) (5) neutral gray (GATF) (6) full tone (ISO) (7) full tone (Kodak) (8) memory color (GATF)

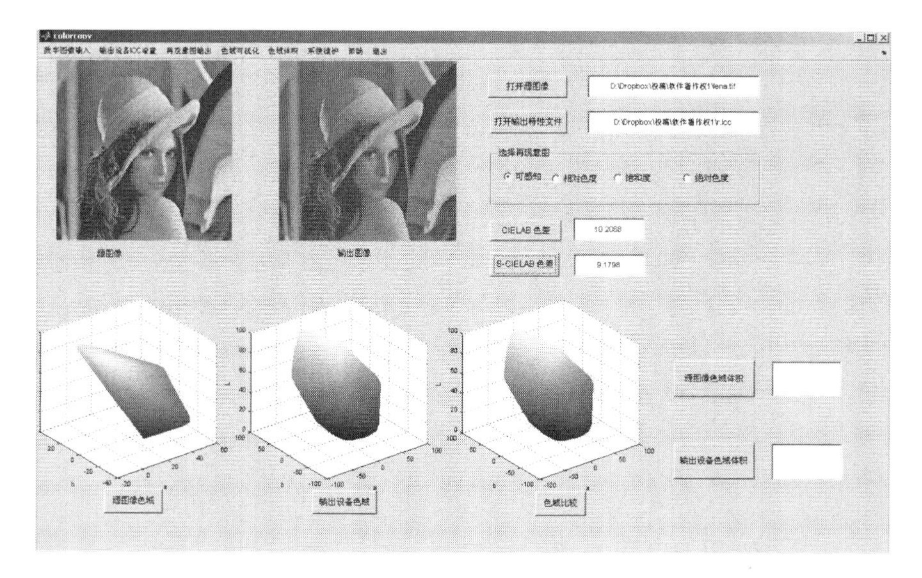

Fig. 71.3 GUI of the softproofing software

Group portrait image. In group portrait pictures, human skin color should be received much attention.

Neutral gray image. In neutral gray images, there still have some warm or cold gray tone besides neutral gray color. The saturated color is a key tone.

Full tonal image. An original full tonal image is usually taken by camera in a close range, as shown in Images 6 and 7 of Fig. 71.2. Both pictures have less object elements, providing a wide range of color and tone.

Memory color image. Image 8 of Fig. 71.2 shows an outstanding outdoor scenery of natural light color. It can arouse human reactions of memory color. Memory colors are processed by the human visual system and stored as memories. When readers see such images, they will first judge whether or not image colors are credible.

71.3.2 Softproofing

After device calibration and characterization, the color conversion will be implemented by a Matlab software for selected testing images. The specific output device profile is chosen to realize color conversions to simulate the actual output effects in a display, which is just so-called softproofing. The graphic user interface (GUI) of the proofing software developed in Matlab 7.0 is shown as in Fig. 71.3.

Three output profiles are used for testing experiments, which are an OKI 9800 digital printer profile, the USWebCoatedSWOP profile, and the JapanWebCoated profile. The developed software can provide proofed images under the four rendering intents. And it also presents the functions of reading and showing images, selecting ICC profiles, showing color gamut of original image and output device, calculating color gamut volumes, selecting rendering intents, as well as computing color differences between original and proofed images.

71.4 Results and Discussion

Table 71.1 gives the calculation results of color difference, as illustrated in Fig. 71.4, where ΔE is color difference under CIE L*a*b* formula, and ΔEs under S-CIE L*a*b* formula.

From results in Table 71.1, we can see that two kinds of color difference values (ΔE) of absolute colorimetry are very small in the four rendering intents for all of

Table 71.1 Color difference under different rendering intents

		Pc		RC		S		AC	
I	P	ΔE	ΔEs	ΔE	ΔEs	ΔE	ΔEs	ΔE	ΔEs
1	O	9.1824	8.4582	6.1809	5.6173	9.1824	8.4582	5.8017	5.1207
	U	11.2083	10.0711	9.98	8.9668	10.8589	9.7939	6.404	5.6574
	J	11.0016	9.9907	8.8864	8.8864	10.5609	9.6557	6.2593	5.5461
2	O	7.7236	7.226	5.0603	4.7205	7.7236	7.226	3.8561	3.471
	U	10.3029	9.3176	9.5364	8.6129	10.1458	9.1847	4.7472	4.257
	J	9.7448	8.935	8.4715	7.7629	9.5454	8.7685	4.5876	4.1535
3	O	9.2044	8.3108	9.4993	8.7942	9.2044	8.3108	4.5012	4.2771
	U	9.7315	8.4912	19.8834	17.7913	19.6872	17.5998	12.6957	11.0366
	J	16.6087	14.9206	16.622	14.9852	16.5207	14.8616	9.7315	8.4912
4	O	8.022	7.1903	7.1535	6.623	8.022	7.1903	4.4031	3.8995
	U	12.6957	12.4205	13.5536	12.2659	13.5833	12.2181	6.1182	5.3619
	J	12.3009	11.0637	11.5951	10.5815	11.9453	10.7998	5.4399	4.7874
5	O	8.9893	8.7623	7.4581	7.2681	8.9893	8.7623	5.9545	5.8116
	U	12.0784	11.2882	11.5081	10.724	12.0716	11.2839	6.2993	6.1302
	J	11.7767	11.2005	11.077	10.5022	11.7513	11.1824	6.7447	6.6214
6	O	8.2202	7.4923	5.9906	5.4734	8.2202	7.4923	5.1105	4.3871
	U	11.7296	10.5297	10.635	9.5469	11.2639	10.1575	5.7357	4.9926
	J	11.0877	10.0266	9.0022	8.1818	10.463	9.5468	5.5072	4.7885
7	O	9.7536	8.2568	8.4941	7.3747	9.7536	8.2568	5.7435	4.7157
	U	14.9363	13.4799	14.614	13.143	14.7822	13.3763	10.3817	9.4473
	J	14.3781	13.0953	13.609	12.3419	14.1692	12.9617	10.5937	9.6664
8	O	10.0666	9.0852	9.0404	8.3656	10.0666	9.0852	6.1823	5.6012
	U	17.4486	15.8903	16.816	15.3668	16.8663	15.4044	13.8966	12.3266
	J	15.6582	14.2787	14.631	13.4386	15.017	13.7611	12.0472	10.7394

I Image, *P* Profile, *Pc* Perceptual, *RC* Relative Colorimetry, *S* Saturation, *AC* Absolute Colorimetry, *O* OKI 9800, *U* USWebCoatedSWOP, *J* JapanWebCoated

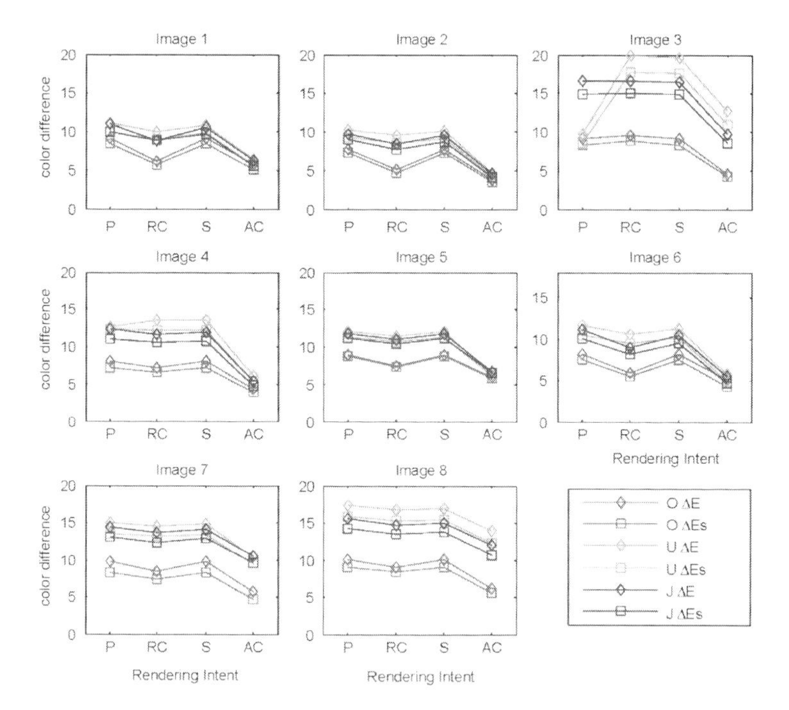

Fig. 71.4 Color difference under different rendering intents

the selected test images. Combined with human visual observation, the conversion effect of absolute colorimetry rendering intent is also the best. All above can prove that the absolute colorimetry rendering intent is suitable for image softproofing.

According to Table 71.1 and Fig. 71.4, the results with perceptual and saturation rendering intents are similar, and these two kinds of color difference values are all somewhat larger; Moreover, color difference values under relative colorimetry and absolute colorimetry rendering intents are somewhat smaller. These can illustrate that absolute colorimetry rendering intent is the best choice when color gamuts of source image and output device are nearly approached.

In addition, for a shadow tone image, the difference of ΔE or ΔEs value between relative colorimetry and absolute colorimetry rendering intents is not very larger. For a light tone image, the ΔEs value among perceptual, relative colorimetry and saturation rendering intents have a little difference. The results of group portrait images, neutral gray images and memory color images are as the similar. For a full tonal images, the ΔEs value of four rendering intents have a large difference, which means the proper selection is the relative colorimetry or absolute colorimetry rendering intent.

71.5 Conclusion

This paper presented results from an experiment of image softproofing under different ICC rendering intents. Through a softproofing software developed in Matlab 7.0, a variety of testing images were used for color difference analysis under CIE L*a*b* and S-CIE L*a*b* formulae. Comparisons and analyses provided useful guidelines for selecting rendering intent of output device profiles in printing processes. Automatic selection for rendering intents is currently under study and progress will be reported in the future.

References

1. Homann, J. P. (2008). *Digital color management: Principles and strategies for the standardized print production* (pp. 10–11). London: Springer. 35–36.
2. Fraser, B., Murphy, C., & Bunting, F. (2005). *Real world color management* (pp. 70–71). Berkeley: Peachpit Press.
3. Morovic, J. (2008). *Color Gamut mapping* (pp. 223–224). London: Wiley.
4. Bakke, A. M., Farup, I., & Hardeberg, J. Y. (2010). Evaluation of algorithms for the determination of color Gamut boundaries. *Journal of Imaging Science and Technology, 54*(5), 2–11.
5. Kulkarni, M. S., Borg, L. U. (2007, May 21). Automatic selection of color conversion method using image state information: US, 8014027.
6. Cai, S. Y., & Liu, R. F. (2004). An idea of automatically selecting rendering intents in color management. *Tian University of Science and Technology, 19*(3), 65–67.
7. Intwala, C., Clara, S. (2009, September 29). Color conversion preserving global and local image characteristics: US, 7965301.
8. Green, P. (2009). *Color management: Understanding and using ICC profiles* (pp. 55–57). London: Wiley. 87–88, 103–104.
9. Zhang, X. M., & Wandell, B. A. (1997). A spatial extension of CIELAB for digital color-image reproduction. *Journal of the Society for Information Display, 5*(1), 61–63.
10. Adams, R. M., & Weisberg, J. B. (2000). *The GATF practical guide to color management* (pp. 44–45). Pittsburgh: GATF Press. 123–125.
11. Sharma, A. (2006). Methodology for evaluating the quality of ICC profiles-scanner, monitor, and printer. *Journal of Imaging Science and Technology, 50*(5), 469–480.

Chapter 72
An Improved Dense Matching Algorithm for Face Based on Region Growing

Xin Xia and Shaoyan Gai

Abstract Traditional dense matching algorithms for face based on region growing have a lot of flaws. To generate a better disparity map, a novel improved method is proposed in this paper. Firstly, scale invariant feature (SIFT) algorithm is adopted to detect feature points for a pair of images, which are taken from two different angles. Secondly, this paper uses normalized cross correlation (NCC) to get match points and uses random sampling consensus (RANSAC) algorithm to eliminate mismatches. Several robust seeds are generated after this step. At last, by using an improved strategy of region growing, in which seeds are evaluated to help determine the locations and sizes of the search windows dynamically, the matching relations of seeds propagate to other parts of images. Experiments show that this method can obtain a good disparity map and has high computation speed.

72.1 Introduction

Three-dimensional information of face is widely used in the area of three-dimensional animations, face recognition, and so on [1]. Passive reconstruction systems for 3D information of face draw a wide attention for its flexible implementation and simple configuration [2]. This method reconstructs the 3D information of the face by using a pair of images taken from two different angles. This paper focuses on the stereo matching algorithm, which is the core of the reconstruction system.

X. Xia
School of Automation, Southeast University, Nanjing 210096, China

S. Gai (✉)
Key Laboratory of Measurement and Control for Complex System of Ministry of Education, School of Automation, Southeast University, Nanjing 210096, China
e-mail: cyoula@gmail.com

W.E. Wong and T. Zhu (eds.), *Computer Engineering and Networking*, Lecture Notes in Electrical Engineering 277, DOI 10.1007/978-3-319-01766-2_72,

Scale invariant feature transform (SIFT) algorithm was proposed by Lowe in 1999 [3]. This method can detect feature points and keep the invariability for scaling, rotation, and several other transformations. Zhou used this algorithm for the research of face matching [4]. Because the SIFT feature points are based on the gradient magnitude calculated by the grayscale value of images, traditional matching algorithm based on SIFT has a poor performance when the difference of light distribution between two images is very large. In some condition, this algorithm cannot find any match points [5].

Image dense stereo matching algorithm based on region growing is a method to match the images by propagating the matching relations to other parts of the images [6]. This algorithm performs well for multi-texture images. Tang adopted this method for image matching [7]. But because this method relies on the amount and accuracy of the seeds, error-matched points would cost more time for the calculating of region growing and lead to the accumulation of mismatch. All these flaws may result in a poor disparity map and make it hard to determine the size of the search window.

Normalized cross correlation (NCC) algorithm is a matching method based on statistical principle [8]. This algorithm calculates the cross-correlation coefficients by using the grayscale values of pixels in the windows of the two images for matching. Compared to SIFT, this method is more stable for the difference of light between images. Random sampling consensus (RANSAC) is an algorithm using random uncertainty [9], which is capable of processing data with a large proportion of error points.

To solve all these problems above, this paper uses a method combining SIFT and NCC algorithms for the first match, then using RANSAC algorithm to eliminate the error-matched points to make the seeds more reliable. In addition, this paper proposes an evaluating system, which optimizes the choice of the size and location of the search window. This system helps to reduce mismatch and to increase the computational efficiency.

72.2 SIFT Feature Points Detecting and NCC First Matching

Detecting of SIFT feature points of each image is necessary for the first match. This method of detecting contains four steps. First of all, Gaussian pyramid and DOG pyramid of each image should be built to detect extreme values and find out the approximate location of feature points by constructing scale space. Secondly, this method fits the local extreme values with 3D quadratic function to determine the location and scale precisely and eliminate points with low contrast on the unstable edge. Thirdly, this method assigns the major orientation of the feature points by using the distribution of the gradient direction in the neighboring windows of the feature points. Lastly, this method describes each feature point with 16 points, each

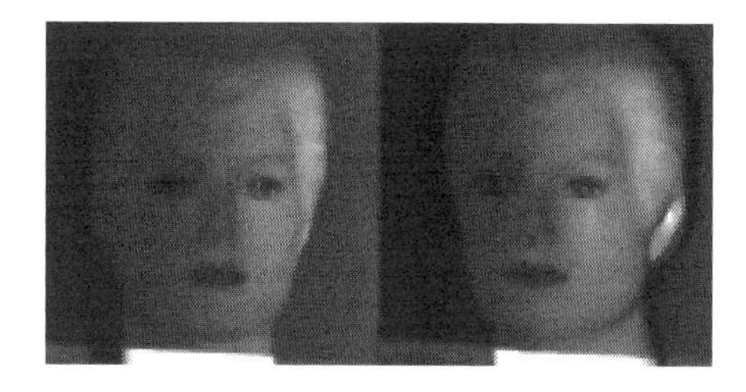

Fig. 72.1 Initial images of Set 1

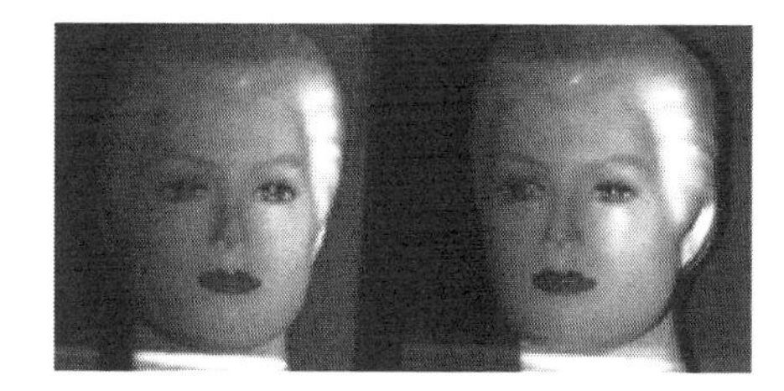

Fig. 72.2 Initial images of Set 2

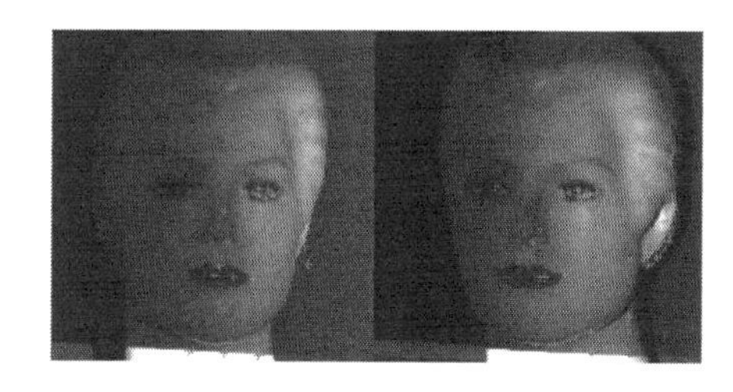

Fig. 72.3 Feature points detected from Set 1

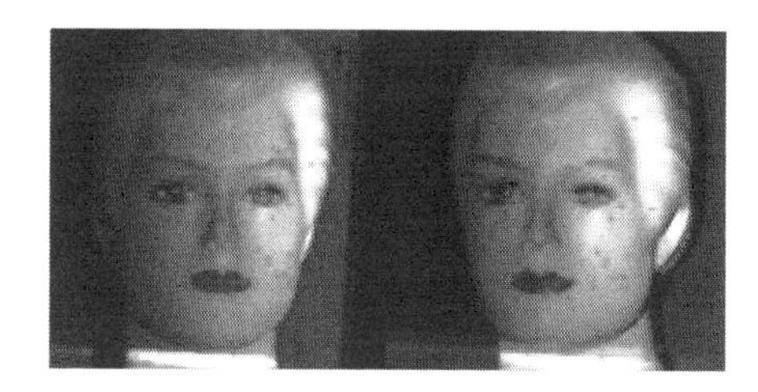

Fig. 72.4 Feature points detected from Set 2

point has information in 8 directions and obtain 128 dimensions feature vectors. Figures 72.1 and 72.2 are two sets of initial images under different lighting conditions. Figures 72.3 and 72.4 show the SIFT feature points detected from the initial images.

Traditional matching algorithms for the SIFT feature points are based on the Euclidean distance. Because the SIFT feature points are based on the gradient magnitude calculated by the grayscale value of images, traditional method has a poor performance when the difference of light distribution between two images is very large.

To solve this problem, this paper combines SIFT and NCC to process the first match. NCC algorithm is a matching method based on statistical principle, which

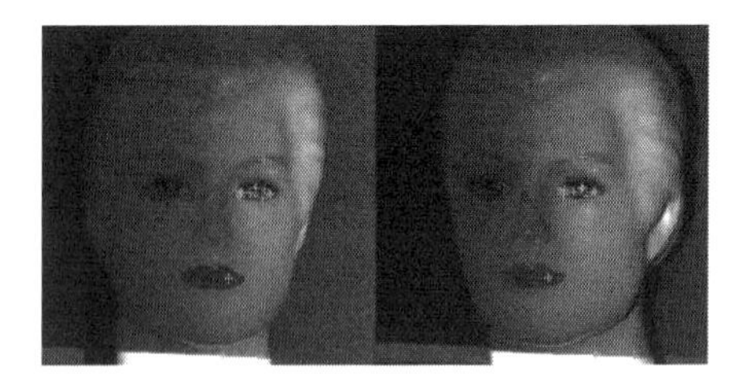

Fig. 72.5 Match points in Set 1 from SIFT

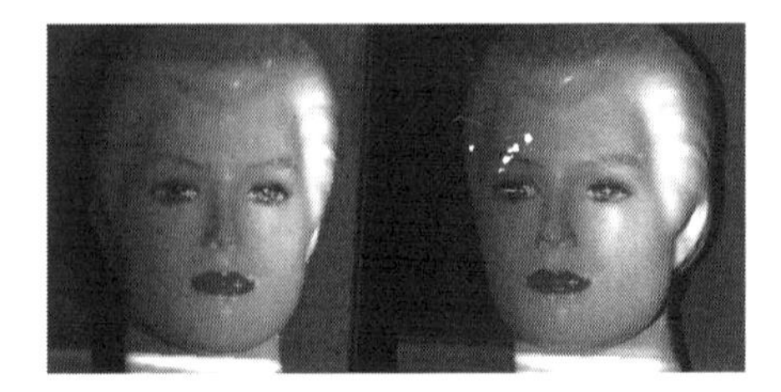

Fig. 72.6 Match points in Set 2 from SIFT

calculates the cross-correlation coefficients by using the grayscale values of pixels in the windows of the two images for matching. In this method, a neighboring window is opened around the feature point, which is to be matched in the lift image. Then windows with same size are opened around every single feature point in the right image. Then the grayscale values of every pixel in those windows are taken into Eq. (72.2) to calculate the cross-correlation coefficients. The point with the largest coefficient and the feature point to be matched are a pair of match point.

$$\rho(i_0, j_0) = \frac{\sum_{(i,j)\in W_{1,2}(i_0,j_0)} \left[G_1(i,j) - \bar{G_1}\right]\left[G_2(i,j) - \bar{G_2}\right]}{\sqrt{\sum_{(i,j)\in W_1(i_0,j_0)} \left[G_1(i,j) - \bar{G_1}\right]^2 \sum_{(i,j)\in W_2(i_0,j_0)} \left[G_2(i,j) - \bar{G_2}\right]^2}} \tag{72.1}$$

In this equation, $W_1(i_0,j_0)$ and $W_2(i_0,j_0)$ are the two neighboring windows around the feature point (i_0,j_0) in the two images. $\rho(i_0,j_0)$ is the cross-correlation coefficient between those two windows. $G_1(i,j)$ and $G_2(i,j)$ are the grayscale values of point (i,j) in two images. $\bar{G_1}$ and $\bar{G_2}$ are the average values of the grayscale values in those two windows.

Compared to the common definitions of NCC coefficients, Eq. (72.1) subtracts the average value of grayscale in single window, which improves the performance of NCC and makes the process of matching not affected by the difference of distribution of light between images. Figure 72.5 shows the 13 pairs of match points in Set 1 computed from SIFT algorithm. Figure 72.6 shows the 22 pairs of match points in Set 2 computed from SIFT algorithm. Figure 72.7 shows the 16 pairs of match points in Set 1 computed from NCC method. Figure 72.8 shows the 28 pairs of match points in Set 2 computed from NCC method.

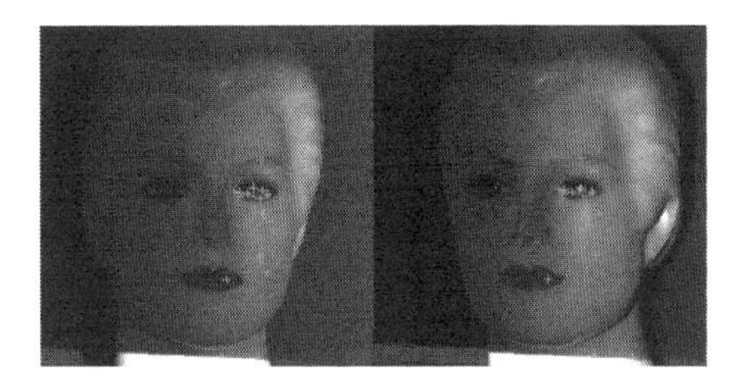

Fig. 72.7 Match points in Set 1 from NCC

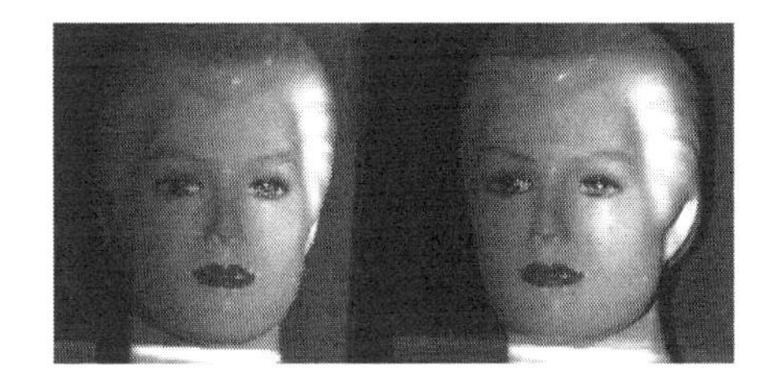

Fig. 72.8 Match points in Set 1 from NCC

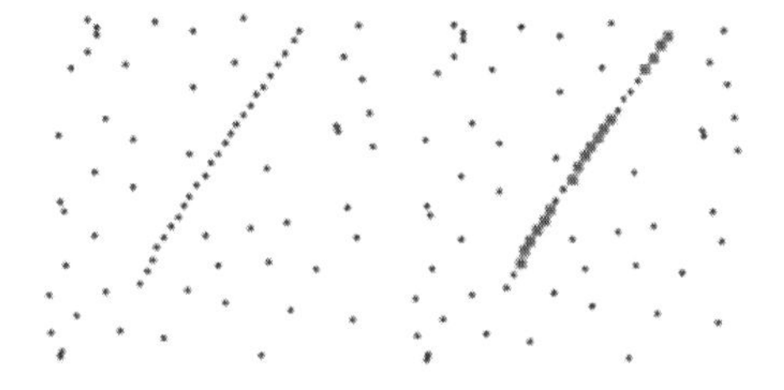

Fig. 72.9 RANSAC experiment

72.3 Eliminating Mismatch by RANSAC

After the first match, several match points are generated while some of them are error matched. In this paper, RANSAC algorithm is used to eliminate the mismatches. RANSAC is an algorithm using random uncertainty [9], which is capable of processing data with a large proportion of error points. Figure 72.9 shows an experiment using RANSAC to find the correct points in a line. There are 25 correct points and 50 error points in the figure, and the error rate is 67 %. After 500 iterations, 17 correct points were found out, and all error points are eliminated.

72.4 Region Growing

Region growing is a process that propagates the matching relations of seeds to other parts of images [10]. The basic strategy of region growing can be shown as Fig. 72.10. In Fig. 72.10, point *La* in left image and point *Ra* in right image are a pair of matched points. Point *Lb* and point *Lc* are around point *La* in the left image,

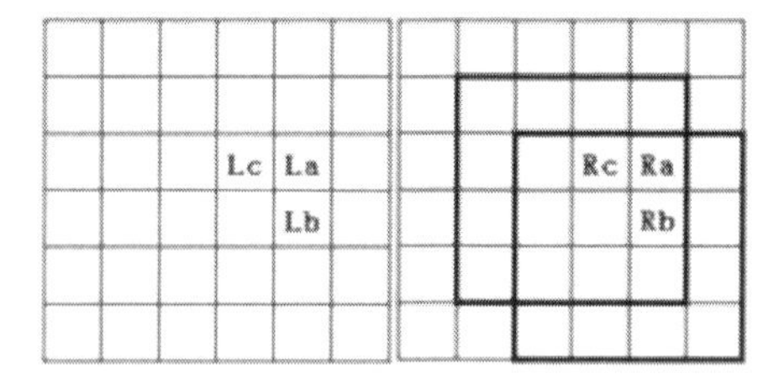

Fig. 72.10 Basic strategy of region growing

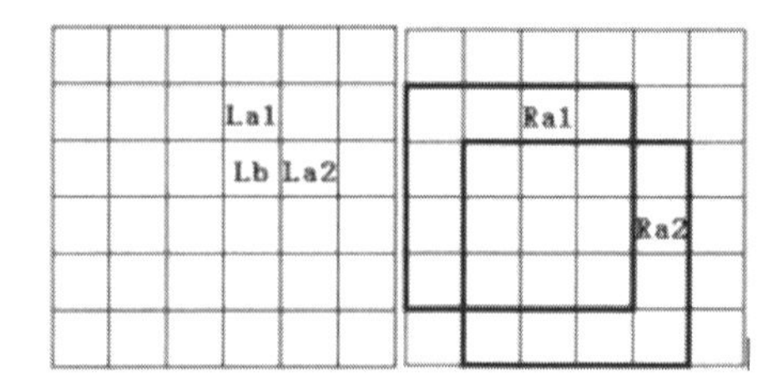

Fig. 72.11 Location of search window

which means point *Rb* and point *Rc*, the match points of point *Lb* and point *Lc*, are around point *Ra* too. So the search areas of those two match points can narrow down to neighboring windows of point *Ra*. With this algorithm, once a pair of robust match points is found, this matching relation can be quickly propagated to other parts of an image.

If two pairs of match points are known, the strategy can be shown as Fig. 72.11. In this figure, point *La*1 with point *Ra*1 and point *La*2 with point *Ra*2 are two pairs of match points. Point *Lb* in the left image is around both point *La*1 and point *Ra*1, which means that point *Rb*, the match point of point *Lb* is around point *Ra*1 and point *Ra*2. The search window is located by both point *Ra*1 and point *Ra*2. Search window of a point can be determined by one pair of match points or several pairs.

But because the degree of accuracy and the computation speed of this method rely on the amount and accuracy of the seeds, to lower the negative effects of seeds with low robustness, this paper proposed an evaluating system for the reliability of the seeds.

$$R(i_0, j_0) = \frac{\sum_{(i,j)\in A_s} \rho(i,j)}{n} \tag{72.2}$$

In Eq. (72.2), $R(i_0,j_0)$ is the evaluation for seed (i_0,j_0). A_s is a neighboring area with the size of 3×3. $\rho(i,j)$ is the NCC coefficient of the seed (i,j) in this area. n is the number of the seeds that in this area. During the process of region growing, seeds with large R will be taken into calculation first. The size of the search windows is inversely proportional to the R. In Eq. (72.3), N is the size of the search window for the next match point. R is the evaluation of the prior seed. λ is a coefficient which determined by the specific conditions.

$$N = \frac{\lambda}{R} \tag{72.3}$$

With this improved strategy of region growing, it can be ensured that growing happens around robust seeds first. Also the adaptive window can help to reduce the computation time and increase the veracity of the algorithm.

72.5 Algorithm Summary

Algorithm in this paper can be summarized as follow:

(a) Detect SIFT feature points of images and use NCC to do the first match.
(b) Eliminate the error-matched points by using RANSAC and obtain robust seeds.
(c) Evaluate all seeds and put them into a sequence in order.
(d) If the sequence is not empty, take out the first pair of seeds in the sequence and determine the size of the search window for the next match points. Calculate the match points of the 8 points around the seed.
(e) Put the new match points into sequence in order.
(f) Repeat step d and step e until the sequence is empty.

72.6 Experiments

Figures 72.12 and 72.14 are the disparity maps by using seeds from Figs. 72.5 and 72.6, which shows the result of the traditional algorithm. Figures 72.13 and 72.15 are the disparity maps by using this paper's method. Match points from Figs. 72.7 and 72.8 are processed with RANSAC and generate seeds for the region growing which uses the improved strategy with the evaluating system. It is obvious that the disparity maps generated by method in this paper are smoother and cover more area of the face. Table 72.1 shows the average computation time for two sets of experiments of both traditional algorithm and method of this paper. This paper's method is more efficient.

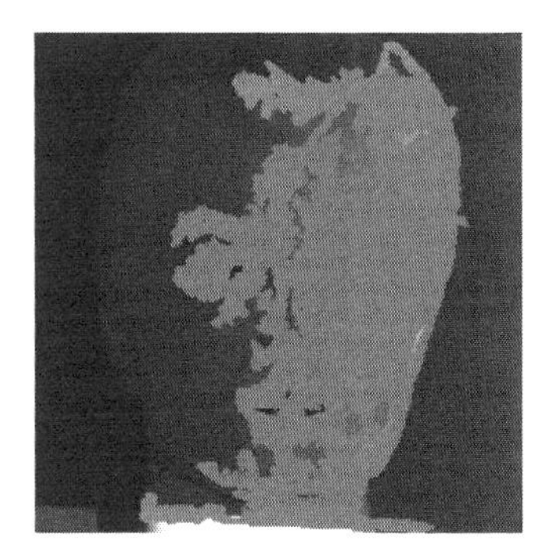

Fig. 72.12 Traditional algorithm for Set 1

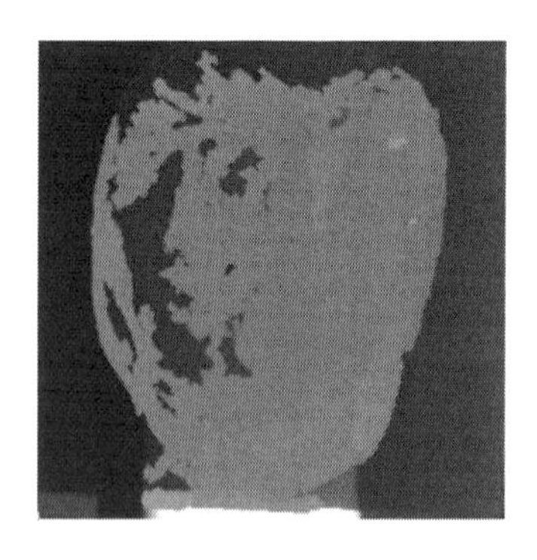

Fig. 72.13 This paper's method for Set 1

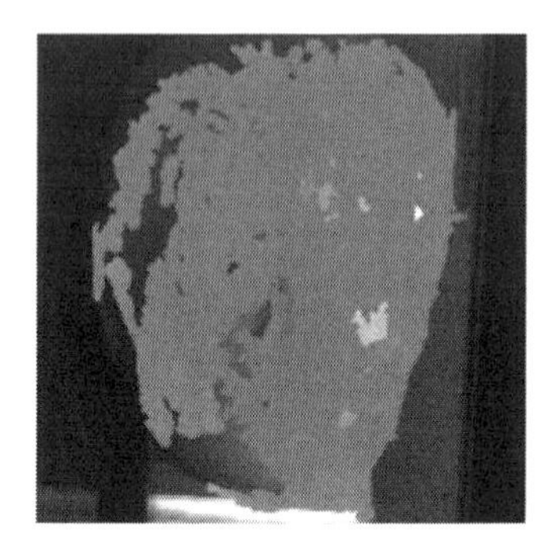

Fig. 72.14 Traditional algorithm for Set 2

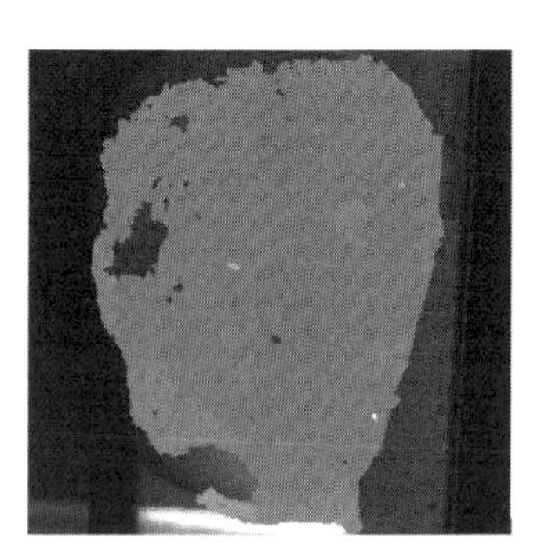

Fig. 72.15 This paper's method for Set 2

Table 72.1 Computation time

Algorithm	Traditional	This paper
Computation time	5.8 s	3.9 s

72.7 Conclusion

This paper proposed an improved dense matching algorithm for face based on region growing. The algorithm combines SIFT and NCC for the initial match, and then uses RANSAC to eliminate the error-matched points to make the seeds more reliable. By using an improved strategy of region growing, in which seeds are evaluated to help determine the locations and sizes of the search windows dynamically, the matching relations of seeds propagate to other parts of images.

From the experimental results, it can be concluded that compared to traditional algorithm, this paper's algorithm can generate better disparity maps, which are smoother and cover more area of the face. Also, this paper's algorithm has a higher computational efficiency.

References

1. Liang, L. H., Ai, H. Z., Xu, G. Y., et al. (2002). A survey of human face detection. *Chinese Journal of Computers-Chinese Edition, 25*(5), 449–458.
2. Zhao, W., Chellappa, R., Phillips, P. J., et al. (2003). Face recognition: A literature survey. *ACM Computing Surveys, 35*(4), 399–458.
3. Lowe, D. G. (1999) Object recognition from local scale-invariant features. Computer vision, 1999. *The Proceedings of the Seventh IEEE International Conference on IEEE, England* 2, pp. 1150–1157
4. Zhou, Z. M., Yu, S. Y., Zhang, R., et al. (2008). An algorithm for face recognition based on SIFT descriptor. *Journal of Image and Graphics, 13*(10), 1882–1885.
5. Liu, X., & You, H. (2009). Matching feature points of multi-temporal space-borne SAR based on SIFT. *Science of Surveying and Mapping, 1*(015), 43–45.
6. Lhuillier, M., Quan, L. (2002). Quasi-dense reconstruction from image sequence. England, Computer Vision—ECCV 2002 (pp. 125–139). Berlin/Heidelberg: Springer
7. Tang, L., Wu, C. K., Liu, S. G., et al. (2004). Image dense stereo matching by technique of region growing. *Chinese Journal of Computers, 27*(7), 936–943.
8. Lewis, J. P. (1995). Fast normalized cross-correlation. *Vision Interface, 10*(1), 120–123.
9. Fischler, M. A., & Bolles, R. C. (1981). Random sample consensus: a paradigm for model fitting with applications to image analysis and automated cartography. *Communications of the ACM, 24*(6), 381–395.
10. Xing, Y. J., Meng, J. J., Sun, J., et al. (2006). An improved region-growth algorithm for dense matching. *Journal of Achievements in Materials and Manufacturing Engineering, 18*(1–2), 323–326.

Chapter 73
An Improved Feature Selection Method for Chinese Short Texts Clustering Based on HowNet

Xin Chen, Yuqing Zhang, Long Cao, and Donghui Li

Abstract Short texts have played an important role in the field of text data mining. Because of the problems arousing from the complexity of Chinese semantics and data sparseness, which is an obvious characteristic of short texts, it is necessary to explore some new semantic-based methods to cluster Chinese short texts. An improved approach of feature selection based on HowNet is applied in this paper to address data sparseness of Chinese short texts. By redefining Vector Space Model in semantic level and merging generalized synonymy features, we present a new feature generation strategy. Experimental results show that by merging semantic similar feature, our method is effective in feature dimension reduction and gets better clustering performance. The proposed HowNet-based feature selection method is suitable for Chinese short texts clustering.

73.1 Introduction

With the popularization of Web2.0, instant message, image captions, twitter etc. which in the form of short texts have played an irreplaceable role in the area of public opinion analysis [1], Social Networking Services (SNS) [2], and topic tracking [3]. The clustering of short texts which is the basis of information analysis, however, presents great challenges. Because of the features of short texts, they cannot provide competent context information or word co-occurrence information for similarity measure [4], which is the basis of clustering methods [5]. Thus, the traditional methods of text clustering are not suitable for short text clustering when they are directly applied [6].

This paper is concerned with improving performance of the Chinese short texts clustering. We propose an effective feature selection method by word semantic

---

X. Chen • Y. Zhang (✉) • L. Cao • D. Li
China University of Geosciences, Beijing 100083, China
e-mail: yqzhang@cugb.edu.cn

W.E. Wong and T. Zhu (eds.), *Computer Engineering and Networking*, Lecture Notes in Electrical Engineering 277, DOI 10.1007/978-3-319-01766-2_73,

similarity calculating based on HowNet [7]. We redefine Vector Space Model (VSM) [8] in semantic level and use a new feature generation strategy that merges generalized synonymy features. The experimental results demonstrate that the method obviously reduces the negative forces of sparse feature and improves the clustering accuracy.

73.2 Related Works

At the present, there are generally three types of approach in short texts clustering. The first proposed by Phan et al. [4] is mining the implicit topic of short texts based on the topic model and improving the co-occurrence probability of the feature terms. Quan et al. [9] also measured similarity between different feature terms in short texts with different topics based on the Latent Dirichlet Allocation (LDA) model. The second is achieving the short texts clustering based on similarity of the feature terms by using search engines by Sahami et al. [10]. The third presented by Gabrilovich et al. [11] is expanding feature terms by world knowledge to improving clustering performance in short texts. Hu et al. [12] using World knowledge such as WordNet or Wikipedia to reduce negative impact of data sparseness.

However, Chinese short texts are quite different from general short texts as a result of the complexity of Chinese semantic. Pu et al. [13] using Latent Semantic Indexing (LSI) for text preprocessing and then using Independent Component Analysis (ICA) to deal with Chinese short texts. The result expresses that the combination of LSI and ICA method is little better than ICA. Zuo et al. [14] improved the performance of Chinese short texts classification by selecting key words from training set. However, it is far from ideal effect. Ning et al. [15] proposed a method by using the domain word ontologies to deal with Chinese short texts. In order to solve the sparse feature keywords in Chinese short texts, Jin et al. [16] expanded semantic features to improve clustering effect. Our method is also committed to reduce the negative forces of sparse feature, but the core is improving feature selection method by word semantic similarity calculating.

73.3 Words Semantic Similarities Calculation

There are many approaches proposed to calculate Chinese words semantic similarities. HowNet as a widely adopted method is used in this paper to address words semantics.

73.3.1 HowNet Overviews

According to the official website of HowNet, the definition of HowNet is an online commonsense knowledge base unveiling inter-conceptual relations and inter-attribute relations of concepts as connoting in lexicons of the Chinese and their English equivalents [7]. The latest version covers over 65,000 concepts in Chinese and close to 75,000 English equivalents, and the relations include hyponymy, synonymy, antonymy, meronymy, attribute-host, material-product, converse, dynamic role, and concept co-occurrence [7]. The knowledge dictionary of HowNet is based on words and their concepts. Concept and sememe are the most important elements in the definition of HowNet. Concept, which is not we usually refer, describes one item of word semantics. Each word includes one or more items, and an item in the HowNet is called as sense corresponding. The concept of HowNet is described using a kind of knowledge representation language, each term used in the knowledge representation language is called as the sememe. A sememe refers to the smallest basic semantic unit. Every concept of a word or phrase and its description form one entry as follow:

NO = word/phrase ID
W_C = Chinese word/phrase form
G_C = Chinese word/phrase syntactic class
E_C = example of Chinese word/phrase usage

73.3.2 Words Semantic Similarities Calculation Based on HowNet

One of the popular methods of Chinese word semantic similarity calculation based on HowNet was proposed by Liu [17]. For two words W_1 and W_2, if W_1 included n concepts C_{11}, C_{12}, ..., C_{1n}, and W_2 included m concept C_{21}, C_{22}, ..., C_{2m}. The similarity between two words W_1 and W_2 defined as follows:

$$Sim(W_1, W_2) = \max_{i=1,2\ldots n, j=1,2\ldots n} Sim(C_{1i}, C_{2j}) \tag{73.1}$$

The similarity between two concepts defined as follows:

$$Sim(C_1, C_2) = \sum_{i=1}^{4} \beta_i \prod_{j=1}^{i} Sim_j(C_{1i}, C_{2j}) \tag{73.2}$$

where $\beta_i (1 \leq i \leq 4)$ means weight parameters of these four parts, respectively, β_i greater than 0.5 in generally, $\beta_1 + \beta_2 + \beta_3 + \beta_4 = 1$ and $\beta_1 \geq \beta_2 \geq \beta_3 \geq \beta_4$. Sim_j represents the similarity of different section of concepts.

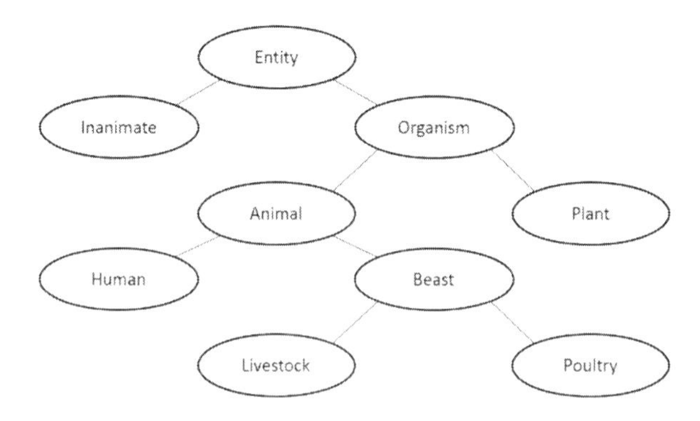

Fig. 73.1 Hierarchy of sememe tree

The similarity between two sememe defined as follows:

$$Sim(P_1, P_2) = \frac{\alpha}{\alpha + dis(P_1, P_2)} \quad (73.3)$$

where $dis(P_1,P_2)$ is the distance between P_1 and P_2 in sememe tree as shown in Fig. 73.1, α is a smoothing factor.

In hierarchy of sememe tree, the depth of generalized sememe is smaller than the depth of specific sememe. Took into account depth information of the sememes and the semantic distance between two sememe nodes, we defined a father node of P_1 and P_2 in sememe tree called Least Common Node $DN(P_1,P_2)$ and used the depth of $DN(P_1,P_2)$ to replace the previous smoothing factor α. Then similarity between two sememe redefined as follows:

$$Sim(P_1, P_2) = \frac{Depth(DN(P_1, P_2))}{Depth(DN(P_1, P_2)) + dis(P_1, P_2)} \quad (73.4)$$

73.4 General Framework for Chinese Short Texts Clustering

In this part, we introduce the presented general framework for Chinese short texts by an improved feature selection method based on HowNet. The framework consists of three phases, including data preprocessing, feature generation and selection, and clustering. As shown in Fig. 73.2, the first step of Chinese short texts clustering is Chinese word automatic segmentation. The matching process includes inserting the separator between Chinese terms according to specific standards, which is different from the space used in English. Moreover, some words are meaningless in Chinese just like "a" and "the" in English, which are called Stop Words. After data preprocessing, the bag of words should be segmented and exclude of Stop Words.

In phases 2, we use HowNet and the method presented in Sect. 73.3.2 to calculate the semantic similarity between each pair of words in bag. Then we get

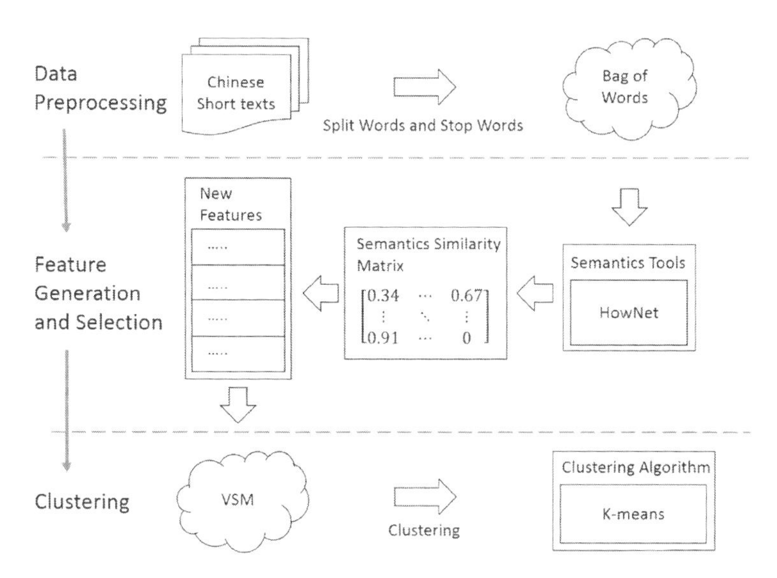

Fig. 73.2 Framework for Chinese short texts clustering

a Semantics Similarity Matrix, which dimension is equal to the number of words in bag. In order to reduce the matrix dimension, a threshold value K in our model is applied. If the semantic similarity between each pair of words is greater than K, we think that these two words are generalized synonymy in semantic. That is to say, their contribution can merge to computation as features. Therefore, the number of new feature will be reduced and the problem of data sparseness will be alleviated. The last step in this workflow is standard clustering algorithm and thus no more introduce.

73.5 Experimental Evaluations

In this section, we first present the experimental data used in our test bed and then introduce the evaluation criteria. Finally, we evaluate the experimental results.

73.5.1 Date Sets and Evaluation Criteria

The microblog which is a typical representative of the Chinese short texts employed in our experiment as the benchmark datasets. We crawled twenty thousand microblogs from Tencent since August to September 2012, and manual tagged them to 15 cluster.

Table 73.1 Average accuracy test condition

	Same class	Different class
Same cluster	*TP* (True positive)	*FN* (False negative)
Different cluster	*FP* (False positive)	*TN* (True negative)

Table 73.2 Results with three feature selection methods

	5 Cluster		15 Cluster	
	F_1-measure	Average accuracy	F_1-measure	Average accuracy
Simple VSM	0.513	0.559	0.489	0.505
Synonymy dictionary	0.604	0.646	0.526	0.553
HowNet (K = 0.6)	0.642	0.677	0.538	0.551

The evaluation standard F_1-measure and Average Accuracy used in our experiments is based on the Confusion Matrix as illustrated in Table 73.1 for Clustering problem.

F_1-measure is a measure of a test's accuracy which considers both precision and recall that measures a cluster contains only objects of a particular class and all objects of that class.

Average Accuracy is a statistical measure defined as follows:

$$Accuracy = \frac{TP + TN}{TP + FP + FN + TN} \tag{73.5}$$

73.5.2 Performance Evaluation

Experimental results of the three feature selection methods on both 5 cluster dataset and 15 cluster dataset using the simple k-means clustering algorithms are reported in Table 73.2. Each result denotes an average of 20 test runs by randomly choosing the initial parameters for k-means algorithm. In this Tables, the feature selection based on HowNet with the threshold value $K = 0.6$ achieves better performance than other two methods except the Average Accuracy in 15 cluster dataset. Moreover, the performance of three methods in 5 cluster is obviously better as compared with them in 15 cluster.

In Fig. 73.3, we note that, in general terms, different threshold values achieve different performance. When $K \leq 0.4$, threshold value is too low that many words which not similarity in semantic merge to one feature, and thus get a lower clustering accuracy. The most appropriate threshold value in our experiment is $K = 0.6$.

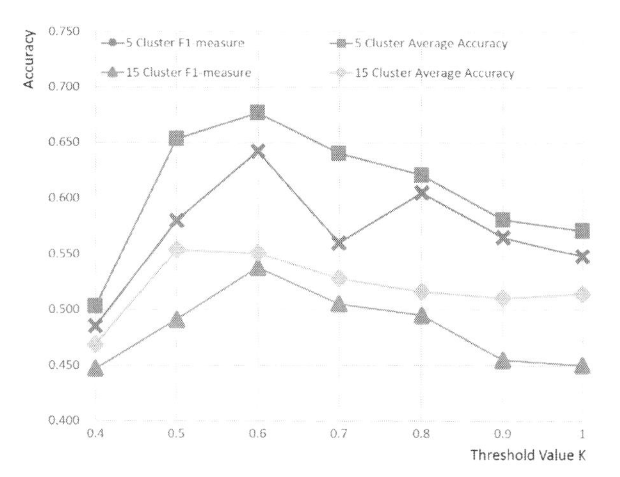

Fig. 73.3 Results with different threshold values in HowNet method

73.6 Conclusion

This paper presented an improved feature selection method to address Chinese short texts. We redefined VSM in semantic level and used a new feature generation strategy that merges generalized synonymy features. Based on this new strategy, we used microblog which is a typical representative of the short text at present to validate the model. The experimental results indicated that the method can effectively reduce the negative forces of sparse feature and improve the clustering accuracy. Due to the various presentations of short texts, however, our experimental dataset was incomplete. Therefore, we should further expand the category of short texts and focus on the applicability of this method.

Acknowledgements Supported by "the Fundamental Research Funds for the Central Universities" of CUGB (2652013102, CUGB), Subject Supportive Project of CUGB (2013) and the Innovative Experiment Plan for College Students of China University of Geosciences, Beijing.

References

1. Yang, Z., Duan, L., & Lai, Y. (2010). Online public opinion hotspot detection and analysis based on short text clustering using string distance. *Journal of Beijing University of Technology, 36*(5), 669–673.
2. Rosso, P., Errecalde, M., & Pinto, D. (2013). Analysis of short texts on the Web: introduction to special issue. *Lang Resources and Evaluation, 47*(1), 123–126.
3. Mo, Y., Liu, S., Liu, Y., & Cheng, X. (2012). An entropy based rule mining algorithm for filtering tweets by topics. *Journal of Chinese Information Processing, 26*(5), 1–6. 39.
4. Phan, X. H., Nguyen, L. M., Horiguchi, S. (2008). Learning to classify short and sparse text & web with hidden topics from large-scale data collections. *In Proceedings of the 17th International Conference on World Wide Web* (pp. 91–100). New York: ACM

5. Hu, J., Fang, L., Cao, Y., Zeng, H., Li, H., Yang, Q., Chen, Z. (2008). Enhancing text clustering by leveraging Wikipedia semantics. *In Proceedings of the 31st ACM SIGIR* (pp. 179–186). New York: ACM
6. Metzler, D., Dumais, S., & Meek, C. (2007). Similarity measures for short segments of text. *Lecture Notes in Computer Science, 4425*, 16–27.
7. Dong, Z., Dong, Q. HowNet knowledge database. http://www.keenage.com/
8. Salton, G., Wong, A., & Yang, C. S. (1975). A vector space model for automatic indexing. *Communications of the ACM, 18*(11), 613–620.
9. Quan, X., Liu, G., Lu, Z., Ni, X., & Liu, W. (2010). Short text similarity based on probabilistic topics. *Knowledge and Information Systems, 25*(3), 473–491.
10. Sahami, M., Heilman, T. D. (2006). A web-based kernel function for measuring the similarity of short text snippets. *In Proceedings of the 15th International Conference on World Wide Web* (pp. 377–386). New York: ACM
11. Gabrilovich, E., Markovitch, S. (2005). Feature generation for text categorization using world knowledge. *In Proceedings of the 19th International Joint Conference on Artificial Intelligence* (pp. 1048–1053). San Francisco CA: Morgan Kaufmann
12. Hu, X., Sun, N., Zhang, C., Chua, T. S. (2009). Exploiting internal and external semantics for the clustering of short texts using world knowledge. *In Proceedings of the 18th ACM Conference on Information and Knowledge Management* (pp. 919–928). New York: ACM
13. Pu, Q., Yang, G. (2006). Short-text classification based on ICA and LSA. *In Proceedings of International Symposium on Neural Networks*. (pp. 265–270). German: Springer
14. Zuo, S., Wu, C., Zhou, Y., & He, H. (2006). Chinese short-text categorization based on the key classification dictionary words. *The Journal of China Universities of Posts and Telecommunications, 13*, 47–49.
15. Ning, Y., Fan, X., Wu, Y., & Text, S. (2009). Classification based on domain word ontology. *Computer Science, 36*(3), 142–145.
16. Jin, C., Zhou, H., & Bai, Q. (2012). Short text clustering algorithm with feature keyword expansion. *Advanced Materials Research, 532–533*, 1716–1720.
17. Liu, Q., & Li, S. (2002). Word similarity computing based on how-net. *Computational Linguistics and Chinese Language Processing, 7*(2), 59–76.

Chapter 74
Internet Worm Detection and Classification Based on Support Vector Machine

Huihui Liang, Min Li, and Jiwen Chai

Abstract This paper proposes a novel Internet worm detection and classification method. The behaviors of worms are different from each other's, and they are also different in terms of the normal Internet activities. So we can detect and classify worms by the extracted features of the network packets. At first, we sniff raw network packets from the local area network (LAN), and extract 13 features from the packet header, and then select 10 important features using the information gain algorithm. With the labeled features, we train Support Vector Machine (SVM) classifiers. The classifiers can classify the behaviors of the worm apart from the normal Internet activities. And this approach can also classify network attacks and Internet worms, although the network attacks and the Internet worms have similar behaviors. In the experiments, this approach performs well in worm detection and classification.

74.1 Introduction

With the increasing popularity of the Internet and the network attack technology continues to change, the network worm has become a critical threat in computer network [1, 2]. By self-propagating and fast spreading, the network worm attacks computer operating system or applications that have specific security vulnerabilities, so as to obtain control of some or all of the computers. Currently, many research works for Internet worm detection have been done [3, 4]. And most of the Internet worm detection methods are based on the intrusion detection system (IDS) [5]. The Internet worm-based IDS mainly has two categories, the network-based Internet worm detection and the host-based Internet worm detection. The host-based Internet worm detection approaches analysis the network packets that already reached the end-host. While the network-based Internet worm detection methods

H. Liang (✉) • M. Li • J. Chai
Sichuan Electric Power Research Institute, Chengdu 610000, China
e-mail: liang_huihui@sohu.com

W.E. Wong and T. Zhu (eds.), *Computer Engineering and Networking*, Lecture Notes in Electrical Engineering 277, DOI 10.1007/978-3-319-01766-2_74,

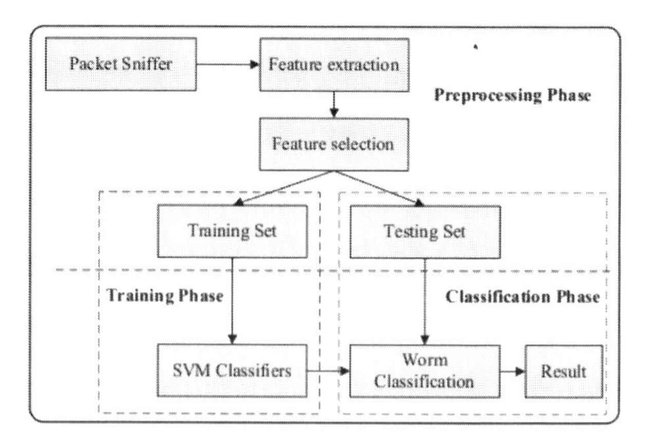

Fig. 74.1 Overview of the framework

consider network packets before they reach an end-host. In this chapter, we detect Internet worms and classify the Internet worms apart from the Internet attacks by analyzing the network packets before they reach an end-host.

In this chapter, we propose a novel network packets based Internet worm detection method. In Fig. 74.1, we give an overview of the framework of our worm detection and classification approach. For worm detection, during the preprocessing phase, we sniffer network packet data from our local area network, and we extract 13 features from each traffic data. And we give each extracted feature vector a label L. If the feature vectors are extracted from a normal traffic data, we set $L = 1$, else, we set $L = c(c = 2, 3, \cdots, C)$ for different Internet worms or Internet attacks, C is the number of the worm and the Internet attacks categories. In order to use some useful features to detect and classify the worms and the Internet attacks. We select ten features from the extracted features using an information gain method. Using this method, we collect our training data and testing data. We train a Support Vector Machine (SVM) classifier for each Internet worm category, each Internet attack category and the normal Internet activity class. During the classification phase, we can detect and classify the Internet worms, the Internet attacks, and the normal Internet activities.

The rest of this chapter is organized as follows. After describe related works of Internet worm detection methods in Sect. 74.2, we give an introduction of the abnormal behaviors of the Internet worms and the Internet attacks of the traffic packets in Sect. 74.3. In Sect. 74.4 we present the technique information of SVM. After introduce the experiment in Sect. 74.5, we conclude this paper in Sect. 74.6.

74.2 Related Works

Through normal network behavior to associate and compare with anomaly behavior of the network traffic, anomaly behavior could be detected. To detect unknown Internet worm, Data mining approach is a good choice. This section provides brief description of these existing detectors.

Wang et al. [6] built a model by training with these features extracted from malware behavior. They presented the results showed that the average detection rate is over 90 %, and the false alarm rate is around 6.67 % with n gram extraction. Siddiqui et al. [7] built a data mining model by train the features extracted from cleaned program and infected program. They presented good results with 95.6 % of detection rate for every end point and with false alarm rate close to 3.8 %.

In this chapter, we propose a novel network packets based Internet worm detection method. Different with most of the previous works, this work uses the one-against-all SVM classification method and trains an SVM classifier for each Internet worm category, each Internet attack category, and the normal Internet activity class. During the classification phase, we can detect and classify the Internet worms, the Internet attacks, and the normal Internet activities. And we get the better performance in our experiment.

74.3 Extract Features of Worms and Attacks

A lot of Internet worms have similar behaviors and payload as the Port Scan and Denial of Service (DoS) attacks. We take Blaster worm, Sobig.E worm, and Forbot-FU as the representation of the common worms. And we take UDP flood and HTTP flood as attack representation of DoS attack. In this chapter, we propose an approach to detect and classify Blaster worm, Sobig.E worm, Forbot-FU worm, Port Scan, and DoS attacks behaviors.

Port Scan is a technique to scan for available port or service from any users, it can be defined as an attack that sends client requests to a range of server port addresses on a host, with the goal of finding an active port and exploiting a known vulnerability of that service.

74.3.1 Feature Selection

In data classification, training set contain a large number of features can produce great computational complexity, and redundant feature vector can reduce the judgment probabilistic. So before we train the SVM classifiers, we select the extracted features. In this section, at first, we introduce the extracted feature records, and then, we select ten features as the classification feature vector using the information gain ratio method.

In order to detect and classify the Internet worm and the Internet attacks, we sniff packets from the local area network. Then we extract a lot of features from the collected packets. We define the 13 features that extracted from all of the packets collected in 1 s as a feature record. For a feature record X_i, we represent each feature component as x_{ij}, $j = (1,2, \cdots,13)$, and the detail information of each feature component is shown as follows:

x_{i1} Represent the number of the source IP address.
x_{i2} Represent the number of the destination IP address.
x_{i3} Represent the number of ICMP packet.
x_{i4} Represent the number of TCP packet.
x_{i5} Represent the number of UDP packet.
x_{i6} Represent the number of the SYN flag.
x_{i7} Represent the number of ACK flag.
x_{i8} Represent the number of RST flag.
x_{i9} Represent the number of source port.
x_{i10} Represent the number of destination port.
x_{i11} Represent the number of different packet size.
x_{i12} Represent $\frac{the\ number\ of\ source\ port}{the\ number\ of\ destination\ port}$.
x_{i13} Represent $\frac{the\ number\ of\ SYN\ flag}{the\ number\ of\ the\ destination\ IP}$.

The purpose of feature selection is in maintaining the accuracy of the premise, as far as possible reduce the number of features. In this paper, we use information gain algorithm to select features. Information gain algorithm is a feature selection methods that use the information gain of feature attributes to decision attributes as the importance measure of feature attribute.

At first gives the training feature record set X, the information entropy of training set X is defined as follows,

$$E(X) = -\sum_{c \in C} \frac{|X_c|}{|X|} \cdot log_2 \frac{|X_c|}{|X|}, \tag{74.1}$$

where C is a subset of X.

Then we reduce the feature j, through measure the degree of entropy reduction (Information Gain) $IG(X,j)$ to evaluate the information content of feature j, the formulation of information gain is shown as follows:

$$IG(X,j) = E(X) - \sum_{v \in V(j)} \frac{|X_v|}{|X|} \cdot E(X_v) \tag{74.2}$$

where set V contains all of the possible values of feature j.

Using this information gain formulation, we evaluate the network features. If the feature has a higher value, it represents this feature is more important than other features that have smaller information gain values. Based on the rank of the information gain values of those features, then we choose the top ten features as a feature vector. At last, based on the ten dimensional feature vectors and its corresponding labels, we train the SVM classifiers, detect and classify the Internet worms and Internet attacks.

74.4 Support Vector Machines

In this paper, we use the support vector machines (SVM) classifiers to classify the Internet worms and Internet attacks. SVM [8, 9] is a pattern recognition method developed from the stoical learning theory, and some works of worm detection that base on SVM have been done [10]. SVM algorithm was originally designed for binary classification problems, when dealing with many problems; it needs to construct a suitable multi-class classifier. Construct the multi-class SVM classifier method mainly has two kinds: the direct method and the indirect method. The direct method directly modifies the objective function, merges the parameter solution problems of multiple classification surface into one optimization problem, by solving the optimization problem to implement "one time" multi-class classification. This method looks like simple, but its computational complexity is very high, and it is difficult to implement. The indirect method achieves multi-class classification by combining multiple classifiers. Common methods include one-against-one and one-against-all two kinds.

The one-against-one SVM method designs an SVM classifier between any two classes of samples. For a training set of C categories, the one-against-one method needs to train $C(C - 1)/2$ classifiers. When classifying an unknown sample, the category has the highest votes is the category of the unknown samples. The one-against-all SVM method needs design C classifiers when solve a C categories classification problem. During training, this method takes the samples of a certain category as one category and takes the other of the rest samples as another category, So that C kinds of samples are constructed C SVM classifiers. During classification, classify the unknown samples as the category that has the maximum value of the classification function.

In this paper, we use the one-against-all method to train SVM classifier. For each classifier, we use the Radial Basis Function (RBF) as a kernel to train the SVM classifier. The formulation of the RBF is shown as follows:

$$K(x, x_i) = exp\left(-\frac{\|x - x_i\|^2}{2\sigma^2}\right), \tag{74.3}$$

where σ is the standard deviation, and $\sigma > 0$. x_i and x are feature vectors.

Based on the radial basis kernel function, using the Lagrange method, the SVM classification decision becomes the problem of finding the solution of the dual optimization problem as follows:

$$max\ L(\alpha) = \sum_{i=1}^{n} \alpha_i - \frac{1}{2} \sum_{i,j=1}^{n} \alpha_i \alpha_j y_i y_j K\left(x_i, y_j\right) \tag{74.4}$$

where $y_i = \pm 1$, $i = 1, \cdots, n$ are class indicator values and α_i, $i = 1, \cdots, n$ are Lagrange multipliers satisfying where C is the regularization parameter.

$$0 \leq \alpha_i \leq C, \quad \sum_{i=1}^{n} \alpha_i y_i = 0$$

The final discriminant function of the SVM classifier of \$j\$ categories is shown as follows:

$$G_j(x) = \sum_{i \in SV} \alpha_i y_i K(x_i, x) + \omega_0 \tag{74.5}$$

where α and ω_0 are the trained parameters, SV is the support vector set. The value of ω is computed by the following function,

$$\omega_0 = \frac{1}{N_{SV'}} \left(\sum_{i \in SV'} y_i - \sum_{i \in SV, j \in SV'} \alpha_i y_i K(x_i, x_j) \right) \tag{74.6}$$

in which SV is the set of support vectors with associated values of α_i, satisfying $< \alpha_i \leq C$ and SV is the set of $N_{SV'}$ support vectors satisfying $0 < \alpha_i < C$.

At last, with the trained SVM classifiers $G_j(x)$, $j = 1, \cdots, C$, we can classify the Internet worms and attacks by the following function.

$$c = arg\max_j G_j(x). \tag{74.7}$$

Where c is the classify category of the unknown sample x.

74.5 Experiment

In this chapter, we set up a local area network and analog the normal working state of the computers. During collect data set, we insert actual blaster worm, Forbot-FU worm, and Sobig.E worm using a reliable on-line source. And generate Port Scan, HTTP flood, and UDP flood into this local area network using the Net Tool at the same time.

74.5.1 Generate Data Sets

In order to collect the experiment data set and the testing data set, we sniffer network packets from the local area network and extract the features from the packet header as describe as in Sect. 74.3. Then we split the collected data set into two parts: the original training set and the original testing set.

Table 74.1 Comparison detection rate and true positive rate

Model	Bayesian network	Decision tree C4.5	Random forest	SVM
Detection rate (%)	96.9	98.6	99.4	99.6
Normal (%)	97.8	98.9	99.6	99.6
Blaster worm (%)	91.2	97.5	99.1	99.4
SoBig.E (%)	94.8	97.5	96.1	97.8
Forbot-FU (%)	98.2	98.2	98.2	98.2
UDP flood (%)	100	100	100	100
HTTP flood (%)	98.0	98.3	98.9	98.8
Port Scan (%)	99.8	99.8	99.8	99.8

The original data set contains redundant feature information. We use information gain method to select ten features from the training set as the final extracted feature vector. And then, we present the training set and the testing set as the ten dimensional feature vectors and its corresponding labels.

At last, we get a training data set of 6,000 feature vectors for normal class, and 1,000 feature vectors for each of the six class Internet worms and the Internet attacks. Totally, the training data set has 12,000 feature vectors. The same size with training data set, the testing data set also have 6,000 feature vectors for normal class, and 6,000 feature vectors for all class of the Internet worms and the Internet attacks. Totally, the testing data set has 12,000 feature vectors too.

74.5.2 Experiment Results

In the experiment, we compute the classification results of the worms and the Internet attacks. In order to evaluate the performance of our detection and classification model, we compare our detection rate, true positive rate, and false alarm rate with other three data mining models.

In Table 74.1, we show the detection rates of our algorithm and make a comparison between our model and other models. Obviously, our algorithm has a better performance. Our detection rate is 99.6 % which is the top score. When the situation is normal or the worm is Blaster worm or SoBig.E worm, we could get a higher rate than other algorithms. When the testing data set is Forbot-Fu worm, UDP Flood, HTTP Flood, or Port Scan, the positive rate is the same with other algorithms. In this case, the positive rate is high even 100 %, so to improve these rate is very difficult. Our algorithm presents a better result in most of the situations, and the same with other algorithms in many cases.

In Table 74.2, we show the false alarm rates of different algorithms. When the testing data is UDP Flood, HTTP Flood, or Port Scan, the false alarm rate is 0. It is the same with other algorithms. The rate of our algorithm is a little higher than Random Forest algorithm and lower than others when the worm is SoBig.E. However, we get the top score when the worm is Blaster Worm or Forbot-FU. In

Table 74.2 Comparisons about false alarm rate

Model	Bayesian network	Decision tree C4.5	Random forest	SVM
Blaster worm (%)	2.1	0.3	0.5	0.3
SoBig.E (%)	4.2	2.1	1.5	1.6
Forbot-FU (%)	0.7	0.6	0.7	0.6
UDP flood (%)	0.0	0.0	0.0	0.0
HTTP flood (%)	0.0	0.0	0.0	0.0
Port Scan (%)	0.0	0.0	0.0	0.0

this case, the Decision Tree C4.5 has the same good performance with ours, and the Random Forest is a little bad. So, the result here is that our algorithm as a whole has a better performance than other algorithms.

74.6 Conclusion

In this paper, a novel method for worm detection and classification was proposed. At first, we sniffered data packets from the local area network, and we extracted a lot of features from the collected traffic data. Then we used information gain algorithm to select useful features. And then by using the selected useful feature vectors, we trained our SVM classifier. At last, we detected and classified the worms and Internet attacks. This algorithm performed well in worm detection and classification. From the experiment results, it can be concluded that our algorithm can correctly detect and classify Blaster worm, SoBig.E worm, Forbot-Fu worm, UDP Flood, HTTP Flood, and Port Scan.

References

1. Magkos, E., Avlonitis, M., Kotzanikolaou, P., et al. (2013). Toward early warning against Internet worms based on critical sized networks. *Security and Communication Networks, 6*(1), 78–88.
2. Julisch, K. (2013). Understanding and overcoming cyber security anti-patterns. *Computer Networks, 57*(10), 2206–2211.
3. Rabinovitch, P., Chow, S. T. H., Abdel-Aziz, B. (2012). Worm detection by trending fan out U.S. Patent 8,095,981[P]. 2012-1-10.
4. Zheng, H., Lifa, W., Huabo, L., et al. (2012). Worm detection and containment in local networks. *Computer Science and Information Processing. Xi'an*: *IEEE*, 595–598.
5. Smith, C., Matrawy, A., Chow, S., & Abdelaziz, B. (2009). Computer worms: Architecture, evasion strategies, and detection mechanisms. *Journal of Information Assurance and Security, 4*, 69–83.
6. Wang, X., Yu, W., Champion, A., Fu, X., & Xuan, D. (2008). *Detecting worms via mining dynamic program execution. Security and privacy in communications networks and the workshops* (pp. 412–421). Nice: IEEE.

7. Siddiqui, M., Wang, M. C., & Lee, J. (2008). Detecting Internet worms using data mining techniques. *Journal of Systemics, Cybernetics and Informatics, 6*(6), 48–53.
8. Bishop, M. (2006). *Pattern recognition and machine learning* (pp. 325–359). New York: Springer-Verlag.
9. Webb, A. (2002). *Statistical pattern recognition* (2nd ed., pp. 123–200). Paris: John Wiley & Sons.
10. Nissim, N., Moskovitch, R., & Rokach, L. (2012). Detecting unknown computer worm activity via support vector machines and active learning. *Pattern Analysis and Applications, 15*(4), 459–475.

Chapter 75
Real-Time Fall Detection Based on Global Orientation and Human Shape

Shuangcheng Wang, Yepeng Guan, and Ruiyue Xu

Abstract Fall detection is an important problem in the research of abnormal behavior recognition. In this chapter, a novel real-time method is proposed to detect human fall with a single uncalibrated camera by the changes of global orientation and human shape. Our algorithm has three basic parts: moving object extraction, fall pre-detection, and fall confirmation. The overall orientations are derived from the combination of Gaussian mixture model and motion history image. The shape deformation is quantified from the silhouettes by an approximated bounding box. Standard deviations of the overall orientation and aspect ratio of bounding box are checked to pre-detect a fall, and a fall is confirmed by unmoving shape of the bounding box with the defined fall angle. Experimental results show that our method can detect all possible types of human fall accurately and successfully.

75.1 Introduction

Intelligent visual surveillance has got more research attention and funding due to increased global security concerns, whose key technology is abnormal behavior recognition. Fall detection is an important problem in the research of abnormal behavior recognition [1], because fall in elderly greatly threatens their health, especially who are living alone.

S. Wang (✉)
School of Communication, Information Engineering, Shanghai University,
Shanghai 201305, China
e-mail: wsc36305@foxmail.com

Y. Guan
Key Laboratory of Advanced Displays and System Application, Shanghai 201305, China
e-mail: ypguan@shu.edu.cn

R. Xu
School of Communication, Information Engineering, Shanghai University,
Shanghai 201305, China

W.E. Wong and T. Zhu (eds.), *Computer Engineering and Networking*, Lecture Notes in Electrical Engineering 277, DOI 10.1007/978-3-319-01766-2_75,

Here, a novel real-time fall detection algorithm is proposed, which detects fall by checking the variation of the global orientation and some changes in human shape with a single fixed uncalibrated camera. We notice that when people walk or run normally, the motion direction varies continuously at small scale, while during a fall, it changes greatly. Moreover, the rebound when the body hits on the ground and the movement of arms and legs can lead to great change in directions. The global orientation performs well on falls parallel or perpendicular to the camera optical axis in our work, which is derived from the combination of Gaussian mixture model (GMM) and motion history image (MHI). Meanwhile, the extracted moving object by GMM is approximated by a bounding box. Firstly, standard deviations of the overall orientation and aspect ratio of bounding box are checked to pre-detect a fall. And then a fall is confirmed by unmoving shape of the bounding box with the defined fall angle.

The organization of this chapter is as follows. Section 75.2 explains the related work on fall detection. Section 75.3 describes the details of the three basic parts of the method. Experimental results are represented in Sect. 75.4, and a conclusion is made at last.

75.2 Related Work

Traditional methods to detect falls use some wearable sensors [2]. However, the problem of such detectors is that people often forget to wear them and it's uncomfortable. Approach based on the computer vision is not only able to overcome these limitations but also capable of getting more information about fall, which may help to analyze the reason of a fall.

There are many related works in the field of fall detection based on the digital video processing techniques. For instance, a simple method was used only based on analyzing aspect ratio of the moving object's bounding box [1]. Some researchers have mounted the camera on the ceiling: Lee [3] detected a fall by analyzing the shape and 2D velocity of the person. Nait-Charif and McKenna [4] tracked the person using an ellipse and analyzed the resulting trajectory to detect inactivity outside the normal zones of inactivity like chairs or sofas. Rougier [5] and Miao Yu [6] both computed a motion coefficient based on the motion history in order to detect a fall. Rougier [5] combined the coefficient with shape descriptor ellipse fitting, while Miao Yu added head tracking using particle filter to find a fall with the speed of head. Other systems used the 3D information to infer events [7], but these mechanisms tend to be more complex, need more additional cost, and are difficult to be applied to real systems.

Unlike the above methods, in our system the global orientation and human shape are adopted to detect a fall. It has an acceptable detection rate, and it can work in real time using a webcam.

75.3 The Details of Our Method

Our algorithm has three basic parts: moving object extraction, fall pre-detection, and fall confirmation.

75.3.1 The Moving Object Extraction

Considering a relatively fixed background in our scene, we adapt an OpenCV algorithm BackgroundSubtractorMOG2 described by Z. Zivkovicin to segment the moving object [8]. In our work, four Gaussian distribution models are defined to describe the state of each pixel [8].

Then, morphological opening-and-closing operation is used to obtain a better result. And a denoising method is applied by means of the edge tracing method. The contours of updated binary foreground image are tracked, and the ones whose circumference is under a certain threshold (in practice, it is set to 80) are eliminated.

75.3.2 Fall Pre-detection

Motion History and Moving Orientation We notice that when people walk or run normally, the motion direction varies continuously at small scale, while during a fall, the direction changes obviously relative to walking or running. Moreover, the rebound when the body hits on the ground and the movement of arms and legs can lead to great changes in directions as it is shown in Fig. 75.1 (the radius indicates the direction). So the global motion orientation is needed.

MHI [9] is a simple and powerful method that records movement information of object by means of time stamp. The MHI is an image where the pixel intensity represents the recency of motion in an image sequence and therefore gives the most recent movement of a person during an action. Contrary to the optical flow [5], which is complex, time-consuming, and bad anti-noise, MHI is brief and real time.

The MHIH_τ(x,y,t) can be computed from an updated function D(x,y,t):

$$H_\tau(x,y,t) = \begin{cases} \tau & , \textit{if } D(x,y,t) = 1 \\ \max\left(0, H_\tau(x,y,t-1) - \delta\right) & , \textit{otherwise} \end{cases}, \tag{75.1}$$

where (x,y) and t show the position and time and D(x,y,t) is a binary sequence of motion regions that is extracted from the original image sequence I(x,y,t), which are already obtained by GMM. The duration τ decides the temporal extent of the movement (normally, in terms of frames), and δ is the decay parameter [9]. The result is a scalar valued image where more recently moving pixels are brighter and vice versa.

Fig. 75.1 Changes in direction when falling. (1) Falling perpendicularly to the camera optical axis, large changes in direction. (**a**) 137th frame, (**b**) 143th frame, (**c**) 152th frame, (**d**) 161th frame. (2) Falling parallelly to the camera optical axis, also large changes in direction. (**e**)120th frame, (**f**) 128th frame, (**g**) 135th frame. (**h**) 142th frame

Since the MHI records silhouette at different times, the overall motion orientation will be obtained by calculating the gradient of MHI. The gradient Orient(x,y) is computed as follows:

$$Orient(x, y) = \arctan\left(D_y(x, y)/D_x(x, y)\right), \quad (75.2)$$

where $D_x(x,y)$ is the difference in x direction and $D_y(x,y)$ is the difference in y direction. The average direction is computed from the weighted orientation histogram, where a recent motion has a larger weight and the motion occurred in the past has a smaller weight, as recorded in MHI.

In the light of above work, we can easily obtain the α standard deviation σ_α from N frames (N is set to 20 in our work). In our work, we set the fixed threshold of σ_α as 1.0 according to our datasets.

Rectangular Approximation for Shape Analysis It is obvious that the human shape changes greatly during a fall. Due to the well performance of global orientation, we simply approximate the human shape with a minimum bounding box [10]. When a person falls, the height and width of bounding box change drastically. We simply defined the aspect ratio ρ as

$$\rho = w/h \quad (75.3)$$

where w is the width of the bounding box and h is the height of the bounding box. We derived its standard deviation σ_ρ from fixed number of frames, which in our work is set to 20. We use σ_ρ to represent the changes in shape. It will stay at a high value when a fall happens, while it will stay at a low value when people walk normally. It's threshold for our datasets is set to1.5.

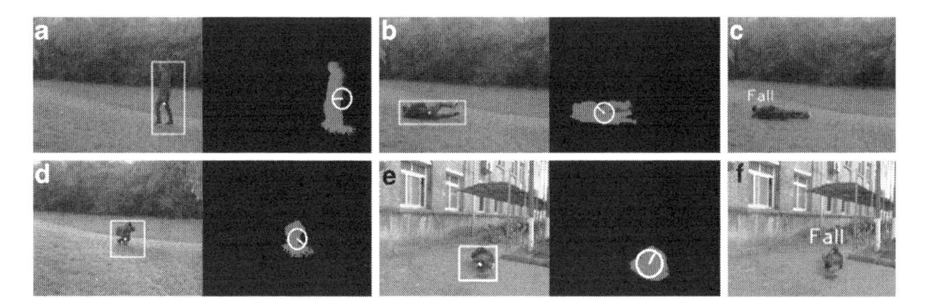

Fig. 75.2 Some examples on various situations (**a**) Walking normally: $\sigma_\alpha = 0.0501$, (**b**) Falling perpendicularly: $\sigma_\alpha = 1.3297$ (**c**) Result $\sigma_\rho = 0.5001$, $\rho = 2.0894$; $\sigma_\rho = 2.3211$, $\sigma_\beta = 0.652$, $\rho = 0.340$ (**d**) Crouch normally: $\sigma_\alpha = 0.7805$, (**e**) Falling parallelly: $\sigma_\alpha = 1.2504$, (**f**) Result $\sigma_\rho = 0.8424$, $\rho = 1.0216$, $\sigma_\beta = 0.9532$; $\sigma_\rho = 1.6033$, $\rho = 0.8598$, $\sigma_\beta = 0.7236$

Since σ_α and σ_ρ need tens of frames to initialize, there may exist false alert at the beginning when a person first appears. Obviously, when normally walking, ρ stays at a low value, while during a fall, ρ stays at a high value. The parameter ρ is used to avoid these faults with a fixed threshold (in our work, we choose 1.0).

75.3.3 Fall Confirmation

The last step of the algorithm is to check whether the person remains motionless on the floor for a few seconds after falling. And the centroid of object [11] is needed.

The center of the bounding box is obtained by computing the coordinates of the center of mass with the first- and zero-order spatial moments [12]. We define fall angle β as the angle of centroid of object with respect to horizontal axis of the bounding box [11]. The unmoving shape of the bounding box is checked by the defined fall angle β. When a person is walking, β varies from 45° to 90° (depends on their style and speed of walking). But after a fall, the β changes very little, so σ_β (standard deviation of tan β) for ten frames will be low.

After a fall is pre-detected, if $\sigma_\beta < 0.7$, a fall is confirmed.

75.4 Experiment Results

Our algorithms implemented in the project are written in C++ using the OpenCV library. Our system is designed to work with a single uncalibrated camera and it can work in real time using a webcam. The camera can be placed with horizontal viewing angle or a wild angle of more than 70°.

Figure 75.2 shows some examples and detection results on various situations. In our testing project, when a fall is confirmed, we put out the text in the frames.

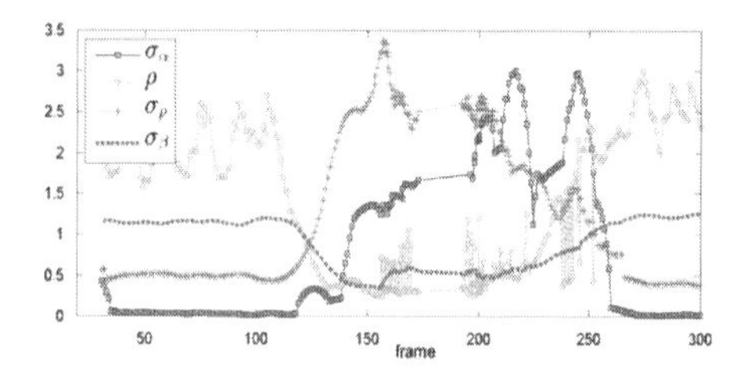

Fig. 75.3 Variation of the coefficients during a fall

Table 75.1 Detected results

	Detected	Not detected	Sensitivity/specificity (%)
Falls (sensitivity)	23	2	92.0
Daily activities (specificity)	4	19	81.8

Table 75.2 Comparison of fall detection approaches in terms of sensitivity and specificity

	Our method (%)	The shape analysis [5] (%)	Posture analysis [13] (%)
Falls (sensitivity)	92.0	84.0	88.0
Daily activities (specificity)	81.8	82.6	78.2

Table 75.3 Comparison of fall detection approaches in terms of time complexity and time consumption

	Time complexity[a]	Time consumption per frame (ms)
Our proposed method	*O(n)*	60
The shape analysis [5]	*O(n)*	80
The posture analysis [13]	*O(n* log *n)*	–

[a] n is the total number of pixels in one frame

In Fig. 75.2a when walking normally, σ_α and σ_ρ are low; meanwhile, ρ is high, so no fall is detected. Figure 75.2b shows a person falling perpendicularly; it is obvious that σ_α and σ_ρ are high and above the fixed thresholds; meanwhile, ρ is low, but σ_β is a little high, so a fall is pre-detected; after some frames σ_β becomes low enough, and a fall is confirmed and it is shown in

Figure 75.2c. A person is crouching normally in Fig. 75.2d; σ_α and σ_ρ are higher than those walking normally, but they are below the fixed thresholds, so no fall is detected. A parallel falling happens in Fig. 75.2e; we notice that σ_α and σ_ρ are above the fixed thresholds and ρ is low, so a fall is pre-detected; and then after some frames σ_β becomes low enough, and a fall is also confirmed, just as Fig. 75.2f shows.

The thresholds we choose are based on practice. Figure 75.3 shows an example of video sequence of a perpendicular fall, which is the same sequence as in Fig. 75.2b. It's obvious that when the person is falling, at about 140th frame, σ_α

and σ_ρ are high and ρ is low. The thresholds of σ_α and σ_ρ are set to 1.0 and 1.5; meanwhile, the threshold of ρ is 1.5. After a fall, σ_β is still low, so a fall is confirmed. 0.7 is chosen as the threshold of σ_β. When the person starts to stand up about the 235th fame, σ_α is still high, while σ_ρ is not as high as that when the person is falling due to its motion intensity, but ρ becomes higher; what's more σ_β is higher than 0.7, so no fall is detected.

To verify the feasibility of the proposed approach, we have used 45 video clips as our test targets, which include 23 daily normal activities (walking, running, crouch, sitting down) and 25 falling video clips (forward falls, backward falls, perpendicular falls, parallel falls). Sensitivity and specificity are obtained according to the method described by B. Töreyin [10].

Table 75.1 shows that our system gives very good results. We get a good detection rate with a sensitivity of 92.0 % and an acceptable rate of false detection with a specificity of 81.8 %. The specificity is not very high due to one brutal sits-down activity on a sofa with rebounding, one brutal crouches-down activity, and one brutal squat-down activity were detected as falls, which have large enough changes in motion detections and box ratio.

We compare our proposed method with the shape analysis approach using ellipse [5] and posture analysis technique [13]. Our proposed method can achieve better performance in terms of sensitivity and time consumption, despite the relatively low specificity, as Tables 75.2 and 75.3 show. The detection rate of the shape analysis [5] is lowest because of the instability of motion coefficient. Posture analysis [13] has high sensitivity, but low specificity, due to the combination of two different analyses used. The posture analysis [13] has the highest time complexity, which means more time consumption, while our method is less time-consuming than the shape analysis approach [5] due to less computation; the data were obtained with a PC of Intel i3 3.3 GHz CPU and 4G RAM.

75.5 Conclusion

In this chapter, we propose a new real-time fall detection approach based on global orientation and human shape. The global orientation performs well in our work. Our approach of fall detection has been proven to be robust and of real time through testing on realistic image sequence of simulated fall and daily activities.

Acknowledgments This work is supported in part by the National Natural Science Foundation of China (Grant no. 11176016, 60872117).

References

1. Popoola, O. P., & Wang, K. (2012). Video-based abnormal human behavior recognition—A review. *Systems, Man, and Cybernetics-Part C: Applications and Reviews, 42*(6), 65–878.
2. Mubashir, M., Shao, L., & Seed, L. (2013). A survey on fall detection: Principles and approaches. *Neurocomputing, 100*(16), 144–152.
3. T. Lee, A. Mihailidis. An intelligent emergency response system: preliminary development and testing of automated fall detection. Journal of telemedicine and telecare. 11(4), 194-198 (2005)
4. Nait-Charif, H., McKenna, S. J. (2004). Activity summarization and fall detection in a supportive home environment. *Proceedings of the 17th International Conference on Pattern Recognition. IEEE Computer Society,Washington DC* (Vol. 4, pp. 323–326).
5. Rougier, C. (2007). Fall detection from human shape and motion history using video surveillance. *Proceedings of 21st International Conference on Advanced Information Networking and Applications Workshops. IEEE Computer Society, Washington DC* (Vol. 2, pp. 875–880)
6. Yu, M., Naqvi, S. M., Chambers, J. (2009). Fall detection in the elderly by head tracking. *Proceedings of IEEE/SP 15th Workshop on Statistical Signal Processing. IEEE, Cardiff Wales* (pp. 357–360).
7. Rougier, C. (2012). 3D head tracking for fall detection using a single calibrated camera. *Image and Vision Computing, 31*(3), 246–254.
8. Zivkovic, Z. (2004). Improved adaptive Gaussian mixture model for background subtraction. *Proceedings of the 17th International Conference on Pattern Recognition. IEEE Computer Society, Washington DC* (Vol. 2, pp. 28–31).
9. Ahad, M. A. R. (2012). Motion history image: Its variants and applications. *Machine Vision and Applications, 23*(2), 255–281.
10. Töreyin, B., Dedeoğlu, Y., & Cetin, A. (2005). HMM based falling person detection using both audio and video. *Computer Vision in Human-Computer Interaction, 3766*, 211–220.
11. Vishwakarma, V., Mandal, C., & Sural, S. (2007). Automatic detection of human fall in video. *Pattern Recognition and Machine Intelligence, 4815*, 616–623.
12. Flusser, J., & Kautsky, J. (2010). Implicit moment invariants. *International Journal of Computer Vision, 86*(1), 72–86.
13. Chen, Y. T., Lin, Y. C., Fang, W. H. (2010). A hybrid human fall detection scheme. *Proceedings of IEEE International Conference on Image Processing. IEEE, Hong Kong.* pp. 3485–348

Chapter 76
The Classification of Synthetic Aperture Radar Oil Spill Images Based on the Texture Features and Deep Belief Network

Xixi Huang and Xiaofeng Wang

Abstract This chapter introduces a new method to classify the SAR oil spill images. That is Deep Belief Network (DBN). Through the experimental certification, it is shown that the SAR images' information extracted by Gray-Level Co-occurrence Matrix (GLCM) can have a better effect in classification then that extracted by Gabor wavelet features. And using DBN to classify 240 samples including oil slick, looks-like oil slick and seawater, we can reach high total classification accuracy up to 91.25 %. Finally, we get a result that the method of DBN with GLCM features can better meet the needs of the SAR oil spill images classification.

76.1 Introduction

The Satellite SAR images of oil spill are rich in texture information that can be used to increase the image classification accuracy. Many researchers use neural network (NN) to classify the SAR images [1–3]. However, the complexity of the NN system and the high theoretical knowledge requirements restrict the widely application of network. It is mainly because of the dimension and structure of the network, the depth of the training set, and some other aspects of NN [4].

In this chapter, we use a new model named Deep Belief Network (DBN) for the classification of sea SAR image. DBN classifier is composed of a number of Restricted Boltzmann Machines (RBMs). After training each group of RBMs from the bottom up to form the hierarchy, DBN can solve the problem that the complex structure in traditional NN and its depth makes it not easy to be trained.

X. Huang (✉) • X. Wang
Information Engineering College, Shanghai Maritime University, Shanghai 201306, China
e-mail: dreamhk5001@gmail.com; xfwang@cie.shmtu.edu.cn

W.E. Wong and T. Zhu (eds.), *Computer Engineering and Networking*, Lecture Notes in Electrical Engineering 277, DOI 10.1007/978-3-319-01766-2_76,

76.2 Experimental Data of SAR Oil Spill Images

In this chapter, the experimental data of SAR oil spill images comes from Infoterra global oil spills database. We select 95-King image of China Sea from ERS-1 and ERS-2 satellite with a resolution of 100 m. As shown in Fig. 76.1, zone 1 and zone 2 are looks-like slick areas; zone 3, zone 4, zone 5, and zone 6 are the real slick contaminated areas.

SAR images in this chapter are a series of preprocessed 600×600 pix SAR oil spill images which have already been corrected, de-noised, and filtered. Because the SAR can cover very wide areas in high resolution, even a large area of sea still looks small in the SAR images. So the SAR images need appropriately and reasonably cut into 40×40 pix images of oil slick, looks-like oil slick and sea water. There are 1,200 samples in all, and each kind of different category has 400 samples. Each image is saved as one-dimensional gray-level image. Then use histogram equalization to enhance the contrast of the gray-level of SAR images in the storing process of the three kind of classified images. In order to ensure that samples are randomly assigned, data are stored randomly, and will be disrupted before selecting the training samples and testing samples.

Before using DBN to classify the SAR images, we need to extract the texture feature information of the preprocessed images. In this chapter, we use Gabor wavelet [5–7] and Gray-Level Co-occurrence Matrix (GLCM) [8, 9] as the auxiliary texture feature information, and then use DBN to classify the feature information extracted from SAR image.

(1) Gabor wavelet feature extraction: Create a Gabor filter with 5 scale by 8 orientation, convolving to 40 transformed images. Decimate the transformed images with the sampling rate at 8, and then we can get a feature matrix with 640-dimensional feature, denoted by F_{gabor}.
(2) GLCM feature extraction: Extract energy, entropy, contrast, correlation, angular second moment, dissimilarity at four directions as 0°, 45°, 90°, 135°. Use the value, mean, and variance of the six features to get a 36-dimensional feature matrix, denoted by F_{glcm}.

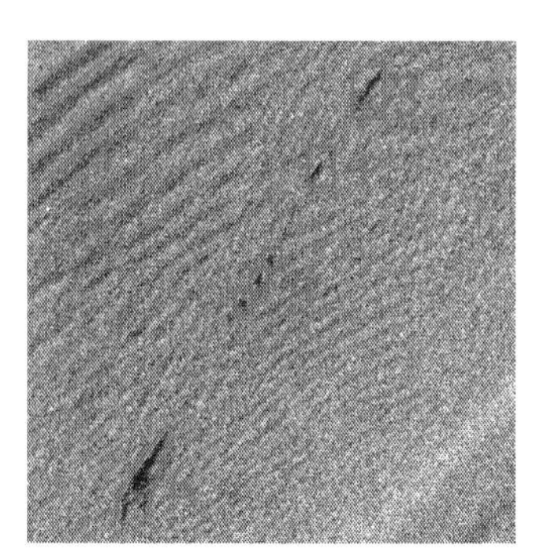

Fig. 76.1 Oil spill image of SAR

Through the experiment we find that using GLCM for classification without setting learning rate can lead to a better result. And momentum has little effect on classification and the variation trend is more stable. When we use Gabor wavelet for classification, it has a bigger impact on learning rate and momentum, the result of classification swings great. So in this chapter, we choose GLCM as the extraction feature for DBN. The follow-up study is based on this.

76.3 Classification Process Based on Texture Features and DBN

76.3.1 DBN

DBN is consisted of several RBMs. Its depth depends on the demand of experiment object on each layer. Actually three-layer DBN is enough, and Fig. 76.2a shows a DBN model. RBM is consisted of Markov Random Field with a visual layer and a hidden layer with no connection between the nodes in the same layer. Figure 76.2b shows a single-layer RBM structure.

The complication of RBM can be achieved by increasing the hidden layer nodes or superimposing a hidden layer [10]. The model can not only simulate original data but also process those representative data after pretreatment, such as the extracted features set.

Construct the training data as RBM network with visual layer and hidden layer. The pixel of image corresponds to the nodes of visual layer in the RBM network [11]. The features will be detected corresponds to the node of hidden layer [12]. A joint configuration, (v, h) of visible and hidden layer has an energy given by:

$$E(v,h) = -\sum_{i\in visible} a_i v_i - \sum_{j\in hidden} b_j h_j - \sum_{i,j} v_i h_j w_{ij} \tag{76.1}$$

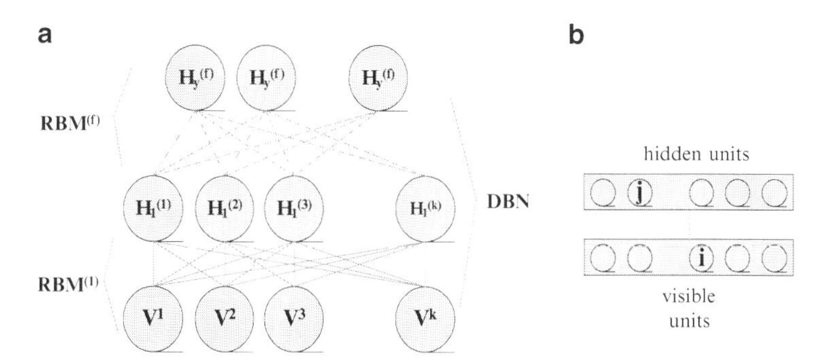

Fig. 76.2 The model of DBN and RBM

where v_i and h_j are the state of visible unit i and hidden unit j, a_i, and b_j are their biases. And w_{ij} is the weight between them. According to the energy formula, the network assigns a probability to every possible pair of visible and hidden layer:

$$p(v, h) = \frac{1}{Z} e^{-E(v,h)} \tag{76.2}$$

where "partial function" Z is the sum of all possible pairs of visible and hidden layers:

$$Z = \sum_{v,h} e^{-E(v,h)} \tag{76.3}$$

Due to the nodes are independent from each other in one layer in RBM, the biases of $<v_i h_j>_{data}$ is easy to calculate. Randomly select a training sample v, the state h_j of each hidden layer is set to 1 with probability:

$$p(h_j = 1|v) = \sigma\left(b_j + \sum_i v_i w_{ij}\right) \tag{76.4}$$

Given a hidden vector, then get an unbiased sample of the state of a visible layer:

$$p(v_i = 1|h) = \sigma\left(a_i + \sum_j h_i w_{ij}\right) \tag{76.5}$$

Where the logistic sigmoid function is as follows:

$$\sigma(x) = \frac{1}{(1 + \exp(-x))} \tag{76.6}$$

The change in a weight is then given by:

$$\Delta w_{ij} = \varepsilon\left(\langle v_i h_j \rangle_{data} - \langle v_i h_j \rangle_{recon}\right) \tag{76.7}$$

Where $<v_i h_j>_{data}$ is model value and $<v_i h_j>_{recon}$ is the "reconstruction" value.

DBN is a generative deep learning structure which is widely used [13]. It can solve the training problems in traditional multilayer neural network: (1) the requirement of huge amount of training set with label; (2) slower convergence rate; and (3) risk of falling into partial optimum due to an improper parameter selection [14].

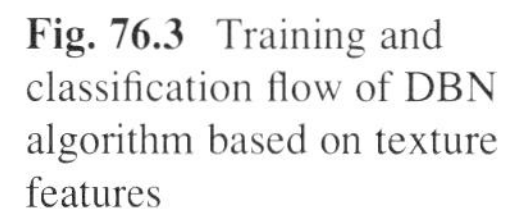

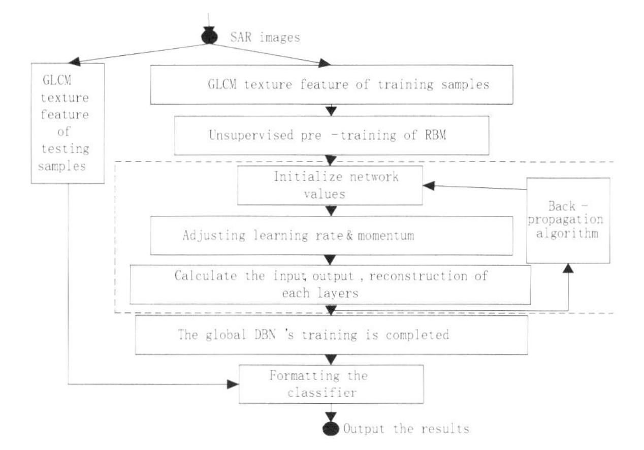

Fig. 76.3 Training and classification flow of DBN algorithm based on texture features

76.3.2 Learning and Classification Process of DBN Algorithm

In this chapter, we use 1,200 samples of SAR images. Choosing the quantity of training samples and test samples is also a part of what we will discuss in this chapter. So the choice of the quantity is gotten after the comparison through the experiment.

During the experiment, we firstly do the unsupervised pretraining on the bottom layer of RBM to get the initial value of the network. Then we use GLCM texture features to adjust the learning rate and momentum parameter of the DBN. Secondly, based on GLCM texture feature, we select an appropriate training number by comparing the results of different training number used in DBN. Finally, we get the most suitable quantity of the training samples by comparing the classification accuracy of DBN on the basis of the number of training samples.

Figure 76.3 describes the training process of DBN algorithm and the flow of how to use the trained DBN for classification.

(1) Use the normalized features as the input data for unit V in the lowest layer and do unsupervised pretraining in RBM module to get the initialize parameter for network. Then adjust the learning rate and momentum.
(2) The output h_i of Group K-1 in RBM's hidden layer (Layer H) becomes the k layer's input v_i in RBM's visible layer (Layer V) and take the training in Layer K. Calculate the input, output, and reconstruction of each intermediate layer. Each hidden layer in RBM model is considered to be a feature detector and will be used to extract the representative features.
(3) Repeat steps (2) to train each RBM group. When the last group of RBM in DBN is finished, do a back-propagation algorithm to fine-tune the overall DBN until getting a global optimal solution for the network. Then the training of DBN is complete and DBN becomes a classifier.
(4) Put the testing GLCM texture features of training set into DBN, and finally calculate the classification results.

76.4 Result Analysis

From Sect. 76.2, we decide to use GLCM extracted from SAR images for DBN's classification. So now we go to analysis the experiment result coming from GLCM and DBN.

We can learn from Fig. 76.4 that with the increase of iteration times, the classification accuracy will be gradually improved and the trend goes more and more smooth. When the number comes up to 500 and 1,000, the average classification accuracy basically remains the same. So if the quantity of training samples is huge enough, we could choose a relatively small number of iterations, in order to reduce the amount of calculation as well as improve the calculation speed.

It can be seen from Fig. 76.5 that with the increase of training sample quantity, the average classification accuracy increases. In this chapter, we made a comparison of 40–80 % sample quantity at different training number. When the sample quantity is certain, and the training number is less than 1,000, the average accuracy

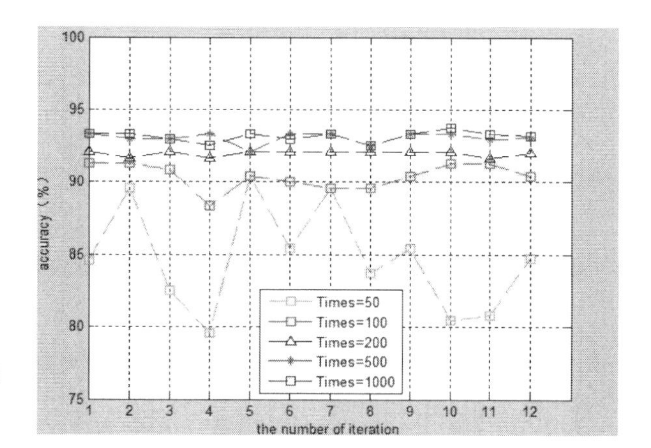

Fig. 76.4 Iterations impact on the result of classification

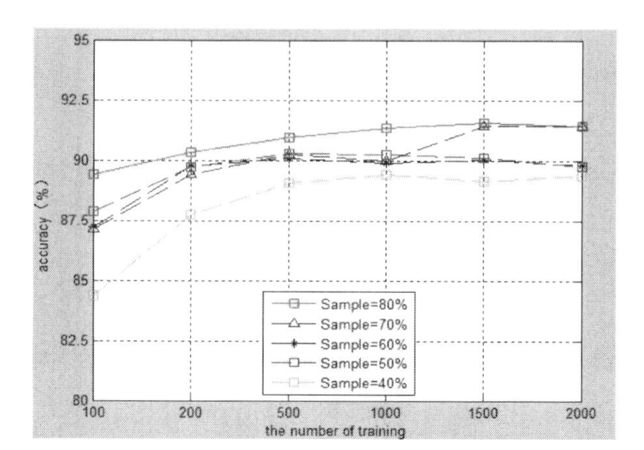

Fig. 76.5 Comparison of classification accuracy under different sample quantity by GLCM

Table 76.1 Total comparison of the classification error statistics of 240 test samples

Category	Oil slick	Looks-like oil slick	Sea water	Total	Classification accuracy (%)
Oil slick	74	6	0	80	92.5
Looks-like oil slick	1	74	5	80	92.5
Sea water	0	9	71	80	88.75
Total	75	89	76	240	91.25

shows a process of surge. With the continued increasing of training time, the average accuracy remains unchanged, even shows a downward trend. When training number is certain, the increasing of training sample quantity will bring a relatively increase of average accuracy. So we come to the result that the effect of classification is most ideal when using 80 % sample quantity with 1,000 times of training for this experiment.

Take the data of oil slick, looks-like oil slick and seawater in Table 76.1 as Matrix A, calculate the total classification accuracy and Kappa coefficient. In the equation, a_{ij} means diagonal elements and correct classification number of A, N means the total amount of all types of samples. $T_{i\bullet}$ and $T_{\bullet j}$ equal to the sum of row i and column j. Total classification accuracy is different from Kappa coefficient in which total classification accuracy only takes correctly classified sample accuracy into account, but Kappa coefficient considers not only the samples which is correctly classified but also the incorrect and missing ones. The result of total classification accuracy (CA) and Kappa coefficient is shown below:

$$\mathrm{CA} = \frac{\sum \mathrm{a}_{u}}{N} = 91.25\%, \mathrm{Kappa} = \frac{N \cdot \sum_{i=1}^{r} a_{ij} - \sum_{j=1}^{r} T_{\bullet j} \cdot T_{i\bullet}}{N^2 - \sum_{i=1}^{r} T_{\bullet j} \cdot T_{i\bullet}} = 0.9205,$$

We can get the total classification accuracy is 91.25 %, respectively, is 92.5 % for oil slick, 92.5 % for looks-like oil slick, and 88.75 % for seawater which are correctly classified. The Kappa coefficient is 0.9205 with a high consistency. Although the low classification accuracy of sea water makes some effect on the final total classification accuracy. But being practical, sea water and looks-like oil slick belongs to nature phenomenon rather than man-made pollution caused by subjective factors in later stage. So we can use it only as a reference data. Thus, we can see that the DBN can meet the needs of the classification of SAR oil spill images in sea oil spill affairs, with a classification rate over 88 %.

76.5 Conclusion

In this chapter, we studied the classification method of SAR oil spill images based on the GLCM and DBN. According to the features of SAR images, we separately extract the GLCM f and Gabor wavelet. Through experiment, we find that GLCM has a better effect in classification. When we use 80 % sample quantity and take the training 1,000 times, the classification of SAR images has a better effect by using GLCM and DBN. The classification accuracy is 4.5 % higher than choosing Gabor wavelet as features. Meanwhile we propose a new deep learning module named DBN. By cutting images into pieces, training batch by batch, each piece is a unit of RBM module, and the depth of learning can be reasonably added according to the actual needs. Eventually, we get total classification accuracy as 91.25 % when using 240 test samples to do the test. The classification accuracy of all category are higher than 88 %, and Kappa coefficient reaches 0.9205, which means DBN shows good classification of oil slick, looks-like oil slick and sea water.

To some extent, the associative level abstraction mode and fast learning method used in DBN are breakthrough of complexity and low computing efficiency of the traditional neural network. But in practices, image processing and recognition is a large-scale project, we still need a lot of further research and exploration to extract an efficient and executable computing ability based on the deep learning.

References

1. Lijian, S., Chaofang, Z., & Peng, L. (2009). Oil spill identification in marine sar images based on texture feature and artificial neural network. *Periodical of Ocean University of China, 39* (6), 1269–1274. In Chinese.
2. Wencheng, X., Chuanqing, W., & Bin, W. (2008). Oil spill detection with SAR in South Korea's oil leak. *Remote Sensing Technology and Application, 23*(4), 410–413. In Chinese.
3. Topouzelis, K., Karathanassi, V., & Pavlakis, P. (2008). Dark formation detection using recurrent neural networks and SAR data. *International Journal of Remote Sensing, 29*(26), 4705–4720.
4. Marghany, M., Hashim, M. (2011). Comparative algorithms for oil spill detection from multi mode RADARSAT-1 SAR satellite data (pp 318–329). Computational Science and Its Applications-ICCSA, Santander, Spain: Springer Link.
5. Luo, X., & Xiaojun, W. (2011). Detection algorithm for infrared small and weak targets based on wavelet transform and Gabor filter. *Infrared and Laser Engineering, 40*(9), 1818–1822. In Chinese.
6. Jin, Q., Tong, X., & Bo, S. (2012). Study of Gabor feature selection for iris recognition. *Computer Engineering and Applications, 48*(19), 201–204. In Chinese.
7. Yi, N., Chen, D., & Yue, W. (2012). Improved Gabor wavelet transformation feature extraction method. *Computer Engineering, 38*(15), 145–147. In Chinese.
8. Zhou, Y., Lin, M., & Ma, T. (2010). SAR feature analysis of oil spill based on GLCM. *Marine Science Bulletin, 29*(4), 455–458. In Chinese.
9. Qing, X., Zheng, J., & Chen, Y. (2011). Application of texture analysis to identification of oil spills in SAR images. *Journal of Hehai University (Natural Sciences), 39*(5), 569–574. In Chinese.

10. Nair, V., & Hinton, G. E. (2009). Implicit mixtures of restricted Boltzmann machines. *Neural Information Processing Systems, 21*(1), 1145–1152.
11. Hinton, G. E. (2010). Learning to represent visual input. *Philosophical Transactions of the Royal Society, 365*(1537), 177–184.
12. Hinton, G. E., Osindero, S., & Teh, Y. W. (2006). A fast learning algorithm for deep belief nets. *Neural Computation, 18*(7), 1527–1554.
13. Sun, Z., Xue, L., & Yangming, X. (2012). Overview of deep learning. *Application on Research of Computers, 29*(8), 2807–2810. In Chinese.
14. Mohamed, A., Dahl, G. E., & Hinton, G. E. (2012). Acoustic modeling using deep belief networks. *IEEE Transactions on Audio, Speech, and Language Processing, 20*(1), 14–22.

Chapter 77
The Ground Objects Identification for Digital Remote Sensing Image Based on the BP Neural Network

Shengkui Cao, Guangchao Cao, Kelong Chen, Chengyong Wu, Tao Zhang, and Jie Yuan

Abstract Spectral information of ground objects target in remote sensing image is complex, more noise, and highly nonlinear. It makes traditional data processing method no longer significant, effective, and efficient. The BP neural network classification-recognition method provides a more ideal solution. Using the TM remote sensing images as the example, this paper experimented the application of the BP neural network to the remote sensing image classification and recognition. Results showed that the classification precision of cultivated land was very low for both the BP neural network and traditional maximum likelihood methods because the spectrum difference between the new cultivated land and the bare land having low plant covered in this area was not significant. Maximum likelihood method wrongly regarded the bare land which had higher soil moisture content by lakeshore as water body. Except the grassland, the classification effect of the BP neural network was superior to maximum likelihood method. The overall classification accuracy by the BP neural network reached 81.79 %; however, the one by the maximum likelihood method was 79.08 %, indicating that the BP neural network classification and recognition was superior to the traditional maximum likelihood method.

77.1 Introduction

Recorded abundant surface spectral information by means of the electromagnetic waves characteristics is reflected or emitted by the ground objects. Remote sensing images provide the appropriate data for quick access to surface features relating to the space scale in human production and living. The remote sensing image represents the differences of the ground objects through pixel high or low value

S. Cao (✉) • G. Cao • K. Chen • C. Wu • T. Zhang • J. Yuan
College of Life and Geography Science, Qinghai Normal University, Xining 810008, China
e-mail: 243263340@qq.com

W.E. Wong and T. Zhu (eds.), *Computer Engineering and Networking*, Lecture Notes in Electrical Engineering 277, DOI 10.1007/978-3-319-01766-2_77,

difference (reflecting spectral information of the ground objects) and spatial variation (reflecting the spatial information of the ground objects). Those are the physical basis to distinguish different images of the ground objects.

The recognition of the remote sensing images is the process of analyzing the spectral and spatial information of various ground objects in remote sensing image, selecting the features, and dividing each pixel in the image into different categories by means of the computer. Then to achieve the classification and recognition of the remote sensing images, we obtain the corresponding information of remote sensing image with the actual ground objects in order [1]. However, due to the complex of spectral information about the characteristics within the internal ground objects target, the effect of traditional data processing methods (such as parallel pipeline algorithm and the maximum likelihood method) is no longer significant [2]. With the diversification of network types, improvement of network structure, as well as perfect network learning algorithm, a parameterized artificial neural networks (ANNs) model family is gradually applied to remote sensing image recognition [3–5]. Moreover, the BP neural network and its variable forms are the most widely applied and successful [6, 7]. This article explored the effectiveness and advantages of the BP neural network classification method through the classification and identification of the land use of a certain local area of Hainan Prefecture in Qing province using the TM images as data, through comparison of the results classified and recognized by the traditional classification method and the BP neural network one.

77.2 Studying Methods

77.2.1 Introduction of the BP Neural Network

The BP (Back Propagation) network was proposed by Rumelhart and McCelland in 1986. It is currently one of the most widely used neural network models. To conduct the error back propagation algorithm as its learning rules that processes the feedforward networks of supervised learning, it requires a considerable number of known samples to learning and training, to find out and remember the relationship between the input sample models and classification categories. It is usually need to be set that the features conditions of the ground objects will be classified, and it acts as a BP network input mode, then, we give the desired output mode, which plays the role in prediction type. Its structure includes input layer, hidden layer, and output one. The neurons in the same layer cannot interconnect, but the neurons in the different layers need full interconnection. The weights of the neural network are connected through a number of neurons (computing elements) by means of the feedforward or feedback. Those neurons lied in the hidden layer, and the input and output layers are connected by them [8].

77.2.2 The BP Neural Network Algorithm

The BP neural network algorithm is described as follows:

1. Initialized the network parameters: learning parameters, such as setting up a network initial weight matrix and learning factor
2. Provided training mode: not training the network until it meets the learning requirements
3. Forward propagation process: for the given training mode input, calculating output mode of the network and comparing with the expected pattern. If there was an error, step (4) would be executed; otherwise, and the program would be returned the step (2)
4. Backward propagation process: calculating the error in the same layer unit and correcting weights and thresholds; back to step (2)

77.2.3 Selecting the Number of Hidden Layer Neurons

When the network mapping is achieved using neural network, the number of hidden layer neurons directly affects the ability of learning and induction of the neural network. The smaller the number of hidden layer neurons, the shorter is each network's learning time. However, because of insufficient learning, the network could not remember all the learning content. When the number of hidden layer neurons is larger, the learning ability strengthened, but along with the longer times learning and the larger storage capacity of the network, resulting in the decline of inductive capacity of the network for the unknown input [9].

77.3 The Implementation of the Remote Sensing Image Classification and Recognition

The BP neural network is used to recognize the feature category of remote sensing image. The main process is divided into sample training and pattern recognition.

77.3.1 Training Sample

Its purpose is to find out certain error conditions of weight matrices [10]. The essence is the input layer of neural network structure. Input items are values of the pixel (R_i, G_i, B_i) that are known classification results. That is usually achieved through selection and drawing the training area of typical ground objects in the remote sensing image. Expected output is the classification-recognition results in correspondence to each pixel, namely, the information categories of the ground objects in the study area.

77.3.2 Pattern Recognition

After the sample training is finished, weight matrix has met the requirements. This moment, put any pixel on the image (R, G, B) value as input, the output vector can be obtained by calculation. Vector component corresponds to probability value of classification types that the pixel appoints every ahead of time. The type corresponding to the biggest probability value is the type belonging to the pixel.

77.4 Application Examples

77.4.1 The Experimental Data and Its Pretreatment

Experimental data chose the TM remote sensing images at the Hainan Prefecture in the southeastern region of Qinghai province in July 2009 (Fig. 77.1). The ground objects intercepted were complex and varied terrain category. The study area where can more clearly reflect the land cover situation is selected. The TM images were selected six bands of TM1, TM2, TM3, TM4, TM5, and TM7, with the spatial resolution of all 30 m. The picture size is 157,994 pixels, and picture contrast is good, without a shadow of a cloud. Data preprocessing mainly included the atmospheric correction, geometric correction, and image enhancement. The atmospheric correction chose the histogram adjustment method; geometric correction used RTK measurement data as control points and used quadratic polynomial to do coordinate transformation and gray resample based on the nearest neighbor method and the precision control of the coordinate transformation within one pixel. Through the result of comparing the land use status, high-resolution remote sensing image, and the visual interpretation, this region can be divided into six types: the cultivated land, grassland, sandy land, bare land, forest land, and water body. To improve the spectral contrast of the land types in the remote sensing image, image enhancement was handled using a fast Fourier transform (FFT) and inverse FFT for weakening the noise in the image (Fig. 77.1).

77.4.2 Sample Selection

On remote sensing image, we set the neurons layer numbers of 6 input layer, 1 hidden layer, and 6 output layer using the means of human–computer interaction

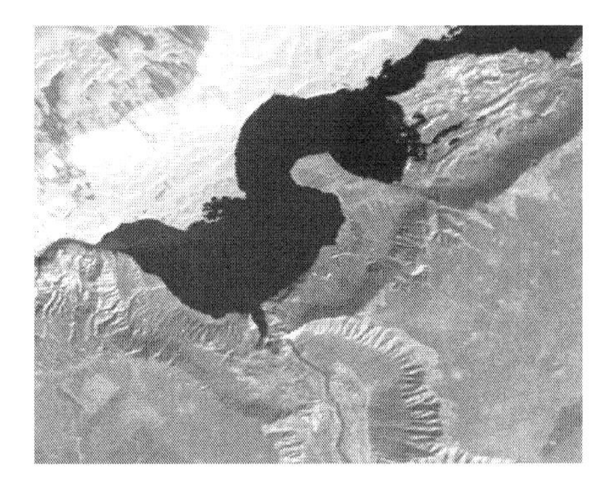

Fig. 77.1 The TM remote sensing image of experimental zone (5/4/3)

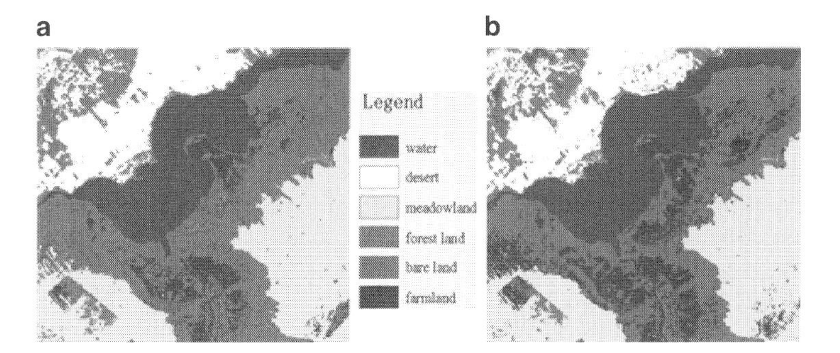

Fig. 77.2 Classification results of different identification methods. (**a**) BP neural network classification. (**b**) Maximum likelihood classification

and the BP neural network algorithm. We selected multiple typical areas having the clear and determined types as the training samples (district), which each kind of sample cannot cross the boundaries of the different categories. According to the characteristics of the land types, eventually 152 cultivated lands, 744 grasslands, 589 sand lands, 600 bare lands, 130 woodlands and 1,052 water-body were selected as training samples. Because the sample quality had greater impact on the recognition accuracy, there required that all land types were homogeneous. Therefore, we need check the sample histogram. If the sample distribution closes to a normal one, it indicates that the sample quality is better. Otherwise, it needs to choose training sample.

The hidden layer number was determined by the constructor method, which validated the model prediction error one by one from the minimum until it reached the maximum value. At last, the hidden layer with the least model error was selected and set the number of 1. Set the training error $E < 0.002$ in the classification. The result of the classification is shown in Fig. 77.2.

Table 77.1 The precision comparing table of different classification methods

	Maximum likelihood method		BP neural network algorithm	
Class	Prod. acc.	Overall accuracy (%)	Prod. acc.	Overall accuracy (%)
Water	80.31	79.08	93.56	81.49
Sandy land	92.67		93.10	
Grassland	92.65		90.10	
Forest land	93.86		94.47	
Bare land	89.24		89.68	
Cultivated land	30.90		40.86	

77.5 Results and Analysis

The time of maximum likelihood classification is 3 s and the time of BP neural network classification is 56 s, which indicated that the time of BP neural network classification is obviously longer. Applying land use map at present as reference diagram, we evaluated the precision of maximum likelihood method and BP neural network classification through overall classification accuracy and mapping accuracy. The overall classification accuracy is equal to the value of correct classified pixel dividing the total pixel number. The mapping accuracy is equal to the value of pixel number of correct classification of each category dividing the number of the corresponding pixel on reference diagram.

Results showed that the classification precision of cultivated land was very low for both methods. The reason is that the spectrum difference between the new cultivated land and the bare land in this area is nonsignificant which leads to big error in classification. The reason for larger error is that the maximum likelihood method regarded the wetter bare land near the lakeshore as water body incorrectly. Except for the grassland, the classification effect of the BP neural network was superior to the maximum likelihood method based on the features of spectral information (Table 77.1). The overall classification accuracy by the BP neural network was higher than the one by the maximum likelihood method.

77.6 Conclusion

As a heuristic algorithm, the BP neural network used to improve the ability of nonlinear approximation. Under normal circumstances, the neural network speed is lower than the traditional remote sensing image classification method. Using neural network algorithm for remote sensing image classification and recognition, it could eliminate the ambiguity and uncertainty caused by the traditional image classification and improve the accuracy of classification. The handle of sample quality before the BP algorithm is very important. Gathering the sample, each ground object must be uniform in distribution in the remote sensing image. There needs to be a gathering of 5–10 training area with at least greater than 100 pixels, and the sample histogram must be close to a normal distribution. Only in this situation, the image classification and identification become more reliable and accurate.

Acknowledgements This paper is cosponsored by National Social Science Foundation (10CJY015), Western Light Project of Chinese Academy of Sciences, Chunhui plan of the Ministry of Education (Z2012092), and Innovation Program of Qinghai Normal University (1294). The authors thank them and the anonymous reviewers.

References

1. Yan, Y., Dong, X. L., & Li, Y. (2011). The Comparative study of remote sensing image supervised classification methods based on ENVI. *Beijing Surveying and Mapping, 3*, 14–16 (In Chinese).
2. Li, L. W., Ma, J. W., Ou, Y. B., & Wen, Q. (2008). High spatial resolution remote sensing image segmentation based on temporal independent PCNN. *Journal of Remote Sensing, 12*(1), 65–69 (In Chinese).
3. Simpson, J. J., & Mcintir, J. T. (2001). A recurrent neural network classifier for improved retrievals of areal extent of snow cover. *IEEE Transactions on Geoscience and Remote Sensing, 39*(10), 2135–2147.
4. Lin, J., Bao, G. S., Jing, R. Z., & Huang, J. X. (2002). A study of FasART neuro-fuzzy networks for supervised classification of remotely sensed images. *Journal of Image and Graphics, 7*(12), 1264–1268 (In Chinese).
5. Han, M., Cheng, L., & Tang, X. L. (2005). Application study of Fuzzy ARTMAP neural network in classification of land cover. *Journal of Image and Graphics, 10*(4), 416–419 (In Chinese).
6. Zhou, T. G., & Su, Y. C. (2004). Artificial neural networks-based study of vegetation's classification for aerial remote sensing image. *Journal of Southwest China Normal University (Natural Science), 29*(6), 1037–1040 (In Chinese).
7. Yang, F. L., Liu, J. G., Zhao, J. H., & Du, Z. X. (2006). Seabed texture classification using BP neural network based on GA. *Science of Surveying and Mapping, 31*(2), 111–115 (In Chinese).
8. Xiu, L. N., & Liu, X. G. (2003). Current status and future direction of the study on artificial neural network classification processing in remote sensing. *Remote Sensing Technology and Application, 18*(5), 339–345 (In Chinese).
9. Lei, J. F., & Sun, J. Y. (2008). Research on image recognition based on artificial neural network. *Modern Electronic Technique, 31*(8), 127–130. 134 (In Chinese).
10. Chen, Y. M. (2002). Application of neural network to classification of remote sensing image. *Journal of Geomatics, 27*(3), 6–8 (In Chinese).

Chapter 78
Detection of Image Forgery Based on Improved PCA-SIFT

Kunlun Li, Hexin Li, Bo Yang, Qi Meng, and Shangzong Luo

Abstract In view of the problem existing in abusive using of image copy-move forgeries, this paper proposes an image forensics algorithm for detecting copy-move forgery based on improved PCA-SIFT. The present method works first by extracting features of an image and then reducing its dimensionality, and the method uses k-nearest neighbor to operate forgery detection. Owing to the similarity between pasted region and copied region, the descriptors are then matched between each other to seek for any possible forgery in images. Extensive experimental results are presented to confirm that the algorithm is able to precisely individuate the tampered image and quantify its robustness and sensitivity to image post-processing and offer a considerable improvement in time efficiency.

78.1 Introduction

With the availability and sophistication of digital imaging technology and image processing tools, it is possible to increasingly easily manipulate the image information. The authenticity and integrity of digital images are experiencing the serious threat and challenge. Therefore, image forensics is one of the key issues to be solved in maintaining information security.

Image forgery forensics techniques are generally divided into two categories: active methods and passive methods [1]. Active methods work when we have some prior information or its source about the image such as digital watermarking or signature. Passive (Blind) methods only make use of characteristics of images to identify image forgery which become the hot topics in the field of image forensics.

K. Li (✉) • H. Li • Q. Meng • S. Luo
College of Electronic and Information Engineering, Hebei University, Baoding 071002, China
e-mail: likunlun@hbu.edu.cn; xinxinlanyi@163.com

B. Yang
College of Mechanical Engineering, Yanshan University, Qinhuangdao 066004, China

W.E. Wong and T. Zhu (eds.), *Computer Engineering and Networking*, Lecture Notes in Electrical Engineering 277, DOI 10.1007/978-3-319-01766-2_78,

Copy-move forgery is one specific type of image forgery where part of an image has been copied and pasted within the same image in an attempt to add or disguise feature in the scene [2]. It is very likely for the duplicated part to be subjected to a variety of post-processing operation. The detecting region duplication principle can be based on the similar characteristic presence of duplicated regions in the image.

There have been disappeared different techniques for detecting copy-move forgery. J. Fridrich [3] proposed an effective approach based on lexicographic sorting of quantized DCT coefficients of image blocks, which has the milestone significance, but the computational complexity is higher. B. Medan et al. [4] yield a fuzzy method to detect tampering based on blur moment invariants, which can resist fuzzy, noise, etc., but fail in the scale and rotated post-processing operation. M.A. Sekeh et al. [5] applied a coarse-to-fine approach to propose an efficient image duplicated region detection model using sequential block clustering, which improves time complexity. H.L. Huang et al. [6] detected image copy-move forgery using SIFT algorithm which suffers from the drawback that the search method is computationally complex, it not only reduces the image matching and retrieval speed but also affect the real-time performance.

In view of the problem existing in detection algorithms of the copy-move image forgeries, we proposed a novel image forensics algorithm for detecting copy-move forgery based on new PCA-SIFT, k-nearest neighbor to operate forgery detection.

The rest of the paper is organized as follows. In Sect. 78.2, we give a description of SIFT and PCA-SIFT algorithms and discuss the proposed method in detail. In Sect. 78.3, we apply our scheme to image forgery detection. Some experimental results and discussions are debated in Sect. 78.4, and conclusions are drawn in Sect. 78.5.

78.2 The Related Work

78.2.1 Review of the SIFT Algorithm

Scale Invariant Feature Transform (SIFT) algorithm [7] is essential to problems including image matching, object recognition. SIFT extracts distinctive features from images which are invariant to image scale and rotation and are robust to distortion, noise, illumination, and viewpoint. The major stages of computation are following.

Scale-space extrema detection: The first step searches over all scales and image locations. Given an input image $I(x,y)$, establish scale space $L(x,y,\sigma)$ through Gaussian function $G(x,y,\sigma)$ and implement the Difference-of-Gaussian (DoG) scale space $D(x,y,\sigma)$ by using DoG function to identify potential interest points that are invariant to scale and orientation. Extrema are detected by comparing each sample point in DoG images to its 26 neighbors in 3×3 regions at the current and adjacent scales in attempt to obtain candidate keypoints.

$$L(x, y, \sigma) = G(x, y, \sigma) * I(x, y) \tag{78.1}$$

$$\begin{aligned} D(x, y, \sigma) &= [G(x, y, k\sigma) - G(x, y, \sigma)] * I(x, y) \\ &= L(x, y, k\sigma) - L(x, y, \sigma) \end{aligned} \tag{78.2}$$

Keypoint localization: The next step performs a detailed model fit to determine location and scale. Candidate keypoints are localized to sub-pixel accuracy, reject keypoints with low contrast and eliminate edge responses to have stable ones.

Orientation assignment: The third step assigns the dominant orientations for each keypoint based on local image gradient directions to achieve invariance to image rotation. The keypoint orientation is calculated from an orientation histogram of local gradients from the closet smoothed image $L(x,y,\sigma)$.The gradient magnitude $m(x,y)$ accumulate to the histogram according to orientation $\theta(x,y)$. An orientation histogram with 36 bins covering the 360° range of orientations is formed from the gradient orientations in the neighborhood of the keypoint. Peaks in the orientation histogram correspond to dominant directions of local gradients.

$$m(x, y) = \sqrt{(L(x+1, y) - L(x-1, y))^2 + (L(x, y+1) - L(x, y-1))^2} \tag{78.3}$$

$$\theta(x, y) = \tan^{-1}\left[\frac{L(x, y+1) - L(x, y-1)}{L(x+1, y) - L(x-1, y)}\right] \tag{78.4}$$

Keypoint descriptor: The final step builds a local image descriptor for each keypoint to guarantee descriptors being distinctive and robust to other variations, such as illuminations and viewpoint. Rotate the coordinate relative to the keypoint orientation to ensure rotation invariance, sample the magnitudes and orientations of the image gradient in a region around the corresponding keypoint, and accumulate orientation histograms with eight bins summarizing the contents over 4×4 subpatches as the local statistics of gradient orientations to capture SIFT descriptor with 128 elements.

78.2.2 Brief of the PCA-SIFT

The algorithm for local descriptors (termed PCA-SIFT) [8] examines the local image descriptor used by SIFT. It can be summarized in the following steps: it consists of the highly restricted set of patches that passed through the first three stages of SIFT discussed above. So it accepts the same input as the standard SIFT: the sub-pixel location, scale, and dominant orientations of the keypoint; it explores alternatives to its local descriptor representation. Like SIFT, the descriptors encode the salient aspects of the image gradient in the feature point's neighborhood; however, instead of using SIFT's smoothed weighted histograms, the method extract the 41×41 patch centered at the keypoint, and rotated to its dominant orientation, concatenate their horizontal and vertical gradient maps, then the descriptor has $2 \times 39 \times 39 = 3042$ elements, and PCA was applied to the normalized gradient patch to 20 elements. Details are given in the paper.

78.2.3 The Proposed Method

A novel method is presented to extract more challenging features. The method extract keypoints of the image based on SIFT to inherit stability of the SIFT, while refusing the traditional PCA-SIFT alternative to describing keypoints, it improves on SIFT by using PCA thoughts in a novel way in which different from traditional PCA-SIFT.

Keypoint detection: Given an image, we extract the keypoints $X = \{x_1, x_2, \cdots, x_n\}$ through three steps of SIFT algorithm. The first step identifies image scale-space extrema which robust to scale and rotation, preliminarily determine the position and the scale of the keypoints. The second step determines the position of the keypoints accurately and removes the low contrast keypoints and the unstable edge response to enhance the matching stability. The third step assigns orientations for each keypoint based on its local image patch. The assigned orientation(s), scale, and location to construct a canonical view for the keypoint that is invariant to similarity transforms.

Keypoint representation: Keypoint description using SIFT is created by sampling the magnitudes and orientations of the image gradient in the patch around the keypoint, and building smoothed orientation histograms to capture the important aspects of the patch. Then image descriptors $D = \{d_1, d_2, \cdots d_n\}$ each with 128 elements are obtained.

The choices behind the SIFT descriptor is complicated, the descriptor matching will have heavy computation based on large eigenspace. Moreover, some characteristic of the eigenspace have very low contribution rate, they are susceptible to external interference, such as noise, blur. Principal Component Analysis (PCA) is used to reduce the eigenspace, simplify the dataset, and retain the identity-related variation while discarding the distortions induced by other effects. For well solve the problems above, it is not unreasonable to believe that the descriptor can be reasonably modeled by applying PCA to accurately represent them with a compact feature representation.

At first calculate the standardized data of input data $d_i = d_i - \frac{1}{n}\sum_{i=1}^{n} d_i$. Then the calculation of the covariance matrix should be done $C = \frac{1}{n}\sum_{i=1}^{n} d_i d_i^T$, and the eigenvalue and eigenvectors of the covariance matrix are extracted. Finally, sort the eigenvalues and select the first r eigenvalues as the main components, the corresponding eigenvectors to makeup matrix U. The transformation of input data that maps the data matrix D in a new matrix Y as $Y = U^T D$. Then the improved PCA-SIFT descriptors with r elements will be done.

78.3 Application to Image Forgery Detection

The proposed method is introduced into image forgery detection. Apply the simple and effective k-nearest neighbor using Euclidean distance as the similarity measure criteria to run image detection. Perform global traversal search to determine the matching decision by taking the ratio of the distance of the closest neighbor to that of the second closest one with respect to a global threshold. Given a keypoint $x \in X$ with a similarity vector $S = \{s_1, s_2, \cdots, s_{n-1}\}$ that represents the distance with respect to the other descriptors. When $s_1/s_2 < T$, the keypoint is matched. Iterating on each keypoint in X, we can obtain the set of matched points which reveal the image forgery.

78.4 Experimental Results

78.4.1 Evaluation Metrics

Precision and accuracy are commonly used in evaluation standards, they are popular in evaluating various classification and detection methods, thereby we use them to evaluate detection performance of the new method, which are defined as follows:

$$Precision = \frac{\#number\ of\ correct\ matches}{\#number\ of\ total\ matches}$$

$$Accuracy = \frac{\#images\ detected\ as\ forged\ being\ forged}{\#forged\ images}$$

78.4.2 Test on a Large Dataset

In order to verify the effectiveness and stability of the proposed algorithm, the method has been tested on various images. Experiments are performed as follows.

Parameter determination: Figure 78.1 demonstrates that the performance of the improved PCA-SIFT as PCA dimension and threshold are various. Increasing the dimensionality of the space has a result in better accuracy, since the representation is able to capture the structure of the patch. Adding the threshold achieves better results too, but once it exceeds a certain size, the precision begins to decline. Note that the requirement for image description improves on image forgery detection. Using fewer components require less storage and faster matching. Thereby weighing the precision and detection time, this article chooses the threshold value T = 0.5, PCA space r = 64.

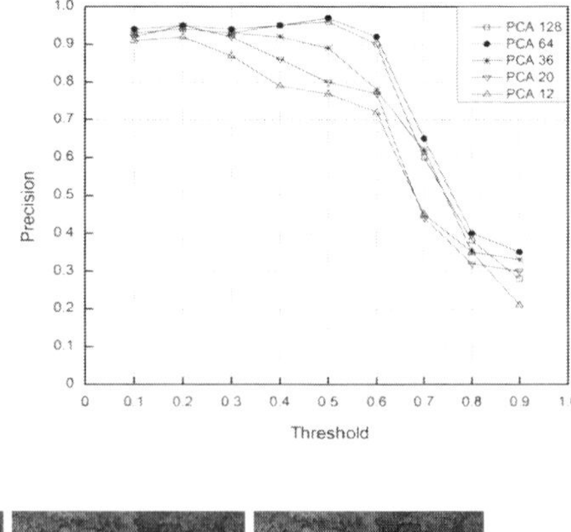

Fig. 78.1 The performance of the improved PCA-SIFT as PCA dimension and threshold changes

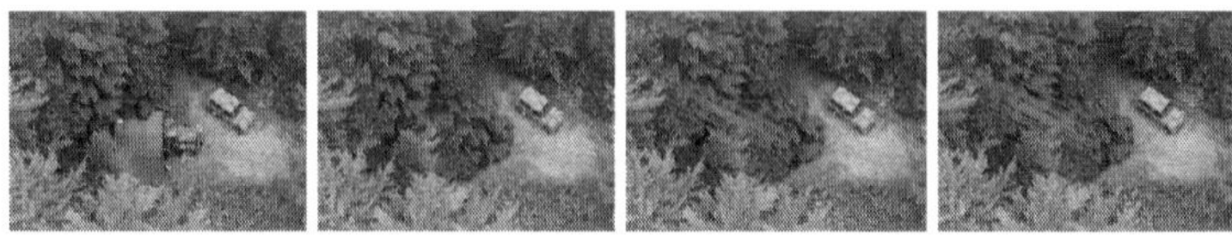

Fig. 78.2 The original image is shown in the fist column; the tampered image is pictured in the second column; the detection results using SIFT and new algorithm are reported in the third and fourth column, respectively

Detection of image forgery with no post-processing: Comparison on the performance of SIFT [7] and new method in copy-move image detection are shown in Fig. 78.2. Threshold is set to 0.5. The new method shows the increased detection accuracy.

Detection of image forgery with post-processing: The dataset is composed by various post-processing in the copy-move forgery. Figure 78.3 shows that the new method can accurately detect and locate forgery region and has resistance ability for post-processing. An interesting phenomenon was found after repeated experiments, when the image is more distinct and the texture is stronger, it may bring about more keypoints and feature matching, the detection results will be better.

Detection accuracy: For statistical detection accuracy of SIFT and new method, each post-processing operations selected 200 tampering pictures including figures, landscapes, buildings, etc., and covering a variety of forgeries such as scale, rotation, and JPEG compression. Figure 78.4 presents that the detection accuracy of the new algorithm is superior to the SIFT algorithm. Furthermore, it greatly reduces the time complexity and improves the detection efficiency.

78.5 Conclusion

This paper proposed a novel algorithm for detecting copy-move forgery using the idea of improved PCA-SIFT. The presented method showed effectiveness with respect to detecting and locating forged images even if the images have undergone

Fig. 78.3 Some examples of original images are shown in the first column; the tampered images (*rotation, sale, blur*) are pictured in the second column; the detection results are reported in the third column

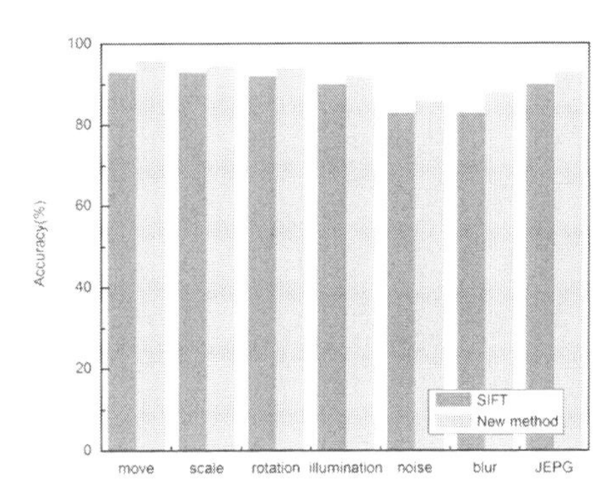

Fig. 78.4 Detection accuracy of SIFT and new method for each different attack (in percentage)

any post-processing. Furthermore, the algorithm also has lower computational complexity and it works well on all the basic image formats. It performed better than some of the previous methods. However, the algorithm still deserves further exploration to more accurately define the tampered region and enhance the robustness of the detection.

Acknowledgements Project supported by the National Science and Technology Support Plan Project (No. 2013BAK07B04) and Natural Science Foundation of Hebei Province. China (No. F2013201170).

References

1. Gavin, L., Shih, F. Y., & Liao, H.-Y. M. (2013). An efficient expanding block algorithm for image copy-move forgery detection. *Information Sciences, 239*, 253–265.
2. Christlein, V., Riess, C., Jordan, J., Riess, C., & Angelopoulou, E. (2012). An evaluation of popular copy-move forgery detection approaches. *Information Forensics and Security, 7*(6), 1841–1854.
3. Jessica, Fridrich., David, Soukal., Jan, Lukas. (2003). Detection of copy–move forgery in digital images. *Proceedings of Digital Forensic Research Workshop (DFRWS'03)* (pp. 55–61). Cleveland, OH: IEEE Computer Society.
4. Babak, M., & Stanislav, S. (2007). Detection of copy–move forgery using a method based on blur moment invariants. *Forensic Science International, 171*(2–3), 180–189.
5. Mohammad Akbarpour, S., Mohd. Aizaini, M., Mohd. Foad, R., & Babak, M. (2013). Efficient image duplicated region detection model using sequential block clustering. *Digital Investigation, 10*(1), 73–84.
6. Hailing, Huang., Weiqiang, Guo., Yu, Zhang. (2008). Detection of copy-move forgery in digital images using SIFT algorithm. *Proceedings of IEEE Pacific-Asia Workshop on Computational Intelligence and Industrial Application (PACIIA'08)* (vol. 2, pp. 272–276). Wuhan: IEEE Computer Society.
7. Lowe, D. G. (2004). Distinctive image features from scale-invariant keypoints. *International Journal of Computer Vision, 60*(2), 91–110.
8. Yan, Ke., Rahul, Sukthankar. (2004). PCA-SIFT: A more distinctive representation for local image descriptors. *Proceedings of 2004 I.E. Computer Society Conference on Computer Vision and Pattern Recognition (CVPR'04)* (vol. 2, pp. 506–513). Washington, DC: IEEE Computer Society.

Chapter 79
A Thinning Model for Handwriting-Like Image Skeleton

Shijiao Zhu, Jun Yang, and Xue-fang Zhu

Abstract In order to solve the letter skeleton problem with handwriting-like attributes, a thinning model is used in this paper. By introducing improved reservation and eliminating produces, additional pixels are constrained by thinning nearest pixels. In the experiment, the proposed method is compared with others in the literature English letters by using the one pass thinning algorithm (OPTA) and Hilditch methods. Empirical results show that the proposed model can thin handwriting-like skeleton in terms of reserving topology and eliminating extra pixels.

79.1 Introduction

Handwriting provides useful personal information that may be used to describe the property of an image [1]. While it is relative easy to describe print character, but difficult for personal handwriting character especially for widely use of handset devices. The first thing of reorganization is to preprocess images with thinning handwriting outlines.

Research in thinning methods has been interest in recent years due to the first step of retrieval features [2, 3]. Some algorithm has been used in some applications [4, 5]. An ideal thinning method has the properties of keeping connections, topology, and relative position. It is best to keep the thinning result outline in middle of handwriting skeleton. Some methods have been proposed, such as Hiditch [6], one pass thinning algorithm (OPTA) [7], and so on.

S. Zhu (✉) • J. Yang
School of Computer and Information Engineering, Shanghai University of Electric Power, Shanghai 200090, China
e-mail: zhusj707@hotmail.com

X. Zhu
Department of Information Management, Nanjing University, Nanjing 210093, China
e-mail: mediate@163.com

W.E. Wong and T. Zhu (eds.), *Computer Engineering and Networking*, Lecture Notes in Electrical Engineering 277, DOI 10.1007/978-3-319-01766-2_79,

But we all known that each method has its better performance based on its characteristic in special field. A widely used method is FAT (fast thinning algorithm) [8] which is based on eight nearest pixels. This method use local pixels properties. From the point compute iterate type, it is classified as serial algorithm and parallel algorithm. Serial algorithm is depended on preprocess order, while parallel algorithm is based on last produce. Therefore parallel algorithm is superior to serial one. For improved contour-based thinning method, it uses shape characteristics of text to get skeleton of nearly same as the true character shape [9]. Our proposed method is based on parallel algorithm.

In the paper that follows, we begin with the definition model in Sect. 79.2. Meanwhile, the produce and algorithm is presented. A framework of the thinning schema is discussed. In Sect. 79.3, we give experimental result and comparison of different algorithm. Section 79.4 briefly summarizes the paper.

79.2 An Improved Model

OPTA algorithm is a typical method for thinning binary image. It uses reservation and elimination templates to apply for an original image.

The whole produce banks of produce of thinning method are described in Fig. 79.1.

For original image, it is made by produce of binary, and then to removing superfluous pixels of skeleton by bank of keeping templates and removing templates.

The whole produce can be described as follow:

(1) For current pixels, find nearest pixels in removing templates, if true then next, otherwise go to step (3);
(2) Make judgment of keeping template, if pattern in keeping method then keep current pixel, otherwise make tag of removing and go to step (1);
(3) No action for current pixel, go to step (1);
(4) Process for next pixel;
(5) Make judgment of the last time parallel produce, if has pixel removed, then continue next loop. Otherwise stop.

For handwriting skeleton, it has some of its own properties. Firstly, it has some subblocks of contiguous regions having similar characteristic such as color, intensity, or edges. Secondly, they are often vertical or horizontal alignment. Thirdly,

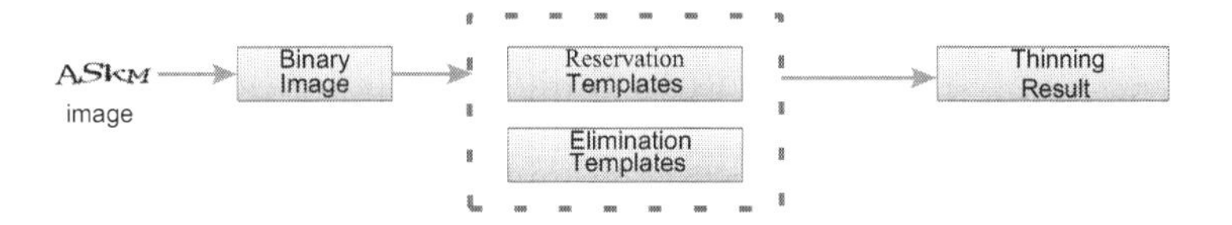

Fig. 79.1 Thinning produce

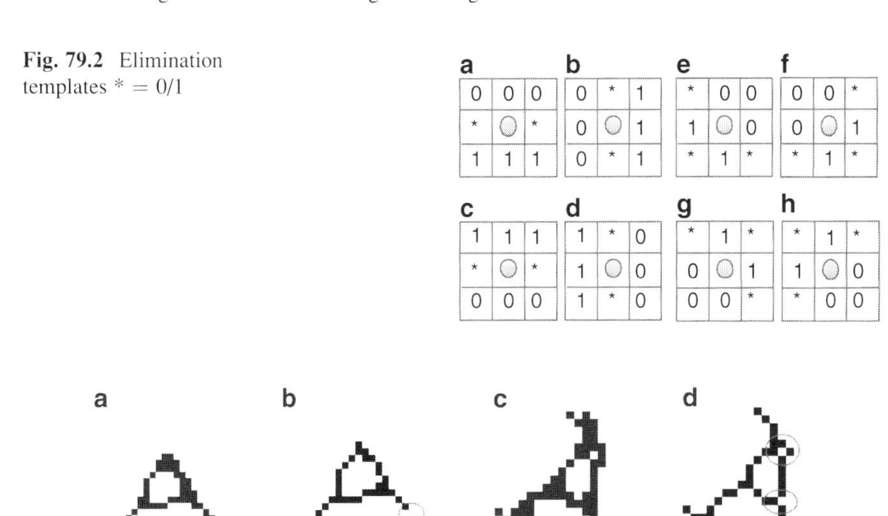

Fig. 79.2 Elimination templates * = 0/1

Fig. 79.3 Two result of OPTA marked by circle. (**a**) Original image. (**b**) Broken line. (**c**) Another original image. (**d**) Close line

Fig. 79.4 Pixel and its near pixels

p1	p2	p3	p4
p5	○	p6	p7
p8	p9	p10	p11
p12	p13	p14	p15

they are often been placed in center of an image. For above-stated reasons, we define some templates as follow (Fig. 79.2).

In paper of the thinning algorithms [2], it gives reservation templates. When pixel and its near pixels compared to the templates, if it meets condition, pixel is kept otherwise it is removed. From experiments, we find it has three negative sides. One is destroy topology structure for some outlines because it uses strict templates for constraint. Next one is invalid for closure structure. The last one is time consume for big size of reservation templates (Fig. 79.3).

In proposed method, we define keeping produce instead of keeping template. Firstly we define its position as symbolized letters as below:

The circle point in Fig. 79.4 is used to represent the location of the current point.

According to elimination templates, model gives candidate pixels with parallel computation. In the next step, reservation templates banks and with option of elimination templates bank was presented (Fig. 79.5).

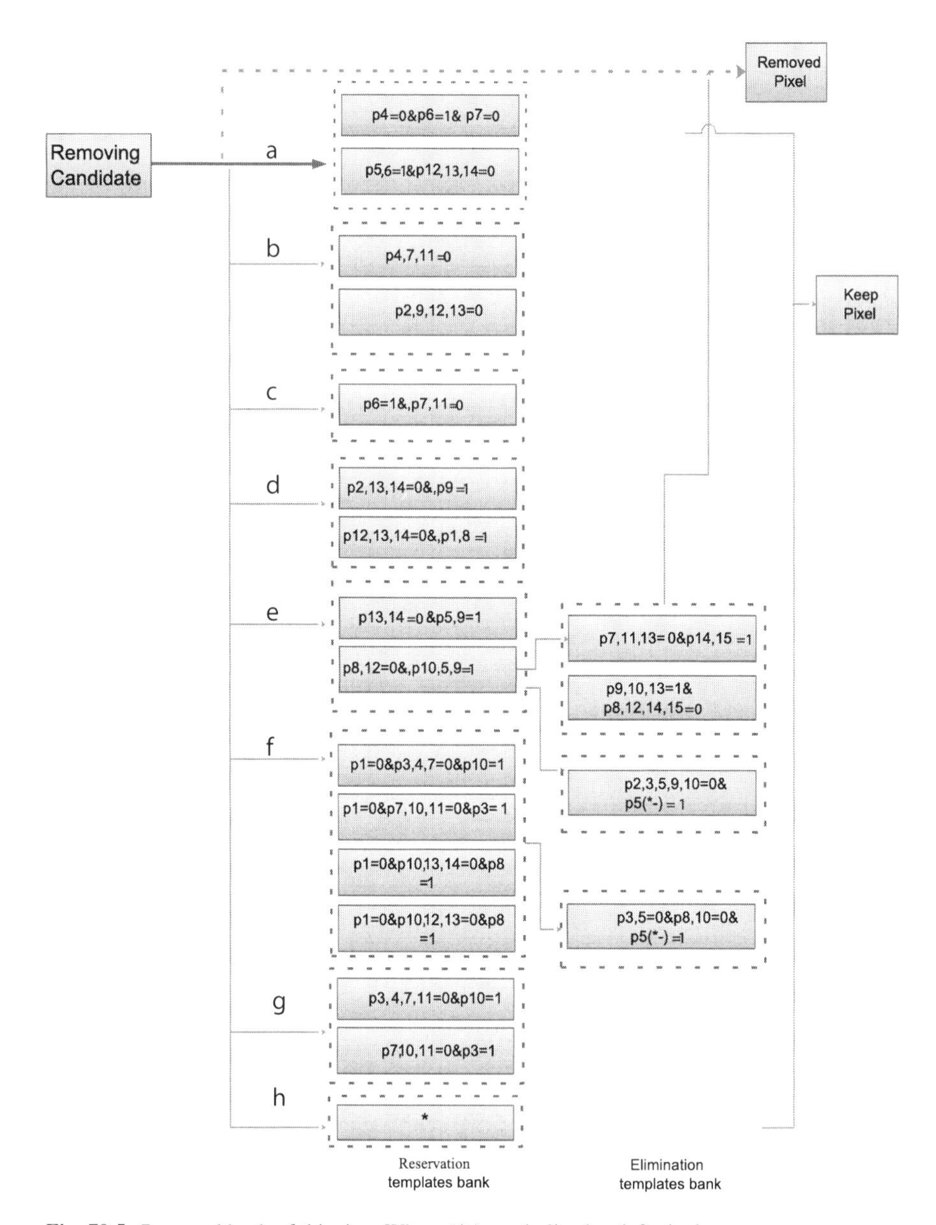

Fig. 79.5 Proposed bank of thinning. Where (*-) symbolized as left pixel

The proposed method can be described as:

Binary image initial from image	
Repeat until done:	
1a.	parallel processing current pixel and its nearest around pixels.
1b.	if judgment can meet removing templates and not meet keeping templates, then it is marked as erased pixel.
2a.	If judgment is empty, exit.
2b.	Make a new binary image based on pre-image and its judgment.
2c.	Use new binary image for next loop.
3a.	Update binary image.

As mentioned, pixels of near candidate pixel are filtered by templates. In our implementation, these points are tagged by temporary array with the ability of parallel computation. The image pixels are, therefore, updated (pixel by pixel) at each iteration of the filtering algorithm.

79.3 Experimental Results

In order to verify and analysis experiment, here we apply our algorithm to a variety of handwriting-like images. In each gray images, their size are 160×40. The computer used in the experiment:

- CPU: Intel Core i5-2430M 2.40 GHz.
- Memory: DDR3 1066 MHz 2 GB.
- Operating System: Windows 7.
- Compiler: Visual C++ 2008.

Where possible, we make comparisons to previously proposed methods. We perform our experiment on the well-known OPTA and Hilditch methods. In experience, we use two category images: one is image with single letter and another one is multi-letters. A single character set is primarily to the validity of algorithm corresponding to the character topology. Multi-letters set is to verify validity of Connectivity between the characters. Figures 79.6 and 79.7 give the result.

From Fig. 79.6, we can notice that the topology of letter can be reserved while comparison of little disconnection by improved OPTA method and many disconnect region by Hilditch method. For multi-letters thinning in Fig. 79.7, Line with more than one pixel was generated in (b) by method of improved OPTA method and disconnection in sub-image with letter M. In (c), connection lines were broken into some regions. In (d), we can notice topology and connection of letters were reserved.

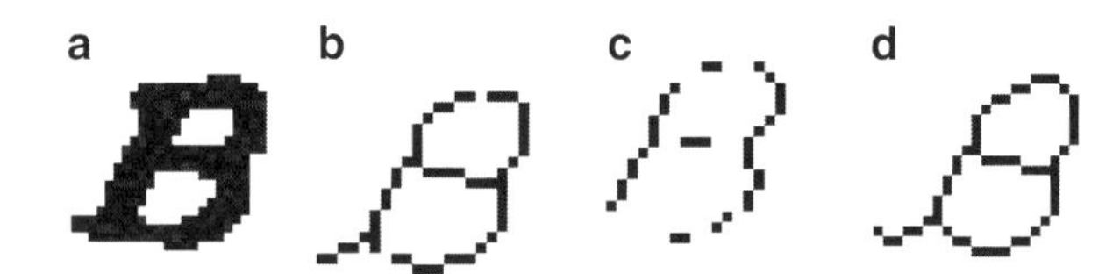

Fig. 79.6 Single Letter result. (**a**) Original single image. (**b**) Improved OPTA method. (**c**) Hilditch method. (**d**) Proposed method

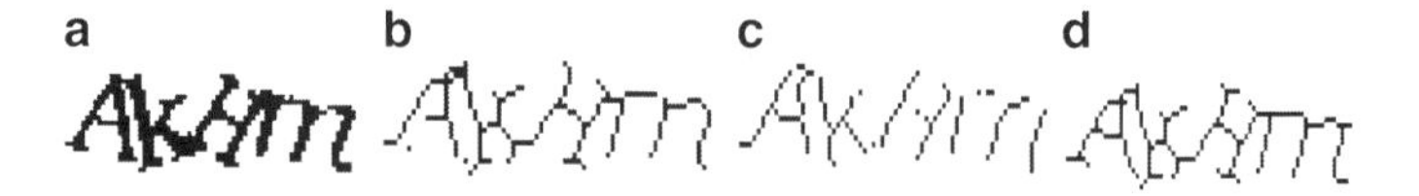

Fig. 79.7 Multi-letters result. (**a**) Original single image. (**b**) Improved OPTA method. (**c**) Hilditch method. (**d**) Proposed method

79.4 Conclusion

In this paper, a model for thinning handwriting-like image was proposed. Based on the proposed model, thinning binary image was generated with better topology structure and connection. It is superior to the OPTA and Hilditch proofed by experiments. What's more, the method is based on produce instead of only templates. It can accelerate the thinning speed.

However, this is not necessarily the best algorithm. Thinning binary image and reserving its topology may be able to further improve the post-process correctness based on keeping handwriting skeleton accuracy. It remains to be confirmed by further studies and experiments.

References

1. Jawahar, C. V., Balasubramanian, A., Million, M., & Namboodiri, A. M. (2009). Retrieval of online handwriting by synthesis and matching. *Pattern Recognition, 42*(7), 1445–1457.
2. Cheng, J., Wang, J., Jiang, S., Zhou, Z.-H., & Hancock, E. (2011). Special edition on semi-supervised learning for visual content analysis and understanding". *Pattern Recognition, 44*(10–11): 2242–2243.
3. Shang, L., Yi, Z., & Ji, L. (2007). Binary image thinning using autowaves generated by PCNN. *Neural Processing Letters, 25*(1), 49–62.
4. Xing-kui, F., Lin-yan, L., & Zu-quan, Y. (1999). A new thinning algorithm for finger print image. *Journal of Image and Graphics, 4*(10), 835–838.
5. Peter I. Rockett, "An Improved Rotation-Invariant Thinning Algorithm," IEEE Transactions on Pattern Analysis and Machine Intelligence, vol. 27, no. 10, pp. 1671–1674, Oct. 2005.
6. Jia, Yu., & Yaqin, Li. (2009). Improving Hilditch thinning algorithms for text image". 2009 International Conference on E-Learning, E-Business. Enterprise information Systems and E-Government. pp. 76–79.

7. Imiya, A., & Saito, M. (2006). Thinning by curvature flow. *Journal of Visual Communication and Image Representation, 17*(1), 27–41.
8. Ravi, J., Raja, K. B., & Venugopal, K. R. (2009). Fingerprint recognition using minutiae score matching. *International Journal of Engineering Science and Technology, 1*, 35–42.
9. Bag, S., & Harit, G. (2011). An improved contour-based thinning method for character images. *Pattern Recognition Letters, 32*(11), 1836–1842.
10. Jinhai, C. (2012). Robust filtering-based thinning algorithm for pattern recognition. *The Computer Journal, 55*(7), 887–896.

Chapter 80
Discrimination of the White Wine Based on Sparse Principal Component Analysis and Support Vector Machine

Rong Wang, Wu Zeng, and Jiao Ming

Abstract In allusion to the urgency of the white wine identification, and the key shortcoming of PCA (Principal Component Analysis), whose all loadings are non-zero, the sparse PCA (SPCA) is employed to enhance the explanatory and remove unnecessary variables. By using elastic net Zou Presented, SPCA can seek sparse factors and explain the maximum amount of variances in the data. For illustration, a comparison of strategy between PCA and SPCA, which combined with the specified categorizer—Support Vector Machine (SVM) and Infrared (IR) Spectroscopy, is utilized. The finally classified result of white wines based on SPCA is up to 93 %, but the PCA's 83 %, and directed categorizer's 80 %. Obviously, the SPCA can extract characteristics more effectively, which benefits the classification accuracy of SVM.

80.1 Introduction

Up to today, white wine, which blended delicately combined with storage perfectly in many years, has become a luxury slowly. The costliness price spilt over into confusion. In consequence, the fraud means emerge in endlessly. Thus, the quality of wine has attracted great importance. As a rapid and non-destructive methodology, NIR technique has been paid much attention recently. Studies have been reported on using NIRS technique to analyze quality of food oil, fruit, and honey [1–3]. IR spectroscopy is absorbed primarily by hydric groups, which is main composition of the wine. To qualitatively detect the category, IR can yet be regarded as a simple but effective method. However, IR acquisition area includes hundreds of wavelength points. The correlation among certain wavelength points surely increases the complexity of the problem, so we must take measures to select

R. Wang (✉) • W. Zeng • J. Ming
Department of Electrical & Electronic Engineering, Wuhan Polytechnic University, Wuhan 430000, China
e-mail: 340689241@qq.com

W.E. Wong and T. Zhu (eds.), *Computer Engineering and Networking*, Lecture Notes in Electrical Engineering 277, DOI 10.1007/978-3-319-01766-2_80,

variable. PCA chances to transform original variables into a few new variables, thus realizes reducing the dimensionality of spectral data in order to extract main information.

PCA seeks linear combinations of the data variables (called principal components PCs) by capturing a maximum amount of variance. However, all PC coefficients (called loadings) are non-zero. This is to say, although PCA concentrates the information in a few key PCs for the convenience of the model interpretation and visualization, the PCs are still constructed by all original variables. To enhance the explanatory and remove unnecessary variables, Zou [4] presented SPCA. In SPCA, we seek a trade-off between the two goals of expressive power (explaining most of the variance or information in the data) and interpretability (making sure that the factors involve only a few coordinate axes or variables) [5].

The outline of the paper is as follows. After brief background introduction, data and methods needed are described in Sects. 80.2 and 80.3, separately. Then in Sect. 80.4, we demonstrate this mixed method and discuss its potential through several experiments. At the end, we conclude and give a general overview to future's work.

80.2 Materials

A total of five bottles of wine samples were bought from a hypermarket in Wuhan, including "guan-gong-fang" wine (ggf), "fen-jiu" aged 10 and 5 years (f10 and f5), "er-guo-tou" aged 8 years, and another "er-guo-tou" (egt8 and egt65d). Just because those wine are famous in china, thus it is more likely to be cheated. Moreover, there are different flavor, alcohol, and year so that selected samples are more typical.

The IR spectra were collected in the reflectance mode using a NEXUS 670 FT-IR spectrophotometer (Nicollet, USA) with an InGaAs detector in the range from 4,000 to 650 cm^{-1}. The sample spectrum was acquired to subtract from a reference spectrum which scanned in an empty cell without sample to remove background noise. The spectrum of each sample was obtained by taking the average reading of 16 scans. In case of abnormal data, more than 50 samples were collected from each bottle. But 50 samples derived from each wine were selected as final experimental data. In multivariate analysis, the wine samples were divided into calibration set and validation set, which was used to train and evaluate the model separately. To testify the classified prediction, the concentrations of each category were arranged in randomly.

80.3 Methods

80.3.1 The Overall Idea of Algorithm

The paper adopts SVM based on the IR spectroscopy technique to realize perfect classification of white wine, whose overall idea is as follows:

1. To remove the abnormal samples;
2. To extract feature used PCA and SPCA;
3. To take the extracted PCs as the input of SVM and build a prediction model;
4. To predict the samples on the model.

80.3.2 PCA

PCA was used here mainly to extract information from the wine spectrum data, and meanwhile to reduce the dimensionality of the spectra. By calculating the eigenvectors of the covariance matrix of normalized calibration data, PCA transforms the original variables into new axes and calculates PCs as new variables to replace the original data [6].

In detail, the data X is an $n \times p$ matrix containing p features for each observation. Without loss of generality, X is normalized. Then the singular value decomposition (SVD) of X is to be:

$$X = UDV' \tag{80.1}$$

So $Z = UD$ is the PCs, and the columns of V are the corresponding loadings.

Notice that PCA has another function to realize clustering. Through observing the score plot, we can find the same kind of data will gather together while deviate far when they are different. Those dates deviated farther are thought to be abnormal. The paper is to use this method to eliminate abnormal samples.

80.3.3 SPCA

As an improve method, SPCA seeks sparse factors or linear combinations of the data variables to explain a maximum amount of variance in the data while having only a limited number of non-zero coefficients [5].

Elastic net, as a regression-type optimization problem, combines Ridge with Lasso, both of which are a penalization technique based on the l_n norm. Through the l_2 norm, Ridge regression contracts coefficients of regression to achieve the purpose of reducing error. But Lasso increases l_1 penalty function based on general linear least square method to make the sum of the absolute coefficient less than a constant.

Due to the constraint of natural property, some coefficients turn to zero. Elastic net integrates both such that removing the limitation of each other, and then simultaneously chooses variables and reduces the model prediction error. Meanwhile, it generally does not compress regression coefficients excessively.

To address these concerns, Elastic net regression attempts to mix the l_1 and l_2 penalties, as shown in (80.2).

$$\hat{\beta} = \underset{\beta}{\text{argmin}} \left(\|Z_i - X\beta\|^2 + \lambda\|\beta\|^2 + \lambda_1\|\beta\|_1 \right) \tag{80.2}$$

The goal in elastic net is to produce a sparse feature space with the l_1 penalty, but improve stability and retain correlated features with the l_2 penalty [7]. Like the lasso this is not a computationally simple problem, but efficient methods for solving it which have been developed. There is also an additional free parameter in α, which determines the relative strength of the penalties. Previous studies have shown this technique to be effective in classification of functional magnetic resonance imaging (fMRI) data [8].

80.3.4 SVM

SVM is a powerful classification technique first proposed by Vapnik, whose theory is based on Vapnik-Chervonenkis theory and structural risk minimization [9]. An optimum network structure and a better generalization performance could be achieved with SVM.

The main idea of SVM is to first map the data points into a high-dimensional feature space by using a kernel function, and then to construct an optimal separating hyperplane between the classes in that space. The primary advantage of SVM over the traditional learning algorithm is that the solution of SVM is always globally optimal and avoids local minima and over-fitting in the training process.

80.4 Results and Discussion

80.4.1 Spectra Investigation

During the process of collecting spectral data, there may be emerge abnormal values, which will affect the classification accuracy. In order to reduce the negative effect, we must take measures to remove these abnormal samples before feature extraction. According to PCA for all data, we find and remove three samples deviating farther, and then arrange new data. Figure 80.1 shows the mean spectral of each class for the original data mapping of five wines. As seen in Fig. 80.1, the

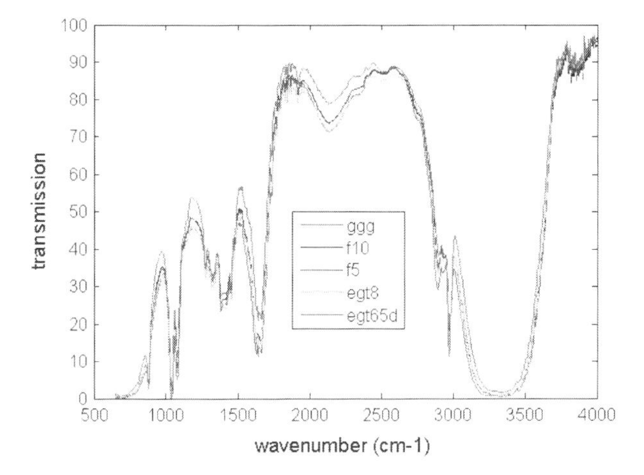

Fig. 80.1 The mean spectral of each class

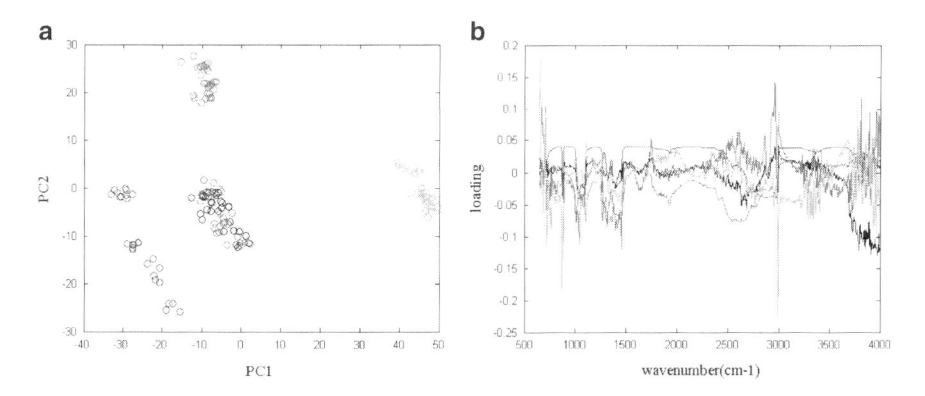

Fig. 80.2 The scores plot of PCA (*left*) and the loading of PCA (*right*)

features of wine spectral were similar generally because of the influence of water in the wine, which affects the result of classification seriously. So we must make measures to select features and improve the classified precision.

80.4.2 PCA and SPCA

Although both of them cannot be used as a classification tool, this behavior may indicate the data trend in visualizing dimension spaces. For visualizing the data trends and the discriminating efficiency, a scores plot using the first two PCs of training samples and the loading of top five PCs were obtained which are showed in Fig. 80.2. As it shows, the scores plot could reflect cluster relation at some degree. Then the sparse feature of SPCA was showed in Fig. 80.3.

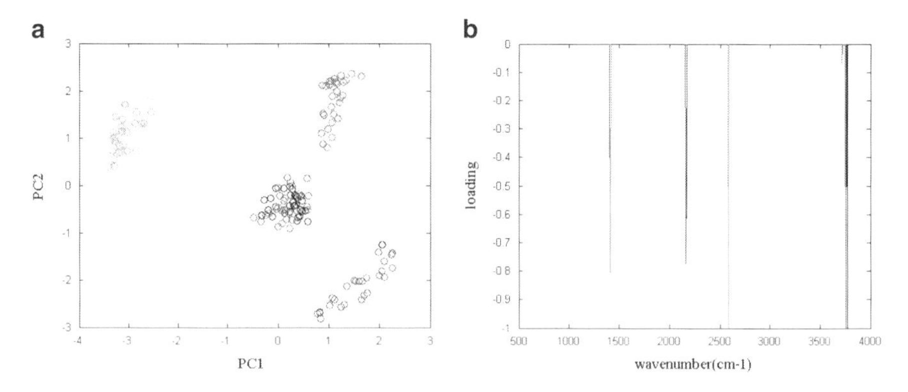

Fig. 80.3 The scores plot of SPCA (*left*) and the loading of SPCA (*right*)

Comparing Figs. 80.2 and 80.3, we can discover that although both scores plot getting into four categories, in which the two of "fenjiu" cannot be separated, the SPCA loadings are sparse enough so that it can clearly point out the strongest relevance between PCs and band. Thus, the property of sparse enhances the explanation and removes the redundance.

80.4.3 SVM

According variance contribution on request, the first extracted 17 PCs by PCA and SPCA, respectively, when training, contained more than 96 % of total variance. So they could almost express the total spectral information. Then the top 17 latent variables were put into the classifiers. As explained by Chih-Wei [10], RBF is the most reasonable choice because of its simplicity and ability to model data of arbitrary complexity. So we entirely focus on the RBF kernel in this paper. To improve SVM predictive performance, we adopt a "heuristic search" on parameters c and g.

After constructing the model, how to predict a new sample is the following work. According to the introduction of PCA and SPCA, we must take validation set to the same mapping and get PCs of it. The best way is to deal with Validation set based on projection matrix of calibration set. Then the obtained PCs of calibration set are taken into the model to predict.

Table 80.1 contains the prediction results for the SVM, PCA + SVM, and SPCA + SVM models. For pure SVM and PCA + SVM classified method, only 80 % and 83 % of true wine were correctly classified, respectively; while the SPCA + SVM models achieved 93 % correct classification. Obviously, the sparse PC as input of SVM could be produced better classification results than others.

Table 80.1 Results for the calibration and validation set using hybrid SVM model

Methods	Data set	c	g	Classification accuracy
SVM	Calibration set	1	0.0313	100 %
	Validation set			80 %
PCA + SVM	Calibration set	4	0.0313	100 %
	Validation set			83 %
SPCA + SVM	Calibration set	32	0.0625	100 %
	Validation set			93 %

80.5 Conclusion

This work tried to improve enforcement of quality of Chinese white wine. The grade and commercial value of five breeds of wine is of great difference. So quickly, efficiently, and correctly identify wine has significant meaning. The study applied NIRS combined with SVM, and extracted optimal PCs using PCA and SPCA as the input of SVM. After comparing the abilities of two kinds of methods, we can conclude that the NIR spectroscopy technique based on SVM has high potential to identify the wine varieties in a rapid and nondestructive way and with a high degree of accuracy.

Although not achieving correct perfectly, the approach using SPCA still indicates that the sparse is effective and promising to classification in some degree. After all the penalty coefficient is selected by experience or man-made, which lack of comprehensive. If we can realize the adaptive selection, the classification accuracy will reach 100 %. So the adaptive selection turns to be the next step before us.

Acknowledgements The authors gratefully acknowledge the laboratory of electronic and information engineering, Huazhong University of Science and Technology (HUST), HuBei. This work has been financially supported by National Science & Technology Pillar Program of China (No. 2012BAK02B06).

References

1. Du, C., & Zhou, J. (2011). Application of infrared photoacoustic spectroscopy in soil analysis. *Applied Spectroscopy Reviews, 46*(5), 405–422.
2. Ying, Y., & Liu, Y. (2008). Nondestructive measurement of internal quality in pear using genetic algorithms and FT-NIR spectroscopy. *Journal of Food Engineering, 84*(2), 206–213.
3. Cozzolino, D., Corbella, E., & Smyth, H. E. (2011). Quality control of honey using infrared spectroscopy: A review. *Applied Spectroscopy Reviews, 46*(7), 523–538.
4. Zou, H., Hastie, T., & Tibshirani, R. (2006). Sparse principal component analysis. *Journal of Computational and Graphical Statistics, 15*(2), 265–286.

5. Luss, R., & d'Aspremont, A. (2010). Clustering and feature selection using sparse principal component analysis. *Optimization and Engineering, 11*(1), 145–157.
6. Yu, H., Lin, H., Xu, H., et al. (2008). Prediction of enological parameters and discrimination of rice wine age using least-squares support vector machines and near infrared spectroscopy. *Journal of Agricultural and Food Chemistry, 56*(2), 307–313.
7. Kelly, J. W., Degenhart, A. D., Siewiorek, D. P., et al. (2012). Sparse linear regression with elastic net regularization for brain-computer interfaces. *Engineering in Medicine and Biology Society (EMBC), 2012 Annual International Conference of the IEEE* (pp. 4275–4278). IEEE.
8. Carroll, M. K., Cecchi, G. A., Rish, I., et al. (2009). Prediction and interpretation of distributed neural activity with sparse models. *NeuroImage, 44*(1), 112–122.
9. Shao, Q., & Feng, C. J. (2012). Pattern recognition of chatter gestation based on hybrid PCA-SVM. *Applied Mechanics and Materials, 120*, 190–194.
10. Hsu, C. W., Chang, C. C., Lin, C. J. (2003). A practical guide to support vector classification. Department of Computer Science, National Taiwan University.

Printed by Publishers' Graphics LLC